袁渭康　院士

袁渭康院士在麻省理工学院旁的查尔斯河边
（1979年，美国波士顿）

袁渭康院士在麻省理工学院做客座研究时与导师韦潜光（J.Wei）教授合影
（1979年，美国波士顿）

袁渭康院士主持第 17 届国际化学反应工程讨论会（ISCRE-17）
（2002 年，中国香港）

ISCRE-17 闭幕式上袁渭康院士把标志下届会议主办权的标牌交给
ISCRE-18 的主席——美国普渡大学的 Varma 教授
（2002 年，中国香港）

袁渭康院士与洛林工大的 Tondeur 教授共同主持 2001 年的
中法化学工程联合实验室年会
（2001 年，法国里昂）

袁渭康院士在第 7 次全球华人化工研讨会上与美国密歇根大学 HenryWang
教授（右）和清华大学费维扬院士（左）愉快交谈
（2014 年，中国香港）

美国弗吉尼亚大学的 Hudson 教授来袁渭康院士家做客
（2003 年，上海）

袁渭康院士陪同德国埃拉根大学 Hofmann 教授（左）参观实验室，
右立者为程振民教授
（20 世纪 90 年代，上海）

袁渭康院士在 ISCRE-23 上与中国科学院李静海副院长（右）愉快交谈

（2014 年，泰国曼谷）

袁渭康院士在 ISCRE-23 上与荷兰德尔夫特大学 Kapteijn 教授（右）、
英国伦敦大学学院 Coppens 教授（中）愉快交谈

（2014 年，泰国曼谷）

袁渭康院士与妻子同摄于埃菲尔铁塔下
(1997年,法国巴黎)

在第7次全球华人化工研讨会期间,袁渭康院士为会议参加者在
《半生行悟》上签字留念
(2014年,中国香港)

中国工程院 院士文集

Collections from Members of the
Chinese Academy of Engineering

袁渭康文集

A Collection from Yuan Weikang

北京
冶金工业出版社
2016

内 容 提 要

本文集收录了袁渭康院士已发表的部分论文,以近二三十年的为主。文集分为中文论著和英文论著两个部分。中文论著部分包括一些评述性的短文及研究论文。英文论著部分收录了他的部分主要研究成果,包括反应器工程、动力学与催化剂、超临界技术、以及一些研究的方法论问题。本书可供我国从事化学工程等研究的科技人员及高等院校相关专业的师生阅读。

图书在版编目(CIP)数据

袁渭康文集/袁渭康著.—北京:冶金工业出版社,2016.1
(中国工程院院士文集)
ISBN 978-7-5024-7092-0

Ⅰ.①袁… Ⅱ.①袁… Ⅲ.①化学工程—文集
Ⅳ.①TQ02-53

中国版本图书馆 CIP 数据核字(2015)第 293243 号

出版 人 谭学余
地　　址　北京市东城区嵩祝院北巷39号　邮编　100009　电话　(010)64027926
网　　址　www.cnmip.com.cn　电子信箱　yjcbs@cnmip.com.cn
策　　划　任静波　责任编辑　李臻　美术编辑　彭子赫
版式设计　孙跃红　责任校对　王永欣　责任印制　牛晓波
ISBN 978-7-5024-7092-0
冶金工业出版社出版发行;各地新华书店经销;北京画中画印刷有限公司印刷
2016年1月第1版,2016年1月第1次印刷
787mm×1092mm　1/16;37.25 印张;4 彩页;909 千字;584 页
179.00 元

冶金工业出版社　投稿电话　(010)64027932　投稿信箱　tougao@cnmip.com.cn
冶金工业出版社营销中心　电话　(010)64044283　传真　(010)64027893
冶金书店　地址　北京市东四西大街46号(100010)　电话　(010)65289081(兼传真)
冶金工业出版社天猫旗舰店　yjgycbs.tmall.com
(本书如有印装质量问题,本社营销中心负责退换)

《中国工程院院士文集》总序

2012年暮秋，中国工程院开始组织并陆续出版《中国工程院院士文集》系列丛书。《中国工程院院士文集》收录了院士的传略、学术论著、中外论文及其目录、讲话文稿与科普作品等。其中，既有院士们早年初涉工程科技领域的学术论文，亦有其成为学科领军人物后，学术观点日趋成熟的思想硕果。卷卷文集在手，众多院士数十载辛勤耕耘的学术人生跃然纸上，透过严谨的工程科技论文，院士笑谈宏论的生动形象历历在目。

中国工程院是中国工程科学技术界的最高荣誉性、咨询性学术机构，由院士组成，致力于促进工程科学技术事业的发展。作为工程科学技术方面的领军人物，院士们在各自的研究领域具有极高的学术造诣，为我国工程科技事业发展做出了重大的、创造性的成就和贡献。《中国工程院院士文集》既是院士们一生事业成果的凝炼，也是他们高尚人格情操的写照。工程院出版史上能够留下这样丰富深刻的一笔，余有荣焉。

我向来认为，为中国工程院院士们组织出版院士文集之意义，贵在"真、善、美"三字。他们脚踏实地，放眼未来，自朴实的工程技术升华至引领学术前沿的至高境界，此谓其"真"；他们热爱祖国，提携后进，具有坚定的理想信念和高尚的人格魅力，此谓其"善"；他们治学严谨，著作等身，求真务实，科学创新，此谓其"美"。《中国工程院院士文集》集真、善、美于一体，辩而不华，质而不俚，既有"居高声自远"之澹泊意蕴，又有"大济于苍生"之战略胸怀，斯人斯事，斯情斯志，令人阅后难忘。

读一本文集，犹如阅读一段院士的"攀登"高峰的人生。让我们翻开《中国工程院院士文集》，进入院士们的学术世界。愿后之览者，亦有感于斯文，体味院士们的学术历程。

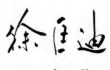

2012年7月

中国工程院 院士文集

前 言

　　本文集收录了我的一些论著，以近二三十年的为主。更早期的论著多散落各处，且历时已久，查找起来颇费时费事，又何况多半学术价值有限。近二三十年的论著可以通过互联网查找和下载，十分方便，因而成了本文集收录的重点。

　　文集分中文论著和英文论著两类。中文论著包括两个部分：一部分是与我从事的化学工程专业有关的一些评述性或综论性的小文，主要是经过多年的研究工作实践后形成的一些认识和观念，并没有什么学术性的价值；另一部分收集了近二三十年来发表的一些研究论文，主要是关于反应工程和催化剂工程的。英文论著部分收录了我的部分研究结果和研究论文。

　　我的专业是化学工程，或者说得更细一些，是化学工程的一个分支——化学反应工程。参加工作数十年来，我基本上没有离开过这个领域。就化学反应工程而言，反应器工程是它的主体部分，也是我研究工作的主要方向。
　　化学工程是一门传统学科，它的发展历史已逾百年。在这漫长的过程中，化学工程学科已形成了比较完整的理论体系和独特的研究方法，但是与其他的传统学科有类似之处，化学工程学科亦面临内涵更新和结构优化的问题。
　　化学反应工程从传统的意义上着眼于反应器的工程放大，也就是把实验室中小型试验的结果在大型反应器中实施，并且维持比较高的操作水平。这就是我们通常说的工程放大，也是我在 20 世纪八九十年代研究工作的主要内容。进入 21 世纪以来，化学工程的更新步伐有所加快。我们在 20 世纪末就在探索新的研究方向，确立了从化学品多层次结构的性能效应，即通常所说的"构效关系"，并导入化学工程方法和手段对化学品多层次结构的效应进行分析，以获取优化的性能。这实际上就是产品工程的核心。
　　从 20 世纪 90 年代起，国际学术界普遍认为化学工程的发展会逐步地从"过程"导向向"产品"导向过渡。所谓过程导向，是指人们把主要目标着眼于过程的放大与优化，而产品导向则是更多地注重产品的结构优化，并由此实现性能优化。

我的研究工作试图与国际学术界的研究方向同步。从最初以过程导向的反应器工程为主，到后来触及从搜索催化剂多层次结构着手的催化剂性能优化，以及用化学工程手段，如超临界技术来优化大宗高分子聚合物的多层次结构，使之成为性能独特的"小众"产品。这也是与我国工业产业结构升级的迫切要求相一致的。我不能说与国际学术界同步得很好，我只是想说明我们是这么去做的。

从20世纪80年代起，我们的研究成果主要是在国际学术刊物上发表。我们也鼓励我们的博士生们在国际上发表他们的结果，让国际同行们来评价他们的研究工作。这就造成了我们的研究工作，凡是比较有价值的，大体都是在国际杂志发表的事实。因此在选择本文集的论著时，我不得不在数百篇英文论文中作一些选择。好在现代的信息技术使得收选工作比较方便。从我校图书馆系统中我找到了1999年以来发表的240篇论文，再通过其他渠道找到了一些更早期的，初步从中选了80篇。再经过两次取舍，最后选定了现在读者看到的32篇。在这些论文中，有一些我的贡献比较多，有一些的主要贡献是我的同事们和学生们，应该承认，我的贡献是很有限的。

在论著的收录过程中，回顾我已走过的岁月，总的来说缺乏重要的成果与贡献。虽说是在国际上这一领域最有声望的学术刊物上发表了多篇论文，但缺乏真正意义上的创新，因而只能说是业绩平平，但我要在此感谢我的同事们和学生们，他们长期和我一起工作，他们对我的帮助是我时刻铭记在心，永难忘怀的。他们中有与我同龄的同事，有后来成为我同事的当年的学生，也有现在尚在为他们的博士论文而努力的学生们。我的知识结构早已老化，我必须向年轻人虚心学习，事实上，我也确实在向他们学习的过程中获益良多。

袁渭康
华东理工大学
2015年7月

中国工程院 院士文集

目 录

中文论著

综论

- 》》《化学工程与技术丛书》序 …………………………………………… 3
- 》》我们的化学工程：关于目标尺度微细化的讨论 …………………… 6
- 》》化学工程：在演化中提升 ……………………………………………… 11
- 》》初议超临界条件下的化学反应 ………………………………………… 16
- 》》《化学反应工程分析》前言——化学反应工程方法论讲座（第一讲） …… 24
- 》》《化学反应工程分析》后语——化学反应工程方法论讲座（第二讲） …… 28
- 》》工业反应过程的开发方法 ……………………………………………… 38
- 》》过程的放大与优化 ……………………………………………………… 41

研究论文

- 》》新型苯加氢反应器的研究 ……………………………………………… 47
- 》》气液向上或向下并流固定床动持液量测定 ………………………… 54
- 》》局部面积搜索法在高压相平衡中的应用 …………………………… 60
- 》》高压脉冲电晕放电等离子体降解废水中苯酚 ……………………… 67
- 》》脉冲电晕放电等离子体降解含4-氯酚废水 ………………………… 72
- 》》密闭体系中乳酸的固相缩聚 …………………………………………… 79
- 》》催化剂活性组成对纳米碳纤维产率和微结构的影响 ……………… 84
- 》》钛硅分子筛/纳米碳纤维催化剂的制备及其液相分离性能 ………… 92
- 》》纳米炭纤维的表面润湿行为 …………………………………………… 96
- 》》超声波对甘氨酸溶析结晶过程的影响 ………………………………… 105
- 》》构造理论在工程领域中的应用研究进展 …………………………… 114
- 》》有机酸杂质对乳酸锌初次成核速率的影响 ………………………… 127

英文论著

研究方法论

» Targeting the Dominating-scale Structure of a Multiscale Complex System: a Methodological Problem ······ 137
» Reactor Engineering: Science, Technology, and Art ······ 155
» An Algorithm for Simultaneous Chemical and Phase Equilibrium Calculation ······ 164
» Intensification of Phase Transition on Multiphase Reactions ······ 180
» Controlling Sandwich-structure of PET Microcellular Foams Using Coupling of CO_2 Diffusion and Induced Crystallization ······ 194

动力学与催化剂

» Modeling Silver Catalyst Sintering and Epoxidation Selectivity Evolution in Ethylene Oxidation ······ 215
» First-principles Calculations of C Diffusion through the Surface and Subsurface of Ag/Ni(100) and Reconstructed Ag/Ni(100) ······ 230
» Catalytic Reduction of Hexaminecobalt(Ⅲ) by Pitch-based Spherical Activated Carbon(PBSAC) ······ 250
» Mechanistic Insight into Size-dependent Activity and Durability in Pt/CNT Catalyzed Hydrolytic Dehydrogenation of Ammonia Borane ······ 263
» Kinetics of Gas-liquid Reaction between NO and $Co(NH_3)_6^{2+}$ ······ 272
» DFT Studies of Dry Reforming of Methane on Ni Catalyst ······ 283
» First-Principles Study of C Adsorption and Diffusion on the Surfaces and in the Subsurfaces of Nonreconstructed and Reconstructed Ni(100) ······ 300
» Diffusion-enhanced Hierarchically Macro-mesoporous Catalyst for Selective Hydrogenation of Pyrolysis Gasoline ······ 315

反应器工程

» Dryout Phenomena in a Three-phase Fixed-bed Reactor ······ 329
» Influence of Hydrodynamic Parameters on Performance of a Multiphase Fixed-bed Reactor under Phase Transition ······ 340
» Redistribution of Adsorbed VOCs in Activated Carbon under Electrothermal Desorption ······ 351
» Process Flow Diagram of an Ammonia Plant as a Complex Network ······ 362
» Determination of Effectiveness Factor of a Partial Internal Wetting Catalyst from Adsorption Measurement ······ 372
» Deep Removal of Sulfur and Aromatics from Diesel through Two-stage Concurrently and Countercurrently Operated Fixed-bed Reactors ······ 384
» A Hybrid Neural Network-first Principles Model for Fixed-bed Reactor ······ 395

- Hydrodynamic Behavior of a Trickle Bed Reactor under "Forced" Pulsing Flow ··· 402
- Practical Studies of the Commercial Flow-reversed SO_2 Converter ················ 413
- Nanomaterials Synthesized by Gas Combustion Flames:
  Morphology and Structure ·· 424

超临界技术

- Co-pyrolysis of Residual Oil and Polyethylene in Sub- and Supercritical Water ··· 437
- Experimental Measurements and Modeling of Solubility and Diffusivity of CO_2 in Polypropylene/Micro- and Nanocalcium Carbonate Composites ················ 452
- Effect of Supercritical Carbon Dioxide-assisted Nano-scale Dispersion of Nucleating Agents on the Crystallization Behavior and Properties of Polypropylene ·············· 471
- CO_2-induced Phase Transition of Isotactic Poly-1-butene with Form Ⅲ upon Heating ·· 489
- CO_2-induced Polymorphous Phase Transition of Isotactic Poly-1-butene with Form Ⅲ upon Annealing ·· 506
- Supercritical Carbon Dioxide-assisted Dispersion of Sodium Benzoate in Polypropylene and Crystallization Behavior of the Resulting Polypropylene ······ 521
- Controlling Crystal Phase Transition from Form Ⅱ to Ⅰ in Isotactic Poly-1-butene Using CO_2 ·· 533
- Effects of Crystal Structure on the Foaming of Isotactic Polypropylene Using Supercritical Carbon Dioxide as a Foaming Agent ······················ 552
- Foaming of Linear Isotactic Polypropylene Based on its Non-isothermal Crystallization Behaviors under Compressed CO_2 ················ 567

中文论著

- 综论
- 研究论文

《化学工程与技术丛书》序*

华东理工大学出版社经过长期调查研究后，决定组织编写并出版一套"化学工程与技术"学科的相关教学用丛书，多位在各自领域学有所长并对研究生培养工作有丰富经验的学者参加编写。

对于华东理工大学出版社的决定，我非常赞同。我本人是研究生毕业，从20世纪70年代末起，一直在指导研究生，包括授课及指导论文，我的主要工作都与研究生培养有关。加上报纸杂志的报道中多认为我国在研究生培养方面尚显不足，我也经常在思考如何提高研究生培养质量的问题。对此常感到有些话想说，只是没有适当的场合去说而已。因此当出版社的编辑们要我为丛书作序时，我便欣然从命，以便借此说上几句。

1 关于教学用书（简称教材）

教材对研究生教学是重要的，好的教材显然十分有利于学生学习和掌握相关的专业知识，此外还可以作为学生在学完课程后的案头参考书。

也有一些非常优秀的教授在教学过程中不规定使用固定教材，他们在课堂上主要讲授思想和方法，或即使使用教材，讲课时也完全不局限于教材内容，然后要求学生在课后通过自学、做习题、讨论、找材料、做笔记等多种形式掌握知识。这种教学方式对于一部分学习主动、基础较好的学生，可能十分有利，但也许会使另一些学生感到困难，甚至抱怨连连。关键是看这些学生是否有克服困难，通过努力争取学习主动的决心。

2 关于例题和习题

例题的重要性丝毫不逊于理论知识。正确的方法应是有目的地讲解例题：一个例题解决一类问题，引导多方面的思路，并培养学生举一反三的能力。

我要特别强调的是习题的作用，使学生巩固、掌握知识和运用学到的方法只是起码的要求，习题的功能应被看做是对学生潜在创造力的培养，以及让学生做好在面对困难时应有的心理准备。这里说的当然不是指我们常见的这些只需稍稍复习就可以依样画瓢式的习题，而是指学生初看会不知如何下手的那一类。当学生要做这类习题时他们不得不去认真复习和思考，相互讨论，查找文献，才能解答。他们会认为这些习题很"难"，但也就是这种"难"，可以培养学生的能力。

我们不是经常在说要培养学生的创新能力么？显然，单靠说是不够的。创新应成为一种习惯，它可能日积月累，潜移默化地养成。做"难"题确实要克服不少困难，

---

* 本文为袁院士为华东理工大学出版社出版的《化学工程与技术丛书》所做的序。

但这正好是一种对今后工作中创新能力的磨炼，也必会形成一种能解决困难问题的心理准备。这是一个师生需要共同认识的问题。对教师，有"敢"或"不敢"布置"难"题的问题；对学生，有对"难"题迎难而上还是尽量规避的不同态度。学生遇到"难"题而又千方百计地去解答，这实际上是在为将来在工作中解决"难"题做准备。

3　关于方法论教学

教师在教学过程中应突出重点、启发式、重概念等，大家都已熟知。但如何具体把握教学中涉及的有限知识，上升到方法论的高度，无疑是十分重要的。

方法论者，简而言之，就是在面对一个具体问题时把所具备的知识实施并应用的理念、思想和方法。教材的性质决定了它能给予你的可能是学科的一般原理，但如何根据你所要解决的问题的具体要求，恰如其分地用好你已具备的知识，尚有相当距离。例如，我们能列举对一个过程产生影响的诸多因素，但对某一特定条件，这些因素中可能只有少数是主导的，其余的可能分别在另一些条件下起主导作用。举例来说，一个多相反应系统，如是反应速率很低的慢反应，那时相际传质可能并不重要，研究工作可不必拘泥于此；反之，对于快速反应，传质可能成为控制因素，反应本身退居次要。你首先应弄清对象特点，根据你的知识，作出正确判别。这样的工作方法越早养成越好，而课程学习正是培养这种能力的良好机会。

4　关于知识的交融

人们解决一个问题，常会说"用化学的知识"，或是"用物理的知识"等。我认为对一位训练有素的专家而言，他可能在解决问题的时候已说不清使用的是哪方面的知识，因为各种知识在他的头脑中已融为一体，不分彼此了。如果在学习某一课程时能广泛联系和思考已学过的课程和已掌握的知识，对于掌握知识和灵活地运用知识必有好处。如果不说得太远，那么在"化学工程与技术"这一学科领域内的各个分支，应尽可能去做到这种知识的交融。下面我举几方面的例子：

（1）一个传热过程，热量从热流体经过器壁传向冷流体。如一侧的传热系数很大，另一侧的很小，在不考虑器壁热阻时要提高总传热系数，必须提高传热系数小一侧流体的传热系数，而进一步提高传热系数大的一侧流体的传热系数，效果必不明显。这一例子对于学过"化工原理"的学生几乎是尽人皆知的。

（2）设计一个由10个部件组成的产品。如在正常使用条件下其中9个部件的使用寿命为2年，一个部件为1年，则在一年后整个产品就不能使用了。解决的办法应是，或把这9个2年部件降级为1年部件，或把这个1年部件用更好的工艺和材料制成2年部件，使整个产品的各部件几乎在同时报废。这一机械设计原则已成为一个设计常识。

（3）一个连续生产流程经过一定时间运行后需要停工"检修"，进行设备清理及更换一些零部件等，再开始下一周期运行。在一个生产周期内，所有的生产环节必然遵循这一运行周期进行规划，催化剂也应在检修期间更换。在生产过程中催化剂会逐步失活，为弥补活性下降，通常的方法是提高反应器操作温度。然而，温度提高更促进了催化剂失活。如在运行周期内由于催化剂失活而温度提高过快，则活性损失过多，可使正常操作难以进行；如在运行周期内温度提高缓慢，在检修时催化剂虽尚有余力，但也不得不更换，这也是另一种浪费。最好的策略是规划好温度提高的序列，使催化剂能力在运行中充分发挥效能，在检修时达到催化剂的经济使用的极限。这实际上是系统学的基本议题。

上述三者涉及对象虽不同，但都说明了一个观点：系统的观点。这一观点显然在任何场合都是适用的，尽管对象可能会有千差万别。

知识的交融需要教师的引导和学生的主动参与。应提倡对所学的各种知识和方法的细嚼慢咽，反复体会，提倡广泛联想，认真论证。现代科学，无处不体现交叉和联想。在学习中千万不应把所学课程孤立起来，在研究生学习阶段注意这一点，必能终身受益。

研究生课程可能是多数学生一生中最后的系统课程学习（博士生的课程往往很少）。学到一些专业知识，通过了考试，得到几个学分，这只是最起码的教学要求。谁都知道，多数学生的目标远非仅此而已。研究生课程的重要性绝不亚于完成论文，应十分珍惜这段时间的学习。利用课程学习更好地掌握知识，需要师生的共同努力。为此，我谨利用为本书写序的机会，奉上这几句可有可无的话，仅供使用本书的师生们参考，也希望本丛书能为同学们在掌握知识、掌握方法方面有所贡献。

我们的化学工程：关于目标尺度微细化的讨论*

摘　要　探讨了化学工程的发展趋势，认为目标尺度不断微细化是化学工程发展的必然趋势。阐述了微结构调控是产品工程的核心内容，而多尺度方法则是联系性能-结构-过程关系的重要途径。说明产品工程是对工业而言的，而多尺度方法则具有学科的内涵，实际上两者之间有相互贯穿相互统一的一面。
关键词　产品工程　多尺度方法　化学工程学科　发展趋势　范畴

化学工程总体来说是一门传统学科。"传统"是相对于"新兴"而言的，它既衬托"新兴"，又与其互补。但是在当今飞速发展的社会，"传统"一词，多少被理解成含有些贬义。这也许是由于化学工程支撑的化学工业、能源工业、石化工业、冶金工业和食品工业等工业部门是以大宗产品的生产为主的，产品的品种比较有限，新产品也不如信息工业的产品那样层出不穷且贴近最终用户。另外，化学工程所支撑的一些工业还面临一些其他的限制，如这些工业大量消耗资源和能源，并往往对环境产生明显的不良影响。

化学工程学科发展的困境不但在中国有，而且在一些发达国家同样存在。但由于中国的传统工业还远没有饱和，技术也远没有成熟，更何况化学工程的作用远不局限于传统工业和传统产品，所以化学工程在中国会有更多的用武之地。

1　过程的开发、放大和优化

自形成学科之日起，化学工程的主要目标就是过程的开发、放大和优化。

常常说化学工程师的任务是把化学家在实验室中取得的研究成果通过各种过程的组合在工业规模上予以实现，获得有竞争力（指质量和成本）的产品。这样一种观念在很长一段时间内被认为是正确的，而且是唯一的。一个重要的原因是，过程工业（化学工业是一种代表性的过程工业）面对的主要是大宗产品生产。大宗产品要求高产量、低消耗和低成本。从产品的结构看，品种和质量的要求并不突出，特别对于短缺经济更是如此。化学工程师与化学家各司其职，各行其是，他们的研究领域通常较少交叉，他们的研究工作也较少融合。

化学工程学科经过 100 多年发展，到 20 世纪末，已逐步形成了完整的理论体系和独特的试验研究方法，并且自 20 世纪五六十年代以来计算机技术的进步极大地充实了化学工程学科。可以说，化学工程发展至今，依靠已有的理论、方法和经验已可借助经过规划的试验研究解决过程工业的绝大多数工程技术问题和过程开发问题。这是就学科整体水平而言的。至于中国作为发展中国家工业过程的开发和实施中存在的技术问题，除化学工程技术自身与国外先进水平有差距外，还常常受制于诸如材料、制造、配套设施等。

---

* 本文原发表于《化工进展》，2004，23（1）：9~11。

但这里要说明的是，化学工程的传统内容主要是针对大宗产品生产的，但随着社会的发展，必然提出新的要求，也必然会滋生新的学科内容。

2 过程与产品

对于一个社会或一个工业部门，产品总是首要的、最有生命力的因素，就像前面提到的信息业之所以兴旺是与其日新月异的产品结构密切相关的。

西方工业国家可能比中国更早地意识到这一点。由于大宗产品的人均产量趋近饱和，传统工业的技术也已十分成熟，他们很早就将目光转向产品的品种、功能和质量，以及为产品实现这些产品生产相应的工业过程。

显然这些产品主要不再是指大宗产品，而是指专用品、功能品等。这些产品的特点是产量相对较小，但有专门的或特殊的功能，而且十分贴近用户。例如，社会的进步必然使一些医药保健用品显得空前重要，对低毒低污染的农用品的要求必然会大幅增加；生活质量的提高还会对食品和轻工业品提出更高的要求。以日常生活中经常碰到的事情为例，人们食用冰淇淋的口感很大程度上取决于冰粒粒度，所用涂料的质量也与涂料中胶乳的粒径分布密切有关。即使冰粒和胶乳的化学成分没有变化，但大小形貌上的差异对冰淇淋和涂料的质量有极其重要的影响，可见产品的介观结构是如何重要。这种所要求的介观结构的形成又与过程如何实施密切相关，而这一类与介观结构有关的问题可能在考虑大宗产品生产时是不予考虑或无需注意的。

关于产品，至少要研究两方面的问题：一是如何根据所需要的产品性能来提出产品应该具有的微观和介观的结构；二是知道了产品的微观和介观结构要求，如何通过过程来保证这些结构的实现。

3 尺度效应和多尺度方法

国际学术界一致认为，化学工程学科发展至今，已经经历了两个里程：第一里程始于19世纪末，是以单元操作为标志的，其考察尺度是宏观的，所采用的基本研究方法是试验数据关联；第二里程大体从20世纪60年代开始，以传递现象为标志，其考察尺度变小，为滴、泡、颗粒等，所采用的研究方法是数学模型法。显然，与第一里程相比，第二里程的考察更能揭示过程的本质。不过目前已经或正在进入第三里程，至于第三里程的标志是什么，至少在目前尚难做出定论。

在20世纪的八九十年代，化学工程学科的发展走入困境，一些大型的过程工业公司纷纷进行重新组合和产品结构调整，这些都促进了对学科前沿的研究和考虑。1996年，Villermaux在第四届世界化学工程会议的大会报告会上就倡议了化学工程应该从过程导向转向产品导向的问题[2]。2001年，Wei在美国一次极有声望的学术报告中提到了产品工程是否是化学工程第三里程的问题[3]。这种所谓的产品工程的核心所在，是指产品的微观和介观结构（从原子、分子尺度结构到纳微米尺度结构）与产品功能的关系，以及其实现这些结构的工程方法。产品工程作为第三里程的说法固然还有争议，但是从化学工程发展的历程可以看出一个明显的趋势，那就是考察的尺度更加微细化。这种微细化的趋势，使人们正视了尺度问题。尺度问题不仅在化学工程学科，在相关的其他学科也受到了前所未有的重视[4~6]。这里所说的尺度并不局限于某一固定的尺度（如纳米尺度），而是对考察对象的物理化学性质起决定作用的尺度，即目标尺度，根据考察对象的不同可能涉及分子、分子簇、纳米、微米等跨度极广的多个尺度。而传统化学工程方法通常只涉及少数几个尺度，如流程尺度、设备尺度、微团尺度等，并且也未得到足够的重视。

随着考察尺度的不断微细化和尺度跨度的不断增加，客观上要求一种新的理论体系和研究方法以建立各种不同尺度现象之间的联系和关系的方法，这就是多尺度法。由于自然界存在空间和时间两种尺度，因此也可以称做时空多尺度方法。目前尚难以明确有某种普遍适用的多尺度法，但这一理念的形成无疑为化学工程研究工作者提供了一个重要的研究指导准则。

4 产品性能-结构-制备的关系

有了对产品性能的要求，就需要确定产品的结构（通常是指微介观结构），就需要有制备这种结构产品的方法——过程。这里要讨论的一个问题是，用什么方法建立这种性能-结构-制备关系。

就目前的学科发展来看，人们还难以形成一个普遍适用的处理性能-结构-制备关系的方法。如果说作为第二里程的主要方法是数学模型法，第三里程的主要方法是分子模型化的推论恐怕还为时过早。传统的数学模型法主要是着眼于一个很小的控制体积（微分体）。把流体（有时甚至是固体颗粒群）看做是连续介质，对这一微分体做物质或能量的衡算，建立起瞬态的或定常态的常/偏微分方程，取适当的求解条件，然后进行计算机模拟。这种已经沿用了数十年的方法无数次被证明是相当有效的。但当物系不能被看做是连续介质时，这种模型法就将失效。由此，人们想到了分子模型化，想到了量子化学计算，想到了深入分子内部，想到了电子云，想到了影响物系性质的"源头"。显然，传统的数学模型法从未考虑过这些。

但是也并不是所有的场合皆源头模型化不可。例如，当考虑流体-颗粒系统时，影响系统行为的最小尺度是单颗粒尺度，那么完全没有必要去做构成颗粒的物质的分子模拟。如果研究的是催化性能，那么构成催化活性中心的分子-原子-电子结构就显得非常必要，要建立性能-结构关系，微细结构的模型化是必不可少的。再举一个宏观的例子，如果研究天体运动行为，那么即使把地球（或其他星球）看成一个最小尺度也未尝不可，大可不必计较地球上的海洋、山脉、建筑和人群了。因此，最小尺度不能一概而论，而应视系统要所研究的性能/行为要求而异。但总的一个必然趋势是，把发展产品的微细结构-性能关系作为研究重点（这一点也是现实的需要），即目标尺度的微细化。

讨论了这些以后，再回过头看产品工程的提法和时空多尺度方法的提法，可以看到两者之间的统一性。产品的性能（通常为宏观尺度）取决于其结构（通常为介观或微观尺度），在这两者之间可能还存在几个过渡的尺度。从产品的角度，这是性能-结构-过程关系；从学科角度，则概括为多尺度方法。

产品的不同结构尺度大体可分为分子结构产品和介观结构产品两类。前者指产品构成分子本身的性质决定着产品的性能和功能，如药物、燃料等；后者则主要是指一些组成相同或相近，但由于在介质尺度上的结构差异，其性能和功能可有很大差别。超细晶钢与组成类似的碳钢相比的优越性能即为一例，类似例子在过程工业产品中不胜枚举。在这两类产品中，一种新的有效的分子结构产品的开发可能需要一个较长的周期，而产品介观结构的研究对于扩充产品品种、提高产品性能也许更有实际意义。可见，研究工作的任务显然就是要解决好以下两类问题（这两类研究问题的相互关系见图1）。

图 1　结构-性能-制备的关系

（1）建立其理性的结构-性能关系，即了解产品功能通过怎样的微、介观结构（分子结构和介观结构）才能实现，这一点可归结为"做什么"。

（2）在了解应具有的结构以后用过程来保证这种结构的实现，也可归纳为"怎么做"，这样所讨论的目标尺度比传统理解的尺度明显地"微细化"了。

5　结语

传统化学工程学科的知识结构显然已经不能适应学科发展的需要，原有的教学研究体制也无疑经受着挑战。总的一个发展趋势是目标尺度的微细化：观察尺度的微细化，结构尺度的微细化和加工目标尺度的微细化。如果主动认识到这一点，无疑会对研究者的工作带来好处。在这样的学科发展背景下，特别需要重视以下几方面的问题：（1）重视与其他学科的交叉，改变原有教学内容的安排；（2）熟悉并充分利用先进的测试方法，如 XRD、FTIR、HRTEM、AFM、XPS、LEED 等；（3）发展微细结构模拟技术，充分利用分子模拟、量子化学计算等模型化和模拟方法。总体而言，化学工程的传统学科性质，并没有阻碍它向新的领域和新的目标发展和渗透，也并没有阻碍人们创立新的学科生长点，关键是需要人们不断地探索、追求和创新。

参 考 文 献

［1］ Amundson N R, et al. Committee on Frontiers in Chemical Engineering: Research Needs and Opportunities, Frontiers in Chemical Engineering ［M］. Washington D C: National Academy Press, 1988.
［2］ Krieger J H. Chemical & Engineering News, 1996, 19: 10~18.
［3］ Wei J. www. princeton. edu/seasweb/ammudson. ppt.
［4］ Lerou J J, Ng K M. Chem. Eng. Sci., 1996, 51: 1595~1614.
［5］ Kwauk M, Li J. Trans. I. Chem. E. Part A, 2002, 80: 699~700.
［6］ 郭慕孙，胡英，王夔，等. 物质转化过程中的多尺度效应 ［M］. 哈尔滨：黑龙江教育出版社，2002.

Miniaturization of Target Scale: Refreshing Chemical Engineering Philosophy

Yuan Weikang

(State Key Laboratory of Chemical Reaction Engineering, East China University of Science and Technology, Shanghai, 200237, China)

Abstract　The author emphasizes that miniaturization of the target scale is an essential tendency that deserves more attention of chemical engineering professionals. Micro/meso structure control is the key of product engineering, and multiscale methodology is applied to establish the property-structure-process relationships. Product engineering is raised from an application-oriented point of views, however, multiscale methodology is to certain extent academic research based. The fact is they two are inter-related.

Key words　　product engineering, multiscale methodology, chemical engineering discipline, trends, paradigm

化学工程：在演化中提升[*]

化学工程是为化学工业和相关过程工业提供工程技术（包括流程、设备、催化剂的设计和优化，过程与设备的操作和优化），并研究其中的工程技术基础，以利用物质的化学（包括生物化学）转化和加工实现工业生产的工程学科。化学工程学科的发展对于我国国民经济的发展具有重要意义。

作为一门工程科学，化学工程与其他工程科学一样，工程背景十分明确，并且以工程应用为宗旨。另外，它是一门科学，以化学、物理和数学等基础科学为基础，同时又有自己系统的理论体系。此外，化学工程还具备一整套颇具特色的实验研究方法。

化学工程的形成大体可以追溯到19世纪末。当时在西方工业国家，化学工业已基本上摆脱了作坊式的生产模式，出现了较大规模的连续生产流程。尽管生产的产品不同，但不同的流程却使用了一些操作特点基本相似的单元操作方式，如蒸馏、吸收等分离过程，传热过程，以及流体流动过程等，这些单元操作就成了早期化学工程的雏形。目前，国际上公认关于这一学科的第一部专著《化学工程手册》[1]就是在那段时间问世的。

化学工程学科经历了百余年的演变和发展，建立了自身的学科体系，服务对象也大为拓展，学科内涵得到了极大的丰富。与传统意义上的化学工程相比，今天的化学工程显然已面目一新。

1 学科的演化

国际学术界公认，化学工程学科的发展已经经历了两个发展阶段（分别以"单元操作"和"传递现象"为标志），目前正处于第三个发展阶段的起始阶段。了解学科发展历程，有助于认识本学科今后的发展趋势，开展前瞻性的基础和应用研究，以在科学和技术方面取得领先优势。

化学工程学科的第一个发展阶段从19世纪末开始一直延续到20世纪50年代末。当时的研究对象主要集中在化学工业中大量出现的与流动、传热和分离有关的单元操作。对反应过程当时研究得很少，并不是因为反应过程不重要，而是因为当时还不具备足够的能力进行深入的研究。在这一段时间里，化学工程的主要研究方法是对实验数据进行归纳和关联（采用无因次数群），以获得实用的经验性规律，如流体流动阻力系数、传热膜系数、传质系数，甚至理论板数和板效率等。以今天的眼光来看，这种研究方法把过程和对象当作黑箱处理，比较粗糙，但在当时的条件下，这样的研究方法显然是可取的，也被证明是十分有效的。

化学工程学科的第二个发展阶段起始于20世纪60年代[2]。当时的化学工程研究特点出现了重要的变化：研究对象不再是在不同工业流程中具有共性的单元操作，而是在不同单元操作中具有共性的传递过程；研究的着眼点从较大的设备尺度集中到了

---

[*] 本文合作者：周兴贵。

湍流尺度,如液滴、气泡、颗粒尺度等;研究方法不仅仅是对已有实验现象和数据的归纳,更重要的是可通过演绎发现新的现象和规律。显然,这比只把对象看做一个整体来认识要更为深入,并且由于演绎法的使用亦提供了过程与设备优化的途径。第二个发展阶段的另一个特点是大量使用数学模型方法及对模型方程的数值分析,使化学工程研究过程中常常遇到的大量非线性方程能够通过数值计算得到定量结果,很明显这是与20世纪60年代计算技术的突飞猛进分不开的。另外值得一提的是差不多在同一时间,开始了对反应过程的系统研究,并很快取得突破性进展,形成了化学反应工程这一化学工程学科的最重要分支,专门研究流动、传热、传质与化学反应的相互作用[3]。

在第二个发展阶段中化学工程的研究对象是不同设备中具有共性的传递过程,研究方法以演绎法为主,而且普遍以严格的数学模型为基础,由此提高了化学工程学科的科学性,深化了化学工程学科基础,丰富了化学工程的学科内涵。这一阶段的研究已摆脱前一阶段把研究对象整体作为黑箱处理的原始状态,将过程和设备进行分解,利用基本的物理/化学定律,通过研究局部的传递和反应行为获得对过程与设备的整体行为的认识,由此获得的研究结论无疑具有更广泛的适用性。这种研究方法也十分有效,有力促进了在20世纪下半叶化学工程学科和化学工业技术的飞速发展。

但值得注意的是,在这一阶段尽管研究尺度从设备尺度过渡到滴、泡、颗粒的尺度,但只关注滴、泡、颗粒的整体行为如相界面积、相间传质系数、反应动力学等,并在建立数学模型时忽略滴、泡、颗粒之间的差异(平均化处理)和内部的差异(均质化处理),将介质进行连续化处理,甚至忽略不同相态的差异(拟均相处理),在此基础上采用控制体积法,即把研究对象分解成微元(微分体或微团),假定每一微元(控制体)中的物性和状态变量均匀,然后对每个控制体应用基本的物理/化学定律(守恒定律、热力学定律、费克定律、傅里叶定律等),建立起数学模型,并对各控制体进行积分以获取过程与设备的宏观行为。采用简化和平均化处理的目的是避开复杂的物理/化学过程,这主要是由于受到当时研究手段、方法和水平的限制。这种简化和平均化处理方法延续了数十年,事实上也被证明十分有效。目前,依托现有的化学工程学科知识,能够解决绝大多数化工过程的设计与操作问题。但这种方法也有严重不足之处。多数(尤其是涉及多相物系的)的化工过程的开发还需要以大量的经过规划的实验为基础,这主要是因为上述简化处理使传递过程参数不准确或不确定,必须对这些传递参数重新进行实验测量或加以实验验证。

化学工程在过去的一个世纪取得了巨大进步和成就,但发展至今遇到了不少问题。一方面现有的化学工程基础能够解决绝大多数传统化工过程与装置的开发,尽管开发代价较大、耗时较长,但目前短时间内难以取得突破性进展;另一方面,非传统化学工业领域中大量的过程和产品设计问题目前得不到解决。这些问题对化学工程学科的发展提出了挑战,但也同时带来重要的发展机遇。正确判断化学工程学科的发展趋势,开展前瞻性学科前沿基础研究无疑对学科的发展和相关行业的技术进步有重要的意义。

2 化学工业发展趋势

技术需求是学科发展的动力。因此,化学工业的发展趋势决定了化学工程学科的发展趋势。

化学工业的主要任务是对石油、煤、天然气、矿物等天然资源进行加工,为其他工业部门提供能源和基本化工原料，因而以"资源加工"为主要特点。化学工业发展初期主要是满足人们衣食住行的基本生活要求，产品量大，使用面广，如合成高分子（包括纤维、塑料和橡胶）。自20世纪末期起，化学工业的发展出现了新的特点，这也是化学工业发展的必然趋势。一方面"资源加工"技术日益成熟，能力接近饱和，利润空间萎缩；另一方面，技术的日新月异和人们生活水平的不断提高，对化学品的功能和性质提出了越来越高的要求。化学工业利润的主要来源在于精细化工品的生产与制造。事实上，20世纪末期就有越来越多的发达国家以及大型化学工业公司将核心产业向精细化工方向转移。精细化工以"产品制造"为特点，是以"资源加工"为特点的化学工业的重要补充，目前已成为世界各大化学工业公司竞争的核心领域。精细化工品的制造技术水平也成为衡量一个公司技术水平和竞争力的重要标志。

对以"资源加工"为特点的化工过程，在满足产量的基础上节能、降耗、提高加工效率是技术进步的驱动力和技术发展的主要目标。与"资源加工"型化学工业生产大宗的、通用的化学品不同，"产品制造"型化学工业生产批量小、附加值高、具有不同特殊功能与用途的产品，也即精细化工品。精细化工品是相对于大宗化工品而言的，是众多专用化学品的总称。精细化工品普遍有附加值高的特点，产品的功效成为技术的关键。

目前，精细化学品的开发基本上还是沿用试错法。典型的如药物与催化剂开发：在一些初步规则的指导下，合成具有不同分子结构的药物或不同配方的催化剂，再通过并行实验比较药效或活性，以从中筛选出最具有价值的产品。这种方式不仅开发时间长、效率低，而且代价巨大。

3 进入第三发展阶段

传统化学工程学科主要服务于以"资源加工"为特色的传统化学工业过程（基本化学工业），内容是三传一反，目的是实现过程（设备）设计，确定设备的型式和尺寸；技术先进性的体现是过程的效能（包括物耗、能耗和废物排放），并集中体现在加工成本上。传统化学工程研究以过程为导向，主要解决"How to do?"的问题。由于传统化学工业技术相对成熟，利润空间较小，化学工程学科面向传统化工领域的主要任务是提供新的学科基础以显著提高过程开发效率并显著降低生产成本。近年来由于世界范围内资源和环境问题的日益严峻，绿色化工过程开发，替代资源的使用和资源综合利用也是当前化学工程学科的重要研究内容。

随着化学工业的重心逐步从基本化工原料生产向精细化工品生产转变，化学工程的角色也必然发生相应的转变，即从"过程导向"转变为"产品导向"。不仅要解决"How to do?"，更重要的是解决"What to do?"的问题。面向以"产品制造"为特色的新化学工业（精细化学工业），化学工程学科的主要研究任务是提供产品设计和可控制备的工程理论基础。

精细化工品是极其个性化的产品，大体上可分为分子产品、配方产品和结构化产品。分子产品是指具有复杂分子结构或特定功能的分子，如（食品）添加剂、染料、医药、植物生长调节剂、合成香料、聚合物、生物分子、超分子等，是纯净物，产品性能由产品的分子结构（包括空间结构，如分子构象和聚集态结构如晶型）唯一确定。配方产品是指多组分产品，如表面活性剂、香水、洗涤剂、农药、墨水、颜料等，主要通过复配而成，组成与比例决定产品的性质。结构化产品是最复杂的一类，产品性能不仅取决于分子结构、化学组成，更取决于物理结构（包括晶体结构和分相结构），如功能材料、催化剂、涂料、化妆品、食品等，涉及胶体、孔液、囊、泡、膏、胶、膜等结构和形态。上述分类只是粗略分类，有些精细化工品可能是分子产品也可能是配方水平，如表面活性剂、调味品、香精、颜料、染料、农药等，也有可能是分子产品但同时又是结构化产品，如分子筛吸附剂、高分子膜、纳米材料等。

决定产品性能的是产品的结构，包括分子结构（如添加剂、表面活性剂）、晶体结构（如药物）、分相结构（如功能材料）等。对于精细化学品工业，需要解决的主要问题是产品的设计，即根据产品性能要求确定产品结构。由于技术成熟，传统化学工程几乎不考虑完成过程与装置设计后的制造加工问题。但对于精细化学工业，确定了产品结构后，还需要考虑在生产过程中如何调控或组装这种结构。因此，化学工程学科在今后的主要任务是研究产品的结构-性能关系及结构调控。

自20世纪90年代开始，随着传统化学工业技术的日益成熟和化学工业重心的转移。国际化学工程界的先驱们就将化学工程学科如何进一步发展作为一个十分重要的议题进行了讨论。1996年，Villermaux在第四届世界化学工程会议的大会报告上就提出化学工程应该从"过程导向"转向"产品导向"的问题。2001年，Wei在一次Ammudson报告中提出"产品工程"可能是化学工程的第三发展阶段的标志，其核心内容是创新和设计满足人们生活需求的产品（innovation and design of useful products that people want）。Charpentier认为[4]，面向功能化学品的过程工程（process engineering adapted to specialty chemicals）是"产品工程"的学科内涵。国内外一些学者认为，"多尺度方法"是化学工程新的发展阶段的主要学科内容。尽管目前对这一发展阶段化学工程学科的内涵和标志尚无统一的看法，但回顾化学工程学科的发展历程和化学工业今后的发展趋势，可基本明确今后化学工程学科研究的若干重要特点：

（1）以产品工程为特点。过去关注过程的效能，以节能降耗减排为目标，今后将重视产品的性能，并以产品结构为核心研究结构-性能关系和结构调控[5]；过去关注反应速率、选择性、反应机理和促进反应速率与提高分子选择性的催化剂设计，今后将更多关注产品结构的涌现和演化过程速率与机制，重点关注对结构演化起促进和选择导向作用的添加剂设计。

（2）研究尺度不断微细化[6]。过去将气泡、液滴、颗粒作为一个整体，采用均质化假定，今后将重视气泡、液滴、颗粒之间的差异和内部状态变量的空间分布，从对更基本的物理化学现象如界面的热质传递、气泡/液滴内部流体的流动、相变（冷凝与蒸发、溶解与结晶）[7]、表面反应动力学等的描述着手，预测设备的整体行为和产品的性能。

（3）多尺度方法为重要手段[8,9]。过去将气泡、液滴、颗粒均质化处理，甚至采用拟均相假定，过程模型涉及的尺度基本上还是设备尺度，一些过程模型保留了非均相特点，此时模型跨越设备和颗粒两个尺度。今后对过程和产品宏观行为和性质的描述需要从分子水平出发，将跨越更多尺度。因此，针对复杂过程模型的多尺度模型化方法和计算方法将成为化学工程的核心研究内容。

（4）化学工程学科的科学性进一步增强。过去化学工程关注的尺度主要是宏观尺度，理论物理、化学关注的是原子和分子尺度，而在其间的介观尺度无论是工程学科还是基础学科都很少涉及。在介观尺度上物质及其转化具有最为丰富的结构和现象，而且绝大多数目前都得不到合理解释。今后，化学工程学科以结构-性能关系及结构调控为研究内容，将在建立新学科体系的同时，也促进物理和化学科学的发展。

参 考 文 献

[1] Davis G E. A Handbook of Chemical Engineering. Manchester：Davis Bros，1901.

[2] Bird R B. Notes on Transport Phenomena. New York：Wiley，1958.

[3] Levenspiel O. Chemical Reaction Engineering. New York：Wiley，1962.

[4] Charpentier J C. The triplet "molecular processes-product-process" engineering：the future of chemical engineering. Chem Eng Sci，2002，57（22）：4667~4690.

[5] Yuan W K. Targeting the dominating-scale structure of a multi-scale complex system：a methodological problem. Chem Eng Sci，2003，58（1）：2702~2705.

[6] 袁渭康. 我们的化学工程：关于目标尺度微细化的讨论. 化工进展，23（1）：9~11.

[7] Zhou Z M，Cheng Z M，Li Z，Yuan W K. Determination of effectiveness factor of a partial internal wetting catalyst from adsorption measurement. Chem Eng Sci，2004，59：4305~4311.

[8] Lerou J L. Chemical reaction engineering：a multiscale approach to a multiobjective task. Chem Eng Sci，1996，51（10）：1595~1614.

[9] Li J，Kwauk M. Complex systems and multi-scale methodology. Chem Eng Sci，2004，59（8~9）：1611~1612.

初议超临界条件下的化学反应*

摘 要 超临界流体有区别于气体和液体的显著性质和能力：扩散系数大于液相，溶解度远大于气体，相对密度介于气体和液体之间，它的电性质可介于"强极性和弱极性"之间。本文讨论了超临界条件对化学反应的影响：可加速反应，提高选择性，抑制催化剂结焦，改善产物品质，实现反应分离一体化等，分析了超临界反应研究尚待解决的一些问题。

关键词 超临界流体　反应过程　临界点　环境友好过程

自从 1869 年安德鲁斯发现临界点，1873 年范德瓦耳斯提出理想气体状态方程以来，对相变的研究已有 100 多年的历史，但对临界现象的广泛研究和应用还只是近 20 多年的事。超临界流体（SCF）所具有的许多专有性质的潜在应用吸引着众多的研究工作者，近年来论文发表数迅速增加。

临界点是指气、液两相共存线的终结点，此时气液两相的相对密度一致，差别消失[1]。在临界温度以上压力不高时与气体性质相近，压力较高时则与液体性质更接近，故称其为"流体"或"密相流体"，由此形成了 SCF 性质介于气液两相之间，并易于随压力调节的特点。主要表现为：有近似于气体的流动行为，黏度小、传质系数大，但其相对密度大，溶解度也比气相大得多，又表现出一定的液体行为。此外，SCF 的介电常数、极化率和分子行为与气液两相均有着明显的差别[2]，见表 1 和表 2。

SCF 的应用研究已涉及化学化工的大部分领域，具体包括超临界条件下的色谱、萃取、反应（包括生物反应、燃烧等）、材料加工、结晶和制细等技术。基础研究包括 SCF 的相行为、物理性质、超临界反应机理和动力学、传递性质和分子间相互作用等。按照操作条件又可分为亚临界、近临界和超临界区域内的研究和应用，而且该三区域内的反应行为差别较大。Brennecke[3]指出在 $1 < T/T_c < 1.1$ 和 $1 < p/p_c < 2$ 超临界范围内的研究是最有价值的，这时 SCF 的专有性质表现突出。但此范围离临界点近，SCF 性质变化大[4]，操作易出现不稳定，状态方程（EOS）在此范围有歧点，造成实验研究和理论计算的困难[5]。

表 1 超临界流体、气体、液体的性质比较

性　质	液体	超临界流体	气体
相对密度/$g \cdot cm^{-3}$	1	0.1~0.5	10^{-3}
黏度/$Pa \cdot s$	10^{-3}	10^{-4}~10^{-5}	10^{-5}
扩散系数/$cm^2 \cdot s^{-1}$	10^{-5}	10^{-3}	10^{-1}

---

* 本文合作者：郭继志。原发表于《化工进展》，2000（3）：8~13。

表 2　水在不同条件下的介电常数值

温度 T/K	压力 p/MPa				
	0.1	1.0	10.0	20.0	40.0
298.15	78.38	78.41	78.74	79.10	79.80
498.15	1.00	1.03	31.11	31.58	32.43
648.15	1.00	1.02	1.28	2.23	13.62
748.15	1.00	1.02	1.19	1.49	3.05

1　超临界反应的基础研究

1.1　超临界流体相行为研究

Subramaniam 等人[6]指出：不弄清研究物系的相行为或相态变化规律就进行所谓 SCF 下的反应研究是不会有多大价值的。只有确保反应是在超临界状态下进行的，才可冠以"超临界反应下的结果"。Brennecke[3]对 SCF 相平衡作了系统的应用分析，提出将 SCF 作为密相气体或膨胀液体处理的模型，并指出状态方程对临界点和临界区计算的局限性，尤其对于不对称混合物组成的物系，难于找到适应性比较好的混合规则。有的作者甚至认为状态方程对临界点的计算是失败的。近来许多作者[3,10]对 SCF 密度、极性、溶解度、相平衡、溶剂相互作用等利用分子动力学[7]和蒙特卡罗[8]等计算机模拟方法[9]作了大量工作，但结果仍难以满足要求。寻求新的和准确的模型方程及计算方法是预测 SCF 相行为和进行 SCF 反应研究的重要保证。

1.2　超临界反应常用的流体介质

SCF 反应可分为两类：一类是反应物系处于超临界状态下[11,12]；另一类是反应混合物处于一种惰性 SCF 介质中。超临界反应介质的选取依据尚处于探索之中，表 3 列出了一些常用的反应介质。其中研究最多的是二氧化碳和水。二氧化碳具有临界温度接近室温，临界压力不高、不活泼、无毒、对环境无污染等特点，是超临界反应的理想介质。水的临界温度和临界压力较高[13]，操作和经济上不如二氧化碳有竞争力。但水从室温到临界状态，介电常数可在 78.38 到 2.23 范围内变化（表2）。因此超临界水可溶解有机物，为有机物和植物的热解、废物和废水的超临界氧化提供了良好的环境，无机盐却因溶解度低而可以自动分离。

表 3　某些反应使用流体的临界性质

流　体	T_c/K	p_c/MPa	流　体	T_c/K	p_c/MPa
乙烯	282.4	5.03	叔丁醇	506.2	3.97
三氟甲烷	299.1	4.87	正己烷	507.4	2.97
二氧化碳	304.2	7.37	丙酮	508.1	4.70
乙烷	305.4	4.88	甲醇	512.6	8.09
丙烷	369.8	4.24	乙醇	516.2	6.38
氨	405.6	11.27	甲苯	591.7	4.11
甲胺	430.0	7.46	水	647.3	22.04
1-己烯	504.0	3.17			

1.3 超临界反应研究的内容和对象

当前,超临界反应的研究基本上是基础理论研究和对具有潜在工业应用前景的反应过程的探索。前者主要是超临界反应的机理和动力学,以及 SCF 的特殊性质(图1和图2)对反应动力学的影响,如描述基元反应的过渡状态原理,超临界反应平衡和选择性,活化体积和压力的动力学关联,介电常数动力学关联,以及溶剂相互作用理论和分子间相互作用理论等。后者则包括超临界下的均相和非均相催化与非催化反应、聚合反应、废物处理、煤和生物高分子转化为燃料和化学品、SCF 的燃烧、材料加工与合成,以及酶催化反应和电化学反应等。

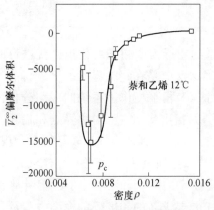
图1 临界点附近的偏摩尔体积变化

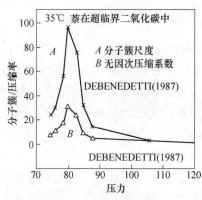
图2 分子簇涨落现象

1.4 超临界条件下的反应动力学

1.4.1 过渡状态原理

通常认为,SCF 在反应过程中不像液体体积基本不变,也不同于气相(密度比气相大得多)。SCF 的性质随压力可调性较大,因此传统的动力学方程中忽略压力影响的假设已难以适应,但如何将压力或其他 SCF 的性质引入动力学方程,成了超临界反应研究的一大难题。过渡状态理论是一个既方便又适应性强的基元反应理论,许多学者利用它描述了超临界反应速率常数和压力、活化体积等因素的关系[14]。并认为活化体积由两类贡献组成:一类是物质分子的力学结构信息,键长和键角;另一类是电伸缩(electrostriction)和其他溶剂效应。Wu[15]发展了这一理论,认为 SCF 的介电常数、扩散、静电相互作用和 SCF 的可压缩性及相行为在近临界点时变化较大,均可认为是表观活化体积的贡献。液相反应的典型活化体积一般不大于 $30 cm^3/mol$,但近临界点反应的表观活化体积可达 $\pm 1000 cm^3/mol$ 的数量级。表观活化体积在一定程度上反映了物质在近临界点时的反应特性,但由于它是一个多因素的综合结果,难以分清究竟是哪些因素起主要作用。

1.4.2 溶剂效应

利用过渡状态原理和分子热力学理论可以建立超临界反应速率常数和活度系数及溶解度参数 δ 的关系。利用溶剂溶解度参数表征溶剂效应,得到了超临界反应速率常数与溶剂溶解度参数呈线性关系的结果,并在超临界水下的热解反应中得到了证实。而溶解度参数随密度变化,因此可以用密度来调节超临界反应。有些研究表明溶剂的介电常

数 ε 与偶极矩 μ 同超临界反应存在一定的关系，尤其对含极性反应物的超临界反应[16]：

$$\ln k = \ln k' - (N_{av}/RT)[(\varepsilon-1)/(2\varepsilon+1)](\mu_A^2/r_A^3 + \mu_B^2/r_B^3 - \mu_{M^*}^2/r_{M^*}^3)$$

式中，k' 是单位介电常数下的速率常数。上式表明，溶剂极性大有利于过渡态产物极性大于反应物极性的反应，由此加盐可改变超临界反应。Yang[17] 研究了溶剂的"笼子"效应：溶剂分子对反应物分子的"包敷"作用。它使高密度高黏度（低扩散）溶剂对超临界反应的影响加剧，尤其对扩散控制反应。

2 超临界流体的物理和化学现象

超临界反应的物理和化学现象，尤其是近临界现象一直是研究的主要对象之一。首先是近临界点"异常"现象问题[4]：Krichevskii[18] 实验表明在近临界区时，反应平衡和动力学有显著变化（图3），温度影响也增强。而 Procaccia[19] 却得出了某些反应的反应速率会在接近临界点时下降的结论，但 Milner[20] 对此提出了怀疑。Randolph[21] 提出对于稀溶液，近临界点区域内的反应活化体积较大，高浓度，远离临界点的区域内较小。Chialvo[22] 对此进行了分子动力学模拟，提出了溶剂和溶质的"吸引"和"排斥"对活化体积的影响理论："分子簇"理论和"笼子效应"，认为由于"分子簇"形成的局部浓度提高和"笼子效应"均会对超临界反应过程发生作用，但对扩散控制的反应例外。这种影响的强弱取决于化学反应的"时间尺度"和"分子簇"的"时间尺度"的相对大小。

超临界条件下的反应选择性。超临界条件下压力和黏度可以影响某些顺-对位反应的选择性（图4），或某些分解反应的途径。SCF 的溶剂效应可以影响异构化反应的机理，还可对某些反应的中间态发生作用（稳定或促进）。Hrnjez[23] 的工作表明，SCF 可以改变化学反应的立体选择性和配位选择性，并认为这是由压力引起的溶剂极性变化所致。还有许多作者利用激光热解双分子反应[24]、电子顺磁共振谱[21] 和分子动力学[25] 对"分子簇"的作用和分子碰撞模型的内在机理作了大量的工作。Kimura[26] 研究了 SCF 的性质对超临界反应平衡的影响。在 2-甲基-2-亚硝基丙烷的二聚可逆反应中，平衡常数在 $\rho_r<0.3$；$0.3<\rho_r<1.4$ 和 $\rho_r>1.4$ 三个密度区域内随 ρ 的增大分别表现为增大、减小、再增大的不同变化，解释为分子间的吸引和剂-质相互作用、多体相互作用和"包敷"作用分别占主导地位所致。Peck[27] 的研究认为对于可逆反应，极性超临界溶剂有利于反应朝极性化合物的方向移动。此外还有用压力调节极性，改变氢键；溶解度与反应平衡常数的关联；以及超临界反应平衡和相平衡方程联立求解等进行研究，在此不作赘述。

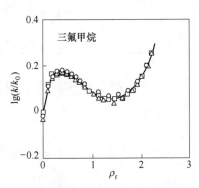

图3 2-甲基-2-亚硝基丙烷在超临界氟仿中反应平衡常数的变化

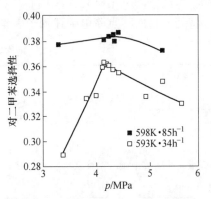

图4 对二甲苯在近临界点的选择性

国家重点反应工程实验室自1989年开始研究超临界苯烃化以来，主要进行了如下工作：超临界天然植物的萃取；超临界活性炭再生；超临界抗溶剂法制备分布较窄的颜料微颗粒；超临界苯烃化系统的临界点测定、近邻界区相行为测定与计算模拟、超临界流体的激光测试以及超临界苯烃化反应等。在超临界苯烃化反应方面得到了如下主要结果：超临界流体可以延缓苯烃化反应催化剂表面的焦前体沉积[28]，从而延长催化剂的寿命，这是充分利用了超临界流体对焦前体的较强溶解能力；利用实验测定的近邻界区苯和乙烯的相行为数据进行了液相、气相、气液共存、超临界和近邻界区条件下的苯烃化反应行为研究（图5是实验操作条件和苯与乙烯的相平衡线的关系分布图）。实验表明各条件下的反应速率差别较大，在近临界区反应速率存在极大值（图6）[29]。速率对操作压力的非常规关系提示我们在考虑影响超临界反应的因素时，应对反应过程中的相行为变化给予足够的重视。利用扫描电镜、X射线衍射等手段考察了不同相区的反应操作条件对催化剂的影响，结果表明近临界区的气液两相操作对催化剂存在较大的破坏作用，压力过高对催化剂也是不利的。苯与丙烯的超临界反应以及相关方面的研究正在进行之中。

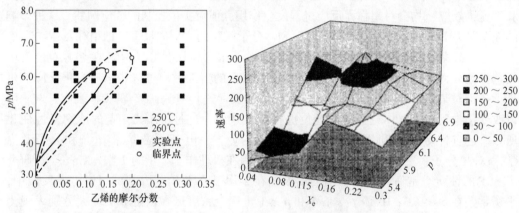

图5　乙烯-苯进料近临界区实验点分布

图6　260℃下的烃化速率$\times 10^2$与压力p（MPa）和乙烯的摩尔分数X_e的关系

3　超临界反应的应用研究

超临界反应的应用研究涉及内容广泛，研究方法和手段以及目的也不同，为简单起见，将其各自的特点简单阐述列于表4中。

表4　超临界反应的可能应用

反应物系	研究对象、内容、分析和结论
Diels-Alder反应	超临界CO_2中异戊二烯与马来酸酐反应。其二阶反应速率常数随压力增大而增大。压力可影响环化加成选择性，压力在近临界点时效应显著[30]
有机金属化合物反应	用SCF（CO_2或X_e）代替水作溶剂。其特点是：SCF可提高有机金属化合物的溶解度，从而提高反应速率和有用产物浓度；SCF使反应物系均相化，使催化剂和产物容易分离；SCF使NMR分析更容易[31]
异构、烷基化反应	SCF可原位萃取焦前物，抑制催化剂结焦。近临界点的反应结果显著不同。SCF还可加速传质，使反应和分离一体化[11,28,29]

续表 4

反应物系	研究对象、内容、分析和结论
F-T 合成反应	利用 SCF 的传热快、溶解度高的特点，可解决原气相反应的热迁移，原液相反应的大分子烃堵催化剂孔的问题，可改善由温度分布引起的选择性问题[32]
酶催化反应	利用 SCF 原位分离出产物；调控优化反应系统含水量，强化酶催化反应或改善立体选择性。密度、溶解度参数和介电常数对酶催化反应的影响显著[33]
其他非均相催化和非均相反应	重油加氢反应中用脂肪族烃作 SCF，比用芳烃为优（耗氢少）；超临界条件下加氢比液相加氢产品的气体少，不易结焦，而且适当提高压力可补偿温度的降低。某些气固反应在 SCF 条件下进行，可加快产物的脱附速率，进而提高反应速率。应用混合 SCF 介质，可由沥青或木沥青生产高质量碳电极[2]
燃料加工	SCF 水对油页岩"湿热解"可降低不饱和烃含量，减少生焦和气体裂解；SCF 可用于煤液化，反应萃取，脱除硫等杂原子。SCF 同煤在升温前或后混合的产物分布不同，混合超临界溶剂效果明显[34]；研究煤的模型化合物表明介电常数大的介质，有利于稳定过渡状态化合物而提高反应速率；超临界水作介质时可用加盐来调控超临界水的极性，有利于亲电和亲核反应，有时比调压更有利
氧化反应	主要是超临界水处理废有机物：高温加速了反应；超临界水能溶解有机物形成均相，加速传质；盐难溶于超临界水可自动析出；可降低某些温度较高的氧化反应温度。但反应器昂贵，易磨损，且反应器的材质和新旧会影响反应结果。超临界氧化燃烧、烷基芳烃和烷烃部分氧化均有很好应用前景[6,13]
生物大分子的转化	用超临界进行脱木质作用、降阶、液化、脱水、反应分离，可提高产率，改善相对分子质量分布[2,35]
材料加工	利用 SCF 可实现多分散聚合物的分级（窄分子量分布），改变聚合物的形态，制造高水平多孔介质和窄的粒子大小分布等。操作条件可在聚合物单体或惰性 SCF 的临界点以上；SCF 的高压低黏有利于加成反应生成大分子产物；溶解度大和易调节性，使脱挥容易；超临界解聚反应和由聚合物再生成聚合物的反应选择性好、产物纯度高，优于纯热解过程；SCF 用作聚合物材料的合成（氢化铁），易于控制粒子大小、结晶结构和产物形态，并可简化流程[36]

4 超临界流体条件下的反应特点

超临界反应研究发展迅猛，涉及领域广泛，是由 SCF 的独特性质所决定的。SCF 性质的压力可调变性，以及加盐对它的电性质的影响，为在超临界反应过程中改变 SCF 性质，实现调控反应的目的奠定了基础，亦为开发超临界反应新技术带来了机遇和挑战。SCF 对反应的作用主要体现在：（1）利用 SCF 中扩散系数介于气液之间，加速液相中的扩散控制反应。（2）利用 SCF 的溶解度比气体大，又具有液体所不具备的溶解选择性，提高反应物浓度，加速反应，使反应均相化或反应分离一体化。（3）SCF 的大部分性质是密度的强函数，而密度又可利用压力（或温度）调节，进而调控反应的速率、平衡和选择性。（4）SCF 的电性和极性与气液差别较大，且可加盐调节，由此可改变某些含极性化合物或过渡中间态生成极性化合物的反应。（5）SCF 的混合物对反应具有协同效应。

5 超临界反应研究可能存在问题和应用前景

尽管超临界反应研究发展迅速，涉及领域广泛，但无论基础研究还是应用研究基本上仍处于对具体过程的探索和积累阶段。1997 年美国 13 届热物理性质会议和 1997 年日本第四届国际 SCF 会议也说明了这一点。传统的反应动力学形式已不再适用，如何建立超临界反应的动力学方程就成了基础研究的重要目标。现有研究中虽然有一些动力学关联式，但适应范围有限，而且对超临界反应的内在机理并不很了解，所以大部分的"异常"现象无法解释，定量描述就更困难。对超临界反应过程的计算模拟，分子设计是一个当前研究的薄弱领域，仅用分子动力学 MD 和蒙特卡罗 MC 模拟，尚不能阐述动力学问题。相行为对超临界反应的影响尚没有引起足够重视[29]。在近临界点附近 SCF 的性质变化大，状态方程适应性降低。目前尚没有针对 SCF 特点的新反应器形式。对非均相超临界催化反应，吸附和脱附的动力学行为，催化剂孔的大小和孔的形状的影响报道也较少。此外超临界反应究竟可以应用在哪些具体的物系，超临界反应器的技术关键在哪里，如何弄清诸因素（包括介电常数、扩散系数、盐效应、溶剂效应、分子几何构型）分别在超临界反应中起什么作用等问题均值得深入研究。

以应用为导向的研究应侧重于：不同反应系统的 SCF 介质的选取方法，SCF 与反应物的作用关系。寻找能更好地利用 SCF 性质特点的反应系统，实现经济效益和社会效益的统一。充分利用 SCF 的密度、溶解度随压力可调的特点，改造使用有毒介质的"三废"严重的工业过程。石油加工中利用超临界反应减少气体量，抑制结焦，使催化剂活性稳定，改变油品的相对分子质量分布，提高油品等级，减少污染，超临界水是一个超临界氧化处理"三废"物质的极有潜力的介质，但氧化机理、工艺条件、经济性等值得深入研究。超临界二氧化碳是使用广泛的环境友好的反应介质，探索其在高分子材料、生物制品、医药、食品等领域的应用新工艺，具有很好前景。除原有两类超临界反应体系外，可考虑在反应混合物体系中加入少量惰性（超临界）介质，以改变混合物的理化性质（如降低反应混合物的临界温度和压力），从而实现调控反应的目的。

致谢

本研究得到国家自然科学基金委和中国石化集团公司支持（合同编号 29792077）。

参 考 文 献

[1] 沈忠耀. 超临界流体. 化工百科全书（第 2 卷），北京：化学工业出版社，1991：325～339.
[2] Savage P E, Gopalan S, Mizan T I, et al. AIChE J, 1995 (41)：1723.
[3] Brennecke J F, Eckert C A. AIChE J, 1989 (35)：1409.
[4] Morrison G. Physical Review A, 1984 (30)：644.
[5] Siepmann J L, Karabornl S, Smit B. Nature, 1993 (365)：330.
[6] Subramaniam B, McHugh M A. Ind Eng Chem Proc Des Dev, 1986 (25)：1.
[7] Cummings P T. Fluid Phase Equilibria, 1996 (116)：237.
[8] Panagiotopoulos A Z. Fluid Phase Equilibria, 1996 (116)：257.
[9] Balbuena P B, Johnston K P, Rossky P J. J Phys Chem, 1995 (99)：1554.
[10] Kontogeorgis G M, Smirlis I, Yakoumis I V, et al. Ind Eng Chem Res, 1997 (36)：4008.
[11] Li F, Nakamura I, Ishida S, et al. Ind Eng Chem Res, 1997 (36)：1458.
[12] 王涛，艾大刚，李成岳. 化工进展，1992 (11)：17.
[13] Holgate H R, Tester J W. J Phys Chem, 1994 (98)：810.
[14] VanELDIK R, Asano T, LeNoble W. J Chem Rev, 1989 (89)：549.

[15] Wu B C, Klein M T, Sandler S I. Ind Eng Chem Res, 1991 (30): 822.
[16] Townsend H S, Abraham M A, Huppert G L, et al. Ind Eng Chem Res, 1988 (27): 143.
[17] Yang H H, Eckert C A. Ind Eng Chem Res, 1988 (27): 2009.
[18] Krichevskii I R, Tsekhanskaya Y F, Rozhnovskaya L N. J Phys Chem, 1969 (43): 1393.
[19] Procaccia I, Gitterman M. Phys Rev A, 1983 (27): 555.
[20] Milner S T, Martin P C. Phys Rev A, 1986 (33): 1996.
[21] Randolph T W, Carlier C. J Phys Chem, 1992 (96): 5146.
[22] Chialvo A A, Debenedetti P G. Ind Eng Chem Res, 1992 (31): 1391.
[23] Hrnjez B J, Mehta A J, Fox M A, et al. J Amer Chem Soc, 1989 (111): 2662.
[24] Roberts C B, Zhang J, Chateauneuf J E, et al. J ACS, 1993 (115): 9576.
[25] Randolph T W, O'Brien J A, Ganapathy S. J Phys Chem, 1994 (98): 4173.
[26] Kimura Y, Yoshimura Y J. Chem Phys, 1992 (96): 3824.
[27] Peck D G, Mehta A J, Johnston K P. J Phys Chem, 1989 (93): 4297.
[28] Gao Y, Liu H Z, Shi Y F, et al. The 4th International Symposium on Supercritical Fluids, Vol. B, May 11-14, Japan, 1997: 531~534.
[29] 石一峰. 近临界区苯与乙烯的烷基化 [D]. 上海: 华东理工大学, 1998.
[30] Ikushima Y, Saito N, Arai M. J Phys Chem, 1992 (96): 2293.
[31] Jessop P G, Ikariya T, Noyori R. Nature, 1994 (368): 231.
[32] 闫世润, 张志新, 周敬来. 天然气化工, 1996 (21): 47.
[33] Kamat S V, Iwaskewycz B, Beckman E J, et al. Proc Nat Acad Sci USA, 1993 (90): 2940.
[34] Adschiri T, Shinji A, Arai K. Proc Int Symp Supercritical Fluids, 1991: 408.
[35] Hirth T, Franck E U, Ber Bunsenges. Phys Chem, 1993 (97): 1091.
[36] Scholsky K M. J Supercrit Fluids, 1993 (6): 103.

《化学反应工程分析》前言*
——化学反应工程方法论讲座（第一讲）

化学反应工程是化学工程学科的一个重要分支。经历了大约半个世纪的发展，化学反应工程已成了四肢健全、羽毛丰满的成熟学科了。

但是"成熟"二字有时也含贬义，即有已近暮年之意。在高科技受到人们特别重视的今天，化学工程被归入传统学科一类。"传统"可理解为是"新兴"的反义词，其含义大体是，经过一段时间发展的传统学科，其前景已远不如那些新兴学科了。

但不论成熟也罢，传统也罢，一门学科一旦形成，它总会顽强地表现出自己的生命力，何况化学反应工程直接关系到用好资源，保护好环境，与国计民生关系密切，又是生物工程和材料科学等的重要基础，据此，我们可以认为化学反应工程还是要发展、要深化的。

1 化学反应工程——化学工程学科发展里程的一个标志

学术界普遍认为，化学工程学科的发展始自 19 世纪末。早期化学工程学科的工业背景是化学工业（可能还有冶金工业），显然，我们今天的理解已绝非仅限于此。今天的化学工程，已成为过程工程的核心。

在发展的早期，化学工程被简单地看做是单元操作，仅限于物理过程，并不包括化学反应工程这一独立的分支。原因显然并不在于化学工业不涉及化学反应，而是由于当时还没有把伴有化学反应的过程提炼为一门学科的科学基础。单元操作就方法而言，主要是归纳法，即把物理量归纳为无因次数，并用实验关联其关系。对伴有化学反应的过程，由于多数具有很强的非线性性质，归纳法已不适用，但当时还缺少知识的积累，也没有进行数值分析的必要工具——计算机，因而不得已将伴有化学反应的过程排除在化学工程学科的范围之外。这一阶段一直延续到大约 20 世纪的五六十年代，现在被学术界公认为化学工程发展的第一个里程。

学术界普遍认为化学工程发展的第二里程始于 20 世纪 60 年代初出现的传递过程作为一个单独的学科分支时。事实上，化学反应工程作为一个重要的学科分支也几乎在同时形成，共同标志着化学工程学科发展进入了第二个里程。第二里程与第一里程的明显区别在于：

（1）观察问题的尺度从单元操作的设备尺度（如塔、热交换器、塔板）转入液滴、气泡、颗粒尺度，也就是更深入到过程的本质行为。

（2）计算技术的迅速普及已使原来无法解决的非线性分析成为可能，从而传统的归纳法已不再是主要可用的研究方法，而代之以演绎法为基础的模型方法。显然，对传质传热和流体流动等基础知识的大量积累在由第一里程向第二里程的过渡中起了重要作用。

---

* 本文为袁院士为《化学反应工程分析》一书所撰写的前言。

化学工程是一门传统学科。化学反应工程虽是一门相对年轻的学科分支，但总体上毫无疑问地应属于传统学科。当然，传统学科仍应深化和发展，包括学科的完善、拓展，以及与其他学科不断的交叉并形成新的研究方向。自从化学工程学科的第二里程形成以来，近半个世纪过去了，学科出现某些老化也在情理之中。关键的问题是如何不断使传统学科萌生新的活力。与新兴的生物、新材料、信息等学科相比，化学工程似乎显得不那么具有活力。但我们还是相信，作为国民经济重要支柱的过程工业的学科基础，作为生物工程和材料科学的支撑学科，化学工程的重要性绝不能低估，其中化学反应工程是一个关键的分支。

国际学术界十分重视化学工程学科的前途。最早系统研究化学工程前沿的，可能是1988年美国国家研究委员会编写的那份研究报告[1]。人们也十分重视什么是化学工程的第三里程，虽然至今尚无定论，但也已提出了一些有前瞻性的观点。"里程"毕竟是有回顾性质的，就像我们当初也不见得已意识到第一里程和第二里程的到来一样。重要的是，我们的同行应对学科的发展保持足够的敏感，尽可能使我们的工作处在前沿。

2 化学反应工程：理论体系和实用化方法

化学反应工程的传统内容包括化学动力学、混合与返混、催化剂外部和内部的传热和扩散、相内外的传递、反应器的稳定性和参数的敏感性，以及后来得到延伸的反应器的控制和优化。复杂反应系统的简化处理方法，如集总方法等，也曾为石油化工这样的复杂对象起过重要作用。

本书作者对化学反应工程的理解是，经过约半个世纪的发展，它已形成了自己完整的学科体系：有特色的理论基础和有效的实验研究方法，以及关于处理复杂实际反应系统的指导思想等，其理论的科学性是毋庸置疑的。论及存在的问题，最突出的可能是化学反应工程的理论几乎都建立在最简单的反应体系的简化假定之上，这些简化假定与复杂的实际过程相去甚远。同行们虽然也为如何处理实际反应体系做了种种努力，但总嫌不足。一方面，我们承认目前还没有足以把实际反应体系进行详尽描述的能力，另一方面，恐怕也应认识到不见得有必要一定要做到这一点，因为正确运用这些在简化假定基础上建立起来的理论，加上我们已掌握了在充分理解理论基础上的判断、鉴别和处理问题的能力、方法和技巧，通过合理有效且有针对性地组织实验工作，是可以解决绝大多数实际问题的。

与任何学科领域一样，化学反应工程作为一门工程学科，其理论体系首先是以建立各种变量之间严格的定量关系为基点的。众多的化学反应工程研究工作者为此做了十分出色的工作，使得我们今天看到的本学科在理论上是如此完善。例如，反应过程的非线性性质是反应工程中表现得非常突出的一个问题。这主要是由反应体系速率与温度之间所存在的十分复杂的非线性关系决定的。正因为过程的非线性性质，系统行为就显得捉摸不定：不但不能外推，甚至连内插都可能会有问题。人们几乎用尽了非线性分析中最有效的方法于化学反应工程，尽管结果令人赞叹不已，但这些理论研究结果的直接应用却有相当困难。理由已如上述，就是我们所遇到的实际过程远较理论研究时所作的假定复杂，或者有些参数的准确确定基本上是不可能的。当然，我们绝不会由此低估理论的意义，相反，理论是如此重要，只有在充分地理解反应工程，有能力灵活地运用理论的前提下，我们才能建立正确的工作方法论，以解决各种千差万别的实际工程问题。

因而,我们曾经把化学反应工程从学科到应用归纳为"科学、技术和艺术"[2]。科学是指从各种尺度研究化学反应和传递过程相互影响所获得的客观规律的认识。技术则是指解决实际工程问题时形成的各种比较成熟的方法,如催化剂制备技术、各种反应动力学实验技术和技巧、冷模实验技术和流体均布技术等。艺术则指在反应工程理论指导下,充分利用具体过程的特殊性,处理千变万化的实际问题的技巧,如简化与分解、一定条件下的线性化、因素的敏感性分析等,正因为这是将复杂问题简化处理却不导致明显偏离过程本质与特征的工作方法,它更加依赖于研究者的经验、才干、素养,甚至灵感,需要大胆的构思和想象[3],有点像艺术家作素描或印象处理时的情景。

回到本书的写作宗旨。正因为上面说到的种种关系,我们不作细微的理论分析和周到的内容罗列,而重"分析",并以此二字定名本书。

3 化学反应工程——深化和延拓

任何学科,不论是传统学科或新兴学科,都有不断深化的需要。任何人都不能说作为传统学科的化学工程已到了不再需要发展的境地。事实上,我们确也碰到过诸多问题,亟须深化认识。以常见的多相反应系为例,至今还有不少问题显得扑朔迷离,难以预测,而这些却是相当有实用价值,且颇为常见的。这些都属于传统化学反应工程的范围。

同样十分重要的是,传统外的技术领域与化学反应工程的关系十分密切,使本学科广有用武之地。且不说早已为人们所知的半导体制备技术中广泛采用的典型化学反应技术——化学气相淀积(CVD)法,化学反应工程的应用,特别是在材料科学和生物工程中的应用不胜枚举,有十分诱人的前景。化学反应工程传统上作为过程工程支撑的另一个话题似乎不得不在此一提。在市场经济迅速发展的今天,"市场"这一名词已越来越多地被人们提及和议论,市场也必然影响到学科方向。市场问题,必然牵涉到消费者。作者认为,信息业近年来受人重视,关键之处就是该行业不断向市场推出新的产品,而这些产品正是消费者所需要的,受最终用户欢迎的。与其相比,应该承认以化学工业为代表的传统过程工业确实望尘莫及。几年前,国际学术界就有人提到化学工程应该从过程导向朝产品导向发展,正是因为看到了这一点。

由此提出的一个问题是,化学反应工程原是一个以过程导向的学科,对产品工程能有多少贡献?如果回过头来再读一遍1988年的那份战略报告[1],我们就会发现,实际上那时已把问题提得很清楚了:我们已不宜坚守传统的过程工程范围,而着眼于一些传统外的问题;化学工程应更多地面向科学(化学),借助科学来找研究方向;与其他学科的交叉领域势必是最有前途的,如界面现象等。我们的理解是:

(1) 随着学科的发展,有一个趋势十分明显,也就是考察现象的尺度必然是越来越小。上面所说化学工程从第一里程进入第二里程的一个标志,就是观察尺度从设备尺度转向滴、泡、颗粒尺度。当我们还在思索第三里程究竟是什么的时候,我们应该注意到观察工具已允许我们着眼于更细小的,更靠近事物本质的尺度——纳米尺度、分子簇尺度,甚至分子原子尺度。事实是我们已得益于此,如在纳米碳纤维制备中的形貌控制正是需要得益于将观察尺度变得很小。高分辨率电镜提供了极好的观察手段。

(2) 只限于对一般的化学反应过程进行研究并得到应用的看法已不能适应,而应有针对性地与一些专门的产品或专门的技术的研究交叉起来，如应特别重视与材料科学、生物工程及环境工程的交叉。可以说这方面我们将会有十分广阔的施展余地。如在极端条件下，利用聚合物性质的变化对其进行化学接枝的技术是对大宗产品利用的一项补充，可以增加有特殊性能的产品品种。

大宗化学品的制备技术历来是化学工程的研究重点，但现在形势显然起了很大的变化。多品种小产量的产品必将越来越受重视。在西方发达国家，很多大宗产品已趋饱和，并且受到了资源和环境的限制，不可能有多大的发展，而各种产量较少的专用品却正在大量发展之中。在未来的年代里，必然会在药物和营养品、农用化学品等方面有大的发展。

学科的界限多是比较模糊的，或互相渗透的，特别是到了高度发展的阶段更是如此。我们已大可不必拘泥于我们的某一项知识是从物理书上来的或是从化学书上来的，重要的是正确理解知识和正确掌握方法。当一位学者的学识和修养已是炉火纯青时，相信他已说不清用来解决问题的某一种方法来自何方，可能是数学的、化学的，也许是经济的，或是艺术的。这意思是说，我们今天在学习某一门知识的时候，更多地应把注意力集中在方法、理念和思想上，而不在这项内容是否属于化学反应工程的范围。

说了这些，再回到本书的写作宗旨。传统上有那么多问题，有的已解决了，有的尚未解决；有那么多内容，在文献上不知还可以找到多少，又有那么多有待发掘、开拓和研究的，我们这本薄薄的书又能给予读者多少？我们只是试图借内容，讲分析；借细节，讲理念；借问题，讲方法。希望读者能在理解内容的基础上，着重了解分析问题的方法。读者如能带着这样的观念去阅读本书，或许不无裨益。

参 考 文 献

[1] National Research Council. Frontiers of chemical Engineering [M]. Washington: National Academy Press, 1988.
[2] 袁渭康，郭慕孙. Reactor Engineering: Science, Technology, and Art [J]. Ind. Eng. Chem. Res., 36 (8): 2910 (1997).
[3] 陈敏恒，袁渭康. 工业反应过程的开发方法 [M]. 北京：化学工业出版社，1985.

《化学反应工程分析》后语[*]
——化学反应工程方法论讲座(第二讲)

本书作者从事化学反应工程的研究和教学工作多年,深感化学反应工程发展至今,已成为一门比较完整的学科,形成了其独特的理论体系。但也应看到,化学反应工程的理论通常是建立在简化的基础上,与实际过程相去甚远。例如,我们在研究一个反应过程时,往往假定是一个简单反应,有时甚至简化到 A→B 这样在实际中几乎不可能遇到的反应。因而当我们在处理远较我们已学过的理论更为复杂的实际问题时,还需要我们具备能联系实际问题的复杂性,并从这些基本的化学反应工程知识找出简便的、可靠的解决问题的途径的能力。这种在掌握理论基础上的解决问题的技巧,就构成了我们常说的"方法论"。

我们的一些学生,虽然完成了化学反应工程方向的博士论文,但往往还会在一些基本问题的处理上束手无策,或犯一些不应犯的错误,然而他们却能通过计算机进行一些复杂的偏微分方程的数值运算。析其原因,还是在于他们在掌握知识的深度方面和熟练运用知识的能力方面尚稍逊一筹。实际上本书的编写意图也正是希望能在解决这一问题方面前进一步。

在写完全书初稿以后,我们还是感到意犹未尽,因而萌生了再写一个"后语"的念头。这个"后语",主要是写一些在前面各章的系统叙述中难以写入,或不宜穿插的带有一些共性的观点,或者说是对一些基本概念的补充:一些理解上和方法上的补充。

1 速率过程和特征时间

化学过程在很多方面表现出速率过程的特点。我们几乎处处时时都在讨论速率,因为没有速率,过程就失去了它的实际意义。

以间壁传热为例,流体从管外向管壁传热,热量通过外壁导向内壁,再传给管内流体。这一过程是由三步传热过程所构成的,因而又称串联过程。对于这样一个串联速率过程,如果是定常的,则每一步传热过程速率相等,没有孰快孰慢之分,只有热阻大小差异。内外流体的温差决定了传热总推动力,而各步所分配到的温差是由热阻大小决定的。热阻大的温差大,热阻小的温差小,热阻大的是传热这一速率过程的控制步骤。我们通常为了简便而说的这一步慢那一步快,至少是一种不严密的说法,因为每一步速率都相等。因此,如果要比较过程速率的快慢,必须是在同样推动力的前提下才是合理的。或可用我们后面讨论的特征时间来表示速率:小的特征时间表示速率高,大的特征时间表示速率低。特征时间基本上排除了强度因素(本书中主要指浓度和温度)的影响;只有对于非线性过程,强度因素才影响特征时间。

1.1 催化反应——一类串联速率过程

在化学反应工程中串联速率过程也十分常见。反应物从流体主体向催化剂外表面

---

[*] 本文为袁院士为《化学反应工程分析》一书所撰写的后语。

传递，然后通过催化剂内部孔道在催化剂内传递，主要在内表面进行催化反应。反应产物则以相反的方向向外传递，到达流体相主体。如果反应伴有热效应，则视反应为吸热或放热，决定传热方向。对于气液反应，或气液固反应，所表现出的串联速率过程从原则上看也是类同的。即使是均相反应，实际上也有类似问题，只是看不到相界面而已。最根本的一点就是各步分配到的推动力是取决于其阻力的。

对于化学工程师来说，重要的是了解串联过程中的哪一步是控制步骤，进而可以设法研究减少这一控制步骤影响的途径，使这一速率过程趋向合理。当然有时我们也会有意识地促成某一步为控制步骤而使其对过程有利，如一些部分氧化反应，氧更有利于副产物的形成，这时控制催化剂表面氧浓度就有利于反应的选择性。

既然是关于速率过程的学科，速率就是一个关键因素。速率必然涉及时间，或者说是在单位时间内传递的热量、物质、消失的反应物、或生成的产物。为了能将上面说到的串联速率的各步骤的相对关系作一个定量的描述，我们常常引入特征时间的概念。对于化学反应工程，有几个特征时间是有特别重要意义的：

扩散时间 $\qquad t_D = \dfrac{L^2}{D_A}$

传热时间 $\qquad t_\lambda = \dfrac{\rho c_p L^2}{\lambda}$

反应时间 $\qquad t_R = \dfrac{c_{A0}}{k c_{A0}^n} = \dfrac{1}{k c_{A0}^{n-1}}$

除此以外尚有物料的停留时间 τ_{res}。它虽不是速率过程的特征时间，但却代表了一种流动特征：

停留时间 $\qquad \tau_{res} = \dfrac{V_R}{q_V}$

上面几个特征时间的物理意义可解释为：t_D 为组分 A 的分子扩散途径 L 长度所需的时间；与扩散时间类似，t_λ 为热量通过传导 L 长度所需的时间；t_R 则代表了反应速率的一个特征。如为简单反应 A→B，反应为 n 级，反应时间 t_R 应看做物系中可能的最高浓度 c_{A0} 与可能的最高反应速率 $k c_{A0}^n$ 之比。显然，特征时间长的过程，就是我们在非严格意义上所说的"慢"过程。

我们从串联速率过程的观点来观察特征时间，问题就更为清楚。以气体在颗粒状固体（为简单计，此处假定为球形颗粒）催化表面进行反应为例，系由催化剂外部传递（以及在外表面的反应）和催化剂内部过程（传递和反应）构成。这两个过程的特征时间可见表1。

表1 催化剂内外传递和反应的特征时间

名 称	催化剂外部	催化剂内部
反应时间	$t_R = \dfrac{1}{k c_{A0}^{n-1}}$	$t_R = \dfrac{1}{k c_{A0}^{n-1}}$
传质（扩散）时间	$t_{De} = \dfrac{1}{k_g a} = \dfrac{L}{k_g}$	$t_{Di} = \dfrac{L^2}{D_e}$
传热时间	$t_{\lambda e} = \dfrac{\rho c_p L}{h_f}$	$t_{\lambda i} = \dfrac{\rho c_p L^2}{\lambda_e}$

1.2 几个有关特征时间的认识

对于一个串联速率过程几个组成步骤的速率，只在同一个推动力的前提下进行比较才有意义，才能辨别其相对大小。特征时间是以过程可能的最大推动力为前提的：反应时间以系统中的最高浓度计（对催化反应为气相主体中的反应物浓度），外部传质时间以外表面浓度为零为基础，内部传质时间以外表面浓度与主体浓度相等为基础。

（1）反应时间的定义为系统中反应物 A 的最高浓度（此处为气体主体中 A 的浓度 c_{A0}）与可能的最高反应速率之比。如后者以幂函数动力学表示，即为 $r_{A0} = kc_{A0}^n$，则 $t_R = \dfrac{1}{kc_{A0}^{n-1}}$。以最高浓度 c_{A0} 作为基准是为了在统一的推动力基础上进行比较。

（2）催化剂外部传质的特征时间定义为 A 的最高浓度 c_{A0} 与可能的最高外部传质速率之比。当催化剂外表面 A 浓度为零时传质速率最高，即为 $k_g a c_{A0}$，因而有 $t_{De} = \dfrac{1}{k_g a}$。$L$ 为催化剂的某一几何特征长度，可为直径或半径；a 为单位催化剂体积 V_P 的外表面积 S_P，无疑为 $a = \dfrac{S_P}{V_P} \propto \dfrac{1}{L}$。如颗粒为球状，则有 $L = \dfrac{R_P}{3}$ 或 $\dfrac{d_P}{6}$。

（3）反应物 A 在催化剂内部的传质应理解为通过一定距离的扩散。因此，内部传质的特征时间应定义为催化剂特征长度（半径或直径）的平方与有效扩散系数 D_e 之比，即 $t_{Di} = \dfrac{L^2}{D_e}$。

（4）催化剂外部与内部传热的特征时间和传质的特征时间完全相对应，所不同的是引入了表征热容的密度与比热容。

（5）可以看到所有的特征时间，除非线性（非一级）反应的反应时间外，均与过程的推动力无关。当 $n = 1$ 时，即线性反应的 $t_R = \dfrac{1}{k}$，与浓度无关。传质与传热为物理过程，均为线性过程，其特征时间与浓度和温度无关。

讨论特征时间的目的是了解串联过程中各步速率的相对关系。表现在几个化学反应过程的重要无因次数实际上只是有关过程特征时间之间的关系。

1.3 特征时间与无因次数

化学工程中无因次数是常用的，因为这些无因次数反映了相互影响因素之间的制约关系，有明确的物理意义，如 Re 为惯性力与黏性力之比，因而其数值也决定了流动阻力系数。下面将要讨论的是化学反应工程中涉及的几个无因次数实际上是几个互为制约的特征时间之比。

（1）以一级反应为例，外部的传质时间与反应时间之比为 $Da = \dfrac{t_D}{t_R} = \dfrac{k}{k_g a}$，这是一个常见的 Damkohler 数。当 $Da \to \infty$，说明传质时间远大于反应时间，催化剂外表面浓度 $c_{AS} = 0$，此为快速反应。当 $Da \to 0$ 时，表面反应时间远大于传质时间，$c_{AS} = c_{A0}$，此为极慢反应。

（2）内部的传质时间与反应时间之比为 $\varphi^2 = \dfrac{t_{Di}}{t_R} = \dfrac{kL^2}{D_e}$，其中 L 常以催化剂颗粒半径 R 代之即为熟悉的 Thiele 数。当 φ 很大时，表明内扩散影响十分显著，内部传质严重影响了表观反应速率。当 $\varphi=0$ 时，表明内扩散影响可以忽略，整个催化剂颗粒内反应物 A 的浓度等于外表面浓度 c_{AS}。

（3）内部传质时间 t_{Di} 与外部传质时间 t_{De} 之比，此即为通常所称的传质 Biot 数以 Bi_m 表示：

$$Bi_m = \frac{k_g L}{D_e}$$

它实际上反映了内部和外部反应物 A 的浓度梯度之比。当 $Bi_m \to \infty$ 时，内部传质极慢，梯度在内部，而外部的梯度几乎为零。要使反应的表观速率提高，关键是要改善内部传质。采用小颗粒会是一个有效手段。反之，$Bi_m = 0$ 则表明外部传质控制。

与其相类似的是内部传热时间 $t_{\lambda i}$ 与外部传热时间 $t_{\lambda e}$ 之比，此即为通常所称的传热 Biot 数：

$$Bi_h = \frac{h_f L}{\lambda_e}$$

当 $Bi_h = 0$，为外部传热控制。

（4）Bi_m 表示内部和外部传质时间的相对关系，Bi_h 则为内外传热时间的相对关系。它们两者之比为：

$$\frac{Bi_m}{Bi_h} = \frac{t_{Di} \cdot t_{\lambda e}}{t_{De} \cdot t_{\lambda i}}$$

上式反映了 t_{Di} 和 $t_{\lambda e}$ 的等位关系。显然，$\frac{Bi_m}{Bi_h}$ 只与系统的物理性质有关，与反应过程没有直接联系。由于传热与传质之间的类似性，可以用 j 因子将 k_g 与 h_f 联系起来：

$$j_D = \frac{k_g}{u} Sc^{2/3} = \frac{h_f}{c_p \rho u} Pr^{2/3} = j_H$$

因而可以得到：

$$\frac{Bi_m}{Bi_h} = \frac{\lambda_e}{De \cdot c_p \rho Le^{2/3}}$$

式中，Le 为 Lewis 数，$Le = \frac{Sc}{Pr}$；Sc 为 Schmidt 数。我们在第四章中已说明，对于某一 $\frac{Bi_m}{Bi_h}$ 值，外部温差与总温差之比在一定条件下亦已确定。在常见条件下，气固相催化反应系的 $\frac{Bi_m}{Bi_h}$ 多见于 $10 \sim 10^4$，由此也可解释气固相催化反应的温差主要是在催化剂外部，而液固体系的 $\frac{Bi_m}{Bi_h}$ 多在 $10^{-4} \sim 10^{-1}$ 范围，其差别甚为明显。

特征时间的概念含义广泛，如气液相反应的八田（Hatta）数 Ha，也反映了液相传质时间 t_D 与反应时间 t_R 之比：

$$Ha = \frac{t_D}{t_R}$$

2　模型方法

数学模型化方法是化学反应工程的主要研究方法，应用模型方法的必要性是由于反应过程所表现出的很强的非线性性质。

模型方法的重要性尽人皆知。人们学得不少，在研究工作和开发工作中用得也不少，但在掌握模型方法应用的技巧方面往往还存在一些问题。这种现状，可能是由对模型方法的理解和观念方面的一些不足所造成的。

2.1 模型方法的适用性和实用性

模型化是一种方法，就像其他方法，如实验方法一样，各有其适用场合，但也绝不是万能的和一成不变的。简言之，模型化是建立一定的数学方程式，用来描述某一种或数种实际过程定态或非定态的行为。模型化的目的大体是用于分析、预测、设计和放大，以及控制等。应该说预测反应过程的行为是模型化的核心。有了预测能力，各种应用的目的大体都可以达到。

模型结构和模型参数为构成模型的两个主要要素。我们常见的模型结构有偏微分方程、常微分方程和代数方程，以及它们的变异形式。在实用中我们常常遵循一些经典的模型。这些模型有时可在文献中找到（如描述轴向扩散的扩散模型，描述反应物在催化剂内部分布的内部扩散-反应模型，描述非均相催化反应动力学的双曲型模型等），并不需要我们另行建立。我们常常遇到的问题是验证或筛选可用的模型，并确定在我们所需条件下的模型参数。在很多场合，模型参数的确定是一个比模型结构更为困难的问题。这一困难常造成模型化方法在应用中的限制。

反应工程领域大体涉及两类模型：反应动力学模型（包括本征和表观）和反应器模型（包括定态和非定态）。这两类模型有一个基本的不同点：反应动力学是化学反应特征的一种表现，理应与反应器的构型和流体的流动等物理因素没有直接联系。反应器内的流动等传递条件只有通过浓度和温度才对反应动力学产生影响。因而可以理解为反应动力学模型不应受到反应器的构型及几何尺寸的直接影响。反应器模型则不同。即使是相同的构型，不同的尺寸大小可能对模型结构提出不同要求；尤其是模型参数，在几何尺寸不同时，可以有很大出入，而这一参数随尺寸的变化通常是难以预计的。例如，小型反应器中的轴向返混可以用扩散模型来描述，其轴向扩散系数可以通过实验测定，但在预测一个放大后反应器的行为时会有明显的不确定性。放大后环流效应可能会使扩散模型是否适用成为疑问，如仍沿用扩散模型，则其放大后的轴向扩散系数的确定就很成问题。

解决这个问题，可以通过建立大型的（尺寸相当于或略小于待定系统，但有模拟意义的）装置，用模拟物系（常用的是水和空气）进行冷模实验。大型冷模实验研究有别于通常所理解的化学反应工程研究，但也需要专门的经验和知识。

读者用数学模型研究化学反应过程，将会对模型方法形成一个清醒的认识。我们必须充分意识到模型方法的独特作用和对反应工程学科的极大的推动，也应看到在不同场合下模型方法的不同的限制。合理的模型对分析过程各有关因素的影响无疑是很有效的。在有实用意义的定量预测方面，数学模型可能会有一些限制。例如，用扩散模型来预测物料的返混就会有一些限制。笼统地说，扩散模型并不很适于长径比不大，或流速不高，也即返混较大的情况，并且其轴向扩散系数的确定也较困难。所以在实用中，很少用扩散模型进行定量预测。与其相反的是催化剂内部扩散-反应模型。有效扩散系数作为一个模型参数是易于确定的，且并不随反应器放大而变。这一模型的实用价值非常明显。我们在这里提出这个问题，是为了提请读者注意形成对模型化方法适用性及局限性的认识。

2.2 模型化方法及关于相应实验的思考

反应器模型可以从不同的视角予以分类：机理和经验，定态和非定态，代数方程型和微分方程型，分布参数型和集中参数型，线性和非线性等。从分类开始认识模型方法无疑是正确的，因为分类不仅仅限于分类学上的意义，更重要的是可以帮助我们了解模型方法的全局。

任何模型都体现了对实际过程的简化，简化是模型建立的基础。有时某种简化孤立地看可能是不合理的，甚至可能是荒唐的，但对于特定的前提，这种简化却是允许的，甚至是必要的。例如，多管壁冷式固定床反应器的一维模型假定床层没有径向温度和浓度分布。这显然是一个与实际情况相去甚远的简化。但是对于一些应用场合，如过程控制，这一简化是常用的。简化到何种程度，则需视模型化的目的和确定参数的可用手段而定，很大程度上是取决于研究人员的经验。

模型建立的另一个基础是过程的分解。分解至少与简化同样重要。事实上，简化和分解是不可分割的。在反应工程中最重要的分解体现在物理过程（流动、传热和传质）与化学反应过程的分解。例如，本书中式（2.38）等号两侧描述的正是分解后的物理过程和化学反应过程。当然在分解后必须综合，否则前者也是残缺的。

模型方法与实验方法同样重要，且关系密切。一位训练有素的研究人员应既懂模型，又懂实验，且能充分融合两者的精髓，以解决研究中的问题。

（1）模型的构思应以实践为基础。实验现象为模型结构提供基本思想。

（2）模型的结构应考虑到与用实验手段获取模型参数的能力相一致。如果建立一个理论上合理的模型，能充分描述实际过程，但它所包含的众多模型参数不能实验获取，则不能认为这是一个适用的模型。反之，如果一个简单的模型，虽然与实际过程有一定程度的"失真"，但参数获取简捷可靠，则这个模型不失是一个实用的模型。模型的建立应充分考虑其参数获取的现实性。

（3）对一个过程可能建立数个竞争模型，因而最后选用哪个模型尚需对这些模型进行实验鉴别，或称筛选。在反应动力学模型化过程中尤常见需用实验筛选模型的情形。可以通过专门的实验进行模型筛选，然后进行参数估计。当定态的实验无法筛选出合适的模型时，可以考虑采用非定态条件下的实验，因为在不少场合，非定态的实验更有利于模型的筛选[1]。

（4）模型参数可分两类，一类有独立的物理-化学意义，可以在实验条件具备时单独通过实验测定，如某气体反应物的分子扩散系数；另一类是在模型建立过程中派生而得的，反映了过程的某些综合的特性，如扩散方程中的轴向扩散系数，它是流动、物系物理性质、设备几何尺寸等多种因素的综合。后一类参数通常不能单独测定，而且脱离了模型，就谈不上什么物理意义。

对于模型中出现的前一类参数，我们应完全利用实测数值（实验测定的或文献中报道他人测定的）。至于后一类参数，其物理意义是相对"模糊"的，一般只能通过一些宏观的实验并通过模型参数估计而得。这类参数是对模型必要的补充，是十分常见而且很有应用价值的。但必须充分认识到这类参数的本质和使用条件，注意避免一般化。

（5）模型参数的确定通常是模型化方法中的大问题，也是妨碍模型方法广泛应用的重要障碍。为了使模型实用化，模型建立应注意参数确定尽可能简便和省时。模型的结构通常极大地影响获取参数的工作量。我们认为，为了使模型实用化，最宜尽可能减少参数，也应使获取的参数尽可能可靠。例如对于一个固体催化的气相反应，我们可以测定其本征动力学（消除内外扩散影响条件下），并根据催化剂粒度及所测定的有效扩散系数等计算效率因子，获得表观动力学。我们也可以根据已选定的催化剂原粒直接测定其表观动力学。我们认为，后者更为直截了当，测定更为省时省钱。当然，两类参数确定都涉及不少实验设计和规划的技巧。

（6）我们常常注意到一些对数学有兴趣的研究工作者对复杂模型的偏爱。他们可能会认为，只有研究复杂的数学模型和先进的数值分析方法才能体现他们的学术水平。但是我们认为这并不是一种值得鼓励的指导思想。因为模型是为了达到某种研究意图采用的方法。以能实现这种意图为前提，模型应该越简单越好。所以在模型研究中，我们并不鼓励过多研究通用的模型（当然如果目的是研究模型的通用理论和一般方法时除外）。我们提倡模型研究的针对性，即尽可能地贴近实际过程和模型化的目的，尽可能地使模型的结构和参数的获取简化（显然这两者是有联系的）。

所谓贴近实际过程，是指利用对象过程的一些特殊性使问题简化。本书作者所在的华东理工大学联合化学反应工程研究所曾有很多成功的实际案例[2]。例如，反应过程本身必然呈现很多非线性性质，但按工业实际，操作只可能是在一个相对狭小的范围内进行。在这一十分有限的范围内，一个总体上看是非线性的问题就可以作为一个线性问题来处理，从而使问题大为简化，就像观察一条曲线在一个小区间内的规律，通常可以近似为一段直线而不导致重大误差。例如，朱中南等[3]曾将一个本应以一组微分方程来表述的反应器模型简化为一组代数方程，参数获取快捷简便，模拟精度很高，使开发过程周期大为缩短。如果要小结一下本节的主要思想，应该说模型研究者更应注意模型的针对性，以及更应注意对象过程的特殊性并予以充分利用。

2.3　模型和参数的不确定性

任何模型必须建立在对实际过程简化的基础上，简化必然带来与实际过程的偏差，所不同的只不过是偏差大小而已。因而任何模型都必然存在对实际过程描述的某些不确定性（正的或负的偏差）。模型参数或从计算获得，或从经验（如操作数据）获得，但主要的是从所规划的实验获得。这种获得的过程也不可避免地导致一些与实际情况的偏差，也就是参数也存在一些不确定性。由此可见，不论研究人员和设计人员如何经验丰富，知识渊博，这些不确定性是不可避免的。所不同的是，有经验的研究人员可以把这种不确定性尽可能地减少。更何况工业生产的现场又会增添一些其他方面的不确定性，如原料的变化，上游生产流程的参数波动，市场和价格因素等，这是一些在原本已有的模型和参数的不确定性之外始料不及的不确定因素。

既然这些不确定性是不可避免的，工业生产的现场操作就需要一些在不同程度上弥补的对策。好在利用化学反应器作为关键生产装置的过程工业的一个特点是存在一些可调的操作变量，如温度、压力、流量等，这给予我们把上述不确定性降低的余地。现场调优是一种常用的弥补方法，即以建成的工业反应器的实际操作状况为基础，以一定的搜索方法进行优化搜索，以获得较好的操作条件。调优并不依赖于曾用于该反应器开发的已有模型（如果有的话），而是通过建立纯属经验性的模型进行调优。调优法是实用的，但调优过程无疑需要一个相当长的时间，且结果不能用于外推，或只能在相当有限的范围内外推。

在控制技术和计算机技术已相当普及的今天，对建成的工业反应器进行在线优化不失为一条现实的途径。在线优化可以采用已用于该反应器开发所用的模型，也可以建立一个新的机理性或经验性的模型，或采用一种把这两者复合起来的模型。在运行过程中不断地更新参数并使得反应器在优化条件下操作。这样既可以部分地弥补模型不确定性和参数不确定性所造成的问题，也可以为始料不及的现场不确定因素的干扰提供适当的对策。

3 原理的应用和应用中的方法论思考

发展至今，化学反应工程学科已形成相当完整的理论体系，包括原理和方法，但这当然并不是说其理论体系已完美无缺，不需要进一步的深化和拓展。然而值得指出的是化学反应工程的理论和方法主要是建立在简单反应的基础上的，但大量的实际问题往往涉及复杂的或十分复杂的反应系统，虽然研究工作者经过长期努力，提出了一些用于复杂反应系统的处理方法，如集总方法等，但毕竟还有不够之处。这是由于实际反应过程实在过于复杂，人们至今所具备的手段尚有欠缺。另外，也应提出一个问题，就我们在现阶段掌握的理论和方法，有没有可能来解决我们面临的复杂的现实反应过程课题。事实是，应用已有的反应工程的原理，是有可能来解决多数复杂的反应过程课题的。这里要强调的是，原理的应用绝不应是生硬的。原理提供的是我们思考的基础，在实用中应十分重视原理应用中的方法论问题。

应用化学反应工程原理无非是解决两类问题，一类是基础性的和探索性的，一类是应用性和开发性的。不论是哪一类问题，盲目地、生硬地搬用反应工程的原理一般是难以奏效，或效果欠佳的。我们认为，特别重要的是首先应对所研究的过程有一个概括的认识（可用计算或实验，或两者并用），再决定系统的研究方案。图1和图2分别概括了作者所在的华东理工大学联合化学反应工程研究所同仁们经常使用的两类方法。

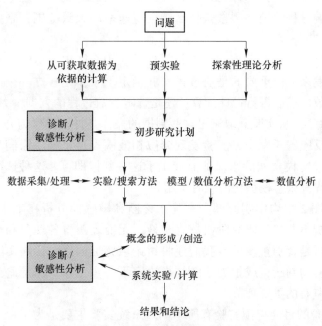

图1　主要用于基础性和探索性研究的工作流程

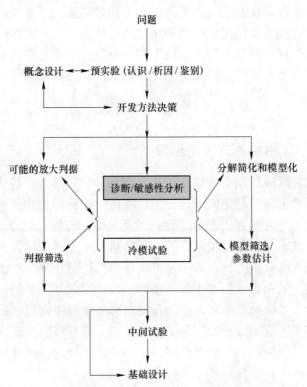

图2 主要用于应用和开发任务的工作流程

这两图所示的工作方法详情似已无进一步解释的必要。我们只想对其要旨作几点补充说明：

（1）课题上手就进行系统实验并不是一种好的工作方法。首先应该是通过一些简便的方法来认识一下所研究的反应过程有什么特征。可以通过查阅文献，经验判断，或通过一些非常简单的实验（称之为预实验）来实现这一步[4]。这些实验可以只是定性的，如鉴别影响过程的控制因素是什么，反应速率的大致范围，副反应大致在何种条件下生成等。

（2）通过预实验，对过程有了一个概括性的认识，则进而拟定研究计划或开发工作的策略。对开发来说，重要的是通过预实验确定下一步是建立一个简化模型，通过实验获取模型参数，然后再行设计计算，还是选定少数过程的判据（判据数越少越好，通常以一个为最好），从实验获取满足这些判据的操作条件，并作为开发依据。例如液液两相搅拌釜式反应器开发中，实验发现分散相液滴直径是唯一需满足的判据。放大过程只需要通过（冷模）实验来保证液滴小于这一判据即可。这种选取判据作为开发依据通常是在过程复杂、模型化困难时采用的。

（3）图1和图2中均用阴影部分突显了诊断和敏感性分析在工作流程中的作用。诊断和敏感性分析不断与各研究阶段信息交换，无疑表明将各现象和结果进行诊断和敏感性因素的分析是特别重要的。我们还将讨论敏感性因素。诊断实际上是对现象和结果所反映的事实的判断，或根据诊断结果决定下一步工作。这是避免我们的工作被一些表面现象所左右的重要环节。

（4）系统实验的目的与预实验有所不同。系统实验主要是为了取得数据。预实验的目的则是：为了认识反应过程的特征，如浓度对速率的大致效应；为了剖析某些现象的生成原因，如某一副产物是通过串联反应抑或通过平行反应产生的；为了对我们已形成的认识作是否的鉴别，如判别传质是否是影响选择性的主要因素等。对于应用和开发工作，如有必要安排中间试验，注意中间试验绝不仅仅是为了观察规模效应，而应该精心策划其试验内容，通过中试验证模型的可靠性和参数的合理性，或是验证所选用判据的平稳性。

上面提到了诊断和敏感性分析问题，按作者的理解，敏感性分析是化学反应工程原理的核心。化学反应工程的原理为分析哪些参数和变量对反应结果有敏感的影响，而哪些则没有提供指导。当然介于其间的还有一些。对敏感的因素我们必须认真对待，对不敏感的则不妨暂忽略之。例如我们研究的反应系统有 A、B 两反应组分，如 B 大大过量，很可能 B 的浓度对反应速率没有影响，我们只需研究 A 浓度的影响即可。至于 B 的浓度是否确实没有影响，我们只需有意识地组织少数几个实验进行验证即可。

敏感性分析可以通过模型计算，也可以通过组织有针对性的实验获得。壁冷多管固定床反应器的参数敏感性是一种典型的敏感性问题。对于强放热反应，进料中反应物浓度和壁温影响都相当敏感，且尤以壁温为甚。

更简便的方法（可能也是更可靠的方法）是用实验搜索敏感性因素的存在范围。以催化反应为例，气体反应物通过固体催化剂层，要观察各种条件对内外扩散的影响。在空速不变的条件下，逐步提高气速并测定反应速率。在起始阶段气速对表观速率有显著影响，然后影响趋于缓和，当气速达到相当数值时，气速的影响几乎可以忽略。这一现象体现了气速从敏感到不敏感的转变。如果保持气速在某一数值以上，它就是一个不敏感因素，或者说，它的影响就可不予考虑。与此类似，为了消除内扩散影响，可以将催化剂颗粒破碎至一定程度。进一步的破碎，已不会对反应结果产生敏感的影响，这时催化剂内部扩散效应基本已经消除，颗粒直径已不是一个敏感因素。

当然可以通过计算来寻找敏感条件，但实验方法无疑是更为可靠和直截了当的。通常人们愿意用实验方法。

诊断在很大程度上依赖敏感性分析的结果。敏感性分析帮助我们去伪存真，去粗存精。

以上我们讨论了三个涉及与化学反应工程各部分都有关系的问题，作为本书的后语和终结，供读者在读完本书各章后参考。我们感到作为后语，篇幅已经太长，但作为横贯各章内容的几个观点，似乎还意犹未尽。我们抱憾地在此搁笔，希望能在其他场合，与读者继续就这些观点作更深入的讨论。

参 考 文 献

[1] Kobayashi H, Kobayashi M. Transient Response Method in Heterogeneous Catalysis [J]. Cat. Rev. Sci. Eng., 1974, 10 (2): 139～176.
[2] 陈敏恒，袁渭康. 工业反应过程的开发方法 I. 开发的方法论问题 [J]. 石油化工, 1994, 23: 170～172.
[3] 朱中南，吴民权，张浩. 工业反应过程的开发方法 I. 负压法乙苯脱氢绝热反应器的开发 [J]. 石油化工, 1994, 23: 173～181.
[4] 陈敏恒，袁渭康. 工业反应过程的开发方法 [M]. 北京：化学工业出版社, 1985.

工业反应过程的开发方法[*]

摘 要 工业反应过程的开发方法是一组叙述利用反应过程开发方法来解决实际问题的若干实例的系列论文。本文是这组文章的首篇。主要阐明工业反应过程开发方法的重要性、主要内容和基本原则,以及如何从实际出发,有效地应用工业反应过程开发的方法。

关键词 工业反应过程 开发 方法论

1 引言

1985 年,我们曾写了《工业反应过程的开发方法》一书[1],专门论述工业反应器开发工作中的一些方法问题。将工业反应器开发的方法作比较系统的讨论,是出于以下原因:工业反应过程的复杂性常使传统的方法归于无效,这是研究方法的必要性;对反应工程理论的掌握提供了讨论开发方法的基础,这是解决问题的可能性。

我们曾多次向工业界人士讲解工业反应过程的开发方法,并辅以多个实例,试图使我国的反应器开发工作水平提高一步。但据反映,在解决实际问题时仍困难重重,不知在实际问题前应该如何起步。然而我们在工作中都深受开发方法之益,恳切希望能与国内同行共享。这就是《工业反应过程的开发方法》一书出版以后旧事重提的原因。

2 要重视反应过程的开发方法

在有些场合,方法的问题比较明确。或者说,有一些现成的方法可以被沿用。因而人们也就在不知不觉中应用了前人提供的成熟方法,似乎也就不存在专门的方法论问题。但对于开发一个工业反应器,就缺乏这种现成的、成熟的方法。

缺乏现成的方法,并不意味着对反应工程的研究尚不充分,或是理论尚不成熟。事实上,反应工程已发展了比较完整的理论体系,也已有大量的专著和文献,详尽地叙述了各种重要原理。但是这些原理和理论在用于实际过程开发时,却显得缺乏活力。这绝不是影射反应工程的理论体系不够完整。我们认为,其中一个重要原因是实际问题过于复杂。

一个工业反应过程体现了传热、传质和反应的结合,远较只进行传热过程的换热设备和进行传热、传质过程的分离设备复杂,更何况三者不是简单的加和,而是融合。反应速率与温度的 Arrehnius 关系,用学术语言来说,反映了过程的强非线性性质,使反应器表现出一些非寻常的、难以捉摸的行为。由于绝大多数反应过程都不仅产生单一的目的产物。而同时产生一些副产品。在开发分离设备时,人们可以假设较大的"安全系数",增大设备体积,充其量多用些材料,多些能耗,似乎总可达到目的。但在开发反应器时,这样做不一定有效,有时可能适得其反。

开发的含义比放大远为广泛。开发包括反应器类型的确定、操作条件的确定,也

---

[*] 本文合作者:陈敏恒。原发表于《石油化工》,1994,23(3):170~172。

包括放大。通过一个或几个中试层次，进行逐级放大，这在今天已被认为是一种无可奈何的方法，因为它耗资费时，而且结果不可靠。于是，有人求助于文献所载的数学模型方法。但是实际上，单纯用数学模型方法能奏效的，即使在工业发达国家也为数极少。其原因大致如下：

（1）模型研究是以单因素为基础的，也即在研究物理因素（如扩散和传热）时，常假定十分简单的化学反应，如 A→B 这样的单组分不可逆反应。实际的反应远较复杂，模型描述极为困难。

（2）在建立模型时，必须对过程作简化。由于实际过程过于复杂，所作的简化假定可能存在相当程度与实际过程的偏离，从而造成模型不足以确切反映实际。

（3）即便模型可靠，但模型涉及的参变量难以准确通过实验测定。以固体催化剂作用下的气相催化反应为例，如用微分反应器或循环式反应器进行测定，催化剂的代表性亦有问题。所测得的动力学关系在放大时可能导致较大误差。

我们并非认为上述两种放大方法是不足取的。我们只是试图说明，任何方法的应用都应该在方法论的指导下视过程的特殊性而定，而不能不加选择地沿用现成的方法。正确的开发方法，首先应揭示过程的特殊性，并利用这种特殊性。在这个基础上进行反应器选型和确定优选的操作条件。然后根据选定的反应器类型进行放大。本文所说的方法，其核心就是利用过程的特殊性，而不是泛泛地沿用现成方法。为了利用，必须先弄清究竟有哪些特殊性。

3 工业反应过程开发的主要原则

工业反应器的开发与反应工程的基础研究有一个基本的不同点。后者作为学科内容，总是从一个具体问题开始认识，逐步推广到一般，希望得到有普遍意义的规律，是一个从特殊到一般的过程。前者则完全不同。在开发一个工业反应过程时，我们应该从所掌握的反应工程的一般原理中，针对本问题进行分析判别，找出其特殊性，是一个从一般到特殊的过程。这个过程，犹如医生根据他的医学知识和病人的症状，并结合必要的检验手段，以诊断病情，对症下药。开发工作者首先要明确的一个思想，即在做开发，不是在做研究。

工业反应过程开发应在反应工程理论的指导下，在正确的试验方法论指导下进行。

（1）组织简单的预试验以充分揭示本反应过程的特殊性。副产物是通过平行反应或串联反应生成的，可以通过改变某一种反应物进入反应器的浓度的简单试验来弄清。该试验虽不能定量得到动力学模型，但对反应器选型却是十分有用的。改变温度，观察活化能的相对大小，有助于选择操作条件。

（2）充分利用过程的特殊性以大幅度地简化试验。过程的特殊性可来自反应和过程自身，也可来自进行反应器开发的工程目的。例如，工程上对某一过程已有了一些约束，如进料配比、操作温度等。测定动力学当然就不必在全程范围内进行，而只需有针对性地测定所需范围的动力学。同样，模型建立和优化也只需在有限的范围内进行。正因为范围有限，近似的线性化就有可能，从而使问题大为简化，可靠性大为提高。

（3）分析特殊问题的各类参变量的敏感性。反应工程问题的特点之一就是影响因素众多。如要全部考虑，则既不可能又不必要。敏感性分析的手段多样，但其实质不外是分清主次，个别对待。对不重要的，或是在一定范围内可退居次要地位的可不予考虑。另一种处理方法是，设计操作点应在参数的不敏感区，从而可以容许一定范围内的设计偏差。

开发方法的手段多样，灵活多变，可能也因为如此，使人们觉得不易掌握。但归结起来，方法论的核心就是以反应工程理论为基础，来判别什么时候做什么或怎样做，及不该做什么。这里所谓该做什么，是指为了满足开发所需的最低限度的工作。怎样做是指最省钱、最省时间的做法。不该做什么是指可做可不做的一律不做。紧紧抓住过程的特殊性。没有充分利用过程特殊性的开发工作必然不是好的开发。

4 从实际出发，主动运用工业反应过程开发方法

实际问题千变万化，扑朔迷离。要面对这些问题，获得一个普遍适用的工作方法，实际上是不可能的。我们庆幸反应工程的理论已为我们运用方法论提供了一个坚实的基础。不少开发工作者正在下意识地运用方法论解决问题。如能在解决问题的过程中主动注意方法论的运用，势必有事半功倍之效。

在《工业反应过程的开发方法》问世以来，我们多次鼓励华东理工大学联合化学反应工程研究所和技术化学物理研究所的研究人员自觉地在解决实际问题过程中运用这些方法，效果十分明显。本系列文章搜集了他们运用该方法解决问题的一些实例，相信读者能从中得到一些启发。不同作者的文体风格各异，我们没有强求一律，还请读者见谅。

参 考 文 献

[1] 陈敏恒，袁渭康. 工业反应过程的开发方法，北京：化学工业出版社，1985.

Methodology for the Development of Industrial Reaction Processes

Chen Minheng, Yuan Weikang

(UNILAB Research Center of Chemical Reaction Engineering, East China University of Science and Technology, Shanghai, 200237, China)

Abstract This is the first one of a series of papers presenting cases of using the methodology for the development of industrial reaction processes as a tool to manipulate various practical reactor development problems. This paper aims at describing why attention should be paid to the methodology, its basic principles and how it is applied to effectively developing industrial reaction processes from a practical point of view.

Key words industrial reaction process, development, methodology

过程的放大与优化

"过程"是制造业普遍应用的一种加工方式。如果将制造业按所采用的主要加工方式加以大略区分,则主要为过程工业和加工工业两个大类。前者通过各类加工过程实现,如冶金和材料工业、化学工业、食品工业等,后者则以各类加工工序为主,如机械工业、电子工业等。

对于加工过程为主的工业,统称为过程工业,其相应技术则为过程工程,如冶金工程、化学工程等的统称。

过程工业与加工工业相比有下列特点:

(1) 过程工业主要通过加工过程来实现工业生产,而不是通过加工工序来实现。如炼铁过程是通过高炉中的反应过程来实现的;石油化学工业中的石油裂解也是通过裂解炉中的反应过程来实现的。与此不同,加工工业则是以机床、焊机等工具的加工工序来实现生产的。

(2) 与之相应,过程工业的大型化主要依靠设备大型化,而主要不是依靠增加设备数来实现。例如大型钢铁厂主要依靠个数极少的大型高炉来体现效益,而绝不能依靠众多的小高炉;同样石油化工厂也往往是依靠一个大型裂解炉。而加工工业的设备除了采用先进技术(如程控技术)外,大型化十分重要的手段是增加设备数,当然也包括同一设备同时进行若干种加工工序的能力。

(3) 过程工业的产品往往以质量或体积计,如以吨、立方米等来计量钢铁、水泥和各种化工产品,而不是如加工工业的产品往往以件数计,小到螺钉,大至汽车、飞机等。

因此,过程工业与加工工业不同,存在一个设备放大问题。现代过程工业的标志之一是设备的大型化,即用一套大型设备实现大规模生产。大型化的合理性十分易于理解,因为大型化可以省投资,省原材料,省能耗,省人力。随着技术的进步,这种符合经济合理性的工业规模还有不断增大的趋势。例如合理规模的单套乙烯装置生产能力从30万吨/年,提高到45万吨/年,又提高到60万吨/年,并有进一步提高的趋势。当然,设备放大以后还必须保证经济上的合理性和各项指标的先进性。但是往往由于放大,有一些指标趋于合理,如能耗一般可以降低。但另一些指标,如反应产物的收率等,由于在大型化以后,操作温度等条件不易控制而往往有所降低,这就是通常所说的"放大效应"。放大效应被认为是一种弊端,似乎总是与大型化伴生的。我们的一个重要任务就是尽可能使这些指标在过程放大后仍保持一个较高水平。另一个现实是,一个实际过程,通常不能处在最优的操作状态下。这是由过程的复杂性和人们的认识能力限制所决定的,何况过程的一些参数会随时间变化(如催化剂的失活)。即使今天找到了最优条件,明天还可能发生变化。因此存在一个寻找最优的操作条件使过程不断优化的问题。这一想法是有可能实现的,因为一般来说,过程的操作条件(如温度、压力、浓度、流量、pH值等)允许在一定范围内调节以便寻优。这是过程工业的又一个特性——可调性。

1 过程普遍存在

过程是普遍存在，广为应用的。例如，旧日上海居民使用的煤球炉中所进行的是一个燃烧过程，属于气固化学反应过程；煤球炉实际上是一种移动床反应器，结构上与高炉有相仿之处。在生物发酵罐中进行微生物发酵过程。大型生物工程生产显然不能依靠增加发酵罐数，而应使发酵罐大型化。上述两类过程虽然表现得完全不同，但从本质上来说，在两种设备中进行的不外乎下面两类过程：

（1）传递过程，包括流体流动过程，传热过程和传质过程，属于没有物系组成变化的物理过程。

（2）化学反应过程，属于有组成变化的化学过程。

很多过程都是这两类过程的不同组合：小至观察尺度为分子级的薄膜镀层（如CVD方法GaAs薄膜制备），大至大气中SO_2的传递。此外还有一些过程，如处理矿物原料的粉碎过程，浮选过程等，以及颗粒状物料的输送过程和分离过程等，这些都属于机械过程。当然现代工业很少是以单个过程来实现生产的，而是由多个过程组合构成过程群或系统来实现。

化学工业是一种典型的过程工业。化工过程也完全是由上述几个过程组成的。化学工业中常见的是将原料预热到一定温度（传热过程），然后在反应器中进行化学反应（化学反应过程），经过反应的产品中必然有所需要的目的产物，以及并非主要目的的副产物和未反应的原料。这类混合物需要经过一个分离器，如精馏塔、吸收塔等进行分离（分离过程或传质过程）。由于分离过程还不能获得纯净的目的产物，有时尚需进一步提纯（又一个分离过程）。与此相似的是，冶金工业也往往是以这类传热、分离和反应过程为主体组成的：高炉可看做是个相当典型的化学反应器。以化学工业为背景，发展了化学工程学科，其主要内容为传热、传质和动量传递（传动）以及化学反应。实际上化学工程的应用范围已远远超越了化学工业，而是成为过程工程的核心。例如火力发电厂排出的大量含SO_2的废气，对大气产生污染。将SO_2从排气中脱除，甚至予以回收利用，也是通过这些化工过程来实现的。然而这些过程有一个共同点，即在实验室中，这些过程的实施多少是比较容易的，或在实验室的理想条件下，可以获得比较理想的指标。另外，实验室的规模和条件，可以使很多过程在间歇条件下实现。但是在工业生产中，这些过程可以比实验室中进行的同一性质过程大数万倍，甚至数十万倍。在小型设备中可不予考虑的温度和浓度不均匀性在大型设备中存在并且影响指标，并且大型过程多数是连续的。因此，将在实验室中所获得的结果在工业规模实施就成了一个完全不同的问题。因此，用最廉价的方法，在最短的时间内，最高质量地将过程大型化，就成为过程工程师的任务。

2 过程的开发和放大

本文所指的放大实际上有两重意义：设备的选型和放大。这一过程也可称作开发。

实验室中的过程通常是在尽可能简单的条件下进行的，并尽可能地排除对过程产生不利影响的因素。例如，一个催化反应需在一定温度下进行。在实验室条件下加热到这一温度是毫无问题的。将催化剂破碎到一定粒度，可消除反应物在催化剂内的扩散阻力以利于传质。将催化剂装入一个细小的反应管内，使其维持等温条件，也毫无困难。这样化学反应就可在所寻找到的优化条件下操作以期得到最好的结果。实验室化学家的任务是制备催化剂，筛选出最好的催化剂，并实验获得反应物浓度、流速和反应温度之间的适宜关系。在化学家的工作基础上，过程工程师的任务是：

（1）选择最适宜的工业反应器类型，或称选型。

选型过程包括对多种因素的综合考虑。例如，所能达到的指标，设备投资，能耗和操作费用，设备制造和材料，环保和安全性，操作和控制以及人员素质等。在权衡了种种得失以后作出的选择反映了工程师本人的知识、经验和思维能力。

（2）根据所选定的反应器类型，通过实验、计算或其他可以利用的一切手段，在最短的时间内，用最少的投资，进行设备的放大，最后提供工业反应器的设计，供设备制造工程师用。此谓放大。

当然这两者是交叉的。最初的选型可以在放大过程中被认为不合适而被放弃，并考虑另一种选型。如此反复，直到获得最好的方案。

放大的方法可以归纳为：

1）逐级经验放大。在确定选型以后，一种最为传统的方法是通过从小型试验，稍大规模的试验，中间试验，扩大中间试验，逐级地实现大型工业生产。这种通过多个试验层次的放大过程必然是耗时费资的。并且，虽然这一过程也可以凝聚着过程工程师的知识和积累，但毕竟由于放大完全是以经验为基础的，必然带有不同程度的盲目性。在过程工业发展的早期，经验放大几乎成了唯一的方法。随着过程工程技术的发展，在今天经验放大显然是并不可取的。但对于一些过于复杂的、人们认识甚少的过程，有时还不得不求助于经验放大。

2）数学模拟法放大。一种近年来被大为提倡的放大方法是建立数学模型（一组数学方程）对过程进行描述，并通过不同规模的实验以确定模型的参数，然后通过计算机模拟过程大型化后的各种行为，以确定放大的准则。这种放大理应是合理的。然而事实表明，单纯地用数学模拟法放大的成功例子甚为罕见。其原因是：

①实际过程通常极为复杂，而人们对它们的认识往往还不够系统和全面，因而为数学模型的建立带来困难。

②即使对复杂的实际过程已完全了解，数学模型的建立必须作出不少简化假定；因而为了便于描述，很可能得到了过度简化的模型。

③实验测定的模型参数的可靠性往往受实验手段的限制和实验过程中噪声的干扰，因此模型参数存在或多或少的不确定性。

由于数学模拟法放大只能适用于人们对过程的认识已相当透彻，参数的测定相当可靠的场合，因而即使是在工业发达国家，完全依靠数学模拟法放大的案例也比较罕见。比较现实的方法是，利用数学模型，但也需依靠一些实验或经验结果以实现放大。

3）以"实验方法论"为基础的放大。以上两种放大方法是就两类极端的类型而论的。经验法主要是用于异常复杂，人们知之甚少的过程；数学模拟则主要用于人们已相当透彻了解的过程。实际情况是，大量的过程介于这两类极端情况之间。对于这种为数众多的中间类型过程，应充分利用已有的理论，并根据实验方法论的原则组织实验，据此进行放大。

实际过程种类繁多，性质各异，难以归纳出可以适用于各种不同过程的通用步骤。实验方法论提供了一些考虑问题的原则和基点。它们包括过程的简化、分解和综合，过程的敏感性分析，工程问题特殊性的分析和利用，冷模实验的组织，预实验和系统实验的组织等。实验方法论包括了利用尽可能简单的实验方法进行预实验（包括认识实验、析因实验和鉴别实验），以获得对过程的初步认识，并以此为据作出采用哪一类放大方法的决策。这里所谓的主要方法包括两大类：以寻找放大判据为主的实验方法，以及以敏感性分析和简化模型为主的简化模型方法。根据不同过程的特殊性以及放大中对过程了解的深化以决定是否需要经过中试以检验放大方法的可靠性，最终再进行放大。

3 两个过程放大的案例

这里简述两个由华东理工大学反应工程国家重点实验室（REL）进行放大的案例。在放大中已初步考虑到优化条件确定，因为这两者是不可分离的。

案例之一：乙苯脱氢反应器的放大。

在铁系催化剂存在的条件下，乙苯可以在550℃上下脱氢生成苯乙烯，这是一种重要的合成材料原料。其主反应为：

$$C_6H_5C_2H_5 \rightleftharpoons C_6H_5C_2H_3 + H_2$$

其他尚有一些副反应，产生一定量的副产物。一个成功的放大应该获得高的产物收率和选择性，能耗应尽可能低。由于这是一个可逆的吸热反应，反应过程中体积增大。为了利用反应混合物自身供热，用高热容的水蒸气稀释反应气体，以维持比较适宜的温度。考虑到反应过程中气体体积增大，操作压力应是一个敏感的条件，负压操作是有利的。

遵循教科书上的方法，则必定先测定反应动力学，建立动力学模型，然后建立反应器模型（微分方程型），通过计算机模拟进行放大。但是这样的方法除了耗资费时外，结果尚不可靠。如果妥善组织实验就可以在很短的时间内成功地完成开发，获得满意的指标。

工业中最简单的实现这类反应过程的反应器是绝热反应器。对于绝热反应器，反应结果将唯一地由进口条件决定：

$$Y = F(T_0, SV, R, P)$$

式中，Y 表示反应结果，如转化率、选择性等；T_0 为进口温度；SV 为空速；R 为浓度（水烃比）；P 为压力。这样就可以免去先测定动力学，然后进行积分。上述关系可简单写作代数型。

从本问题的工程约束来看，T_0 不能过高，一般在630℃以下，R 不能太大，一般在1.5上下，转化率至少应在50%以上，选择性应在92%C以上。由于这些工程上的约束，这一系统的操作只能在一个相对狭窄的范围内。因而一个原来是非线性的问题可简化为一个线性问题。上述关系可简化为一个线性代数型方程。通过最普通的实验设计（如正交设计）即可获取优化条件。

大型装置的实现需保证气体均布。本案例为实现负压低压降操作，采用径向反应器，通过大型冷模研究气体的均布方法以完成反应器的放大。

放大后的工业规模反应器，完全实现了小试的指标。这体现了一个通过工程特征分析建立简化模型进行放大的实例。

案例之二：丙烯氯醇化过程的放大。

丙烯和氯在水中反应生成氯丙醇是环氧丙烷生产中的一个反应过程。其大致的反应步骤为：氯溶于水生成次氯酸，丙烯与次氯酸反应生成氯丙醇（简称醇）。主要副反应是丙烯直接与氯反应生成二氯丙烷（简称烷），另一个副反应是氯丙醇与丙烯继续反应生成二氯异丙基醚（简称醚）。合理的反应器应有高的醇选择性，低的烷和醚的选择性。

（1）小试发现，气相丙烯和气相氯极易产生烷。因此大装置应避免丙烯和氯直接接触。溶氯应与反应分离。

（2）溶氯是一个快速过程，因而其速率取决于气-液接触界面积。醚的生成是一个相对缓慢的反应，生成量与反应器体积有关。因而溶氯器应具有大的气液接触面积，小的体积。

（3）为避免过量醚的生成，应保持醇的低浓度，因而多级反应器应是理想的选择。通过实验寻找放大判据，主要是实验获得一个合理的喷射式溶氯器的放大依据，成功地实现了放大。这体现了一个通过寻找放大判据进行放大的实例。

4 过程优化

优化显然指对一定的目标函数的优化。这些目标函数一般是指过程的某项重要经济指标，或经简化后得到的一些关键操作指标，如产品产量、纯度、产物收率、能耗等。过程的优化包括两类：离线优化和在线优化。前者一般可理解为先验优化，绝大多数为定态优化，即在实现过程以前已先验地设计好优化条件，然后在过程实施中予以实现。在线优化是在过程进行之中，经过对过程行为的观察和了解，然后逐步地进行优化。在线优化是目前公认的一种先进的动态优化方法，即利用计算机在线地采集动态操作数据，进行系统的状态和参数的辨识，作为优化的依据。离线优化是一种比较早期的方法，它的效果对于大系统和大型过程甚为明显。主要原因是，对于大型过程，即使是一个相当小的百分比的优化，对于降低原料消耗和能耗的意义却是巨大的。另外，现代过程通常都不是单一的，而是由多个过程所组成的系统或大系统。系统有如下特性：

（1）对于大系统（大量的单元过程，性质相同的和性质不同的过程的组合），通常是难以通过人们的经验和知识判断如何进行优化组合的。用计算机进行组合过程的模拟（或称流程模拟），是一种很有效的定量方法。目前已有各类商业化的流程模拟系统，如 ASPEN PLUS 和 PROCESS 等，可以提供离线优化的基础。

（2）组成系统的各类单元过程有各自的优化问题。但是单个过程的优化往往以系统优化为前提，即系统的优化并不意味着每一个单元过程都处于优化状态。因此单元过程的优化应受到系统优化的制约；在单元过程的优化时不能脱离系统优化的前提。

（3）系统和单元过程的优化都是以模型为基础的，以一个单元过程构成一个"模块"，每一模块可以包含为数众多的方程。如一个精馏塔的模块就包含数十以至数百个代数方程，用以描述精馏塔的定态行为。单元过程优化以模块为基础。

（4）以模块为基础的单元过程优化，构成了系统优化的序贯法和联立模块法。但系统优化是以系统为基础的，重点并不在于单元过程的优化，当然很大量的系统优化问题并不与单元过程优化相矛盾。因此，理想的系统优化方法是联立方程法。联立方程法可以跨越单元化的局限。

离线优化需要单元过程模型和大量的计算，从已有的软件和现有的计算机水平来说，这已不是什么困难。但现有的通用商业软件一般并不提供一些特殊或专用过程的模块。有关这些模块需使用者自行开发。尽管离线定态优化有其非常可观的实用价值，但是也存在一些比较明显的不足之处，这通常是由于一些过程的不确定性质引起的：

（1）先验的优化是以模型为基础的。由于对过程了解存在或多或少的不足，先验的优化往往不能反映出真实的优化状态。这主要是由于用于先验优化的模型往往只是实际过程的过度简化。由于过度简化，模型对实际过程的描述能力显然受到限制。这是模型不确定性的结果。

（2）模型中必然包括一些参数，或由文献和数据库获得，或需由实验测定，或由现场采集的数据关联而得。这些参数包含了测定中的误差和噪声干扰。这是参数的不确定性。利用这些参数进行优化也必然带来一些不确定的结果。

（3）有些过程有时变性。因此过程虽为连续，但随时间会产生一些或快或慢的变化。例如化学反应器中催化剂的失活和粉化，传热装置中传热面的结垢，原料和产品的价格变化等，都会使已完全实现离线优化的过程偏离已设定的优化条件。此外，还有一些大家都已知道的时变因素，如生产负荷的变化、操作的变化等。实现了离线定态优化的系统显然并不能适应这些时变因素，因此理想的优化方法应是在线优化。

在线优化也有不同的含义：

（1）以在线辨识所得的动态模型为基础，对动态模型的定态形式实现优化，此为在线定态优化，对慢时变过程相当有效。

（2）在线辨识得到包括过程时变性的模型，以此动态模型为基础实现长时间范围内的优化。此类优化对间歇过程（如发酵）尤为适用。

在线优化显然比离线优化合理，但也远为复杂和困难。在线优化需要实时地决定优化策略，因此需要快速的计算机模拟，甚至像 CRAY 那样的公司也参与了过程的在线优化工作。但是目前看来比较现实的和有效的仍是单元过程的在线优化。单元过程的作用虽然是局部的，但对于一些关键的单元过程，对全系统有至关重要的影响，特别是一些反应过程（如裂解、冶炼、发酵等），其作用直接影响到全局。这些过程的优化是支配性的；其他过程的优化可以作为优化的第二层次。

在线优化属于先进过程控制的范围。将在线优化过程分为两个相：优化相和控制相。控制设定值由优化相根据过程动态模型的定态形式优化获得。控制相的一个关键部分是预测模型状态估计和在线参数辨识，并通过这一模型实现优化。这一预测模型可以是机理性的，也可以是经验性的，主要视其是否具备在动态操作中的学习能力。

正在 REL 开发的新颖的预测模型的结构特点是利用 Karhunen-Loeve（K-L）展开法压缩动态数据求取过程的特征函数以改造传统的神经元网络模型，使后者反映出过程的动态特征。所开发的软件通过采集动态数据（为了在工业上实施方便，测量点数应尽可能少）进行学习，实时地提供优化设定，并进行在线优化。整个优化工作通过一个微机进行，并可以与已有的 DCS 系统兼容。

我们的初步研究结果表明，通过在线优化，已可以使一个催化反应过程的收率比通过常规方法优化的结果高 1% ~ 3%。

5　结束语

过程的特殊性决定了过程工程的方法。可以看到放大和优化都在相当程度上考虑到过程的可调性，即过程的操作变量允许在一定的范围内波动。

过程放大最重要是分析和利用过程的特殊性，这种特殊性可通过简单实验定性（或半定量）地获得。实验揭示操作变量范围对结果的敏感性影响常可获得放大的依据。但是不论某一过程的放大的依据是否可靠，在放大中应充分考虑到放大以后进行优化（经验的调优或动态的优化）的余地。

新型苯加氢反应器的研究*

摘　要　文章提出了依靠相变蒸发吸收反应热,同时解决移热和加速反应等问题的新型化学反应器并开展了实验研究。实验在加压下进行,催化剂床层高度为 1.0m,反应器内径为 0.02m,苯加氢生成环己烷为研究体系,实验中采用气液并流向上流动的方式通过催化剂床层。在操作条件:$p = 0.5 \sim 3.0$ MPa、$\theta = 150$℃、液体空速 $= 1.1 \sim 6.9 h^{-1}$、氢油体积比 $= 300 \sim 1910$ 下,苯的转化率及环己烷的收率均可达 99.9% 以上。为了防止反应器内发生飞温,实验采用苯与环己烷的混合物作为原料,其中苯的质量分数为 15% ~ 25%。随着反应物料不断进入反应器,液相物料吸收反应放出的热量而蒸发,因此在适当的操作条件下床层内可同时存在液相反应区、气液两相区和气相反应区。

关键词　相变反应器　苯　加氢　环己烷

环己烷是生产聚酰胺纤维的主要中间体之一。此外,环己烷还用于尼龙 6 和尼龙 66 的生产,制备增塑剂和聚氨酯等材料,以及作为性能优良的溶剂等[1]。由苯加氢生产环己烷的工艺路线较多,根据反应条件的不同,可分为气相法、液相法和气液两段加氢法三大类[2,3]。加氢过程可在固定床反应器或液相循环反应器(IFP 技术)中进行。在各种生产方法中,如何从反应器中移出大量的反应热,并保证环己烷产品的纯度,是需要解决的关键问题。

本文所研究的新型连续相变苯加氢反应工艺过程流程简单,可连续操作,液相反应和气相反应可同时在一个反应器内进行,为简化、改善与优化苯加氢制环己烷的工艺迈出重要一步[4]。

1 实验部分

实验流程见图 1。反应器内径为 20mm,采用 $\phi 2 \sim 3$ mmDG 钯催化剂。催化剂装填高度为 1.0m,装填量为 316g,床层两端装填与催化剂同样大小的玻璃球,高度均为 30cm,因此反应器总高为 1.6m。反应器进口装有不锈钢液体分布器,以保证液体分布均匀。利用出口气液分离器后面的压力调节器来控制反应器的操作压力。氢气和液体原料(苯和环己烷混合物)分别用压缩机和液体泵连续输送至气液混合预热器,加热至设定温度后从底部进入反应器。利用绕在反应器壁上的加热带先将反应器壁温度升高,并保持与进口温度相同。实验时先设定压力和进口温度,调节所需的氢气和液体流量,当进口温度达到设定值后开始记录温度随床层高度的变化。出口温度稳定后(一般需 20min)取样,通过色谱分析各组分浓度。

气相流率和液相流率分别用气相和液相高压流量计进行测量;催化剂床层轴向温度则由伸入到反应器中心的热电偶检测,以计算机进行在线显示和记录。实验原料为

---

* 本文合作者:A. M. Anter、程振民、肖琼、胡劲松。原发表于《中国工程科学》,2000,2(6):59 ~ 63。

纯度 99.96% 的环己烷，99.97% 的苯，纯度为 99.9% 的氢气。通过调配使其中苯的质量分数为 $w_B = 15\% \sim 25\%$。

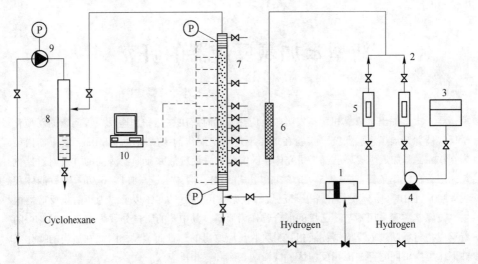

图 1　实验流程简化示意图

Fig. 1　Flow sheet of the experimental apparatus

1—Compressor；2—Liquid flow meter；3—Liquid tank；
4—Liquid pump；5—Hydrogen flow meter；6—Preheater；7—Reactor；
8—Gas liquid separator；9—Back pressure control valve；10—Computer；
✕—Valve；Ⓟ—Pressure

2　实验结果与讨论

2.1　反应器内流体相变分析

作者采用催化剂床层内同时具有液相反应区、混合相变反应区和气相反应区的气液并流向上流动新工艺研究了苯加氢生成环己烷的过程。该新技术不但可以通过改变环己烷和苯的配比，也可利用环己烷部分循环来控制苯加氢过程所产生的热量。该技术与其他过程相比具有可以减少反应器数量、简化流程和产品无需分离等优点。该反应器的相态分布如图 2 所示。反应物料以液体方式进入反应器底部进行液相反应，然后随着反应放出热量，液相不断蒸发汽化。同时床层温度逐渐升高，反应器内物料相态由液相经过气液两相并存进入气相状态。在这样的过程中，反应器床层下方催化剂为完全润湿状态，中部的催化剂为部分润湿状态，上部的催化剂为全干状态，从而实现了在单个催化剂床层内同时存在液相和气相反应。

在该过程中，液相和气相连续并流由反应器的底部进入催化剂床层发生反应，最后以气相状态从反应

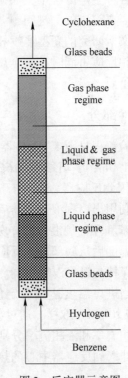

图 2　反应器示意图

Fig. 2　Schematic diagram of reactor

器（催化剂床层）的顶部排出。催化剂床层的典型轴向温度分布如图3所示，从图中可看出温度沿流动方向升高。由于在催化剂床层底部发生液相反应，反应速率较低（苯的转化率较小），反应放出的热量较少，另外反应放出的热量大部分被液相在蒸发汽化时所吸收，因此温升较小。当部分液相开始汽化（气液两相反应区内）和全部汽化（纯气相反应区内）时，反应速率加快，反应放出大量热量，造成床层温度升高。

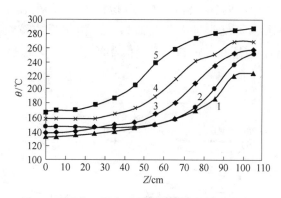

图3 不同进口温度下反应器温度的轴向分布图

Fig. 3 Bed axial temperature profiles corresponding to different inlet temperatures

（压力 $p = 1.0$MPa；催化剂床层总高 $h = 100$cm；液相流率 $L = 1.23$kg/h；氢气流率 $G = 15.5$L/min；$w_{B,i} = 16.6\%$；$\theta_{i,1} = 130$℃；$\theta_{i,2} = 140$℃；$\theta_{i,3} = 150$℃；$\theta_{i,4} = 160$℃；$\theta_{i,5} = 170$℃）

2.2 稳态操作下床层轴向温度分布

图4列出了稳态操作时反应器床层轴向温度随时间的变化。从图中不难看出在所研究的实验条件下，当反应器进口温度稳定在设定值上后，轴向各点温度较稳定，温度变化幅度最大不超过3～6℃。这也说明反应器的稳定操作特性。

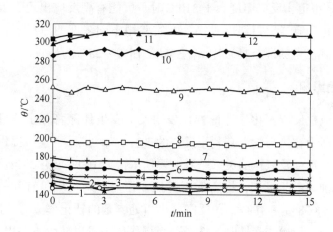

图4 稳态操作时轴向温度随时间的变化图

Fig. 4 Transient axial temperature profiles under steady-state operation

（$p = 1.0$MPa；$\theta_i = 150$℃；$L = 1.23$kg/h；$G = 15.5$L/min；$w_{B,i} = 36.4\%$；
$Z_1 = 0$cm；$Z_2 = 5$cm；$Z_3 = 15$cm；$Z_4 = 25$cm；$Z_5 = 35$cm；$Z_6 = 45$cm；$Z_7 = 55$cm；
$Z_8 = 65$cm；$Z_9 = 75$cm；$Z_{10} = 85$cm；$Z_{11} = 95$cm；$Z_{12} = 105$cm）

2.3 动态操作下床层轴向温度随时间的变化

为了观测操作条件改变时床层温度随时间的变化,实验采用 w_B 分别为 16.6% 和 36.4% 的混合液相物料。在液体流量为 1.26kg/h 下切换苯的进料浓度,得到当浓度周期变化时的床层轴向温度曲线,如图 5 所示。从图中可看出,在靠近反应器进口的区域中,由于液相的存在,反应速率较慢,另外液相汽化时将吸收反应放出的热量,因此,液相区和气液两相区内各处温度对进口物料中 w_B 的变化不敏感,各处温度较为平稳,变化幅度不大。当反应进入气相阶段时,由于反应较快,反应放出的热量将被气相吸收,因此各处温度对进口物料中 w_B 的变化较为敏感,而且呈现规律性变化。

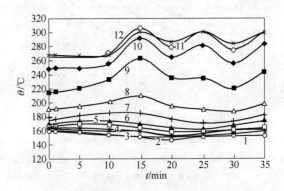

图 5 动态操作时床层各处温度随时间的变化

Fig. 5 Transient axial temperature profiles under dynamic operation

(Switch Time: 5min; $\theta_i = 160℃$; $p = 1.0MPa$;

$Z_1 = 0cm$; $Z_2 = 5cm$; $Z_3 = 15cm$; $Z_4 = 25cm$; $Z_5 = 35cm$; $Z_6 = 45cm$; $Z_7 = 55cm$;

$Z_8 = 65cm$; $Z_9 = 75cm$; $Z_{10} = 85cm$; $Z_{11} = 95cm$; $Z_{12} = 105cm$)

据文献 [2] 报道:反应温度大于 310℃ 时,可能有副反应出现。因此为了控制反应器的温度、防止反应器出口温度过高,本实验用不同浓度的原料进行实验,结果表明:物料(苯和环己烷的混合物)中 w_B 控制在 15% ~ 25% 为最宜。

2.4 反应压力的影响

苯加氢是强放热反应,化学计量方程表明反应发生伴随着分子数目的减少,因而降低温度、提高压力在热力学上有利于反应的进行。如文献 [5] 所提供的苯在平衡状态下的质量分数与温度及氢压的关系来看,要使苯接近 100% 转化,当反应温度升高时压力必须相应增加。由此可知,提高氢压(操作压力)对反应是有利的。在本实验的操作压力范围内 (0.85 ~ 3MPa),在反应器出口处,苯的转化率 x 均可达 99.99%。因此在设计反应器设备及操作运转时,选择适当操作压力可以减少设备以及操作的费用,从而提高经济效益。

表 1 的数据提供了在出口温度小于 280℃ 时,操作条件与反应器进、出口物流中苯的质量分数或转化率(用色谱分析得到数据)及反应器效率之间的关系(表中下标 i、e 分别代表进、出口)。

表1 操作条件与转化率的关系

Table 1 Relation between operation conditions with benzene conversion

p/MPa	G/L·min^{-1}	L/kg·h^{-1}	θ_i/℃	$w_{B,i}$/%	$w_{B,e}$/%	$x[100\times(w_{B,i}-w_{B,e})/w_{B,i}]$
0.85	17.90	1.20	150	23.69	0.00	100
1.00	10.80	1.30	135	24.10	0.00	100
1.00	10.80	0.80	150	24.10	0.00	100
1.00	20.40	0.64	144	24.54	0.00	100
2.00	15.50	1.25	150	20.31	0.00	100
1.00	15.50	1.25	150	25.10	0.00	100
3.00	15.50	1.24	150	25.10	0.00	100
0.50	15.50	0.48	150	12.24	0.00	100
1.00	15.50	1.24	150	36.50	11.45	68.63

2.5 液体空速对加氢效果的影响

在入口温度为（150±2）℃、压力为1.0MPa、液相中 w_B 为22%，氢气流率 $G = V^e/t$ 为10.8L/min 的操作条件下，改变液体空速进行实验，其床层温度曲线如图6所示。大的空速将使催化剂完全被液体覆盖，反应器床层处于液相操作，导致苯的转化率降低。因为反应热被液体吸收，所以床层的温升随着液体空速的增加而减小。

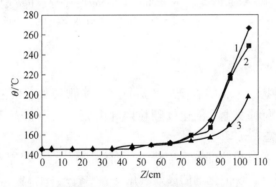

图6 不同液体流量下的床层温度分布

Fig. 6 Axial temperature profiles at deferent liquid flow rates

（LHSV$_1$ = 4.75；LHSV$_2$ = 5.70；LHSV$_3$ = 6.97（LHSV 为液体空速））

2.6 氢气流率对转化率的影响

实验固定压力为1.0MPa、液体流率为0.8kg/h，液相中 w_B 为22%，将不同的氢气流率下的出口转化率列于表2。从表中的数据可看出小氢气流率时，转化率随着此值的增加而增加，而大氢气流率时转化率就不受氢气流率的影响。从化学计量上分析，氢、苯摩尔比要等于或大于3时，才有可能使苯完全反应。在实验条件下，由于苯和氢在床层内处于流动状态，它们与催化剂的接触时间较短，当氢、苯摩尔比接近12时，苯转化率可达99.99%。实验还观测到在一定液体流率下，氢气流率较低（或在一定氢气流率下，液相流率较高），会使反应器处于全液相操作，即反应为液相反应，出口处苯的转化率变小。

表2 氢气流率对转化率的影响

Table 2 Effects of hydrogen flow rate on outlet conversion

G/L·min^{-1}	5.30	7.80	10.80	15.50	20.40
x/%	57.3	77.27	95.16	100	100

氢气的流量大小会影响到主体物料在床层内的停留时间,从而影响到反应结果和床层的温升。图7是在上述实验中的床层温度分布,图中的曲线表明,氢气流量对床层下半部分影响不大,而床层上半部分中,随着氢气流量的增加,液体的汽化量增加,从而引起床层的温升增大。

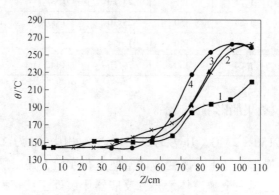

图7 不同氢气流量下的床层温度分布

Fig. 7 Axial temperature profile at deferent hydrogen flow rates

1—G = 5.3L/min; 2—G = 7.8L/min; 3—G = 10.8L/min; 4—G = 15.7L/min; Z—床层高

3 结论

(1) 在液相进料状态下,随着反应的进行,可实现液相、气液两相和气相反应在反应器内的不同部位同时发生,催化剂颗粒内部由完全润湿状态经部分润湿状态连续过渡至完全变干的状态。

(2) 利用液相蒸发可吸收部分反应热,从而解决强放热反应(例如苯加氢制环己烷)的移热问题,同时实现催化剂的高效利用。

(3) 连续相变过程可实现液相与气相操作在空间上的统一,而且能获得高转化率,从而大大地简化苯加氢制环己烷的工业过程。

致谢

本题目得到上海市科委新材料研究中心和国家重点化学工程联合实验室资助,谨致谢忱。

参考文献

[1] Pierlig A, et al. Hydrogen spillover effects in hydrogenation of benzene over P_t/γ-Al_2O_3 catalysts [J]. J. Catal., 1982, 75: 140~150.

[2] 庞先荣,陈立斌,易建峰,等. 在加压积分反应器上气相苯加氢 Pt/Al_2O_3 催化剂的活性评价 [J]. 化学反应工程与工艺, 1990, 6 (4): 1~7.

[3] Hydro. Proc. 1979, 58 (11): 149.

[4] 程振民,袁渭康,A. M. Anter,等. 相变法苯加氢制备环己烷工艺 [P]. 国家发明专利 (CNP) 申请号 00111762.9 (2000年3月2日).

[5] 勒巴日 J F,等. 接触催化 [M]. 北京: 石油工业出版社, 1984: 322~323.

A Novel Benzene Hydrogenation Reactor

Anter A. M., Cheng Zhenmin, Xiao Qiong, Hu Jinsong, Yuan Weikang

(State Key Laboratory of Chemical Reaction Engineering, East China University of Science & Technology, Shanghai, 200237, China)

Abstract In this work, the evaporation of the liquid cyclohexane is used to remove the heat of reaction in a novel phase transition reactor for hydrogenation of benzene to form cyclohexane under elevated pressure up to 3.0MPa. The reactor is of fixed-bed with Pd catalyst, 20mm in diameter and 100cm in height. The operation conditions are 150℃ in inlet temperature. 1.1~6.9h^{-1} in LHSV and 330~1910 in volume ratio of hydrogen to liquid. The experimental results show that the conversion of benzene is over than 99.99% and the yield of cyclohexane also over than 99.9%.

A mixture of benzene (accounting for 15%~25%) and cyclohexane is used as feed to prevent the abrupt temperature rise in the bed. The bed axial temperature distribution indicates the phase transition behavior of the liquid feed which varies from the liquid state, via the mixed phase, and eventually to the gaseous state. So there exists a three phase's regime in the catalyst bed.

Key words phase transition reactor, benzene, hydrogenation, cyclohexane

气液向上或向下并流固定床动持液量测定*

关键词　固定床　黏度　动持液量　关联

1　引言

固定床也可用作气-液-固三相反应器,其中气体和液体可并流向下或并流向上通过催化剂颗粒床层而发生化学反应。它广泛应用于各种化工过程[1]。床层中的动持液量是一个估算液体平均停留时间和液膜厚度必不可少的参数,也是反应器内液-固接触效率的一个量度,它直接影响催化剂的润湿程度和反应器的性能,因此持液量常是关联固定床(或滴流床)反应器内传质系数和传热系数的变量之一。随着化工工业的发展,往往要进行原料脱杂质处理过程(一般是加氢过程),需要足够的催化剂床层持液量。工业上采取催化剂颗粒小型化、异形化和脉动操作[2,3]等办法提高催化剂床层中的持液量。为了改善滴流床反应器床层中的持液量,还提出采用气体和液体并流向上流动通过催化剂床层的操作方式,可以满足某些工业过程需要高持液量的需求。

固定床中较高压力下气液并流向上和并流向下流动这两种操作方式的床层中动持液量测定和比较在文献中很少看到。本文的工作就是补充这方面的实验和研究,并提供有参考价值的数据。

本文在内径为 20mm、填料床层高为 1.6m 的固定床内,采用空气-不同黏度的水溶液(本文用 CMC 溶质来增加水的黏度)－ϕ2~3mm 和 5mm 的瓷球填料体系,在 15~20℃和 0.5~5.0MPa 的压力下进行实验,并考察了两种操作方式、液体黏度、气液流率以及填料大小对持液量的影响。

2　实验流程与设备

实验流程见图 1。固定床反应器内径为 20mm,高为 1.60m。反应器进口(塔顶和塔底)分别装有一个有 13 个 ϕ3mm 孔的不锈钢盘作为液体分布器,以保证液体分布均匀。利用出口分离器的压力调节器控制反应器进口的压力保持在设定值上。空气和水(或水溶液)分别用气体压缩机和液体泵输送混合后进入塔顶(或塔底)。气液两相在不同流速下通过填料层,待稳定后(一般为 10min),取其流量读数。同时关闭反应器气体和液体的进出口电磁阀,然后收集塔底流出的液体(实验中一般收集液体 30min,这时液体几乎不再流出)。再把装有填料的塔体拆下来,连接塔顶和塔底的法兰,并在相同的操作条件下进行端效应测定实验,扣除端效应后,即可得到动持液量值(重复实验之间的误差均低于 5%)。

---

* 本文合作者:A. M. Anter、肖琼、程振民。原发表于《化工学报》,2000,51(4):535~539。

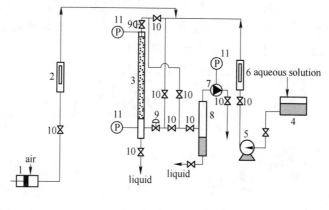

Fig. 1 Experimental flow sheet

1—Air compressor; 2—Air flowmeter; 3—Reactor; 4—Liquid tank;
5—Liquid pump; 6—Liquid flowmeter; 7—System back pressure valve;
8—Gas-liquid separator; 9—Electronic valve; 10—Valve; 11—Pressure meter

3 实验测量与结果

3.1 气体、液体流率对动持液量的影响

动持液量是指单位体积床层内的液体量。通常用两种表示方法：一种是以液相所占的床层体积分率 h 表示。另一种是以床层空隙内液相所占的体积分率 β_d 表示。本文采用第 2 种表示方法。

动持液量随液体流率变化关系见图 2a 和 b。从图 2 可看出，当气体流量一定时，气液两相并流向上或向下流动两种操作方式下，动持液量都随着液体流率的增加而增加，而且气液两相向上流动时的动持液量比向下流动时大。气液两相向上流动时动持液量随填料直径增加而增加。气液两相向上流动时，有利于液体在填料间的均匀分布、填料的润湿和液体充满填料中间的空隙（或流体通道），从而使得液体在填料中的流动接近平推流，使得动持液量比并流往下流大。

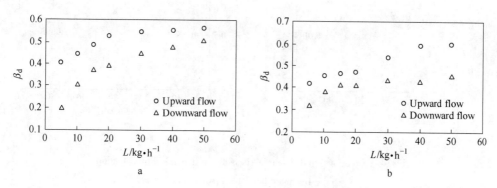

Fig. 2 Effect of liquid flowrates on dynamic liquid holdup

a—$d_p = 2 \sim 3$mm glass beads; b—$d_p = 5$mm glass beads

($G = 30$L/min, $p = 2.0$MPa, $\mu_1 = 1.08 \times 10^{-3}$Pa·s)

动持液量随气体流率变化关系见图 3a 和 b。图中的曲线表明，随着气体流率的增

加,动持液量减少,特别是气液两相向上流动时在较低的气体流率条件下,曲线的下降趋势较快。从图3还可看出,当气体流率较小时,气液两相向上流动的动持液量远高于两相向下流动时的动持液量,而且可以充满填料间空隙的90%以上。当气体流率较大时,向上流动与向下流动时的动持液量比较接近。

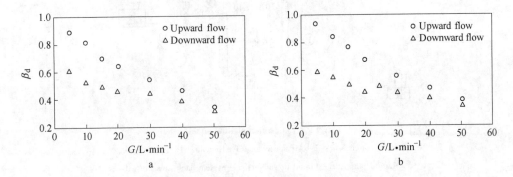

Fig. 3　Effect of gas flowrates on dynamic liquid holdup
a—d_p = 2~3mm glass beads; b—d_p = 5mm glass beads
(L = 30kg/h, p = 2.0MPa, μ_l = 1.08 × 10^{-3}Pa·s)

3.2　液体黏度对动持液量的影响

对3mm的瓷球填料床（高1.6m,ε = 0.3563）进行了动持液量与液体黏度关系的测定,测定结果见图4a和b。从图中看出,液体流率较低时动持液量明显随着液体流率的增大而增大,而液体流率高时这种变化超缓,甚至开始变小,另外随着黏度的增加动持液量的减少并不十分明显。由于两相并流流动方向不同（向上或向下）,两相在填料中的流动通道和分布的影响也就不同,因此当其他条件相同时,图4b的动持液量要大于图4a的动持液量。

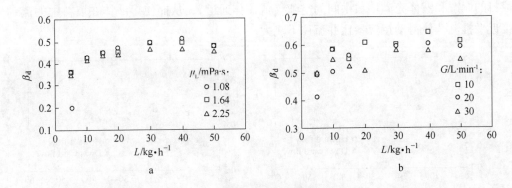

Fig. 4　Effect of liquid dynamic viscosity on dynamic liquid holdup
a—Downward flow; b—Upward flow
(G = 30L/min, p = 2.0MPa)

3.3　操作压力对动持液量的影响

压力的增加影响气相流速,从而影响流体的流动和分布,同时加压体系中水溶液的物性和颗粒表面液膜形态开始变化,而影响动持液量。实验测定的压力对动持液量的影响如图5a和b所示。虽然图5a和b的条件不同,但动持液量都是随着压力的增加而增加,当压力较高时动持液量经过一个峰值开始下降。

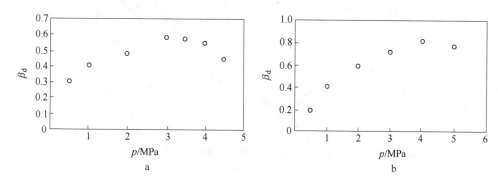

Fig. 5 Effect of operation pressure on dynamic liquid holdup
a—Downward flow ($H = 1.2$m; $G = 40$L/min; $L = 25$kg/h);
b—Upward flow ($H = 1.6$m; $G = 30$L/min; $L = 30$kg/h)

由上述实验结果（图2～图5）可看出,固定床中两相向上流动时的流动特性和液体在空隙里的分布以及液体与填料之间的黏结力等和两相向下流动时不同,两相往上流动时的动持液量比两相向下流动时大。

3.4 两种操作方式动持液量的关联式

由上可见,影响动持液量的主要因素是液体和气体的流速,但填料（或催化剂）颗粒大小、床层高度以及流体性质等因素对其也有影响。本文在15～20℃和2.0MPa压力下对以3mm瓷球为填料,床层高度1.60m（空隙率$\varepsilon = 0.3563$）,以纯水（$\mu = 1.08 \times 10^{-3}$Pa·s）和空气分别作为液相和气相的系统进行动持液量的测定。实验时,气体操作流率为10L/min、20L/min和30L/min,液体流率变化范围为5～50kg/h。选用Speechia和Baldi[4]以及Wammes[5～7]等提出的关联式对持液量数据进行关联,得到结果如下。

两相向上流时:
$$\beta_d = 0.4525 Re_L^{0.1552} Ga^{-0.0314} (a_V \cdot d_p/\varepsilon)^{-0.0018} \tag{1}$$

两相向下流时:
$$\beta_d = 0.3745 Re_L^{0.1931} Ga^{-0.0476} (a_V \cdot d_p/\varepsilon)^{-0.0022} \tag{2}$$

关联式的计算值与实验值的最大相对误差分别为13.5%和14.1%。与文献不同的是,文献[4]～[9]中关联式用床层压力降数据计算修正Galileo数,而本文用气体和液体Reynolds数之比代替压力降（因为当其他条件如操作压力、床层高度等不变时,影响压力降的主要因素是气液两相流动情况（流速）、流体特性以及床层空隙率等）。上述的关联式形式简单,避免了在固定床中高压下测定床层压力降所需要的高精密度法和高精度传感器问题,具有预测性很好、误差较小、关联式包含了床层的特性和流体的物性数据（如密度、黏度）等优点。部分关联式的计算值、实验值见图6a和图6b。从图中可看出,实验值除了个别点外,其他点的误差都低于10%。而实验的重复性比较好,3次实验的相对误差在5%以内。

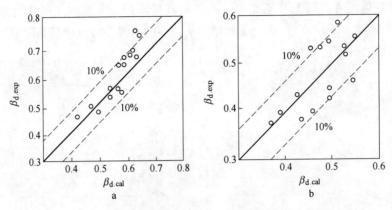

Fig. 6 Comparison of experimental and calculated dynamic liquid holdup
a—Upward flow; b—Downward flow

4 结论

(1) 固定床层中,气液两相并流向下或并流向上流动,在加压系统中床层的动持液量随液体流率的增加而增加,而随气体流率的增加而减少。液相黏度的增加对动持液量影响不大,两相并流向上流动的操作方式的动持液量总是比向下流动的动持液量大。常压或低压下动持液量随压力的增加而增加,但压力更高时则变小。

(2) 在本文实验操作范围内用式 (1) 和式 (2) 能对固定床中两种操作方式的动持液量进行预测和计算。

符号说明

a_V	单位床层体积的填料表面积,m^2	
d_p	颗粒直径,m	
G	气体体积流率,L/min	
Ga	修正 Galileo 数	
	($Ga = d_p^3 \rho_L [\rho_L g + \mu_L U_L Re_g/(Re_L H^2)]/\mu_L^2$)	
g	重力加速度,m/s^2	
H	床层(填料)高度,m	
L	液体质量流率,kg/min	
p	压力,Pa	
Re	Reynolds 数 ($Re = d_p U \rho / \mu$)	
U	空塔流速,m/s	
β_d	动持液量	
ε	填料在床层上的空隙率	
μ	黏度,Pa·s	
ρ	密度,kg/m^3	

下角标

g	气体
L	液体

参 考 文 献

[1] Larachi F, Laurent A, Midoux N, Wild G. Chem. Eng. Sci., 1991, 46 (5/6): 1233~1246.
[2] Li Shunfen (李顺芬), Zhao Yulong (赵玉龙). Journal of Chemical Reaction Engineering and Technology (化学反应工程与工艺), 1994, 10 (2): 169~178.
[3] Ge Shiying (葛世英), Cai Yunsheng (蔡云生), Li Pansheng (李盘生). Journal of Chemical Engineering of Chinese Universities (高校化学工程学报), 1991, 5 (3): 219~224.
[4] Speechia V, Baldi G. Chem. Eng. Sci., 1977, 32 (5): 515~523.
[5] Wammes W J A, Mechielsen S J, Westerterp K R. Chem. Eng. Sci., 1991, 46 (2): 409~417.
[6] Wammes W J A, Mechielsen S J, Westerterp K R. Chem. Eng. Sci., 1990, 45 (10): 3149~3158.
[7] Wammes W J A, Westerterp K, R. Chem. Eng. Sci., 1990, 45 (8): 2247~2254.
[8] AL-dahhan M, Dudukovic M P. Chem. Eng. Sci., 1994, 49 (24B): 5681~5698.
[9] Wammes W J A, Midelkamp J, Huisman W J, DeBaas C M, Westerterp K R. AIChE J., 1991, 37 (12): 1849~1861.

Liquid Holdup Measurements in Downward Flow and Upward Flow Fixed-bed Reactors

Anter A. M., Xiao Qiong, Cheng Zhenmin, Yuan Weikang

(State Key Laboratory of Chemical Reaction Engineering, East China University of Science and Technology, Shanghai, 200237, China)

Abstract In this work, at a pressure of 2.0MPa and room temperature, the effects of gas and liquid flowrates, liquid viscosity and operation pressure on the dynamic liquid holdup were investigated in a downward flow and upward flow fixed-bed reactor. With an air-water-solid packing system, experimental results show that the dynamic liquid holdup increased with an increases of liquid mass flowrates. It decreased with an increasing of gas volume flowrates and porosity of packed bed. It was also shown that there was no large effect on dynamic liquid holdup with an increase of liquid dynamic viscosity. The correlations of dynamic liquid holdup was also given under these two kinds of operation.

Key words fixed-bed, viscosity, dynamic liquid holdup, correlation

局部面积搜索法在高压相平衡中的应用*

摘　要　用最小Gibbs能曲线积分的正面积最大方法（AM）进行相平衡的计算和预测，可以比传统的闪蒸计算法更好地计算高压下和近临界区的相平衡组成。但是这一方法由于使用了在全组成范围的"穷举"方法搜索最大正面积对应的相平衡组成，使得收敛速度慢，计算和搜索时间过长；将其修改成多次局部面积搜索法（LSAM），极大提高了收敛速度和减少了计算时间，并利用所测定的苯与丙烯高压平衡数据对其进行了有效性检验。

关键词　吉布斯能　高压气液平衡　相平衡计算方法　苯　丙烯

1 引言

超临界流体技术作为近来发展迅速的一个全新的领域，不仅得到了全世界的广泛重视，而且其在工业中的应用也初显端倪[1]，被认为是21世纪化学领域中一项具有极大应用前景的新兴绿色化学技术。超临界技术的研究和应用主要是研究超临界流体和超临界环境所具有的性质，因此正确认识超临界流体和超临界条件是进行研究的前提。为此，高压相平衡的实验测定和计算模拟就成了超临界研究的关键，尤其是对近临界点区域的范围[2]。

Ammar(1987)[3]将相平衡计算方法分为两类：平衡闪蒸法和吉布斯能最小法。闪蒸法是将物料平衡和化学位相等作为建立方程的准则，对建立的方程利用迭代、牛顿和鲍威尔等优化算法进行这些高度非线性方程组的求解得到相平衡组成。但是闪蒸法使用的相平衡条件只是必要条件，而非充分条件。得到的"平衡解"可能是局部最小吉布斯能而非整体最小吉布斯能，因此可能不是实际的平衡组成。特别是在临界点附近由于状态方程存在歧点[4]，使得相平衡的计算难以收敛。此方法需要求导数，对初值的要求一般也较高。Cairns(1990)[5]对吉布斯能最小方法作了详细介绍。然而，一般的吉布斯能方法虽然改善了平衡组成的可信度，但仍然存在需要比较精确的初值，需要对计算方程求导数等不足。Eubank(1992)[6]提出了最小吉布斯能的最大正面积搜索方法（AM）。此方法保证了搜索条件是充分必要条件，不仅不需要初值和导数，而且由于使用了"穷举"搜索法，可以完全有效地搜索整体最小吉布斯能对应的两相平衡组成或所有存在的多相平衡组成。但是正如Eubank在AM方法[6]一文中所阐明的，获得正确的平衡组成是首先要考虑的问题，而计算时间的长短是第二位的，所以Eubank采用了"穷举"搜索法，致使搜索次数巨大，计算时间太长。虽说可以使用Eubank随后提出的二次搜索方法（先用低精度搜索得到近似的平衡组成，舍去平衡组成以外的部分再用高精度进行搜索）减少搜索次数，但节约的时间很有限（只有百分之几十）。尤其是设定计算精度较高时，计算时间并没有本质上的改变。如果用于分离过程的计

---

* 本文合作者：郭继志、刘涛。原发表于《中国工程科学》，2001，3（3）：56~60。

算机模拟，这一计算速度问题就会显得较为突出。文章利用多次局部面积搜索法（LSAM）对其进行改进，即使在设定精度较高的情况下也会大大减少计算相平衡的时间。换一种方式说：即使增加设定精度，而计算时间却无明显增加。作者利用实验测定的苯和丙烯二元系的高压平衡数据❶进行了两种方法的计算对比。

2 局部面积搜索法（LSAM）

2.1 AM 方法

图 1 给出了二元系两相共存时的气液吉布斯能曲线（$\Delta_m G/RT \equiv \Phi$）。给定温度和压力，依据所设定的精度对（0,1）区间进行 N 等分（N 是正整数，等于精度的倒数），得到 $N+1$ 个节点，然后针对任意两个节点 $a, b [a, b \in (0,1)]$ 利用式（1）计算积分面积 A，求得与正值的最大面积相对应的 a, b 两点即为气液平衡组成，如图 1 中的 A 和 B 两点。

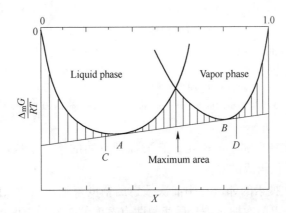

图 1　气液两相的 $\Delta_m G/RT$ 曲线

（中间三线围成的区域为正值最大面积 A，B 是正值最大面积
对应的平衡组成竖线表示对（0,1）区域的等分）

Fig. 1　$\Delta_m G/RT$ curve for a binary mixture forming two phases at fixed T and P

$$A(a,b) = \left| [\Phi(b) + \Phi(a)] \frac{(b-a)}{2} \right| - \left| \int_a^b \Phi(x) dx \right| \tag{1}$$

通常如果采用 Eubank 的二次搜索法，即在一次搜索的基础上缩小区间进行二次搜索。如图 1 所示，$[C, D]$ 区间就是经过一次搜索后得到的二次搜索区间。搜索方法同一次搜索。

2.2 多次局部面积搜索法（LSAM）

类似 AM 方法，如图 2 所示，首先选定一个合适的预设精度对（0,1）区间进行首次正值最大面积搜索，然后对得到的近似平衡组成进行二次或多次搜索。采用这一方法的关键问题在于：

（1）首次等分的精度选择，因为它不但直接决定着整个计算的时间，而且关系到

---

❶ Guo Jizhi, Liu Tao, Dai Yingchun, et al. Vapor-liquid equi. libria of benzene and propylene at elevated temperature and pressure, Submitted to J Chem Eng Data.

是否能搜索到平衡组成。首次精度取得过大会使计算时间过长，过小可能会搜索不到平衡组成。尤其是在接近临界点时，精度过小会使搜索越过平衡组成而导致搜索失败，并可能误认为处于单相区。

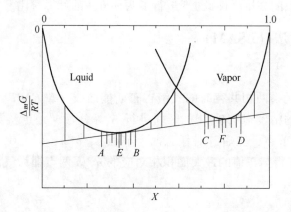

图 2　气液两相的 $\Delta_m G/RT$ 曲线

（中间三线围成的区域为正值最大面积 E，F 是首次搜索正值最大面积对应的平衡组成。间距大的竖线表示对（0，1）区域的首次等分，$[A, B]$ 和 $[C, D]$ 是二次等分的邻域，它们中间的小间距竖线是对相应邻域的二次等分）

Fig. 2　$\Delta_m G/RT$ curve for a binary mixture forming two phases at fixed T and P

（2）二次或多次搜索邻域的确定，它要保证真实的平衡组成落在选取的邻域中，否则就难以搜索到平衡组成的真值。

（3）二次或多次搜索的邻域利用前次搜索的精度来确定。即以前次搜索得到的近似平衡组成为中心，以前次搜索等分后相邻两点宽度的 2 倍为长度形成新的搜索邻域。如图 2 中，E、F 是首次搜索后得到的近似平衡组成，$[A, B]$ 和 $[C, D]$ 是首次搜索后得到的再次搜索邻域。同样的可以在 $[A, B]$ 和 $[C, D]$ 中产生三次，四次或更多次的搜索邻域。

（4）图 3 给出了多次局部搜索计算程序。

3　结果讨论

3.1　AM 和 LSAM 方法在面积搜索次数上的比较

依据 AM[6] 方法的计算原理，它的正值最大面积的搜索次数（NT）可以由组合理论计算。AM 法将（0，1）组成区间分成 N 等分得到 $N+1$ 个节点，在去掉 0 和 1 点（平凡点）的 N-1 个节点中搜索。N-1 个节点中取任意两点构成积分区间，利用式（1）计算面积。然后在所有的面积中求出正值最大面积对应的两点即是平衡组成。所以面积的计算次数（或称为搜索次数）由组合理论可得：

$$NT = C_{N-1}^2 = \frac{N(N-1)}{2} \tag{2}$$

通常等分节点数 N 可由所设定的精度 EPS 计算，不失一般性的取为 $N = 1/\text{EPS}$（EPS = 10^{-1}，10^{-2}，…，10^{-m}，…）（以后的分析和讨论都使用这样的精度）。代入式（2）可以得到不同精度下的面积搜索次数，而对应相邻两个精度的搜索次数之比为：

$$\frac{NT(m+1)}{NT(m)} = \frac{\dfrac{10^{m+1}(10^{m+1}-1)}{2}}{\dfrac{10^m(10^m-1)}{2}} = \frac{10^{m+1}(10^{m+1}-1)}{10^m(10^m-1)} \approx 10^2 = 10^{2\Delta m} \quad (3)$$

即相邻两次面积搜索的次数相差 100 倍，所以随着精度的提高，所需的搜索次数成幂级数增长。如果将精度由 0.01 提高到 0.0001，搜索次数就会增加 10000 倍。这显然是一般计算机所难以承受的。

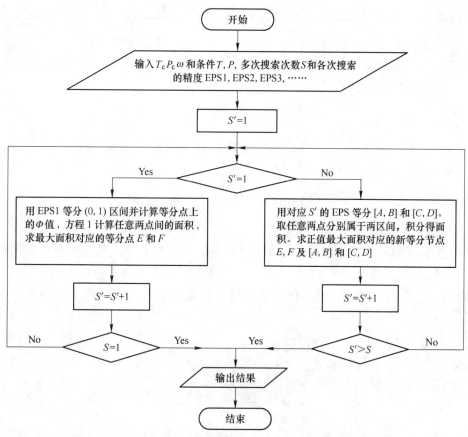

图 3　多次局部面积搜索法（LSAM）计算相平衡数据程序图

Fig. 3　Flow diagram of multi local search area-method for calculation of the VLE compositions

LSAM 方法是在一定精度下进行首次等分和搜索之后，再进行逐步提高精度的逐次局部搜索。由图 3 不难看出：如果每次只增加一位精度，则对围绕上次得到的近似平衡组成构成的单个邻域的等分就只有 21 个节点，气液两相相加就有 42 个节点（如果是两相共存）。对这 42 个节点的搜索次数是 21×21 即 441 次，与上次的搜索次数无关，而且一般比首次的搜索次数会小得多。例如首次搜索精度为 10^{-3}，则搜索次数为 $1000 \times 999 \approx 10^6$，远大于二次局部搜索的次数 441。即使将精度由首次搜索的 10^{-3} 提高到 10^{-6}，依据 LSAM 方法，首次搜索后再需三次局部搜索的次数的总和也只有 1323 次，与首次的搜索次数 10^6 相比也是微不足道的。如果与 AM 方法的 10^{12} 次数（AM 方法达 10^{-6} 精度所需要的搜索次数）相比，就更加微小了。由此说明，LSAM 方法的计算时间主要取决于首次搜索，提高设定精度（表 1）对计算时间几乎是无影响的。

表 1 AM 和 LSAM 方法用于回归气液平衡数据的计算时间对比

Table 1 Comparison of calculation times of regressions of the experimental VLE data by the two method AM and LSAM

名 称	计算机型号	首次精度	设定精度	计算时间 t/h		
				250℃	260℃	270℃
AM 方法	PⅡ350		10^{-4}	215.0	198.5	176.0
LSAM 方法	PⅡ350	10^{-3}	10^{-4}	4.5	4.2	3.8
LSAM 方法	PⅡ350	10^{-2}	10^{-4}	0.05167	0.04667	0.03833

不仅如此，依据 LSAM 方法的搜索过程，在二次或多次局部搜索中，可以利用分段积分避免对首次搜索后的 [B, C] 区间（图 2 中）内的节点的吉布斯能值 Φ 的重复计算，从而进一步加快计算速率，节约计算时间。

3.2 计算实例

应用测定的苯和丙烯二元系相平衡数据❶对 AM 和 LSAM 方法进行了考核计算和比较。其中二元系吉布斯自由能的计算见式 (4)[7]：

$$\Phi = \sum_{1}^{2} x_i \ln x_i + \sum_{1}^{2} x_i \ln \phi_i - \sum_{1}^{2} x_i \ln \phi_i^* \tag{4}$$

逸度系数 ϕ 由 Peng-Robinson 状态方程采用传统的混合规则计算[8]。利用图 3 的平衡组成计算程序 (LSAM) 与优化方法结合可以对测定的苯和丙烯二元系平衡数据进行回归。结果见图 4，计算值和实验值具有较好的一致性，而且基本上可以得到封闭的计算相平衡线。这也说明 LSAM 方法在计算临界点附近的平衡组成时比平衡闪蒸法容易收敛。

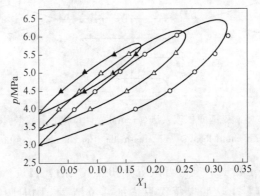

图 4 丙烯 (1) + 苯 (2) 二元系高压下不同温度的气液平衡数据❶
○—523.15K；△—533.15K；▲—543.15K；实线—PR EOS

Fig. 4 Comparison between the experimentally determined vapor—liquid equilibria for the propylene (1) + benzene (2) system with the calculated results❶
○—523.15K；△—533.15K；▲—543.15K；Solid line—PR EOS

在 PⅡ350 计算机上对 LSAM 方法与 AM 方法所需的计算时间作比较，结果示于表

---

❶ Guo Jizhi, Liu Tao, Dai Yingchun, et al. Vapor-liquid equilibria of benzene and propylene at elevated temperature and pressure, Submitted to J Chem Eng Data.

1。显然 LSAM 方法所需的时间只是 AM 方法的几十分之一至几千分之一。尤其是首次设定精度越小，所需时间就越少，因而比 Eubank[6] 的二次搜索方法更加有效，而 Eubank 的二次搜索方法只能节约百分之几十的时间。由图 1 也可以看到这一点。二次搜索的区间 [C, D] 的搜索节点数一般只比首次搜索节点数少百分之几十。只有在接近临界点的平衡组成时，由于两相的平衡组成比较接近，搜索节点数才会相差较大。

3.3 LSAM 的有效性

LSAM 方法的有效性是指两个方面。一是指确保搜索得到的正值最大面积是整体最大，而非局部最大。由于状态方程除了在临界点存在奇异性外，其他点基本上是连续和光滑的。如图 1 所示，利用状态方程计算得到的吉布斯能曲线无强烈振荡现象，由此从数学理论上能够说明取首次搜索得到的近似平衡组成点的邻域做再次搜索的区间的方法（LSAM）可以保证得到的是整体最大。但是在临界点附近由于气液两相的平衡组成相距越来越近，如果首次选用的精度值大于气液平衡组成的距离，就会导致首次搜索失败，或者因面积计算的误差较大也会导致搜索失败。这一点可以通过选取适当的首次搜索精度来克服，只是首次设定的搜索精度太高会增加计算所需的时间。二是 LSAM 法和 AM 法计算的结果在同样的精度下应该是完全一致的。由于采用的首次精度是 10^{-m}，二次搜索的邻域是以上次搜索得到的近似平衡组成为中心和上次等分间距的两倍为宽度所形成的。这一邻域再用 $N = 2 \times 10$ 等分。所以这些邻域的新的等分点刚好是 AM 法中用精度 $10^{-(m+1)}$ 等分的点。多次等分依次类推。因此两种方法所搜索的点本质上是相同的，其搜索的结果应该一致。本文对上述实验数据使用两种方法进行计算，得到了完全一致的结果，也证明了这一结论的正确性。

4 结论

文章提出的多次局部搜索方法，既保持了 AM 方法计算平衡组成的整体有效性与无需初值和导数的优点，又大大提高了收敛速率和缩短了计算时间。其次，对 LSAM 方法的有效性和整体性做了理论分析，并利用实验数据对两种方法在微机上进行了统一性和一致性检验。LSAM 方法的建立为许多化工过程，尤其是高温高压的化工过程的在线或离线优化提供了计算平衡数据的有效和快捷的方法，对化工过程的控制和优化具有现实的意义。

致谢

本文由国家自然科学基金委员会和中国石化集团公司支持。

参 考 文 献

[1] 郭继志，袁渭康. 初议超临界条件下的化学反应 [J]. 化工进展，2000, 19 (3)：8~13.

[2] Brennecke J F, Eckert C A Phase equilibria for supercritical fluid process design [J]. AIChE J 1989, 35 (9)：1409~1427.

[3] Ammar M N, Renon H The isothermal flash problem: New methods for phase split calculations [J]. AIChE J 1987, 33 (6)：926~939.

[4] Siepmann J L, Karabornl S, Smit B. Simulating the critical behavior of complex fluids [J]. Nature, 1993, 365 (23)：330~332.

[5] Cairns B P, Furzer I A. Multicomponent three-phase azeotropic distillation. Part Ⅱ. Phase-stability and phase-splitting algorithms [J]. Ind Eng Chem Res. 1990, 29 (7): 1364~1382.

[6] Eubank P T, Elhaasan A E, Barrufet M A. Area method for prediction of fluid-phase eequilibria [J]. Ind Eng Chem Res, 1992, 31 (3): 942~949.

[7] 郭天民. 多元汽—液平衡和精馏 [M]. 北京: 化学工业出版社, 1983.

[8] 童景山, 李敬. 流体热物理性质的计算 [M]. 北京: 清华大学出版社, 1982.

The Application of Local Search for Maximum Area to High Pressure Phase Equilibrium

Guo Jizhi, Liu Tao, Yuan Weikang

(State Key Laboratory of Chemical Reaction Engineering, East China University of Science and Technology, Shanghai, 200237, China)

Abstract The area method (AM) that minimizes the Gibbs energy by integrating, instead of differentiating, the Gibbs energy curve provides a sufficient condition for global Gibbs energy minimization rather than only a necessary condition provided by the flash equilibrium calculation method. However, it uses the method of exhaustion to search the positive maximum area, therefore requires a long calculation time. Here this method is modified as LSAM (local search area method), which is a repeated search shrinking vicinities of the liquid and gas equilibrium composition points located by previous rough search. The calculation time decreases to one over several thousands of AM's or less. The experimental equilibrium data of benzene and propylene are used to demonstrate the merit of this method.

Key words Gibbs energy, VLE high pressure, VLE calculation method, benzene, propylene

高压脉冲电晕放电等离子体降解废水中苯酚*

摘　要　考察了多种因素对高压脉冲电晕放电低温等离子体法降解水中苯酚效果的影响。提高脉冲电压峰值和放电频率、延长放电时间等均可大大提高降解效果。自由基清除剂（如正丁醇）和缓冲剂（如硼酸盐）的存在均会显著降低降解效果。100mg/L 苯酚废水溶液放电处理 180min，最高降解率达 67.3%。当放电处理 420min 时，废水的 TOC 下降 83.8%。

关键词　高压脉冲放电　等离子体　降解　苯酚　废水处理

废水的毒性成分有重金属离子和毒性有机物等。重金属离子可通过电沉积法有效回收[1]，而废水中溶解性毒性有机物，如酚用常规工艺如生物处理法难以将其有效降解。在难降解有机废水处理技术中，近年来较快地发展了以产生氧化自由基为主的高级氧化技术。利用高活性自由基进攻有机物分子可破坏其结构和去除有机物。根据产生自由基的方式和反应条件的不同，高级氧化技术可分为湿式空气（催化）氧化法、超临界水（催化）氧化法、光化学（催化）氧化法、低温等离子体氧化法等[2~4]。本文利用高压脉冲电晕放电产生低温等离子体[5,6]降解以苯酚为代表的水中酚类化合物。

1　实验方法

用分析纯苯酚和超纯水配制模拟废水。实验装置由脉冲电源、放电反应器和辅助设备组成（图1）。脉冲电压波形如图2所示。放电反应器材质为有机玻璃（直径50mm，高200mm）。放电电极为不锈钢注射针头，接地电极为直径40mm的不锈钢圆板。废水体积为150mL，温度为室温。外通氧气流量为200mg/L。苯酚初始浓度为100mg/L。放电间距为3.0cm。脉冲电源参数为：脉冲电压峰值≤50kV、脉冲宽度≤400ns、脉冲前沿上升时间≤50ns、放电频率≤250Hz。

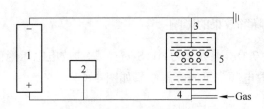

图1　高压脉冲放电降解水中有机物示意图

Fig. 1　Schematic diagram of pulsed high-voltage discharge degradation of organic compounds in aqueous solution

电压测量系统采用美国泰克公司的 P6105A 高压探头和 100MHz 示波器。溶液的 pH 值和电导率分别采用 PHB-1 型酸度计和 DDS-11C 型电导仪测量。用分光光度法（GB 7490—87）分析废水中苯酚的浓度。降解率（η）的定义为放电前后苯酚的浓度变化率。

---

* 本文合作者：陈银生、张新胜。原发表于《环境科学学报》，2002，22（5）：566~569。

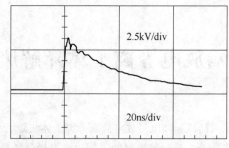

图 2　峰压为 12kV 的脉冲电压波形图

Fig. 2　An example of the waveform of the voltage

2　结果与讨论

2.1　脉冲电压峰值对苯酚降解效果的影响

废水初始 pH 值为 7.0，初始电导率为 80μS/cm，放电频率为 60Hz，放电电极直径为 0.9mm。不同电压峰值下的降解结果如图 3 所示。

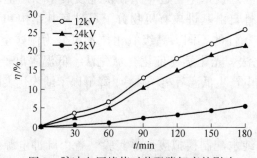

图 3　脉冲电压峰值对苯酚降解率的影响

Fig. 3　Effect of peak voltage on degradation of phenol

放电间距一定时，提高脉冲电压峰值可以增强电极间的电场强度，因而可以提高自由电子能量和速度，导致由电子轰击产生的各种自由基和臭氧等氧化性粒子含量增加以及产生的紫外光强度增强。所有这些现象都将促进苯酚的降解。

2.2　放电频率对苯酚降解效果的影响

废水初始 pH 值为 7.0、电导率为 80μS/cm，脉冲放电电压峰值为 32kV，放电电极直径为 0.9mm。不同放电频率下的降解结果如图 4 所示。

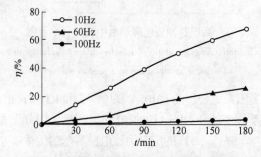

图 4　脉冲放电频率对苯酚降解率的影响

Fig. 4　Effect of discharge frequency on degradation of phenol

放电频率决定了单位时间内向废水体系中注入能量的多少。提高放电频率使得单位时间内产生的高能电子数量增加，因而产生的氢氧自由基、活性氧、臭氧等强氧化剂数量增加，导致废水中苯酚被氧化降解的速率提高。但由于放电产生的各种氧化剂数量并不与放电频率成线性关系，而且这些氧化剂之间还在水分等作用下相互转化，因此，在相同的放电时间内苯酚的降解率并不与放电频率成正比。另外，高压脉冲放电等离子体通道内的高温、高压和由此产生的紫外光也会促进有机物的降解[7]。而低频率放电产生的高温、高压和紫外光降解效应远远低于高频率放电产生的上述降解效应。因此在本实验中100Hz时的苯酚降解率远远高于60Hz和10Hz时的降解率。

图3和图4表明，提高脉冲峰压和放电频率可以显著提高有机物的降解率。但是放电行为受到溶液的性质（如电导率和温度）、脉冲电源参数（如脉冲峰压和放电频率）以及放电电极直径和放电间距等条件的影响。在上述实验条件下，32kV和100Hz是保持脉冲电晕放电所能达到的最高峰压和最大放电频率，再提高脉冲峰压和放电频率将会出现火花放电和溶液被击穿的现象。在脉冲宽度和脉冲前沿上升时间等电源参数不变和放电方式不变的前提下，选择最佳脉冲峰压和放电频率的原则是有机物的降解量与消耗的能量比达到最大。

2.3 自由基清除剂和缓冲剂对降解效果的影响

醇是一种有效的自由基清除剂[8]，而硼酸盐是常用的缓冲剂。本文考察了正丁醇和硼酸钠的存在对放电降解过程的影响。脉冲电压峰值为32kV，脉冲频率为60Hz，电极直径为0.6mm。当考察正丁醇对降解效果的影响时，废水初始pH=7.0、电导率为80μS/cm；当考察硼酸钠对苯酚降解效果的影响时，废水初始pH=10.0、电导率为1000μS/cm。苯酚降解结果如图5和图6所示。

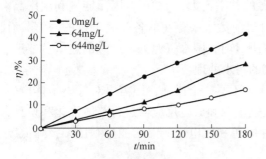 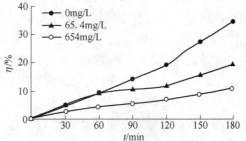

图5　正丁醇含量对苯酚降解率的影响　　　图6　硼酸钠含量对苯酚降解率的影响
Fig. 5　Effect of n-butanol　　　　　　　Fig. 6　Effect of sodium borate
　　on degradation of phenol　　　　　　　　　on degradation of phenol

正丁醇与·OH反应速率常数大于等于10^8L/(mol·s)[8]，正丁醇的存在能明显地抑制苯酚的降解。当硼酸钠含量很低时就可显著地抑制苯酚的降解。缓冲剂中的阴离子是·OH的淬灭剂[9]。·OH含量的下降，降低了苯酚降解速率。这说明放电产生的·OH对苯酚的降解起重要作用。

2.4 废水TOC的变化

当脉冲电压峰值为32kV，脉冲频率为60Hz，电极直径为0.6mm，废水初始pH值和电导率分别为7.0和80μS/cm时，苯酚废水溶液的总有机碳含量与放电时间的关系如图 7 所示。用 $Ca(OH)_2$ 可以检测到放电过程中有 CO_2 生成。

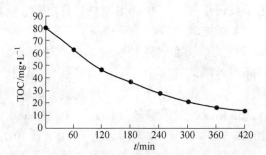

图 7 TOC 与降解时间的关系

Fig. 7 TOC vs discharge time

2.5 脉冲放电法的应用前景和存在的主要问题

脉冲放电等离子体法是一种兼具高能电子辐射、臭氧氧化和光化学氧化三种作用于一体的全新概念的废水处理技术。因而处理效果优于上述三种方法单独使用时的效果[10]。在其他高级氧化技术的特点中，湿式氧化法和超临界氧化法的反应速率和反应程度高，但要求的温度和压力都极高，对反应器材质要求高，能量消耗也很大，因而其推广应用受到一定程度的限制。光化学氧化法的条件温和、氧化能力强，但光能利用率低，消耗的时间太长。寻找稳定、高效和宽波长的光催化剂成为该法实现工业化的关键。利用超声空化效应的声化学氧化法能够使几乎任何污染物完全氧化降解。但超声波发生装置复杂，而且空化高温使废水温度升高导致能量利用率下降。本文的实验数据表明，脉冲放电法处理污染物效果好。与上述其他高级氧化技术相比，由于脉宽和脉冲前沿上升时间均为 ns 级，脉冲电源的载空比大于 $1:10^4$，因而能量利用率高。而且该技术要求的条件温和，操作简便，对目前污水中存在的各类有机物的降解具有广泛的适用性，不产生二次污染，因而是解决目前我国水资源污染严重的造纸、酿造、制药和印染等工业废水问题的一项前景广阔的污水处理技术。

但就应用技术本身来讲，脉冲放电法目前还不完全成熟，还需要对等离子体产生机理和有机物降解机理进行深入研究，对放电电源和放电反应器进行整体优化和设计，从而进一步提高脉冲电源的稳定性以及脉冲电源和放电反应器的能量利用率，降低废水处理成本。

3 结论

（1）提高脉冲电压峰值和放电频率均可明显地加快废水中苯酚的降解。

（2）自由基清除剂正丁醇和缓冲剂硼酸钠可以与氢氧自由基作用，因而显著地降低苯酚的降解速率。

（3）废水的 TOC 不断下降和 CO_2 的生成表明废水中的有机物不断被矿化。

参 考 文 献

[1] 郑远扬. 污水的电化学处理 [J]. 化工进展，1990，2：34～42.

[2] 齐军，顾温国，李劲. 水中难降解有机物氧化处理技术的研究现状和发展趋势 [J]. 环境保护，2000，3：17～19.

[3] 韦朝海,侯轶.难降解毒性有机污染物废水高级氧化技术[J].环境保护,1998,11:29~31.

[4] 徐中其,戴航,陆晓华.难降解有机废水处理新技术[J].江苏环境科技,2000,13(1):32~35.

[5] 李胜利,李劲,王泽文,等.用高压脉冲放电等离子体处理印染废水的研究[J].中国环境科学,1996,16(1):73~76.

[6] 文岳中,姜玄珍,吴墨.高压脉冲放电降解水中苯乙酮的研究[J].中国环境科学,1999,19(5):406~409.

[7] 苏建龙,黄卫东.液电脉冲等离子降解高浓度有机废液的机理研究[J].环境科学动态,1997,2:13~16.

[8] Sato M, Ohgiyama T, Clements J S. Formation of chemical species and their effects on microorganisms using a pulsed high voltage discharge in water [J]. IEEE Transactions on Industry Application, 1996, 32(1):106~112.

[9] Sharma A K, Locke B R, Arce P, et al. A preliminary study of pulsed streamer corona discharge for the degradation of phenol in agueous solutions [J]. Hazardous Waste & Hazardous Materials. 1993, 10 (2):209~219.

[10] 孙亚兵,任兆杏.低温等离子体技术在三废处理中的应用与研究[J].环境保护科学,1997,23(5):45~48.

A Preliminary Study of Pulsed High-voltage Corona Discharge Plasma for the Degradation of Phenol in Agueous Solution

Chen Yinsheng, Zhang Xinsheng, Yuan Weikang

(State Key Laboratory of Chemical Reaction Engineering, East China University of Science and Technology, Shanghai, 200237, China)

Abstract Effects of some factors were studied on the degradation of phenol in aqueous solution with pulsed high-voltage corona discharge plasma. The degradation of phenol could be raised considerably by increasing the peak voltage and discharge frequency and by prolonging the discharge time. The existence of radical scavenger such as n-butanol and buffer such as borate could decrease the degradation of phenol. The maximum degradation of 100mg/L phenol solution was 67.3% after discharging for 180 minutes. The TOC of wastewater decreased by 83.8% when discharge time was 420 minutes.

Key words pulsed high-voltage discharge, plasma, degradation, phenol, wastewater treatment

脉冲电晕放电等离子体降解含 4-氯酚废水*

摘 要 考察了多种因素对高压脉冲电晕放电等离子体降解废水中 4-氯酚效果的影响,同时对 4-氯酚降解过程的动力学进行了研究。提高脉冲电压峰值和气体的流量以及降低废水溶液的电导率均可提高 4-氯酚的降解效果,而醇类化合物的存在将明显降低 4-氯酚的降解率。4-氯酚的降解过程符合一级反应,降解速率常数与降解温度的关系符合 Arrhenius 公式。当废水的初始 pH 值为 7.0、电导率为 $80\mu S/cm$、脉冲电压峰值为 30kV、放电频率为 60Hz、放电电极直径为 0.6mm、放电距离为 3.0cm 时,指前因子 $A = 1.365 \times 10^{-2} min^{-1}$,实验活化能 $E_a = 5.129 kJ/mol$。得到了降解速率常数与脉冲电压峰值、放电频率、放电距离和初始氧气流量的关系。

关键词 脉冲电晕放电 等离子体 4-氯酚 降解 动力学 废水处理

1 引言

水体中存在一些可溶性毒性有机物,用常规工艺很难将其有效除去[1]。针对难降解有机废水问题,近年来较快地发展了以产生氧化自由基为主的废水处理技术,如湿式氧化法、超临界氧化法、光化学氧化法、等离子体氧化法等[2~4]。利用在水溶液中进行直接高压脉冲电晕放电产生低温等离子体降解以 4-氯酚为代表的含难降解有机化合物的废水也有报道[5~9]。本文研究了该技术多种工艺参数对 4-氯酚降解效果的影响,测定了 4-氯酚降解过程的速率常数以及速率常数与降解温度、脉冲峰压、放电频率、放电距离和氧气流量等的关系。

2 实验方法

用化学纯 4-氯酚和超纯水配制模拟废水。实验装置由脉冲电源(大连理工大学制造)、放电反应器(自制)和辅助设备组成(图1)。脉冲电压波形如图2所示。放电反应器材质为有机玻璃($\phi 50mm \times 200mm$)。放电电极为不锈钢注射针头,接地电极为 $\phi 40mm$ 的不锈钢圆板。脉冲电源参数为:脉冲电压峰值 $\leq 50kV$,脉冲宽度 $\leq 400ns$,脉冲前沿上升时间 $\leq 50ns$,放电频率 $\leq 250Hz$。除特殊说明外,本文所用的工艺条件如下:$c_0 = 100mg/L$,$V_p = 30kV$,$f = 60Hz$,$d = 0.6mm$,$l = 3.0cm$,废水初始 pH 值 7.0,$\kappa = 80\mu S/cm$,$Q_0 = 200mL/min$。每次实验用废水量为 180mL,模拟废水的温度控制在 $25℃ \pm 1℃$。

电压测量系统采用美国泰克公司的 P6105A 高压探头和 100MHz 示波器。溶液的 pH 值和 k 分别采用 PHB-1 型酸度计和 DDS-11C 型电导率仪测量。用 HP1100 型高效液相色谱仪分析废水中 4-氯酚的浓度。η 定义为放电前后 4-氯酚的浓度变化率。

---

* 本文合作者:陈银生、张新胜、戴迎春。原发表于《化工学报》,2003,54(9):1269~1273。

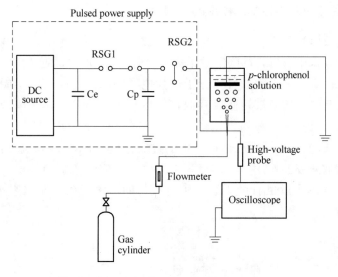

Fig. 1　Schematic diagram of experimental setup

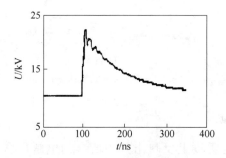

Fig. 2　Example of voltage waveform of pulsed discharge

3　结果与讨论

3.1　脉冲电压峰值对 4-氯酚降解率的影响

脉冲电压峰值是脉冲放电的主要参数，电压峰值对 4-氯酚降解率的影响如图 3 所示。放电距离一定时，提高脉冲电压峰值可以增强电极间的电场强度，因而可以提高自由电子能量和速度，导致放电过程中产生的臭氧量上升、电子轰击产生的各种氧化剂含量增加以及产生的紫外光强度增强，使 4-氯酚的降解率上升。

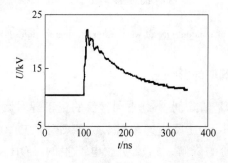

Fig. 3　Effects of peak voltage on degradation of 4-chlorophenol

3.2 气体种类和流量对4-氯酚降解率的影响

在高压脉冲放电过程中，气体放电比液体放电更容易发生。从放电电极通入的气体种类和流量对废水中有机物的降解效果影响显著。O_2 和 N_2 的通入对4-氯酚降解率的影响如图4所示。鼓入废水中的气体分子被强电场中的高能高速电子轰击成为各种自由基，这些自由基再直接或被转变成其他粒子后与4-氯酚分子作用以破坏有机物。鼓入 O_2 比鼓入 N_2 更能使4-氯酚被氧化降解。这是因为 O_2 放电可产生 O_3，而 N_2 放电却不能产生 O_3[8]。这表明 O_3 以及由 O_3 产生的过氧化氢（H_2O_2）和氢氧自由基（·OH）对4-氯酚的降解起重要的作用。气体流量上升时，由放电产生的各种自由基密度也上升[8]，因此提高气体流量也有助于废水中有机物的降解。在工业化废水处理过程中，可用空气替代 O_2。

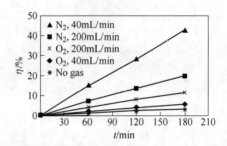

Fig. 4　Effects of gas bubbling in wastewater on degradation of 4-chlorophenol

3.3 废水电导率对4-氯酚降解率的影响

废水的电导率在一定程度上反映出其中所含的电解质含量。在实验过程中，用 K_2SO_4 调节废水溶液的电导率。不同电导率下的4-氯酚降解结果如图5所示。高电导率将抑制放电过程中流光（streamer）的产生[7]，因此可以认为高电导率废水溶液中由放电产生的各种自由基减少，从而导致4-氯酚降解率下降。

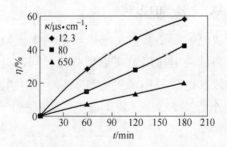

Fig. 5　Effects of conductivity of wastewater on degradation of 4-chlorophenol

3.4 正丁醇对4-氯酚降解率的影响

醇是一种有效的自由基清除剂[10]。不同含量的正丁醇对废水中4-氯酚降解率的影响如图6所示。正丁醇与·OH反应速率常数大于等于 $10^8 L/(mol·s)$[8]，正丁醇的存在能明显地抑制4-氯酚的降解。这说明放电产生的·OH对4-氯酚的降解起重要作用。

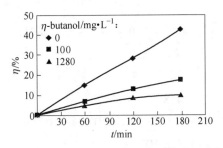

Fig. 6 Effects of addition of *n*-butanol on degradation of 4-chlorophenol

3.5 4-氯酚降解过程动力学研究

3.5.1 有机物降解机理

高压脉冲电晕放电过程中有机物降解位置有空化泡内、空化泡与本体溶液之间的气液界面以及本体溶液内。其中空化泡内存在高温热解和 O_3 氧化，空化泡气液界面上存在 O_3 氧化，本体溶液中存在 ·OH 氧化、H_2O_2 氧化和 O_3 氧化以及热解，另外还有紫外光氧化等，可见反应机制相当复杂。一般脉冲放电等离子体降解有机物速率方程可表述为：

$$r_T \approx r_{g,py} + r_{g,oz} + r_{surf,oz} + r_{b,·OH} + r_{b,hp} + r_{b,oz} + r_{b,py} + r_{uv} \tag{1}$$

由式（1）可见，很难给出研究对象的精确动力学方程式。为了得出有机物降解动力学模型，很有必要针对污染物和废水的特性以及具体的降解条件对降解过程进行合理的简化。

3.5.2 4-氯酚降解机理的简化

由于 4-氯酚是亲水性和难挥发性有机物，而且空化泡气液界面为疏水性[11]，4-氯酚很难扩散进入空化泡内。与本体溶液中大量 4-氯酚被降解相比，4-氯酚在空化泡内和空化泡气液界面上的降解量估计都可以忽略。因此，4-氯酚降解反应可视为主要发生在本体溶液中。3.2、3.3 和 3.4 节的工艺实验表明，在放电过程中，O_3、由 O_3 与 H_2O 作用产生的 H_2O_2 和 ·OH 以及由高能电子辐射水分子直接产生的 ·OH 对 4-氯酚的降解起主要作用。因此：

$$r_T \approx r_{b,·OH} + r_{b,hp} + r_{b,oz}$$

由于放电过程中 O_3、H_2O_2 和 ·OH 等氧化性粒子不仅与有机物反应，而且它们之间也相互作用和相互转化，因此脉冲放电降解有机物动力学模型的建立还依赖实验测定。

3.5.3 4-氯酚降解反应级数和降解速率常数的测定

废水中 4-氯酚的初始浓度 c_0 分别为 101.8mg/L、104.6mg/L、118.1mg/L、110.2mg/L 和 107.6mg/L。对实验数据进行分析发现，285.15K、295.15K、305.15K、315.15K 和 324.15K 下 4-氯酚浓度的对数值 $\ln c$ 与放电时间 t 均成线性关系，而且线性相关系数 R^2 均大于等于 0.9873（图 7）。因此，可认为高压脉冲放电降解 4-氯酚过程符合一级反应，即：

$$c = c_0 e^{-kt}$$

由于降解速率常数的对数值 $\ln k$ 与降解温度的倒数 $1/T$ 的线性相关系数接近 0.990（图 8），可认为 k 与 T 的关系符合 Arrhenius 公式，即：

$$k = Ae^{-E_a/(RT)}$$

式中，$A=1.365\times10^{-2}\text{min}^{-1}$；$E_a=5.129\times10^{3}\text{J/mol}$。同时还研究了 V_p、f、l 和 Q_0 对 k 的影响，在一定的放电条件下有如下关系：

$$k = 10^{-8}V_p^{4.221}$$

$$k = 4.872\times10^{-6}f^{1.806}$$

$$k = -1.186\times10^{-2}\ln l + 0.02156$$

$$k = 5\times10^{-5}Q_0^{1.0026}$$

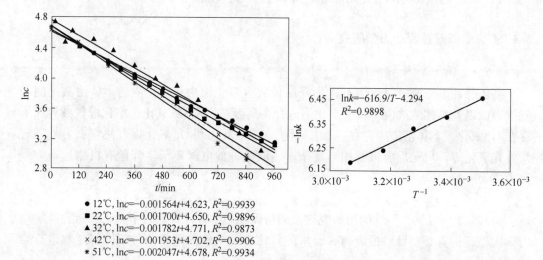

lnc
t/min
● 12℃, lnc=−0.001564t+4.623, R²=0.9939
■ 22℃, lnc=−0.001700t+4.650, R²=0.9896
▲ 32℃, lnc=−0.001782t+4.771, R²=0.9873
× 42℃, lnc=−0.001953t+4.702, R²=0.9906
＊ 51℃, lnc=−0.002047t+4.678, R²=0.9934

Fig. 7 Correlation between lnc and discharge time

lnk=−616.9/T−4.294
R²=0.9898
−lnk
T⁻¹

Fig. 8 Correlation between rate constant and temperature

4 结论

（1）提高脉冲电压峰值和所通气体的流量以及降低废水溶液的电导率均可提高4-氯酚的降解效果。

（2）醇类化合物的存在明显降低4-氯酚的降解率。

（3）4-氯酚的降解主要是通过放电产生的臭氧、过氧化氢和氢氧自由基的氧化实现的。

（4）4-氯酚的降解过程符合一级反应：$c=c_0e^{-kt}$。降解速率常数 k 与降解温度 T 的关系符合 Arrhenius 公式：$k=Ae^{-E_a/(RT)}$。

符号说明

c　4-氯酚浓度，mg/L

d　放电电极直径，mm

f　放电频率，Hz

l　放电距离，cm

Q　气体流量，mL/min

r　4-氯酚降解速率，mg/(L·min)

t　降解时间，min

V_p 脉冲峰压，kV
κ 溶液电导率，μS/cm
η 4-氯酚降解率（$\eta = 1 - c/c_0$）

下角标
b 本体溶液
g 空化泡内
hp 过氧化氢
O 氧气
oz 臭氧
py 热解
surf 空化泡气液界面
T 总和
uv 紫外光
0 初始

参 考 文 献

[1] Shi Hanchang（施汉昌），Zhao Yinhui（赵胤慧），Ji Jingping（冀静平）. Study and Progress of Biotreatment of Chlorophenol in Wastewater. Chemistry（化学通报），1998（8）：1～5.

[2] Qi Jun（齐军），Gu Wenguo（顾温国），Li Jin（李劲）. Current Research of Oxidation Technology of Nondegradable Organics in Water and Its Trend. Environmental Protection（环境保护），2000（3）：17～19.

[3] Wei Zhaohai（韦朝海），Hou Yi（侯轶）. Advanced Oxidation Techniques on the Treatment of Refractory Toxic Organic Contaminations in Wastewater. Environmental Protection（环境保护），1998（11）：29～31.

[4] Xu Zhongqi（徐中其），Dai Hang（戴航），Lu Xiaohua（陆晓华）. New Techniques on the Treatment of Nondegradable Organics in Wastewater. Jiangsu Environmental Science and Technology（江苏环境科技），2000，13（1）：32～35.

[5] Li Shengli（李胜利），Li Jin（李劲），Wang Zewen（王泽文），Yao Honglin（姚宏霖），Gao Qiuhua（高秋华），Yin Xiaogen（尹小根）. Study on Treatment of Dyeing Wastewater by High Voltage Pulse Discharge Plasma. China Environmental Science（中国环境科学），1996，16（1）：73～76.

[6] Wen Yuezhong（文岳中），Jiang Xuanzhen（姜玄珍），Wu Mo（吴墨）. Preliminary Study of Destruction of Acetophenone in Waster by High Voltage Pulse Discharges. China Environmental Science（中国环境科学），1999，19（5）：406～409.

[7] Sharma A K，Locke B R，Arce P，Finney W C. A Preliminary Study of Pulsed Streamer Corona Discharge for the Degradation of Phenol in Aqueous Solutions. Hazardous Waste & Hazardous Materials，1993，10（2）：209～219.

[8] Sun B，Sato M，Clements J S. Use of a Pulsed High Voltage Discharge for Removal of Organic Compounds in Aqueous Solution. J. Phys. D：Appl. Phys.，1999，32：1908～1915.

[9] Wen Yuezhong（文岳中），Jiang Xuanzhen（姜玄珍），Liu Weiping（刘维屏）. Degradation of 4-Chlorophenol in Aqueous Solution by High-voltage Pulsed Discharge-ozone Technology. Environmental Science（环境科学），2002，23（2）：73～76.

[10] Sato M，Ohgiyama T，Clements J S. Formation of Chemical Species and Their Effects on Microorganisms Using a Pulsed High Voltage Discharge in Water. IEEE Transactions on Industry Application，

[11] Riesz P, Kondok, Krishna C M. Sonochemistry of Volatile and Non-volatile Solutes in Aqueous Solutions: e. p. r. and Spin Trapping Studies. Ultrasonics, 1990, 28: 295~303.

Degradation of 4-chlorophenol in Aqueous Solution with Pulsed High-voltage Discharge Plasma

Chen Yinsheng, Zhang Xinsheng, Dai Yingchun, Yuan Weikang

(State Key Laboratory of Chemical Reaction Engineering, East China University of Science and Technology, Shanghai, 200237, China)

Abstract The effects of some factors on degradation of 4-chlorophenol in waste water with pulsed high-voltage discharge plasma were studied. Furthermore, the kinetics of degradation of 4-chlorophenol was also studied. The degradation efficiency of 4-chlorophenol could be raised considerably by increasing peak voltage, bubbling rate of gas and by decreasing electric conductivity of waste water. The presence of alcohol would decrease the degradation efficiency of 4-chlorophenol. The degradation of 4-chlorophenol in waste water with pulsed high-voltage plasma was a first-order reaction. The correlation between degradation rate constant k and reaction temperature T could be expressed by Arrhenius eguation. At initial pH 7.0, electric conductivity of waste water 80μS/cm, peak voltage 30kV, discharge freguency 60Hz, the diameter of discharge electrode 0.6mm, and discharge distance were 3.0cm, A was $1.365 \times 10^{-2} min^{-1}$ and experimental activation energy E_a was 5.129kJ/mol. The degradation rate constant k could also be expressed.

Key words pulsed high-voltage discharge, plasma, 4-chlorophenol, degradation, kinetics, waste water treatment

密闭体系中乳酸的固相缩聚[*]

摘　要　研究在真空密闭、存在脱水剂的条件下聚乳酸固相缩聚工艺的可行性，得到了相对分子质量达 25 万的聚乳酸；实验研究了催化剂浓度、预聚体粒度和反应时间对缩聚过程的影响，表明降解副反应仍然是控制聚乳酸最终分子量的重要因素。
关键词　聚乳酸　固相缩聚　脱水剂　密闭

聚乳酸（PLLA）是一种无毒、无刺激性，具有良好生物相容性和生物降解性能的合成高分子[1]。作为一种很有希望的生物医用材料和环保型高分子材料，正越来越受到人们的关注。目前，聚乳酸的制备方法主要有两种：开环聚合和直接缩聚。其中开环聚合法是目前聚乳酸工业生产的主要工艺，和直接法相比较容易获得高相对分子质量的聚乳酸。但该工艺的主要难点在于丙交酯的制备和提纯，不仅流程长，而且能耗大，使聚乳酸的生产成本居高不下，妨碍了它的大规模生产应用。因此，许多研究者都将目光投向了直接法制备高分子量的聚乳酸[2~5]。

乳酸在聚合过程中涉及两个主要的反应平衡，即脱水平衡和降解平衡。

（1）脱水平衡：

$$H{\vphantom{)}}{\left(O-CH(CH_3)-CO\right)}_x OH + H{\left(O-CH(CH_3)-CO\right)}_y OH \rightleftharpoons H{\left(O-CH(CH_3)-CO\right)}_{x+y} OH + H_2O \quad (1)$$

（2）降解平衡：

$$H{\left(O-CH(CH_3)-CO\right)}_n OH \rightleftharpoons H{\left(O-CH(CH_3)-CO\right)}_{n-2} OH + \text{丙交酯} \quad (2)$$

为了得到高相对分子质量的聚乳酸，所采用的工艺条件必须有利于脱水反应的进行，并同时抑制降解副反应。为此，我们提出了在密闭环境中利用吸附剂选择性脱除水的聚乳酸固相缩聚新工艺[6]。该工艺不仅充分利用了固相缩聚的优点，而且避免了连续抽真空。由于反应体系中能保持较高的丙交酯分压，因此可在一定程度上抑制热降解副反应的发生。

本文首次验证了该工艺的可行性，并通过实验研究了工艺条件对聚乳酸相对分子质量的影响。

1　实验部分

实验的主要原料乳酸为荷兰 PURAC 公司产品（PURAC(r) HS88），其中含有 12%

---

[*]　本文合作者：钱刚、朱凌波、周兴贵。原发表于《高分子材料科学与工程》，2004, 20 (2)：51~53, 56。

的水，乳酸中L-乳酸含量大于95%。脱水剂选用CaO，不仅因为CaO的相对分子质量小，而且作为脱水剂使用时可在反应温度下维持很低的水分压；产物Ca(OH)$_2$也易于处理。

多种质子酸和路易斯酸都可作为乳酸缩聚催化剂。研究表明强质子酸能够阻止丙交酯的形成，对热降解有一定的抑制作用。Moon等[4]研究了SnCl$_2$作为催化剂，不同质子酸作为共催化剂对乳酸直接缩聚的影响，发现对甲苯磺酸（TSA）能够有效地抑制降解副反应，而且能够抑制外消旋和变色等副反应，有利于获得较高相对分子质量的聚乳酸。因此，本实验采用SnCl$_2$·2H$_2$O和TSA作为聚乳酸合成催化剂。实验中所用的SnCl$_2$·2H$_2$O和TSA都为分析纯，分别为上海试四赫维化工有限公司和上海凌峰化学试剂有限公司产品。

预缩聚（熔融缩聚）：预缩聚在三口烧瓶中进行，油浴加热。首先，控制油浴温度为110℃，在常压下脱水4h，再等摩尔添加催化剂SnCl$_2$·2H$_2$O和TSA。抽真空，同时提高油浴温度，在170℃反应8h。

终缩聚（固相缩聚）：预聚物经冷却、粉碎并筛分后得到不同目数的样品。样品在105℃经结晶预处理后分装于一系列内径为8mm、长约10cm、一端封闭的玻璃管中。为了避免脱水剂与聚乳酸直接接触，氧化钙粉末用铝箔疏松裹装后再置于玻璃管中。玻璃管另一端在抽真空条件下封闭，之后平放于恒温烘箱中于150℃进行固相缩聚。定期取出一根玻璃管，分析产品的相对分子质量。本文中聚乳酸相对分子质量为黏均相对分子质量。黏度测量在37℃进行，以四氢呋喃为溶剂。通过$[\eta]=1.04\times10^{-4}M_v^{0.75}$[7]计算黏均相对分子质量。

2 结果与讨论

首先，为了确定脱水剂CaO对缩聚过程的促进作用，分别在有和无脱水剂存在的条件下进行固相缩聚。表1列出了黏均相对分子质量随时间的变化。可见，存在脱水剂时，聚乳酸的相对分子质量迅速上升。脱水剂对缩聚过程的促进作用显而易见。

Table 1 The effect of the adsorbent CaO on the molecular weight of PLLA

Time/h	10	15	20
M_v (Adsorbent)	69000	105300	149700
M_v (no adsorbent)	7403	7166	15000

Polymerization conditions: $m_{SnCl_2\cdot 2H_2O}/m_{OLLA}=1.0\%$, $n_{SnCl_2\cdot 2H_2O}/n_{TSA}=1:1$, 150℃, 40~60mesh.

影响固相缩聚过程的因素众多，除预聚物的相对分子质量与分布外，反应温度、催化剂浓度、预聚物粒度和反应时间都对聚乳酸的相对分子质量有重要的影响。

固相缩聚温度不仅影响反应的速率，还影响小分子的扩散速率。如温度过低，反应速率太慢，达到规定的相对分子质量所需要的时间就很长；相反，如果温度过高，又可能因热降解副反应的加剧得不到高相对分子质量聚乳酸，而且也易于使聚合物变色。此外，过高的温度将使聚合物颗粒相互黏结。一般固相缩聚反应温度控制在聚合物熔点以下10~40℃。本文将固相缩聚温度固定在150℃。

催化剂是在熔融缩聚过程中加入的。由于乳酸直接熔融缩聚过程中存在着乳酸、水和聚乳酸的化学平衡及伴随着丙交酯的形成,这使得直接法制备高相对分子质量聚乳酸变得困难。实验发现催化剂质量分数分别为0.5%,1.0%和1.5%时,预聚物的相对分子质量分别为1.7万、1.5万和1.8万,相当接近。

提高催化剂浓度可促进固相缩聚反应速率,同时也促进了降解等副反应的发生。但在预聚反应中由于聚乳酸的相对分子质量较低,降解对相对分子质量大小影响不显著。而在固相缩聚过程中,由于相对分子质量较大,预聚物中残留的催化剂对聚乳酸相对分子质量的变化有重要的影响。

实验结果表明,适当提高催化剂的质量分数(从0.5%提高到1.0%)将在固相缩聚初期促进聚乳酸相对分子质量的增长,但后期因降解副反应占优,相对分子质量反而降低,见图1。而如果催化剂的质量分数太高,从一开始聚乳酸的相对分子质量就上升很慢,并且所能达到的最高相对分子质量也要低得多。

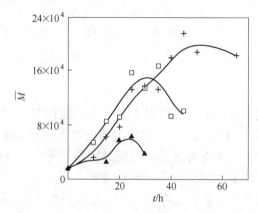

Fig. 1 The effect of mass fracture of catalyst (w) on the solid-state polycondensation
(polymerization conditions: $n_{SnCl_2 \cdot 2H_2O}/n_{TSA} = 1:1$, 150℃, 40~60mesh;
+—$w = 0.5\%$; □—$w = 1.0\%$; ▲—$w = 1.5\%$)

此外从图1还可看出反应时间对聚乳酸相对分子质量的影响。在缩聚过程初期,低聚物的浓度较大,链增长反应起主导作用,因此相对分子质量不断上升。而在反应后期,随着端基浓度的不断降低,链增长速率随之下降。与此同时,热降解反应却一直在进行,当聚乳酸的相对分子质量达到一定值时,降解反应占了主导地位,这使聚乳酸的相对分子质量反而下降了。在密闭环境中进行固相缩聚,尽管可维持气相较高的丙交酯分压,但如果聚合物中丙交酯浓度始终低于平衡浓度,就不能完全抑制生成丙交酯的副反应。另外,热降解副反应是一类反应,除反应(2)外,还包含生成其他相对分子质量大于丙交酯的环状物的降解反应。对于这些反应,密闭的作用更小。

图2为不同粒度预聚体对固相缩聚过程的影响。粒度将影响小分子(如水、丙交酯、乳酸等)的扩散速率。但是,由于固相缩聚温度较低,反应速率较慢,缩聚过程将由动力学控制。而图2中却清楚地显示,粒度较大时,在缩聚反应初期,相对分子质量上升更快。一种可能的解释是,小分子产物(水和乳酸)在聚合过程中起催化作用。而在缩聚后期,水浓度因脱水剂的作用降低,端基浓度随链增长而不断下降,降解成为控制相对分子质量的主要因素。此时,粒度大小对聚乳酸相对分子质量的变化几乎无影响。

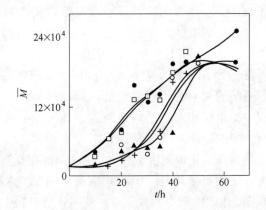

Fig. 2 The effect of particle size on the solid-state polycondensation
(polymerization conditions, $m_{SnCl_2 \cdot 2H_2O}/m_{OLLA} = 0.5\%$, $n_{SnCl_2 \cdot 2H_2O}/n_{TSA} = 1:1$, 150℃；
●—20~40mesh；□—40~60mesh；+—60~80mesh；▲—80~100mesh；○—100~120mesh）

3 结论

本文验证了在密闭和有脱水剂的情况下进行聚乳酸固相缩聚的可行性，得到了最高相对分子质量为 25 万的聚乳酸。并通过实验研究了催化剂浓度、预聚体粒度和反应时间对缩聚过程的影响，表明降解副反应仍然是控制聚乳酸最终相对分子质量的主要因素。但本工艺勿需连续抽真空，可降低操作成本，还能避免因丙交酯的损失而对收率的影响。同现有直接缩聚工艺相比有一定的优越性，因而有良好的工业化前景。

固相缩聚过程中聚乳酸的相对分子质量大小一直是链增长和降解反应相互竞争的结果。在缩聚反应初期因相对分子质量较低、降解不显著，如采用较高的反应温度无疑可促进聚乳酸相对分子质量的快速增长，缩短反应时间。但也由于固相缩聚反应初期相对分子质量较小，聚乳酸熔点较低，较高的温度将引起聚乳酸黏结。因此，在固相缩聚过程中，在保证聚乳酸颗粒不黏结的情况下，如果随相对分子质量的上升相应地提高反应温度，就有可能在较短的时间内获得较高的聚乳酸相对分子质量，以提高产率。这部分研究工作正在进行中。

参 考 文 献

[1] 李孝红 (LI Xiao-hong)，袁明龙 (YUAN Ming-long)，熊成东 (XIONG Cheng-dong). 高分子通报 (Polymer Bulletin)，1999 (1)：24~32.

[2] Fukushima T, Sumihiro Y, Koyanagi K, et al. Intern. Polymer Processing, 2000, 15 (4)：380~385.

[3] Ajioka M, Enomoto K, Suzuki K, et al. Bull. Chem. Soc. Jpn., 1995, 68 (8)：2125~2131.

[4] Moon S I, Lee C W, Miyamoto M, et al. J. Polym. Sci. Part A, Polym. Chem., 2000, 38：1673~1679.

[5] Moon S I, Lee C W, Taniguchi I, et al. Polymer, 2001, 42：5059~5062.

[6] 周兴贵 (ZHOU Xing-gui)，朱凌波 (ZHU Ling-bo)，袁渭康 (YUAN Wei-kang). 中国专利 (Zhongguo Zhuanli)，01113146.2 (2001).

[7] 孙俊全 (SUN Jun-quan)，崔力强 (CUI Li-qiang)，吴兰亭 (WU Lan-ting). 功能高分子学报 (Journal of Functional Polymers)，1996, 9 (2)：252~256.

Solid-state Polycondensation of Poly (*L*-lactic Acid) in a Closed System

Qian Gang, Zhu Lingbo, Zhou Xinggui, Yuan Weikang

(State Key Laboratory of Chemical Reaction Engineering, East China University of Science and Technology, Shanghai, 200237, China)

Abstract An innovative process for the production of high molecular weight poly (*L*-lactic acid) (PLLA) was presented, featuring solid-state polycondensation of *L*-lactic acid (LA) in a closed system with a dehydrant to remove water selectively. The final PLLA obtained has a molecular weight of 250000. The influences of different operating variables on the changes of molecular weight, as well as the highest molecular weight attainable were investigated experimentally, which reveals that thermal degradation is still the most important factor that controls the molecular weights of PLLA products.

Key words PLLA, solid-state polycondensation, dehydrant, closed system

催化剂活性组成对纳米碳纤维产率和微结构的影响[*]

摘　要　采用沉积-沉淀法制备了活性组成可控的镍铁系列催化剂，以 CO/H_2 为碳源，在 600℃下进行了纳米碳纤维的催化生长，考察了催化剂活性组成变化对纳米碳纤维产率和微结构的影响。结果表明，镍铁系列催化剂具有较好的催化活性，在反应 36h 内没有失活，催化剂中 Fe 的存在有利于提高纳米碳纤维的产率。表征结果表明，纳米碳纤维的直径分布较为均匀，在 20～50nm 之间，比表面积为 130～200m^2/g；纳米碳纤维中石墨层与轴之间夹角随催化剂中 Fe 含量的增大而增大；TPO 结果表明，Fe 的存在提高了纳米碳纤维的石墨化程度。结合纳米碳纤维的生长机理，认为活性组成的变化影响了 CO 与催化剂表面的反应和碳在金属中的扩散，进而影响纳米碳纤维的产率和微结构。

关键词　纳米碳纤维　催化生长　微结构　活性金属组成　沉积-沉淀法

纳米碳纤维因其优异的物理化学性质而在催化剂载体、电极材料、高效吸附剂（特别是储氢材料）和结构增强添加剂等方面具有较好的应用前景[1~7]。利用含碳气体在过渡金属（如 Ni，Fe 和 Co 等）催化剂上催化热解制得的纳米碳纤维具有纯度高、结构可调和可实现大规模生产等优点[1,3,8]。纳米碳纤维的微结构（包括石墨层空间堆积方式，石墨层间距和直径等）与其生长条件（如催化剂活性组成与制备方法、含碳气体种类和温度等因素）有关[9,10]。根据纳米碳纤维中石墨层的平面截面结构（以石墨层与中心轴的夹角 θ 为基准），可将其分为管式（$\theta=0°$）、鱼骨式（$0°<\theta<90°$）和片状纳米碳纤维（$\theta=90°$）。研究者对纳米碳纤维制备过程进行了大量的研究，但至今仍不能确定其微结构与制备条件的关系，对纳米碳纤维的微结构控制尚需进一步研究[8~11]。本文采用沉积-沉淀法制备了活性组成可控的系列镍铁催化剂，研究了 CO 在镍铁系列催化剂上分解形成纳米碳纤维的过程，考察了活性组成对纳米碳纤维的产率、纳米碳纤维的直径、石墨化程度和石墨层堆积方式的影响。

1　实验部分

1.1　镍铁系列催化剂的制备

采用沉积-沉淀法[12,13]，通过控制硝酸镍和铁氰化物的含量制备了以 Al_2O_3 为载体，金属活性组成不同的镍铁系列催化剂。其中，催化剂还原后的金属活性组分的负载量为 20%，不同 Ni/Fe 比例的沉淀配合物组成列于表 1 中。催化剂的制备过程如下：取 4.0g 磨碎至 200 目以上的活性 Al_2O_3 与 500mL 蒸馏水混合成悬浮液，用硝酸调节 pH 值至 5.0，搅拌 2h 后加入一定量的 $Ni(NO_3)_2·6H_2O$，继续搅拌，用计量泵在 2h 内缓慢加入 $Na_2Fe(CN)_5NO$ 溶液，室温下搅拌过夜，过滤后多次洗涤，所得滤饼在 120℃下干燥，在空气（流量 100mL/min）中升温至 600℃，焙烧 3h，再用 25% H_2/Ar（流量

---

[*] 本文合作者：赵铁均、朱贻安、李平、De Chen、戴迎春、Anders Holmen。原发表于《催化学报》，2004，25（10）：829~833。

160mL/min）于600℃还原3h，冷却至室温，用4% O_2/N_2 钝化2h，密封备用。

1.2 纳米碳纤维的催化生长与提纯

纳米碳纤维的生长在放置有石英管反应器的卧式电热炉中进行。装有1.0g催化剂的石英舟置于反应器（外径50mm，长度800mm）中部，进口气体流量通过质量流量计控制，预混后通入反应管。催化剂用25% H_2/Ar（流量160mL/min）于600℃还原3h，切换为80% CO/H_2（流量100mL/min），保持一定时间后停止加热和通气，冷却至室温并收集产品，称量分析。为比较纳米碳纤维的石墨化程度，将制得的纳米碳纤维在2mol/L硝酸中浸泡3d后过滤，在120℃下干燥、密封备用。

1.3 样品的表征

催化剂的XRD测试在日本Rigaku D/Max 2550VB/PC型X射线衍射仪上进行，Cu K_α 射线，$\lambda = 0.15405$nm，电压40kV，电流100mA，扫描范围20°~80°，扫描速率8°/min。N_2 吸附在Micromeritics ASAP2010型吸附仪上进行。纳米碳纤维样品在190℃和133Pa下真空脱气处理6h，在液氮温度下进行 N_2 吸附实验。TEM测试在日本JOEL JEM2010型透射电子显微镜上进行，首先将样品制成纳米碳纤维-乙醇悬浮液，经超声分散后滴加到碳筛上，再进行TEM分析。纳米碳纤维的程序升温氧化（TPO）实验在瑞士Mettler-Toledo STA热重分析仪上进行，CO_2 为氧化气体，流量60mL/min，以10℃/min的速率升温至1200℃。

2 结果与讨论

2.1 催化剂的XRD结果

图1为还原后不同催化剂的XRD谱。可以看出，Ni/Al_2O_3 催化剂中金属态的Ni为

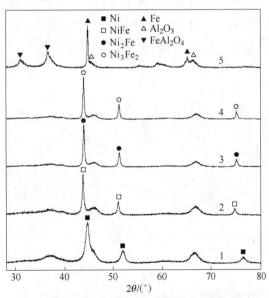

图1 还原后不同催化剂的XRD谱

Fig.1 XRD patterns of different reduced catalysts

1—Ni/Al_2O_3；2—NiFe/Al_2O_3；3—Ni_2Fe/Al_2O_3；4—Ni_3Fe_2/Al_2O_3；5—Fe/Al_2O_3

(The metal loading of the reduced catalyst is 20%)

fcc 晶体结构；Fe/Al$_2$O$_3$ 中 Fe 为 bcc 结构；Ni$_2$Fe/Al$_2$O$_3$，Ni$_3$Fe$_2$/Al$_2$O$_3$ 和 NiFe/Al$_2$O$_3$ 中的合金相都呈 fcc 结构，且三者的 XRD 谱非常相似。文献［14］报道，当 Ni-Fe 合金中的 Fe 含量增大时，可以转变为 bcc 结构。

通过对催化剂中过渡金属（合金）晶相的（111）面衍射峰进行分析，可以得到不同催化剂中金属或者合金的晶面间距和晶粒尺寸（Fe/Al$_2$O$_3$ 为（110）面），结果如表 1 所示。可以看出，不同 Ni/Fe 配比的双金属催化剂中晶面间距和晶粒大小都比较接近；Ni/Al$_2$O$_3$ 中金属相晶粒的平均直径最小。在 NiFe/Al$_2$O$_3$ 催化剂中，没有发现 Fe 和 Ni 的特征峰；而在 Fe/Al$_2$O$_3$ 中，发现了少量尖晶石结构的 FeAl$_2$O$_4$ 晶相。实验发现，NiFe/Al$_2$O$_3$ 较 Ni/Al$_2$O$_3$ 和 Fe/Al$_2$O$_3$ 容易还原，表明催化剂的活性组成不是两种活性金属的简单加合，这可能与 Ni 和 Fe 之间存在电子传递有关[14,15]。

表 1　不同催化剂的物化性质
Table 1　Physicochemical properties of different catalysts

Catalyst	Stoichiometric complex	Lattice spacing (111)/nm	$\bar{d}$/nm
Ni/Al$_2$O$_3$	—①	0.2035	23.1
NiFe/Al$_2$O$_3$	NiFe(CN)$_5$NO	0.2069	28.0
Ni$_2$Fe/Al$_2$O$_3$	Ni$_2$Fe(CN)$_6$	0.2062	32.0
Ni$_3$Fe$_2$/Al$_2$O$_3$	Ni$_3$[Fe(CN)$_6$]$_2$	0.2064	27.2
Fe/Al$_2$O$_3$	FeFe(CN)$_5$NO	0.2027(110)	29.0

①Using ammonia solution as precipitant.

2.2　催化剂活性组成对纳米碳纤维产率的影响

图 2 给出了不同催化剂上纳米碳纤维产率与生长时间的关系。可以看出，四种含 Fe 催化剂上纳米碳纤维的产率比较接近，生长 36h 后产率大于 25g/g，催化剂没有明显的失活；Ni/Al$_2$O$_3$ 催化剂在 16h 内没有发现明显的失活现象，但随着时间的延长，纳米碳纤维的生长速率缓慢降低，至 36h 时产率为 18g/g 左右。

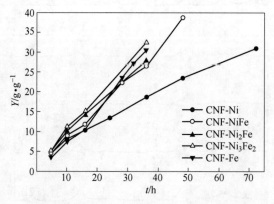

图 2　不同催化剂上纳米碳纤维的产率与生长时间的关系
Fig. 2　Relationship between carbon nanofiber (CNF) yield and growth time on different catalysts
(CNF-Ni (NiFe, Ni$_2$Fe, Ni$_3$Fe$_2$ and Fe) stands for the CNF grown on the Ni/Al$_2$O$_3$, NiFe/Al$_2$O$_3$, Ni$_2$Fe/Al$_2$O$_3$, Ni$_3$Fe$_2$/Al$_2$O$_3$ and Fe/Al$_2$O$_3$, respectively; Y—Yield of CNF (g) on per gram of catalyst)

纳米碳纤维的生长与含碳气体在金属催化剂上的吸附/解离、碳在金属颗粒中的扩散以及溶解的碳在特定金属面上形成石墨层等过程有关[1]。纳米碳纤维的稳定生长需要上述三个过程存在动态平衡。如果 CO 在金属表面的吸附/解离速率大于碳在金属体相中的扩散速率，表面碳的浓度就可能过高，导致碳在过渡金属表面聚合形成包覆层，包覆碳层的存在抑制了含碳气体在金属表面的解离，引起催化剂的失活，因此碳在金属表面的浓度存在一个与催化剂活性金属组成有关的最佳值。

纳米碳纤维的产率与反应过程中产生的 CO_2 有关。通过 CO 分解生成的 CO_2 可促使金属表面沉积的碳气化，从而避免催化剂表面存在过高浓度的碳。Pinheiro 等[15]发现，在 CO 中加入 CO_2 可大大延长 Fe-Co 催化剂的寿命。Park 等[9]报道，当反应温度由 600℃升高到 750℃时，Ni-Fe 合金上沉积的纳米碳纤维含量明显降低，这与高温下 CO_2 气化能力增大有关。他们还发现，纳米碳纤维产率与 NiFe 合金的组成有关，当 Fe < 25% 时，纳米碳纤维的产率明显降低；当 Fe > 30% 时，产率与 Fe 含量的关系不大。这说明催化剂的活性组成存在一个敏感区，在此区间内组成变化对纳米碳纤维产率的影响显著，这是由于不同组成的 Ni-Fe 合金的结构不同。Rao 等[14]在对 Ni-Fe/SiO_2 进行研究时也发现了合金结构与金属组成的关系，发现 Ni/Fe 比影响催化剂的催化性能。本文中纳米碳纤维的产率与催化剂的 Fe 含量没有明显的对应关系，这可能是由于催化剂活性组成不处于敏感区。

2.3 催化剂活性组成对纳米碳纤维微结构的影响

2.3.1 纳米碳纤维的直径与分布

不同催化剂上生长的纳米碳纤维的 TEM 照片如图 3 所示。可以看出，纳米碳纤维的直径为 20～50nm，分布较窄，长度可达到几十微米，这可能是由于采用沉积-沉淀法可以制得高负载量和高分散度的催化剂[12]。

2.3.2 催化剂活性组成对石墨层与轴之间夹角的影响

图 4 示出了不同催化剂上生长的纳米碳纤维的 HRTEM 照片。可以看出，纳米碳纤维都具有中空的结构，而且纳米碳纤维中石墨层与轴之间的夹角随活性组成变化而变化。在 Ni/Al_2O_3 催化剂上，纳米碳纤维的石墨层与轴平行，$\theta = 0°$；在三种 NiFe/Al_2O_3 上，石墨层与轴间的夹角 θ 为 5°～10°；在 Fe/Al_2O_3 上，相应的 θ 值为 20°～30°。

一般认为，石墨层堆积方向是含碳气体在金属表面的吸附/解离，碳在金属表面或体相传递及金属本身结构等因素共同作用的结果。对上述实验现象的可能解释是，当金属表面解离的碳原子浓度不足以促进体相扩散时，趋向于形成管式纳米碳纤维；当碳原子趋向于在金属颗粒的体相进行扩散时，则形成鱼骨式纳米碳纤维。文献 [17]、[18] 研究了不同配比的镍铁系列金属催化剂上 CO 的解离，发现催化剂中 Fe 的比例大于 0.3 时有利于 CO 解离。这表明当 Fe 金属与 CO 接触时，CO 吸附/解离速度较快，在 Fe 金属表面形成的碳浓度高，因此与金属体相之间存在较大的浓度梯度，趋向于体相扩散，促进了鱼骨式纳米碳纤维的形成。

2.3.3 催化剂活性组成对纳米碳纤维织构的影响

不同催化剂上生长的纳米碳纤维的织构性质如表 2 所示。可以看出，纳米碳纤维的比表面积为 130～200m^2/g，微孔很少，主要由中孔和大孔组成。这是因为纳米碳纤维外表面本身没有孔，孔隙结构是由纤维的相互缠绕而形成的[19]。不同催化剂上生长

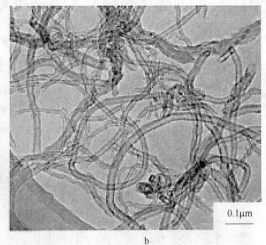
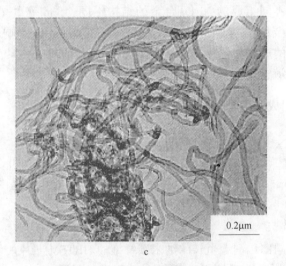

图3 不同催化剂上生长的纳米碳纤维的TEM照片

Fig. 3 TEM images of CNF grown on different catalysts

a—CNF-Ni; b—CNF-NiFe; c—CNF-Fe

的纳米碳纤维比表面积的大小顺序为 CNF-Ni < CNF-NiFe(CNF-Ni_2Fe, CNF-Ni_3Fe_2) < CNF-Fe。由于 Fe/Al_2O_3 上生长的纳米碳纤维具有更多的石墨边界,因而具有相对较高的比表面积。

表2 不同催化剂上生长的纳米碳纤维的织构性质

Table 2 Textural properties of CNF grown on different catalysts

CNF	Surface area/$m^2 \cdot g^{-1}$	Pore volume/$cm^3 \cdot g^{-1}$	Micropore/$cm^3 \cdot g^{-1}$
CNF-Ni	139.4	0.3785	0.003
CNF-NiFe	165.8	0.3891	0.005
CNF-Ni_2Fe	160.4	0.4211	0.006
CNF-Ni_3Fe_2	172.3	0.4102	0.005
CNF-Fe	193.7	0.4600	0.009

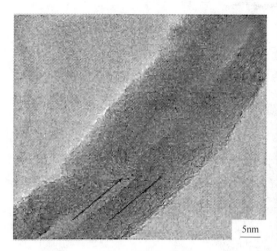
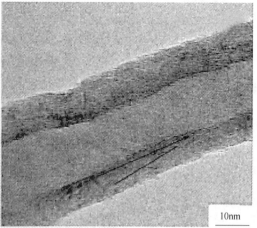
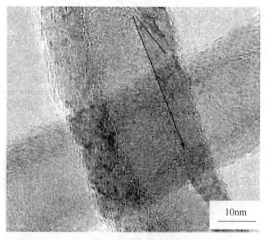

图4 不同催化剂上生长的纳米碳纤维的 HRTEM 照片
Fig. 4 HRTEM images of CNF grown on different catalysts
a—CNF-Ni; b—CNF-NiFe; c—CNF-Fe

2.3.4 催化剂活性组成对纳米碳纤维石墨化程度的影响

利用 TPO 实验可考察石墨类材料在氧化气氛中的稳定性,从而可了解其石墨化程度。图5示出了不同催化剂上生长的纳米碳纤维的 TPO 谱。可以看出,在 CO_2 气氛中,无论是否经过硝酸处理,在 Fe/Al_2O_3 上生长的纳米碳纤维 CNF-Fe 的稳定性最好,即其石墨化程度最高;在合金催化剂上生长的纳米碳纤维石墨化程度相近;在 Ni/Al_2O_3 上生长的纳米碳纤维的石墨化程度最低。这可能与石墨层形成过程中金属与石墨层间的界面行为有关。Park 等[9]发现,相比于 Ni 催化剂,Fe 催化剂与石墨层之间有更好的浸润性,可形成更完整的石墨层结构。此外,石墨化程度还与石墨层的空间堆积方式有关,Ni/Al_2O_3 上生长的管式纳米碳纤维石墨层弯曲程度更大,石墨层能量更高,因而其石墨化程度较低。

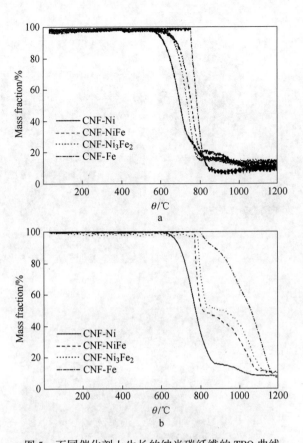

图 5 不同催化剂上生长的纳米碳纤维的 TPO 曲线
Fig. 5 TPO profiles of CNF grown on different catalysts
a—Untreated; b—Treated with HNO_3
(The oxidant in TPO was CO_2 with flow rate of 60mL/min)

参 考 文 献

[1] De Jong K P, Geus J W. Catal Rev-Sci Eng, 2000, 42 (4): 481.

[2] Baker R T K. Carbon, 1989, 27 (3): 315.

[3] Rodriguez N M. J Mater Res, 1993, 8 (12): 3233.

[4] Ledoux M J, Vieira R, Pham-Huu C, Keller N. J Catal, 2003, 216 (1-2): 333.

[5] Lozano K, Barrera E V. J Appl Polymer Sci, 2001, 79: 125.

[6] Dai H, Hafner J H, Rinzler A G, Colbert D T, Smalley R E. Nature, 1996, 384 (6605): 147.

[7] Bessel C A, Laubernds K, Rodriguez N M, Baker R T K. J Phys chem B, 2001, 105 (6): 1115.

[8] Rodriguez N M, Chambers A, Baker R T K. Langmuir, 1995, 11 (10): 3862.

[9] Park C, Rodriguez N M, Baker R T K. J Catal, 1997, 169 (1): 212.

[10] Chen P, Zhang H-B, Lin G-D, Hong Q, Tsai K R. Carbon, 1997, 35 (10~11): 1495.

[11] Li Y D, Chen J L, Liu Ch, Qin Y N. J Catal, 1998, 178 (1): 76.

[12] Boellaard E, van der Kraan A M, Geus J W. Stud Surf Sci Catal, 1995, 91: 931.

[13] Boellaard E, van der Kraan A M, Sommen A B P, Hoebink J H B J, Marin G B, Geus J W. Appl Catal A, 1999, 179 (1~2): 175.

[14] Rao C N R, Kulkarni G U, Kannan K R, Chaturvedi S. J Phys Chem, 1992, 96 (18): 7379.

[15] 赵铁均. [博士学位论文]. 上海: 华东理工大学, 2004 (Zhao T J. [PhD Dissertation]. Shang-

hai: East China Univ Sci Technol, 2004).
[16] Pinheiro J P, Gadelle P. J Phys Chem Solids, 2001, 62 (6): 1015.
[17] Boellaard E, Vreeburg R J, Gijzeman O L J, Geus J W. J Mol Catal, 1994, 92 (3): 299.
[18] Vreeburg R J, Van de Loosdrecht J, Gijzemen O L J, Geus J W. Catal Today, 1991, 10 (3): 329.
[19] 成会明. 纳米碳管制备、结构、物性及应用. 北京: 化学工业出版社 (Cheng H M. Preparation, Structure, Physical Properties and Application of Carbon Nanotubes. Beijing: Chem Ind Press), 2002, 176.

Effect of Active Metal Composition on the Yield and Microstructure of Carbon Nanofiber

Zhao Tiejun[1], Zhu Yian[1], Li Ping[1], De Chen[2], Dai Yingchun[1], Yuan Weikang[1], Anders Holmen[2]

(1. State Key Laboratory of Chemical Reaction Engineering, East China University of Science and Technology, Shanghai, 200237, China; 2. Department of Chemical Engineering, Norwegian University of Science and Technology, N-7491 Trondheim, Norway)

Abstract A series of alumina supported nickel/iron catalysts with different active metal compositions were prepared by the deposition-precipitation method, and the carbon nanofiber (CNF) was catalyticly grown on the above catalysts at 600℃ by using CO/H_2 mixture as a carbon source. The catalytic activity was stable after 36 hours on stream, and the presence of Fe in the catalyst promoted the growth of CNF. The formed CNF diameter distribution was narrow in the range of 20 ~ 50nm. The HRTEM result showed that the angle between the CNF axis and the graphite layer increased with the increase of Fe content. The TPO result indicated that the graphitization degree of the CNF depended on the active metal composition. The yield and microstructure of CNF changed with the active metal composition, which was attributed to the surface reaction between CO and active metal (alloy) together with the carbon diffusion mode on/in the metal particle.

Key words carbon nanofiber, catalytic grown, microstructure, active metal composition, deposition-precipitation method

钛硅分子筛/纳米碳纤维催化剂的制备及其液相分离性能[*]

摘　要　通过在纳米碳纤维（CNF）上原位生长钛硅分子筛（TS-1）的方式制备了负载型 TS-1/CNF 催化剂。对该催化剂进行 TS-1 和 CNF 的结合强度测试的结果表明，TS-1 与 CNF 表面之间具有较强的结合程度。负载型 TS-1/CNF 催化剂和单分散 TS-1 的分离性能对比显示，前者易于分离，可显著降低 TS-1 应用中的分离难度。

关键词　钛硅分子筛　纳米碳纤维　负载　分离性能

钛硅分子筛（TS-1）在以过氧化氢为氧化剂的选择性氧化反应中有着优异的催化性能[1]。但是在实际应用中，由于催化活性较好的 TS-1 晶粒度分布一般为 100~200nm[2]，其较难从液相反应体系中分离和回收。进行 TS-1 的负载化可以增加催化剂的特征尺寸，从而改善反应产物和催化剂之间的分离问题。

纳米碳纤维（CNF）是一种新型碳材料，具有优异的物理化学性质，如特殊的长纤维结构、较大的比表面积、很高的机械强度、较好的化学稳定性等。另外，纳米碳纤维结构可控、均一。因而其作为性能优良的催化剂载体受到日益广泛的关注[3]。

笔者在 CNF 上原位生长了 TS-1，对制备的 TS-1/CNF 催化剂进行了结合强度测试，并将其和纯 TS-1 的分离性能进行了对比。

1　实验部分

1.1　TS-1/CNF 合成方法

将四丙基氢氧化铵、蒸馏水和异丙醇配成混合溶液，对 CNF 进行预浸渍。在剧烈搅拌条件下将硅酸四乙酯加入上述混合物中，1h 后，将溶于异丙醇的钛酸四丁酯滴加进去，得到 $n(SiO_2):n(TiO_2):n(TPAOH):n(H_2O) = 1:0.025:0.2:20$ 的合成 TS-1 用胶液。室温下继续搅拌 3h，将混合物移入带有聚四氟乙烯内衬的不锈钢反应釜内，80℃老化 2d，175℃晶化 3d。产物经冷却、分离、洗涤和干燥，并在 550℃氩气气氛中焙烧 6h 后待用。

TS-1 合成参考 TS-1/CNF 的合成方法，不添加 CNF。

1.2　样品分散和分离方法

将 2g 待测样品和 50mL 蒸馏水，加入置于磁力搅拌器上的三口瓶中，控制一定转速搅拌 14h。将分散后的混合液放入离心试管，在离心分离机中控制一定转速离心分离，分离液相与固体样品。

---

[*] 本文合作者：赵茜、李平、张京纬、胡喜军。原发表于《石油学报（石油加工）》，2006（增刊）：221~223。

1.3 表征

采用日本 JOEL JSM3360LV 扫描电子显微镜（SEM）观察样品的外貌和粒径分布。采用日本 Rigaku X 射线衍射仪测定样品的物相结构。采用英国 Mastersizer 2000 激光粒度分析仪测定粒度分布。

2 结果与讨论

2.1 催化剂的表征

图 1 为 TS-1、CNF 和 TS-1/CNF 的 XRD 谱图。由图 1 可知，TS-1 具有典型的 MFI 型结构。CNF 的 XRD 谱图中，$2\theta = 26.2°$ 处出现的峰归属于 CNF 石墨结构的衍射峰，30°~40°之间出现了残存催化剂形成的 Fe_3C 衍射峰。对比可知，TS-1/CNF 的谱图中出现了所有 TS-1 和 CNF 的特征峰，说明 TS-1/CNF 具有完整的晶型结构。

图 2 为合成样品的扫描电镜图。由图 2a 可知，TS-1 为均匀的球形颗粒，粒度分布在 100~150nm；由图 2b 可知，对于 TS-1/CNF，球形 TS-1 生长在 CNF 表面，粒度在 100~150nm。

另外，合成样品已通过 HRTEM/EDS 测试[4]，进一步证明所得晶粒为 MFI 型 TS-1。

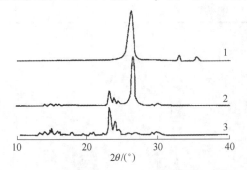

图 1 TS-1，CNF 和 TS-1/CNF 的 XRD 谱图
Fig. 1 XRD patterns of TS-1，CNF and TS-1/CNF
1—CNF；2—TS-1/CNF；3—TS-1

图 2 合成样品的扫描电镜图
Fig. 2 SEM photos of synthesized samples
a—TS-1；b—TS-1/CNF

2.2 样品的分离性能

为了考察不同物质的分离性能，在相同条件下对 TS-1、CNF、TS-1/CNF 进行分散

处理，均可得到悬浮溶液，然后对其进行离心分离，结果如表1所示。由表1可知，经长时间分散后，TS-1在水溶液中以高分散形式存在，以4000r/min离心分离40min，仍很难完全分离。少量TS-1残留在液相中形成白色悬浊液，对其进行粒度分析，结果如图3所示。进一步证明，颗粒度在150nm左右的TS-1样品在该条件下很难完全分离。而CNF具有较大的特征尺寸，在较低的离心转速（2000r/min）和较短时间（2min）下TS-1/CNF即可从水中完全分离，液相呈透明状，不存在残余物。也说明经长时间剧烈搅拌，TS-1/CNF中TS-1未发生脱落现象，TS-1与载体CNF具有较大的结合强度。通过将TS-1负载到CNF上可明显降低TS-1在实际应用中的分离难度。相对TS-1而言，TS-1/CNF的分离难度大大降低。

表1 不同样品的分离性能

Table 1 Separation properties of different samples

Sample	Contrifuging speed/r·min^{-1}	Centrifuging time/min	Status of aqueous solution
TS-1	4000	40	White and turbid
TS-1	2000	2	White and turbid
CNF	2000	2	Clear
TS-1/CNF	2000	2	Clear

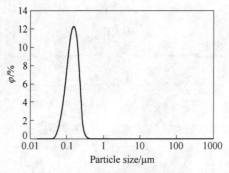

图3 TS-1悬浮液的粒度分布

Fig. 3 Particle size distribution of TS-1 suspension

3 结论

（1）合成的负载型TS-1/CNF催化剂中活性组分TS-1与载体CNF之间具有较强的结合力，在剧烈搅拌条件下TS-1不会出现脱落现象。

（2）负载型催化剂TS-1/CNF易于分离，可显著降低TS-1应用中的分离难度。

参 考 文 献

[1] Notari B. Catal Today, 1993, 18: 163~172.

[2] 周继承，王祥生. 化工进展, 1998. 10 (4): 381~394.

[3] 赵铁均，朱贻安，李平. 等. 催化学报, 2004, 25 (10): 829~833.

[4] Zhao Q, Li P. Li D Q, et al. 4th Asia Pacific Congress on Catalysis, Singapore, 2006, Accepted.

Synthesis and Separation Properties of Titanium Silicate Supported on Carbon Nanofiber

Zhao Qian[1,2], Li Ping[1], Zhang Jingwei[1], Yuan Weikang[1], Hu Xijun[3]

(1. State Key Laboratory of Chemical Engineering, East China University of Science and Technology, Shanghai, 200237, China;
2. Department of Chemical Engineering and Technology, Hebei University of Technology, Tianjin, 300130, China;
3. Department of Chemical Engineering, The Hong Kong University of Science and Technology, Hong Kong, China)

Abstract A novel catalyst consisting of *in-situ* crystallized titanium silicate (TS-1) supported on carbon nanofiber (CNF) was synthesized. The binding strength between TS-1 and CNF was investigated. The results showed that the TS-1 was anchored firmly on CNF surface. The comparison of separation properties between TS-1/CNF And TS-1 showed that the former could be easily separated from liquid system.

Key words titanosiliacate (TS-1), carbon nanofiber (CNF), support, separation properties

纳米炭纤维的表面润湿行为*

摘 要 采用动态渗透法研究了不同结构的纳米炭纤维（Carbon nanofibers，CNF）的表面润湿性以及在不同溶剂中润湿性的变化和表面改性对其润湿性的影响。结果表明，生长条件如催化剂组成、碳源等对 CNF 的表面性质有显著影响，并最终决定其在溶剂中的润湿能力。以 $Fe/\gamma-Al_2O_3$ 为催化剂，以 C_2H_4 为碳源得到的 CNF 在水中的润湿性能最差；而以 $Ni/\gamma-Al_2O_3$ 为催化剂，CO 为碳源得到的 CNF 在环己烷中的润湿性能最好。CNF 在不同溶剂中的相对接触角测定表明 CNF 是一种表面非极性较强的材料。设 CNF 在润湿性能最好的环己烷中的接触角为 0°，则 CNF 在水、丙酮、乙醇中的相对接触角分别为 81.6°、45.2°、24.8°。不同的表面改性手段可对 CNF 的表面性质进行调变以控制其在不同溶剂中的润湿行为。在浓硝酸中液相氧化可提高其对水溶液和环己烷的润湿性能；在氩气中的高温热处理可提高其对水溶液的润湿性能，但降低了对环己烷的润湿能力；而在过氧化氢溶液中的处理则同时降低了对水溶液以及环己烷的润湿能力。

关键词 纳米炭纤维　润湿　动态法　相对接触角

1 前言

纳米炭纤维（CNF）作为一种新型炭结构，由于其特殊的物理化学性能，在催化材料、电极材料、高效吸附剂和结构增强剂等方面具有较好的应用前景，成为近十年来的研究热点[1~3]。CNF 有序排列的石墨边缘结构，独特的中孔结构，可调控的微结构等，与组成结构和性质最接近的活性炭相比，有较多的优势，负载金属后对催化剂的活性、选择性、寿命等产生了积极的影响[4~6]。

炭材料作为催化剂载体用于负载活性金属时，其与溶剂的亲和性，对溶剂的吸附能力，均与其表面性质、表面组成密切相关。尤其是对于贵金属催化剂而言，通常采用浸渍的方式来制备，因此炭材料的表面性质及其与溶剂之间的相互亲和力，不仅影响活性金属在载体表面的分布过程，而且最终将影响催化剂的活性[7,8]。

CNF 作为催化剂载体负载贵金属时，炭纤维粉末与浸渍液之间的润湿行为直接反映了两者之间的亲和性和相互作用，本文主要报道动态渗透法结合 Washburn 方程考察 CNF 的润湿过程。研究了不同微结构的 CNF 表面润湿行为以及在不同溶剂中润湿行为的变化和表面改性对其润湿性的影响，以期对其作为催化剂载体在负载活性金属的过程中提供有益的指导和帮助。

2 原理及实验

2.1 纳米炭纤维的制备

本研究所用 CNF 纤维采用化学气相沉积法（CCVD）在实验室制备，通过在不同

---

* 本文合作者：周静红、隋志军、李平、戴迎春。原发表于《新型炭材料》，2006，21（4）：331~336。

的活性金属上催化裂解不同的含碳气体,得到了具有不同微结构以及表面润湿性能的 CNF。考察了 γ-Al_2O_3 负载 Ni、Fe 以及 NiFe 合金系列催化剂,以 CH_4、C_2H_4、CO 为碳源制备的 CNFs 的润湿行为。生长 CNF 的镍铁系列催化剂的制备以及生长过程参见文献 [9]。生长过程简述如下:

CNF 的生长在水平放置的石英管式反应炉中进行。气体流量通过质量流量计控制。将催化剂 1.0g 平铺于石英舟中,置石英舟于反应器中部恒温段。在 Ar: H_2 = 120mL/min: 40mL/min 的混合气流中程序升温至 600℃,并保持 3h,将催化剂还原。而后,切换为反应气体。反应气体流量控制为:C_2H_4: H_2 = 40mL/min: 20mL/min,CH_4: H_2 = 80mL/min: 10mL/min,CO: H_2 = 80mL/min: 20mL/min。生长过程持续 24h。使用的气体全部购自于雷磁-创益公司,纯度为:Ar > 99.9%,H_2 > 99.99%,CO > 99.99%,CH_4 > 98%,C_2H_4 > 98%。

2.2 修正 Washburn 法原理

Washburn 法将多孔阻塞视作一束平均半径为 r 的毛细管,将液体在孔内渗透高度 h 的平方和时间 t 作图可得直线关系[10]。K. Grundke 等人推导出以质量替代高度的修正 Washburn 方程,如式(1)所示,该方法较透过高度法更为精确和方便,因此许多商用的仪器采用该原理测定。

$$w^2 = \frac{C\gamma\rho^2\cos\theta}{\eta} \times t \quad (1)$$

由式(1)可知,以 w^2 和 t 作图可得到一条直线,并可从图中求得直线段斜率 k,$k = \cos\theta\rho^2\gamma C/\eta$。因此,该斜率的大小直接反映了粉末样品在测定液体中的润湿能力,斜率越大,表明两者的润湿性越好。

还可利用以下公式计算粉末的接触角:

$$\cos\theta = \frac{w^2}{t} \times \frac{\eta}{\rho^2\gamma C} \quad (2)$$

式中,w 为吸附的液体质量;γ 为液体的表面张力;ρ 为液体的密度;η 为液体的黏度;θ 为润湿接触角;t 为时间;C 为常数,由粉末柱的有效毛细管半径以及粉末特性等决定。一般用一种对样品润湿接触角为零的液体,先确定出 C 值,再测定其他同等实验条件下液体的接触角。

2.3 相对润湿接触角

事实上,要找到使 CNF 完全润湿的液体几乎不可能,这样式(1)中的 C 值就无法确定,也就难以得到粉末材料与不同液体之间的接触角。但是,正如前言中所述及的,研究者关注的往往是不同粉体对相同液体的润湿能力,或者同种粉体对不同液体的润湿能力。参照文献 [11] 研究结果,采用相对润湿接触角(relative wetting contact angle,RWCA)来半定量表示粉末对不同液体的润湿能力。

令样品在液体 L 和润湿性更好的液体 L_0 中的接触角分别为 θ 和 θ_0,根据实验测得的 $w^2 \sim t$ 的近似线性直线的斜率分别为 k 和 k_0,同一样品在 L 和 L_0 中的 C 相同,$\theta > \theta_0$,令 $\theta_0 = 0°$,则 $C = k_0\eta_0/(\rho_0^2\gamma_0)$,由式(2)可得到相对接触角为:

$$\theta_{RWCA} = \arccos[k\eta\rho_0^2\gamma_0/(\rho^2\gamma k_0\eta_0)] \quad (3)$$

式中,γ、η、ρ 和 γ_0、η_0、ρ_0 分别为液体 L 和润湿性更好的液体 L_0 的表面张力、黏度

和密度。式（3）所确定的接触角只具有相对的意义。对于评价同一种样品或相似的润湿性能的样品在不同液体中的润湿能力，能采用定量化的研究，具有较好的对比意义和实际应用价值。

2.4 试验仪器和方法

润湿动力学采用 Thermo Cahn 公司的 Radian Series 3000 型表面张力仪完成，其原理如图 1 所示。

实验步骤如下：

（1）将 2.1 节制备的不同种类 CNF 经研磨至 120 目至 180 目，准确称取质量后装入测量管并用一定质量的砝码压实。

（2）将测量管小心悬挂于天平的挂钩，平衡后使天平复零。

（3）采用仪器自带软件控制升降台上升直到液体刚刚接触石英玻璃管。这时升降台自动停止移动，并开始记录时间和质量的变化曲线。

（4）当质量的变化速率低于给定值时，仪器自动停止测量。

（5）通过仪器自带软件和其他软件计算 $w^2 \sim t$ 直线段的斜率。

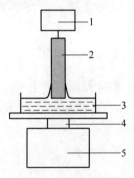

图 1　润湿动力学测试装置示意图

1—精密电子天平；2—底端带烧结板的石英玻璃管（$D8 \times 36mm$）；
3—待测溶剂；4—升降台；5—自动马达

Fig. 1　Experimental setup for wicking measurements

1—Electronic balance；2—Silex glass tube with sinter at one end（$D8 \times 36mm$）；
3—Test liquid；4—Lifting stage；5—Auto motor

3　结果与讨论

3.1　催化剂活性组成对 CNF 润湿行为的影响

研究表明[12,13]，制备工艺如催化剂活性、气相流动状态等对 CNF 的微结构和表面性质均有影响。CNF 的微结构（包括石墨层的空间堆积方式、直径等）可以通过控制其生长条件（如催化剂活性组成、含碳气体种类、生长温度等）来进行调节以得到具有不同的表面性质的材料[14,15]。因此当制备 CNF 所采用的催化剂改变时，所获 CNF 的润湿能力必然也随之改变。图 2 给出了不同组成的催化剂制备的 CNF 在水中润湿过程中 $w^2 \sim t$ 的关系图。可以看出，随着催化剂活性组分的变化，得到的 CNF 对水的润湿性有明显的区别。对于镍铁系列催化剂而言，镍的含量越高，$w^2 \sim t$ 图中的斜率越大，这意味着所制得 CNF 的亲水性越强。随着活性金属中铁的含量逐渐加大，直至活性组分为纯铁时，其亲水性降至最低。研究表明[14]，对于活性组分类似的催化剂如镍铁合金制备得到的 CNF，其直径和宏观的比表面积以及孔径分布等差别不大，但其石墨层的空间堆积方式发生了变化。当以纯镍作为活性组分时，石墨层与纤维的轴向平行，夹角成0°。随着活性组分中加入铁，石墨层与纤维轴向的夹角逐渐增大，当以纯铁作为活性组分时，夹角增大为30°。另一方面，由于在石墨层堆积方式发生变化的同时，纤维的石墨化程度不一，从而使得暴露在空气中的表面基团有了明显的变化。正是这两方面的原因，使 CNF 表面与水之间相互作用发生了变化，并导致润湿行为的差异。由此认为，当 CNF 应用于催化领域负载活性金属时，活性金属的浸渍工艺应随 CNF 的制备工艺调整，以期获得优良的催化性能。

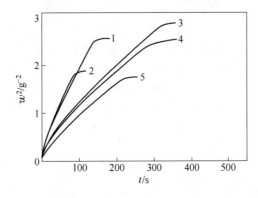

图 2　催化剂活性组成对 CNF 在水中的润湿行为的影响

催化剂活性组成：1—Ni/γ-Al$_2$O$_3$；2—Ni$_3$Fe$_2$/γ-Al$_2$O$_3$；

3—Ni$_2$Fe/γ-Al$_2$O$_3$；4—NiFe/γ-Al$_2$O$_3$；5—Fe/γ-Al$_2$O$_3$

（碳源全部为：CO）

Fig. 2　Influence of catalyst composition on wettability of CNF

Catalyst composition：1—Ni/γ-Al$_2$O$_3$；2—Ni$_3$Fe$_2$/γ-Al$_2$O$_3$；

3—Ni$_2$Fe/γ-Al$_2$O$_3$；4—NiFe/γ-Al$_2$O$_3$；5—Fe/γ-Al$_2$O$_3$

（Carbon source：CO）

3.2　碳源对 CNF 润湿行为的影响

图3给出了采用不同碳源在 Ni$_3$Fe$_2$/γ-Al$_2$O$_3$ 催化剂表面热解得到的 CNF 在水中（a）以及环己烷中（b）的润湿过程的 $w^2 \sim t$ 图。与图2比较可知，碳源变化对 CNF 润湿性能的影响比催化剂活性组成显著得多。不同碳源得到的 CNF 润湿速率差别明显，在水溶液中的润湿速率排序为 $CH_4 > CO > C_2H_4$；而在环己烷中的润湿速率排序则为 $CO > CH_4 > C_2H_4$。水溶液中以甲烷为碳源得到的 CNF 达到渗透平衡时的吸水率达 3.20g/g，远远高于以乙烯为碳源得到的 CNF 的吸水率 0.30g/g。在环己烷中，以甲烷为碳源得到的 CNF 达渗透平衡时溶剂渗透率为 2.22g/g，而以一氧化碳为碳源得到的 CNF 渗透的环己烷仅为 1.10g/g。这可能是由于采用不同的碳源时，炭纤维的石墨层排列角度有所变化，因此表面粗糙度以及表面基团—CH$_3$、—CH$_2$—等的数量均有显著变化。另外，当碳源中含有氧原子时，CNF 表面含有较多数量的含氧基团。而固相表面的基团种类及数量均是影响其对不同溶剂的亲和力的关键因素之一[16,18]，因此不同碳源得到的 CNF 在不同溶剂中的润湿性能有显著的差异。其表面化学组成对炭纤维与溶

剂间相互作用的影响尚在进一步研究之中。

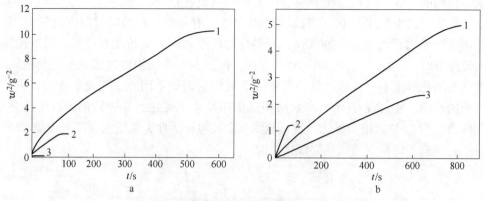

图 3　碳源对 CNF 的润湿行为的影响
a—CNF 在水溶液的润湿行为；b—CNF 在环己烷中的润湿行为
（催化剂活性组分采用 $Ni_3Fe_2/\gamma\text{-}Al_2O_3$；碳源分别为：1—$CH_4$；2—CO；3—$C_2H_4$）

Fig. 3　Influence of carbon source on wettability of CNF
a—Wettability of CNFs in water；b—Wettability of CNFs in cyclohexane
(Catalyst composition：$Ni_3Fe_2/\gamma\text{-}Al_2O_3$；Carbon source：1—$CH_4$；2—CO；3—$C_2H_4$)

3.3　CNF 在不同溶剂中的润湿行为

图 4 给出了以 20% $Fe/\gamma\text{-}Al_2O_3$ 为催化剂，CO 为碳源气相生长得到的 CNF 在水、乙醇、丙酮、环己烷四种不同溶剂中的动态润湿过程。各种溶剂的主要物性数据列于表 1。从图 4 表观看，符合 Washburn 方程的直线段斜率在不同溶剂中的大小依次为：丙酮＞环己烷＞乙醇＞水，根据式（2），CNF 与溶剂之间的接触角与溶剂的物性密切相关，由于丙酮与其他的溶剂相比，黏度小得多，因而溶剂在 CNF 堆积形成的毛细孔中扩散阻力也小得多，表现出来就是 $w^2 \sim t$ 图的斜率最大。由此可推断，润湿最好的溶剂为环己烷。根据式（2），假定 CNF 在润湿性能最好的环己烷中的接触角为 0°，则利用图 4 可以得到 CNF 在其他三种不同溶剂中的相对接触角分别为 81.6°、24.8°、45.2°（表 1）。从中可以看到，CNF 在不同的溶剂中接触角差别显著。对活性炭和石墨的表面润湿性能的研究表明[17]，活性炭或石墨的表面极性越大，与极性溶剂的润湿性越好，而非极性固体在极性溶剂中的润湿热较低，即非极性固体与极性溶剂的亲和力较小，润湿程度差。结合表 1 和图 4 结果，可以认为 CNF 是一种表面非极性较强的材料，与非极性溶剂的润湿能力远远高于与极性溶剂的润湿能力，这是由于 CNF 是一种高度石墨化的炭材料，其表面较活性炭规整，主要的表面基团为非极性基团，润湿热也相应较小。

表 1　不同溶剂的物性数据
Table 1　Parameters of different solvents

Solvent	Surface tension γ /mN·m^{-1}	Density ρ /kg·m^{-3}	Viscosity η /mPa·s	Dielectric constant ε_0	RWCA θ /(°)
Water	72.8	998.2	1.005	80.000	81.6
Ethannol	22.8	789.0	1.150	24.300	24.8
Acetone	23.7	792.0	0.320	20.700	45.2
Cyclohexane	25.3	778.6	0.970	2.016	0.0

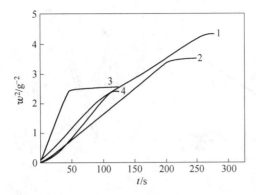

图4 CNF 在不同溶剂中的润湿行为
溶剂分别为：1—H_2O；2—CH_3CH_2OH；3—CH_3COCH_3；4—C_6H_{12}
(CNF 生长条件：催化剂组成，Fe/γ-Al_2O_3；碳源：CO)

Fig. 4 Wettability property of CNF in different solvents
Different solvents：1—H_2O；2—CH_3CH_2OH；3—CH_3COCH_3；4—C_6H_{12}
(CNF growth condition：catalyst composition，Fe/γ-Al_2O_3；Carbon source：CO)

3.4 表面改性对 CNF 润湿行为的影响

CNF 在不同溶剂中的润湿试验结果表明 CNF 是一种表面极性很弱的材料，这一性质对其作为催化剂载体而言有时不利，尤其当其应用在极性催化反应体系时。为此笔者采用不同的表面处理方法对 CNF 进行表面改性，以期调变 CNF 作为催化剂载体在不同的催化反应中的适用性。

图 5a 和 5b 分别给出了 CNF 经过硝酸液相氧化处理，过氧化氢液相氧化处理，高温氩气处理等不同表面改性处理后在水及环己烷中的动态润湿过程。从中可以看到，CNF 经过不同的表面改性后，润湿性能得到显著的调变。当 CNF 在浓硝酸中回流 0.5h 后，在水溶液以及环己烷中的润湿性都有明显改善，尤其在水溶液中的润湿能力提高更为明显。这可能是由于在硝酸中回流处理增加了 CNF 表面的含氧基团的数量，尤其是羧基的数量[19,20]，而含氧基团易与极性溶剂之间形成更强的作用，从而使得 CNF 对水溶液的润湿性更好。当 CNF 在过氧化氢溶液中处理 24h 后，在水溶液和环己烷中的润湿性却明显减弱；而当 CNF 在惰性气体氩气中 900℃高温处理 4h 后，在水溶液中的润湿性能有所增加，而在环己烷中的润湿性则减弱。通常认为高温热处理过程中 CNF 表面的含氧基团数量减少，羰基等碱性基团增加[21]，因此表面极性增强，所以在强极性的水溶液中的润湿性增强，而在弱极性的环己烷中润湿性降低。

4 结论

CNF 的生长工艺条件如催化剂的活性组成和碳源，尤其是碳源对 CNF 的润湿性能有显著的影响，对 CNF 负载活性金属用于催化产生较大作用。采用动态渗透法利用 Washburn 方程可以有效地表征 CNF 的表面润湿性能，并比较 CNF 在不同溶剂中的相对接触角。采用不同的表面改性手段可以调变 CNF 在不同溶剂中的润湿性能，并应用于不同的催化反应体系。

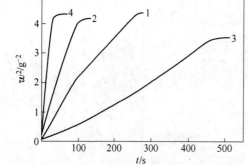
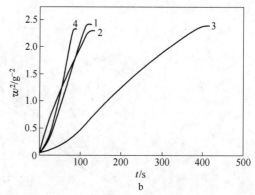

图 5　表面改性对 CNF 润湿行为的影响

a—CNF 在水中的润湿行为；b—CNF 在环己烷中的润湿行为

1—生长得到的 CNF；2—在 900℃氩气氛中处理 4h 的 CNF；3—在 H_2O_2 中处理 24h 的 CNF；

4—在浓硝酸中回流处理 0.5h 的 CNF

（催化剂组成：Fe/γ-Al_2O_3；碳源：CO）

Fig. 5　Influence of surface modification on wettability of CNFs

a—Wettability in water；b—Wettability in cyclohexane

1—As-grown CNF；2—CNF after heated in argon at 900℃ for 4h；

3—CNF after treatment in H_2O_2 for 24h；4—CNF after refluxed in concentrated HNO_3 for 0.5 h

（Catatlyst composition：Fe/γ-Al_2O_3；Carbon source：CO）

参 考 文 献

［1］ 杨全红. "纳米"-碳质材料研究的新视角-Carbon2004 参会有感 ［J］. 新型炭材料，2004，19 (3)：161～165.
（YANG Quan-hong. New insights into carbon research-A brief report on Carbon 2004 ［J］. New Carbon Material，2004，19 (3)：161～165）.

［2］ Ledoux M J，Vieira R，Pham-Huu C，et al. New catalytic phenomena on nanostructured (Fibers and Tubes) catalysts ［J］. J Catal，2003，216 (1～2)：333～342.

［3］ Serp P，Corrias M，Kalack P. Carbon nanotubes and nanofibers in catalysis ［J］. Appl Catal A，2003，253 (2)：337～358.

［4］ Esther Ochoa Fernandez. Carbon nanofibers supported Ni catalyst ［D］. Norway：Norwegian University of Science and Technology，2003.

［5］ Marjolein Toebes. Carbon nanofibers as catalyst support for noble metals ［D］. Netherland：University of Utrecht，2004.

［6］ Cuong Pham-Huua，Nicolas Keller a，Gabrielle Ehret c，et al. Carbon nanofiber supported palladium catalyst for liquid-phase reactions：An active and selective catalyst for hydrogenation of cinnamaldehyde into hydrocinnamaldehyde ［J］. Journal of Molecular Catalysis A：Chemical，2001，170：155～163.

［7］ Toebes M L，van Dillen J A，de Jong K P. Synthesis of supported palladium catalysts ［J］. J Mole Catal A，2001，173：75～98.

［8］ 赵江红，刘振宇. 载体炭对 CuO/AC (F) 催化—吸附剂干法催化氧化苯酚的影响 ［J］. 新型炭材料，2005，20 (2)：115～120.
（ZHAO Jiang-hong，LIU Zhen-Yu. Effect of carbon support on CuO/AC (F) catalyst-sorbents used for catalytic dry oxidation of phenol ［J］. New Carbon Material，2005，20 (2)：115～120）.

[9] 赵铁均．纳米炭纤维的微观结构调控及相关催化性能研究［D］．上海：华东理工大学，2004．
(ZHAO Tie-jun. Carbon nanofiber with well-controlled microstructure as catalyst material [D]. Shanghai: East China University of Science & Technology, 2004).

[10] Grundke K, Boerner M, Jacobasch H J. Characterization of fillers and fibres by wetting and electrokinetic measurements [J]. Colloids and Surfaces, 1991, 58: 47~59.

[11] 艾德生，李庆丰，戴遐明，等．用透过高度法测定粉体的润湿接触角［A］．理化检验-物理分册［C］．2001，37（3）：110~112．
(AI De-sheng, LI Qing-feng, DAI Xia-ming, et al. Measurement of wetting contact angle of powder by permeating height method [A]. PTCA (Part A: Physical Testing) [C]. 2001, 37 (3): 110~112).

[12] 雷中兴，刘静，王建波，等．催化剂结构与形态对碳纳米管生长的影响［J］．新型炭材料，2003，18（4）：271~276．
(LEI Zhong-xing, LIU Jing, WANG Jian-bo, et al. The effects of catalyst structure and morphology on the growth of carbon nanotubes [J]. New Carbon Material, 2003, 18 (4): 271~276).

[13] 侯鹏翔，白朔，范月英，等．气体流动状态对纳米炭纤维制备的影响［J］．新型炭材料，2000，15（4）：17~20．
(HOU Peng-xiang, BAI Shuo, FAN Yue-ying, et al. Effect of gas flowing state on the preparation of carbon nanofibers [J]. New Carbon Material, 2000, 15 (4): 17~20).

[14] 赵铁均，朱贻安，李平，等．催化剂活性组成对纳米炭纤维产率和微观结构的影响［J］．催化学报，2004，25（10）：829~833．
(ZHAO Tie-jun, ZHU Yi-an, LI Ping, et al. Effect of active metal composition on the yield and microstructure of carbon nanofiber [J]. Chinese Journal of Catalysis, 2004, 25 (10): 829~833).

[15] Reshetenko T V, Avdeeva L B, Ismagilov Z R, et al. Catalytic filamentous carbon structural and textural properties [J]. Carbon, 2003, 41: 1605~1615.

[16] 顾惕人，李外郎，马季铭，等．表面化学［M］．北京：科学出版社，2003：359~388．
(GU Ti-ren, LI Wai-lang, MA Ji-ming, et al. Surface Chemistry [M]. Beijing: Science Press, 2003: 359~388).

[17] 任兰正，王道宏，王日杰，等．炭黑粉末润湿性质的研究［J］．化学工业与工程，2003，20（4）：200~204．
(REN Lan-zheng, WANG Dao-hong, WANG Ri-jie, et al. Study on wettability of carbon black powder [J]. Chemical Industry and Engineering, 2003, 20 (4): 200~204).

[18] Alexander Bismarck, M. Emin Kumru, Jürgen Springer. Influence of oxygen plasma treatment of PAN-based Carbon fibers on their electrokinetic and wetting property [J]. Journal of Colloid and Interface Science, 1999, 210: 60~72.

[19] 商红岩，刘晨光，徐永强，等．碳纳米管的表面修饰对CoMo催化剂HDS性能影响的研究［J］．新型炭材料，2004，19（2）：129~136．
(SHANG Hong-yan, LIU Chen-guang, XU Yong-qiang, et al. Effect of the surface modification of multi-walled carbon nanotubes (MWCNTs) on hydrodesulfurization activity of Co-Mo/MWCNTs catalysts [J]. New Carbon Material, 2004, 19 (2): 129~136).

[20] 邱军，王国建，屈泽华，等．氧化处理方法与多壁碳纳米管表面羧基含量的关系［J］．新型炭材料，2006，21（3）：269~272．
(QIU Jun, WANG Guo-jian, QU Ze-hua, et al. Relationship between oxidation treatment method and carboxylic group content on the surface of MWCNTs [J]. New Carbon Materials, 2006, 21 (3): 269~272).

[21] Montes-Moran M A, Suarez D, Menendez J A, et al. On the nature of basic sites on carbon surfaces: An overview [J]. Carbon, 2004, 42 (7): 1219~1225.

The Wettability of Carbon Nanofibers

Zhou Jinghong, Sui Zhijun, Li Ping, Dai Yingchun, Yuan Weikang

(State Key Laboratory of Chemical Engineering, East China University of Science and Technology, Shanghai, 200237, China)

Abstract The surface chemical properties and wettability of carbon nanofibers (CNFs) synthesized by CVD were studied by wicking kinetic measurements. The influence of the microstructure, preparation conditions and the surface modification of the CNFs on their wettability in different solvents were investigated. Results indicated that the preparation conditions such as catalyst composition, and especially the carbon source, have an obvious effect on the surface properties and wettability of CNFs. CNFs from $Fe/\gamma\text{-}Al_2O_3$, and C_2H_4 has the worst wettability in water, while CNFs from $Ni/\gamma\text{-}Al_2O_3$ and CO has the best wettability in cyclohexane. Measurement of contact angle in cyclohexane, acetone, ethanol and water showed that interaction between the CNF surface and the solvent changed significantly with the solvent used. The contact angles of CNFs from $Fe/\gamma\text{-}Al_2O_3$ and CO in water, acetone and ethanol relative to that in cyclohexane (in which the contact angle is arbitrarily set to 0°), are 81.6°、45.2°、24.8° respectively. It is believed that a small polarity of the CNF surface leads to its higher wettability in a non-polar solvent such as cyclohexane than in polar solvent such as water. The nature of the CNF surface groups and the wettability of CNFs in different solvents could be tailored by different surface modifications. Treatment in concentrated nitric acid increased the CNF wettability both in water and in cyclohexane. while treatment in hydrogen peroxide decreased the CNF wettability in both solvents. Interestingly, thermal treatment in argon increased the CNF wettability in water, but decreased it in cyclohexane.

Key words carbon nanofiber (CNF), wettability, wicking test, relative wetting contact angle

超声波对甘氨酸溶析结晶过程的影响*

摘 要 以甘氨酸水溶液的丙酮溶析结晶为对象,探讨了超声波对结晶过程的影响。在超声波作用下,结晶过程经历空泡形成、超声波诱导成核、二次成核多个阶段;在不同的阶段施加超声波,或在相同时刻引入超声波但持续不同的时间,都可能影响晶体的粒径大小和分布。在自然均相成核点之前施加超声波并持续较短时间,使晶核以超声波诱导成核为主时,可获得较大颗粒的晶体;在接近均相成核点处施加超声波,将产生更多的晶核,使晶体平均粒径降低。在晶体生长过程中继续使用超声波,超声波的破碎效应也将降低晶体的平均粒径。

关键词 溶析结晶 超声波 甘氨酸 成核 生长

1 前言

结晶单元操作是制药和生物工程中最重要的分离过程之一,对药物和生物产品的质量,包括化学纯度及晶系、晶习、晶体大小及分布等有重要的影响,从而直接影响产品的后加工性(如过滤性、流动性)、储存稳定性(如货架期)使用效果(如药效)等。在结晶过程中,需要控制结晶温度、压力、进料,并采用外加晶种、清液、分级排料及增加机械刺激等手段达到控制晶体形成的目的[1]。近年来,随着超声波发生设备的日益普及,超声波作为辅助结晶手段,引起了人们的广泛关注。

超声波作为过程强化的一种手段可用于促进反应和传递,在有机合成[2~7]和功能材料制备[8~11]中得到广泛的应用,但文献中却少有超声波对结晶过程影响的报道。

超声波结晶可替代传统的加入晶种的操作,缩短成核诱导时间,阻碍晶体的团聚,对产品的进一步加工非常有益[12]。Guo等[13]研究了超声波对Roxithromycin溶析结晶的影响,表明超声波能显著缩短诱导时间,降低介稳区的宽度,阻止晶体的团聚,并改变晶习。Kim等[14]在晶浆的后结晶过程中,利用超声波的粉碎作用产生细小晶体,避免在后期进行干研磨。Li等[15]指出,超声波可促进成核,改善晶体产品的粒晶分布,得到较完美的晶体,避免了由于在高过饱和度状态下结晶而产生的迸发成核。另外,超声波可改变晶体形貌、尺寸[14,16],从而改变晶体尤其是药物晶体的物理性能,提高晶体质量。

超声波对结晶过程的影响非常复杂,对连续流体的最基本的影响是对其施加交变的压强[17]。在低强度下,这种压强可促进流体局部流动和混合;在高强度下,流体中局部压强可降至流体的蒸汽压以下,生成一些小泡和空穴,当流体中的局部暂时负压增大时,其中的小泡就会长大,在到达一定程度时突然塌陷,形成更多的空穴,即空穴现象,为晶核形成提供特殊的空间和能量,促使晶体的形成。超声波同样对溶液有显著的热效应和混合效应[18],从而对结晶过程产生重要影响。

本工作以甘氨酸水溶液的丙酮溶析结晶为对象,研究了结晶过程中引入超声波后

---

* 本文合作者:周甜、钱刚、周兴贵。原发表于《过程工程学报》,2007,7(4):728~732。

出现的特殊现象，并探讨了超声波施加时刻和持续时间对晶体成核、生长及晶体粒径分布的影响。

2 实验

实验装置如图1所示，使用低温恒温槽（上海衡平仪器仪表厂，DC-2006）为控温系统，蠕动泵（保定兰格恒流泵有限公司，BT100-1J/YZ1515X）为溶剂和溶析剂的加料泵，恒温磁力搅拌器（上海司乐仪器厂，85-2型）搅拌，采用氦氖激光发射器（上海激光技术研究所，632.8nm）和计算机录入模块（上海诺达佳工业控制技术有限公司，I-7521/牛顿-7018）测量溶液的混浊度。

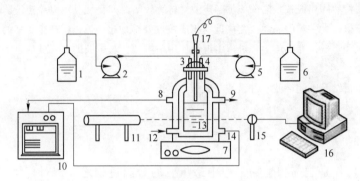

图1 超声波结晶实验装置图

Fig. 1 Schematic diagram of experimental apparatus for ultrasound-aided crystallization

1—Acetone tank；2—Acetone pump；3—Acetone inlet；4—Water inlet；
5—Water pump；6—Water tank；7—Magnetic stirring；8, 9—Cooling water out；
10—Water bath；11—He-Ne laser emitter；12, 14—Cooling water in；
13—Stirrer；15—Laser receiver；16—Computer；17—Ultrasound probe

结晶在三层夹套结晶器中进行，体积为250mL，内径8cm，最内层为母液结晶区，次外层通低温恒温槽的循环水，控制结晶温度，最外层夹套通常温水以避免空气中水分在玻璃外壁凝结而影响激光透过。

采用 PVM（Particle Vision Measurement）探头（梅特勒-托利多仪器（上海）有限公司，700型）、光学显微镜（上海彼爱姆光学仪器制造有限公司，BM-14）和数码相机（尼康映像仪器销售（中国）有限公司，COOLPIX4500）拍摄晶体的形貌。晶体的粒度由马尔文粒度分析仪（Mastersizer 2000）测定。使用日本理学 Rigaku D/MAXRB 型 X 射线衍射仪（XRD）测定晶型，入射光源为 Cu K_α 靶，管电流100mA，管电压40kV，扫描范围 $2\theta = 3°\sim 60°$，扫描速率 $2°/min$。

本工作选取甘氨酸-水-丙酮为结晶体系。实验时先配制一定浓度的溶液，开启低温恒温槽，降温至所需温度，恒定一段时间后，将一定浓度的溶液加入结晶器中。开启激光信号发射器、信号接收器，稳定3~4h。以恒定速率和既定顺序滴加丙酮或水，超声波按照相应程序和功率开启，同时计算机开始记录。

3 结果与分析

3.1 结晶过程

所有实验在20℃下进行。首先在60mL 13.1g/mL 甘氨酸水溶液中以 3.11mL/min

的恒定速率加入丙酮，出现大量固体颗粒后，停止加入丙酮，再以恒定速率加入水。由于纯溶剂（丙酮或水）加入后完全分散并与混合溶剂互溶需要一定时间，而丙酮与水的折光率不同，因此激光透过率出现波动。此外，由于日光的影响，即使激光完全不能通过结晶釜，信号接收器仍有响应。激光主要用于跟踪晶体（或气泡）出现与消失的时刻，因此光电信号的绝对值并不重要。激光透过强度与时间的关系见图2a。

当无超声波时，随溶析剂的加入，溶液逐渐变浑浊；此后加入主溶剂，溶液逐渐变澄清。在 A 点前，透光率几乎不变，A 点后因晶核出现，透光率直线下降，因此 A 点对应超溶解度；$C \sim D$ 点间晶体不断消失，透光率上升，D 点后透光率维持恒定，因此 D 点对应溶解度。在结晶过程中引入超声波后，溶液的透光率经历复杂的变化过程，如图 2b 所示。一旦引入超声波，透光率即开始下降（$A \sim B$），之后经历相对稳定期（$B \sim C$），后迅速下降（$C \sim D$）。向晶浆中加水至一定程度后，晶体逐渐溶解，直至溶液澄清。

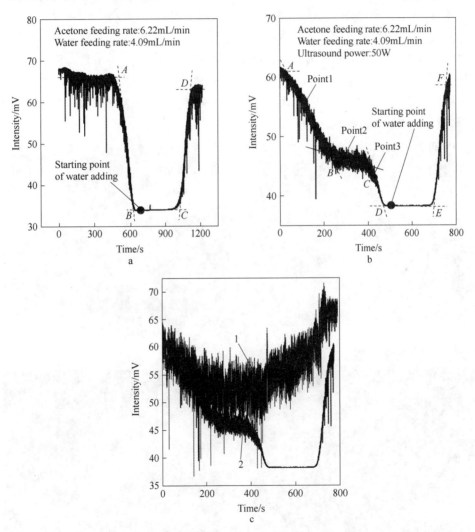

图2　有无超声波作用下结晶过程激光信号曲线及超声波作用下的激光透过对比

Fig. 2　Laser signals in crystallization with and without ultrasound and comparison of laser signals with ultrasound

a—Without ultrasound；b—With ultrasound；c—Comparison of laser signals

1—Without glycin；2—With glycin

为了解释超声波的影响,在无溶质的纯溶剂中重复上述实验,见图2c。由于超声波的空化作用,溶液中产生大量的小气泡并开始浑浊,透光率下降,说明空泡浓度受溶剂配比(表面张力)的影响。停止加入丙酮并开始加水后,溶液的表面张力向相反方向变化,气泡浓度降低,透光率上升。当溶剂中有一定浓度的甘氨酸时,由于物系物理性质与混合溶剂不同,其中发生的物理过程也有所差异。在 $A \sim B$ 段,因超声波的空化作用产生大量气泡,透光率下降,之后透光率变化出现明显转折,转折点的出现可能是因为空泡浓度已达到一平衡值,但更可能是发生了其他物理过程。从 $B \sim C$ 到 $C \sim D$ 段透光率变化又出现了一次转折,源于丙酮的不断加入,溶液浓度超过了超溶解度,出现大量成核。在 $E \sim F$ 段,随着水的加入,晶体不断溶解,直至完全消失。

图2b 的 $A \sim B$ 与 $B \sim C$ 段溶液外观类似,为白色浑浊溶液。在 A 点处停止施加超声波,迅速移走超声波探头并将PVM探头插入溶液,没有观察到颗粒(图3a);按照同样方法,在图2b 的 B 点处可观测到大量几十微米的气泡(图3b),这些气泡可能由尺寸更小的气泡聚并而成,源于超声波的空化作用,使溶液中交替出现气泡的生长和湮灭。PVM观测结果说明在超声波存在的条件下,气泡可能稳定相当长一段时间。在 C 点处用同样的手段可观察到大量细晶(图3c),说明晶核已经产生,但此时观察不到气泡。在此阶段,空泡的湮灭促进了成核,成核也加速了空泡的湮灭。图3d 显示 D 点时结晶器内已经出现大量晶体。

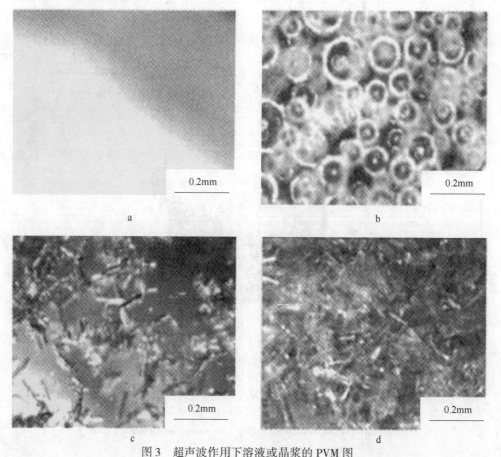

图3 超声波作用下溶液或晶浆的 PVM 图

Fig. 3 PVM observation of the solution at the different stages in Fig. 2b

a—Photo in point A; b—Photo in point B; c—Photo in point C; d—Photo in point D

上述结果表明，在超声波作用下，溶液中出现两次成核，一次为超声波诱导成核，始于 B 点（图2b），另一次（二次成核）始于 C 点。从透光率的变化幅度可知超声波诱导成核速率较慢，而二次成核速率较快。二次成核主要以均相成核为主，但不排除超声波和溶液中已有晶体的诱导作用。比较图2a 和图2b 可以发现，在超声波的作用下，无论是一次还是二次成核，都较无超声波条件下的均相成核提前，因此介稳区宽度缩小。

超声波作为结晶过程操作的一种辅助手段，主要目的是控制晶体产品的质量，包括晶习、粒径分布等，超声波的施加时刻和持续时间是关键操作变量。

3.2 超声波施加时刻的影响

选择从结晶过程开始后的100s、280s、430s 时刻（分别对应图2b 中 AB、BC 和 CD 段点1、2、3处）开始以50W 功率连续施加超声波100s，在结晶过程中，丙酮以3.11mL/min 的速率连续加入，600s 后结束结晶过程，将产品过滤、干燥，在显微镜下观测晶体形貌，并用马尔文粒度分析仪分析粒径大小（d）及分布，结果见图4 和图5。

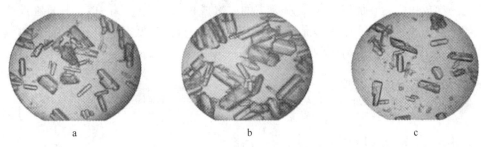

图4　超声波不同施加时刻的晶体形貌图
Fig. 4　Microscopic views of the crystals vs. different ultrasound intervention times （120×）
a—100s；b—280s；c—430s

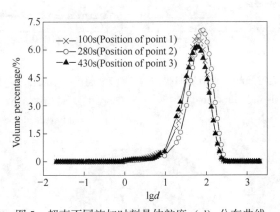

图5　超声不同施加时刻晶体粒度（d）分布曲线
Fig. 5　Crystal size（d）distributions vs. different ultrasound intervention times

在 AB 段施加超声波，在溶液中将产生大量空泡，但也可能在后期诱导少量晶核生成。停止施加超声波后，空泡将消失，晶核仍将保留。超声波诱导产生的晶核数量较少，过饱和度降低也很小。随着丙酮的继续加入，出现二次成核，爆发大量晶核，最终得到的晶体平均粒径为 62.8μm。在 BC 段施加超声波，将诱导和促进大量成核，使溶液过饱和度下降，也使之后的二次成核延迟，同时由于溶液绝对浓度下降，均相成核数量下降。因而在 BC 段施加超声波产生的晶核数量总体上比在 AB 段少，最终得到的晶体粒径也较大，为 80.9μm。

在 CD 段施加超声波时，由于均相成核已经发生，超声波一方面可能促进混合传递，利于晶体生长，但也可能因混合过于强烈对晶体起破碎作用，在均相成核的基础上促进二次成核，使晶体粒径变小，为 61.1μm，分布变宽。从实验结果看，本工作使用的超声波功率对晶体有较强的破碎作用。

3.3 超声波持续时间的影响

结晶过程中连续施加 50W 超声波，在 60mL 13.1g/mL 甘氨酸水溶液中以 6.22mL/min 的速率连续加入丙酮，得到激光透过值与时间的关系，见图 6。在相同的实验条件下，初始时刻不施加超声波，选择图 6 中位置 1 处对应的时刻开始施加超声波，分别持续 100s、200s、300s，获得的粒径大小分布见图 7，晶体形貌见图 8。对比图 6，如果超声波仅仅持续 100s，晶核的产生主要源于超声波的诱导作用，因此晶核数量较小，最终得到的晶体粒径较大，为 84.5μm。如果超声波施加时间较长（200s），将在由超声波诱导的一次成核之后叠加以均相成核为主的二次成核，产生较多晶核，并且在此之后由于超声波的粉碎作用产生更多晶核，由此晶体平均粒径降低至 75.8μm。如果继续延长超声波作用时间，最终得到的晶体粒径进一步降低到 57.2μm。

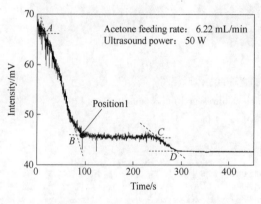

图 6　超声波作用下结晶过程的激光信号曲线
Fig. 6　Laser signals with ultrasound

图 7　超声波不同持续时间时晶体粒度分布曲线
Fig. 7　Crystal size distributions vs. different ultrasound durations

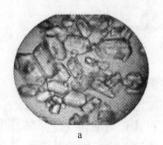
a

b

c

图 8　超声波持续不同时间的晶体形貌图
Fig. 8　Microscopic views of the crystals vs. different ultrasound durations（120×）
a—100s；b—200s；c—300s

3.4 超声波对晶体形貌和晶型的影响

分别采用连续施加超声波（50W，从始至终）和不加超声波两种方式进行结晶，获得的晶体形貌见图9，对应的平均粒径分别为 57.3μm 和 79.6μm。连续施加超声波，一方面由于超声波对晶体的粉碎作用，产生细晶，降低平均粒径，另一方面也因为超声波对混合的促进作用增加了晶体的长径比。晶体各晶面由于表面能的不同，生长速率存在差异。如果晶体生长受外扩散控制，生长速率差异将变小。因此，利用超声波促进外扩散，晶体的形貌将更接近表面能差异确定的形貌。

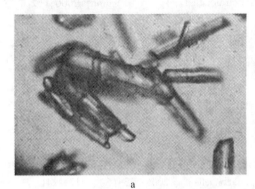 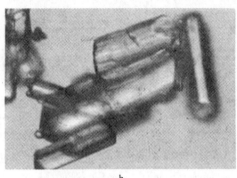

a　　　　　　　　　　　　　　　b

图9　有无超声波时晶体形貌的比较

Fig. 9　Comparison of crystallization habits with and without ultrasound （400×）

a—With ultrasound；b—Without ultrasound

据文献 [13]、[19] 报道，超声波可能对晶体的结构产生影响。本研究体系和操作条件下，XRD 结果表明晶体结构未受到超声波的影响。

4　结论

探讨了超声波对溶析结晶过程的影响，实验结果表明：

（1）在超声波作用下，结晶过程经历空泡形成、超声波诱导成核、二次成核多个阶段。

（2）采用不同的超声波施加时刻和持续时间，可获得比不施加超声波更大或更小的晶体粒径。在自然均相成核点之前施加超声波并持续较短的时间，晶核的产生以超声波诱导成核为主，可获得较大颗粒的晶体；在接近均相成核点处施加超声波将产生更多的晶核，晶体平均粒径降低；在晶体生长过程中继续使用超声波，由于破碎效应，降低了晶体的平均粒径。

超声波的确定作用有空化、搅拌混合、粉碎和致热。对工业结晶过程，能够利用的超声波作用主要是前三种。超声波的空化作用能诱导成核，缩短诱导时间，或降低超溶解度，搅拌混合作用可促进热质传递，提高晶体生长速率，并调节晶形，而粉碎作用能够获得粒度更小的晶体。

参 考 文 献

[1] 时钧. 化学工程手册 [M]. 2版. 北京：化学工业出版社，1996：34~35.

[2] Peng Y, Dou R, Song G, et al. Dramatically Accelerated Synthesis of β-Aminoketones via Aqueous

Mannich Reaction under Combined Microwave and Ultrasound Irradiation [J]. Synlett, 2005, 14: 2245~2247.

[3] Nulty J, Steere J A, Wolf S. The Ultrasound Promoted Knoevenagel Condensation of Aromatic Aldehydes [J]. Tetrahedron Lett., 1998, 39: 8013~8016.

[4] Jin T S, Zhang J S, Wang A Q, et al. Ultrasound-assisted Synthesis of 1, 8-Dioxo-octahydroxanthene Derivatives Catalyzed by p-Dodecylbenzenesulfonic Acid in Aqueous Media [J]. Ultrason. Sonochem., 2006, 13 (3): 220~224.

[5] Wang S Y, Ji S J. Facile Synthesis of 3, 3-Di (heteroaryl) indoline-2-one Derivatives Catalyzed by Ceric Ammonium Nitrate (CAN) under Ultrasound Irradiation [J]. Tetrahedron, 2006, 62 (7): 1527~1535.

[6] Cravotto G, Cintas P. Power of Ultrasound in Organic Synthesis: Moving Cavitational Chemistry from Academia to Innovative and Large-scale Applications [J]. Chem. Soc. Rev., 2006, 35 (2): 180~196.

[7] Li J T, Dai H G, Xu W Z, et al. An Efficient and Practical Synthesis of Bis- (indolyl) -methanes Catalyzed by Aminosulfonic Acid under Ultrasound [J]. Ultrason. Sonochem., 2006, 13 (1): 24~27.

[8] Park J E, Mahito A, Toshio F. Synthesis of Multiple Shapes of Gold Nanoparticles with Controlled Sizes in Aqueous Solution Using Ultrasound [J]. Ultrason. Sonochem., 2006, 13 (3): 237~241.

[9] Stefani H A, Oliveira C B, Almeida R B, et al. Dihydropyrimidin- (2H) -ones Obtained by Ultrasound Irradiation: A New Class of Potential Antioxidant Agents [J]. Eur. J. Med. Chem., 2006, 41 (4): 513~518.

[10] Zhou W J, Ji S J, Shen Z L. An Efficient Synthesis of Ferrocenyl Substituted 3-Cyanopyridine Derivatives under Ultrasound Irradiation [J]. J. Organomet. Chem., 2006, 691 (7): 1356~1360.

[11] Okitsu K, Yue A, Tanabe S, et al. Formation of Colloidal Gold Nanoparticles in an Ultrasonic Field: Control of Rate of Gold (III) Reduction and Size of Formed Gold Particles [J]. Langmuir, 2001, 17: 7717~7720.

[12] McCausland L, Cains P. Ultrasound to Make Crystals [J]. Chem. Ind., 2003, 5: 15~17.

[13] Guo Z, Zhang M, Li H, et al. Effect of Ultrasound on Anti-solvent Crystallization Process [J]. J. Cryst. Growth, 2005, 273: 555~563.

[14] Kim S, Wei C, Kiang S. Crystallization Process Development of an Active Pharmaceutical Ingredient and Particle Engineering via the Use of Ultrasonics and Temperature Cycling [J]. Org. Process Res. Dev., 2003, 7: 997~1001.

[15] Li H, Wang J K, Bao Y, et al. Rapid Sonocrystallization in the Salting-out Process [J]. J. Cryst. Growth, 2003, 247: 192~198.

[16] Li Z W, Tao X J, Chen Y M, et al. A Simple and Rapid Method for Preparing Indium Nanoparticles from Bulk Indium via Ultrasound Irradiation [J]. Mater. Sci. Eng., A, 2005, 407: 7~10.

[17] McCausland L J, Cains P W, Martin P D. Use the Power of Sonocrystallization for Improved Properties [J]. Chem. Eng. Process., 2001, 6: 56~61.

[18] Nishida I. Precipitation Calcium Carbonate by Ultrasonic Irradiation [J]. Ultrason. Sonochem., 2004, 11 (6): 423~428.

[19] 孟明, Stievano L, Lambert J F. 甘氨酸在 SiO_2 表面的吸附及热缩合反应 [J]. 催化学报, 2005, 26 (5): 393~398.

Effect of Ultrasound on Anti-solvent Crystallization Process of Glycin

Zhou Tian, Qian Gang, Zhou Xinggui, Yuan Weikang

(State Key Laboratory of Chemical Engineering, East China University of Science and Technology, Shanghai, 200237, China)

Abstract Crystallization of aqueous glycin solution with acetone as the anti-solvent is studied to examine the effects of ultrasound on crystal size and habit. With ultrasound excitations, the solution experiences cavitation, first nucleation, second nucleation, crystal growth, fragmentation, etc. Introducing ultrasound at different times or for different durations will have different influences on the crystallization distribution. If ultrasound is introduced for a short time before the onset of homogeneous nucleation to induce nucleation, one will obtain larger crystals, whereas if the ultrasound is introduced at or after the onset of homogeneous nucleation, smaller crystals will be obtained. If ultrasound is used in the growth stage, the crystals will also be small because the crystals will be fragmented by ultrasound.

Key words anti-solvent crystallization, ultrasound, glycin, nucleation, growth

构造理论在工程领域中的应用研究进展*

摘　要　从工程设计与优化的角度，对构造理论在不同工程领域的应用进行了总结，重点介绍了构造理论在传热和流体分布等单元设备设计和优化中的应用，展现了构造理论（Constructal theory）在工程应用中的意义和重要价值。构造理论作为相对较新的过程系统设计理论，还需要继续发展和完善。对构造理论在化工过程及设备的设计和优化中的应用前景进行了展望。

关键词　构造理论　工程设计　工程优化　多尺度

1　前言

在进行过程系统的工程设计时，都希望能在达到设计目标的同时使系统效能最佳。如对热水供给系统的设计，要求在将热水送达用户（设计目标）的同时，使输送耗能和热量损失最小（最佳效能）。但要实现最优设计，必须首先明确系统结构和效能之间的关系，这种关系在绝大多数情况下是定性的、模糊的，工程师通常只有依靠经验进行设计，尽管能够达到设计目标，但却难以达到最佳效能。构造理论的出现为摆脱这种局面提供了一种新方法。

构造理论是由美籍罗马尼亚科学家 Bejan[1] 在 20 世纪 90 年代中期首先提出的。构造理论是一种最优设计方法，它从系统的基本单元结构开始优化，之后再将这些经过优化的最小单元结构通过优化逐级组合起来，一直到满足设计要求。在构造理论中，系统效能和结构之间的关系由物理和数学方程描述，系统结构通过优化方法确定。因此，利用构造理论进行系统设计在某种意义上是最优设计。

构造理论的提出引起科技人员的极大兴趣，目前已被应用于众多不同的研究领域，如电网[2,3]、热水供给系统[4,5]、冷却系统[6]的构建、建筑墙体[7~9]和电磁体的结构设计[10]，以及生命体结构、地球气候、土地的风化龟裂等众多自然现象的诠释等[11~14]。

构造理论其实是 20 多年来热力学系统优化的一种延续[15]。到目前为止，对构造理论关注较多的主要是美国、法国以及罗马尼亚。在我国，陈林根等[16,17]最早在介绍热力学优化理论时提到 Constructal theory，并将其翻译为构形理论，以后程新广等[18]将其应用于构造导热通道，周圣兵等[19]从广义热力学优化的角度对构造理论进行了介绍，罗灵爱等[20]着重介绍了构造理论在流体分布多尺度优化中的应用。

本工作在简单介绍构造理论基本思想的基础上，综述了构造理论在众多领域，特别是在传热和流体分布系统工程的设计和优化中的应用，并对构造理论今后在化学工程等应用领域的发展进行了展望。

2　构造理论的基本思想

Bejan 在研究城市街道网络时发现，众多城市的街道网络具有非常相似的树状结

---

*　本文合作者：范志伟、周兴贵。原发表于《过程工程学报》，2007，7（4）：832~839。

构。城市的主干街道较宽，上面行驶的多是速度较快的机动车辆，且流量较大，而行人则多在较窄的小巷中穿行，流量较小。他认为城市街道网络的树状结构是城市长时间发展和进化的结果，它使居民在耗能（花费）最小的情况下，出行速度最快，因此具有最佳效能。为此，Bejan按照逐级优化的方法建立了一个街道网络模型，准确地描述了实际情况。在此基础上，Bejan[21~23]进一步研究了传热和流体流动系统的设计和优化，最终发展形成了构造理论（Constructal theory）[24,25]。

下面用一个简单的例子说明构造理论方法。在城市建设中，合理的街道网络可使居民出行更加高效、快捷。将此问题简化为构建从面（A）上任意一点到达点（M）时间最短的路径，如图1所示。假设体系中有不同速度的运动方式：慢速（步行，速度V_0），快速（车辆，速度$V_1 < V_2 < V_3$）。设计从选取基本单元A_1开始，其中A_1的面积（$A_1 = H_1 L_1$）是固定的，但其形状（H_1/L_1）是可变的，而A_1（最小街区）的大小与当地的自然状况和居民的习惯有关。

通过分析找到影响从A_1上各点到边界点M_1间速度的结构参数，其中包括A_1的形状参数H_1/L_1、快速通道（速度V_1）的位置、步行（速度V_0）的角度α_1。当速度V_1远大于V_0时，经过分析优化可确定这个矩形的最佳几何参数为快速通道位于长轴上，$H_1/L_1 = 2V_0/V_1$，$\alpha_1 = 0$。

下一步是数个基本单元A_1优化组合到一起。如图1所示，一级组合A_2就是由N_2个基本单元A_1构成的，速度为$V_2(>V_1)$的路径用来连接速度为V_1的路径和所有的A_1。以A_2与点M_2之间行进速度最快为目标进行分析优化，可以推导出街道（速度V_2）的位置（区域A_2的长轴）、A_2的形状参数H_2/L_2和角度α_2。

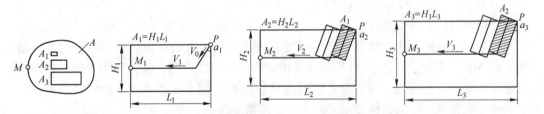

图1　运用构造理论构建点面间行进速度最快的路径

Fig. 1　Construction of high speed paths between points and areas with constructal theory

通过重复相同的步骤就可得到更高级、尺寸更大的组合结构，如A_3，并直到所得组合区域的大小满足设计要求。在整个结构中，每个几何结构参数都是为达到同一目标（行进速度最快）优化的结果。在此过程中，没有任何假设条件，也没有受到任何自然结构的影响。

以上只是构造理论工程应用的一个小例子，而在其他与流体相关的工程领域中，构造理论方法同样适用。工程师只需在应用相关流体方程的同时，遵循一条简单的、唯一的构造理论法则：从小到大，用几何学的方法，对系统的不完美特性进行最优的配置。如在解决一点与无数点（体点或面点）之间的流体流动问题时，构造理论先是构造一个最小单元结构（其大小与制造工艺水平有关），然后再将一定数量的最小单元组合成更大的单元结构，并按此步骤逐级继续，直到得到符合要求的流体路径结构。在整个构造过程中，每次组合的结构参数，包括单元组合结构的几何尺寸、组合单元的数量等，都是以优化流体流动为目标的分析计算结果。在构造结构中，传输阻力大的流动形式，如慢速流动、高耗能流动、扩散、步行等，被限制在每个最小单元结构内,而传输阻力较小的流动形式,如快速流动、低能耗流动、强对流、运输工具的行驶等,则出现在各个组合之间,用来连接体点或面点间的流动,最后全部构建单元组成一个有机的整体,形成树状结构的流动路径,其中每层结构的几何结构参数都是优化所得。

按构造理论得到的系统结构与分形结构看上去非常相似,但事实上构造理论与分形论有着本质区别。分形论[26]是应用分形几何学对自然界中的各种几何图形和结构如树木、河流等进行研究的方法。由于分形具有扩展对称性的自相似结构,其部分结构在形式上与其整体结构本身相似。分形作为一种从大到小的无限过程,可以非常近似地描述自然界中的许多几何形状和结构。分形论也被用于流体分布器、混合器[27]和传热系统[28]的设计优化中,并起到提高系统性能的作用。但需要强调的是,分形是一个从大到小的无限过程,它将系统不断分割一直到无限小。分形的过程在时间方向上从大到小,是描述性和非决定性的,而构造理论却正好与其"相反",在进行系统结构设计时,从有限小的结构单元优化开始,将小的单元结构不断组合为更大的结构,整个过程通过一级一级的构建来实现。构造理论在时间方向上是从小到大的,是可预测的和决定性的[24]。

3 构造理论的发展应用

构造理论提出之初就被用于高导热材料优化配置强化传热的研究中,其后又被用于对流换热强化。与此同时,研究者还将构造理论应用于流体分布、传质过程强化等。

3.1 构造理论优化系统传热

从 Bejan[21] 运用构造理论解决电子元件中高热导材料的布置以强化热传导开始,这方面的研究就一直备受关注。与 Bejan 的最小化热阻不同,Dan 等[29]研究了如何在最短时间内将一定体积内的热量导出的问题,将研究从二维的点面扩展到三维的点体范围;而 Almogbel 等[30]提出两种提高二维树状结构导热性能的方法,可用于解决任意形状电子元件的冷却问题。按此方法得到的导热路径(图2)具有典型的树状结构,而且通过比较发现,增加结构复杂度有利于提高系统性能。

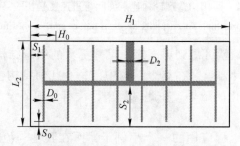

图 2 内含 2 个一级高导热材料组装结构的二级热传导结构[30]

Fig. 2 Second-assembly configuration containing two first assemblies[30]

Almogbel 等[31]在研究过程中发现,如放松对基本单元结构的尺寸限制,提高结构自由度,能够提高系统的热传导性能,并可得到与自然界中真实结构更加相似的树状结构。Rocha 等[32]运用构造理论设计出圆形电子元件上高热导材料的最优分布(图3),以将电子元件产生的热量从圆心导出(图中从圆心出发的黑线代表高热导材料),最后得到辐射状布置结构,其中扇形面积(α)的大小、辐射路径长度(L_0)以及分叉数目都是优化得到的结果。

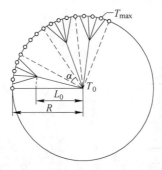

图3 扇形区域内具有分叉状冷却结构的圆碟型发热体[32]

Fig. 3 Disc-shaped body cooled by a structure consisting of several branched sectors[32]

Ghodoossi 等[33,34]研究了由两种基本单元（矩形和三角形）组合成不同形式（矩形和三角形）的导热路径，并比较了这些组合的性能。组合方式包括矩形组合成矩形（RR），矩形组合成三角形（RT），三角形组合成三角形（TT）以及三角形组合成矩形（TR），如图 4 所示，结果发现当冷却区域面积相同时（$A_2 = L_2H_2$），矩形组合成三角形的组合方式性能最好。另外，他们还探讨了三角形电子元件的传热冷却问题[35]。

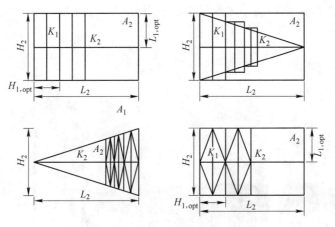

图4　各种二级组合[33,34]

Fig. 4　Second RR, RT, TT, and TR assembled configurations
(R for rectangular, and T for triangular, from left to right and top to bottom)[33,34]

除了强化热传导之外，构造理论在对流换热强化方面也有许多应用。Bejan 等[36]运用构造理论提出了一种多尺度板片组合结构以达到强化层流换热的目的（图5）。他们发现在大的板片（长 L_0，间距 D_0）中间插入较小的板片（$L_2 < L_1 < L_0$），可以改善层流换热时大板片边界层之间部分的传热状况，从而提高对流换热能力。其中，最佳的插入位置 D_1、D_2 和相应的板片长度 L_1、L_2 都可由优化计算确定。

随后 Bello-Ochende 等[37~39]将这种方法用于强制对流换热强化研究，提出了多尺度管路组合结构（图6）以强化传热性能，优化参数包括各尺度管尺寸（D_0，D_1，D_2）的选择以及各尺度管在组合中的位置安排（S_0，S_2）。

Silva 等[40,41]则研究了自然对流换热下多尺度结构的设计方法和如何在含有热源的封装结构中构造多尺度结构，以达到最大传热性能。

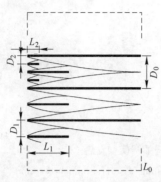
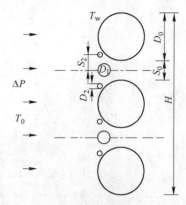

图 5　优化的多尺度平行板片组合[36]

Fig. 5　Optimal multi-scale package of parallel plates[36]

图 6　三种不同尺寸管路组合结构[38]

Fig. 6　Row of cylinders with three sizes[38]

从图 7 可看出插入小的板片可强化系统换热（红色区域越大、蓝色区域越小代表传热性能越好）。从左到右，随着新板片的插入，系统的传热性能也越来越强。而 Silva 等[42,43]在其他关于对流换热强化的研究中主要研究自然层流和强制对流冷却的壁面上分散热源的安置问题。此外他们还求得了自然对流冷却情况下，垂直开放通道里离散热源的最优分布（图 8 中 S_0，S_1，S_2，S_3）和其几何尺寸（D_0，y_0）[44]，并对这些研究进行了分析总结[45]。

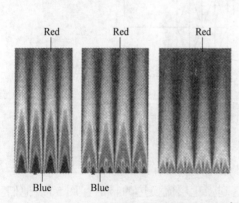
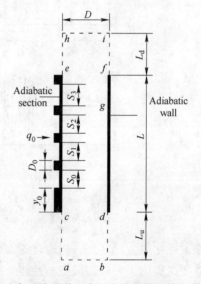

图 7　具一、二、三级尺度板片结构的温度分布[40]

Fig. 7　The temperature distributions in packages with one, two and three length scales[40]

图 8　垂直开放通道上多尺度热源的不均匀分布[44]

Fig. 8　Multiple length scales of the non-uniform distribution of finite-size heat sources on a vertical channel[44]

构造理论另一常见的强化传热应用是通过优化翅片形状及结构参数以达到翅片热阻最小和传热能力最大等目标。Bejan 等[46]研究了在翅片材料总量和总体积一定的条件下（图 9）如何通过优化 T 状翅片的几何形状（L_1/L_0，t_1/t_0）达到热传导性能最大的目的，并将这种 T 状翅片与末端向下弯曲的 Tau 状翅片进行了比较。

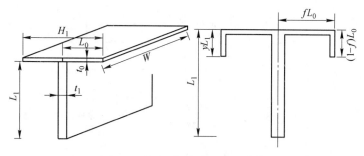

图 9 T 状和 Tau 状翅片结构[46]

Fig. 9 T-shaped (left) and Tau-shaped (right) assembly of plate fins[46]

Almogbel 等[47]研究了圆柱体和辐射状翅片组合结构的外部形状、主干和翅片直径比及翅片形状对组合结构传热性能的影响。在此基础上加入新的分支翅片，可得到传热表面更大、结构更加紧凑、传热性能更强的新结构[48]。Silva 等[49]研究了在材料总量一定的条件下，碟状发热体中矩形高导翅片的分布情况，以使发热体的热阻最小。Matos 等[50]运用模拟和实验相结合的方法研究了翅片和椭圆形管道组合结构在体积一定和外部流动恒定时，各几何结构参数（管与管之间的空间距离（L）、翅片间距（δ）椭圆形管的偏心率）对结构传热效能的影响，并求得其最佳值（图10）。这一研究引入了实验工作，这在构造理论应用探索研究中为数不多，因而具有重要的现实意义。

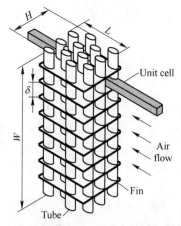

图 10 椭圆管道和翅片的组合结构以及用于模拟计算的三维区域[50]

Fig. 10 Arrangement of finned elliptic tubes and three-dimensional computational domain[50]

换热器是重要的化工系统单元设备，研究者通过设计新的结构和优化结构参数来提高其换热性能。由 Bejan[51]设计的树枝形换热器具有多尺度内部流体通道结构，并以消除层流流动中纵向温差、强化换热为目的，对换热器基本单元结构进行优化，同时通过调整基本单元之间的空隙和流体通道的几何尺寸，降低换热器内流体的流动耗能。而由 Bonjour 等[52]设计的圆柱形双流体共轴换热器（图11）则通过在环形空间中插入翅片和优化翅片几何结构达到强化换热的目的。研究发现最佳翅片组合设计形式为发射状还是分枝状与换热器横截面的大小（R）密切相关。

最近几年，关于微型换热器在过程强化方面的研究和应用受到很大关注，出现了很多新颖的设备结构形式。Muzychka[53]在微型换热器的构形设计中引入了构造理论，进行体积大小和压降固定条件下微通道列管式换热器的设计，考察了不同截面形状

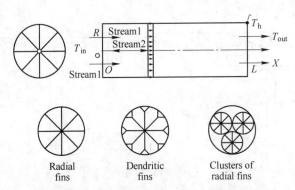

图 11 具有发散状翅片的共轴双流体换热器和三类翅片组合设计形式[52]

Fig. 11 Coaxial two-stream heat exchanger with radial plate fins (left) and three competing configurations in the design of fins for coaxial heat exchanger (right)[52]

（平板式、矩形、椭圆形、圆形以及三角形等）和尺寸的微通道及其组合对换热器性能的影响（图12），发现正方形和等腰三角形微通道换热器的换热性能最好。

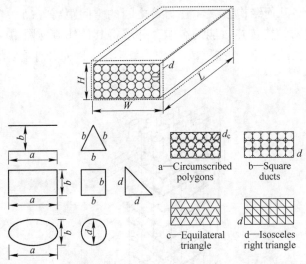

图 12 微型换热器（上）、各种通道截面形式（左下）和其组合形式（右下）[53]

Fig. 12 Micro heat exchanger (top) elemental geometries being considered (left bottom) and some possible packing arrangements (right bottom)[53]

以上介绍的只是构造理论在强化传热方面的部分应用，其实在工程应用中，它还有很大的拓展空间，如具有翅片的冷却或加热装置，包括快速吸热的气化器、乙烯裂解装置中的核心部件——裂解管的结构设计和优化等。

3.2 构造理论优化系统流体分配

构造理论除了可以用来构造强化热传导的流体通道之外，对其他流体通道路径的构造也同样适用，如构造耗能最小的流体分布网络等。以下将主要介绍构造理论在流体分配中的应用。

燃料电池作为新型洁净能源装置，受到能源工程界的广泛关注。作为反应器，其内部结构的合理设计对于提高燃料电池的性能非常重要。Vargas 等[54~56]通过优化燃料电池的单元外形和内部结构,在达到电力输出最大的同时,将供给反应物的耗能降到最低。Senn 等[57]设计和优化了 DMFC 燃料电池单元内部的多尺度流体通道和电流收集结构的几何尺寸 (L_k,h_k,$b_k - h_k$),得到比传统单一通道燃料电池更大的输出电力和更高的电池效率(图13)。

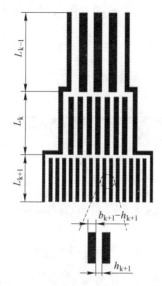

图 13　金字塔状流体分布几何结构[57]

Fig. 13　Geometric structure of the pyramidal tree network fluid delivery system including channels (white) and current collector shoulders (black)[57]

在列管式反应器、换热器等化工设备的设计中,经常会遇到流体均布的问题。在寻找最佳解决方法的过程中,Tondeur 等[58]提出了多尺度流体分布器的概念(图14),它可以将速率一定的流体均匀地分布到一组规则分布的孔道中,并同时达到停留时间最小和耗能最少的目的。值得一提的是,他们用立体激光印刷术制造了第 1 个构造理论的实体部件——分枝结构流体分布器。Luo 等[59]对分枝结构流体分布器进行了进一步发展和完善,提出了新的四分叉树状结构以降低流体能耗,并用高速摄像机观察记录了分布器中流体的均布情况,定性地表明了分布器的流体均布能力。他们还将流体分布器与一个微型换热器组合在一起,通过实验考察了不同分布器对换热器性能的影响。

  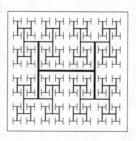 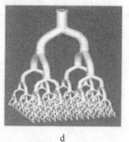

a　　　　　　　　b　　　　　　　　c　　　　　　　　d

图 14　具有 8 级多尺度、256 个出口的流体分布器[58]

Fig. 14　Constructal fluid distributor with 8 generations and 256 outlets[58]

a—Outline; b—Inner frame; c—Projection of pore network on base plane;
d—Pore network with smooth direction changes

此外，还有构造理论用于其他一些流体分布的研究，如点到面、点到圆周和点到直线等[60~64]其间流体路径的构建（图15），以及包含 T 状或 Y 状接管的管路网络的构建[65,66]，虽不如上面介绍的流体分布器有典型的可被应用于化工过程的特征，但在化工管路网络的设计和优化研究中仍具有巨大的潜在价值。而 Alvarez 等[67]将构造理论用于吸附-脱附过程时空结构的优化，又显示了构造理论在优化质量交换结构时的优越性。

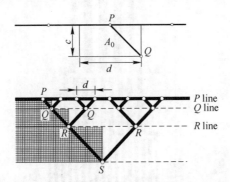

图15　具有一条流体通道的基元系统（上）和点线间长度最小的树状结构（下）[62]

Fig. 15　Elemental system with one flow segment (top) and the construction of the minimal-length tree between a line and a point (bottom)[62]

除了以上构造理论在传热和流体分布等系统工程优化方面的应用，还有研究者将构造理论运用到自然界结构经济性最优的分析和证明[68,69]，对一些电学理论如欧姆定律的证明[70]及飞行物体自身质量和能耗之间关系的最优分析[71]之中，显示了构造理论的普适性。

4　总结和展望

正如 Bejan 构造理论专著[25]的书名（从工程到自然）所暗示的那样，构造理论开始应用于工程设计和优化，之后又应用于自然界中一些结构和现象的分析和解释。构造理论不仅被用于优化流体流动单元外部形态和内部构造，也被用来解释一些自然规律（生物结构是生物体自身以性能最优为目标不断进化的结果）。对自然规律的成功解释反过来也证明了构造理论的合理性和正确性，由此进一步拓展了其在系统结构的设计、分析和预测中的应用。

构造理论自身有着鲜明的特点，与现代物理学从大到小步步简化的研究方法相反，构造理论则是按照从小到大、层层优化组合的方式，将基本单元结构组合构建成优化的整体宏观结构。构造理论本身也不是完美无缺的，需要我们在认识、运用的同时，加以发展和完善。构造理论作为一种相对较新的过程系统设计理论，目前了解、接受和应用的研究者还不多。但随着时间的推移，越来越多的研究者会认识构造理论。构造理论提供了工程设计与优化创新的新方法，在与不同工程领域相结合的过程中，将产生丰富多彩的创新设计思想，并最终变为现实，应用于各工程领域。

化学工程是跨度最广的工程领域，构造理论在化工过程及设备的工程设计和优化中大有用武之地。目前，化工过程强化是化学工程的重点发展领域，而化工流体力学设备（用于进行混合、均布、反应、分离、换热等）的强化是过程强化的主要内容。对这些设备进行强化的根本途径是对流体进行结构化，而为实现流体结构化必须对设备内部进行结构化。目前对流体进行结构化的主要方式是利用特殊设计的内构件（如导流筒、挡板、盘管、花板等）和规整填料（用于催化反应器、精馏塔、静态混合器等）。不可否认，这些内构件和填料的设计在很大程度上依赖经验，与最优设计有一定距离。这些手段尽管能够实现过程强化，有时甚至效果十分显著，但仍有潜力可挖。因此，构造理论不仅如前所述能够提供新的设备强化思路，也为现有过程强化手段的挖潜和优化提供了重要的理论基础。

参 考 文 献

[1] Bejan A. Street Network Theory of Organization in Nature [J]. J. Adv. Transp., 1996, 30 (2)：85~107.

[2] Valentin A, Alexandru C, Bejan A. Constructal Tree Shaped Networks for the Distribution of Electrical Power [J]. Energy Convers. Manage., 2003, 44 (6)：867~891.

[3] Valentin A, Alexandru C, Bejan A. Integral Measures of Electric Power Distribution Networks：Load-Length Curves and Line-network Multipliers [J]. Energy Convers. Manage., 2003, 44 (7)：1039~1051.

[4] Wechsatol W, Lorent S, Bejan A. Tree-shaped Insulated Designs for the Uniform Distribution of Hot Water over an Area [J]. Int. J. Heat Mass Transfer, 2001, 44 (16)：3111~3123.

[5] Wechsatol W, Lorent S, Bejan A. Development of Tree-shaped Flows by Adding New Users to Existing Networks of Hot Water Pipes [J]. Int. J. Heat Mass Transfer, 2002, 45 (4)：723~733.

[6] Zamfirescu C, Bejan A. Tree-shaped Structures for Cold Storage [J]. Int. J. Refrig., 2005, 28 (2)：231~241.

[7] Lorente S, Bejan A. Combined "Flow and Strength" Geometric Optimization：Internal Structure in a Vertical Insulating Wall with Air Cavities and Prescribed Strength [J]. Int. J. Heat Mass Transfer, 2002, 45 (16)：3313~3320.

[8] Gosselin L, Lorent S, Bejan A. Combined "Heat Flow and Strength" Optimization of Geometry：Mechanical Structures Most Resistant to Thermal Attack [J]. Int. J. Heat Mass Transfer, 2004, 47 (14/16)：3477~3489.

[9] Rocha L, Lorenzini E, Biserni C. Geometric Optimization of Shapes on the Basis of Bejan's Constructal Theory [J]. Int. Commun. Heat Mass Transfer, 2005, 32 (10)：1281~1288.

[10] Gosselin L, Bejan A. Constructal Thermal Optimization of an Electromagnet [J]. Int. J. Therm. Sci., 2004, 43 (4)：331~338.

[11] Bejan A, Ikegami Y, Ledezma G A. Constructal Theory of Natural Crack Pattern Formation for Fastest Cooling [J]. Int. J. Heat Mass Transfer, 1998, 41 (13)：1945~1954.

[12] Bejan A. The Constructal Law of Organization in Nature：Tree-shaped Flows and Body Size [J]. J. Exp. Biol., 2005, 208 (9)：1677~1686.

[13] Reis H, Bejan A. Constructal Theory of Global Circulation and Climate [J]. Int. J. Heat Mass Transfer, 2006, 49 (11/12)：1857~1875.

[14] Bejan A. How Nature Takes Shape：Extensions of Constructal Theory to Ducts, Rivers, Turbulence, Cracks, Dendritic Crystals and Spatial Economics [J]. Int. J. Therm. Sci., 1999, 38 (8)：653~663.

[15] Poirier H. Une Théorie Explique l'intelligence de la Nature [J]. Science & Vie, 2003, 1034 (1)：44~63.

[16] 陈林根, 孙丰瑞. 有限时间热力学理论和应用的发展现状 [J]. 物理学进展, 1998, 18 (4)：395~422.

[17] 陈林根, 孙丰瑞. 热力学优化理论的研究进展 [J]. 武汉化工学院学报, 2002, 24 (1)：81~85.

[18] 程新广, 李志信, 过增元. 基于仿生优化的高效导热通道的构造 [J]. 中国科学 E 辑, 2003,

33（3）：251~256.

[19] 周圣兵，陈林根，孙丰瑞. 构形理论：广义热力学优化的新方向之一 [J]. 热科学与技术，2004，3（4）：283~292.

[20] Luo L A, Tondeur D. Multiscale Optimisation of Flow Distribution by Constructal Approach [J]. China Particuology, 2005, 3（6）：329~336.

[21] Bejan A. Constructal-theory Network of Conducting Paths for Cooling a Heat Generating Volume [J]. Int. J. Heat Mass Transfer, 1997, 40（4）：799~816.

[22] Bejan A. Constructal Tree Network for Fluid Flow between a Finite-size Volume and One Source or Sink [J]. Rev. Gén. Therm., 1997, 36（8）：592~604.

[23] Bejan A. Theory of Organization in Nature: Pulsating Physiological Processes [J]. Int. J. Heat Mass Transfer, 1997, 40（9）：2097~2104.

[24] Bejan A. Constructal Theory: From Thermodynamic and Geometric Optimization to Predicting Shape in Nature [J]. Energy Convers. Manage., 1998, 39（16/18）：1705~1718.

[25] Bejan A. Shape and Structure, from Engineering to Nature [M]. Cambridge: Cambridge University Press, 2000：5~19.

[26] Mandelrot B. The Fractal Geometry of Nature [M]. New York: WH Freeman, 1982：9~33.

[27] Mike M K. Engineered Fractals Enhance Process Applications [J]. Chem. Eng. Prog., 2000, 96（12）：61~68.

[28] Chen Y P, Cheng P. Heat Transfer and Pressure Drop in Fractal Tree-like Microchannel Nets [J]. Int. J. Heat Mass Transfer, 2002, 45（13）：2643~2648.

[29] Dan N, Bejan A. Constructal Tree Networks for the Time-dependent Discharge of a Finite-size Volume to One Point [J]. J. Appl. Phys., 1998, 84（6）：3042~3050.

[30] Almogbel M, Bejan A. Conduction Trees with Spacings at the Tips [J]. Int. J. Heat Mass Transfer, 1999, 42（20）：3739~3756.

[31] Almogbel M, Bejan A. Constructal Optimization of Nonuniformly Distributed Tree-shaped Flow Structure for Conduction [J]. Int. J. Heat Mass Transfer, 2001, 44（22）：4185~4194.

[32] Rocha L, Lorent S, Bejan A. Constructal Design for Cooling a Disc-shaped Area by Conduction [J]. Int. J. Heat Mass Transfer, 2002, 45（8）：1643~1652.

[33] Ghodoossi L, Egrican N. Flow Area Optimization in Point to Area or Area to Point Flows [J]. Energy Convers. Manage., 2003, 44（16）：2589~2608.

[34] Ghodoossi L, Egrican N. Flow Area Structure Generation in Point to Area or Area to Point Flows [J]. Energy Convers. Manage., 2003, 44（16）：2609~2623.

[35] Ghodoossi L, Egrican N. Conductive Cooling of Triangular Shaped Electronics Using Constructal Theory [J]. Energy Convers. Manage., 2004, 45（6）：811~828.

[36] Bejan A, Fautrelle Y. Constructal Multi-scale Structure for Maximal Heat Transfer Density [J]. Acta Mechania, 2003, 163（1）：39~49.

[37] Bello-Ochende T, Bejan A. Maximal Heat Transfer Density: Plates with Multiple Lengths in Forced Convection [J]. Int. J. Therm. Sci., 2004, 43（12）：1181~1186.

[38] Bello-Ochende T, Bejan A. Constructal Multi-scale Cylinders in Cross-flow [J]. Int. J. Heat Mass Transfer, 2005, 48（7）：1373~1383.

[39] Bello-Ochende T, Bejan A. Constructal Multi-scale Cylinders with Natural Convection [J]. Int. J. Heat Mass Transfer, 2005, 48（21/22）：4300~4306.

[40] Silva A K, Bejan A. Constructal Multi-scale Structure for Maximal Heat Transfer Density in Natural Convection [J]. Int. J. Heat Fluid Flow, 2005, 26（1）：34~44.

[41] Silva A K, Lorente S, Bejan A. Constructal Multi-scale Structures with Asymmetric Heat Sources of Fi-

nite Thickness [J]. Int. J. Heat Mass Transfer, 2005, 48 (13): 2662~2672.

[42] Silva A K, Lorente S, Bejan A. Optimal Distribution of Discrete Heat Sources on a Plate with Laminar Forced Convection [J]. Int. J. Heat Mass Transfer, 2004, 47 (10/11): 2139~2148.

[43] Silva A K, Lorente S, Bejan A. Optimal Distribution of Discrete Heat Sources on a Wall with Natural Convection [J]. Int. J. Heat Mass Transfer, 2004, 47 (2): 203~214.

[44] Silva A K, Lorenzini G, Bejan A. Distribution of Heat Sources in Vertical Open Channels with Natural Convection [J]. Int. J. Heat Mass Transfer, 2005, 48 (8): 1462~1469.

[45] Silva A K, Lorente S, Bejan A. Constructal Multi-scale Structures for Maximal Heat Transfer Density [J]. Energy, 2006, 31 (5): 620~635.

[46] Bejan A, Almogbel M. Constructal T-shaped Fins [J]. Int. J. Heat Mass Transfer, 2000, 43 (12): 2101~2115.

[47] Almogbel M, Bejan A. Cylindrical Trees of Pin Fins [J]. Int. J. Heat Mass Transfer, 2000, 43 (23): 4285~4297.

[48] Almogbel M. Constructal Tree-shaped Fins [J]. Int. J. Therm. Sci., 2005, 44 (4): 342~348.

[49] Silva A K, Vasile C, Bejan A. Disc Cooled with High-conductivity Inserts that Extend inward from the Perimeter [J]. Int. J. Heat Mass Transfer, 2004, 47 (19/20): 4257~4263.

[50] Matos R S, Laursen T A, Vargas J, et al. Three-dimensional Optimization of Staggered Finned Circular and Elliptic Tubes in Forced Convection [J]. Int. J. Therm. Sci., 2004, 43 (5): 477~487.

[51] Bejan A. Dendritic Constructal Heat Exchanger with Small-scale Crossflows and Larger-scales Counter Flows [J]. Int. J. Heat Mass Transfer, 2002, 45 (23): 4607~4620.

[52] Bonjour J, Rocha L, Bejan A, et al. Dendritic Fins Optimization for a Coaxial Two-stream Heat Exchanger [J]. Int. J. Heat Mass Transfer, 2003, 47 (1): 111~124.

[53] Muzychka Y S. Constructal Design of Forced Convection Cooled Microchannel Heat Sinks and Heat Exchangers [J]. Int. J. Heat Mass Transfer, 2005, 48 (15): 3119~3127.

[54] Vargas J, Ordonez J C, Bejan A. Constructal Flow Structure for a PEM Fuel Cell [J]. Int. J. Heat Mass Transfer, 2004, 47 (19/20): 4177~4193.

[55] Vargas J, Bejan A. Thermodynamic Optimization of Internal Structure in a Fuel Cell [J]. Int. J. Energy Res., 2004, 28 (4): 319~339.

[56] Vargas J, Ordonez J C, Bejan A. Constructal PEM Fuel Cell Stack Design [J]. Int. J. Heat Mass Transfer, 2005, 48 (21/22): 4410~4427.

[57] Senn S M, Poulikakos D. Pyramidal Direct Methanol Fuel Cells [J]. Int. J. Heat Mass Transfer, 2006, 49 (7/8): 1516~1528.

[58] Tondeur D, Luo L A. Design and Scaling Laws of Ramified Fluid Distributors by the Constructal Approach [J]. Chem. Eng. Sci., 2004, 59 (8/9): 1799~1813.

[59] Luo L A, Tondeur D. Optimal Distribution of Viscous Dissipation in a Multi-scale Branched Fluid Distributor [J]. Int. J. Therm. Sci., 2005, 44 (12): 1131~1141.

[60] Ledezma G A, Bejan A, Errera M R. Constructal Tree Networks for Heat Transfer [J]. J. Appl. Phys., 1997, 82 (1): 89~100.

[61] Wechsatol W, Lorente S, Bejan A. Optimal Tree-shaped Networks for Fluid Flow in a Disc-shaped Body [J]. Int. J. Heat Mass Transfer, 2002, 45 (25): 4911~4924.

[62] Lorente S, Wechsatol W, Bejan A. Tree-shaped Flow Structures Designed by Minimizing Path Lengths [J]. Int. J. Heat Mass Transfer, 2002, 45 (16): 3299~3312.

[63] Gosselin L, Bejan A. Tree Networks for Minimal Pumping Power [J]. Int. J. Therm. Sci., 2005, 44 (1): 53~63.

[64] Gosselin L. Minimum Pumping Power Fluid Tree Networks without a Priori Flow Regime Assumption

[J]. Int. J. Heat Mass Transfer, 2005, 48 (11): 2159~2171.

[65] Bejan A, Rocha L, Lorente S. Thermodynamic Optimization of Geometry: T- and Y-shaped Constructs of Fluid Streams [J]. Int. J. Therm. Sci., 2000, 39 (9/11): 949~960.

[66] Zimparov V D, Silva A, Bejan A. Thermodynamic Optimization of Tree-shaped Flow Geometries [J]. Int. J. Heat Mass Transfer, 2006, 49 (9/10): 1619~1630.

[67] Rivera-Alvarez A, Bejan A. Constructal Geometry and Operation of Adsorption Processes [J]. Int. J. Therm. Sci., 2003, 42 (10): 983~994.

[68] Bejan A, Badescu V, De Vos A. Constructal Theory of Economics [J]. Appl. Energy, 2000, 67 (1/2): 37~60.

[69] Bejan A, Badescu V, De Vos A. Constructal Theory of Economics: Structure Generation in Space and Time [J]. Energy Convers. Manage., 2000, 41 (13): 1429~1451.

[70] Lewins J. Bejan's Constructal Theory of Equal Potential Distribution [J]. Int. J. Heat Mass Transfer, 2003, 46 (9): 1541~1543.

[71] Ordonez J C, Bejan A. System-level Optimization of the Sizes of Organs for Heat and Fluid Flow Systems [J]. Int. J. Therm. Sci., 2003, 42 (4): 335~342.

Advances of Constructal Theory in Engineering Applications

Fan Zhiwei, Zhou Xinggui, Yuan Weikang

(State Key Laboratory of Chemical Engineering, East China University of Science & Technology, Shanghai, 200237, China)

Abstract In view of engineering design and optimization, applications of constructal theory in several engineering areas are introduced, especially in design and optimization of heat transfer and fluid distribution devices, the significance and potential value of constructal theory in engineering applications are highlighted. As a novel process system designing theory, constructal theory needs to be developed and improved. The application prospect of constructal theory in chemical engineering process/devices design and optimization is also discussed.

Key words constructal theory, engineering design, engineering optimization, multi-scale approach

有机酸杂质对乳酸锌初次成核速率的影响*

关键词 成核 乳酸锌 杂质 初次成核动力学

1 引言

溶液结晶作为一种高效、低能耗、少污染的分离和提纯技术,在医药、食品、生物等工业生产中的应用非常广泛。特别在医药行业,溶液结晶是主要的单元操作,85%以上的医药产品的生产过程中含有结晶单元[1]。结晶过程涉及成核和生长两个阶段,成核和生长的热力学和动力学决定最终晶体产品的粒度分布以及产品质量,是进行结晶器分析、设计、操作和优化的主要依据[2]。

工业结晶过程中不可避免有杂质存在。通常情况下,溶液中存在少量的杂质即可能对晶体的成核、生长动力学和晶习产生显著的影响。杂质对结晶过程的影响机理十分复杂,不同的杂质对结晶过程有不同的影响。杂质的存在将改变介稳区宽度,增加或降低成核能垒[3,4],从而影响初次成核;杂质也可能吸附在晶体表面缺陷位并增大表面裂缝,使晶体易碎而产生二次晶核[5]。此外,Kubota等[6]通过研究发现,当溶液浓度高于某一临界浓度时,晶体表面生长控速步骤由表面生长控制转为溶质扩散控制,杂质对结晶过程的影响消失。溶液结晶时杂质的存在也可能改变固-液边界层的特性,从而影响生长单元的聚集[7,8]。杂质本身也可以选择性地吸附在晶体的不同晶面或晶面的不同活性位从而促进或抑制特定晶面的生长,有些杂质还能嵌入晶格中,并影响晶体的空间结构[9,10]。

成核动力学与生长动力学是结晶过程模拟、设计与控制的基础。初次成核动力学可通过直接测定初次晶核的生成速率来确定,如Garten等[11,12]利用晶体成核时的发光现象来研究成核速率。由于晶体成核瞬间(小于10^{-7}s)发生,直接测量晶核的生成速率对仪器要求极高,不易实现。初始成核动力学也可间接地通过测量介稳区宽度加以确定,这种方法由Nývlt[13]提出。由于该法简便易行,在工业中得到了广泛的应用[14,15]。

本文提出了一种杂质存在条件下的晶体初次成核动力学研究方法,用于研究杂质(苹果酸、丁二酸或草酸钠)对乳酸锌初次成核动力学的影响。该方法以Nývlt成核动力学模型[13]为基础,适用于间歇降温结晶过程。在工业结晶中,初次成核速率一般采用简单的经验关联式表示:

$$B = K_n \Delta c^n \tag{1}$$

式中,B为初次成核速率,$mol/((kgH_2O) \cdot s)$;K_n为初次成核速率常数;$\Delta c = c - c^*$为溶液过饱和度,$mol/(kgH_2O)$;n为成核级数。

根据Nývlt提出的成核动力学模型,极限过饱和度时的成核速率(初次成核速率)

---

* 本文合作者:张相洋、周兴贵、钱刚。原发表于《化工学报》,2011,62(5):1302~1307。

可以采用以下方程表示：

$$B = b\frac{dc^*}{dT}\bigg|_{T=T_n} \tag{2}$$

式中，b 为降温速率，K/s；T_n 为成核发生时的温度，K；$\frac{dc^*}{dT}\big|_{T=T_n}$ 为溶解度曲线在 T_n 处斜率。合并式（1）、式（2）可得：

$$\lg b = n\lg\Delta c + \lg K_n - \lg\frac{dc^*}{dT}\bigg|_{T=T_n} \tag{3}$$

因此，如能获得不同条件下的溶解度曲线，并在不同的降温速率 b 下确定 T_n，就可确定初次成核速率常数（K_n）和成核级数 n。

需要注意的是采用测量介稳区宽度法获得的初次成核速率 B 单位为 $mol/((kgH_2O)\cdot s)$，而成核速率 B_n 的常用单位为 $No./(m^3\cdot s)$。在忽略加入溶质后引起的溶液体积变化时，两者之间存在如下关系：

$$B_n = BN\rho/\omega \tag{4}$$

式中，N 为阿伏伽德罗常数；ρ 为溶剂密度，kg/m^3；ω 为单个晶核所包含的分子个数，可通过式（5）进行计算得到：

$$\omega = \frac{V}{v} = \frac{\frac{4}{3}\pi r_c^3}{v} \tag{5}$$

式中，V 为临界晶核体积；r_c 为临界晶核的半径；v 为单个分子体积。

临界晶核的半径可通过式（6）确定[16]：

$$r_c = 2\gamma v/kT\ln S \tag{6}$$

式中，γ 为表面自由能，J/m^2；k 为玻耳兹曼常数，$1.38\times10^{-23} J/K$；T 为温度；S 为溶液过饱和度。

2 实验方法

所有实验均在图 1 所示的装置中进行。该结晶釜体积为 200mL，温度控制精度为 ±0.1K，搅拌速率控制精度为 ±0.5r/min。在测定某种溶质在一种溶剂中的溶解度时，常用的方法是将确定量的溶质加入溶剂中，形成溶质过量、有固体悬浮的饱和溶液，之后逐渐提高饱和溶液温度，直至固体颗粒最终溶解消失。由于激光法具有响应快、灵敏度高、准确性好等优点，因此本文采用该法确定溶质溶解时刻以及过饱和溶液中首批晶核出现的时机。激光监视装置由激光发生器、接收器、光电转换器和光强显示仪组成。实验中采用 He-Ne 激光发射器，通过观察透过被测溶液的激光强度对应的电压（mV）随时间变化情况来确定成核温度和溶解温度。图 2 为溶液在以一定速率降温和以一定速率升温过程中透过溶液的激光强度对应的电压随时间的变化。从电压变化曲线的转折点可分别从图中确定成核温度 T_n 和溶解温度 T_s。

利用这种方法确定热力学溶解温度时，悬浮液升温的速率必须尽可能小，Mullin[16] 建议控制在 2K/h 以下，因此真正地取决于热力学的溶解温度的确定很耗时。在一定的升温速率下测定的固体颗粒的消失温度由于受传递速率的影响，与热力学溶解温度有所不同，升温速率越大，差距越大。

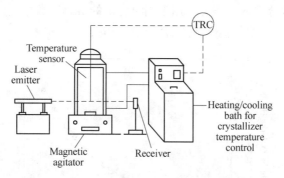

图1 降温结晶实验装置图

Fig. 1 Schematic diagram of experimental apparatus for crystallization

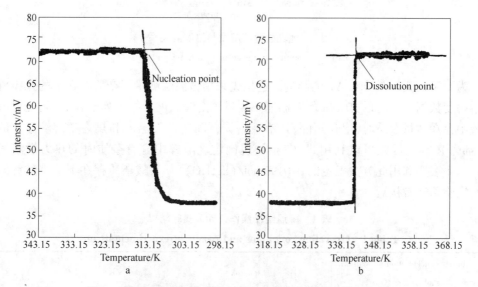

图2 降温和升温过程中激光强度对应的电压随时间的变化

Fig. 2 Relation between laser beam intensity and time

a—Cooling crystallization; b—Heating dissolution

为了提高实验效率，本文采用了较大的升温速率（20K/h）先确定固体的消失温度T_{diss}，之后通过校正升温速率的影响来确定热力学溶解温度T_s。对于溶解度已知的纯乳酸锌溶液（溶解度采用恒温法测量[17]），可确定在此升温速率下T_s与T_{diss}之差ΔT；对于溶解度未知的含杂质的乳酸锌溶液，由于杂质浓度较低，认为ΔT几乎不变。因此，在杂质存在时的热力学溶解温度T_s可以直接根据实验测量到的固体颗粒消失温度T_{diss}估算，即：

$$T_s = T_{diss} - \Delta T \tag{7}$$

3 结果与讨论

3.1 杂质存在时的乳酸锌溶解度

溶解度与温度的关系通常可用式（8）关联[18]：

$$\ln c^* = \alpha + \frac{\beta}{T} + \gamma \ln T \tag{8}$$

式中，α、β、γ为拟合常数，需要利用不同温度下的溶解度数据确定。图3为纯乳酸锌溶液溶解度（点）和采用式（8）的拟合结果（虚线）。

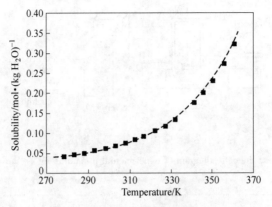

图3　纯乳酸锌溶解度与温度的关系

Fig. 3　Solubility curve of pure zinc lactate

为了验证微量杂质对 ΔT 的影响，这里先采用恒温法测量了杂质存在下的几组乳酸锌溶解度数据，之后以20K/h的升温速率测量了溶解温度，结果列于表1。可见乳酸锌溶液中存在微量杂质时，与纯乳酸锌溶液一样，固体消失温度和热力学溶解温度的差 ΔT 都为1.2K。因此可以利用式（7）快速确定乳酸锌中含有杂质时的热力学溶解温度。图4表明利用此方法确定的0.0199mol/(kgH$_2$O) 苹果酸杂质存在下的乳酸锌溶解度与实验值非常接近。

表1　降温结晶操作条件及实验结果

Table 1　Operating conditions and results of experimental runs

No.	Initial concentration of solution /mol·(kgH$_2$O)$^{-1}$				T_s/K	T_{diss}/K	T_n/K					
	Zinc lactate	Malic acid	Succinic acid	Sodium oxalate			b = 0.001389 K/s	b = 0.00283 K/s	b = 0.002778 K/s	b = 0.003472 K/s	b = 0.004167 K/s	b = 0.005556 K/s
1	0.2070				346.1	347.3						
2	0.1967				344.3	345.5						
3	0.1956				344.1	345.3	334.5	331.9	330.0	328.3	327.1	324.5
4	0.1956	0.0199			339.7	340.9		325.5	323.6	321.8	320.3	
5	0.1956	0.0149			340.7	341.9			325.3	323.5	322.0	
6	0.1956	0.0099			342.4	343.6			327.7	325.9	324.5	
7	0.1956	0.0049			343.2	344.4			328.7	326.9	325.5	
8	0.1956		0.0112		342.9	344.1		329.3	327.1	325.5	323.6	
9	0.1956			0.0025	344.5	345.7		332.9	330.5	328.8	326.9	

如表1所示，当乳酸锌浓度为0.1956mol/(kgH$_2$O) 时，0.0199mol/kg 的苹果酸使乳酸锌溶液热力学溶解温度下降4.4K，0.0112mol/(kgH$_2$O) 的丁二酸使热力学温度下降1.2K，而0.0025mol/(kgH$_2$O) 的草酸钠则使热力学溶解温度升高0.4K。杂质改变了乳酸锌在溶液中的复杂有机酸电离平衡和有机酸根与锌离子的络合平衡，由此改变乳酸锌的溶解度。

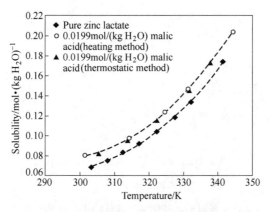

图4 乳酸锌溶解度与温度的关系
Fig. 4 Solubility curve of zinc lactate

3.2 杂质存在时的乳酸锌初次成核动力学

3.2.1 不同杂质存在下的乳酸锌初次成核动力学

表1中也列出了通过激光法测量得到的不同降温速率下的成核温度。在本文的操作条件下，纯乳酸锌溶液的介稳区宽度在9.8～20.3K之间，且随着降温速率的增加而增加。当体系中有杂质存在时，乳酸锌溶液介稳区的位置和宽度均发生了变化。其中苹果酸和丁二酸存在时，介稳区上移，且宽度增加值在0.2～2.4K之间。而草酸钠杂质的引入则起到了相反的作用。对于介稳区位置的改变可以解释为杂质对溶解度的改变。溶解度增加后，介稳区位置必然发生上移，反之亦然。介稳区宽度的变化则是由成核能垒的变化引起的：当体系中存在杂质时，杂质分子与溶质分子之间的相互作用通常会引起成核能垒的改变，并由此改变介稳区宽度。

根据式（3）对杂质存在时的乳酸锌初次成核动力学参数进行拟合，结果如表2所示。

表2 杂质存在时的乳酸锌初次成核动力学参数
Table 2 Primary nucleation parameters in presence of impurities

Characteristics of aqueous solution	K_n	n	R^2
Pure zinc lactate solution	3.8×10^3	2.28	0.9973
Zinc lactate solution added with 0.0199mol/(kgH$_2$O) malic acid	6.3×10^3	2.58	0.9993
Zinc lactate solution added with 0.0112mol/(kgH$_2$O) succinic acid	3.2×10^3	2.30	0.9976
Zinc lactate solution added with 0.0025mol/(kgH$_2$O) sodium oxalate	1.2×10^3	1.83	0.9948

苹果酸和丁二酸杂质存在时的乳酸锌初次成核级数分别为2.58和2.30，均高于纯乳酸锌的成核级数。而草酸钠杂质的成核级数低于纯乳酸锌，为1.83。图5为乳酸锌溶液的过饱和度对初次成核速率的影响。由图可知，苹果酸和丁二酸杂质的存在抑制了成核的发生。此外，在相同过饱和度时，苹果酸和丁二酸对乳酸锌成核的抑制能力比较接近；而草酸钠在低过饱和度时对成核起到了促进作用，只有在高过饱和度时才对乳酸锌成核起抑制作用。

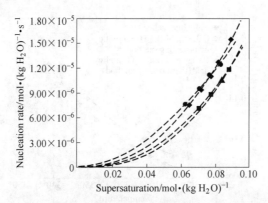

图 5 杂质存在时的乳酸锌初次成核速率与过饱和度的关系
Fig. 5 Fitted primary nucleation kinetic of zinc lactate by the least square method

◆—Pure zinc lactate; ▲—With 0.0199mol·$(kgH_2O)^{-1}$ malic acid; ■—With 0.0112mol·$(kgH_2O)^{-1}$ succinic acid;
●—With 0.0025mol·$(kgH_2O)^{-1}$ sodium oxalate; ---fitted by $K_n\Delta c^n$

草酸钠对乳酸锌初次成核速率的影响随着溶液过饱和度发生转变是因为影响成核的控制因素在不同过饱和度时发生了变化。经典成核理论认为，晶核的形成涉及分子间相互碰撞和相互结合两个过程。因此，成核过程受两方面因素的控制：（1）溶液浓度，代表分子间相互碰撞的概率；（2）结合能，代表分子间结合能力。在相同的乳酸锌浓度下，草酸的存在增加了过饱和度，或在相同的过饱和度情况下，有草酸时的乳酸锌的浓度较低。草酸钠杂质的存在改变了乳酸锌的溶解度，同时也因参与成核过程改变了成核能垒。低过饱和度时，成核速率主要受结合能控制；草酸根的存在增加了参与成核分子之间（包括草酸锌之间和草酸与乳酸锌之间）的结合能力，由此提高了成核速率。而过饱和度较高时，成核速率主要受溶液浓度影响；在过饱和度相同时，有草酸时的乳酸锌溶液的浓度比纯乳酸锌溶液的浓度更低，因此晶体成核速率也降低。

3.2.2 不同苹果酸杂质浓度下的乳酸锌初次成核动力学

根据表 1 中的成核温度和在该温度下的溶解度曲线斜率，利用式（3）回归杂质存在时的乳酸锌初次成核动力学参数，结果列于表 3。乳酸锌初次成核级数随着苹果酸浓度的增加而增加，且均高于纯乳酸锌的成核级数。为便于比较苹果酸浓度变化对乳酸锌初次成核动力学的影响，参考式（1）将乳酸锌初次成核速率与过饱和度之间的关系绘于图 6。可见任意浓度条件下的苹果酸均对乳酸锌成核起到了抑制作用，且这种作用随着苹果酸杂质浓度的增加而增加。苹果酸（包括丁二酸）的引入提高了乳酸锌的溶解度，在成核发生时乳酸锌有更高的浓度，原则上应该有更大的成核速率。但苹果酸的存在削弱了参与成核的分子（乳酸锌和苹果酸锌）之间的结合能力，由此降低成核速率。

表 3 不同苹果酸浓度条件下的乳酸锌初次成核动力学参数
Table 3 Values of nucleation parameters in the presence of the malic acid

Characteristics of aqueous solution	K_n	n	R^2
Zinc lactate solution added with 0.0199mol/(kgH_2O) malic acid	6.3×10^3	2.58	0.9993
Zinc lactate solution added with 0.0149mol/(kgH_2O) malic acid	4.1×10^3	2.38	0.9994
Zinc lactate solution added with 0.0099mol/(kgH_2O) malic acid	4.1×10^3	2.34	0.9974
Zinc lactate solution added with 0.0049mol/(kgH_2O) malic acid	3.9×10^3	2.31	0.9973
Pure zinc lactate solution	3.8×10^3	2.28	0.9973

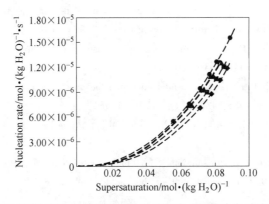

图 6 不同苹果酸浓度条件下的乳酸锌初次成核速率与浓度的关系

Fig. 6 Malic acid effect on primary nucleation at different concentrations

●—Pure zinc lactate；▼—With 0.0049mol·(kgH$_2$O)$^{-1}$ malic acid；▲—With 0.0099mol·(kgH$_2$O)$^{-1}$ malic acid；
■—With 0.0149mol·(kgH$_2$O)$^{-1}$ malic acid；◆—With 0.0199mol·(kgH$_2$O)$^{-1}$ malic acid；- - -Fitted by $K_n \Delta c^n$

4 结论

本文在确定较高的升温速率下，利用溶液升温时的溶解温度 T_{diss} 与理论溶解温度 T_n 之间的差值（$\Delta T = T_{diss} - T_n$）保持不变这一特点来估计微量杂质存在条件下的理论溶解度。实验表明，该方法准确可靠，能显著提高实验效率。根据 Nývlt 成核动力学模型提出了一种通过匀速降温实验快速确定溶液冷却结晶初次成核动力学的方法。并以此考察了苹果酸、丁二酸和草酸钠杂质对乳酸锌初次成核速率的影响。研究表明，苹果酸和丁二酸都对乳酸锌成核起抑制作用；草酸钠在低过饱和度时促进乳酸锌成核，只有在高过饱和度时才对乳酸锌成核起抑制作用。

参 考 文 献

[1] Wang Jingkang（王静康）. Present and future of industrial crystallization [J]. Chemical Engineering（化学工程），1992，20（2）：57~63.

[2] Myerson A S. Handbook of Industrial Crystallization [M]. 2nd ed. Amsterdam：Elsevier Science & Technology Books，2001：44~53.

[3] Gibbs J W. Thermodynamic [M]. New Haven：Yale University Press，1948：237.

[4] Volmer M. Kinetics of Phase Formation [M]. Steinkopff：Dresden，1939：146.

[5] Serig S，Mullin J W. The size reduction of crystals in slurries by the use of habit modifiers [J]. Industrial and Engineering Chemistry Process Design and Development，1980，19（3）：490~494.

[6] Kubota N，Ito K. Shimizu K. Chromium ion effect on contact nucleation of ammonium sulphate [J]. Journal of Crystal Growth，1986，76（2）：272~278.

[7] Rashkovich L N，Kronsky N V. Influence of Fe^{3+} and Al^{3+} ions on the kinetics of steps on the {100} faces of KDP [J]. Journal of Crystal Growth，1997，182（3/4）：434~441.

[8] Dugua J，Simon B. Crystallization of sodium perborate from aqueous solutions（Ⅱ）：Growth kinetics of different faces in pure solution and in the presence of a surfactant [J]. Journal of Crystal Growth，1978，44（3）：280~286.

[9] Thomas B R，Chernov A A，Vekilov P G. Distribution coefficients of protein impurities in ferritin and lysozyme crystals self-purification in microgravity [J]. Journal of Crystal Growth，2000，211（1~4）：149~156.

[10] Valery T, Chani I, Inoue K, Shimamura K. Segregation coefficients in β-Ga$_2$O$_3$: Cr crystals grown from a B$_2$O$_3$ based flux [J]. Journal of Crystal Growth, 1993, 132 (1/2): 335~336.

[11] Garten V A, Head R B. Crystalloluminescence and the nature of the critical nucleus [J]. Philosophical Magazine, 1963, 95 (8): 1793~1803.

[12] Garten V A, Head R B. Homogeneous nucleation and the phenomenon of crystalloluminescence [J]. Philosophical Magazine. 1966, 132 (14): 1243~1253.

[13] Nývlt J. The Kinetics of Industrial Crystallization [M]. Amsterdam: Elsevier Science & Technology Books, 1985.

[14] Nývlt J. Kinetics of nucleation in solutions [J]. Journal Cryst. Growth, 1968, 7 (3/4): 377~383.

[15] Bravi M. Mazzarotta B. Primary nucleation of citric acid monohydrate [J]. Chemical Engineering Journal, 1998, 70 (3): 197~202.

[16] Mullin J W. Crystallization [M]. 2nd ed. London: Butterworth, 1972.

[17] Zhang Xiangyang (张相洋), Zhong Liang (钟亮), Qian Gang (钱刚), Zhou Xinggui (周兴贵), Yuan Weikang (袁渭康). Purification of zinc lactate by crystallization in presence of malic acid [J]. CIESC Journal (化工学报), 2010, 61 (7): 1815~1820.

[18] Williamson A T. The exact calculation of heats of solution from solubility data [J]. Trans. Faraday Soc., 1944, 40: 421.

Effect of Impurity on Primary Nucleation Kinetics of Zinc Lactate

Zhang Xiangyang, Zhou Xinggui, Qian Gang, Yuan Weikang

(State Key Laboratory of Chemical Engineering, East China University of Science and Technology, Shanghai, 200237, China)

Abstract An approach was proposed to determine the solubility of pure zinc lactate at a high heating rate (20K/h), which significantly improves the efficiency of solubility measurement. The results showed a good agreement with the solubility data obtained through isothermal processes. This approach was extended to measure the solubility of zinc lactate in the presence of small amount of impurity. The influences of impurities, including malic acid, succinic acid and sodium oxalate, on the primary nucleation kinetics of zinc lactate in cooling crystallization were investigated, and a method was presented to estimate the primary nucleation kinetics based on Nývlt theory by measuring the metastable zone width of zinc lactate at specific cooling rate in the solution. The results showed that the rate of primary nucleation of zinc lactate decreased in the presence of malic acid or succinic acid. However, when sodium oxalate was introduced in the solution, the nucleation of zinc lactate was accelerated at low supersaturation but suppressed at high supersaturation.

Key words nucleation, zinc lactate, impurity, primary nucleation kinetics

英文论著

- 研究方法论
- 动力学与催化剂
- 反应器工程
- 超临界技术

Targeting the Dominating-scale Structure of a Multiscale Complex System: a Methodological Problem[*]

Abstract It has long been a problem of finding out a proper methodology for dealing with multiscale complex systems, including processes and products. However, it is seemingly impractical to start with the most fundamental structure of the system, and then to extend step by step to obtain its holistic behaviors. The author's idea is to study the system only to set up the ties between some key structure, or dominating-scale structure, and certain holistic performance of the system that interests people, and then to manipulate the found dominating-scale structure to achieve our target. Several cases from the author's laboratory have been quoted for elucidation. Identification of the dominating-scale structure is made on the system individuality base, rather than on some generalized principle. The same viewpoint is suggested when manipulating the dominating-scale structure.

Key words   multiscale, complex system, dominating scale, structure

To extract physical knowledge from a complex system, one must focus on the right level of description.

Every good model starts from a question. The modeler should always choose the correct level of detail to answer the question.

(Goldenfeld, N., Kadanoff, L. P., Simple lessons from complexity. Science 1999, 284, 87-89)

1 Introduction: Process- and Product-oriented Chemical Engineering

Chemical engineering underwent two development stages in the past century. As commonly understood, it is currently marching forward in its third stage.

The first stage scanned a period from the late 19th century till the 1960s of the past century, with unit operations as a mark of the research interest of this period. Experimental result correlation with various dimensionless groups was the basic method dealing with the relations among influencing variables and parameters. Defining and predicting the holistic performances and behaviors of process units and equipment was seemingly the main focus of the chemical engineering community. Obviously, from the updated viewpoint the involved scale of concern of that period could be the process-unit/equipment scale, including the component scale, such as a plate in a distillation column, or a tube in a heat exchanger.

Starting from the late 1950s and the early 1960s, transport phenomena (Bird et al., 1960) rose from the horizon and rapidly became a more and more important branch of chemical engineering, symbolizing initiation of the second stage. Transport phenomena in fluid flow and those related to turbulence, droplets, bubbles, and pellets became more and more attractive because

---

[*] Reprinted from Chemical Engineering Science, 2007, 62: 3335-3345.

they had helped people to get insight into the process and they had proved themselves closely related to the holistic process performances that needed to be modified and optimized. Compared with the process unit scale, transport phenomena studies were based on a smaller, or even a much smaller scale. The fundamental method of this stage was mathematical modeling, mainly controlled volume modeling, which was entirely different from the conventional experimentation-correlation, and it has been widely applied since then, aided by the significant progress in computer science that made numerical solution of nonlinear model equations possible.

When approaching the end of the past century, people noticed that a new paradigm was being uncovered though its essence was not sufficiently clear. With the aid of advanced laboratory instruments and of *ab initio* computation techniques, people were able to observe and to understand structures of much smaller scales of the process, such as molecule structure, and structures of colloid, cluster, turbulence, and pore levels, etc. Structures of these levels actually directly affect the properties and performances of a process or a product, particularly a solid-state product. As an example for a process, the active components of a catalyst that are of molecular cluster scale determine the intrinsic rates of the reaction, but the pore structure affects the reactant diffusion. The apparent rates which are observed as a holistic property of the catalyst is affected at least by its structure of the two scales. An example for a product with desired holistic properties is also clear: we need to understand its proper structures from the molecule level to several larger levels to meet our target, and to master means of manufacturing the product with the desired structures. The problem can then be categorized as process-oriented and product-oriented. The latter has become extensively important because more and more focuses have been shifted from commodity products to fine chemical, functional, and high value-added specialty products whose properties are closely tied to their various scale structures.

Although connotation of the third stage is still under discussion, there is an obvious tendency that, with deepening of the chemical engineering discipline, the focusing scale goes down from macroscopic to smaller, and even to microscopic, as is shown in Table 1.

Table 1 A summary of three development stages

Development stage	Major focusing concern, or topic	Scale of major interest	Major methodology
First stage (late 19th century till 1960s)	Unit operations	Process-unit/equipment/component	Dimensionless group correlation (inductive)
Second stage (1960s till late 20th century)	Transport phenomena, reaction engineering	Turbulence, droplet, bubble, pellet	Mathematical modeling (deductive)
Third stage (starting from late 20th century)	Product engineering? Complex systems? Multiscale structure/property relations?	Tinier, mesoscopic, or microscopic	Advanced computation, advanced synthesis, modification and assembly

In the past century, the horizon of process engineering offered two very different categories of products. On the one hand, there were those such as ethylene, gasoline, and commodity polymers, etc., for which more mature technologies had been applied. These products are undifferentiated with similar properties and functions, adding only a small value. The target of chemical engineering was to develop processes that manufactured the products under optimized conditions and to ensure minimum cost, consistent with safety, health, and environmental concerns. Normally, product manufacturing of this kind came to depend on large volume, continuous processes with well designed and controlled equipment. This is what we called process-oriented.

The other category of process industry supplied comparatively high value-added specialty products and functional products, such as health-care agents, special-quality composite materials, and specialty chemicals, where the innovation rate was most important and where the product suppliers competed on product functions(Cussler and Wei, 2003). Manufacturing could be realized with batch processes. The key was perhaps no longer low cost, or even proper operation, but product quality and speed to market. This is what we call product-oriented industry, which obviously symbolizes a trend of the latest development.

Both process-oriented engineering and product-oriented engineering are facing complex systems, which are normally multiscale as well. To deal with problem like this, people naturally think of interactions between structures of the neighboring levels, or scales. People also consider that the systems need to be reduced to its most fundamental structure, the molecular scale(if sub-molecular scale structures are not considered in engineering). People then move forward from scale to scale, starting with the molecular scale, up to the one they need, followed by integration of the decoupled scale structures and then to obtain the holistic properties. This methodology, which is similar to what is called "reductionism" in physics, is likely reasonable. However, for complex systems, it is almost impractical, at least at the present stage. Is there a more straightforward means which is practical and acceptable?

2 The Concept: Identifying and Manipulating the Dominating-scale Structure

Whether a system is regarded as simple or complex first depends on the observability of its various level structures.

Examining a catalyst pellet we notice that there exist at least three scales: molecule scale, pore scale, and pellet scale. However, looking back to some 80 years ago, people might just take it as a simple black cylindrical pellet with some catalytic activity. With the aid of devices such as HRTEM, XRD, and BET, we have learnt its complexity. We certainly can go deeper inside into the pellet and explore its sub-molecular structures when necessary. If our goal is very limited, for example, to the apparent rate of a given reaction under some defined operating condition, what we only need to do is to determine the rate through a laboratory packed-bed reactor without making clear the complex aspect of the catalyst. For this purpose, to deal with the catalyst as simple pellets may be adequate, even though its complexity has been known. However, for those beyond, thorough understanding of other scale structures would be necessary. If the reaction rate is very low, internal diffusion might be neglected. With a higher reaction rate, the internal diffusion begins to play a role, therefore the porous scale structure needs to be taken care of. At a much higher reaction rate, the external diffusion might dominate the process, so the pellet size and morphology affect the apparent rate. This is to explain that the dominating scale shifts as the condition varies, and it is likely that, under a certain condition, there exists only one(occasionally together with a neighboring one) principal structure which is of real significance.

Similar situation is encountered in product-oriented processes. Chemical composition, or molecular structure is the foundation of a product. However, more and more facts have revealed, in addition to chemical composition, the micro/nano structure greatly affects properties of a product. Countless examples can be given to show importance of the structure-based effects. Carbon materials and composite materials with diverse performances and properties are well known. An excellent example can be found in developing a high-porosity whiffle aerosol for pulmonary drug delivery(Edwards et al., 1997). The specially manipulated morphology and structure enable the large aerosol particles to slowly deliver the drug and to pass through the narrow lung passages without being blocked. This technical breakthrough makes possible an effective spray drug of this kind. This example also shows that, in case of the same drug, the effectiveness and functions could be entirely different according to its particle structure. Looking back to what had been foreseen by Richard Feynman, the 1965 Noble laureate in physics, "When we have some control of the arrangement of things on a small scale, we will get an enormously greater range of possible properties that substance can have, and of different things we can do." (Feynman, 1959). We chemical engineers would be fond of working in an area to control things arranged on a small scale.

For both process-oriented and product-oriented issues, people need to raise an idea and to create a methodology. We have mentioned that it is neither practical nor necessary to decouple or to reduce the problem to the most fundamental structures, and then to integrate or reconstruct the structures of various scales to obtain the holistic natures of a process or a product. In physics, after a long-term discussion and controversy, it is likely the scientists tend to apply the concept of "emergent" property of a system, raised by Philip Anderson, the 1977 Nobel laureate in physics, instead of the conventional "reductionism" (Anderson, 1972, 1994). As we can understand, it implies that a stride from one scale to the other, emergence of a much different, or perhaps entirely new property might appear. The key is to extract the most significant structure of a complex system. In genetics studies, it has been well established that the sequence-wise DNA structure is the biological basis of inheritance and variations. The DNA molecule itself is like all other organic substances, built up from the elements C, H, O, N, P, but the study of genetics is never reduced to that level(i.e., the atoms)—as long as DNA is concerned as a genetic material, only the basepair arrangement is relevant. Therefore, the biological study of DNA has always been focused to looking for connections between the DNA sequence(genotype) and the phenotypes, rather than to reduce the subject to a lower level of structure. Similar cases can be found anywhere. In semiconductor studies, people have found that conductivity is only based on the electron energy band structure of the solid, but not on the nucleus or the quark.

What counts is to explore the right scale of which its structure dominates a certain property of interest; what follows is to realize the desired structure applying all kinds of possible chemical engineering measures. As soon as the desired structure is achieved, by and large the expected goal can be attained. We therefore can summarize the research methodology schematically with

Fig. 1. From the desired performances (of a process) or the desired properties (of a product), we need to determine the dominating-scale structure. The two-way arrows imply that the knowledge feedback and back-and-forth studies are always required. Understandably, various chemical engineering manipulations can be applied only aiming at realizing the dominating-scale structure. We thus can reduce the discussing complex system to a relatively "simple" one as soon as the dominating-scale structure is found. What follows is to take advantage of all possible chemical engineering means to manipulate the dominating-scale structure so as to obtain the desired performances or properties. The author would like to further discuss this idea through several research cases that have been carried out in his laboratory.

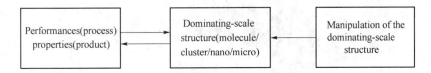

Fig. 1　Schematic for dominating-scale structure manipulation

3　Several Case Studies Representing Manipulation of Dominating-scale Structure

Case 1　Shifting of the dominating-scale structure in a multiphase reactor with continuous phase transition

Consider a packed-bed multiphase reactor in which gas and liquid go concurrently upward through the packed catalyst bed. A reactor like this is most suitable for highly exothermic reactions when high reactant conversion is required (Cheng et al., 2001a, b, 2003a). An excellent example is hydrogenation of benzene using Pd/γ-Al$_2$O$_3$ as catalyst (Cheng et al., 2003b; Zhou, 2004). Benzene conversion must be higher than 99.5% since its separation from the product cyclohexane is almost impractical.

As the reactor is properly operated, three sections of entirely different flow modes spontaneously appear, due to vaporization of the liquid phase caused by reaction exothermicity. Lengths of the three sections can be conveniently adjusted through slightly changing the operating conditions such as pressure, temperature, and/or feeding rate (Zhou et al., 2005). The three sections include (Fig. 2):

(1) The lower section, where exterior surface of the catalyst pellets remains wet, though the liquid content in the bulk gradually reduces with the bed height.

(2) The middle section, where the exterior surface is "dry", but the interior surface remains more or less "wet", according to the capillary effect of the pores.

(3) The upper section, both exterior and interior of the catalyst pellets are dry, where a "pure" gas-phase catalytic reaction occurs, as to guarantee a high conversion.

Understandably, the liquid reactant feed (benzene (B. P. 80.1 ℃) + some recycled product cyclohexane (B. P. 80.7 ℃)) should be kept at a temperature lower than its boiling point. For example, when the reactor is operated at a pressure of 1 MPa, the boiling point of the reactant is

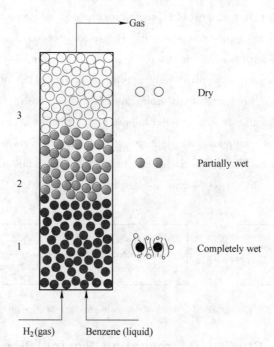

Fig. 2 Diagram of fixed-bed reactor with continuous phase transition

about 180℃. Fig. 3 gives a measured temperature profile within a laboratory reactor of 1.0 m high, in which the temperature raises from below the B. P. up to a higher level, within an allowable limit of the catalyst(Zhou,2004;Zhou et al.,2005). This is why part of the product needs to be recycled to dilute the feed.

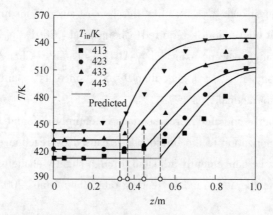

Fig. 3 Effect of inlet temperature on the temperature profile and conversion(at 1.0MPa)

In the lower section, the reaction is greatly affected by the interior diffusion due to the relatively slow diffusion in the liquid phase. However, due to high reactant concentration, the reaction rate is significant. In the upper section, also called finishing section, although the reactant concentration is low, the higher temperature and fast interior diffusion make high conversion possible. The middle section, which links the lower and the upper, is the key of the reactor. In this section, the interior surface of the catalyst pores shifts from completely liquid filled to entirely dry, depending on its pore diameter, temperature, pressure, and vapor composition.

Capillary effect may cause catalyst pores filled with liquid though exterior surface of the catalyst is dry. At a low relative pressure p/p_0, e. g. ,0. 05-0. 1, adsorbed adsorbent molecules form a monomolecular layer on the pore wall. With increase of p/p_0, a multimolecular layer of thickness t appears, which causes the pore to become even smaller; an actual core radius of $r = r_0 - t$. The thickness t can be determined using relations such as that suggested by Halsey(1948), or its modified form(Androutsopoulos and Salmas,2000;Zhou et al.,2004):

$$t = \left(\frac{M_w}{a_m N_A \rho}\right)\left[\frac{5}{\ln\left(\frac{p}{p_0}\right)}\right]^{\frac{1}{3}} \left(\frac{p}{p_0}\right)^m \qquad (1)$$

where, p_0 is the saturation vapor pressure of the adsorbate; M_w is the molecular weight; a_m is the molecular area of the adsorbate; N_A is Avogadro constant; ρ is the liquid density; m is a parameter to be experimentally determined. It is well-known that the classical Kelvin equation can be applied to calculate the critical radius r_k. The pores whose radii are smaller than r_k must be filled with liquid:

$$r_k = \begin{cases} t + \dfrac{2M_w \sigma}{R_g T \rho \ln(p_0/p)} & \text{(desorption or evaporation)} \\ t + \dfrac{M_w \sigma}{R_g T \rho \ln(p_0/p)} & \text{(adsorption or condensation)} \end{cases} \qquad (2)$$

where, σ is the surface tension of the adsorbate; T is the temperature; R_g is the universal gas constant. These three regimes can be schematically elucidated in Fig. 4. In fact, some difference exists between evaporation and condensation. The temperature difference between the catalyst and the bulk can also conveniently be calculated through reaction, heat and phase equilibria.

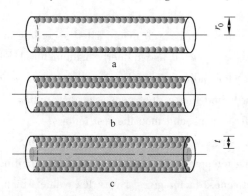

Fig. 4 Schematic of adsorption of adsorbate molecules in a pore

The BET measured pore size distribution of the catalyst we used is presented in Fig. 5, and is expressed in a log-normal form(Zhou et al.,2004):

$$f(r) = \frac{V_0}{\sqrt{2\pi}r\omega}\exp\left[-\left(\ln\frac{r}{r_a}\right)^2/2\omega^2\right] \qquad (3)$$

where, $f(r)$ is the pore size distribution density at radius r; V_0 is the total pore volume of the catalyst. The model parameters used are regressed by fitting Eq. 3 with the experimental data, $V_0 = 0.47 \text{cm}^3/\text{g}, \omega = 0.376$, and $r_a = 3.546 \text{nm}$.

Understandably, the second section starts from the bottom of the section with a gradual decrease of the liquid content in the catalyst pellet, down to zero at the upper margin. Composition of the gas bulk varies accordingly.

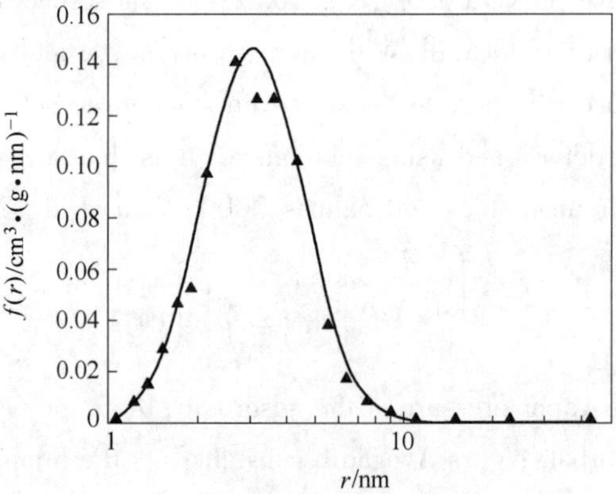

Fig. 5　Pore size distribution of catalyst pellet

We can imagine the largest pore becomes dry first, or becomes "open", then again the next largest, and so on, right up to all at the top of the section where the smallest pore eventually goes dry. It should be pointed out that whether a pore is wet or dry also depends on the bulk concentration and temperature. Pore size distribution dominates this continuous phase transition process. Pellets are actually partially "open", since each of them involves size diversified pores. From the observed reaction rate point of view, there are only two kinds of pores: "open" and "close". The former includes mono/multimolecular layer covered, while the latter implies the liquid-filled ones. The reaction carried out in the liquid-filled pores is generally negligible, compared with that in the "open" pores, where the reaction can be regarded as a regular gas-solid catalytic one.

In practice, the pore structure can be well manipulated according to the reaction characteristics. The catalyst can then be designed and properly manufactured. For example, the pore size distribution is a means for adjusting and controlling the pellet temperature as desired. Sometimes people will find that a dual peak size distribution is good in this regard, because a portion of pores first opens, and then the rest follows.

We can thus make clear entirely different reaction characteristics in the three sections. In the lower section, the operating temperature is actually equal to the boiling point of the liquid mixture. The calculated effectiveness factor gives a very low value which implies that the interior diffusion resistance is considerable. In the upper section, the reaction turns to external diffusion control, due to the high temperature and low interior diffusion resistance. The middle section provides a transition from the interior diffusion control to external control. Catalyst design, such as its activity and its pore size distribution, determines smoothness of the transition and the length of the section as well. An improper design of the catalyst may cause some unexpected situation since problems like local overheating are still possible. In this case the capillary/pore scale structure dominates the process. This structure can effectively be manipulated according to the researcher's purpose.

Case 2　Optimizing time-scale structure of electroreduction accompanied by a homogenous consecutive reaction

Nitrobenzene(NB) is reduced on an electrode to prepare the desired product p-aminophenol

(PAP). A simplified mechanism(Fig. 6)shows that NB is reduced electrochemically to an intermediate phenylhydroxylamine(PHA) which undergoes further electrochemical reduction to aniline(AN). PHA can also undergo a homogeneous chemical reaction in the liquid electrolyte bulk, through the Bamberger rearrangement, to produce PAP(Nolen and Fedkiw,1990).

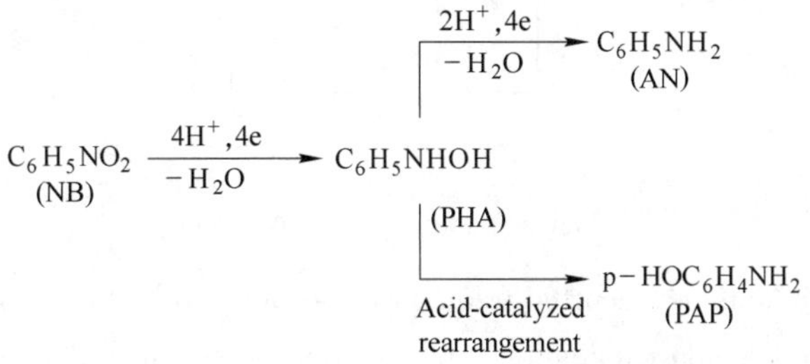

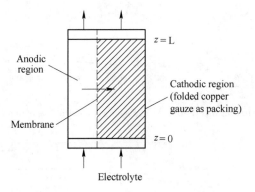

Fig. 6 Simplified mechanism for nitrobenzene reduction to p-aminophenol

To maintain a high PAP selectivity, the electrode potential needs to be kept at a lower level which however reduces the reaction rate. Therefore, a packed-bed electrode(Yang,1993) has been considered for practical purpose, in addition to the earlier rotating disk electrode(RDE) studies(Nolen and Fedkiw,1990) on the same system. The packed-bed electrode is made up with a feeding electrode and a folded copper gauze which greatly raises the electrode surface area, as shown schematically in Fig. 7.

Fig. 7 Schematic of the packed-bed electrode

From the reaction scheme one judges that for reactions of this sort, a controlled periodic forced oscillating current would benefit the PAP selectivity, compared with the conventional direct current (dc) operation. The problem is how to optimize the time-scale structure of the periodic current operation in terms of kinetics and mass transfer based on the same average current density. Fig. 8 presents the waveform for periodic current control(rectangular wave) of which at least three parameters have to be considered: the frequency $f = 1/\tau$, the duty cycle t_{high}/τ, and the average current density $<i> = [i_{high} \cdot t_{high} + i_{low}(\tau - t_{high})]/\tau$.

Modeling of this process can be done based on the measured kinetics and mass transfer results under some reasonable assumptions, e. g., uniformity of liquid flow, etc. The packed-bed electrode can be assumed to be one-dimensional(1-D)(Yang,1993) or three-dimensional(3-D)(Zhang et al.,1999) according to the required accuracy. In the 1-D modeling only axial gradients are to be taken care of, but in many cases it is likely that the 1-D modeling is fairly suffi-

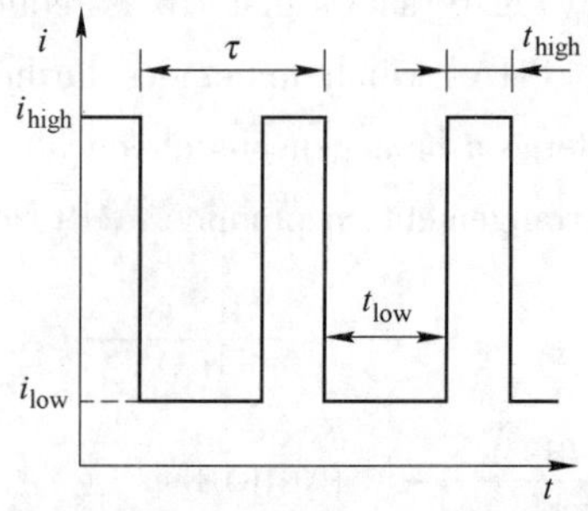

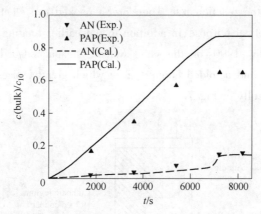

Fig. 8 Diagram of the waveform

cient to set up the relations between the frequency and duty cycle, the conversion and selectivity. Both differential-conversion calculations and integral-conversion calculations are considered, the latter being compared with the experiments. Presented in Fig. 9 and Fig. 10 are some of our experiments and calculations.

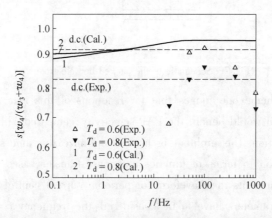

Fig. 9 Concentration distribution in homogenous electrolysis under periodic control

Fig. 10 Comparison of the selectivity between calculated results and experimental data

Like the significance of the spatial scale structure planning in steady-state systems, the time-scale structure is as important in the time-dependent systems. Obviously, a proper time-scale structure needs to be determined based on the process characteristics. Our experience shows that unsteady-state modeling is usually very effective for this purpose. In fact, conventional process control and on-line process optimization belong to the same area of study. The author would like to briefly quote this example, only to explain that a proper time-scale structure is also of advantage to the time-dependent process.

Case 3 Manipulating multiscale structure of carbon nanofiber as catalyst

As catalyst or catalyst support, three types of carbon nanofibers (CNFs) have been prepared using catalytic chemical vapor deposition (CCVD): platelet, tubular, and herring-bone, as shown in Fig. 11. The category of CNF produced depends on the catalyst used for CNF growth, the carbon source, and the CNF growth conditions, as exemplified in Table 2 (Zhao, 2004; Zhou, 2006).

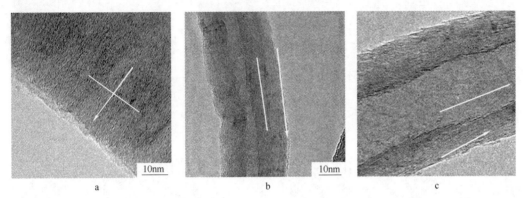

Fig. 11 HRTEM images of CNF with different microstructures
a—Platelet; b—Tubular; c—Herring-bone

Table 2 Synthetic conditions and CNF microstructure

Sample	Catalyst	Reactant(sccm)	Temperature/℃	Microstructure
CNF1	Fe powder	$CO/H_2 = 80:20$	600	Platelet
CNF2	$Ni/\gamma-Al_2O_3$	$CO/H_2 = 80:20$	600	Tubular
CNF3	$NiFe/\gamma-Al_2O_3$[①]	$CO/H_2 = 80:20$	600	Herring-bone

①The molar ratio of Ni to Fe is 1:1.

Different CNF morphologies and structures can meet different reaction requirements. The CNF diameter changes from about 15 to 200nm, which is more or less controllable. The carbon layer orientation determines the ratio between the edge plane and the basal plane. In some cases, the edge plane benefits the reaction, while a certain percentage of basal plane helps the others, depending on the reaction characteristics. In addition to the fiber structure, the surface structure, which implies the molecule/molecular group scale structure, can be modified to better meet special requirements of each individual reaction, as shown in the following two applications of CNF.

3.1 Reduction with loaded active metal over the CNF surface

Active metal Pd should be loaded over the CNF as a Pd/C catalyst for selective hydrogenation of 4-carboxybenzaldehyde (4-CBA) in purification of terephthalic acid (TA). Our experiments have shown that platelet CNF (P-CNF) is most suitable for the 4-CBA hydrogenation, compared with other CNFs (Zhou, 2006). However, when loading the metal, it has been found that Pd dispersion strongly depends on the surface structure. Surface treatment and modification cause the surface to contain a number of oxygen-containing complexes, e. g., strong acidic complexes, weak acidic complexes, neutral complexes, and basic oxides. TPD-MS can be used to determine the relative amount of these complexes on the modified CNF surface. Fig. 12 gives some of our characterization results(Zhou, 2006). The conditions for CNF treatment are presented in Table 3. Compared with our mass titration, we have found that there exist some well-recognized relations between the pHpzc (pH at zero charge) value and the characterized amount of basic oxides, as presented in Fig. 13.

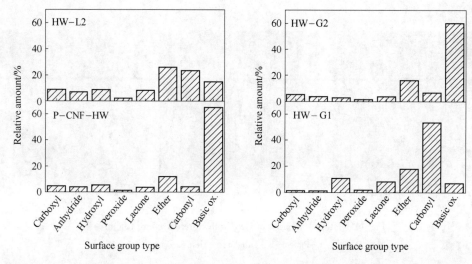

Fig. 12　Effect of surface modification on the relative amount of oxygen complexes on P-CNF surface

Table 3　Some treatment conditions for P-CNF surface modification, corresponding to Fig. 12

P-CNF code	Surface modification methods
P-CNF-HW	P-CNF stirred in 4mol/L HCl(P-CNF: HCl = 1:20) at 60℃ for 1h, filtered and washed by deionized water. Repeated 3-5 times till the Fe residue lowers than 0.1%
HW-L2	Purified P-CNF stirred in 30% H_2O_2 (P-CNF: H_2O_2 = 1:20) at room temperature for 24h
HW-G1	Purified P-CNF oxidized in stagnant air(in an oven) at 450℃ for 4h
HW-G2	Purified P-CNF treated in Ar at 900℃ (120mL/min) for 4h

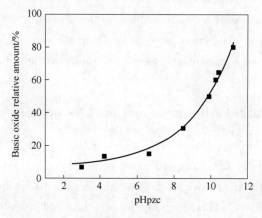

Fig. 13　Plot of pHpzc and basic oxide concentration

It is explained that when preparing the catalyst applying incipient wetness impregnation, the metal precursor can anchor on the active adsorption sites of the support surface through the forces such as electrostatic forces. The basic oxides are of higher thermal stability, so as to keep the anchored precursor stable in the following processes. The result is, the stronger the surface basicity(high pHpzc), the higher the Pd dispersion over the CNF surface, as shown in Fig. 14 (all based on a 0.5% Pd content).

It should be pointed out that there are point 1 and point 2 that deviate obviously from the straight line. Point 2 shows the CNF underwent 3h annealing under 1700℃, and its fiber edge is of some looped, or U-shaped structure (Fig. 15) (Zhou, 2006). This structure is likely to have provided more basal plane surface which actually reduces the desired edge of the routine platelet structure. Point 1 is considered to have strayed from the straight line because the air oxidation has made the treated P-CNF a larger pore volume(Zhou et al. ,2006b).

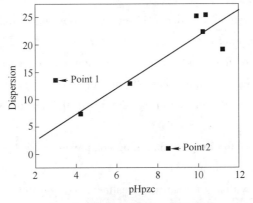

Fig. 14 Plot of metal dispersion vs. pHpzc

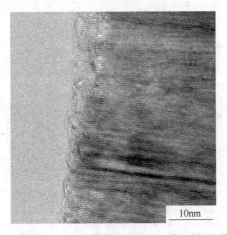

Fig. 15 TEM image of P-CNF after annealing(P-CNF-HT)

The catalyst samples have been applied for 4-CBA reduction in an agitated batch reactor (Zhou et al. ,2006a). Some results of a 0.5h operation are expressed in Fig. 16.

3.2 Oxidation with the modified or surface-treated CNF without extra active component loading

Different means of surface treatment induce dissimilar oxygen-containing complexes that form

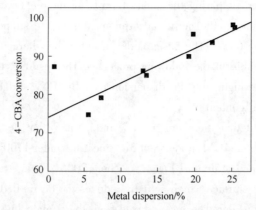

Fig. 16 Plot of metal dispersion vs. 4-CBA conversion
(Operating conditions:Temperature—280℃;Hydrogen pressure—0.7MPa)

active sites for various oxidation reactions. However, reaction individuality always raises its own requirements toward the catalyst.

Propane oxidation to produce propene is used as a sample reaction, targeting high activity and high C_3H_6 selectivity. From our experiments, the herring-bone CNF(H-CNF) is likely to be the most suitable for propane oxidation(Sui et al., 2005). In this regard, we need at least to make clear the following:

(1) Required category of the oxygen-containing complexes;
(2) Suitable amount of the complexes;
(3) Proper method for realizing the surface treatment requirement.

Generally, surface treatment is made by first removing the metal particles contained during the CNF growth, followed by treatment at high temperature(1700℃) in an inert gas, or in air under various temperatures, or in a mixture of HNO_3/H_2SO_4, or even in an H_2/Ar mixture, etc., as is given in Table 4(Sui, 2005).

Table 4 Some treatment conditions for H-CNF surface modification, corresponding to Table 5

H-CNF code	Metal removal	Surface modification
3B-G 400	Alternative acid and base treatment	Stagnant air, 400℃, 2h
3B-G 600	Alternative acid and base treatment	Stagnant air, 600℃, 2h
3B-G 700	Alternative acid and base treatment	Stagnant air, 700℃, 2h
3B-L	Alternative acid and base treatment	HNO_3/H_2SO_4 mixture, 2h
HT-G 700	1700℃, Ar, 12h	Stagnant air, 700℃, 2h
3B-H	Alternative acid and base treatment	H_2/Ar mixture, 900℃, 2h

TPD-MS experiments have been carried out to define the oxygen-containing complexes for all the H-CNF samples. As expected, no oxygen-containing complexes are found with the 3B-H coded fiber. Further studies are made to explore the functions of various oxygen-containing complexes applying TPSR. The samples are treated in advance in an Ar atmosphere at a temperature rising rate of 10℃/min, up to 550℃, under which carbonyl groups and basic oxides are not to be affected. The temperature is then lowered down to 50℃ to start the TPSR runs in a

C_3H_8 (8%)/Ar mixture up to 600 ℃. Fig. 17 shows the result of a TPSR experiment for the 3B-L coding H-CNF. Table 5 briefs on some of the TPSR results of interest of selected samples for comparison. By integrating the area under the desorption curves, we achieve the desorbed C_3H_6, CO, and CO_2. Obviously, C_3H_6 is produced through oxidation of C_3H_8 by the oxygen-containing complexes on the catalyst surface. It is noticed from Table 5, the higher the temperature for surface treatment, the more amount of surface carbonyl groups produced, and the more total desorbed agents (C_3H_6, CO, and CO_2) released. Clearly, this fact implies that with raising the treatment temperatures, the catalyst becomes more active; however, more undesired product CO_x is generated caused by some parallel reactions. Therefore, there actually exists a certain amount of carbonyl groups, which may achieve an optimized result, according to the reaction individuality, or to the researcher's aim. One may also notice that the acid-treated H-CNF(HT-L) seemingly provides more optimistic results. The liquid-phase surface treatment, as generally acknowledged, creates the oxygen-containing complexes on both CNF edge plane and basal plane, and makes their distribution more uniform.

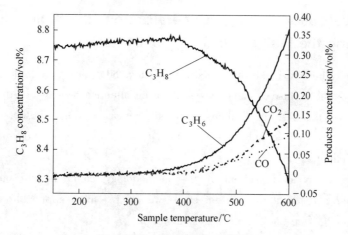

Fig. 17 Propane TPSR on 3B-L

Table 5 The impact of surface treatment methods on the TPSR results

Sample	Carbonyl /mmol·g⁻¹	Basic oxides /mmol·g⁻¹	C_3H_6 desorption /mmol·g⁻¹	Total desorption /mmol·g⁻¹	C_3H_6 selectivity
3B-G 400	0.19	0.07	0.085	0.123	87.03
3B-G 600	0.84	0.04	0.110	0.326	60.44
3B-G 700	1.00	0.06	0.080	0.469	38.16
3B-L	0.73	0.07	0.296	0.505	80.95
HT-G 700	0.068	0.02	0.059	0.069	94.66
3B-H	—	—	—	—	—

So far we have focused on the molecular scale structure of the CNF catalyst. This structure obviously dominates the activity, based on the individual CNF. However, pelletization of the CNF is sometimes inevitable, for example, when it is applied in a packed-bed reactor. When the CNF is pelletized via some adhesion agent, the key is to keep sufficient mesosized pores with enough mechanical strength, and without lowering its original activity (Li et al., 2005). The problem then shifts from the CNF's activity to the pellet's nanoscale structure, which becomes the dominating-scale structure needs to be studied.

Different from Case 1 and Case 2, both presenting process studies, Case 3 is concerned with CNF as catalyst, which is a product of special properties. Surface of the catalyst is the only place where real catalytic reactions are carried out, so surface structure must be of top concern. Following Fig. 1, one needs to identify the required surface characteristics. Here methods such as TPD-MS and TPSR are applied, together with the first-principles computations, since microstructures are not directly observable. The identified dominating structure serves as a guideline for the following processes for manipulating the desired surface. For the discussed CNF, the exterior surface of individual fibers is the only accessible surface, because the interlayer distance which is of 0.3-0.4nm makes the inner surface unavailable to most molecules. However, after palletizing, the interior of the pellet is composed by the exterior surface of the CNF; therefore, the rate becomes diffusion affected.

4 Individualization Versus Generalization: the Philosophy of Approaching the Multiscale Complex System

Scientists naturally tend to make their theories and methodologies more generally applicable, when approaching the multiscale complex system. Their attention may also be drawn to development of generalized theories. People then develop their ideas starting from the study of the most fundamental structure such as sub-molecular, then to the upper scale, and so forth, similar to the concept of reductionism.

This would certainly be excellent if the methodology could handle our problems in practice. However, at least at the present stage, we need a more applicable way of thinking which can help us to face the complexity of our tasks.

The present author suggests to deal with our complex systems individually, but not with a generalized methodology. The individuality of a specified system needs to be explored and identified to determine the dominating-scale structure of the system, or to make it an "emergent" structure worth studying, on which further experimental studies and modeling work are to be done. Unfortunately, no ready-made experimental method and theory have been worked out hitherto to create a unified basis.

We would emphasize the importance of understanding the individuality of a complex system. For this purpose, whatever method may be considered: theoretical, computational, and experimental. As the dominating-scale structure is identified, the problem can be accurately and straightforwardly solved.

The author has quoted several cases from his laboratory in the above sections, only intends to elucidate, though in a very limited sense, that exploration of the dominating-scale structure is actually the key for finally solving the whole problems.

5 Concluding Remarks: Simplicity from Complexity

When dealing with a complex system, one may need to extract from a large number of experimental results of the system the most significant knowledge and the most influential factor. This problem can actually turn to that of searching for a dominating structure of a certain level, or scale. A suggested idea is to find out the relations between the performances and properties of the system, and its multiscale structures. These relations provide a scientific base to identify the dominating structure. Extraction of the dominating structure is a highly individualized approach. After exploring the dominating-scale structure, we will be able to manipulate the structure applying all our chemical engineers' expertise, in achieving the desired system characteristics. Besides, we should be fully aware of the possible sensitivity when moving our research conditions. Within a certain range, the sensitivity is low, and people do not have to question about reliability of the conclusion. However, frequent checking of effects of the research conditions on the system is recommended to assure reliable results.

Acknowledgements

This paper is written under the support of the major project on multiscale methodology by NSFC/PetroChina(Grant No. 20490204).

References

Anderson, P. W., 1972. More is different. Science 177, 393-396.

Anderson, P. W., 1994. Physics: the opening to complexity. Preface to the NAS Proceedings of the Colloquium on Physics, June 27-28, Irvine, CA, USA.

Androutsopoulos, G. P., Salmas, C. E., 2000. A new model for capillary condensation-evaporation hysteresis based on a random corrugated pore structure concept: prediction of intrinsic pore size distribution. 2. Model application. Industrial Engineering and Chemical Research 39, 3764-3777.

Bird, R. B., Stewart, W. E., Lightfoot, E. N., 1960. Transport Phenomena. Wiley, New York.

Cheng, Z. M., Anter, A. M., Khalifa, G. A., Hu, J. S., Dai, Y. C., Yuan, W. K., 2001a. An innovation reaction heat offset operation for a multiphase fixed bed reactor with volatile compounds. Chemical Engineering Science 56, 6025-6030.

Cheng, Z. M., Anter, A. M., Yuan, W. K., 2001b. Intensification of phase transition on multiphase reactions. A. I. Ch. E. Journal 47(5), 1185-1192.

Cheng, Z. M., Yuan, W. K., Bhatia, S. K., 2003a. Unconventional operation of three-phase fixed-bed reactors: a perspective trend in reaction engineering. Trends in Chemical Engineering 8, 35-59.

Cheng, Z. M., Anter, A. M., Fang, X. C., Xiao, Q., Yuan, W. K., Bhatia, S. K., 2003b. Dryout phenomena in a three-phase fixed-bed reactor. A. I. Ch. E. Journal 49(1), 225-231.

Cussler, E. L., Wei, J., 2003. Chemical product engineering. A. I. Ch. E. Journal 49(5), 1072-1075.

Edwards, D. A., Hanes, J., Caponetti, G., Hrkach, J., Ben-Jebria, A., Eskew, M. L., Mintzes, J., Deaver, D., Lotan, N., Langer, R., 1997. Large porous particles for pulmonary drug delivery. Science 276, 1868-1871.

Feynman, R., 1959. There is plenty of room at the bottom. A speech at the Annual Meeting of the American Physical Society, December 29, Pasadana, CA.

Halsey, G., 1948. Physical adsorption on non-uniform surfaces. Journal of Chemical Physics 16(10), 931-937.

Li, P., Zhao, T. J., Zhou, J. H., Sui, Z. J., Dai, Y. C., Yuan, W. K., 2005. Characterization of carbon nanofiber composites synthesized by shaping process. Carbon 43, 2701-2710.

Nolen, T. R., Fedkiw, P. S., 1990. Reaction selectivity enhancement under periodic-current control: the reduction of nitrobenzene on the rotating disk electrode. Journal of the Electrochemical Society 137(9),

2726-2735.

Sui, Z. J. , 2005. Catalytic oxidative dehydrogenation of propane over carbon nanofibers. Ph. D. Dissertation, East China University of Science and Technology.

Sui, Z. J. , Zhou, J. H. , Dai, Y. C. , Yuan, W. K. , 2005. Oxidative dehydrogenation of propane over catalysts based on carbon nanofibers. Catalysis Today 106(1-4) ,90-94.

Yang, X. Q. , 1993. Studies on unsteady operation of a packed bed electrode reactor. M. S. Thesis, East China University of Science and Technology.

Zhang, X. S. , Wu, G. B. , Ding, P. , Zhou, X. G. , Yuan, W. K. , 1999. Studies on paired packed-bed electrode reactor: modeling and experiments. Chemical Engineering Science 54 ,2969-2978.

Zhao, T. J. , 2004. Microstructure control of CNF and its catalytic performance. Ph. D. Dissertation, East China University of Science and Technology.

Zhou, J. H. , 2006. Design and characterization of a novel Pd/CNF catalyst for TA hydropurification. Ph. D. Dissertation, East China University of Science and Technology.

Zhou, Z. M. , 2004. Kinetics behavior of the partially wetted catalyst in the fixed-bed reactor with continuous phase transition. Ph. D. Dissertation, East China University of Science and Technology.

Zhou, Z. M. , Cheng, Z. M. , Li, Z. , Yuan, W. K. , 2004. Determination of effectiveness factor of a partial internal wetting catalyst from adsorption measurement. Chemical Engineering Science 59 ,4305-4311.

Zhou, Z. M. , Cheng, Z. M. , Yuan, W. K. , 2005. Simulating a multiphase reactor with continuous phase transition. Chemical Engineering Science 60 ,3207-3215.

Zhou, J. H. , Shen, G. Z. , Zhu, J. , Dai, Y. C. , Yuan, W. K. , 2006a. Terephthalic acid hydropurification over Pd/C catalyst: mechanism and kinetics. Studies in Surface Science and Catalysis 159 ,293-296.

Zhou, J. H. , Cui, Y. , Zhu, J. , Dai, Y. C. , Yuan, W. K. , 2006b. A novel catalyst for PTA manufacture: carbon nanofibers supported palladium. Studies in Surface Science and Catalysis 159 ,753-756.

Reactor Engineering: Science, Technology, and Art[*]

Abstract As we are approaching the threshold of the 21st century, the reaction engineer can contribute more to the profession if he plays the triple role of an engineering scientist, an engineering technologist, and an engineering artist: he analyzes, models, and computes as a scientist, he uses and devises new processes and new equipment as a technologist, and he has to master, above all, the art of knowledge synthesis and redevelopment including that of his own creation and what was foreign to him.

Throughout the long time prior to the first European Symposium on Chemical Reaction Engineering(ESCRE,1957)(Rietema,1957), which initiated a field of study leading to the present-day reactor engineering, many chemical reactors were built in the chemical and allied industries. Engineers developed, scaled up, and designed chemical reactors, though they did not know sufficiently the real happenings in these reactors. What they did depended greatly on their knowledge and experience in the unit operations, in addition to some odd scraps of information regarding reactors. Reactors were scaled up mainly through multiscale tests, and reactor design was basically empirical, aided by such simple calculations as were necessary to estimate reactor volume, coolant temperature, heat transfer area, etc. The then reactor engineer was not acquainted with such phenomena as multiple steady states, micromixing effect, and so on. But they were ingenious enough to take advantage of related technologies to develop efficient catalysts and skillful enough to integrate their knowledge to create such epoch-making processes as fluid catalytic cracking(FCC) as early as in the 1950s. However, in the framework of today's reactor engineering, their understanding of reactors was perceptual and meagre.

Compared to the unit operations, development of reactor engineering has depended to a far greater extent on computers because the combination of chemical reaction with the transport processes, more often than not, creates nonlinear problems that are harder to solve (Seider et al., 1991). This might explain why chemical reaction engineering has grown dramatically since the 1950s-side by side with computers and computation techniques. In this sense, a novel area has been opened to chemical engineers through the amalgamation of concepts, theories, and methodologies in solving problems for the design, scale-up, and operation of chemical reactors.

It appears, therefore, that reactor engineering(RE) can be resolved into its three major component parts: science, with its fundamental concepts, theories, and methodologies, technology, not only of its own but as a result of interaction with and incorporation of related technologies, and art, consisting of the skill and ability of applying and integrating the available theories, methods, and experiences to create innovations. It is not intended to make this paper highly

---

[*] Coauthor: Mooson Kwauk. Reprinted from *Ind. Eng. Chem. Res.*, 1997, 36: 2910-2914.

comprehensive but merely to discuss through some examples the essence and methodology of reactor engineering from a generalized viewpoint.

1 Creating a Scientific Basis for Reactor Engineering

As a discipline, reactor engineering has its own scientific basis. During its early stage of development, a number of basic problems in reactor engineering drew most researchers' attention: coupling of chemical reaction with heat, mass, and momentum transfer; back-mixing and residence time distribution; micromixing; parametric sensitivity and multiplicity.

With progress made in these areas, many conceptual conclusions were obtained to bring maturity to reactor engineering. Some new subjects have been provoked, e. g., unsteady-state operations of reactors (e. g., forced inlet reactant concentration oscillation (Bailey, 1977)) and the methodology for treating complex reaction systems (Wei and Kuo, 1969). Modeling of these processes is quite different from that used in the early days for steady-state and simple reactions. People noticed with interest an expanding world in reactor engineering. They started to sense that certain problems of generalized implications were of particular significance, such as nonlinearity (Villermaux, 1993) as one of the most popular characteristics existing in almost all reaction systems due to the generally strong temperature effect of most chemical reactions. Modeling has therefore been regarded as the major method for handling this characteristic and has become the core of reactor engineering, particularly because modeling has been successfully applied to the above mentioned areas. Multiphase flow is probably another example which has drawn wide attention (Krishna and Sie, 1994). Generalization of reactor engineering problems was very similar to what happened in the 1950s when transport phenomena were regarded as to identify a more generalized area derived from the unit operations.

People then focused more and more on the methodology dealing with the nonlinear properties of reactors. Several methods have been created for this problem and have been proven quite effective. Online estimation, a prerequisite of online optimization, is one of them (Windes et al., 1989). If a reactor engineer follows the catalytic reaction engineering textbooks, very possibly he will first determine the reaction kinetics (intrinsic or observed) to set up a kinetic model for a reaction system under development. The next step is to establish a reactor model to study the transport behaviors in the reactor for determining the parameters involved. Computer simulation will then be carried out to complete the reactor design and to find out the optimized conditions.

These procedures are traditional and well-known, but obviously certain problems need further investigation, for instance: (1) uncertainty of the kinetic determination (with an isothermal or a gradientless reactor) and kinetic model; (2) oversimplification of the reactor model; (3) uncertainty of parameter determination and/or estimation; (4) time variation of model parameters in the process, e. g., deactivation, or disintegration of the catalyst. Model-based a priori optimization is actually not reliable. A methodology which needs discussion is the on-line optimization strategy through which the model is refined by online parameter estimation in order to minimize its deviation from the real process. The state estimation technique can also be incorporated to account for the inaccuracy of the model and the noises in the measurements. For simple, straightforward and effective implementation of the online optimization, it has to be as follows:

(1) For easy realization in industry, the fewer sampling points the better.

(2) The model should be refined to adapt it to oversimplification of the model and uncertainty of model parameters.

(3) Noises in the measurements should be properly handled. Since the models, basically mechanistic, are inevitably oversimplified and operating data have to be used to refine the parameters, one would rather use a case-independent model structure, e. g., neural network, to approximate the underlying functional relationship between the inputs and the outputs of a real process. This idea is advantageous in that no deep understanding of the process is needed, and therefore, the methodology can be easily transplanted.

With recent developments in measurement techniques, profuse dynamic data are generally available from operating reactors. These data must be simply handled and compressed. For this purpose, neural networks can be combined with Karhunen-Loeve (K-L) expansion (Tan et al., 1994), or proper orthogonal decomposition (POD), in modeling, for instance, a fixed-bed reactor with oxidation of benzene as a working system (Zhou et al., 1996). Here POD is used as a preprocessor to compress the original data most effectively, while reserving as much information in the data as possible. The size of the networks can thus be greatly reduced to make on-line optimization possible. Fig. 1 shows the prediction scheme of the KL-NN model.

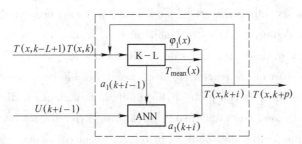

Fig. 1 p-step ahead prediction scheme ($i = 1, 2, \cdots, p$)

Referring to this figure, suppose two-step-ahead temperatures of a wall-cooled fixed-bed reactor are to be predicted, i. e., $p = 2$. First, the average temperature $T_{\text{mean}}(x)$ and eigenfunction $\varphi_1(x)$ are determined by the K-L expansion using the last L measurements, and coefficient $a_1(k)$, by the last measurement; $a_1(k+1)$ is generated by the network with inputs $a_1(k)$ and the operating conditions $U(k)$, i. e., the inlet benzene concentration, the wall temperature, and the gas flow rate. It is assumed that the eigenfunction $\varphi_1(x)$ and the average temperature in the time domain $(k-L+1, k)$ are the same as those in $(k-L+2, k+1)$. Then the temperature $T(x, k+1)$ is predicted from the average temperature, the eigenfunction, and the coefficient generated by the network. With $T(x, k+1)$ available by prediction, the K-L expansion is updated and a new K-L coefficient is available. Thus, a two-step prediction is obtained. A p-step-ahead prediction can be obtained by repeating these procedures p times. Fig. 2 shows a five-step-ahead prediction (18s step length) as compared with experimental data. Curves numbered from 1 to 5 correspond to the thermocouples positioned at 0.2m, 0.3m, 0.4m, 0.5m, and 0.6m, respectively, from the top of the reactor, at a wall temperature of 350℃, a gas flow rate of 25L(STP)/min, and an inlet pseudorandom binary signal (PRBS) of benzene concentration.

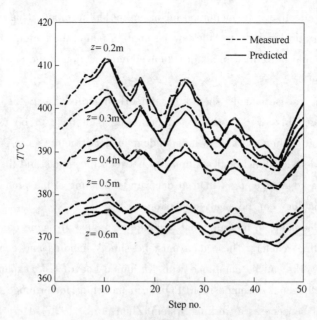

Fig. 2　Five-step-ahead prediction of bed temperatures

Although there are limitations for the mechanistic models as mentioned above, it was not implied that modeling is to be negated. On the contrary, modeling is by all means extremely important for reactor engineers both in theoretical studies and in the application of theories as well. In addition to reactor modeling, it is also very important for reactor engineers to master different methods, to invent new methods, and to create an all-round scientific basis for reactor engineering. Thus, the science part of the reactor engineer consists of his perception in seeing the controlling factors of a chemical reactor problem, his ability in formulating mathematically the problem in terms of these factors (now quantified), his skill in solving the mathematical formulation, and his acumen in interpreting the solutions in relation to physical reality, particularly in discovering the new and innovating the existent as a result of such quantitative analyses.

Although theories derived from such quantified analysis are themselves rigorous, and modeling does play a significant role in theoretical studies, it is by no means infallible and all-encompassing. We have yet to confess that after having done so much research, we are still seldom able to design a reactor, even the conventional fixed-bed reactor, completely from modeling using only kinetic and transport data. What is often needed is to find through experiments the so-called "windows" which show the proper operating conditions (Schmidt et al., 1994). But compared to the early days of chemical reaction engineering, much more has been learned, quantitatively, regarding reactor behavior by applying existing theories and methods.

2　Benefiting from Related Technologies for Reactor Engineering

Technologies related to reactor engineering include mechanical (e.g., mixers, fluid distributors, etc.), material (e.g., corrosion resistance, membrane, etc.), structural (e.g., bed grids, heat exchangers, etc.), and instrumental (instruments, controllers, computers, etc.), and so on. Some of these technologies may appear to the theorists as tricks or techniques. Analytically they might not appear sophisticated, yet they might be equally ingenious and highly effective in practice and contribute as much, if not less, to improvement of reactor performance. These technologies are developed along professional expertise and intellectual standards other than those for solving PDEs.

Catalyst, for example, is a problem of great concern to reactor engineers. Reactor engineers need catalysts with the following desirable characteristics: (1) high activity; (2) high product selectivity; (3) toxic resistant and long operating time; (4) attrition resistant and mechanically strong; (5) low cost and easy to prepare. Since the past decade, in pursuit of a rational basis for making catalysts, people have kept talking about catalyst design. A well-known aspect of catalyst design is the spatial distribution of active species within a catalyst pellet. A conventional policy is to uniformly load the catalyst support pellet with catalytically active species, in order to maximize catalytic activity per unit reactor volume. If the effectiveness factor is significantly low, however, the active species can be saved by leaving the inner core of the pellet free without losing much of its overall activity. This applies also to the case for a higher desired product selectivity, or when the reaction is consecutive and the desired product is an intermediate compound. Such a catalyst is called the "egg-shell" type. Nonuniform activity distribution possesses some particular advantage when an irreversible poisoning reaction occurs, especially when it is diffusion limited; that is, poisoning appears essentially in the outer shell of the pellet. In such a case, the bare support can be relied upon to absorb the poison in the outer shell, and the active species needs to be impregnated only in an inner layer or in the core of the pellet. We thus have the "egg-white"- and "egg-yolk"-type catalysts. Understandably, the major consideration to determine the suitable activity distribution of the active species comes from the classical simultaneous reaction-diffusion concepts.

More and more attention has been paid to combining reaction and separation into a single step, so that the reaction product is continuously removed to facilitate the reaction to proceed, thus circumventing stalling the reaction in view of chemical equilibrium due to product accumulation, e. g. , reactive distillation columns and membrane reactors. Another interesting example is a modified zeolite which is capable of promoting p-xylene selectivities to values far higher than that for equilibrium when alkylating toluene with methane. When equilibrium distribution of xylene isomers occurs at the active sites, p-xylene molecules tend to leave much faster since their kinetic diameter is approximately 0.03nm smaller than those of the other two isomers (Chen et al., 1989). With control of the pore size of zeolite, the effective diffusivity can be made 3 orders of magnitude higher than those of the other two isomers. The key technique is how to prepare the controlled-pore-size zeolite pellets.

There are other types of combined reaction-separation technologies, such as the incorporation of a solid adsorbent, packed together with a catalyst in a fixed-bed reactor, to reduce a designated product in a reversible reaction so as to enhance conversion. This technology has been tested for the dehydrogenation of methylcyclohexane to toluene over a Pt-Al_2O_3 catalyst at temperatures as low as 450 K. Experiments have been carried out to scan the high temperature (400-700K) adsorption properties of methylcyclohexane, toluene, and hydrogen on various adsorbents (Alpay and Chatsiriwech, 1994). A clay-based adsorbent was found to be particularly suitable for this case. Clearly, to develop a process like this, we need a low-temperature catalyst and a high-temperature adsorbent, possessing appropriate reaction and adsorption properties, respectively.

Another combined reaction-separation technology consists of the use of membranes to remove one of the reaction products in order to induce a reaction to proceed or to raise its selectivity beyond equilibrium. Membranes can also be used for staging reactant feeds to reactors in order to optimize the reactant concentration profiles for higher product yields(Veldsink et al., 1992).

Another technique of interest to the reactor engineer is the control of catalyst strength versus its surface-to-volume ratio(S/V) and of bed pressure drop versus S/V. To make a catalyst more attrition-resistant, often a special technique needs to be developed. For instance, vanadium phosphate catalyst has been found to be too weak to withstand the mechanical forces in a fluidized bed. Spray-drying vanadium phosphate solution to-gether with an added silica hydrogel allows the silica to migrate to the outer region of the pellets, thus encapsulating the active core in a porous yet strong silica shell(Contractor and Bergema, 1987), which retains sufficient free openings for reactant and product molecules to diffuse in and out.

In the field of solid processing in fluidized-bed reactors, the bubbling species of fluidization was in vogue for many years: the Winkler coal gasifier followed by the bubbling fluid-bed coal combustor, the bubbling fluid-bed pyrite roaster, etc. Improved modes of fluid-solids processing have resulted in the adoption of the circulating fluid-bed combustor, although pyrite roasting seems to have continued its conventional practice, apparently unaware of advancement in the technology of fluidization. Recent emphasis on bubbleless fluidization(Kwauk, 1992) has identified a new class of alternate species of fluidization other than bubbling, promising better fluid-solids contacting, lower pressure drop, and smaller equipment. Furthermore, studies based on multi-scale-energy-minimization modeling(Li and Kwauk, 1994) has differentiated the bubbling-type and the non-bubbling-type particle-fluid two-phase flow systems and uncovered a continuous range of transition states, forecasting future trends of transplanting, in practice, the advantages of the one to the other(Kwauk et al., 1996). Such in-depth investigations provide the reactor engineer with sufficient derivative technologies to improve his work.

The reactor engineer needs to be constantly aware of what is new in neighboring technologies which might be of use and to be interested in learning enough of the ABC's of these technologies to be conversant with their experts, in order to extend the frontiers of his own profession and realize innovation and improvement in his own work. Compared to decades ago, the current rapidly changing environment is throwing out more new problems that challenge reactor engineers(Wei, 1990).

3 Mastering the Art of Knowledge Synthesis for Reactor Engineering

There are indeed many ways to synthesize or to integrate(and to apply) existing concepts, theories, methods, and technologies so as to design novel reactors or reaction systems, just as a painter integrates colors and shapes or a composer combines notes and tempos. Colors and notes are all well-known, but the paintings and musical compositions synthesized therefrom are all unique by themselves.

Let us examine a typical example of combining traditional autothermal reactors (cf. Westerterp, 1992). The classical method of utilizing the heat released in an exothermic reaction is to heat up the incoming reactant gas by the outgoing hot product stream through a heat exchanger. For a reversible reaction, some appropriate temperature profile should be maintained, at the same time, in order to guarantee high conversion. The shortages of the classical method are a large heat transfer surface area is needed and, more seriously, the pressure drop through the reacting and heat exchanging system is high. For reversible reactions, such as SO_2 conversion in making sulfuric acid, a multistage reactor with interstage heat exchangers needs an even higher pressure drop. In the 1980s Matros suggested a reversing flow SO_2 reactor (Matros, 1985), using the thermal capacity of the catalyst bed for heat regeneration, which dispenses with the need of a heat exchanger and thus consumes less metal and energy. The reversing flow SO_2 reactor is somewhat analogous to the intermittent water-gas generator of yore and presents a synthesis and modification of some conventional methods. Matros' contribution was an ingenious combination of un-steady-state operation with adiabatic chemical reaction to create a preferred moving temperature profile and, as such, constitutes a noteworthy technological breakthrough. A large amount of research then followed to extend the application of the Matros concept. For example, by integrating with one or several heat exchangers, this reactor configuration can hopefully be used for endothermic reactions, such as ethylbenzene dehydrogenation (Haynes et al., 1992). Reactors with flow reversal can be used to treat waste gas streams containing combustible components (Nieken et al., 1994). Based on the flow reversal operation, a few reactor configurations have been derived, for instance, hot gas injection (Nieken et al., 1994), central cooling (Nieken et al., 1994), interstage quenching (Xiao and Yuan, 1994), and hot gas withdrawal (Nieken et al., 1994), so as to better control the temperature profile. The Matros scheme, original as well as derived, has naturally generated quite a number of fundamental research projects, including those on process optimization.

Another example, though yet somewhat speculative, may show how a reactor engineer might use his imagination. Sometimes a reversible gas/solid reaction may be used to advantage by decoupling it into two reactions working against each other, such as in metals recovery and separation from complex ores. For instance, in chlorinating roasting of nonferrous ores, while the nonferrous metals are converted into their chlorides, iron is chlorinated at the same time. Since iron is often present in such ores at much higher concentration than the nonferrous metals, it consumes several times more chlorine than is needed for the nonferrous metals, and downstream disposal of large amounts of iron chloride also spells an abominable problem. However, it is highly possible, in thermodynamic principle, to utilize the reverse reaction to chlorination to regenerate iron oxide, thus avoiding the production of iron chloride and recovering chlorine at the same time (Academic Division of Chemistry, 1995):

$$2Fe_2O_3 + 6Cl_2 \underset{\text{ca. } 700^\circ C}{\overset{\text{ca. } 1000^\circ C}{\rightleftharpoons}} 4FeCl_3(g) + 3O_2$$

To go another step further in imagination, one might even cogitate whether or not it is possible to separate metal elements as some suitable compounds in the vapor phase through chemical transport as has been conventional in ore dissolution in the molten state as in pyrometallurgy and in aqueous solution as in hydro-metallurgy.

The reactor engineer really needs to play with concepts, theories, methods, and technologies (Krishna and Sie, 1994), just as the artist plays with colors and forms and the composer with notes and tempos, to assert his skill and ability, not only in science or in transplanting readily available technologies but in synthesis and integration of even more basic components. Vested in him is the very attribute of the "engineering artist", who is skillful at integrating well-known principles in the engineering disciplines together with well-known know-hows in related areas, to make his work ever more creative.

4 Final Remarks on Science, Technology, and Art

We visualize that the reactor engineer is an engineering scientist, an engineering technologist, and an engineering artist, three in one. The functional attributes of this triple personality are briefly

Science: phenomenon, concept, prediction

Technology: planning, experiment, know-how

Art: integration, configuration, realization

The relations among these attributes are interactive rather than mutually dominative. We hope this portrayal of the reactor engineer is not too demanding, when we think that he is already at the threshold of converting the heritage of the 20th century into the fresh missions of the 21st.

References

Academic Division of Chemistry, CAE. Bull. Chin. Acad. Sci. 1995, 9, 310.

Alpay, E.; Chatsiriwech, D. Combined Reaction and Separation in Pressure Swing Processes. Chem. Eng. Sci. 1994, 49, 5845.

Bailey, J. E., in Lapidus and Amundson, N. R. Chemical Reaction Engineering: A Review; Prentice-Hall: Englewood Cliffs, NJ, 1977.

Chen, N. Y.; Garwood, W. E.; Dwyer, F. G. Shape Selective Catalysis in Industrial Applications; Marcel Dekker: New York, 1989.

Contractor, R. M.; Bergema, H. E. Butane Oxidation to Maleic Anhydride over Vanadim Phosphate Catalysts. Catal. Today 1987, 1, 49.

Haynes, T. N.; Geogakis, C.; Caram, H S. The Application of Reverse Flow Reactors to Endothermic Reactions. Chem. Eng. Sci. 1992, 47, 2927.

Krishina, R.; Sie, S. T. Strategies for Multiphase Reactor Selection. Chem. Eng. Sci. 1994, 49, 4029.

Kwauk, M. FLUIDIZATION, Idealized and Bubbleless; Science Press: Beijing and Ellis Horwood: Chichester, 1992.

Kwauk, M.; Li, J.; Liu, D. Particulate and Aggregative Fluidization-50 years in Retrospect. World Congr. Chem. Eng., 5th 1996.

Li, J.; Kwauk, M. Particle Fluid Two-Phase Flow: The Energy-minimization Multi Scale Method; Metallurgical Industry Press: Beijing, 1994.

Matros, Y. Unsteady Processes in Catalytic Reactors; Elsevier: Amsterdam, 1985.

Nieken, U.; Kolios, G.; Eigenberger, G. Control of the Ignited Steady State in Autothermal Fixed-Bed Reactors for Catalytic Combustion. Chem. Eng. Sci. 1994, 49, 5507.

Rietema, K. Chemical Reaction Engineering; Pergamon: Oxford, 1957.

Schmidt, L. D.; Huff, M.; Bharadwaj, S. S. Catalytic Partial Oxidation Reactions and Reactors.

Chem. Eng. Sci. 1994,49,3981.

Seider, W. D. ; Brengel, D. ; Widagado, S. Nonlinear Analysis in Process Design. AIChE J. 1991,37,1.

Tan, H. ; Dai, Y. C. ; Yuan, W. K. A Novel Kinetic Model Discrimination Method: Eigenfunction Distribution via Forced Concentration Oscillation. Chem. Eng. Sci. 1994,49,5563.

Veldsink, J. W. ; Van Damme, R. M. J. ; Versteeg, G. F. ; Van Swaaij, W. P. M. A Catalytically Active Membrane Reactors for Fast Exothermic Heterogeneously Catalyzed Reactions. Chem. Eng. Sci. 1992,47,2939.

Villermaux, J. Future Challenges for Basic Research in Chemical Engineering. Chem. Eng. Sci. 1993,48,2525.

Wei, J. New Horizons in Reaction Engineering. Chem. Eng. Sci. 1990,45,1947.

Wei, J. ; Kuo, J. A Lumping Analysis in Monomolecular Reaction systems. Ind. Eng. Chem. Fundam. 1969, 8,114.

Westerterp, K. Multifunctional Reactors. Chem. Eng. Sci. 1992,47,2195.

Windes, L. C. ; Schwedock, M. J. ; Ray, W. H. Steady State and Dynamic Modeling of a Packed Bed Reactor for the Partial Oxidation of Methanol to Formaldehyde. Chem. Eng. Commun. 1989,47,1.

Xiao, W. D. ; Yuan, W. K. Modeling and Simulation for Adiabatic Fixed-Bed Reactors with flow reversal. Chem. Eng. Sci. 1994,49,3631.

Zhou, X. G. ; Liu, L. H. ; Yuan, W. K. ; Hudson, J. L. Modeling of a Fixed-Bed Reactor using the K-L Expansion and Neural Networks. Chem. Eng. Sci. 1996,51,2179.

An Algorithm for Simultaneous Chemical and Phase Equilibrium Calculation*

Abstract A new generalized algorithm has been developed for solving simultaneous chemical and phase equilibrium equations of two-phase systems. By rearranging convergence loops of the algorithm of Sanderson and Chien (1973), this new approach shows improved robustness and speed. Solutions to the material balance and chemical equilibrium equations are obtained with an improved Marquardt method (Zhan, 1976). An esterification reaction and dissociation reactions of electrolytes in phase equilibrium are presented as examples.

1 Introduction

In chemical, petrochemical and metallurgical industries, more and more processes use operations with simultaneous chemical reaction and mass transfer in the same vessel, such as reactive distillation and chemical absorption, to meet energy conservation and environmental protection requirements. Chemical and phase equilibrium calculations are often necessary for designing these processes. Therefore, in flowsheeting systems, a simultaneous chemical and phase equilibrium module is usually considered to be essential.

Strong interactions among components, phases, and reactions have made the chemical and phase equilibrium computation very difficult. Not until the early 1970's, after the computer power became sufficient, did this probram attract interest. In 1972, Dluzniewski and Adler (1972) extended the minimum Gibb's free energy method for calculating the chemical equilibrium compositions of an ideal gas mixture by White et al. (1958) to vapor-liquid and liquid-liquid systems, thus creating a generalized method known as RAND. During the last decade, RAND has been much improved. It was enhanced to automatically determine phase regions (Gautam and Seider, 1979a) for electrolyte systems (Gautam and Seider, 1979b), for slow reactions (White Ⅲ and Seider, 1981) and for alternative (i. e., other than temperature and pressure) specifications (Gautam and Wareck, 1986). Several other algorithms based on the minimum Gibb's free energy had been published (George et al., 1976; Gautam and Seider, 1979c; Casittilo and Grossmann, 1981; Zhou and Xu, 1987). Hitherto, methods of this kind have found broad applications and have been generally favored by metallurgists, chemists, and chemical engineers (Mather, 1986).

This problem can also be stated in an alternative formulation. The equilibrium compositions and phase numbers of a system can be obtained by solving a set of mass balance, phase and chemical equilibrium equations for specified components and reactions. This is called the k-val-

---

* Coauthors: Xiao Wende, Zhu Kaihong, Henry Hung-yeh Chien. Reprinted from *AIChE Journal*, 1989, 35(11): 1813-1820.

ue method. For those cases that only certain fast reactions reach equilibrium (Seider et al., 1980) or only the activity coefficient model is available for the phase equilibrium, the minimum energy method is not easy to apply but the k-value method remains effective.

In early 1973, Sanderson and Chien published a paper on a k-value method which was perhaps the first rigorous method for multiphase systems (henceforth referred to as the S-C algorithm). Electrolyte systems are invariably solved in a k-value formulation (Zemaitis and Rafal, 1975). This paper presents another k-value method based on "effective" chemical equilibrium constants, K_z's, for equivalent single-phase compositions, z_i's (henceforth referred to as the KZ algorithm). It reverses the order of chemical reaction and phase equilibrium calculation loops in the S-C algorithm. This strategy is analogous to the inside-out idea of Boston and Britt (1978) for flash calculations. For many problems, the new algorithm can be faster and yet more stable than that of Sanderson and Chien.

Combining the idea of Sanderson and Chien in calculating chemical equilibrium and that of Boston and Britt in calculating phase equilibrium, the KZ algorithm shows improved robustness and speed when compared to the S-C algorithm. The average number of k-value iterations is only one fourth to one tenth that of the S-C algorithm with little increase in iterations of extents.

The KZ algorithm is very flexible. It can easily be extended to solve problems with alternative specifications. With the final amounts of species being iteration variables, it can accurately calculate compositions of trace species.

Finally, it should be noted that the KZ algorithm has the same requirement as others that an accurate description of the thermodynamic characteristics of the system is critical for a valid solution.

2 Problem Formulation and the S-C Algorithm

For a two-phase system involving N components and M reactions:

$$\sum_{i=1}^{N} \gamma_{ij} A_i = 0; \quad j = 1, \cdots, M \tag{1}$$

where, i and j are the component and reaction indices, respectively; γ is the stoichiometric coefficients; A_i is the chemical species. The reaction equilibrium constants are defined by:

$$\prod_{i=1}^{N} (\gamma_i x_i)^{\gamma_{ij}} = \prod_{i=1}^{N} \left(\frac{p_i \phi_i}{f_i^o}\right)^{\gamma_{ij}} = K_j; \quad j = 1, \cdots, M \tag{2}$$

The equilibrium constants, K_j, may be determined experimentally or calculated by:

$$K_j = \exp\left(\frac{-\Delta G_j^o}{RT}\right)$$

where

$$\Delta G_j^o = \sum_{i=1}^{N} \gamma_{ij} G_i^o$$

Symbols are defined in the notation section. Eq. 2 are valid for systems with a vapor phase and one stable liquid phase.

The material balance equations for the reactions are:

$$F_{outi} = F_{ini} + \sum_{j=1}^{M} \gamma_{ij} \xi_j; \quad i = 1, \cdots, N \tag{3}$$

The phase equilibrium equations and the flash equations are:

$$z_i = \frac{F_{outi}}{\sum_{i=1}^{N} F_{outi}}; \quad i = 1, \cdots, N \tag{4}$$

$$z_i = \alpha y_i + (1 - \alpha) x_i; \quad i = 1, \cdots, N \tag{5}$$

$$y_i = \frac{f_i^o \gamma_i}{P \phi_i} x_i = k_i x_i; \quad i = 1, \cdots, N \tag{6}$$

$$\sum_{i=1}^{N} (y_i - x_i) = 0 \tag{7}$$

The liquid activity coefficients, γ_i, may be correlated by any excess Gibbs energy model: e. g., Wilson, NRTL, or UNIQUAC. The vapor fugacity coefficients, ϕ_i, may be calculated using any equation of state: e. g., Soave-Redlich-Kwong or Peng-Robinson. An accurate description of the mixture nonideality is important for a valid solution to the problem. However, the numerical method described in this paper is independent of the correlation or the equation of state used.

For a system with fixed temperature, pressure and feed compositions, Eq. 2-Eq. 7 are solved iteratively for the $4N + M + 1$ unknowns:

$$F_{outi} \quad i = 1, 2, \cdots, N$$
$$x_i, y_i, z_i \quad i = 1, 2, \cdots, N$$
$$\xi_j \quad j = 1, 2, \cdots, M$$
$$\text{and } \alpha$$

An algorithm using only extents as the iterative variables must search for a solution that satisfies the following set of linear constraints:

$$F_{outi} \geq 0; \quad i = 1, \cdots, N \tag{8}$$

or

$$\sum_{j=1}^{M} \gamma_{ij} \xi_j \geq -F_{ini}; \quad i = 1, \cdots, N \tag{9}$$

Sanderson and Chien (1973) stated that generalized methods for constrained problems produced disappointing results when applied to this problem because of its nonlinearity. They developed an algorithm that searches for ξ_i's and F_{outi}'s to satisfy Eq. 2 and Eq. 3 in its outer loop; x_i's, y_i's and α to satisfy Eq. 5 to Eq. 7 in its inner loop. Their approach allows constraint equations (Eq. 9) to be violated during the search and thus avoids the typical complications for these problems when solutions are close to the constraints. It was shown to be effective in obtaining solutions for difficult problems.

The S-C algorithm is as follows:

(1) Estimate ξ_i's and F_{outi}'s.
(2) Perform flash calculation on F_{outi}'s to get x_i's and y_i's and α by solving Eq. 5 to Eq. 7.
(3) Repeat above steps if Eq. 2 and Eq. 3 are not satisfied.

The simultaneous chemical reaction and phase equilibrium calculations are done in two loops: the former is in the outer loop and the latter in the inner loop. Theoretically, the S-C algorithm should perform well for systems with simple phase equilibrium models but with complex reactions. If the phase equilibrium calculation in the inner loop is slow, however, this algorithm may consume too much CPU time. Chen (1981) suggested the use of this algorithm for constant k values that are updated in yet another outer loop for reduced k-value evaluations with improved speed and reliability.

(1) Estimate k values.

(2) Estimate ξ_i's and F_{outi}'s.

(3) Perform flash with constant k values on F_{outi}'s to get x_i's and y_i's and α by solving Eq. 5 to Eq. 7.

(4) Return to step 2 if Eq. 2 and Eq. 3 are not satisfied.

(5) Return to step 1 until k values are converged.

This idea was originally used by Boston and Britt (1978) for flash calculations. It reduces the number of k-value evaluations to improve speed and stabilizes the highly sensitive k values to improve reliability. However, moving the k-value calculations outside represents only a minor modification to the S-C algorithm. It is possible to move the entire flash loop to the outside as described in the following section.

In calculations involving phase equilibria, it has been reported that the computation of composition ratios, k_i's, requires as much as 80% of the total CPU time (Westerberg et al., 1980; Chimowitz et al., 1984). Therefore, it is efficient for an algorithm to reduce the number of k-value evaluations as much as possible. A different approach to reduce these evaluations is used by the new algorithm. The outer loop of this algorithm is used for phase equilibrium calculations with constant k values, and the inner loop is for chemical equilibrium computations. For systems with simple chemical reactions, the new method can be extremely effective.

3 A New Formulation

To present the KZ algorithm, Eq. 2-Eq. 7 are rearranged as follows.

From Eq. 5 and Eq. 6, the liquid composition can be expressed as:

$$x_i = \frac{z_i}{(k_i - 1)\alpha + 1}; \quad i = 1, \cdots, N \tag{10}$$

Eq. 7 can be rewritten as:

$$S_\alpha \equiv \sum_{i=1}^{N} \frac{(k_i - 1)z_i}{(k_i - 1)\alpha + 1} = 0 \tag{11}$$

when z_i's are given, solving Eq. 6, Eq. 10, Eq. 11 is equivalent to the conventional isothermal flash calculation.

Substituting Eq. 10 into Eq. 2 yields:

$$\prod_{i=1}^{N} (z_i)^{\gamma_{ij}} = K_{zj}; \quad j = 1, \cdots, M \tag{12}$$

where

$$K_{zj} \equiv K_j \prod_{i=1}^{N} \left[\frac{\gamma_i}{(k_i - 1)\alpha + 1} \right]^{-\gamma_{ij}}; \quad j = 1, \cdots, M \tag{13}$$

In this formulation, effects of multiple phases and their nonideality on the reaction equilibria have been merged into a set of new parameters, $K_{zj}; j = 1, \cdots, M$. Given these parameters, solving Eq. 3, Eq. 4 and Eq. 12 is equivalent to the chemical equilibrium calculation of a single-phase ideal mixture.

In order to avoid division by zero and the nonlinearity of division, Eq. 12 are rewritten into a kinetic form:

$$S_j \equiv K_{zj}\left[\prod_{i=1}^{N}(z_i)^{-\gamma_{ij}}\right]_{\gamma_{ij}<0} - \left[\prod_{i=1}^{N}(z_i)^{\gamma_{ij}}\right]_{\gamma_{ij}>0} ; \quad j = 1,\cdots,M \tag{14a}$$

for $K_{zj} \geq 1$ and

$$S_j \equiv \left[\prod_{i=1}^{N}(z_i)^{-\gamma_{ij}}\right]_{\gamma_{ij}<0} - \frac{1}{K_{zj}}\left[\prod_{i=1}^{N}(z_i)^{\gamma_{ij}}\right]_{\gamma_{ij}>0} ; \quad j = 1,\cdots,M \tag{14b}$$

for $K_{zj} < 1$.

The material balance equations (Eq. 3) are redefined to be:

$$S_{M+i} \equiv F_{\text{outi}} - F_{\text{ini}} - \sum_{j=1}^{M}\gamma_{ij}\xi_j ; \quad i = 1,\cdots,N \tag{15}$$

At the solution, the functions $S_t; t = 1,\cdots,N+M$, are to be forced to zero.

4 A New Algorithm

The overall "basic" algorithm proceeds as follows:

(1) Estimate $k_i; i = 1,\cdots,N$.

(2) Estimate α.

(3) Calculate $K_{zj}; j = 1,\cdots,M$ with Eq. 13.

(4) Estimate $\xi_j; j = 1,\cdots,M$ and $F_{\text{outi}}; i = 1,\cdots,N$.

(5) Return to step 4 with new estimates from the modified Marquardt method until Eq. 14 and Eq. 15 are satisfied.

(6) Return to step 2 until Eq. 11 is satisfied.

(7) Return to step 2 with new estimates of k_i until $E \equiv \sum(k_{\text{inew}}/k_i - 1)^2 < 10^{-6}$.

The phase equilibrium calculation (in step 2 and step 6) is performed in the outer loop by solving for α to satisfy Eq. 11 with a typical algorithm for isothermal flash. The k-value loop (in step 1 and step 7) is often enhanced with a bounded Wegstein method. In this work, however, we used a simple direct iteration method. The chemical equilibrium calculation of this algorithm (in step 4 and step 5) is described in detail in the next section.

In the fine tuning of this "basic" algorithm, we found that in the initial phase of the search when k values are not yet close to their final values, there is little justification to force convergence of the inner loops, i. e., the α and extents loops. By taking the extents loop out of the α loop, we saved many steps in the search of extents at the cost of extra k-value evaluations. This is accomplished by replacing step 6 with: calculate α to satisfy Eq. 11.

The recommended algorithm uses this "preliminary" algorithm when $E > 10^{-4}$ and switch to the "basic" algorithm afterwards. In example 1 of this paper, we show a comparison of these variations.

5 Chemical Equilibrium Calculation

When $K_{zj}(j = 1,\cdots,M)$ are given, solving Eq. 4, Eq. 14 and Eq. 15 (in step 4 and step 5) is equivalent to the chemical equilibrium calculation of a single-phase ideal mixture. This calculation is placed in the inner loop of the KZ algorithm.

In the S-C algorithm, F'_{outi}s are included as iterative variables in addition to the extents and Eq. 15 as additional objective functions to eliminate linear constraints 8 or 9 in order to obtain consistent solutions in the presence of nonlinearity. Without the nonlinearity in the S-C approach, the new set of reaction equilibrium equations may be solved by using only extents as iterative variables. However, we found that Eq. 3 are very susceptible to round-off errors in extents. When certain reactions are directed by equilibria to approach completion, the residue trace quantities of reactants cannot be accurately evaluated by the differences of extents expressed in finite digits. This problem can be solved by including F'_{outi}s as iterative variables.

To illustrate this problem with an example, Xiao (1988) used the steam shift reaction of methane at 1400K and 1 atm with 1mol methane and 5mol steam in the feed. At such a high temperature, the reaction is in vapor phase and methane is almost completely converted. The residue methane (at 0.565×10^{-7} mol) if calculated from the difference of the initial methane (at 1mol) and the reaction extent (at 0.9999999435) would have required that the extent be accurate to 10 digits. If the final amount of methane is also a search variable, the error introduced to the objective function by the material balance equation (Eq. 15) for methane would be very small when the extent is accurate only to 5 digits.

Eq. 4, Eq. 14 and Eq. 15 are solved for extents and F_{outi}'s by a modified Marquardt method using the following Jacobian.

For $s = 1, \cdots, M$:

$$\frac{\partial S_s}{\partial \xi_j} = 0; \quad j = 1, \cdots, M \tag{16}$$

$$\frac{\partial S_s}{\partial F_{outi}} = \sum_{i=1}^{N} \frac{\partial S_s}{\partial z_i} \cdot \frac{\partial z_i}{\partial F_{outi}}; \quad i = 1, \cdots, N \tag{17}$$

For $s = M+1, \cdots, M+N$:

$$\frac{\partial S_s}{\partial \xi_j} = -\gamma_{s-M,j}; \quad j = 1, \cdots, M \tag{18}$$

$$\frac{\partial S_s}{\partial F_{outi}} = \delta_{s-M,i}; \quad i = 1, \cdots, N \tag{19}$$

For $K_{zj} \geqslant 1$:

$$\frac{\partial S_j}{\partial z_i} = -K_{zj} \frac{\gamma_{ij}}{z_i} \Big[\prod_{i=1}^{N} (z_i)^{-\gamma_{ij}} \Big]_{\gamma_{ij}<0} \tag{20}$$

or

$$\frac{\partial S_j}{\partial z_i} = -\frac{\gamma_{ij}}{z_i} \Big[\prod_{i=1}^{N} (z_i)^{\gamma_{ij}} \Big]_{\gamma_{ij}>0} \tag{21}$$

For $K_{zj} > 1$:

$$\frac{\partial S_j}{\partial z_i} = -\frac{\gamma_{ij}}{z_i} \Big[\prod_{i=1}^{N} (z_i)^{-\gamma_{ij}} \Big]_{\gamma_{ij}<0} \tag{22}$$

or

$$\frac{\partial S_j}{\partial z_i} = -\frac{\gamma_{ij}}{K_{zj} z_i} \Big[\prod_{i=1}^{N} (z_i)^{\gamma_{ij}} \Big]_{\gamma_{ij}>0} \tag{23}$$

Also

$$\frac{\partial z_i}{\partial F_{outi}} = \frac{\delta_{ti} - z_i}{\sum_{s=1}^{N} F_{outs}}; \quad t = 1, \cdots, N; \quad i = 1, \cdots, N \qquad (24)$$

The modified Marquardt method (Zhan, 1976) is used to minimize the objective function:

$$\sum_{i=1}^{M+N} S_i^2$$

It calculates the search vector with the following equation:

$$L(D + \lambda I)L^T \Delta x = -g \qquad (25)$$
$$H = LDL^T = J^T J \qquad (26)$$
$$g = J^T S \qquad (27)$$
$$\Delta x = (\Delta \xi, \Delta F_{out}) \qquad (28)$$

Eq. 25 are valid in that the Hessian Matrix, H, is positive definite. This method was shown to be very efficient since it avoids the repeated operations on H and g. Moreover, it replaces the unity matrix, I, which in the classic Marquardt method (1963) improves only the diagonal elements of matrix H with a positive definite matrix, LL^T, to improve the entire H.

To enhance the performance of the Marquardt method, objective functions equations (Eq. 14 and Eq. 15) are scaled by dividing each equation by the sum of the absolute values of all terms on its righthand side plus

$$\frac{1}{\sum_{i=1}^{N} F_{outi}}$$

These scale factors are updated at each new base point.

In the implementation of the modified Marquardt method, if the calculated new objective function is greater than that at the base point, the value of λ is increased by a factor of 4 and the calculation of a new objective function is repeated; if the new function is smaller, λ is decreased by a factor of 10 and the base point is updated. In either case, the iteration counter is advanced by one.

6 Comparisons between the KZ and the S-C Algorithm

Two systems have been included in this paper as examples to compare these two methods. One is the esterification reaction of ethanol and acetic acid, and the other is an electrolyte system with complex dissociation reactions. Both problems are initialized from:

$$\gamma_i = 1; \quad k_i = p_i^o/P; \quad i = 1, \cdots, N$$
$$\xi_j = 0; \quad j = 1, \cdots, M$$
$$F_{outi} = F_{ini}; \quad i = 1, \cdots, N$$

and

$$\alpha = 0$$

Convergence criteria are as follows.

For the outer loop:

$$\sum_{i=1}^{N} \left[\frac{k_{inew}}{k_i} - 1 \right]^2 < 10^{-6}$$

For the inner loop:

$$\sum_{j=1}^{M}\left[\frac{1}{K_{zj}}\prod_{i=1}^{N}(z_i)^{\gamma_{ij}}-1\right]^2+\sum_{i=1}^{N}S_{M+i}^2<10^{-6}$$

and

$$|S_\alpha|<10^{-5}$$

In both methods, direct iteration is used for the k-value convergence and the modified Marquardt method is used for the reaction extent.

6.1 Example 1

The esterification reaction of ethanol and acetic acid:

$$\text{EtOH} + \text{HAC} \Longleftrightarrow \text{EtAC} + \text{H}_2\text{O}$$

approaches equilibrium in the presence of acid catalyst. It has often been used to check algorithms for simultaneous chemical and phase equilibria (Sanderson and Chien, 1973; George et al., 1976; Gautam and Seider, 1979c; Casittilo and Grossmann, 1981; Zhou and Xu, 1987). Since its chemical equilibrium can be explicitly solved as a quadratic equation, it is ideally suited for the KZ algorithm.

Its chemical equilibrium constant, based on vapor compositions, can be expressed as:

$$\ln(K_y) = -20.718 + 2518.399/T + 3.08\ln T - 0.0031T + 1.2\times 10^{-6}T^2$$

by correlating the data from Reid et al. (1976). The K in Eq. 2 is evaluated with:

$$K = K_y \frac{p_1^s p_2^s}{p_3^s p_4^s}$$

The vapor pressures, p_i^s's, are calculated by the Antoine equation using constants in Table 1. For convenience, we also show K_y and K at different temperatures in Table 2.

Table 1   Antoine constants for EtOH-HAC-EtAC-H_2O[①]

Comps.	a	b	c
EtOH	9.95614	1440.52	-60.44
HAC	9.6845	1644.05	-39.63
EtAC	9.22298	1238.71	-56.15
H_2O	10.09171	1668.21	-45.14
	$\log_{10}p^s(\text{N/m}^2) = a - b/(T,K+c)$		

①Suzuki et al. (1970).

Table 2   Chemical system equilibrium constants of EtOH + HAC $\Longleftrightarrow$ EtAC + H_2O

T/K	K_y	K
345	41.154	20.505
350	35.743	19.537
355	33.546	18.644
358	32.223	18.142

The liquid activity coefficients for this system are computed by the UNIQUAC equation, with its parameters, which are shown in Table 3, generated by the UNIFAC model (Fredenslund et al., 1977). The vapor-phase mixture is assumed to be ideal, and the polymerization of acetic acid in the vapor phase has also been ignored.

Table 3 UNIQUAC parameters for system EtOH-HAC-EtAC-H$_2$O

Volume and surface area parameters				
Comps.	EtOH	HAC	EtAC	H$_2$O
q	1.972	2.092	3.116	1.400
R	2.1055	2.2024	3.4786	0.920

Binary interaction parameters ($\tau_{ij} = A_{ij} + B_{ij}T$)				
System	A_{12}	A_{21}	B_{12}	B_{21}
EtOH-HAC	−0.213	4.250	1.37×10^{-3}	-5.11×10^{-3}
EtOH-EtAC	1.340	0.289	-4.05×10^{-4}	5.57×10^{-4}
EtOH-H$_2$O	0.098	2.140	9.61×10^{-4}	-1.66×10^{-3}
HAC-EtAC	2.760	−0.076	-2.64×10^{-3}	1.07×10^{-3}
HAC-H$_2$O	3.140	−0.073	-3.90×10^{-3}	1.29×10^{-3}
EtAC-H$_2$O	0.223	1.040	-1.03×10^{-4}	2.08×10^{-5}

$$\ln\gamma_k = \ln\gamma_k^c + \ln\gamma_k^R$$

$$\ln\gamma_k^c = \ln\frac{\phi_k}{x_k} + \frac{Z}{2}q_k\ln\frac{\phi_k}{x_k} + l_k - \frac{\phi_k}{x_k}\sum_{i=1}^{N} x_i l_i$$

$$l_k = \frac{Z}{2}(R_k - q_k) - (R_k - 1); \qquad Z = 10$$

$$\theta_k = \frac{q_k x_k}{\sum q_i x_i}; \qquad \phi_k = \frac{R_k x_k}{\sum R_i x_i}$$

$$\ln\gamma_k^R = q_k\left[1 - \ln(\sum \theta_i \tau_{ij}) - \sum_{i=1}^{N}\frac{\theta_i \tau_{ik}}{\sum \theta_j \tau_{jk}}\right]$$

Table 4 lists four cases to compare the KZ algorithm with the S-C algorithm using the system EtOH-HAC-EtAC-H$_2$O. The number of iterative calculations of k_i's of the S-C method is about four times that of the KZ method, and the number for extents of the KZ algorithm doubles that of S-C's.

Table 4 Comparison between two algorithms with system EtOH-HAC-EtAC-H$_2$O

Temp. /K	α	No. of k_i's		No. of ξ_j's	
		S-C	KZ	S-C	KZ
358	1.000	35	10	7	9
355	0.877	42	9	10	23
350	0.413	23	6	10	21
345	0.000	5	3	5	11

As stated before, the extent for this system can be easily calculated using the explicit solution of a quadratic equation (in step 4 and step 5). This is possible since Eq. 14 are identical to those for a single-phase ideal-solution problem. Inasmuch as the quadratic solution requires almost zero time and produces results reliably, it is preferred over any search method. The S-C algorithm, with its flash loop inside the reaction loop, invariably must use a search method for the reaction calculation even for the simplest of reactions. The number of extents trials for the KZ method as shown in Table 4 reflects only the use of a generalized approach.

Table 5 and Table 6 summarize the iterative steps of these two algorithms for the case at 355K. Note that the S-C method moved very slowly at the beginning. The movements in the extent were generated by the Marquardt method. These changes were large enough to have required large numbers of updates of k values in the inner loop. In the KZ algorithm, the chemical equilibrium is solved in the inner loop. Consequently, the reaction extent moved very quickly to the solution.

Table 5 Iteration summary at 355K and 1 atm of the S-C method

Outer loop counter	k values inner loop	α	K_z	Reaction extent	Objective function
0	2	0.000	4.958	0.000	1.537
1	1	0.000	4.958	2.3×10^{-4}	1.537
2	2	0.000	4.856	0.953	0.983
3	4	0.334	8.006	5.369	0.128
4	5	0.652	14.524	7.052	3.66×10^{-2}
5	6	0.790	19.725	7.819	6.68×10^{-3}
6	7	0.846	22.692	8.141	9.29×10^{-4}
7	6	0.867	23.962	8.262	1.08×10^{-4}
8	5	0.874	24.421	8.303	1.16×10^{-5}
9	3	0.874	24.543	8.317	1.18×10^{-6}
10	1	0.877	24.583	8.321	1.14×10^{-7}
Total	42				

Note: 1 atm = 101.325kPa.

Table 6 Iteration summary at 355K and 1 atm by the KZ method

Outer loop counter	Iterations inner loop	α	K_z	Reaction extent
0	6	0.000	18.644	8.120
1	5	0.888	21.350	8.221
2	4	0.884	24.201	8.311
3	2	0.879	23.858	8.301
4	1	0.875	24.023	8.306
5	2	0.876	24.224	8.311
6	1	0.876	24.378	8.315
7	1	0.877	24.472	8.320
8	1	0.877	24.538	8.321
9	0	0.877	24.581	8.321
Total	23			

Using the "basic" algorithm, i.e., forcing convergence of extents inside the α loop, this problem took 35 calculations of extents and seven sets of k values. Using the S-C algorithm as modified by Chen, i.e., moving k values to the outer most loop, this problem took 40 extents calculations and seven sets of k values. As described earlier, not forcing convergence in the inner loop of the recommended algorithm reduced the number of extents calculations but increased the number of k values calculated slightly.

Table 7 gives results of the equilibrium calculation at 355K and 1 atm. Initially, the system contains 10kg · mol of ethanol and 10kg · mol of acetic acid. The calculated extent of reaction is 8.321kg · mol, and the flash ratio is 0.877.

Table 7 Numerical results of esterification equilibrium[①]

Comps.	mol%		
	Feed stream	Liquid stream	Vapor stream
EtOH	50.0	6.86	8.62
HAc	50.0	23.76	6.24
EtAc	0.0	13.73	45.51
H_2O	0.0	55.65	39.63

①Conditions: temperature, 355K; pressure, 1 atm; conversion, 83.21%; flash ratio 87.7%.

6.2 Example 2

In processes for purifying plant effluent streams, one often encounters aqueous volatile weak electrolyte systems, such as NH_3-CO_2-H_2O. There have been many articles about its thermodynamics (Edwards et al., 1975, 1978; Beutier and Renon, 1978; Chen et al., 1979; Zhou and Zu, 1983) and equilibrium calculations (Zemaitis and Rafal, 1975; Gautam and Seider, 1979b). This paper uses it as an example for comparing the KZ algorithm with the S-C algorithm. The results should be of general interest due to its complex reactions and strong nonideality.

There are five reactions in the system NH_3-CO_2-H_2O:

$$NH_3 + H_2O \Longleftrightarrow NH_4^+ + OH^- \tag{29}$$

$$CO_2 + H_2O \Longleftrightarrow HCO_3^- + H^+ \tag{30}$$

$$HCO_3^- \Longleftrightarrow CO_3^{2-} + H^+ \tag{31}$$

$$H^+ + OH^- \Longleftrightarrow H_2O \tag{32}$$

$$NH_3 + HCO_3^- \Longleftrightarrow NH_2COO^- + H_2O \tag{33}$$

At 373K, the equilibrium constants (defined in Eq. 2) for the five reactions are 1.45×10^{-5}, 3.96×10^{-7}, 7.08×10^{-11}, 1.79×10^{12} and 0.44, respectively (Edwards et al., 1978). The liquid activity coefficients of the system are also based on the equations and parameters developed by them. Their equations, which are best suited for aqueous volatile weak electrolyte systems are based on work published by Zhou and Xu (1983) and can be applied to systems with high ionic strength comparable to equations developed by Chen et al. (1979). Vapor fugacity coefficients of the system are computed with the equation of state published by Nakamura et al. (1976).

An iteration summary of the S-C algorithm for a system with total molalities of NH_3 and CO_2 at 0.05 and 0.1, respectively, is given in Table 8. Because of the low electrolyte concentrations, the liquid nonideality is minimal and the number of k value updates in the inner loop is small. Yet the total number by the S-C algorithm is 15 in contrast to only three iterations of k values required by the KZ algorithm as shown in Table 9. There is at least one k value update for each iteration of extents. For very dilute systems, this is apparently not efficient.

Table 8　Iteration summary of the S-C algorithm for system NH_3-CO_2-H_2O[①]

Outer loop counter	k value calc.	K_z (Eq. 25)	Ionic strength	Objective function
0	1	24.407	2.507×10^{-8}	0.9945
1	2	25.403	4.964×10^{-3}	1.32×10^7
2	1	27.960	4.440×10^{-3}	1.95×10^6
3	1	26.006	5.640×10^{-3}	6.61×10^4
4	1	24.604	1.126×10^{-2}	2.18×10^3
5	1	24.230	2.302×10^{-2}	4.36×10^1
6	1	23.903	4.036×10^{-2}	0.2657
7	1	23.717	5.265×10^{-2}	0.9944
8	1	23.710	5.251×10^{-2}	3.23×10^{-2}
9	1	23.694	5.157×10^{-2}	8.54×10^{-3}
10	1	23.759	4.995×10^{-2}	1.23×10^{-4}
11	1	23.779	4.850×10^{-2}	2.31×10^{-5}
12	1	23.784	4.809×10^{-2}	4.57×10^{-6}
13	1	23.784	4.808×10^{-2}	1.11×10^{-8}
Total	15			

①Conditions: $T = 373K$; $P = 10atm$; F_{in}, $H_2O = 1kg$; $NH_3 = 0.05mol$; $CO_2 = 0.1mol$.

Table 9　Iteration summary of the KZ algorithm for system NH_3-CO_2-H_2O conditions[①]

Outer loop counter	Iterations inner loop	K_z (Eq. 25)	Ionic strength
0	14	0.5058	3.330×10^{-2}
1	6	19.886	4.778×10^{-2}
2	4	23.787	4.807×10^{-2}
3	2	23.784	4.808×10^{-2}
Total	26		

①See Table 8.

When the total molalities of NH_3 and CO_2 reach 2.9 and 1.45, respectively, the nonideality becomes strong. The S-C algorithm, as shown in Table 10, did not reach the solution smoothly. Instead, the "objective function" increased many times, at first, 5th, 9th and 12th iterations, indicating frequent increases of λ in the Marquardt method. This phenomenon is certainly attributable to the strong nonideality. Yet, the path by the KZ algorithm, which is shown in Table 11, was very stable and fast. The numbers of k-value evaluations by the two algorithms are 47 and 5, respectively, and those of extents are both 38.

Table 10　Iteration summary of the S-C algorithm for system NH_3-CO_2-H_2O[①]

Outer loop counter	Iterations inner loop	K_z (Eq. 25)	Ionic strength	Objective function
0	1	19.0125	2.79×10^{-8}	0.8624
1	2	30.1505	4.70×10^{-3}	1.606×10^7
2	1	38.8530	3.50×10^{-3}	2.451×10^6

Outer loop counter	Iterations inner loop	K_z (Eq. 25)	Ionic strength	Objective function
4	2	31.0743	2.03×10^{-2}	8.168×10^2
5	1	29.2720	0.2162	2.534×10^3
8	5	14.0150	1.5370	4.930×10^1
9	2	18.2190	0.7887	9.900×10^1
12	4	17.4550	1.0030	1.263×10^2
13	1	16.8970	1.0300	1.074×10^2
18	9	12.4000	1.6642	5.210×10^{-5}
27	9	12.3674	1.7613	8.579×10^{-7}
36	10	12.3595	1.7855	8.076×10^{-8}
38	2	12.3587	1.7878	5.142×10^{-8}
Total	47			

①Conditions: $T = 373$ K; $P = 10$ atm; F_{in}, $H_2O = 1$ kg; $NH_3 = 2.9$ mol; $CO_2 = 1.45$ mol.

Table 11   Iteration summary of the KZ algorithm for system NH_3-CO_2-H_2O conditions①

Outer loop counter	Iterations inner loop	K_z (Eq. 25)	Ionic strength
0	16	0.4925	0.9879
1	12	12.2968	1.5530
2	4	12.6277	1.7224
3	3	12.4392	1.7723
4	2	12.3653	1.7852
5	1	12.3578	1.7880
Total	38		

①See Table 10.

Table 12 gives the results of equilibrium calculation for the system NH_3-CO_2-H_2O-CH_4-C_2H_6, in which CH_4 and C_2H_6 are inert with respect to the chemical equilibrium and assumed to be insoluble in water. All ions are assigned small k values, e.g., 10^{-10}, and CH_4 and C_2H_6 large k values, e.g., 10^{10}. The calculated fractions of NH_3 and CO_2 absorbed in water are 97.10% and 66.00%, respectively. The ionic strength is 2.404 M.

Table 12   Calculated results of system NH_3-CO_2-H_2O-CH_4-$C_2H_6$①

Species	Molalities/mol·kg^{-1}	Activity coeff.	Partial pres.	Fugacity coeff.
NH_3	0.6151	1.010	0.1543	0.9761
CO_2	0.0183	1.014	1.8080	0.9793
CH_4			3.5457	0.9937
C_2H_6			3.5457	0.9691
H_2O	55.555	1.035	0.9462	0.9651
NH_4^+	2.2256	0.188		

Species	Molalities/mol·kg^{-1}	Activity coeff.	Partial pres.	Fugacity coeff.
HCO_3^-	1.6712	0.162		
CO_3^-	0.1780	0.004		
NH_2COO^-	0.1984	0.392		
H^+	5.34×10^{-8}	0.483		
OH^-	4.61×10^{-5}	0.445		

①Conditions: $T = 373K$; $P = 10$atm; F_{in}, $H_2O = 1$kg; $NH_3 = 3$mol; $CO_2 = 3$mol; $CH_4 = 2$mol; $C_2H_6 = 2$mol.

Acknowledgements

This research was supported by the National Natural Science Foundation, China, Grant Number 9287001-05.

Notation

A_i chemical species i

D decomposed diagonal matrix of H (Eq. 25)

f_i fugacity of species i, atm

F_{ini} initial amount of species i, mol

F_{outi} final amount of species i (or estimate), mol

G_i^o standard Gibb's free energy of formation of species i, J

ΔG_j^o standard Gibb's free energy of reaction for reaction j, J

g gradient vector of equation (Eq. 27)

H Hessian matrix of equation (Eq. 26)

I unity matrix

J Jacobian matrix

K_j chemical equilibrium constant for reaction j (Eq. 2)

K_{zj} constant for reaction j defined by Eq. 13

k_i phase equilibrium ratio for species i ($= y_i/x_i$)

L decomposed lower triangle matrix of H (Eq. 26)

M number of reactions

N number of components

P system pressure, atm

p_i partial pressure of species i, atm

R universal gas constant

S_i error in i, described in Eq. 14, Eq. 15

S_α error in flash equation (Eq. 11)

T system temperature, K

x_i mole fraction of species i in liquid phase

y_i mole fraction of species i in vapor phase

z_i mole fraction of species i in total system including liquid and vapor

Greek letters

α vapor to liquid mole ratio
δ Kronecker delta
γ_i activity coefficient of species i
λ Marquardt parameter
γ_{ij} stoichiometric coefficient of species i in reaction j
ξ_j extent of reaction j, mol
$\hat{\phi}_i$ fugacity coefficient of species i in the mixture

Subscripts

i species number
j reaction number

Superscripts

o standard state (at 1 atm and system temperature)
s saturation
T transpose

References

Beutier, D., and H. Renon, "Representation of NH_3-H_2S-H_2O, NH_3-CO_2-H_2O and NH_3-SO_2-H_2O Vapor-Liquid Equilibrium," Ind. Eng. Chem. Process Des. Dev., 17, 220 (1978).

Boston, J. F., and H. I. Britt, "A Radically Different Formulation and Solution of the Single-Stage Flash Problem," Comp. and Chem. Eng., 2, 109 (1978).

Casittilo, J., and I. E. Grossmann, "Computation of Phase and Chemical Equilibria," Comp. and Chem. Eng. 5, 99 (1981).

Chen, C. C., private communication (1981).

Chen, C. C., H. I. Britt, J. F. Boston, and L. B. Evens, "Extension and Application of the Pitzer Equation for Vapor-Liquid Equilibrium of Aqueous Electrolyte Systems with Molecular Solutes," 25(5), 820 (1979).

Chimowitz, E. H., S. Macchietto, T. F. Anderson, and L. F. Stutzman, "Local Model for Representing Phase Equilibria in Multicomponent Non-ideal Vapor-Liquid and Liquid-Liquid Systems: 2. Application to Process Design," Ind. Eng. Chem. Process Des. Dev., 23, 609 (1984).

Dluzniewski, J. H., and S. B. Adler, "Calculation of Complex Reaction and/or Phase Equilibria Problems," Int. Chem. Eng. Symp. Ser. 5, Instn. Chem., Egrs., London, 4, 21 (1972).

Edwards, T. J., J. Newman, and J. M. Prausnitz, "Thermodynamics of Aqueous Solutions Containing Volatile Weak Electrolytes," AIChE J., 21, 248 (1975).

Edwards, T. J., G. Maurer, J. Newman, and J. M. Prausnitz, "Vapor-Liquid Equilibria in Multicomponent Aqueous Solutions of Volatile Weak Electrolytes," AIChE J., 24, 966 (1978).

Fredenslund, A., J. Gmelring, and P. Rasmussen, Vapor-Liquid Equilibria Using UNIFAC A Group-Contribution Method, Elsevier, New York (1977).

Gautam, R., and W. D. Seider, "Calculation of Phase and Chemical Equilibrium. Part I: Local and Constrained Minima in Gibb's Free Energy," AIChE J., 25(6), 990 (1979a).

"Calculation of Phase and Chemical Equilibrium. Part II: Phase-Splitting," AIChE J., 25(6), 999 (1979b).

"Calculation of Phase and Chemical Equilibrium: III. Electrolytic Solutions," AIChE J., 25(6), 1008 (1979c).

Gautam, R., and J. S. Wareck, "Computation of Physical and Chemical Equilibria—Alternate Specifications," Comp. and Chem. Eng., 10(2), 143 (1986).

George, B. L., L. P. Brown, C. W. Farmer, P. Buthod, and P. S. Manning, "Computation of Multicomponent, Multiphase Equilibrium," Ind. Eng. Chem. Process Des. Devl., 15(3), 373 (1976).

Mather, A. E., "Phase Equilibrium and Chemical Reaction," Fluid Phase Equilib., 30, 83 (1986).

Marquardt, D. W., "An Algorithm for Least-Square Estimation of Non-linear Parameters," J. Soc. Ind. Appl. Math., 11, 431 (1963).

Nakamura, R. G., J. F. Gerrit Breedveld, and J. M. Prausnitz, "Thermodynamic Properties of Gas Mixtures Containing Common Polar and Nonpolar Components," Ind. Eng. Chem. Process Des. Dev., 15, 557 (1976).

Reid, R. C., J. M. Prausnitz, and T. K. Sherwood, The Properties of Gases and Liquids, 3rd ed., McGraw-Hill, New York (1976).

Sanderson, R. V., and H. H. Y. Chien, "Simultaneous Chemical and Phase Equilibrium Calculation," Ind. Eng. Chem. Process Des. Dev., 12(1), 80 (1973).

Seider, W. D., R. Gautam, and C. W. White III, "Computation of Phase and Chemical Equilibrium: A Review," Comp. Applic. to Chem. Eng., R. G. Squires and G. V. Reklaitis, eds., ACS Symp. Ser., 124, 115 (1980).

Suzuki, I., H. Komarsu, and M. Hirata, "Formulation and Prediction of Quaternary Vapor-Liquid Equilibrium Accompanied by Esterification," J. Chem. Eng. Japan, 3, 152 (1970).

Westerberg, A. W., H. P. Hutchinson, R. L. Motard, and P. Winter, Process Flowsheeting, 1st ed., Cambridge University Press, Cambridge (1979).

White, W. B., S. M. Johson, and G. B. Dantzig, "Chemical Equilibrium in Complex Mixtures," J. Chem. Phys., 28, 751 (1958).

White III, C. W., and W. D. Seider, "Computation of Phase and Chemical Equilibrium: IV. Approach to Chemical Equilibrium," AIChE J., 27(3), 466 (1981).

Xiao, W., M. Sc. Thesis, East China Univ. Chem. Tech., Shanghai, China (1988).

Zemaitis, J. F., and M. Rafal, "ECES—A Computer System to Predict the Equilibrium Compositions of Electrolyte Solutions," AIChE Meeting, Los Angeles (1975).

Zhan, C., "An Modified Marquardt Algorithm for Least Squires Estimation of Non-linear Parameters," J. Northwest Univ., China, 2(3), 79 (1976).

Zhou, J., and Z. Xu, "Vapor-Liquid Equilibrium Study on Ternary Aqueous Solutions of Volatile Weak Electrolytes—NH_3-CO_2-H_2O-NH_3-H_2S-H_2O and NH_3-SO_2-H_2O Systems," J. Chem. Ind. and Eng., China, 3, 234 (1983).

Zhou, W., and Z. Xu, "Calculation of Multicomponent Multiphases Equilibria (I): Modified Algorithm M-SVMP Method," J. Chem. Ind. and Eng. China, 1, 39 (1987a).

Zhou, W., and Z. Xu, "Calculation of Multicomponent Multiphase Equilibria: II. A New General Method GCG," J. Chem. Ind. and Eng. China, 1, 49 (1987).

Intensification of Phase Transition on Multiphase Reactions[*]

Abstract Chemical reactions, generally conducted under full gas or liquid phases, are problematic for volatile liquid reactants. For such reactants, the presence of phase transition can be favorable, since evaporation of the liquid could not only balance the reaction heat but improve the effectiveness factor of the porous catalyst. This principle was applied to engineering applications. Experiment was carried out to investigate effects of catalyst activities, flow directions, operation pressures, gas and liquid flow rates, and reactant concentrations on the reactor behavior. Quench operation with a cold-injection side stream was initiated to prevent an excessive temperature rise, which was shown to be effective and flexible. With this novel optimizing method, the reactor temperature could be kept around 270℃ under 1.0MPa, even with a benzene concentration of 35%. Research on phase transition of benzene hydrogenation proved successful and could be extended to reaction systems with a similar range of process intensification.

1 Introduction

Chemical reactors are generally operated under constant phase conditions without or with only a little phase transition, but this may no longer be appropriate for liquid reactants with substantial volatility, because if the reaction is carried out in a fixed bed for gas operation, the reactant has to be vaporized before it is fed into the reactor. Nevertheless, a great deal of cooling energy should be supplied to remove the reaction heat, and this energy consumption is unreasonable. On the other hand, if the reaction is carried out in a trickle bed under liquid condition, the reaction rate would unfortunately be largely lowered, since the catalyst interior is completely filled with liquid; however, removal of reaction heat cannot be avoided if the reaction is exothermic.

The purpose of this work is to investigate the underlying principles in scope of intensification of multiphase reactions through phase transition. However, previously this has been considered dangerous, since vaporization of the liquid phase usually leads to partial wetting of the catalyst pellets and thus the inception of gas-phase operation, with a possible large rise in temperature and even damage to the catalyst (Eigenberger and Wegerle, 1982; Hanika et al., 1986).

In spite of the large amount of research that has been done on catalyst efficiency under partial wetting conditions, research on phase transition is still scarce. Important contributions in this latter field, for example, are those of Kim and Kim (1981a,b), Hu and Ho (1987), Bhatia (1988,1989), Jaguste and Bhatia (1991), and Waston and Harold (1993,1994). These researchers have only investigated the hysteresis phenomenon on the pellet scale, and an impor-

---

[*] Coauthors: Cheng Zhenmin, Abdulhakeim M. Anter. Reprinted from *AIChE Journal*, 2001, 47(5): 1185-1192.

tant conclusion from their work is that hysteresis is a route-dependent phenomenon—that is, the internal wetting condition is not exclusively determined by the temperature of the catalyst—that also depends on the catalyst being heated or cooled, as described by the Kelvin or the Cohan equations. Studies on reactor scale were also made by only a few investigators, under narrow operating conditions (such as Satterfield and Ozel, 1973; Sedriks and Kenney, 1973; and Hanika et al., 1975, 1976, 1977). These experiments were all conducted under 1 atm in a glass column under the trickling flow condition. The researchers found that phase transition was very dangerous, as it could cause reactor runaway. Theoretical analysis of phase transition on the reactor scale was given by Hanika et al. (1986), Kheshgi et al. (1992), and Khadilkar et al. (1999), who simulated the transition and hysteresis phenomena observed by Hanika et al. (1975). It seems more experimental work using close to industrial conditions as possible should be conducted to fulfill the gap between engineering application and theoretical understanding on the phase transition.

Regardless of the possibility of runaway reactors, phase transition nevertheless could provide an alternative to process intensification, since it has been observed that the reaction rate under gaseous condition is seven times as large as that under liquid conditions (Sedriks and Kenney, 1973; Hanika et al., 1975). To prevent the danger posed by phase transition, a concentration shift operation has been proposed in the literature (Lange et al., 1994; Hanika, 1999). It should be noted that concentration oscillation would inevitably lead to operational complexity, and even worse, to unstable product quality.

In this work, the hydrogenation of benzene to cyclohexane is carried out as the working system because of both its industrial importance and the inherent academic significance of the outcome. Additionally, the hydrogenation product, cyclohexane, has physical properties similar to benzene and is stable during the reaction, so that benzene hydrogenation can be treated as a single-component system in order to simplify the theoretical analysis.

2 Principles

2.1 Thermodynamics of the reaction system

Hydrogenation of benzene is a complex process, as is indicated by the following network:

$$\bigcirc + 3H_2 \rightleftharpoons \bigcirc + 51.62 \text{kcal/mol} \qquad (1)$$

$$\bigcirc + 3H_2 \longrightarrow 3C + 3CH_4 + 75.44 \text{kcal/mol} \qquad (2)$$

$$\bigcirc \rightleftharpoons \bigcirc^{-CH_4} - 3.98 \text{kcal/mol} \qquad (3)$$

$$\bigcirc + 6H_2 \longrightarrow 6CH_4 + 81.72 \text{kcal/mol} \qquad (4)$$

In the above equations, the reaction heat is given at 500K. Eq. 1 is called the principal reaction in producing cyclohexane, and Eq. 2 accounts for the hydrocracking of benzene, with carbon and methane as the final products. Eq. 1 and Eq. 2 are parallel for the conversion of benzene.

Because of the high reaction heat, a substantial temperature rise will occur in the reactor. The temperature effects on the four reactions are different, as can be seen by the equilibrium constants listed in Table 1. The equilibrium constant for the principal reaction is found to decrease substantially under high temperatures, and it seems that the temperature suitable for benzene conversion cannot exceed 600K. At temperatures lower than this, the benzene conversion will be seriously restricted, thus leaving much of the benzene unconverted. However, the coexistence of benzene hydrocracking provides additional conversion because of its high equilibrium constant. Additionally, high temperatures also favor the conversion of cyclohexane to other by-products. Therefore, the temperature should not be too high to guarantee the conversion of benzene into cyclohexane.

Table 1 Equilibrium constants of the benzene hydrogenation network

T/K	Reaction, log K_P			
	1	2	3	4
300	16.932	49.092	−0.686	58.629
400	7.842	35.632	0.036	44.296
500	2.259	27.448	0.472	35.485
600	−1.527	21.916	0.759	29.458
700	−4.257	17.922	0.957	25.050
800	−6.311	14.896	1.097	21.669
900	−7.910	12.518	1.198	18.976
1000	−9.183	10.605	1.270	16.785

To provide the most available reaction rate, the reaction temperature can approach the equilibrium value, as is determined from the following assumptions:

(1) The conversion of benzene to cyclohexane is as high as 99.95%.

(2) The side reactions are assumed negligible under a wide temperature range.

Assumption 1 is also a constraint that has to be satisfied for industrial operation, since the boiling point of cyclohexane approaches that of benzene; otherwise, the separation of these two components will be very difficult. The second assumption is made to simplify the theoretical analysis, and should be expected in view of the rapid development of new catalysts.

A thermodynamic diagram based on these assumptions is shown in Fig. 1. Specifically, in order to obtain 99.95% conversion from benzene to cyclohexane under an operation pressure of 1.0MPa at a molar ratio of $H_2/C_6H_6 = 9$, the operation temperature should not exceed 270℃.

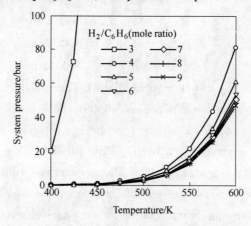

Fig. 1 Thermodynamic conditions for benzene conversion to 99.95% (1 bar = 10^5 Pa)

2.2 Gas-liquid equilibrium inside of the catalyst interior

The liquid filling condition in the catalyst interior is closely related to the effectiveness factor of the porous catalyst. According to the capillary effect, as described by the Kelvin equation (Eq. 5), the gas-liquid equilibrium in the catalyst interior will be much different from the bulk phase:

$$\ln \frac{p_i^0}{p^0} = \frac{2\sigma M}{RT\rho} \left(\frac{-1}{\hat{r}_i} \right) \tag{5}$$

In Eq. 5, $\hat{r}_i$ is the average gas-liquid interface radius, and is expressed by:

$$\frac{2}{\hat{r}_i} = \frac{1}{r_1} + \frac{1}{r_2} \tag{6}$$

where, $\hat{r}_i$ depends on the system being heated or cooled, and is obtained according to Jaguste and Bhatia(1991):

$$\hat{r}_i = \begin{cases} r_i, & \text{where } r_1 = r = r_i \text{ for evaporation} \\ 2r_i, & \text{where } r_1 = r_i, r_2 = \infty \text{ for condensation} \end{cases} \tag{7}$$

The gas-liquid equilibrium data within the capillary is shown in Table 2, where pressure p_i^0 indicates the equilibrium vapor pressure of the liquid in the pores under bulk temperature, T^0. Here T_i^e and T_i^c correspond, respectively, to the bulk phase temperatures, under which the liquid in the pores starts to evaporate or to the vapor in the bulk phase starts to condense within the pores. The difference between T_i^e and T_i^c forms a hysteresis cycle from evaporation to condensation.

Table 2 Capillary effects on phase transition within catalyst pores[①]

d_i/Å	Evaporation		Condensation	
	p_i^0/bar	T_i^e/℃	p_i^0/bar	T_i^c/℃
30	7.65	197	8.85	189
50	8.60	190	9.39	185
120	9.53	184	9.89	182

①The bulk phase condition is $T^0 = 180℃$; $P^0 = 1.025$MPa.

3 Experimental Studies

According to the hydrodynamic analysis of Ng(1986) and Cheng and Yuan(1999), liquid maldistribution has been one of the major problems in the trickle-bed operation, especially prior to the inception of the pulsing flow. In the literature, maldistribution has been considered one of the origins of operational uncertainty (Stanek et al., 1981; Funk et al., 1990). To get a uniform liquid distribution without the tedious catalyst dilution procedure, we prefer the concurrent upward flow of the gas-liquid.

To get a steady hydrogen flow rate over a long period, the process is facilitated with a G2V-5/20 diaphragm compressor(20.0MPa and 5Nm³/h) for hydrogen recycling and making up. The reactor, which is made of a stainless steel pipe, has a wall thickness of 2.5mm, is 1.6m in length, and has an inside diameter of 20mm. Twelve thermocouples, 10cm distant from each other, are inserted across the reactor wall and are connected to a data-acquisition computer. Side-stream concentration analysis and quench operation are also facilitated. The reactor wall is wrapped with a 1.5kW electrical heating band and is insulated by a thick layer of glass wool. A flow sheet diagram of the whole process is shown in Fig. 2.

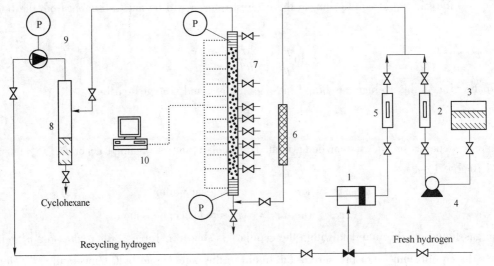

Fig. 2 Experimental flow sheet

1—Compressor;2—Liquid flowmeter;3—Liquid tank;4—Liquid pump;5—Hydrogen flowmeter;
6—Preheater;7—Reactor;8—Gas-liquid separator;9—Back-pressure control valve;
10—Computer;⋈—Valve;Ⓟ—Pressure gauge

Two different catalysts, 0.5% Pd over Al_2O_3 (catalyst A) and amorphous Ni-B alloy catalyst (catalyst B), are employed in the present work. It is assumed that catalyst B is several times more active than catalyst A. The catalysts are 2-3mm in diameter, and are packed to a height of 1.0m between two inert packing layers 30cm on both ends, in order to offer uniform distribution and saturation of the reaction mixture by hydrogen, and eventually the temperature is leveled off to satisfy the Danckwerts' boundary condition. The reactor is designed to operate simultaneously downward and upward for the two fluids. A two-stage gas-liquid separator is installed to accomplish a complete phase separation. The reactor pressure is controlled through a back-pressure regulator installed next to the first-stage separator.

4 Results and Discussion

To conduct the benzene hydrogenation effectively, the choice of catalyst and flow direction should be the fundamental task. A comparison of catalyst activity is shown in Fig. 3, where the reactor temperature increases for catalyst A from 150℃ to 291℃ and to 316℃ for catalyst B. It is also found, under the same operating conditions, the hot spot for catalyst B is about 50cm ahead of that for catalyst A.

The existence of phase transition can be verified in Fig. 3 from the boiling point of benzene/cyclohexane mixture, which is 180℃ at 1.0MPa. The liquid will therefore not exist above this temperature in the bulk phase in the back of the reactor. It should be noted that, due to the capillary effect, the liquid may possibly exist in the catalyst pores. From Table 2, it is expected that the liquid can be stored in the pores at temperatures between 184℃ and 197℃ as the pore diameter decreases from 12nm to 3nm.

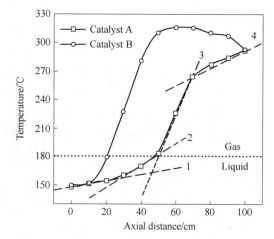

Fig. 3 Effect of catalyst activity on the reactor temperature profile under upward flow
($P = 1.0\text{MPa}; T_{in} = T_w = 150℃; C_B^0 = 25\%; L = 1.24\text{kg/h}; G = 15.5\text{nL/min}; \text{Conversion}: \bigcirc, \square\text{—}100\%$)

A plot of temperature against reactor length in Fig. 3 shows four different slopes. Slope 1 denotes the reaction rate at the reactor inlet under the liquid condition. With some liquid evaporation, the catalyst's external surface will be partially dried and the reaction rate increases to slope 2. Above the boiling point of 180℃, the catalyst becomes partially wetted internally and leads to a stepwise increment in the reaction rate, as denoted by slope 3. Slope 4 corresponds to a reduced reaction rate, since at the end of the reactor, the reactant concentration is much lower, which causes a drop in reaction rate. Different temperature profiles are obtained for catalysts A and B under the same 100% reactant conversion, and this should be attributed to the side reaction of benzene hydrocracking, for providing additional reaction heat. Such an explanation could be verified if carbon deposits were found on the catalysts.

As Fig. 4 shows, a phase-transition diagram is outlined according to the preceding analysis, with five progressive reaction regions classified along the flow direction.

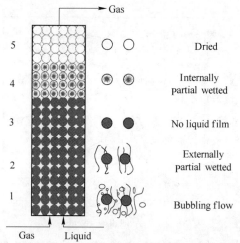

Fig. 4 Phase-transition diagram with vaporization of the liquid component
Catalyst states: 1, 2, 3—fully wetted; 4—partially wetted; 5—dried

The effect of flow direction is shown in Fig. 5. It is likely that the downward flow leads to a reduced utilization of catalysts, since it is seen that the conversion of benzene is even lower than that for catalyst A under the upward flow condition. The reduced conversion under downward flow comes from the lower liquid holdup. The flow rates of liquid and gas under the experimental condition just mentioned ($T = 150\,^\circ\text{C}$ and $P = 1.0\,\text{MPa}$) are $u_L = 1.25 \times 10^{-3}\,\text{m/s}$ and $u_G = 0.127\,\text{m/s}$, which suggests that the flow condition is bubbling flow or trickling flow for the upward and downward flow patterns (Charpentier and Favier, 1975; Shah, 1978; Cheng and Yuan, 1999). The dimensionless liquid holdups are estimated to be 0.33 and 0.25, respectively, according to Turpin and Huntington (1967) and Ellman et al. (1990).

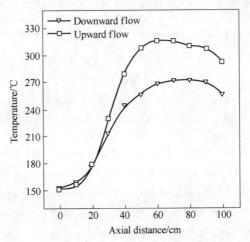

Fig. 5 Effect of flow direction on the reactor behavior for catalyst B
($P = 1.0\,\text{MPa}; T_{in} = T_w = 150\,^\circ\text{C}; C_B^0 = 31.5\%; L = 1.24\,\text{kg/h}; G = 15.5\,\text{nL/min}; \text{Conversion}: \triangledown\text{—}69\%; \square\text{—}100\%$)

4.1 Operations under concurrent upflow

In the following paragraphs, the reactor is operated under upward flow and packed with catalyst B to have a higher catalyst activity and uniform liquid distribution.

No concentration effect is observed in Fig. 6 under liquid-phase condition, although the reactant

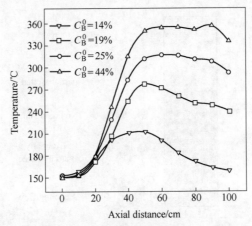

Fig. 6 Effect of reactant concentration on the bed temperature profile
($P = 1.0\,\text{MPa}; G = 15.5\,\text{nL/min}; L = 1.24\,\text{kg/h}; T_{in} = T_w = 150\,^\circ\text{C}; \text{Conversion}: \circ, \square, \triangledown, \triangle\text{—}100\%$)

concentration varies from 14% to 44%. This means that the reaction order to benzene in the liquid phase is almost 0, which could be explained by the large excess of benzene relative to hydrogen in the liquid phase. The concentration effect appears at above 180℃, which indicates that the reaction order to benzene in the gas phase is greater than 0, and this is ascribed to the shortage of benzene compared with hydrogen in this case.

The operation pressure influences both the reaction rate and phase transition. Since the operation pressure represents a total of the hydrogen and benzene, an increase in this variable increases the concentration of the reactants, and thus the reaction rate. In addition, under a fixed mass flow rate, the increase in operation pressure would lead to a longer residence time and higher conversion. These two effects are combined to give different temperature profiles, as is observed in Fig. 7.

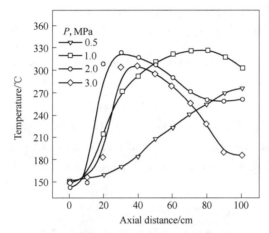

Fig. 7 Effect of operation pressure on phase transition

($C_B^0 = 25\%$, $T_{in} = T_w = 150℃$, $G = 15.5$ nL/min; $L = 12.4$-1.30 kg/h; Conversion: ▽—73.8%; ○,□,◇—100%)

In spite of the similarities in the temperature profiles in the front of the reactor at 1.0MPa, 2.0MPa, and 3.0MPa, a large difference is observed downstream, as is indicated by the significant temperature declines at 2.0MPa and 3.0MPa. The reason for this is that at high pressures, significant condensation of the gas phase, and thus increased heat loss, will occur because of the large thermal conductivity of the liquid phase.

The effect of hydrogen flow rates is shown in Fig. 8. Under a hydrogen flow rate of 5.4nL/min, the conversion is only 56%, which is almost half of the other two cases. The low conversion comes from the insufficiency in hydrogen, since the actual flow rate is less than the stoichiometric value of 5.6nL/min. The reactant conversion under 5.4nL/min indicates that the hydrogen converted in the reaction is only half the stoichiometric value. In this proportion, the complete conversion of benzene requires twice the stoichiometric value for hydrogen, and it was verified by the experiment with $G = 10.5$ nL/min. It seems that, in view of the equilibrium limitation, extra hydrogen is needed to ensure high benzene conversion.

Fig. 9 depicts the effect of the liquid flow rate on the temperature profiles. At a liquid flow rate of 4.74kg/h, the maximum temperature is 180℃, which means that the reaction is carried out under full liquid condition. At a liquid flow rate of 3.24kg/h, a critical temperature profile

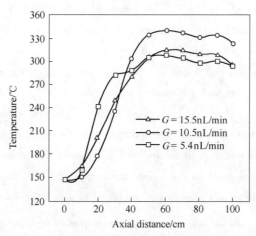

Fig. 8 Effect of hydrogen flow rate on the bed temperature profile
($P = 1.0\text{MPa}$; $C_B^0 = 31.5\%$; $T_{in} = T_w = 150\text{°C}$; $L = 1.24\text{kg/h}$; Conversion: △,○—100%; □—56.5%)

is observed which indicates the transition from liquid-phase operation to gas phase. At liquid flow rates of 2.16kg/h and 1.24kg/h, the operations shift rapidly to gas and exhibit a similar temperature trend. At a liquid flow rate of 2.16kg/h, however, the conversion is only 53.6%, although the temperature profile is similar to that at 1.24kg/h. The reason for this is that benzene hydrogenation is inhibited by equilibrium at high temperatures, and the high temperature region contributes only a little to the conversion of benzene.

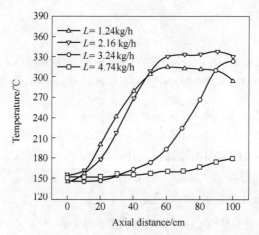

Fig. 9 Effect of liquid flow rate on phase transition
($P = 1.0\text{MPa}$; $C_B^0 = 31.5\%$; $T_{in} = T_w = 150\text{°C}$; $G = 15.5\text{nL/min}$; Conversion:
△—100%; ▽—53.6%; ○—41%; □—14%)

The relationship between the liquid flow rate and conversion is plotted in Fig. 10. It is estimated from the extrapolations that the reaction rate under liquid phase is one half that with a substantial phase transition, which implies the contribution of the gas-phase reaction to the conversion of benzene.

4.2 Optimization with quench operation

To achieve a conversion from benzene to cyclohexane of 99.95%, the temperature should not

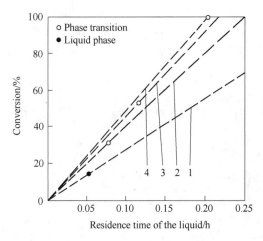

Fig. 10 Effect of phase transition on conversion of the reactant
($P = 1.0\text{MPa}; C_B^0 = 31.5\%; T_{in} = T_w = 150\text{°C}; G = 15.5\text{nL/min}$; Liquid flow
rates(kg/h):1—4.74;2—3.24;3—2.16;4—1.24)

exceed 270 °C, provided the operation pressure is 1.0 MPa. Above this temperature, benzene will be converted to C and CH_4 through hydrocracking, although the conversion still reaches 100%. To control the reaction temperature within 270 °C, it is found from the preceding paragraphs and as depicted in Fig. 6 that the reactant concentration should not exceed 18%. Nevertheless, this would reduce the processing capacity of benzene. It is believed that this problem arises from the low utilization of the reactor space. From Fig. 6 it can be concluded that the reactions only take place in the front of the reactor, with almost a half of the reactor space not used.

To produce a uniform temperature profile over the reactor length, quench operation is employed (Cheng et al., 2000). It is found from Fig. 11 that the operation with two injection points is more flexible than with only one point as shown in Fig. 12. The local maximum temperatures in the operation with two injection points are 226 °C, 269 °C, and 272 °C, all of them approaching the thermodynamic temperature limit of 270 °C. The perfect temperature control has provided

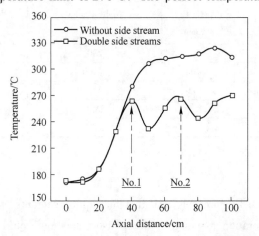

Fig. 11 Side-stream optimization with two injection points
($P = 1.0\text{MPa}; C_B^0 = 31.5\%; T_{in} = T_w = 150\text{°C}; L = 1.24\text{kg/h}; G = 15.5\text{nL/min}$;
○—$L_1 = L_2 = 0, X = 100\%$; □—$L_1 = L_2 = 6\text{mL/min}, X = 100\%$)

promising product quality and benzene conversion. From the GC analysis, it is found that the product purity in cyclohexane is greater than 99.98% and that the conversion of benzene approaches 100%. Additionally, with the temperature control satisfied, no carbon deposits have been found on the catalyst surface.

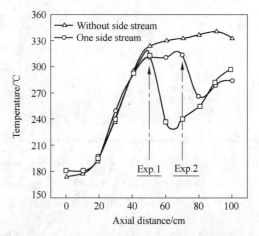

Fig. 12 Side-stream optimization with one injection point
($P = 1.0$ MPa; $C_B^0 = 31.5\%$; $T_{in} = T_w = 150$ ℃; $L = 1.24$ kg/h; $G = 15.5$ nL/min; △—$L_1 = 0$, $X = 100\%$; ○—$L_1 = 6.7$ mL/min, $X = 100\%$; □—$L_1 = 18.4$ mL/min, $X = 82.7\%$)

The quench operation has also made the operation available at the high benzene concentration of 35%. The reduction in reaction rate due to the liquid entering the catalysts is not expected around the injection points, since the local temperature minima are 234 ℃ and 246 ℃, both are greater than the values for vapor condensation, as shown in Table 2. The side-stream operation was proved successful, as can be seen by a comparison with other technologies shown in Table 3 (Cheng et al., 2000).

Table 3 Benzene hydrogenation among different technologies

Parameter	Trickle bed	Magnetically suspended slurry	Phase transition operation	
Catalyst	Ni-Cu-Cr 2-3mm	Ni-Re-P alloy 0.2mm	Pd/Al$_2$O$_3$ 2-3mm	Ni-B alloy 2-3mm
Inlet temp. /℃	150	230	150	150
Hydrogen vs. benzene by mol	>100	>100	6-9	6-9
Liquid space vel. of benzene/h^{-1}	1.62	3.2	1.6	3.4
Benzene in feedstock/vol%	21.5	25.1	16-40	16-40
Benzene in product/vol%	<1	0.73	<0.01	<0.01

5 Conclusions

It may be concluded from the present work that for liquid reactants with substantial volatility such as benzene, an operation associated with phase transition would exhibit multifunctional effects. Advantages due to this new development are summarized as follows:

(1) The energy requirement is reduced, since the reaction heat is to a certain extent counterbalanced by the liquid evaporation.

(2) Phase transition improved the overall reaction rate by 100% in comparison with the full liquid-phase operation.

(3) The reactant concentration can be much higher than in the conventional gas-phase operation and with no risk of hot-spot formation.

(4) The reactor operation is flexible and could be optimized through side-stream injection.

The research work initiated on benzene hydrogenation proved successful in applying the phase transition principle to the intensification of multiphase reactions.

Acknowledgements

The present work is performed under support from the New Materials Center of Shanghai Scientific and Technical Committee and the SINOPEC. The authors are indebted to Professor Jing-Fa Deng of Fundan University for supplying the amorphous alloy catalyst.

Notation

C_B^0	inlet benzene concentration in dilution by cyclohexane, vol%
d_i	catalyst pore diameter, m
G	gas flow rate, nL/min
K_P	equilibrium constant
L	liquid flow rate, kg/h
M	molecular weight
L_1	liquid flow rate of side stream at position 1, kg/h
L_2	liquid flow rate of side stream at position 2, kg/h
P	operation pressure, bar
p^0	bulk phase pressure, bar
p_i^0	vapor pressure of the liquid inside the catalyst, bar
r_1, r_2	principal radii of the gas-liquid interface, m
$\hat{r}_i$	average radius of the gas-liquid interface, m
T	temperature, ℃
T^0	bulk-phase temperature, ℃
T_i^c	condensation temperature of vapor inside of the catalyst, ℃
T_i^e	evaporation temperature of liquid inside of the catalyst, ℃
T_{in}	inlet temperature of the reactant mixture, ℃
T_w	reactor wall temperature, ℃
u_G	linear velocity of the gas, m/s
u_L	linear velocity of the liquid, m/s
X	conversion of the reactant

Greek letters

σ	surface tension, N/m
ρ	density of the liquid, kg/m³

References

Bhatia, S. K., "Steady State Multiplicity and Partial Internal Wetting of Catalyst Particles," AIChE J., 34, 969 (1988).

Bhatia, S. K., "Partial Internal Wetting of Catalyst Particles with a Distribution of Pore Size," AIChE J., 35, 1337(1989).

Charpentier, J. C., and M. Favier, "Some Liquid Holdup Experimental Data in Trickle-Bed Reactors for Foaming and Nonfoaming Hydrocarbons," AIChE J., 21, 1213(1975).

Cheng, Z. M., and W. K. Yuan, "Necessary Condition for Pulsing Flow Inception in a Trickle Bed," AIChE J., 45, 1394(1999).

Cheng, Z. M., W. K. Yuan, A. M. Anter, and J. S. Hu, "A New Technology for Benzene Hydrogenation into Cyclohexane via Phase Transition," Chinese Patent Application, No. 00111762.9(March 2, 2000).

Eigenberger, G., and U. Wegerle, "Runaway in an Industrial Hydrogenation Reactor," Proc. Int. Symp. on Chemical Reaction Engineering, Boston(1982).

Ellman, M. J., N. Midoux, G. Wild, A. Laurent, and J. C. Charpentier, "A New, Improved Liquid Hold-Up Correlation for Trickle-Bed Reactors," Chem. Eng. Sci., 45, 1677(1990).

Funk, G. A., M. P. Harold, and K. M. Ng, "A Novel Model for Reaction in Trickle Beds with Flow Maldistribution," Ind. Eng. Chem. Res., 29, 738(1990).

Hanika, J., K. Sporka, V. Ruzicka, and J. Krausova, "Qualitative Observations of Heat and Mass Transfer Effects on the Behaviour of a Trickle Bed Reactor," Chem. Eng. Commun., 2, 19(1975).

Hanika, J., K. Sporka, V. Ruzicka, and J. Hrstka, "Measurement of Axial Temperature Profiles in an Adiabatic Trickle Bed Reactor," Chem. Eng. J., 12, 193(1976).

Hanika, J., K. Sporka, V. Ruzicka, and R. Pistek, "Dynamic Behavior of an Adiabatic Trickle Bed Reactor," Chem. Eng. Sci., 32, 525(1977).

Hanika, J., B. N. Lukjanov, V. A. Kirillov, and V. Stanek, "Hydrogenation of 1,5-Cyclooctadiene in a Trickle Bed Reactor Accompanied by Phase Transition," Chem. Eng. Commun., 40, 183(1986).

Hanika, J., "Safe Operation and Control of Trickle-Bed Reactor," Chem. Eng. Sci., 54, 4653(1999).

Hu, R., and T. C. Ho, "Steady State Multiplicity in an Incompletely Wetted Catalyst Particle," Chem. Eng. Sci., 23, 1239(1987).

Jaguste, D. N., and S. K. Bhatia, "Partial Internal Wetting of Catalyst Particles: Hysteresis Effects," AIChE J., 37, 650(1991).

Khadilkar, M. R., P. L. Mills, and M. P. Dudukovic, "Trickle-Bed Reactor Models for Systems with a Volatile Liquid Phase," Chem. Eng. Sci., 54, 2421(1999).

Kheshgi, H. S., S. C. Reyes, R. Hu, and T. C. Ho, "Phase Transition and Steady-State Multiplicity in a Trickle-Bed Reactor," Chem. Eng. Sci., 47, 1771(1992).

Kim, D. H., and Y. G. Kim, "An Experimental Study of Multiple Steady States in a Porous Catalyst due to Phase Transition," J. Chem. Eng. Jpn., 14, 311(1981a).

Kim, D. H., and Y. G. Kim, "Simulation of Multiple Steady States in a Porous Catalyst due to Phase Transition," J. Chem. Eng. Jpn., 14, 318(1981b).

Lange, R., J. Hanika, D. Stradiotto, R. R. Hudgins, and P. L. Silveston, "Investigations of Periodically Operated Trickle-Bed Reactors," Chem. Eng. Sci., 49, 5615(1994).

Ng, K. M., "A Model for Flow Regime Transitions in Concurrent Down-Flow Trickle-Bed Reactors," AIChE J., 32, 115(1986).

Satterfield, C. N., and F. Ozel, "Direct Solid-Catalyzed Reaction of a Vapor in an Apparently Completely Wetted Trickle Bed Reactor," AIChE J., 19(6), 1259(1973).

Sedriks, W., and C. N. Kenney, "Partial Wetting in Trickle Bed Reactors—The Reduction of Crotonaldehyde over a Palladium Catalyst," Chem. Eng. Sci., 28, 559(1973).

Shah, Y. T., Gas-Liquid-Solid Reactor Design, McGraw-Hill, New York(1979).

Stanek, V., J. Hanika, V. Hlavacek, and O. Trnka, "The Effect of Liquid Flow Distribution on the Behavior of a Trickle Bed Reactor," Chem. Eng. Sci., 36, 1045(1981).

Turpin, J. L., and R. L. Huntington, "Prediction of Pressure Drop for Two-Phase, Two-Component Concurrent Flow in Packed Beds," AIChE J., 13, 1196(1967).

Watson, P. C., and M. P. Harold, "Dynamic Effects of Vaporization with Exothermic Reaction in a Porous Catalyst Pellet," AIChE J., 39, 989(1993).

Watson, P. C., and M. P. Harold, "Rate Enhancement and Multiplicity in a Partially Wetted and Filled Pellet: Experimental Study," AIChE J., 40, 97(1994).

Controlling Sandwich-structure of PET Microcellular Foams Using Coupling of CO_2 Diffusion and Induced Crystallization*

Abstract Controlling sandwich-structure of poly(ethylene terephthalate) (PET) microcellular foams using coupling of CO_2 diffusion and CO_2-induced crystallization is presented in this article. The intrinsic kinetics of CO_2-induced crystallization of amorphous PET at 25℃ and different CO_2 pressures were detected using *in situ* high-pressure Fourier transform infrared spectroscopy and correlated by Avrami equation. Sorption of CO_2 in PET was measured using magnetic suspension balance and the diffusivity determined by Fick's second law. A model coupling CO_2 diffusion in and CO_2-induced crystallization of PET was proposed to calculate the CO_2 concentration as well as crystallinity distributions in PET sheet at different saturation times. It was revealed that a sandwich crystallization structure could be built in PET sheet, based on which a solid-state foaming process was used to manipulate the sandwich-structure of PET microcellular foams with two microcellular or even ultra-microcellular foamed crystalline layers outside and a microcellular foamed amorphous layer inside.

Key words PET microcellular foams, CO_2 diffusion, CO_2-induced crystallization, coupling, sandwich-structure

1 Introduction

Semicrystalline poly(ethylene terephthalate) (PET) is a low-cost engineering thermoplastics with good mechanical and thermal characteristics such as high elastic modulus, relatively high glass transition temperature (T_g) and good solvent resistance[1]. Besides for the production of fibers, films, trays, and bottles, PET was also applied for mechanical components and in some cases for replacement of commodity metals such as steel and aluminum[2]. In response to an industry challenge to reduce the amount of material used in plastic productions without sacrificing desirable mechanical properties, Suh et al.[3,4] proposed the concept of microcellular foam. The microcellular foams are defined as foams with cell size smaller than 10μm. Unfortunately, the conventional melt process for producing microcellular foams, such as microcellular injection molding and extrusion foaming, are inadaptable for fabrication of PET foams from ordinary semi-crystalline PET resins due to their low melt strength at processing temperatures. That is why in the past two decades, researches about preparation of microcellular foams focused mainly on amorphous polymers such as polystyrene[5,6], poly(methyl methacrylate)[7], polysulfone[8,9], and poly(ether imide)[10,11]. Only a few studies have been conducted on the microcellular foaming of PET, and most of them are focused on the solid-state foaming, i.e., the foaming temperature is lower than the melting temperature of specific PET resin[12-21].

As carbon dioxide (CO_2) has many unique properties such as nonflammable, nontoxic, rela-

---

* Coauthors: Li Dachao, Liu Tao, Zhao Ling. Reprinted from *AIChE Journal*, 2012, 58(8): 2512-2523.

tively inexpensive, easy to reach a supercritical state (critical temperature, 31.1℃, and critical pressure, 7.38MPa), and relatively large solubility in polymers[22,23], it has been used as a popular physical foaming agent in many foaming applications[14,23-27]. Dissolving CO_2 into polymers will affect their properties in both melt and solid states due to the so-called plasticization effect. It depresses the glass-transition temperature[28-30] and the crystallization temperature[31], and changes the crystallization kinetics of several semicrystalline polymers[31-33]. Several studies of CO_2-induced crystallization have been reported for poly(aryl ether ether ketone), polycarbonate, poly(phenylene sulfide) and PET as well. Mizoguchi et al.[33] and Lambert and Paulaitis[32] investigated the CO_2-induced crystallization of amorphous PET at 35-80℃ and 4-6MPa, and compared it with thermal crystallization. They found that the crystallization took place even at temperatures below T_g measured in air due to the sorption of CO_2, and the crystallization rate at temperatures above T_g was greatly increased. However, the thermal properties from these studies were not *in situ* ones under high-pressure CO_2. The method adopted in these studies was to subject the polymer to a delay time between thermal characterization and pressurization, during which the polymer specimen was first enclosed in a high-pressure CO_2 chamber for a period of time to reach sorption equilibrium. Then, the specimen was taken out for further characterization after the CO_2 was released. Zhang and Handa[30] presented the *in situ* studies of PET thermal transitions under high-pressure CO_2 using high-pressure differential scanning calorimeter(DSC) and found that the absorbed CO_2 enhanced the mobility of the chain segments, depressed the crystallization temperature, and caused a lower T_g than room-temperature. However, it has been still impossible to study the kinetics of CO_2-induced crystallization of PET using this method due to the relatively slow crystallization rate of PET and poor signal-to-noise ratio of DSC in high-pressure CO_2 environment. In fact, the basic natures of CO_2-induced crystallization of semicrystalline PET at room-temperature, are far more complicated than we consider[34] because of the coupling effect of CO_2 diffusion and induced crystallization[32]. On the one hand, the presence of compressed CO_2 in PET matrix can enhance or increase the rate of crystallization. On the other hand, the rate of gas sorption in polymers can be coupled to additional kinetic phenomena, including the rates and extent of polymer swelling[22], stress relaxation, and crystallization as well. The extent of crystallization in PET will affect gas permeability by reducing both the equilibrium solubility[13,32] and the diffusivity of gas in the polymer[13,35]. Hence, the kinetics of CO_2 sorption and induced crystallization are coupled with each other[36].

The solid-state batch foaming process for producing PET microcellular foams was conducted by Baldwin et al.[13-15] The PET specimen was placed in a high-pressure vessel charged with CO_2, to a constant saturation pressure at room-temperature. Once the required saturation time was reached, the vessel was discharged. Then, the PET specimens were foamed, unconstrained, in a glycerin bath with certain temperature for the desired foaming time, and the specimens were quenched in a water bath to vitrify the microcellular structure. The cell morphology of the PET foams was found to be relevant to four major processing variables, i.e., gas saturation time, gas saturation pressure, foaming time, and temperature[13]. Crystallinity was found[13,37] to play a major role in the microcellular processing on (1) cell nucleation mechanisms resulting in larger cell densities due to heterogeneous nucleation at the amorphous/crystalline boundaries and (2) cell growth mechanisms resulting in smaller cell sizes due to the increased stiffness of the semi-crystalline matrix. Kumar[12] used the solid-state foaming process combining with the CO_2-induced crystallization to produce PET foams with integral crystalline skin, i. e., a foamed core layer with unfoamed crystalline skins. This structure exhibited enhanced physical properties compared to conventional PET foams. The largest drawback of this batch foaming method was time consuming due to the long saturation time. Kumar et al.[16,17] successfully converted this batch process into a semicontinuous process which allowed essentially continuous production of microcellular foams. PET foams, nowadays, have been successfully commercialized and used in applications such as packaging, thermal insulation, optical reflection, and preferable sandwich materials in wind energy, marine and transportation[12,38-41].

In this work, we decoupled the complicated relationships between CO_2 diffusion in and CO_2-induced crystallization of amorphous PET. The intrinsic kinetics of CO_2-induced crystallization of amorphous thin PET films ((15 ± 2) μm) were detected at 25℃ and different CO_2 pressures using *in situ* high-pressure Fourier transform infrared spectroscopy (FTIR). Magnetic suspension balance (MSB) was used to determine the solubility and diffusivity of CO_2 in PET matrix. A model coupling the CO_2 diffusion and induced crystallization was subsequently proposed to correlate and predict the CO_2 concentration distribution as well as the crystallinity distribution in the PET matrix at different saturation time. On the basis of the modeling results, a solid-state foaming process was used to manipulate sandwich-structure of PET microcellular foams with two microcellular or even ultra-microcellular foamed crystalline layers outside and a microcellular foamed amorphous layer inside. The thickness of the foamed crystalline layer agreed well with that calculated by the model.

2 Experimental

2.1 Materials

PET(BRH-400) with intrinsic viscosity of 1.0dL/g (corresponding viscosity average molecular weight $\overline{M}_v = 42000$g/mol) was kindly provided by Shanghai Petrochemical Co. The crystallinity and melting temperature of the PET were 32.8% and 255℃, which were determined by DSC (NETZSCH DSC 204 HP, Germany) under atmospheric N_2. Before being used, the PET pellets were dried in a vacuum oven at 80℃ for 8h to eliminate moisture. CO_2 (purity: 99.9%, w/w) was purchased from Air Products Co. (Shanghai, China).

2.2 *In situ* high-pressure FTIR

The amorphous PET films were prepared from pellets using a hot press at 280℃ and 10MPa for 5min and rapidly quenched by plunging into cold water. The film thickness measured by micrometer caliper was (15 ± 2) μm. Compared with crystallization time, the CO_2 diffusion time (or saturation time) in such thin film could be negligible. Therefore, it was assumed that the film was saturated as soon as the high-pressure CO_2 was applied, and the obtained crystallization kinetics should be intrinsic kinetics. The intrinsic kinetics of the CO_2-induced crystallization of the amorphous PET films at 25℃ and high CO_2 pressures was investigated using *in situ* FTIR of type Bruker Equinox-55 equipped with a Harrick high-pressure demountable cell, the details of which had been described elsewhere[42]. The FTIR spectra were recorded at a resolution of 2.0cm^{-1} and a rate of one spectrum per 10min. The IR intensities refer to the peak height. The scanned wave number was in the range of 4000-400cm^{-1}.

2.3 Magnetic suspension balance

Melting of PET was performed in a Haake Minilab system (Thermo Electron Co.) under 0.6MPa nitrogen atmosphere at a temperature of 280℃ and a screw speed of 50 rotations/min. The melts were then delivered into a Haake Minijet system (Thermo Electron Co.) and molded into PET sheets with geometry of 30mm × 15mm × 1.2mm at a pressure of 650bar and mold temperature of 50℃. The molded amorphous PET sheets were used for the solubility and diffusivity measurements and foaming experiments.

The solubility and diffusivity of CO_2 in PET was measured using MSB (Rubotherm Prazisionsmesstechnik GmbH, Germany). The MSB has an electronically controlled magnetic suspension coupling that transmits the weight of the sample in a pressure vessel to a microbalance outside of the cell. The MSB can be used at pressures up to 35MPa and temperatures up to 523K. Resolution and accuracy of the microbalance (Mettler AT261, Switzerland) are 0.01mg and 0.002%, respectively. The system temperature and pressure was controlled at the accuracy of ±0.2℃ and ±0.05MPa, respectively. Density of carbon dioxide needed for buoyancy correction was measured simultaneously by MSB. The most important advantage of MSB method was that it could accurately detect the mass variation of polymer sample during gas sorption process. The original sample volume was determined by a blank test with Helium and used to correct the gas solubility by considering gas buoyancy acting on the polymer. Details of the MSB apparatus and experimental procedure used in this work have been described in previous publications[43-46].

2.4 Foaming process

A so-called temperature rising foaming process was used to fabricate PET microcellular foams. The amorphous PET sheets were placed in a high-pressure vessel. The latter was then sealed, carefully washed with low-pressure CO_2 and charged with 6MPa CO_2. Thereafter, the vessel was immersed into a water bath with a constant temperature of 25℃. Different saturation time ranging from a few hours to as long as 15 days was applied to the PET sheets. After that, the PET sample was taken out and immediately immersed, without constraint, into a high-temperature silicone oil bath with a temperature of 235℃ to induce bubble nucleation and growth. After a foaming time of 10s, the samples were quenched in an ice-water bath to vitrify the foam structure.

2.5 Foam characterization

The cell morphologies of the PET foams were characterized by a JSM-6360LV (JEOL-Tokyo, Japan) scanning electron microscopy (SEM). The samples were immersed in liquid nitrogen for 10min and then fractured. The SEM scanned fractured surfaces with Pd (palladium) coating. The average cell size was obtained through the analysis of the SEM photographs by the software of Image-Pro Plus (Media Cybernetics, Silver Spring, MD). The number average diameter of all the cells in the micrograph, d, was calculated using the following equation:

$$d = \frac{\sum d_i n_i}{\sum n_i} \tag{1}$$

where, n_i is the number of cells with a perimeter-equivalent diameter of d_i. The mass densities of foamed PET samples ρ_f were measured according to ASTM D792-00, involving weighing polymer foam in was using a sinker. ρ_f was calculated as follows:

$$\rho_f = \frac{a}{a + w - b} \rho_{water} \tag{2}$$

where, a is the apparent mass of specimen in air without sinker; b is the apparent mass of specimen and sinker completely immersed in water; w is the apparent mass of the totally immersed sinker. The volume expansion ratio of the PET foams, R_v, is defined as the ratio of the bulk density of the virgin PET(ρ_0) to that of the foamed one(ρ_f):

$$R_v = \frac{\rho_0}{\rho_f} \tag{3}$$

The cell density (N_0), the number of cells per cubic centimeter of foamed PET was determined from the equation:

$$N_0 = \left(\frac{nM^2}{A}\right)^{3/2} \tag{4}$$

where, n is the number of cells in the SEM micrograph; M is the magnification factor; A is the area of the micrograph (in cm^2).

3 Results and Discussion

3.1 Intrinsic kinetics of CO_2-induced PET crystallization

A number of studies have been done on the IR spectroscopy characterization of contributing conformers of PET chains[47-49]. The —O—CH_2—CH_2—O— moiety of a PET chain shows gauche and trans conformers through the internal rotation of the COC bond. In the crystalline phase, the —O—CH_2—CH_2—O— moiety adopts a trans conformation. Whereas in the amorphous phase, it is mainly in the gauche conformation with some small contribution of the trans conformation[49]. Therefore, the crystallinity of PET can be estimated. In the IR spectrum of PET, the 1340cm^{-1} and 1370cm^{-1} band are assigned to the CH_2 wagging mode in the trans and gauche conformers, respectively. As the intensities of these two bands are comparable in the spectrum of amorphous PET, we chose these two bands as the key bands for determining the relative conformational population. The crystallinity, I, of PET can be estimated as follows:

$$I = A_{1340}/(A_{1340} + 6.6 \times A_{1370}) \tag{5}$$

where, A_{1340} and A_{1370} were the integral absorbance of the 1340cm^{-1} and 1370cm^{-1} bands, respectively. The factor 6.6, describing the absorption coefficient ratio of 1340cm^{-1} and 1370cm^{-1} [50], had been obtained from the slope of a plot of the integral absorbance of the

1340cm^{-1} band vs. that of the 1370cm^{-1} band according to the spectra of PET samples with different crystallinity. This method had been used in the crystallinity determination of bulk PET[40] and PET thin film[49].

The *in situ* high-pressure FTIR spectra of the PET film during isothermal crystallization induced by CO_2 at 25℃ and different pressures are shown in Fig. 1. As the crystallization time was prolonged, the absorbance of the 1340cm^{-1} band grew, whereas the absorbance reduction of the 1370cm^{-1} band was relatively small. This was attributed to the difference in the absorption coefficients of these two bands. Fig. 2 shows the variation of the crystalliniy I of the PET films with the crystallization time under different pressure CO_2 during isothermal crystallization at 25℃. It was shown that no crystallinity increase could be detected under 4.5MPa CO_2 even after 2000min, while a very long crystallization period of more than 1500min was observed under 5.0MPa CO_2. These results indicated that the least critical pressure, corresponding to the least critical CO_2 concentration in the PET, for CO_2-induced crystallization of PET was 5.0MPa at the temperature of 25℃. The crystallization rate of PET film increased rapidly with increasing saturation CO_2 pressure, and the final crystallinity was also a little bit higher at the higher saturation CO_2 pressure.

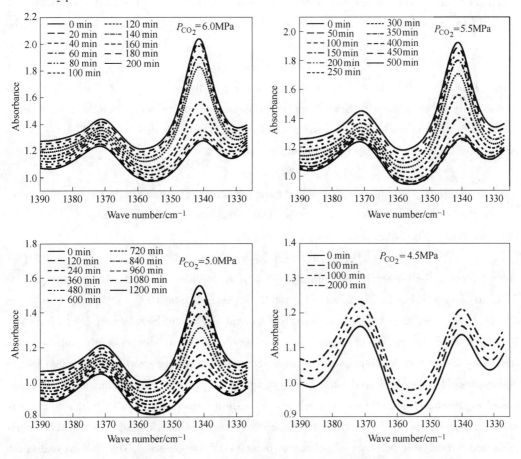

Fig. 1 High-pressure FTIR spectra in the 1320-1390cm^{-1} region of the PET film at different crystallization times during isothermal crystallization at 25℃ and different CO_2 pressures

(Color figure can be viewed in the online issue, which is available at wileyonlinelibrary.com.)

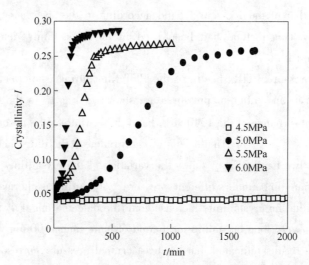

Fig. 2 Changes in crystallinity of PET films during isothermal crystallization at 25℃ and different CO_2 pressures

The well-known Avrami equation[51-53], which provided an insight into the progress of nucleation and crystal growth that occurred during isothermal crystallization, was used to study the isothermal CO_2-induced crystallization kinetics of PET films[49,54-56]. The Avrami equation is in the form as:

$$1 - X_t/X_\infty = \exp(-Kt^n) \qquad (6)$$

where, X_t is the crystallinity of the sample at time t; X_∞ is the crystallinity at which a further increase in the crystallinity with time is imperceptible; K is a constant that includes the rate constants of growth and nucleation; n is the Avrami exponent with a value between 1 and 4. We adopted the Avrami equation to fit our results as follows:

$$X_t/X_\infty = (I_t - I_0)/(I_\infty - I_0) \qquad (7)$$

where, I_0, I_t, and I_∞ are the fractions of the trans conformers at the beginning of crystallization, at time t, and at later periods when a further increase is imperceptible, respectively.

Taking double logarithms, Eq. 6 then becomes:

$$\ln[-\ln(1-X_t)] = n\ln t + \ln K \qquad (8)$$

Avrami plots of the isothermal crystallization of PET films induced by different pressure CO_2 at 25℃ are shown in Fig. 3. The relatively crystallinity ranging from 10% to 95%, i.e., $\ln[-\ln(1-X_t)]$ from -2.2 to 1.1, was selected for the analysis. When the relatively crystallinity was lower than 95%, the growth of spherulites could be considered as independently and there was barely any interaction or contact between adjacent spherulite during growth. In this case, Avrami equation could perfectly describe the growth of spherulite and the linear behavior could be obtained as shown in Fig. 3. When the relatively crystallinity was larger than 95%, due to the limitation of growing space, the contact and interaction between adjacent spherulite would be inevitable and a nonlinear behavior or inflection point would be observed. The half-crystallization time, $t_{1/2}$, defined as the time at which the relative crystallinity is 50wt%, can be determined either from the experiment directly or from the obtained crystallization kinetic parameters as follows:

$$t_{1/2} = \left(\frac{\ln 2}{K}\right)^{1/n} \tag{9}$$

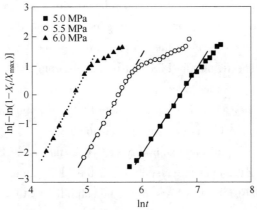

Fig. 3 Avrami plots of the isothermal crystallization of PET films induced by different pressure CO_2 at 25℃

The calculated Avrami exponent, n, and crystallization rate constant, K, from Eq. 8, as well as both calculated and experimental half-crystallization time, $t_{1/2}$, at different CO_2 pressures were given in Table 1. The physical interpretation of the Avrami exponent, n, is the dimension of crystal growth. For homogeneous nucleation, like in our case, the value of n equal to 2, 3, or 4 represents the one-, two-, or three-dimension (3-D) growth of crystals, respectively. As the saturation pressure varied from 5 to 6 MPa, the value of n increased from 2.94 to 3.71, which suggested that the way of crystal growth converted from 2-D dominated to 3-D dominated with increasing CO_2 pressure. This observation was reasonable because the molecular chain of the PET could have more mobility under higher-pressure CO_2 due to the plasticization effect. The crystallization rate constant, K, at 6 MPa was almost one magnitude higher that of 5 MPa, which significantly reduced the half crystallization time from 720 to 109 min. The consistency between calculated and experimental half-crystallization time suggested that the Avrami equation could well describe the CO_2-induced crystallization behavior of PET.

Table 1 n and K values, and half crystallization time of the PET film at different CO_2 pressures

T/℃	P/MPa	n	K	$t_{1/2}$(Experimental)/min	$t_{1/2}$(Calculated)/min
25	5.0	2.94	2.81×10^{-9}	720	715
	5.5	3.32	8.96×10^{-9}	238	246
	6.0	3.71	2.07×10^{-8}	109	107

3.2 Solubility and diffusivity of CO_2 in the PET at 25℃

The MSB method was adopted to investigate the solubility and diffusivity of CO_2 in the PET sheet at 25℃ and 6 MPa. Fig. 4a shows the CO_2 mass uptake in the PET sheet at 25℃ and 6 MPa after buoyancy correction of MSB method. Details about buoyancy correction of MSB method had been described elsewhere[43-46]. As CO_2 diffused into the PET matrix, distribution of CO_2 concentration was founded and varied with saturation time. The CO_2-induced crystallization took place where the CO_2 concentration was high enough. Therefore, the layers near the sheet surface crystallized earlier. The formation of crystals would subsequently reject CO_2 and decrease the CO_2 content in the PET matrix. Thus, there were two main factors affecting the solubility: one was the CO_2 diffused in the PET matrix before the equilibrium state reached, which would increase the CO_2 content; the other was CO_2-induced crystallization, which would reject CO_2. Before the sorption curve reached the "peak" value of the knee, the sorption speed was larger than the rejection speed, and the PET matrix was absorbing CO_2 as a whole. As the sorption curve passed through the knee (i. e., peak value), the sorption speed was passed over by the rejection speed, and the CO_2 content in the whole PET matrix began to decrease and finally reached an equilibrium value when crystallization completed.

As shown in Fig. 4a, the equilibrium solubility of CO_2 in the whole PET sheet, S, was determined to be 78.7g/kg PET, while the crystallinity of the PET sample after MSB measurement was determined to be 28.6% by DSC. Assuming the solubility of CO_2 in crystalline regions of polymer is zero, which has been widely accepted[22,44], the saturation concentration, C_0, of CO_2 in amorphous regions of the PET at 25℃ and 6MPa can be determined to be 11.0wt% with the following equation:

$$C_0 = \frac{S/1000}{1 - X_\infty} \times 100\% \qquad (10)$$

where, S is the equilibrium solubility of CO_2 in the whole PET specimen at 25℃ and 6MPa; X_∞ is crystallinity of the PET sample after MSB measurement.

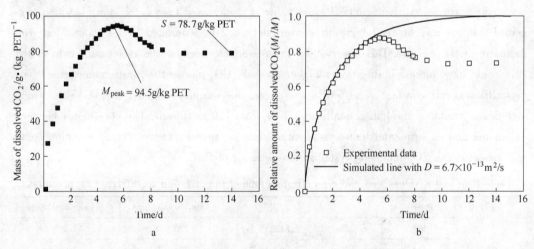

Fig. 4 Sorption profiles for CO_2 in PET at 25℃ and 6MPa

The least critical CO_2 concentration, $C_{critical}$, i. e., equilibrium concentration of CO_2 in amorphous PET under 5MPa CO_2, for induced crystallization of PET at 25℃ was estimated to be 9.17wt% assuming the sorption of CO_2 could be expressed by the Henry's law at relatively low CO_2 pressure.

Assuming that the diffusion could be expressed by Fick's second law:

$$\frac{\partial C}{\partial t} = D \frac{\partial^2 C}{\partial x^2} \qquad (11)$$

Because the other 2-D of the PET sheet were more than 10 times larger than the thickness, CO_2 entered effectively through the plane faces and a negligible amount through the edges. Thus, the diffusion process could approximately be treated as 1-D diffusion in an infinite sheet with constant surrounding concentration, i.e., constant surface concentration. The correspondingly initial and boundary conditions in this case were:

$$C = 0, \text{at} -l < x < l, t = 0$$
$$C = C_0, \text{at } x = l, t \geq 0$$
$$C = C_0, \text{at } x = -l, t \geq 0$$

where, C is the concentration of CO_2 in PET; $2l$ is the thickness of PET sheets; D is the diffusion coefficient of CO_2 in the polymer, which was treated as being independent on the gas concentration during gas dissolution. The appropriate solution of the diffusion equation had been given by Crank[57]:

$$\frac{M_t}{M_\infty} = 1 - \sum_{n=0}^{\infty} \frac{8}{(2n+1)^2 \pi^2} \exp\left[\frac{-D(2n+1)^2 \pi^2 t}{4l^2}\right] \quad (12)$$

where, M_t and M_∞ are the mass of dissolved gas in the polymer at $t = t$ and $t = \infty$, respectively.

In fact, the obtained "knee-like" sorption profile was not a typical Fick diffusion. Thus, Eq. 12 could not be used to describe the whole sorption process. As stated above, this sorption process could be divided into sorption dominated step and crystallization dominated step. Before the sorption curve reached the peak value of the knee, the diffusion process was dominated by the sorption of CO_2 into PET matrix. Thus, the sorption data from $t = 0$ to $t = t_{peak}$, the time sorption reached the peak value, was adopted to correlate the diffusion coefficient with Eq. 12. M_∞ in Eq. 12 was determined by the solubility of CO_2 in amorphous PET at this condition, i.e., C_0. As shown in Fig. 4b, the obtained diffusion coefficient at the experimental condition was $6.7 \times 10^{-13} \text{m}^2/\text{s}$ with an average relative deviation of 1.8%. Note that this value was six to eight times larger than those typically reported for the diffusivity of lower-pressure (no more than 1MPa) CO_2 in PET matrix[58,59]. This was because the T_g of PET under low-pressure CO_2 was still higher than the experimental temperature, i.e., 25℃, the molecular chain of PET could not move and the specimen could not be swollen by CO_2. While in our case, the T_g of PET under 6MPa CO_2 was lower than 25℃, otherwise the induced crystallization could not happen, and the PET sample could be swollen by CO_2. It had been reported that the dissolution of CO_2 in polymer matrix would induce the polymer swelling and increase the free volume so as to lead a dramatically increase in diffusion coefficient[13,22,43].

3.3 Modeling of the coupling of CO_2 diffusion in and induced crystallization of PET

What we concerned about was: after a certain saturation time, the CO_2 concentration and crystallinity distributions in the PET sheet. To investigate the coupling process of CO_2 diffusion in and induced crystallization of the PET sheet, a model was proposed as shown in Fig. 5. Assuming that the PET sheet was consisted of N well-contacted layers in the thickness direction, the thickness of each layer was equal and so small that the CO_2 concentration distribution was uniform in each individual layer. A few assumptions were further made:

(1) The diffusion of CO_2 in the PET sheet could be expressed by Fick's second law;
(2) The CO_2 concentration distribution was uniform in each individual thin layer;
(3) The CO_2 concentration and the rate of CO_2 induced crystallization in an individual thin

layer were constant in a short time (e.g., 1 min);

(4) Crystallinity of PET sheet had no effect on the diffusion coefficient.

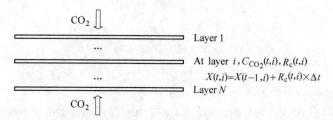

Fig. 5 Schematic diagram of the proposed model

(Color figure can be viewed in the online issue, which is available at wileyonlinelibrary.com.)

In such case, the concentration of CO_2 in each layer could be determined by the following equation[57]:

$$\frac{C(t,i)}{C_0} = 1 - \frac{4}{\pi}\sum_{n=0}^{\infty}\frac{(-1)^n}{2n+1}\exp\left[\frac{-D(2n+1)^2\pi^2 t}{4l^2}\right]\times \cos\frac{(2n+1)\pi x(i)}{2l} \quad (13)$$

where, $C(t,i)$ is the concentration of CO_2 at saturation time t in layer i; $x(i)$ is the location of layer i in the thickness direction of the PET sheet; D is the diffusion coefficient, which had been determined to be $6.7\times 10^{-13}\,\mathrm{m^2/s}$.

The crystallization rate, R_c, in each individual layer could be determined by the differential form of Eq. 6:

$$R_c = \frac{\mathrm{d}X_t}{\mathrm{d}t} = Kt^{n-1}\exp(-Kt^n) \quad (14)$$

where, n and K are the Avrami exponent and crystallization rate constant, respectively. The flow chart of the simulation process was shown in Fig. 6.

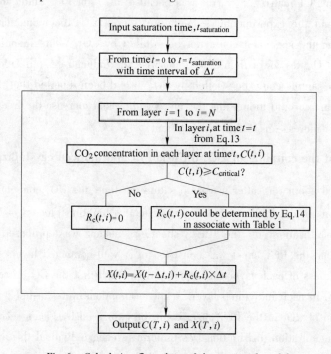

Fig. 6 Calculation flow chart of the proposed model

Evolution of CO_2 concentration distribution and CO_2-induced crystallinity distribution against saturation time were shown in Fig. 7 and Fig. 8, respectively. Because of the relatively small diffusion coefficient of CO_2 in the PET sheet at the experimental condition, the saturation process should take as long as 15 days to approach the equilibrium state (99% saturation). CO_2-induced crystallization could only occur at those positions where the CO_2 concentration was higher than $C_{critical}$, i.e., $C/C_0 \geq C_{critical}/C_0 = 0.833$. While after CO_2 concentration exceeded $C_{critical}$, it would still take about 1 day to complete the crystallization process, which was determined by the crystallization kinetics discussed above. Thus, in the first few days of saturation, the evolution of CO_2-induced crystallization was pretty slow. For example, after 4 days' saturation, the thickness of crystallized layer was still no more than 0.15mm. As shown in Fig. 7, the CO_2 concentration was higher than $C_{critical}$ in most part of the PET sheet after 5 days' saturation, and higher than $C_{critical}$ in the whole PET after 6 days. Corresponding evolution rate of crystallinity from day 5 to 6, as shown in Fig. 8, was much faster than that in the first 4 days. The crystallization process in the whole PET sheet completed after 7 days' saturation.

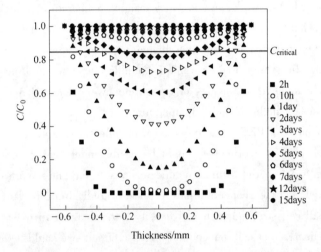

Fig. 7 CO_2 concentration distribution in the PET sheet at different saturation time under 6MPa CO_2

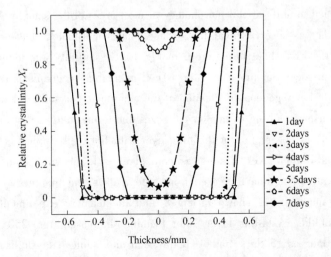

Fig. 8   Crystallinity distribution in the PET sheet at different saturation time under 6MPa CO_2

From the crystallinity distribution profile after different saturation time, as shown in Fig. 8, it was interesting to find that a "sandwich" crystallization structure could be built after an appropriate saturation time, e. g. ,4 or 5 days. This structure was resulted from the rate difference between CO_2 diffusion and induced crystallization, i. e. , the coupling effect of them. Because of the different foaming behaviors of amorphous PET and crystalline PET[13-15,18], it was highly possible to control a sandwich foaming morphology using the obtained sandwich crystallization structure.

3.4 Controlling of sandwich-structure of PET microcellular foams

A so-called temperature rising foaming process was conducted to fabricate PET microcellular foams with sandwich structure using the coupling effect of CO_2 diffusion and induced crystallization of PET. Kumar et al. [17] used the "temperature rising" foaming process to produce high relative density PET microcellular foams using CO_2 as a blowing agent. Their samples were saturated at room-temperature and elevated CO_2 pressures, and then foamed at temperatures ranging from 50 to 90 ℃. In this work, we adopted a foaming temperature as high as 235 ℃, which was just 20 ℃ lower than the melting temperature of the PET. There are two reasons for choosing this high foaming temperature: one was that higher foaming temperature would create larger thermodynamic instability, which was favorable for bubble nucleation and the other was that the difference between foaming morphologies of crystalline layer PET and amorphous layer PET could be more obvious at higher foaming temperature.

The overall morphology of PET foams prepared with different saturation times were shown in Fig. 9. The evolution of foam morphology could be easily understood if considered combining with Fig. 7 and Fig. 8. When the saturation time was just 1 day, most parts of the PET sheet were still unsaturated, and CO_2 concentration decreased rapidly from the surface to the central area. Relatively small cell size and high cell density were obtained in the areas near the surface, while a sharp decrease of cell density and increase of cell size could be observed in the areas near the center, and the central part of the PET sheet was unfoamed. After 5 days' saturation, as shown in Fig. 7, the distribution of CO_2 concentration had been relatively uniform in the PET sheet, which led to uniform cell morphology in most areas of the sample. Amazingly, a thin layer with fine cells was observed near the surface. Foams morphology between the thin layer and inner part of the PET sheet were shown in Fig. 10. Crystallinity was found[13,37] to play a major role in the microcellular processing on (1) cell nucleation mechanisms resulting in larger cell densities due to heterogeneous nucleation at the amorphous/crystalline boundaries and (2) cell growth mechanisms resulting in smaller cell sizes due to the increased stiffness of the semicrystalline matrix. Thus, observation of the thin layer with fine cells indicated the formation of CO_2-induced crystallization layer, which was coincident with the simulation result shown in Fig. 8. When the saturation time increased from 5 to 7 days, the thickness of the crystalline layer increased dramatically as shown in Fig. 8, and the foaming experiment also confirmed it. The thickness of the crystalline layer (after foaming) increased from 250 to 900 μm. After saturation time as long as 15 days, both CO_2 diffusion and induced crystallization completed, and the foams' morphology was quite uniform all through the PET sheet.

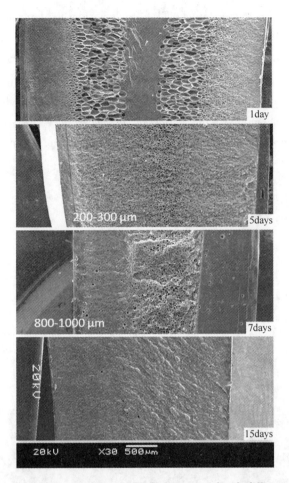

Fig. 9 The overall bubble morphology of PET foams prepared with different saturation times
(Color figure can be viewed in the online issue, which is available at wileyonlinelibrary. com.)

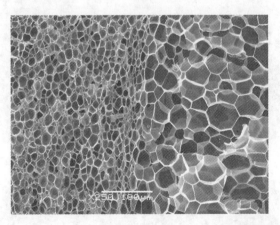

Fig. 10 Typical boundary between the crystalline layer and the amorphous layer of the PET foam

Details about the evolution of the crystalline layer against saturation time were shown in Fig. 11. No crystalline layer could be observed after 1 day's saturation, because barely any crystalline layer could form during such a short saturation time, as shown in Fig. 8. When the saturation time was longer than 2 days, a foamed layer with smaller bubble size and larger bubble density could be observed near the skin of PET foam, indicating the formation of crystalline layer. The boundary of crystalline layer and amorphous layer after foaming was also clearly shown in Fig. 11. Coincident with Fig. 8, the thickness of crystalline layer increased with saturation time. When the saturation time was longer than 7 days, due to completion of the CO_2-induced crystallization, the foaming morphology became uniform. Interestingly, the saturation time as long as 15 days further improved the perfection of CO_2-induced crystallization of the PET sheet, and an ultra-microcellular structure with cell size distribution as small as 100-300nm was obtained.

Time	Panoramic	Amorphous layer	Boundary	Crystalline layer
1 day	Fig. 11-1a	Fig. 11-1b	No	No
2 days	Fig. 11-2a	Fig. 11-2b	Fig. 11-2c	Fig. 11-2d
3 days	Fig. 11-3a	Fig. 11-3b	Fig. 11-3c	Fig. 11-3d
5 days	Fig. 11-4a	Fig. 11-4b	Fig. 11-4c	Fig. 11-4d
7 days	Fig. 11-5a	Fig. 11-5b	Fig. 11-5c	Fig. 11-5d
12 days	Fig. 11-6a	Fig. 11-6b	No	Fig. 11-6d
15 days	Fig. 11-7a	No	No	Fig. 11-7d

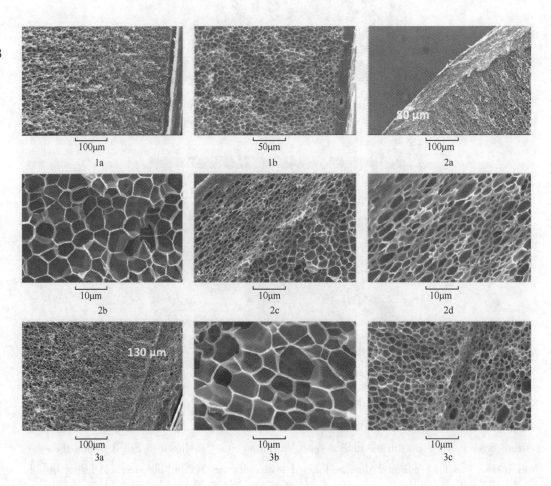

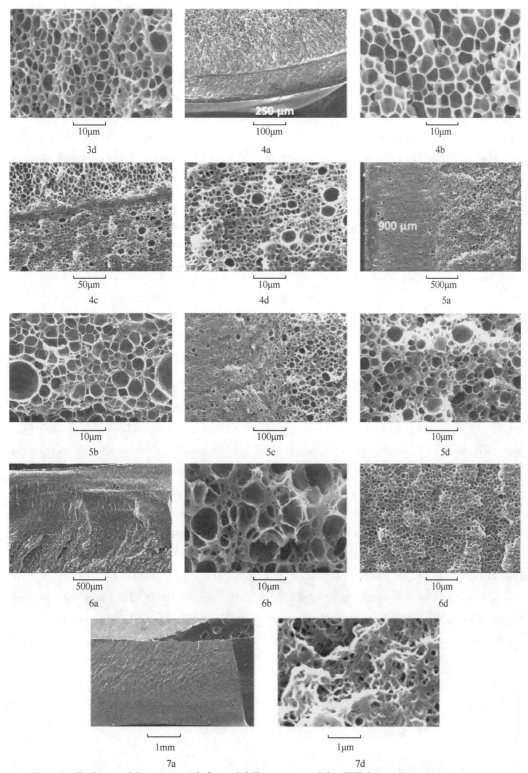

Fig. 11　Evolution of foaming morphology of different areas of the PET foams against saturation time
(Color figure can be viewed in the online issue, which is available at wileyonlinelibrary.com.)

The thicknesses of the crystalline layers calculated using the model coupling CO_2 diffusion and induced crystallization, as well as those measured from the PET foams, at different saturation time were shown in Table 2. Basically, the model could well predict the evolution of crystalline layer against different saturation time. When the saturation time was relatively short, i. e., short than 3 days, the thickness of crystalline layers in the PET foams was slightly larger than that predicted from the model, which was due to the expansion of crystalline layer during foaming experiment. While at relatively long saturation time, e. g., 5 or 7 days, the evolution of crystalline layer in the PET foams was slower than that from the modeling. In the model assumption, it was assumed that the crystallization of PET had no effect on the diffusion coefficient, but in fact, CO_2-induced crystallization of PET would reduce the free volume of samples and extend the path of CO_2 diffusion. These effects would decrease the diffusion coefficient of CO_2 to some extent and delay the evolution of crystalline layer.

Table 2 Comparison of the evolution of crystalline layer thickness calculated using the model coupling CO_2 diffusion and induced crystallization and in PET foams

Saturation time/d	1	2	3	5	7
Thickness of crystalline layer/μm					
In PET foams	0①	80	130	250	900
Modeling	20	50	100	300	Complete

①The crystalline layer was not detected in the PET foams due to the overlap of crystalline and the un-foamed skin layer of the PET specimen.

The evolution of bubble size and bubble density of both crystalline and amorphous layers, and expansion ratio in the PET foams against saturation time is shown in Fig. 12. Because of the increase of CO_2 concentration with increasing saturation time in the amorphous layer of the PET specimen, more CO_2 was available to support the bubble growth, which led to an increase in both bubble size and expansion ratio within the saturation time of 5 days. When the saturation time was longer than 7 days, as revealed by the model, the crystal structure of the crystalline became perfect, and small amount of crystals could have formed in the amorphous layer. Existence of crystal regions in PET matrix restricted the bubble growth, and changed the bubble nucleation mechanism from homogeneous nucleation to heterogeneous one, which dramatically reduced the activation energy of bubble nucleation and greatly increased the bubble density of PET foams. Especially at the saturation time of 15 days, ultra-microcellular PET foams with average bubble size as small as 193nm and bubble density as large as 3.37×10^{13} were obtained.

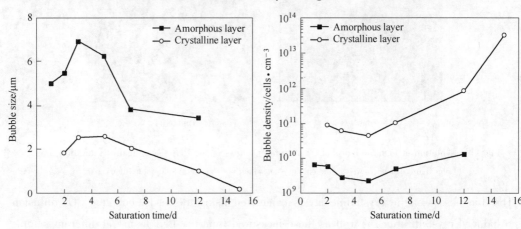

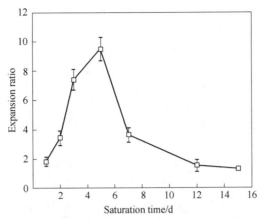

Fig. 12 Characterization of the PET foams obtained at different saturation time
(The lines only indicate the trend)

4 Conclusion

In this work, the intrinsic kinetics of CO_2-induced crystallization of amorphous PET at 25℃ and different CO_2 pressures were detected using *in situ* high-pressure FTIR and correlated by Avrami equation. The least critical CO_2 pressure, at which CO_2-induced crystallization could occur, was determined to be 5MPa at 25℃. Sorption of CO_2 in PET was measured using MSB and the diffusivity determined by Fick's second law. The diffusion coefficient of CO_2 in PET was determined to be $6.7 \times 10^{-13} m^2/s$ at 25℃ and the CO_2 pressure of 6MPa. The least critical CO_2 concentration for CO_2-induced crystallization was 9.17wt% in PET at 25℃. A model coupling CO_2 diffusion in and CO_2-induced crystallization of PET was proposed to calculate the CO_2 concentration as well as crystallinity distributions in PET at different saturation times. A sandwich crystallization structure was found in the PET sheet at an appropriate saturation time, based on which a solid-state foaming process was used to control sandwich-structure of PET foams with two microcellular or ultra-microcellular foamed crystalline layers outside and a microcellular foamed amorphous layer inside. The thickness of the foamed-crystalline layer agreed well with that calculated by the model. At the saturation time as long as possible, PET ultra-microcellular foams with average bubble size of 193nm and bubble density of 3.37×10^{13} could be fabricated.

Acknowledgements

The authors are grateful to the National Natural Science Foundation of China (Grant No. 20976045 and 20976046), Shanghai Shuguang Project(08SG28), Program for New Century Excellent Talents in University(NCET-09-0348), Innovation Program of Shanghai Municipal Education Commission(11CXY23), Program for Changjiang Scholars and Innovative Research Team in University(IRT0721), the 111 Project(B08021), and the support of "the Fundamental Research Funds for the Central Universities".

References

[1] Sorrentino L, Di Maio E, Iannace S. Poly(ethylene terephthalate) foams: correlation between the polymer

properties and the foaming process. J Appl Polym Sci. 2010;116:27-35.

[2] Mark JE. Polymer Data Handbook. London:Oxford University Press,1999.

[3] Martini JE. The Production and Analysis of Microcellular Foam(Master Thesis). Department of Mechanical Engineering, Massachusetts Institute of Technology, Cambridge, MA, 1981.

[4] Martini JE, Suh NP, Waldman FA. Microcellular closed cell foams and their method of manufacture. 1982. US Patent 4,473,665.

[5] Shen J, Han X, Lee LJ. Nanoscaled reinforcement of polystyrene foams using carbon nanofibers. J Cell Plast. 2006;42:105-126.

[6] Han X, Koelling KW, Tomasko DL, Lee LJ. Continuous microcellular polystyrene foam extrusion with supercritical CO_2. Polym Eng Sci. 2002;42:2094-2106.

[7] Krause B, Mettinkhof R, Van der Vegt NFA, Wessling M. Microcellular foaming of amorphous high-T_g polymers using carbon dioxide. Macromolecules. 2001;34:874-884.

[8] Sun H, Mark JE. Preparation, characterization, and mechanical properties of some microcellular polysulfone foams. J Appl Polym Sci. 2002;86:1692-1701.

[9] Huang Q, Klötzer R, Seibig B, Paul D. Extrusion of microcellular polysulfone using chemical blowing agents. J Appl Polym Sci. 1998;69:1753-1760.

[10] Krause B, Sijbesma HJP, Münüklü P, Van der Vegt NFA, Wessling M. Bicontinuous nanoporous polymers by carbon dioxide foaming. Macromolecules. 2001;34:8792-8801.

[11] Krause B, Diekmann K, Van der Vegt NFA, Wessling M. Open nanoporous morphologies from polymeric blends by carbon dioxide foaming. Macromolecules. 2002;35:1738-1745.

[12] Kumar V. Polyethylene terephthalate foams with integral crystalline skins. 1993. US Patent 5,223,545.

[13] Baldwin DF, Shimbo M, Suh NP. The Role of gas dissolution and induced crystallization during microcellular polymer processing—a study of poly(ethylene-terephthalate) and carbon-dioxide systems. J Eng Mater Technol. 1995;117:62-74.

[14] Baldwin DF, Park CB, Suh NP. A microcellular processing study of poly(ethylene terephthalate) in the amorphous and semicrystalline states. 1. Microcell nucleation. Polym Eng Sci. 1996;36:1437-1445.

[15] Baldwin DF, Park CB, Suh NP. A microcellular processing study of poly(ethylene terephthalate) in the amorphous and semicrystalline states. 2. Cell growth and process design. Polym Eng Sci. 1996; 36: 1446-1453.

[16] Kumar V, Schirmer H. Semi-continuous production of solid state polymeric foams. 1997. US patent 5,684,055.

[17] Kumar V, Juntunen RP, Barlow C. Impact strength of high relative density-microcellular CPET foams produced by the solid-state process using CO_2 as a blowing agent. Foams'99: First International Conference on Thermoplastic Foam, New Brunswick, NJ, 1999:157-164.

[18] Kumar V, Juntunen RP, Barlow C. Impact strength of high relative density solid state carbon dioxide blown crystallizable poly(ethylene terephthalate) microcellular foams. Cell Polym. 2000;19:25-37.

[19] Guan R, Wang BQ, Lu DP. Preparation of microcellular poly(ethylene terephthalate) and its properties. J Appl Polym Sci. 2003;88:1956-1962.

[20] Guan R, Wang BQ, Lu DP, Fang Q, Xiang BL. Microcellular thin PET sheet foam preparation by compression molding. J Appl Polym Sci. 2004;93:1698-1704.

[21] Guan R, Xiang BL, Xiao ZX, Li YL, Lu DP, Song GW. The processing-structure relationships in thin microcellular PET sheet prepared by compression molding. Eur Polym J. 2006;42:1022-1032.

[22] Li DC, Liu T, Zhao L, Yuan WK. Solubility and diffusivity of carbon dioxide in solid-state isotactic polypropylene by the pressure decay method. Ind Eng Chem Res. 2009;48:7117-7124.

[23] Tomasko DL, Li H, Liu D, Han X, Wingert MJ, Lee LJ, Koelling KW. A review of CO_2 applications in the

processing of polymers. Ind Eng Chem Res. 2003;42:6431-6456.

[24] Li Y, Xiang B, Liu J, Guan R. Morphology and qualitative analysis of mechanism of microcellular PET by compression moulding. Mater Sci Tech-Lond. 2010;26:981-987.

[25] Jacobs LJM, Kemmere MF, Keurentjes JTF, Mantelis CA, Meyer T. Temperature-induced morphology control in the polymer-foaming process. AIChE J. 2007;53:2651-2658.

[26] Goel SK, Beckman EJ. Nucleation and growth in microcellular materials—supercritical CO_2 as foaming agent. AIChE J. 1995;41:357-367.

[27] Li DC, Liu T, Zhao L, Lian XS, Yuan WK. Foaming of poly(lactic acid) based on its nonisothermal crystallization behavior under compressed carbon dioxide. Ind Eng Chem Res. 2011;50:1997-2007.

[28] Chow TS. Molecular interpretation of the glass transition temperature of polymer-diluent systems. Macromolecules. 1980;13:362-364.

[29] Handa YP, Zhang Z, Wong B. Effect of compressed CO_2 on phase transitions and polymorphism in syndiotactic polystyrene. Macromolecules. 1997;30:8499-8504.

[30] Zhang Z, Handa YP. An *in situ* study of plasticization of polymers by high-pressure gases. J Polym Sci Part B: Polym Phys. 1998;36:977-982.

[31] Takada M, Hasegawa S, Ohshima M. Crystallization kinetics of poly(L-lactide) in contact with pressurized CO_2. Polym Eng Sci. 2004;44:186-196.

[32] Lambert SM, Paulaitis ME. Crystallization of poly(ethylene terephthalate) induced by carbon dioxide sorption at elevated pressures. J Supercrit Fluid. 1991;4:15-23.

[33] Mizoguchi K, Hirose T, Naito Y, Kamiya Y. CO_2-induced crystallization of poly(ethylene terephthalate). Polymer. 1987;28:1298-1302.

[34] Eaves D. Handbook of Polymer Foams. Shrewsbury: Rapra Technology Limited, 2004.

[35] Brantley NH, Kazarian SG, Eckert CA. *In situ* FTIR measurement of carbon dioxide sorption into poly(ethylene terephthalate) at elevated pressures. J Appl Polym Sci. 2000;77:764-775.

[36] Kalospiros NS, Astarita G, Paulaitis ME. Coupled diffusion and morphological change in solid polymers. Chem Eng Sci. 1993;48:23-40.

[37] Jiang XL, Liu T, Xu ZM, Zhao L, Hu GH, Yuan WK. Effects of crystal structure on the foaming of isotactic polypropylene using supercritical carbon dioxide as a foaming agent. J Supercrit Fluid. 2009;48:167-175.

[38] Muschiatti LC. High melt strength PET polymers for foam application and methods relating thereto. 1993. US Patent 5,229,432.

[39] Subramanian PM, Tice CL. Melt fabrication of foam articles. 1992. US Patent 5,128,202.

[40] 3Acomposites. http://www.corematerials.3Acomposites.com/america.html. Accessed on September 6, 2011.

[41] Armacell. http://www.armacell-core-foams.com/ Accessed on September 6, 2011.

[42] Li L, Liu T, Zhao L, Yuan WK. CO_2-induced crystal phase transition from Form II to I in isotactic poly-1-butene. Macromolecules. 2009;42:2286-2290.

[43] Sato Y, Takikawa T, Takishima S, Masuoka H. Solubilities and diffusion coefficients of carbon dioxide in poly(vinyl acetate) and polystyrene. J Supercrit Fluid. 2001;19:187-198.

[44] Lei ZG, Ohyabu H, Sato Y, Inomata H, Smith RL Jr. Solubility, swelling degree and crystallinity of carbon dioxide-polypropylene system. J Supercrit Fluid. 2007;40:452-461.

[45] Sato Y, Takikawa T, Sorakubo A, Takishima S, Masuoka H, Imaizumi M. Solubility and diffusion coefficient of carbon dioxide in biodegradable polymers. Ind Eng Chem Res. 2000;39:4813-4819.

[46] Sato Y, Takikawa T, Yamane M, Takishima S, Masuoka H. Solubility of carbon dioxide in PPO and PPO/PS blends. Fluid Phase Equilib. 2002;194-197:847-858.

[47] Lee WH, Ouyang H, Shih MC, Wu MH. Kinetics of solvent-induced crystallization of poly(ethylene terephthalate) at the final stage. J Polym Res-Taiwan. 2003;10:133-137.

[48] Keum JK, Kim J, Lee SM, Song HH, Son YK, Choi JI, Im SS. Crystallization and transient mesophase structure in cold-drawn PET fibers. Macromolecules. 2003;36:9873-9878.

[49] Zhang Y, Lu YL, Duan YX, Zhang JM, Yan SK, Shen DY. Reflection-absorption infrared spectroscopy investigation of the crystallization kinetics of poly(ethylene terephthalate) ultrathin films. J Polym Sci Part B:Polym Phys. 2004;42:4440-4447.

[50] Qian R, Shen D, Sun F, Wu L. The effects of physical ageing on conformational changes of poly(ethylene terephthalate) in the glass transition region. Macromol Chem Phys. 1996;197:1485-1493.

[51] Avrami M. Kinetics of phase change. I. General theory. J Chem Phys. 1939;7:1103-1112.

[52] Avrami M. Kinetics of phase change. II. Transformation-time relations for random distribution of nuclei. J Chem Phys. 1940;8:212-224.

[53] Avrami M. Kinetics of phase change. III. Granulation, phase change, and microstructure. J Chem Phys. 1941;9:177-184.

[54] Run M, Hao Y, Yao C. Melt-crystallization behavior and isothermal crystallization kinetics of crystalline/crystalline blends of poly(ethylene terephthalate)/poly(trimethylene terephthalate). Thermochim Acta. 2009;495:51-56.

[55] Wellen RMR, Rabello MS. Antinucleating action of polystyrene on the isothermal cold crystallization of poly(ethylene terephthalate). J Appl Polym Sci. 2009;114:1884-1895.

[56] Deshpande VD, Jape S. Isothermal crystallization kinetics of anhydrous sodium acetate nucleated poly(ethylene terephtalate). J Appl Polym Sci. 2010;116:3541-3554.

[57] Crank J. The Mathsmatics of Diffusion, 2nd ed. London:Oxford University Press, 1975.

[58] McGonigle EA, Liggat JJ, Pethrick RA, Jenkins SD, Daly JH, Hayward D. Permeability of N_2, Ar, He, O_2, and CO_2 through biaxially oriented polyester films—dependence on free volume. Polymer. 2001;42:2413-2426.

[59] Lewis ELV, Duckett RA, Ward IM, Fairclough JPA, Ryan AJ. The barrier properties of polyethylene terephthalate to mixtures of oxygen, carbon dioxide, and nitrogen. Polymer. 2003;44:1631-1640.

Modeling Silver Catalyst Sintering and Epoxidation Selectivity Evolution in Ethylene Oxidation[*]

Abstract Catalyst sintering is a slow complex process which may last for years but is contributed by physical processes of atomic femtosecond scale. In this article, sintering caused silver particle growth is considered to be the result of random migration and coalescence, on the basis of which a random walk model is proposed. Simulation results show that the particle size evolution can be well fitted by generalized power law expression, indicating that the random walk model is valid for sintering processes with different sintering rates and time scales. With the assumption of identical grain sizes, changes in the specific surface intergrain boundary length during particle sintering are predicted by the random walk model and the epoxidation rate is related to the specific surface intergrain boundary length in terms of a selectivity factor. The results elucidate quite well the phenomenon of nonmonotonic change of the selectivity during silver catalyst sintering.

Key words ethylene epoxidation, silver catalysts, sintering, intergrain boundary, random walk

1 Introduction

Ethylene epoxidation is a very important catalytic reaction in the chemical industry. Silver is still the only known catalyst with sufficient activity and selectivity and fixed-bed reactor is the only used reactor in the industry. Despite its simple configuration, the fixed-bed reactor for catalytic heterogeneous reaction is indeed a complex system, the performance of which depends not only on the large number of macroscopic factors in reactor design and operation, but also on the time-dependent catalyst surface structure of nano/microscale. For quite a long time the reaction mechanism and reactor design and modeling have been the subject of considerable study by both chemists and chemical engineers for the development of higher selectivity catalysts and more efficient reactors and great improvement have been made in the past few decades. However, they work quite independently, and little work has been done to relate the phenomena on nano/micro to the performance of the reactor on macro-scale.

Though ethylene epoxidation was first discovered early in 1931 and commercialized in 1938, the mechanism of this reaction is still not completely understood and a universally accepted kinetics for industrial use is not available. As a heterogeneous catalytic reaction, it involves adsorption, surface reaction and desorption, and a number of adsorbates besides transition state intermediates that are usually very difficult to identify. The electronic structure that determines how the reaction proceeds is, on the other hand, determined by the property of bulk and surface characteristics of the catalyst. The sizes and morphology of catalyst particles and surface structure (defects of different sizes and dimensions), which are usually on nano/microscale, play de-

---

[*] Coauthor: Zhou X. G. Reprinted from *Chemical Engineering Science*, 2004, 59: 1723-1731.

cisive roles in the reaction and is affected by impurities and promoters in the catalyst and the procedure for catalyst preparation.

Advances in instrument analysis and molecular modeling in the latest few years have greatly improved the understanding of the mechanism of ethylene epoxidation on atomic and molecular scale. The energetic profile of the reaction is now partially understood based on which reaction coordinate can be followed(Nakatsuji et al.,1997;Avdeev et al.,2000). However,it is still not possible to build a microkinetic model that is free of fitting parameters,because silver catalyst is structure sensitive, and oxygen and chlorine can induce surface reconstruction. Moreover, it is difficult to incorporate the adsorbate-adsorbate (O_2-C_2H_4-CO_2) interactions into the model (Broadbelt and Snurr,2000). More intensive study is needed to know about the microstructure evolution and the non-uniform characteristics on the catalyst surface. In practice, chemical engineers still have to turn to macroscopic rate equations by neglecting surface reaction details(Borman and Westerterp,1995).

Catalyst deactivation or decay during operation is associated with almost all heterogeneous catalysts and is to be considered in process design and optimization. Coke formation, poison deposition and solid-state transformation are among the main causes for activity loss(Bartholomew,2001). For silver catalyst,both poisoning by sulfur and metal sintering are responsible for deactivation. In chemical engineering deactivation by coke formation and poison are considered by introducing an activity factor a that is assumed to follow the power low(Levenspiel,1999):

$$-\frac{d\alpha}{dt} = k_d c_w^u \alpha^v$$

where, w is the component that causes loss in activity. Combination of macroscopic rate equations and activity changes with time constitutes the chemical kinetics of a catalytic process, which is the basis for process development and operation.

Solid-state transformation involves sintering, i.e. crystallite growth and support loss, and chemical transformation of catalytic phases to non-catalytic phases as a result of solid-solid reaction. Sintering data are usually fitted to the generalized power law expression(GPLE)(Bartholomew,1997)of the form:

$$-\frac{d(D/D_0)}{dt} = k_d \left(\frac{D}{D_0} - \frac{D_{eq}}{D_0}\right)^m$$

where, D_0 is the initial dispersion; D_{eq} is the limit dispersion at infinite time; m is the sintering order, which is usually two. D is proportional to the total surface of the particles and is therefore an indication of the catalyst activity.

Molecular dynamic(MD) techniques are proposed to describe grain growth(Zeng et al., 1998). However, days of computational time are elapsed only to simulate less than 100 ps of the real process. Catalyst sintering is a very slow process and may last for a few years, a time scale many, many orders of magnitude beyond the current capabilities offered by the most powerful computers. Monte Carlo(MC) procedure(Sault and Tikare,2002) can simulate a real process of 1μs, which is much longer but still far below the time scale over which typical sintering occurs.

Previous modeling approaches mainly focus on the evolution with time of the dispersion of particles during sintering. Activity losses are contributed to the decrease of catalyst surface area, by poisoning or sintering, and a monotonic decline in selectivity is implicated as the temperature is increased to obtain a desired total reaction rate. However, evidence exists showing that the reaction rate of ethylene epoxidation is strongly related to the number of microcrystalline domains in the silver particles(Tsybulya et al., 1995), indicating the dependence of selectivity not only on temperature but also on particle size(Goncharova et al., 1995). Tsybulya et al. (1995) proposed that the changes of intergrain boundaries adjacent to the particle surface are responsible for selectivity changing with particle size. The intergrain boundary provides a low-diffusion resistance path for oxygen(Nagy et al., 1998), resulting more subsurface oxygen(van den Hoek et al., 1989) for selective epoxidation.

In this preliminary report, we would provide a randomwalking algorithm to simulate particle sintering, on the basis of which a specific surface intergrain boundary length, which is defined as the ratio of intergrain boundary length on a particle to its surface area, is calculated. By assigning the intergrain boundary a higher activity for selective epoxidation, the change of selectivity during sintering can be qualitatively predicted. Although the proposed approach is quite simple, it provides a modeling approach at meso-length scale for a satisfactory description of particle growth and nonmonotonic changes in selectivity as the particle grows.

2   Modeling of Silver Particle Sintering

Three principle mechanisms of metal particle growth have been advanced (Bartholomew, 2001): (1) particle migration and coalescence; (2) atom emission and recapture(Ostwald ripening); (3) vapor-phase transport. In the first mechanism, the particle moves over the support surface followed by collision and coalescence. The second mechanism involves detachment of metal atoms from the metal particle, migration of these atoms over the support surface and ultimately captured by larger particles. The third mechanism, vapor-phase transport, takes place at high temperatures with atom transport via vapor phase from the smaller particles to the larger ones. For all these mechanisms, the driving force for particles growth is the same. As indicated by Gibbs-Thomson equation, a smaller particle has higher chemical potential and has a tendency to grow to lower its chemical potential.

Vapor-phase transport is of relatively small significance in sintering of silver catalyst becuase the catalyst is usually used below 250℃, much lower than the melting point of silver (961.93℃). An early investigation(Montrasi et al., 1983) showed that after 6 years of industrial use, the loss of silver is 8% which is quite small. Currently silver catalysts are designed to operate at much lower temperatures and silver loss by vapor-phase transport will be much smaller. Therefore, contribution of vapor-phase transport to silver sintering can be neglected.

Ostwald ripening of the supported Ag particles is most likely to be diffusion controlled. If the growth kinetics is the first-order, then the change in particle radius with time can be formulated (Mullin, 1993):

$$\frac{dr}{dt} = \frac{D_e v}{r}\left[(c - c^*) - \frac{2\delta v c^*}{\gamma R_g T}\right] \tag{1}$$

where, v is the molar volume of the crystal; γ is the interfacial tension; D_e is the diffusion coefficient; c is the equilibrium surface concentration on the support for a small particle of size r; c^* is that for larger particles ($r \to \infty$). The change of particle sizes with time can thus be predicted by a population model (Sivakumar et al., 2001).

Particle migration over the support is a random behavior. When two particles collide, bridging occurs and then the particles fuse into one. Several factors may participate in particle migration, such as shear stress, gravity, and surface energy. The role the shear stress and gravity play in silver catalyst sintering is minor because accumulation of metal on the lower side or at the outlet of the packing was not observed. On the other hand, surface energy would be the most important drive for particle migration. Therefore, particle migration may not be an independent mechanism of sintering, but an integration of Ostwald ripening, vapor-phase transport, particle spreading or support wetting, etc. If the behavior of particle migration is described, the most important feature of catalyst sintering can be captured without considering the underlying governing physics. Only when particle migration is geometrically limited, may Ostwald ripening and vapor-phase transport be taken in to account as independent mechanisms of sintering.

Ag particles on $\alpha\text{-}Al_2O_3$ can be assumed as spheres, with large contact angles and small contact areas, as evidenced by scanning electron spectroscopy (Hoflund and Minahan, 1996a). The smaller the particle, the larger the distance it travels. Hence, it is reasonable to define the traveling distance as:

$$d = \frac{d_0}{1 + \kappa r} \quad (2)$$

where, d_0 is the traveling distance of an infinitely small particle; κ is a scaling factor of the traveling distance d changing with particle size. d_0 is temperature dependent:

$$d_0 = k_0 \exp\left(-\frac{E_m}{R_g T}\right) \quad (3)$$

The new position a particle migrate to is:

$$x = x_0 + d\sin\theta$$

where, θ is a random value uniformly distributed over the interval $(0, 2\pi)$. When two particles collide, i.e.:

$$\sum_{i=1}^{2} [x_1(i) - x_2(i)]^2 \leq (r_1 + r_2)^2 \quad (4)$$

they fuse and become one particle locating at:

$$x = \frac{r_1 x_1 + r_2 x_2}{r_1 + r_2} \quad (5)$$

a place closer to the bigger one.

3 Modeling of Intergrain Boundary Evolution

Intergrain boundary is a two-dimensional defect in a crystal, which is the irregular junction of two adjacent grains with orderly arranged pattern. The narrow zone at the intergrain boundary corresponds to the transition from one crystallographic orientation to another, thus separating one grain from another. Intergrain boundary can be formed during crystallization, or as a result of coalescence. Since the grain shapes are irregular and of different sizes, as illustrated in

Fig. 1, the intergrain boundaries they constitute are also irregular. A known fact about the intergrain boundaries is that three-fold junctions (Fig. 1) are the dominating junctions, and the dihedral angles should be equal to 120℃ if the boundary energy is uniform.

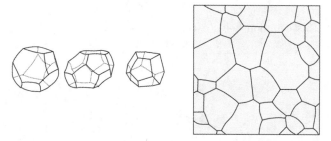

Fig. 1 Illustration of the irregular shape and size of grains and the triple junction of intergrain boundary

As there is large excess surface free energy associated with the boundary, there is a tendency of the boundary to migrate towards its center of curvature to reduce its local surface area. Consequently, the average grain of crystal size increases with time, which occurs through a competition in which some grains shrink and disappear, while other grains grow to fill the volume formerly occupied by the shrinking grains. However, this growing process is usually slow and is remarkable only at high temperatures. If the boundaries are trapped or pinned, grain growth will be significantly retarded. Modeling grain growth is a hot research area and has been studied widely, see Frost and Thompson (1996) and references therein.

Since only the outermost of the Ag crystal, including only a few layers of Ag atoms, is associated in catalytic reaction, we are concerned with only the intergrain boundaries at the surfaces of Ag particles. By assuming that the grain size is the same for all grains and remains unchanged after coalescence, changes in the boundaries on the particle surface as a function of particle can be calculated if the shape of the grain is specified. However, simple geometries do not result in close packing with three-fold junctions and 120℃ dihedral angles. For example, seamless packing with regular dodecahedrons is impossible, while packing with hexahedron will produce cross-junctions, which are unstable and will easily split into two three-fold junctions with a reduction in free energy.

To resolve this geometrical problem, we first make do with hexahedrons of unit length. For closest packing that gives the smallest surface area, the total length of intergrain boundary on the outer surface and surface area can be easily calculated, as listed in Table 1. Also listed is the specific surface intergrain boundary length defined as the ratio of boundary length to surface.

Table 1 Surface intergrain boundary length, surface area and specific surface intergrain boundary length constructed by different number of hexahedrons

Number of hexahedrons	1	2	3	4	5	6	7	8	27	64
Surface boundary length	0	4	7	12	14	18	21	24	72	144
Surface area	10	10	14	16	20	22	24	24	54	96
Specific boundary length	0	0.40	0.50	0.75	0.70	0.82	0.86	1	1.33	1.5

The specific surface intergrain boundary length as a function of number of hexahedrons can be well approximated by:

$$L_s = \frac{2(\sqrt[3]{n} - 1)}{\sqrt[3]{n}} \tag{6}$$

as shown in Fig. 2. This equation holds exactly for $n = 1^3, 2^3, 3^3, \cdots$.

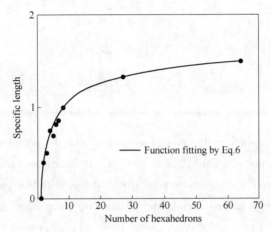

Fig. 2 Fitting of the specific surface intergrain boundary length by Eq. 6

For an infinite n, $L_{s,n\to\infty} = 2$, and the outer surface consists of an infinite number of square cells. However, since hexahedrons constitute unstable cross-junctions, grains are less likely in hexahedrons. To have more stable three-fold junctions, a different grain geometry, which is approximately dodecahedron, is needed. How the grains are packed in the inter part is not our concern, but on the surface the intergrain boundaries must constitute an infinite number of hexagons if the number of grains are infinite. Therefore $L_{s,n\to\infty} = 2/\sqrt{3}$ for unit edge length and Eq. 6 is revised as:

$$L_s = \frac{2}{\sqrt{3}} \cdot \frac{\sqrt[3]{n} - 1}{\sqrt[3]{n}} \tag{7}$$

It can be shown that packing with dodecahedrons resulting in a slightly larger specific surface intergrain boundary length than with hexahedrons and the ratio is constant. Therefore, one can use $(\sqrt[3]{n} - 1)/\sqrt[3]{n}$ to calculate the specific surface boundary length and for grains of size r_g, the specific surface intergrain boundary length

$$L_s = \frac{1}{r_g} \cdot \frac{\sqrt[3]{n} - 1}{\sqrt[3]{n}} \tag{8}$$

and the total surface intergrain boundary length is:

$$L_{sb} = \frac{1}{r_g} \sum_i \frac{\sqrt[3]{n_i} - 1}{\sqrt[3]{n_i}} r_{pi}^2 \tag{9}$$

where, r_p and r_g are the sizes of particle and grain.

Fig. 2 indicates that L_s increase sharply as the size of silver particle increases, which is in accordance with experimental observations made by Tsybulya et al. (1995) and Goncharova et al. (1995) that the epoxidation rate increases monotonously with particle size and the change is much more significant when the particle size is small.

Denote the reaction rates of epoxidation and total combustion on the unit regular surface as R_1 and R_2, and R_1' as the epoxidation rate contributed by unit surface intergrain boundary length. CO_2 formation is assumed not to be affected by intergrain boundary because a visible particle size effect on deep oxidation has not been observed. The instantaneous selectivity is therefore:

$$\frac{R_1 S + R_1' L_{sb}}{(R_1 + R_2) S + R_1' L_{sb}} \tag{10}$$

Assume by kinetic study that r_{EO} and r_{CO_2} are determined to be the EO and CO_2 formation rate on unit weight fresh catalyst; A_0 and A are the specific surface areas (per unit weight) for fresh and aged catalysts, respectively. Then the EO and CO_2 formation rates on unit catalyst surface become αr_{EO} and αr_{CO_2}, with $\alpha = A/A_0$.

Since the fresh catalysts have already possessed a certain amount of intergrain boundary, the kinetics has included the influence of surface intergrain boundary on the initial particles. Hence the total EO formation rate is revised as:

$$R_{EO} = \alpha r_{EO} + R_1'(\Lambda_{sb} - \Lambda_{sb0}) = \alpha(1 + k\beta) r_{EO} \tag{11}$$

where

$$\beta = \frac{1}{\alpha}(\gamma - 1) \tag{11a}$$

$$k = \frac{R_1'}{r_{EO}} \Lambda_{sb0} \tag{11b}$$

$$\gamma = \frac{\Lambda_{sb0}}{\Lambda_{sb}} \tag{11c}$$

Λ_{sb0} and Λ_{sb} are the surface intergrain boundary lengths on silver particles in unit weight fresh and aged catalysts. Eq. 11 relates the EO formation rate with catalyst surface and intergrain boundary in terms of activation factor α and selectivity factor β, which is actually the specific surface intergrain boundary length in dimensionless form. The instantaneous selectivity of ethylene epoxidation is:

$$\vartheta = \frac{(1 + k\beta) r_{EO}}{(1 + k\beta) r_{EO} + 2 r_{CO_2}} \tag{12}$$

4 Simulation Results

First, suppose the temperature of catalyst sintering is constant. Then there are only two parameters in the random walking model, i. e. the traveling distance of an infinite small particle and the size scaling factor.

To start with, 100 particles are randomly generated with a mean size of $0.03 \mu m$ and a squared variance of $1 \times 10^{-4} \mu m^2$ and are equally distributed in a $1 \times 1 \mu m^2$ square. In one cycle of random walking, each particle migrates in a random direction and moves to a new position d away from its original position. Particles coalescence up contacting and as a result the averaged size increases and the total surface area decreases. Fig. 3 shows the changes of averaged particle size as particles migrate and coalescence. The size scaling factor κ used for simulation is currently set to be $100 \mu m^{-1}$.

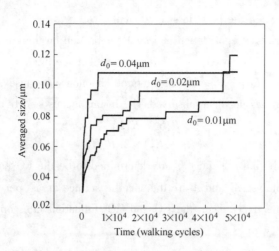

Fig. 3 Dependence of sintering on migration distance d_0 ($\kappa = 100\mu m^{-1}$)

Migration distance is a kinetic parameter in the random walk model. Larger migration distance results in a rapid growth in particle size. When the number of particles becomes very small, coalescence is less likely to happen and further increase of the averaged size is slow and by chance due to the random nature of the sinter process. As more time is allowed for sintering, ultimately only one particle is left and its size reaches maximum.

The size scaling factor also imposes a definite influence on particle growth, as is seen in Fig. 4. If the scaling factor is large, big particles will be slow to move, leading to a sluggish increase in particle size. In order to determine whether our simulations are consistent with the GPLE, data shown in Fig. 3 and Fig. 4 are fitted by the GPLE:

$$\frac{D}{D_0} = \frac{1}{k_d t + 1/(1 - D_{eq}/D_0)} + \frac{D_{eq}}{D_0} \qquad (13)$$

For the same pool of particles, D_{eq}/D_0 remains unchanged regardless of sintering rates. So for all the fitting curves, only k_d is different. Fig. 5a and b show that the data produced by random walking model can be well fitted by GPLE with a fixed D_{eq}/D_0, which indicates the trustfulness of the proposed model.

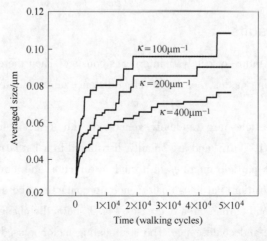

Fig. 4 Dependence of sintering on scaling factor κ ($d_0 = 0.02\mu m$)

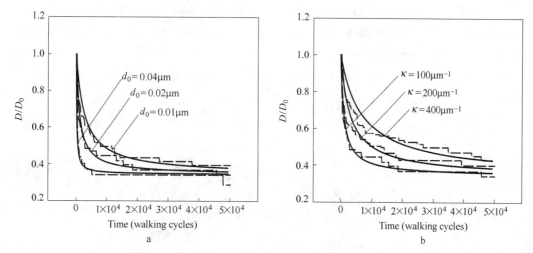

Fig. 5 Fitting of particle dispersion by GPLE
a—$\kappa = 100 \mu m^{-1}$; b—$d_0 = 0.02 \mu m$

It was found by high-resolution electron microscopic image that relatively large silver particles (more than 40nm) possessed a micrograined structure, which was not observed for Ag microcrystals with sizes in the range from 20 to 30nm (Tsybulya et al., 1995). Hence we define a minimum grain volume

$$V_g = \frac{4}{3}\pi r_g^3 \tag{14}$$

and assume r_g is 25nm. The specific intergrain boundary length of a particle is calculated according to Eq. 6 with the number of grains calculated by:

$$n = \frac{r_p^3}{r_g^3} \tag{15}$$

If $r_p^2 < 2r_g^3$, $n = 1$, and the particle consists of only one grain and has no intergrain boundary. For $n = 2$, the particle consists of two grains and has one intergrain boundary. Actually, the specific surface intergrain boundary length changes with the difference between the volumes of the two grains. However, for a large number of particles, the volume difference between two grains in a particle is random and the surface intergrain boundary length will also be a random number. On average n can be any value greater than two and the specific surface intergrain boundary length is determined solely by particle size from Eq. 8. The selectivity factor is calculated by:

$$\beta = \frac{1}{r_g} \cdot \frac{\sum r_0^2}{\sum\limits_{i=0,1,\cdots} r_{pi}^2} \sum\limits_{i=0,1,\cdots}\left(\frac{\sqrt[3]{n}-1}{\sqrt[3]{n}}r_{pi}^2 - \frac{\sqrt[3]{n_0}-1}{\sqrt[3]{n_0}}r_{p0}^2\right), n \geq 2 \tag{16}$$

Fig. 6 shows how the selectivity factor changes with particle size when the particles in the fresh catalyst contain a certain averaged number of grains. This figure is obtained by simulating the random walking model, with different initial particle sizes (0.025μm, 0.0286μm, 0.0315μm, 0.0360μm, and 0.0427μm for $n_0 = 1, 1.5, 2, 3$ and 5, respectively) and a squared variance of $1 \times 10^{-4} \mu m^2$. If in the fresh catalyst the silver particles are small and have fewer intergrain boundaries, the selectivity will increase first as particle size grows by sintering, and then decrease in the long run; if the initial particles are bigger, the selectivity will decrease monotonously. These are the consequences of opposite evolution of the specific intergrain boundary length and the total surface area in catalyst sintering. Fig. 7 shows the evolution of the selectivity factor with time (walking cycles) with migration distance of 0.01 μm and size scaling factor of 100 μm^{-1}.

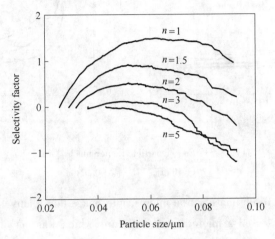

Fig. 6 Change of selectivity factor (with scalable unit) with particle size

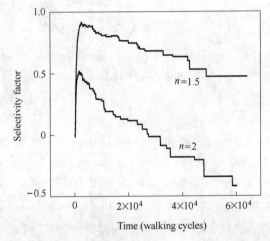

Fig. 7 Change of selectivity factor (with scalable unit) with time

In a differential reactor one would expect from Eq. 10 the selectivity undergoes a change indicated in Fig. 6 in catalyst aging at a constant temperature. In integral reactors such as CSTR or fixed-bed reactor, or if the temperature changes during sintering, the changes of selectivity will be influenced by the conversion and the reaction temperature (Hoflund and Minahan, 1996b) and will not be straightforward to predict.

As an illustration, ethylene epoxidation is carried in a fixed-bed reactor, which is 31.3mm in inner diameter and packed with silver catalyst to a height of 8.2m at a packing density of 590kg/m^3. The reaction kinetics is adapted from Gu et al. (2003) as:

$$r_{EO} = e^{10.30-8358.5/T} p_{ET} p_{O_2}^{0.75}/(1+Dn) \tag{17a}$$

$$r_{CO_2} = e^{12.54-9835.2/T} p_{ET} p_{O_2}/(1+Dn) \tag{17b}$$

$$Dn = 1 + e^{9612.8/T - 21.68} p_{CO_2} + e^{16129.2/T - 34.58} p_{O_2}^{0.5} p_{H_2O} \qquad (17c)$$

A one-dimensional pseudo-homogeneous model (Froment and Bischoff, 1990) is used for this reactor, with the overall heat transfer parameter being $600J/(m^2 \cdot s \cdot K)$. Inlet concentrations are 28.0%, 7.5%, 5.0% and 59.5% for ethylene, oxygen, carbon dioxide and balance methane, respectively. The reactor is operated at a GHSV of $5500h^{-1}$, pressure of 2.1MPa and inlet temperature of 180℃. To remain a constant throughput, the outlet EO concentration is fixed at 1.8%.

Simulation begins with 400 particles randomly generated in a $2 \times 2\mu m^2$ square, with a mean size of $0.0286\mu m$ and a squared variance of $0.1 \times 10^{-4} \mu m^2$. During sintering the migration distance is assumed to be temperature dependent, $d_0 = 1.75e^{-50 \times 10^3/R_g T}$ and the size scaling factor κ fixed at $100\mu m^{-1}$.

The changes of epoxidation selectivity with time are shown in Fig. 8. As a result of sintering, the catalyst activity decreases and the coolant temperature (also drawn in Fig. 8) has to be raised to attain the required outlet EO concentration. Since the activation energy for deep oxidation is higher than for partial oxidation, more CO_2 will be produced at higher reaction temperature. Therefore the selectivity decreases rapidly for aged catalyst, despite the sluggish decrease of selectivity factor in the second stage of sintering shown in Fig. 7 for this case.

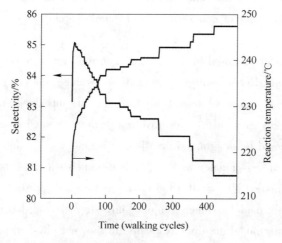

Fig. 8 Simulated change of reaction temperature and epoxidation selectivity with time

Similar trends, as shown in Fig. 9, of the epoxidation selectivity are observed for a commercial silver catalyst (YS, SINOPEC) used in an industrial wall-cooled tubular reactor in North China, which underwent a rapid increase in the first stage, a rapid decrease in the second stage and a slow decline in the last stage. What is different for the industrial catalyst is that the selectivity oscillated obviously, which can be explained by the periodical change (as opposed to the continuous increase in simulation) of reaction temperature, by the irregular variation in inhibitor concentrations, or probably by grain growth not considered in this preliminary study. Grain growth results in a decreased specific surface intergrain boundary and offsets partially the increase of specific surface intergrain boundary by particle coalescence.

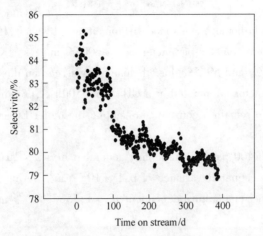

Fig. 9   Changes in selectivity of industrial catalyst with time on stream

5 Conclusion and Discussion

As mentioned previously in this article, though sintering is a process taking years, the basic mechanism of catalyst sintering is an atomic process over femtosecond time scale. Modeling of catalyst sintering can be started from the very bottom, with regard to the femtosecond atomic-scale events, by MD or MC simulation. However, the process that can be simulated by MD or MC techniques is of the order of nanoseconds and microseconds, while sintering is far beyond this time scale.

The random walk model proposed in this article is on the particle level. Actually it is a technique of coarse graining. By adjusting the migration distance, one can simulate a sintering process lasting minutes or years. Although two parameters are introduced in the random walking model, they are physically meaningful. The traveling distance of an infinite small particle (one atom) on the surface could be determined by MD or dynamic MC simulations, while κ could also be theoretically predicted or determined by well designed experiments. Compared with molecular modeling techniques (MD or MC) and chemical kinetics approaches (Sivakumar et al., 2001), the random walk model is much simpler and extremely rapid in computation. The parameters in random walk model are directly related to the real physical process and simulation can be accelerated for processes from microseconds to years. Consequently it provides a useful tool for complex system modeling.

By now knowledge on the mechanism of epoxidation on silver catalyst is not sufficient for the development of microkinetics with zero-adjusting parameters. Moreover, the practical use of microkinetics will be very limited even if it is easily available, because it involves a large number of species and energy terms, some of which are surface structure and coverage dependent. Most probably some of the structure parameters and energy terms are difficult to determine or inaccurate to use and have to be adjusted by experiments. To this regard, macroscopic rate equations are competitive and will dominate in engineering practice. Nevertheless, as macroscopic rate equations are determined by correlation among apparent observations, microscopic phenomena are frequently overlooked as a result of ignorance, inaccessibility by experiments or poor theoretical understanding.

As evidenced by experimental observation, there exists a definite relation between the microstructure, i.e. intergrain boundary, and the rate of epoxidation. On this basis we proposed a simple mechanism to relate the specific surface intergrain boundary length to particle size and subsequently to the epoxidation selectivity. Although we have assumed grains with a regular geometry and omitted the growth of the intergrain boundaries during sintering, the results are qualitatively in good accordance with experimental observations. However, considerable study is needed to know about the role of the intergrain boundary in epoxidation, and the effect of intergrain boundary migration and particle morphology evolution. As particle sintering, intergrain boundary migration and particle morphology evolution are also femtosecond atomic events and last for years, modeling approaches with regard to mechanisms at atomic scale and effective enough to capture the overall behavior over large time scale is highly necessitated.

Notation

A	surface area, m^2
c	concentration, mol/m^3
d	migration distance, μm
D	dispersion of silver particles, m^{-1}
D_e	diffusion coefficient, m^2/s
Dn	expression defined by Eq. 17c
k	constant
k_d	rate constant for catalyst deactivation
L	total surface intergrain boundary length, m
L_s	specific surface intergrain boundary length, μm^{-1}
L_{sb}	total surface intergrain boundary length, μm
m	sintering order
n	number of grains in a silver particle
p	particle pressure, MPa
r	particle size, μm
R_1	epoxidation rate, $mol/(s \cdot m^2)$ (silver surface)
R_1'	epoxidation rate, $mol/(s \cdot m)$ (surface intergrain boundary length)
R_2	combustion rate, $mol/(s \cdot m^2)$ (silver surface)
r_{CO_2}	epoxidation rate, $kmol/(s \cdot kg)$ (catalyst)
r_{EO}	epoxidation rate, $kmol/(s \cdot kg)$ (catalyst)
R_g	gas constant, $J/(mol \cdot K)$
S	total silver surface, m^2
t	time
T	temperature, K
u	order of deactivation with respect to activation factor
v	order of deactivation with respect to w

V volume of a silver particle, m^3

x position of a particle, vector

Greek letters

α activation factor

β selectivity factor

γ ratio of surface boundary lengths defined by Eq. 11c

δ interfacial tension, N/m

κ size scaling factor, μm^{-1}

Λ_{sb} total surface intergrain boundary lengths on silver particles in unit weight catalyst, mm/kg(catalyst)

ϑ instantaneous selectivity

θ angle, radian

υ molar volume of a crystal, m^3/mol

Subscript

0   initial or fresh

eq   limit value

g   grain

p   particle

Superscript

*   limit value

Acknowledgements

This work is financially supported by "863" project (2002AA412120) and by special funding for outstanding doctoral dissertation winners by the Ministry of Education of China.

References

Avdeev, V. I., Boronin, A. I., Koscheev, S. V., Zhidomirov, G. M., 2000. Quasimolecular stable forms of oxygen on silver surface. Theoretical analysis by the density functional theory method. Journal of Molecular Catalysis A 154, 257-270.

Bartholomew, C. H., 1997. Sintering and redispersion of supported metals: perspectives from the literature of the past decade. In: Bartholomew, C. H., Fuentes, G. A. (Eds), Catalyst Deactivation. Elsevier, Amsterdam.

Bartholomew, C. H., 2001. Mechanisms of catalyst deactivation. Applied Catalysis A 212, 17-60.

Borman, P. C., Westerterp, K. R., 1995. An experimental study of the kinetics of the selective oxidation of ethene over a silver on a-alumina catalyst. Industrial Engineering Chemical Research 34, 49-58.

Broadbelt, L. J., Snurr, R. Q., 2000. Applications of molecular modeling in heterogeneous catalysis research. Applied Catalysis A 200, 23-46.

Froment, G. F., Bischoff, K. B., 1990. Chemical Reaction Analysis and Design, 2nd Edition. Wiley, New York.

Frost, H. J., Thompson, C. V., 1996. Computer simulation of grain growth. Current Opinion in Solid State and Materials Science 1, 361-368.

Goncharova, S. N., Paukshtis, E. A., Bal'Zhinimaev, B. S., 1995. Size effects in ethylene oxidation on silver catalysts. Influence of support and Cs promoter. Applied Catalysis A 126, 67-84.

Gu, Y. L., Gao, Z., Jin, J. Q., 2003. Kinetic models for ethylene oxidation to ethylene oxide. Petrochemical Technology 32(Suppl.), 838-840.

Hoflund, G. B., Minahan, D. M., 1996a. Ion-Beam characterization of alumina-supported silver catalysts used for ethylene epoxidation. Nuclear Instruments and Method B 118, 517-521.

Hoflund, G. B., Minahan, D. M., 1996b. Study of Cs-promoted, α-alumina-supported silver, ethylene-epoxidation catalysts. Journal of Catalysis 162, 48-53.

Levenspiel, O., 1999. Chemical Reaction Engineering, 3rd Edition. Wiley, New York.

Montrasi, G. L., Tauszik, G. R., Solari, M., Leofanti, G., 1983. Oxidation of ethylene to ethylene oxide: catalyst deactivation in an industrial run. Applied Catalysis 5, 359-369.

Mullin, J. W., 1993. Crystallization, 3rd Edition. Butterworth-Heinemann, London.

Nagy, A., Mestl, G., Ruhle, T., Weinberg, G., Schlogl, R., 1998. The dynamic restructuring of Electrolytic silver during the formaldehyde synthesis reaction. Journal of Catalysis 179, 548-559.

Nakatsuji, H., Nakai, H., Ikeda, K., Yamamoto, Y., 1997. Mechanism of the partial oxidation of ethylene on an Ag surface: dipped adcluster model study. Surface Science 384, 315-333.

Sault, A. G., Tikare, V., 2002. A new Monte Carlo model for supported-catalyst sintering. Journal of Catalysis 211, 19-32.

Sivakumar, S., Subbanna, M., Sahay, S. S., Ramakrishnan, V., Kapur, P. C., Pradip, Malghan, S. G., 2001. Population balance model for solid state sintering II. Grain growth. Ceramics International 27, 63-71.

Tsybulya, S. V., Kryukova, G. N., Goncharova, S. N., Shmakov, A. N., Bal'zhinimaev, B. S., 1995. Study of the real structure of silver supported catalysts of different dispersity. Journal of Catalysis 154, 194-200.

van den Hoek, P. J., Baerends, E. J., van Santen, R. A., 1989. Ethylene epoxidation on Ag(110): the role of subsurface oxygen. Journal of Physical Chemistry 93, 6469-6475.

Zeng, P., Zajac, S., Clapp, P. C., Rifkin, J. A., 1998. Nanoparticle sintering simulations. Materials Science and Engineering A 252, 301-306.

First-principles Calculations of C Diffusion through the Surface and Subsurface of Ag/Ni(100) and Reconstructed Ag/Ni(100) *

Abstract Density functional theory calculations are performed to investigate the C diffusion through the surface and subsurface of Ag/Ni(100) and reconstructed Ag/Ni(100). The calculated geometric parameters indicate the center of doped Ag is located above the Ni(100) surface owing to the size mismatch. The C binding on the alloy surface is substantially weakened, arising from the less attractive interaction between C and Ag atoms, while in the subsurface, the C adsorption is promoted as the Ag coverage is increased. The effect of substitutional Ag on the adsorption property of Ni(100) is rather short-range, which agrees well with the analysis of the projected density of states. Seven pathways are constructed to explore the C diffusion behavior on the bimetallic surface. Along the most kinetically favorable pathway, a C atom hops between two fourfold hollow sites via an adjacent octahedral site in the subsurface of reconstructed Ag/Ni(100). The "clock" reconstruction which tends to improve the surface mobility, is more favorable on the alloy surface because the $c(2 \times 2)$ symmetry is inherently broken by the Ag impurity. As a consequence, the local lattice strain induced by the C transport is effectively relieved by the Ag-enhanced surface mobility and the C diffusion barrier is lowered from 1.16 to 0.76eV.

Key words DFT, nickel, silver, adsorption, diffusion

1 Introduction

Steam reforming of hydrocarbon is the major industrial route to produce hydrogen and syngas. The cheap supported Ni catalyst is regarded as the most attractive catalyst for this process, though some noble metals such as Ru and Rh are found to be more reactive [1]. However, the main issue associated with Ni catalyst is the catalyst deactivation owing to carbon deposition [2].

Various approaches have been used to minimize coke formation and growth of carbon nanofibers [3-8]. The first method is to limit the ensemble size of active sites, and subsequently control the selectivity of the reaction to favor steam reforming rather than coke formation. As suggested by Rostrup-Nielsen and Alstrup [9], the ensemble required by coke formation is larger than that required by steam reforming. The second way is based on the information that coke formation is initiated at the step site (defect) [4,10]. Blocking the step site can substantially suppress the nucleation of graphite. The third one is to balance the C diffusion rate with the generation rate of C atoms so as to avoid the C accumulation on catalyst surfaces and the C dissolution into bulk metal. Accordingly, the resistance to coke formation and growth of carbon nanofibers is inherently achieved, promoting the long-term stability of catalysts.

---

* Coauthors: Zhu Yian, Chen De, Zhou Xinggui, Per-Olof Åstrand. Reprinted from *Surface Science*, 2010, 604: 186-195.

It has long been known that the addition of a group I B metal to group VIII metals can lead to an increased selectivity of catalysts and to a suppression of coke formation. Besenbacher et al. used Au to modify the Ni catalyst for steam reforming of methane[3]. They found the negative impact of Au on C nucleation is considerably stronger than that on CH_4 dissociation, and therefore the Au/Ni surface alloy is a less reactive, but more robust, steam reforming catalyst. In their subsequent work, density functional theory (DFT) calculations revealed that Au atoms prefer to replace the Ni atoms on the step edge or block the step site to eliminate the nucleation site of graphite formation[4]. More recently, the effect of Ag loading on the catalytic behavior of Al_2O_3-supported Ni catalyst has been investigated by Parizotto et al. [6] They reported that the alloy catalysts with the Ag loading more than 0.3wt% show high resistance to coke deposition in steam methane reforming. However, the experimental results from the same group indicated the coking in steam reforming of ethanol cannot be suppressed by the introduction of Ag[7]. They explained the activation of ethanol is easier than that of methane and independent on the coordination of the active Ni sites. Consequently, the activity of Ni catalyst is not significantly influenced by the Ag-induced blockage of the step sites and the generated CH_x species remain abundant on the alloy surfaces, the latter being the driving force for the accumulation of carbonaceous compounds. However, the role of Ag impurity in the resistance to graphite nucleation is neglected in their statement. Thus, the knowledge of the C adsorption and diffusion on Ag/Ni alloy surfaces is of vital importance.

As presented in our previous calculations[11,12], the energy barrier for the C diffusion on Ni(111) is much lower than that on Ni(100), and therefore the C diffusion rate on Ni(100) determines the overall rate of the C surface transport. In the past two decades, some experimental and theoretical work has been devoted to investigating the C adsorption and diffusion on Ni(100)[12-16]. Klink et al. studied the C binding to Ni(100) using scanning tunneling microscopy (STM) in the early time[13]. The realspace information indicated that at low coverages less than 0.2ML, C atoms are located at the fourfold hollow sites, while above a local critical coverage, the surface is driven into a $(2 \times 2)p4g$ structure by disorder-order phase transition, which is known as "clock" reconstruction. They suggested that the driving force for the reconstruction is the preference for C atoms to be five-coordinated. In order to explain the origin of the C-induced relaxation on Ni(100), Stolbov et al. conducted a first-principles study to reproduce this surface reconstruction[16]. Through the calculation of the geometric and electronic structures, they found two important factors that determines the reconstruction: (1) the formation of covalent bonds between the C atoms and their nearest neighbor (nn) Ni atoms in the second layers substantially reduces the total energy of the system and (2) the reduction of local electronic charges on the C atoms prevents the strong electron-electron repulsion. Recently, we have performed detailed DFT calculations to depict the C diffusion through the surface and subsurface of Ni(100) and reconstructed Ni(100). In that study[12], it was revealed that C atoms tend to accumulate on the Ni surface to a higher level owing to the relatively slow diffusion rate at low coverages. When the critical C coverage is attained, the surface reconstruction takes place and promotes the C diffusion dramatically.

In this contribution, the C adsorption and diffusion on the surface and in the subsurface of

Ag/Ni(100) and reconstructed Ag/Ni(100) are investigated by means of first-principles calculations. Firstly, based on the experimental information, models are constructed for the Ag/Ni(100) surface and the electronic structure is calculated to analyze the variation in the surface properties arising from the introduction of Ag. Secondly, we calculate the adsorption energies and geometric parameters as C atoms bind to Ag/Ni(100), and evaluate the probability of C-induced surface reconstruction. At last, the climbing-image nudged elastic band (CI-NEB) method is used to explore the minimum energy path (MEP) for C diffusion through the surface and subsurface of Ag/Ni(100) and reconstructed Ag/Ni(100). Based on the comparison of the energy barriers for seven diffusion pathways, we conclude by discussing the implication of our results for understanding the effect of Ag introduction and surface reconstruction upon the C diffusion behavior on Ni(100).

2 Computational Details

The first-principles calculations performed in this study are based on spin-polarized density functional theory. The Vienna *ab initio* simulation package (VASP)[17-20] is used, in which the interactions between valence electrons and ion cores are described by pseudopotentials and the electronic wavefunctions at each k-point are expanded in terms of a discrete plane-wave basis set. Here we use Blöchl's all-electron-like projector augmented wave (PAW) method[21] as implemented by Kresse and Joubert[22], which regards the 4s 3d states as the valence configuration for Ni, 4d 5s states for Ag, 2s 2p states for C. The PAW method is necessary for accurate calculations of certain transition metals which are sometimes poorly described by the ultrasoft pseudopotentials. Exchange and correlation of the Kohn-Sham theory are treated with the generalized gradient approximation (GGA) in the formulation of PW91 functional[23]. A plane wave energy cutoff of 370eV is used in the present calculation and all geometries are optimized using a velocity quench algorithm until the forces acting on each atom are converged better than 0.1eV/nm. Brillouin zone sampling is performed using a Monkhorst-Pack grid[24] and electronic occupancies are determined according to a Methfessel-Paxton scheme[25] with an energy smearing of 0.2eV. Because there is a magnetic element (Ni) involved in the system, spin-polarized effect has been considered. The calculations performed by Kresse and Hafner showed that surface magnetism is essential for an accurate quantitative description of adsorption energy[26].

The Ni(100) surfaces is represented as a five-layer slab with $p(2 \times 2)$ and $p(3 \times 3)$ supercells, and then one (or two) surface Ni atom(s) per supercell is replaced with one (or two) Ag atom(s) to form the Ag/Ni(100) alloy surface. As for the reconstructed Ag/Ni(100) surface, we use a five-layer (4×4) supercell with the quasi-$(2 \times 2)p4g$ geometry, consisting of 16 metal atoms per layer and eight (or seven) C atoms on the surface. In all models, the bottom two layers of the slab are constrained to their crystal lattice positions. The neighboring slabs are separated in the direction perpendicular to the surface by a vacuum region of 1nm. A preliminary test calculation is performed to show that the relaxation of deeper layers has only a very small effect on the surface reaction since the forces on these layers are negligible[12]. The first Brillouin zone of the $p(2\times2)$, $p(3\times3)$ and (4×4) supercell is sampled with a $5\times5\times1$, $3\times3\times1$ and $2\times2\times1$ k-point grid, respectively, which is evidenced to be sufficient for each supercell[12,16].

The CI-NEB method[27-29], an improved version of the NEB method, is used to locate the MEPs and transition states for C diffusion. A set of intermediate images are constructed along the diffusion path between the energetically favorable reactant and product. A spring interaction between adjacent images is added to ensure continuity of the path, thus mimicking an elastic band. With the two endpoints fixed, the intermediates are optimized using a force-based conjugate-gradient method[30] until the maximum force in every degree of freedom is less than 0.1eV/nm. In the current work, seven pathways are constructed to explore the C diffusion behavior on the Ag/Ni alloy surface. Then, four and nine system images are interpolated between the initial and final state for PATH1-PATH2 and much longer PATH3-PATH7, respectively, to evaluate the energy barriers.

3 Results and Discussion

3.1 Ag/Ni(100) surface

It was reported by Christensen et al. that the segregation energy of Ag atoms from bulk Ni to Ni surface is negative (-0.58eV)[31], indicating that the deposited Ag prefers to stay on the Ni surface without the formation of subsurface alloy. On the other hand, they also calculated the surface mixing energy, which is defined as the energy difference between the phase-separated and the surface alloy phase, and found Ag and Ni atoms do not form separate islands but mix on the Ag/Ni surface.

To our knowledge, the geometry of the Ag-doped Ni(100) surface has not yet been depicted using first-principles calculations. Based on the aforementioned information, the Ag/Ni(100) surface is constructed by substituting Ag atoms for surface Ni atoms, achieving the Ag coverages of 1/9, 1/4 and 1/2ML. The geometric parameters of the Ag/Ni(100) surface is defined in Fig. 1 and the comparison is summarized in Table 1. Because of the size mismatch, the center of the substitutional Ag atom is located above the Ni surface by 0.052~0.064nm, and the outermost and second Ni layers are pushed towards bulk metal, accompanied by the contraction of the interplanar spacing between them. For the $Ag_{1/2}$/Ni(100) surface ($Ag_{1/2}$/Ni(100) denotes the Ag/Ni(100) surface at the Ag coverage of 1/2ML), two models are available to represent the structure. One model is illustrated in Fig. 1c where the nearest neighbors of a substitutional Ag atom are four surface Ni atoms, while in the other one, every substitutional Ag atom has two nearest neighboring Ag atoms. We have calculated the total energies of these two models and find the latter is 0.21eV/cell higher in energy, which agrees well with the prediction using the surface mixing energy by Christensen et al.[31] The reason for the instability of the Ag-Ag geometry is that the Ag atoms are large relative to their spacing, and consequently repel each other.

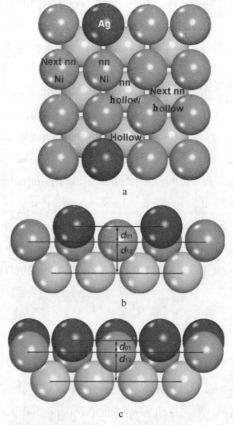

Fig. 1 Top and side views of the structural models for the Ag/Ni(100) surface at the Ag coverages of 1/9ML(a), 1/4ML(b), and 1/2ML(c)

Table 1 Calculated geometric parameters of the Ag/Ni(100) surface

Ag coverage/ML	d_{01}[①]/Å	d_{12}[②]/Å	Δz_1[③]/Å	Δz_2[④]/Å	Δd_{12}[⑤]/Å
1/9	0.52	1.68	-0.02	-0.01	-0.01
1/4	0.59	1.67	-0.05	-0.02	-0.03
1/2	0.64	1.62	-0.11	-0.03	-0.08

Note: 1Å = 0.1nm.

① d_{01} Denotes the interlayer spacing perpendicular to the surface between the substitutional Ag atom and the outermost Ni layer.

② d_{12} Denotes the interlayer spacing perpendicular to the surface between the outermost and second Ni layer.

③ Δz_1 Denotes the relaxation of the outermost Ni surface with respect to the fully relaxed pure Ni surface.

④ Δz_2 Denotes the relaxation of the second Ni surface with respect to the fully relaxed pure Ni surface.

⑤ Δd_{12} Denotes the change of the interlayer spacing between the outermost and second Ni layer with respect to the fully relaxed pure Ni surface.

To investigate the effect of doped Ag on the catalytic behavior of Ni(100), the C adsorption energies on $Ag_{1/9}$/Ni(100) are calculated and compared with those on pure Ni(100). The definition of the C adsorption energies will be detailed in Section 3.2 and the binding sites considered are shown in Fig. 1a, including the hollow site, the nn hollow site of atomic Ag, and the next nn hollow site. As indicated in Table 2, the adsorption energy at the hollow site is reduced by up to 0.66eV since the interaction between C and Ag is much less attractive. The generally accepted reason for the weaker C—Ag bond is that the hybridized antibonding C—Ni d states are empty, while the corresponding states are filled on Ag[32]. That is to say the antibonding states for the C—Ni and C—Ag interactions are located above and below the Fermi level, respectively, the latter giving rise to a repulsion between the C-induced level and Ag d band. Furthermore, the orthogonalization energy cost arising from the overlap of the electronic states is dramatically driven up with the introduction of Ag, as the size of the coupling matrix element between the s and d state of Ag is much larger than that of Ni. At the nn hollow site sharing two Ni atoms with Ag, the C binding to the Ni surface is less affected and the adsorption energy is slightly decreased by 0.05 eV. At the next nn hollow site far away from the doped Ag, the C adsorption energy is unchanged, as compared to that on pure Ni(100), indicating the influence of the substitutional Ag on the adsorption property of the Ni surface is rather short-range. This is in agreement with what is observed by Termentzidis et al. on the Au-doped Ni(111) surface[33].

Table 2 Calculated C adsorption energies and local valence charge around Ni and Ag atoms on pure Ni(100) and $Ag_{1/9}$/Ni(100) alloy surfaces

Sites	Pure Ni(100) hollow	$Ag_{1/9}$/Ni(100)		
		Hollow	nn hollow	Next nn hollow
C adsorption energy/eV	-8.22	-7.56	-8.17	-8.22
Atoms	Ni	Ag	nn Ni	Next nn Ni
Local charge Q	10.01	11.06	10.01	10.01

In order to facilitate the interpretation of the reduction of the C adsorption energy by analyzing the electronic structure, local valence charges Q around surface Ni and Ag atoms and projected density of states (PDOS) on the d band of surface Ni atoms are calculated, as shown in Table 2 and Fig. 2, respectively. The local charge is calculated using the Bader's analysis suggested by Bader and implemented by Henkelmann et al.[34-36] In this method, the so-called "zero flux" surfaces are used to divide the system into atoms. A zero flux surface is a two dimensional surface where the gradient of the charge density is zero along the surface normal. In practice, the total charge, the sum of the core and valence charge, is firstly calculated and used as a reference to define the Bader's volume of atoms, and then the valence charge is integrated in these volumes. In order to reproduce the correct core charge, the FFT grid is increased twice as large as that used for geometry optimization calculations. From Table 2, one can see the introduction of Ag does not change the valence charge of surface Ni atoms, while the substitutional Ag accepts only 0.06 electrons from the deeper Ni layers. This negligible charge transfer is due to the close electronegativity (Ni 1.91, Ag 1.93[37]) of these two elements.

As shown in Fig. 2, the PDOS on the d band of the nn Ni positioned at about -4.2 eV to -5 eV below the Fermi level is enhanced and the d-band width is broadened because of the hybridization of Ni 3d with Ag 4d states. Simultaneously, the d-band center of the nn Ni is calculated to be shifted to a lower energy though the energy shift is only 10 or more meV, as compared to that of Ni atoms on pure surface. As for the next nn Ni of atomic Ag, the PDOS on the d band is almost unchanged. That is why the doped Ag plays a negligible role in the C adsorption energy at the next nn hollow site. Hence, the electronic structure calculations also indicate that only the local area involving Ag impurity and its nn Ni atoms is influenced by the atomic replacement.

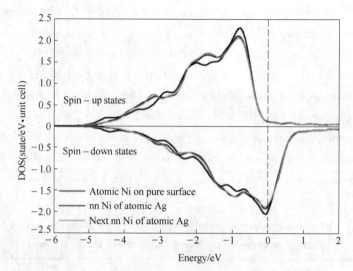

Fig. 2 PDOS on the d band of surface Ni atoms on the Ni(100) and $Ag_{1/9}/Ni(100)$ alloy surfaces. The zero energy refers to the Fermi level

3.2 C adsorption on the surface and in the subsurface of Ag/Ni(100)

The C adsorption on Ni(100) with the C coverages of 1/4 ML and 1/9 ML has been investigated detailedly by Stolbov et al.[16] and in our previous work[12], respectively. Here, we recalculate the C adsorption energies on Ni(100) and compare them with those on Ag/Ni(100) by using the same computational method, as shown in Table 3. As for the definition of the adsorption sites on pure Ni(100), readers can refer to our previous work[12]. The high-symmetry sites for the C adsorption on $Ag_{1/4}/Ni(100)$ and $Ag_{1/2}/Ni(100)$ are illustrated in Fig. 3, and the adsorption energies at these sites are calculated as:

$$\Delta E_{ads} = E_{C+surface} - E_{surface} - E_C \quad (1)$$

The first term on the right-hand side is the total energy of the surface with one C atom adsorbed; the second term is the total energy of surface. The first two terms are calculated by applying the same parameters (k-point sampling, energy cutoff, etc.). The third term is the total energy of an isolated C atom in its 3P ground state, which is estimated by putting a C atom in a box with dimensions of 1.5 nm × 1.55 nm × 1.6 nm and carrying out a spin-polarized Γ-point calculation. With this definition, a more negative value of ΔE_{ads} corresponds to a stronger binding between C and surface.

As seen from Table 3, the fourfold hollow site is the most stable binding site for the C adsorption on the surface of Ni(100) and Ag/Ni(100) with the highest binding energy. In particular, the C atoms placed initially at the Bridge1 and Atop1 sites are relaxed to the hollow site at the Ag coverage of 1/9 ML. On the other hand, the introduction of Ag atoms on Ni(100) weakens the C binding dramatically and the adsorption energy becomes less negative as the C coverage is increased.

Table 3 C adsorption energies (ΔE_{ads}) on the surface and in the subsurface of Ni(100) and Ag/Ni(100) at the C coverages of 1/9ML and 1/4ML (eV)

Ag coverage /ML	C coverage /ML	Hollow	Bridge1	Bridge2	Atop1	Atop2	Sub-O1	Sub-O2	Sub-T1	Sub-T2
0	1/9	-8.22	-6.02	—	-4.50	—	-7.07	—	-6.27	—
1/9	1/9	-7.56	-7.56 (hollow)[①]	-4.80	-7.56 (hollow)	-1.97	-7.19	-6.73	-8.17	-7.19 (Sub-O1)
0	1/4	-8.13	-5.99	—	-4.50	—	-7.06	—	-5.94	—
1/4	1/4	-7.48	-6.01	-4.79	-4.47	-1.94	-7.25	-6.79	-7.29	-7.26
1/2	1/4	-6.96	—	-4.97	-4.93	-2.00	-7.53	-6.89	—	-5.94

① Sites in parentheses denote the final positions of C atoms after relaxation.

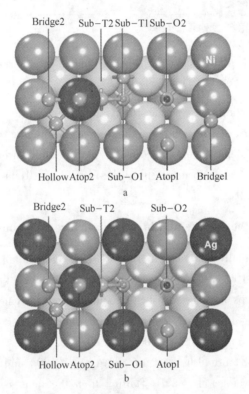

Fig. 3 Schematic representations of high-symmetry binding sites for C adsorption on the surface and in the subsurface of Ag/Ni(100)
a—1/4ML Ag coverage; b—1/2ML Ag coverage
(For clarity, the model size of some metal atoms in the outmost layer is reduced)

In the subsurface of Ni(100) and $Ag_{1/2}$/Ni(100), the subsurface octahedral site under a surface Ni atom (Sub-O1) is most energetically favorable for the C adsorption. However, in the case of $Ag_{1/4}$/Ni(100), C atoms bind more strongly to the Sub-T1 and Sub-T2 site since a C-induced surface reconstruction occurs, as shown in Fig. 4. At the Sub-T1 site, the C atom is relaxed to a subsurface quasi-octahedral site and coordinated to two surface Ni atoms as well as four Ni atoms in the second layer. Simultaneously, the Ag/Ni(100) surface is relaxed to the distinct fcc(111) geometry. At the Sub-T2 site, the C atom moves towards the adjacent Sub-O1 site, accompanied by a weaker reconstruction in which the substitutional Ag and the surface Ni atom over the adsorbed C atom are displaced laterally by 0.051nm. We find these two reconstructed structures are stable even if the converge tolerance of geometry optimization is decreased to 0.01eV/nm. For $Ag_{1/9}$/Ni(100), the C atom positioned initially at the Sub-T1 site is relaxed to the nn hollow site involving four Ni atoms, and therefore in Table 3, the C adsorption energy at the Sub-T1 site is even higher than that at the hollow site.

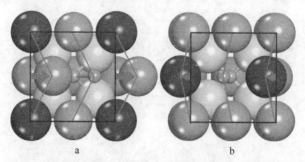

Fig. 4 Illustrations of the C-induced surface reconstruction at the Sub-T1 (a) and Sub-T2 (b) site of $Ag_{1/4}$/Ni(100)

In contrast to the C binding to the surface, the C adsorption in the subsurface is strengthened with the Ag introduction, as depicted by the increased binding energies at the Sub-O1 site. Moreover, though C atoms are predicted to stay on the alloy surface at low Ag coverages, the C diffusion from surface to subsurface turns thermodynamically favorable when the Ag coverage is increased to 1/2ML.

3.3 C adsorption on the surface and in the subsurface of reconstructed Ag/Ni(100)

It was reported experimentally that when a critical C coverage is attained, the Ni(100) surface is reconstructed to a Ni(100)-(2×2)$p4g$-C structure[13], which is known as the "clock" reconstruction. Several groups have previously reproduced this phenomenon using DFT calculations and a $C_{1/2}$/Ni(100) model[12,15,16]. Therefore, it is interesting to explore whether the C-induced reconstruction occurs on Ag/Ni(100) as well.

The C adsorption at the fourfold hollow sites of $Ag_{1/4}$/Ni(100) and $Ag_{1/2}$/Ni(100) is firstly investigated at the C coverage of 1/2ML. Starting from the quasi-c(2×2) symmetry, the geometry of the Ag/Ni alloy surface is optimized to an equilibrium quasi-(2×2)$p4g$ structure. The lateral displacements of the surface Ni atoms are calculated to be 0.05nm and 0.046nm on the $Ag_{1/4}$/Ni(100) and $Ag_{1/2}$/Ni(100) surface, respectively, as shown in Fig. 5. This situation seems different from what was observed on pure Ni(100) by Kirsch and Harris[15] and Stolbov et al.[16] In their studies, it was declared that the system does not undergo a c(2×2)-$p4g$ transformation using the simple quasi-Newton or conjugate-gradient optimization method. The reconstruction is achieved only after rotating surface Ni atoms slightly around the adsorbed C atoms. In fact, the observation in the present study confirms once the c(2×2) symmetry is inherently broken by the Ag introduction, the transformation is spontaneous. Consequently, there is no energy barrier for the "clock" reconstruction, supporting the statements by Kirsch and Harris[15] and Stolbov et al.[16]

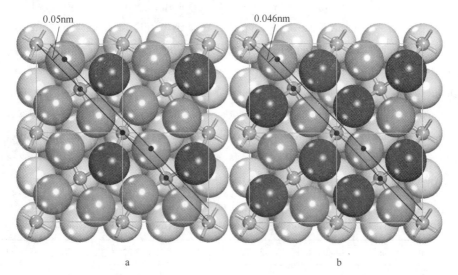

Fig. 5 Schematic representations of C-induced "clock" reconstruction on the $Ag_{1/4}/Ni(100)$ (a) and $Ag_{1/2}/Ni(100)$ (b) surface at the C coverage of 1/2ML

On the surface and in the subsurface of reconstructed $Ag_{1/4}/Ni(100)$ and $Ag_{1/2}/Ni(100)$, 13 and four binding sites are considered for the C adsorption, respectively, as shown in Fig. 6. The adsorption energies at these sites are calculated as

$$\Delta E_{ads} = E_{surface+8C} - E_{surface+7C} - E_C \qquad (2)$$

The first term on the right-hand side is the total energy of the slab consisting of alloy surface and eight C atoms (seven C atoms are adsorbed at the 4-hollow sites and one C atom is adsorbed at the investigated site); the second term is the total energy of the slab consisting of alloy surface and seven C atoms adsorbed at the 4-hollow sites. The third term is the same as that

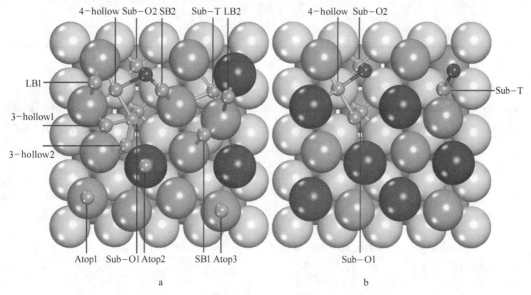

Fig. 6 Schematic representations of high-symmetry binding sites for C adsorption on the surface and in the subsurface of reconstructed Ag/Ni(100)

a—1/4ML Ag coverage; b—1/2ML Ag coverage

in Eq. 1. After geometry optimization, we find the 4-hollow, 3-hollow1, Sub-O1, Sub-O2 and Sub-T sites are stable for the C adsorption on the $Ag_{1/4}$/Ni(100) surface. The C atoms placed initially at the other surface binding sites are predicted to be relaxed to the adjacent 4-hollow or 3-hollow1 site. The calculated C adsorption energies on the surface and in the subsurface of reconstructed $Ag_{1/4}$/Ni(100) are listed in Table 4. From the table, it can be seen that the doped Ag atom weakens the C adsorption on the surface, while it enhances the binding at the Sub-O1 site, which is similar to the situation on nonreconstructed Ag/Ni(100). The only difference, however, is that the C dissolution into subsurface of reconstructed Ag/Ni(100) has already turned favorable at the Ag coverage of 1/4ML.

Table 4　C adsorption energies (ΔE_{ads}) on the surface and in the subsurface of reconstructed Ni(100) and Ag/Ni(100) at the C coverages of 1/2ML (eV)

Ag coverage/ML	4-hollow	3-hollow1	3-hollow2	LB1	LB2	SB1	SB2
0[12]	-7.98	-5.84	—	-6.12	—	-5.84 (3-hollow1)	—
1/4	-6.73	-5.88	-5.88 (3-hollow1)①	-6.73 (4-hollow)	-6.73 (4-hollow)	-5.88 (3-hollow1)	-5.88 (3-hollow1)
1/2	-6.05	—	—	—	—	—	—

Ag coverage/ML	Atop1	Atop2	Atop3	Sub-O1	Sub-O2	Sub-T
0[12]	-7.98 (hollow)	—	—	-7.23	—	-7.23 (Sub-O1)
1/4	-6.73 (4-hollow)	-6.73 (4-hollow)	-6.73 (4-hollow)	-7.35	-6.98	-6.50
1/2	—	—	—	-7.72	-7.52	-7.52 (Sub-O2)

① Sites in parentheses denote the final positions of C atoms after relaxation.

Based on the aforementioned information, the 4-hollow, Sub-O1, Sub-O2 and Sub-T sites are selected to investigate the C adsorption on reconstructed $Ag_{1/2}$/Ni(100) in order to save computational efforts. As seen from the binding energies at the Sub-O1 and Sub-O2 (an octahedral site under a surface Ag) sites, the subsurface is further favored by C atoms with the Ag coverage increased from 1/4 to 1/2ML. The Sub-T site is unstable for the C adsorption and C atoms are predicted to be relaxed to the adjacent Sub-O2 site. The same trend has ever been observed on reconstructed pure Ni(100)[12].

3.4　C diffusion through the surface of Ag/Ni(100) at the C coverage of 1/4ML and 1/9ML

It is assumed that the C diffusion may be described by the simple version of quantum corrected harmonic transition state theory (TST), which predicts the rate constant k through an energy barrier E_a:

$$k = \frac{k_B T}{h} \exp\left(\frac{-E_a}{k_B T}\right) \tag{3}$$

where, k_B is the Boltzmann constant; h is the Planck constant. This approach neglects contributions to the local partition functions from vibrational levels other than the ground state. These contributions can be included as yielding an effective activation energy that is slightly temperature dependent and approaches the classical result at high temperatures. Now that the most stable C adsorption sites on the surface and in the subsurface of Ag/Ni(100) and reconstructed Ag/Ni(100) are determined, the CI – NEB method is used to assess the diffusion barriers between the fourfold hollow sites.

Two C diffusion pathways are explored through the surface of $Ag_{1/4}$/Ni(100) and $Ag_{1/9}$/Ni(100), as shown in Fig. 7. Along PATH1, a C atom diffuses between two hollow sites via an adjacent Bridge1 site involving two Ni atoms, while along PATH2, a Bridge2 site consisting of one Ni atom and one Ag atom is passed through. The converged MEPs and the configurations of the transition states for PATH1 and PATH2 are schematically represented in Fig. 8. As one can see, the rate of the C diffusion along PATH1 is higher than that through the surface of Ni(100). This is because the starting point for PATH1 is much less stable and it is easier for C atoms to leave the hollow sites of the alloy surfaces, while the geometries of the transition states are rather similar. To further investigate the interaction between C and Ag atoms by analyzing the electronic structure, charge density difference ($\Delta\rho$) upon the C adsorption at the hollow sites of pure Ni(100) and $Ag_{1/4}$/Ni(100) is calculated relative to the sum of the isolated C atomic charge density and the charge density of the metal surfaces, as plotted in Fig. 9. It is found that strong Ni—C covalent bonds are formed on both the surfaces and there is a net charge transfer to the adsorbed C atom, arising from the charge depletion of the adjacent Ni d_{z^2} orbital. However, the Ag—C covalent bond almost disappears as little charge transfer is observed in Fig. 9b. Furthermore, the substitutional Ag pushes the atomic C toward the opposite surface Ni by about 0.011nm to reduce the overlap between the C – induced and Ag d states, and therefore the degree of filling of the antibonding states is reduced[32].

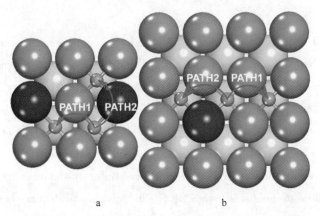

Fig. 7 Schematic representations of C diffusion through the surface of Ag/Ni(100) along two different pathways

a—$Ag_{1/4}$/Ni(100); b—$Ag_{1/9}$/Ni(100)

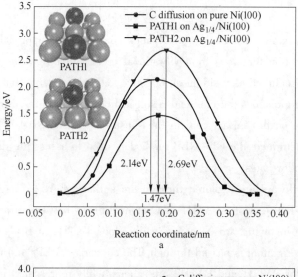

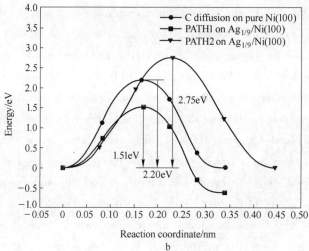

Fig. 8 MEPs for C diffusion on Ni(100) and Ag/Ni(100) between hollow sites through surface bridge sites and the geometries of the transition states for PATH1 and PATH2

a—$Ag_{1/4}$/Ni(100); b—$Ag_{1/9}$/Ni(100)

On $Ag_{1/4}$/Ni(100), the saddle point for PATH2 is 1.22 eV higher in energy than that for PATH1. Once one Ni atom is replaced with Ag at the Bridge2 site, the hopping C atom is displaced towards vacuum by about 0.032 nm, as shown in Fig. 8. The overall C diffusion rate through the surface of $Ag_{1/4}$/Ni(100) is much lower than that on the pure Ni(100) surface because the slower C transport along PATH2 is inevitable at the Ag coverage of 1/4 ML. As the C and Ag coverages are reduced to 1/9 ML, the total energies of the initial states for both PATH1 and PATH2 are lowered owing to the release of lattice strain. On the other hand, the geometries of the transition states are unaffected since no C-induced lateral displacement of surface Ni is found, and consequently the energy barriers are slightly increased. However, the 1/9 ML Ag is predicted to have a minor effect on the C diffusion rate because sufficient surface sites are available for C atoms to bypass the Bridge2 site to lower the barrier.

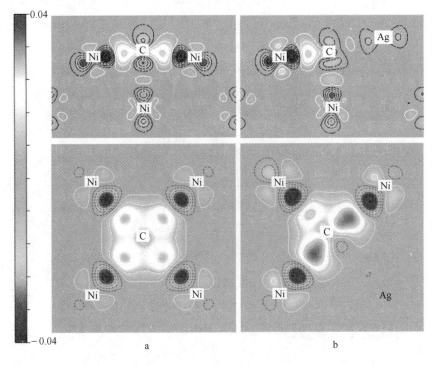

Fig. 9 Charge density difference($\Delta\rho = \rho(\text{surface} + \text{C}) - \rho(\text{surface}) - \rho(\text{C})$) upon the C adsorption at the hollow sites of the Ni(100) (a) and $Ag_{1/4}$/Ni(100) (b) surface, plotted perpendicular (upper panel) and parallel (lower panel) to the (100) surface with the C atoms involved (White (black) contour and red (blue) shading indicate the charge accumulation (depletion), relative to the sum of the isolated C atomic charge density and the charge density of the metal surfaces (For interpretation of the references to colour in this figure legend, the reader is referred to the web version of this article))

3.5 C diffusion through the subsurface of $Ag_{1/9}$/Ni(100) at the C coverage of 1/9ML

To evaluate the contribution of the subsurface migration mechanism to the C transport on the alloy surface, two C diffusion pathways named PATH3 and PATH4 are constructed, as shown in Fig. 10. Along PATH3 and PATH4, a C atom hops from a hollow site, passing through a Sub-O1 and Sub-O2 site, respectively, and finally arrives at a neighboring hollow site. Fig. 11 shows the potential energy diagram of the C diffusion along these pathways as well as the atomic arrangements of the corresponding transition states. Comparing the energy barriers with that for the C diffusion through the subsurface of Ni(100), one can see the C transport is dramatically promoted by the introduction of Ag. In the transition state for PATH3, the C atom is located at a quasi-Sub-T1 site and its two adjacent surface Ni atoms (the Ni1 and Ni2 atoms in the left inset of Fig. 11) are pushed apart, together with the displacement of the Ni2 and Ag atoms along the direction of the marked red arrows. On Ni(100), however, the energy barrier for the C hopping from the hollow to subsurface tetrahedral site is calculated to be 1.97eV[12], reflecting the fact that the substitutional Ag tends to enhance the mobility of the Ni surface to minimize the local lattice strain. The intermediate for PATH3 is 0.37eV higher in energy than the initial state, in which the C atom is adsorbed at a Sub-O1 site and the surface metal atoms are relaxed approximately back to their initial positions except that the Ni2 atom is pushed towards vacuum by 0.014nm.

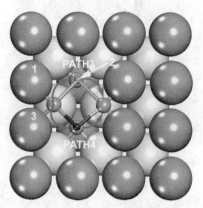

Fig. 10 Schematic representation of C diffusion through the subsurface of $Ag_{1/9}$/Ni(100) along two different pathways at the C coverage of 1/9ML

(The surface Ni atoms coordinated to the hopping C atoms are labeled by numbers)

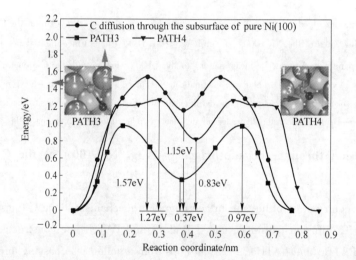

Fig. 11 MEPs for C diffusion on Ni(100) and $Ag_{1/9}$/Ni(100) between hollow sites through subsurface octahedral sites and the geometries of the transition states for PATH3 and PATH4

As for PATH4, the converged MEP shows a symmetrical structure with double maxima at 1.27eV and one local minimum at 0.83eV. Unlike PATH3, the C atom diffuses to a quasi-Sub-O2 site to overcome the energy barrier, without passing through the quasi-tetrahedral site. At the saddle point, the Ag atom is not only pushed towards vacuum by 0.040nm but also driven along the direction of the marked yellow arrow by 0.044nm, which subsequently leads to the displacement of the Ni2 and Ni3 atoms. In the intermediate, the C atom moves to the Sub-O2 site and the position of the Ag atom is 0.016nm higher than its image in the initial state. As we note that the center of the Ag atom has originally been located above the Ni surface by 0.057nm, the further protrusion is attributed to the repulsion between the Ag and C atoms. Based on the energy barriers calculated above, it is believed that the subsurface transport mechanism is more likely on Ag/Ni(100).

3.6 C diffusion through the surface of reconstructed $Ag_{1/4}$/Ni(100) at the C coverage of 7/16ML

As investigated in Section 3.3, the alloy surface is reconstructed to the "clock" geometry when a critical C coverage is attained. Then, we continue to evaluate the effect of the reconstruction on the C diffusion barriers through the surface and subsurface of $Ag_{1/4}$/Ni(100). In this section, the C coverage is decreased from 1/2ML to 7/16ML so as to leave sufficient surface sites for the C transport, which is found to be high enough to drive the alloy surface to undergo the transformation. Using this model, the diffusion of a single C atom between two 4-hollow sites is explored with the remaining C atoms binding to the corresponding 4-hollow sites throughout the process. As for the surface migration, one pathway named PATH5 is constructed in Fig. 12, along which a 3-hollow1 site is passed through. Fig. 13 shows the potential energy diagram and the configurations of the two saddle points. Along the unsymmetrical MEP, the C atom is located at a LB1 and LB2 site in the first and second transition state, respectively. These two transition states are very similar in geometry to those for PATH1 and PATH2. However, the surface is partially relaxed to a closer (111) geometry, resulting in larger interstitial that can be contracted. On the other hand, the reconstruction is more favorable on the alloy surface owing to the missing of the $c(2 \times 2)$ symmetry. Therefore, the mobility of the metal atoms on reconstructed Ag/Ni(100) is dramatically enhanced, which subsequently makes it easier to relieve the lattice strain induced by the C diffusion, as predicted by the lowered energy barrier in Fig. 13.

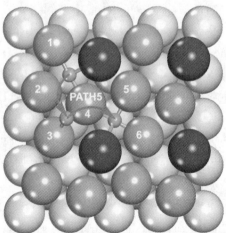

Fig. 12 Schematic representation of C diffusion through the surface of reconstructed $Ag_{1/4}$/Ni(100) along two different pathways at the C coverage of 7/16ML

(For clarity, the immovable C atoms are not displayed in the figure and the surface Ni atoms coordinated to the hopping C atoms are labeled by numbers)

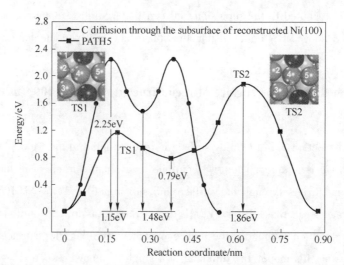

Fig. 13 MEPs for C diffusion on reconstructed Ni(100) and $Ag_{1/4}$/Ni(100) between 4-hollow sites through surface bridge sites and the geometries of the transition states for PATH5

3.7 C diffusion through the subsurface of reconstructed $Ag_{1/4}$/Ni(100) at the C coverage of 7/16ML

As indicated in our previous calculations[12], the C diffusion through the subsurface of reconstructed Ni(100) is most kinetically favorable. Thus, PATH6 and PATH7 are constructed to explore the C diffusion through the subsurface of reconstructed $Ag_{1/4}$/Ni(100), as shown in Fig. 14. Along these pathways, the C atom diffuses between two nearest neighboring 4-hollow sites, passing through a Sub-O1 or Sub-O2 site. The symmetrical potential energy diagrams of PATH6 and PATH7 are shown in Fig. 15, and the corresponding energy barriers are lowered to 1.14 and 0.76eV. Therefore, the subsurface migration mechanism is more preferred by the C transport on reconstructed Ag/Ni(100) as well. One and two saddle points are found along each half of PATH6 and PATH7, respectively. In the transition state for PATH6, the C atom is located at a quasi-Sub-O1 site and coordinated to five Ni atoms (two on the surface and three in the subsurface) with the average bond length of 0.196nm. At the first saddle point for PATH7, the C atom diffuses to a quasi-Sub-O2 site and forms five Ni—C covalent bonds with the similar lengths to those in the transition state for PATH6, as shown in Table 5. Consequently, the potential energies of the TS1 for PATH7 and the TS for PATH6 take almost the same value. The second transition state for PATH7 is quite close to the intermediate on the potential energy surface, leading to a very low energy barrier of 0.22eV. In conclusion, the C transport on the Ag/Ni alloy surface are predicted to follow PATH7 dominantly, and the diffusion rate is predicted to be higher than that on Ni(100).

Table 5 Comparison of the geometries of the transition states for PATH6 and PATH7

Transition state	d_{Ni-C}/nm
PATH6	0.177, 0.179, 0.190, 0.199, 0.237
First transition state for PATH7	0.182, 0.189, 0.191, 0.208, 0.223

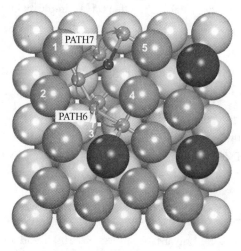

Fig. 14 Schematic representation of C diffusion through the subsurface of reconstructed $Ag_{1/4}/Ni(100)$ along two different pathways at the C coverage of 7/16ML

(For clarity, the immovable C atoms are not displayed in the figure and the surface Ni atoms coordinated to the hopping C atoms are labeled by numbers)

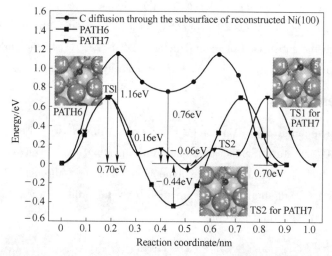

Fig. 15 MEPs for C diffusion on reconstructed Ni(100) and $Ag_{1/4}/Ni(100)$ between 4-hollow sites through subsurface octahedral sites and the geometries of the transition states for PATH6 and PATH7

4 Conclusion

Spin-polarized DFT-GGA calculations are performed to evaluate the effect of substitutional Ag on the geometry and property of Ni(100). It is found that the center of the doped Ag atom is located above the Ni surface by 0.052-0.064nm owing to the size mismatch. On the other hand, the Ni surface is pushed towards bulk metal, accompanied by the contraction of the interplanar spacing. A strong reduction of the C binding strength at the hollow site is observed, arising from the less attractive interaction between Ag and C atoms. However, the C adsorption energy at the next nn hollow site is unchanged, which agrees well with the results of electronic structure analysis that the PDOS on the d band of the next nn Ni atom is almost the same as that of atomic Ni on pure surface. Hence, the effect of Ag impurity on the adsorption property of Ni(100) is rather short-range.

Unlike the situation on the surface, the C adsorption in the subsurface is gradually enhanced as the Ag concentration is increased. When the Ag coverage of 1/2ML is attained, the C diffusion from surface to subsurface turns energetically favorable. At the C coverage of 1/2ML, the Ag/Ni alloy surface is reconstructed to the equilibrium quasi-$(2\times2)p4g$ structure without any activation energy. The 4-hollow and Sub-O1 sites are most preferred by the C adsorption on the surface and in the subsurface of reconstructed Ag/Ni(100), respectively. In this case, the C dissolution into subsurface has already become energetically favorable at the Ag coverage of 1/4ML.

Then, seven pathways between fourfold hollow sites are constructed to explore the C diffusion behavior through the surface and subsurface of Ag/Ni(100) and reconstructed Ag/Ni(100) using the CI-NEB method. The subsurface migration mechanism is expected to be more likely owing to the higher diffusion rate. On the other hand, it is found previously that the reconstruction can promote the C transport in the subsurface of pure Ni(100) because the local lattice strain induced by the C diffusion is effectively relieved, and the lowest diffusion barrier is predicted to be 1.16eV[12]. As for the alloy surface, the "clock" reconstruction is more favorable because of the missing of the $c(2\times2)$ symmetry. As a consequence, the energy barrier for the C diffusion through the subsurface of reconstructed Ag/Ni(100) is further lowered to 0.76eV as large space is achieved by the enhanced surface mobility to hold hopping C atoms.

Acknowledgements

This work is supported by Doctoral Fund of Ministry of Education of China (No. 200802511007), Natural Science Foundation of Shanghai (No. 08ZR1406300) and sponsored by Shanghai Educational Development Foundation through Chenguang plan (No. 2007CG41). The computational time provided by Notur project (www.notur.no) is highly acknowledged (nn2920k and nn4685k).

References

[1] G. Jones, J. G. Jakobsen, S. S. Shim, J. Kleis, M. P. Andersson, J. Rossmeisl, F. Abild-Pedersen, T. Bligaard, S. Helveg, B. Hinnemann, J. R. Rostrup-Nielsen, I. Chorkendorff, J. Sehested, J. K. Nørskov, J. Catal. 259 (2008) 147.

[2] D. L. Trimm, Catal. Today 49 (1999) 3.

[3] F. Besenbacher, I. Chorkendorff, B. S. Clausen, B. Hammer, A. M. Molenbroek, J. K. Nørskov, I. Stensgaard, Science 279 (1998) 1913.

[4] H. S. Bengaard, J. K. Nørskov, J. Sehested, B. S. Clausen, L. P. Nielsen, A. M. Molenbroek, J. R. Rostrup-Nielsen, J. Catal. 209 (2002) 365.

[5] E. Nikolla, J. Schwank, S. Linic, J. Catal. 250 (2007) 85.

[6] N. V. Parizotto, K. O. Rocha, S. Damyanova, F. B. Passos, D. Zanchet, C. M. P. Marques, J. M. C. Bueno, Appl. Catal. A-Gen. 330 (2007) 12.

[7] J. W. C. Liberatori, R. U. Ribeiro, D. Zanchet, F. B. Noronha, J. M. C. Bueno, Appl. Catal. A-Gen. 327 (2007) 197.

[8] J. C. S. Wu, H. C. Chou, Chem. Eng. J. 148 (2009) 539.

[9] J. R. Rostrup-Nielsen, I. Alstrup, Catal. Today 53 (1999) 311.
[10] S. Helveg, C. Lopez-Cartes, J. Sehested, P. L. Hansen, B. S. Clausen, J. R. Rostrup-Nielsen, F. Abild-Pedersen, J. K. Nørskov, Nature 427 (2004) 426.
[11] Y. A. Zhu, Y. C. Dai, D. Chen, W. K. Yuan, Surf. Sci. 601 (2007) 1319.
[12] Y. A. Zhu, X. G. Zhou, D. Chen, W. K. Yuan, J. Phys. Chem. C 111 (2007) 3447.
[13] C. Klink, L. Olesen, F. Besenbacher, I. Stensgaard, E. Laegsgaard, N. D. Lang, Phys. Rev. Lett. 71 (1993) 4350.
[14] R. Terborg, J. T. Hoeft, M. Polcik, R. Lindsay, O. Schaff, A. M. Bradshaw, R. L. Toomes, N. A. Booth, D. P. Woodruff, E. Rotenberg, J. Denlinger, Surf. Sci. 446 (2000) 301.
[15] J. E. Kirsch, S. Harris, Surf. Sci. 522 (2003) 125.
[16] S. Stolbov, S. Hong, A. Kara, T. S. Rahman, Phys. Rev. B 72 (2005) 155423.
[17] G. Kresse, J. Hafner, Phys. Rev. B 47 (1993) 558.
[18] G. Kresse, J. Hafner, Phys. Rev. B 49 (1994) 14251.
[19] G. Kresse, J. Furthmüller, Comp. Mater. Sci. 6 (1996) 15.
[20] G. Kresse, J. Furthmüller, Phys. Rev. B 54 (1996) 11169.
[21] P. E. Blöchl, Phys. Rev. B 50 (1994) 17953.
[22] G. Kresse, D. Joubert, Phys. Rev. B 59 (1999) 1758.
[23] J. P. Perdew, K. Burke, M. Ernzerhof, Phys. Rev. Lett. 77 (1996) 3865.
[24] H. J. Monkhorst, J. D. Pack, Phys. Rev. B 13 (1976) 5188.
[25] M. Methfessel, A. T. Paxton, Phys. Rev. B 40 (1989) 3616.
[26] G. Kresse, J. Hafner, Surf. Sci. 459 (2000) 287.
[27] H. Jónsson, G. Mills, K. W. Jacobsen, in: B. J. Beren, G. Ciccitti, D. F. Coker (Eds.), Classical and Quantum Dynamics in Condensed Phase Simulations, World Scientific Publishing Co., Singapore, 1998, p. 385.
[28] G. Henkelman, B. P. Uberuaga, H. Jonsson, J. Chem. Phys. 113 (2000) 9901.
[29] G. Henkelman, H. Jonsson, J. Chem. Phys. 113 (2000) 9978.
[30] D. Sheppard, R. Terrell, G. Henkelman, J Chem. Phys. 128 (2008) 10.
[31] A. Christensen, A. V. Ruban, P. Stoltze, K. W. Jacobsen, H. L. Skriver, J. K. Nørskov, F. Besenbacher, Phys. Rev. B 56 (1997) 5822.
[32] B. Hammer, J. K. Nørskov, Nature 376 (1995) 238.
[33] K. Termentzidis, J. Hafner, F. Mittendorfer, J. Phys. Condes. Matter. 18 (2006) 10825.
[34] G. Henkelman, A. Arnaldsson, H. Jonsson, Comput. Mater. Sci. 36 (2006) 354.
[35] E. Sanville, S. D. Kenny, R. Smith, G. Henkelman, J. Comput. Chem. 28 (2007) 899.
[36] W. Tang, E. Sanville, G. Henkelman, J. Phys. Condes. Matter. 21 (2009) 7.
[37] D. L. Lide, CRC Handbook of Physics and Chemistry, Academic Press, 1997.

Catalytic Reduction of Hexaminecobalt(Ⅲ) by Pitch-based Spherical Activated Carbon(PBSAC)*

Abstract The wet ammonia (NH$_3$) desulfurization process can be retrofitted to remove nitric oxide (NO) and sulfur dioxide (SO$_2$) simultaneously by adding soluble cobalt(Ⅱ) salt into the aqueous ammonia solution. Activated carbon is used as a catalyst to regenerate hexaminecobalt(Ⅱ), Co(NH$_3$)$_6^{2+}$, so that NO removal efficiency can be maintained at a high level for a long time. In this study, the catalytic performance of pitch-based spherical activated carbon(PBSAC) in the simultaneous removal of NO and SO$_2$ with this wet ammonia scrubbing process has been studied systematically. Experiments have been performed in a batch stirred cell to test the catalytic characteristics of PBSAC in the catalytic reduction of hexaminecobalt(Ⅲ), Co(NH$_3$)$_6^{3+}$. The experimental results show that PBSAC is a much better catalyst in the catalytic reduction of Co(NH$_3$)$_6^{3+}$ than palm shell activated carbon (PSAC). The Co(NH$_3$)$_6^{3+}$ reduction reaction rate increases with PBSAC when the PBSAC dose is below 7.5g/L. The Co(NH$_3$)$_6^{3+}$ reduction rate increases with its initial concentration. Best Co(NH$_3$)$_6^{3+}$ conversion is gained at a pH range of 2.0-6.0. A high temperature is favorable to such reaction. The intrinsic activation energy of 51.00kJ/mol for the Co(NH$_3$)$_6^{3+}$ reduction catalyzed by PBSAC has been obtained. The experiments manifest that the simultaneous elimination of NO and SO$_2$ by the hexaminecobalt solution coupled with catalytic regeneration of hexaminecobalt(Ⅱ) can maintain a NO removal efficiency of 90% for a long time.

Key words hexaminecobalt ion, nitric oxide, pitch-based spherical activated carbon, reduction, sulfur dioxide

1 Introduction

Power plant flue gas frequently contains NO$_x$ and sulfur dioxide (SO$_2$) pollutants that cause acid rain and urban smog. The removal of these contaminants to comply with environmental emission standard is of imperative necessity. Nitric oxide(NO) is 90%-95% of the NO$_x$ present in typical flue gas streams[1]. Traditionally, various wet and dry processes have been put forward. It has been agreed that using chemical additives to an aqueous scrubbing solution could have a significant impact on control strategies[2,3]. Several methods have been developed to enhance NO absorption, including the use of oxidants to oxidize NO to the more soluble NO$_2$ and the addition of various iron(Ⅱ) chelates to bind and activate NO[4,5]. So far, none of these methods have been put into commercial application.

Long et al.[6,7] put forward a novel technique for the simultaneous elimination of NO and SO$_2$ from the flue gas by adding soluble cobalt(Ⅱ) salt into aqueous ammonia (NH$_3$) solution. The mechanism for the simultaneous removal of NO and SO$_2$ with this technique has been reported in these two literatures. The Co(NH$_3$)$_6^{2+}$ formed by ammonia binding with cobalt(Ⅱ) is the

---

* Coauthors: Chen Yu, Mao Yanpeng, Zhu Haisong, Cheng Jingyi, Long Xiangli. Reprinted from *Clean-Soil, Air, Water*, 2010,38(7):601-607.

active constituent of scrubbing NO from the flue gas streams. Dissolved oxygen, in equilibrium with the residual oxygen in the flue gas, is the oxidant. NO is converted into nitrate and nitrite. However, $Co(NH_3)_6^{2+}$ may also be oxidized to $Co(NH_3)_6^{3+}$ by the dissolved oxygen, which is unable to activate oxygen molecules or bind NO. The NO removal efficiency will decrease as the reaction proceeds. In order to regenerate the active constituent $Co(NH_3)_6^{2+}$ to maintain the capability of removing NO from the gas streams, water is used to reduce cobalt (Ⅲ) to cobalt(Ⅱ) under the catalysis of activated carbon. The reaction scheme for the catalytic reduction of $Co(NH_3)_6^{3+}$ with activated carbon catalyst can be illustrated as follows:

$Co(NH_3)_6^{3+}$ may be absorbed on the surface of activated carbon quickly and form a transition complex $AC\cdots Co(NH_3)_6^{3+}$:

$$AC + Co(NH_3)_6^{3+} \rightleftharpoons AC\cdots Co(NH_3)_6^{3+} \qquad (1)$$

Ammonia molecules may react with the acidic part of carbonyl groups and phenolic hydroxyl groups to form $CO\text{-}(NH_4)^+$ complexes on carbon surface[8]. The formation of these complexes may speed up the $Co(NH_3)_6^{3+}$ ions' dissociation to cobalt(Ⅲ):

$$AC\cdots Co(NH_3)_6^{3+} \rightleftharpoons Co^{3+} + 6NH_3 + AC \qquad (2)$$

The equilibrium between NH_3 and NH_4^+ in the aqueous solution can be illustrated as follow:

$$NH_3 + H^+ \rightleftharpoons NH_4^+ \qquad (3)$$

Electrochemical half-cell reduction potential of Co^{3+}/Co^{2+} (1.82V) shows that cobalt(Ⅲ) ions are strong oxidants and can be reduced to Co^{2+} easily. On the other hand, the basicity of an activated carbon is due to the presence of basic oxygen-containing functional group (e.g., pyrones or chromenes) and/or graphene layers acting as Lewis bases and forming electron donor-acceptor (EDA) complexes with H_2O molecules. Montes-Morán et al.[9] explored the causes of the basic nature of carbonaceous materials and concluded that an unconventional bond may be established between the H_3O^+ ions and the cloud of π electrons of the aromatic rings of the activated carbon. These latter basic sites are located at π-electron-rich regions within the basal planes of carbon crystallites away from the crystallite edges[10]. This delocalized π electron system can act as a Lewis base in aqueous solution[10]:

$$-C\pi + 2H_2O \rightleftharpoons C\pi-H_3O^+ + OH^- \qquad (4)$$

This delocalized π-electron system ($-C\pi$) can act as the reduction center of Co^{3+}. The electrochemical half-cell reduction potentials of $E_{O_2/OH^-} = 0.401V$ and $E_{O_2/H_2O} = 1.229V$ show that hydroxyl ions are more liable to be oxidized to O_2 than H_2O molecules by cobalt(Ⅲ) ions. The reduction of cobalt(Ⅲ) is realized as the reaction (5) takes place:

$$Co^{3+} + OH^- \longrightarrow Co^{2+} + \frac{1}{2}H_2O + \frac{1}{4}O_2 \qquad (5)$$

The net reaction for the regeneration of cobalt(Ⅱ) ions can be written as follows:

$$[Co(NH_3)_6]^{3+} + 0.5H_2O + 5H^+ \xrightarrow{AC} 6NH_4^+ + Co^{2+} + \frac{1}{4}O_2 \qquad (6)$$

Cobalt(Ⅱ) ions may combine with ammonia in the solution:

$$Co^{2+} + 6NH_3 \longrightarrow Co(NH_3)_6^{2+} \qquad (7)$$

As a result, the regeneration of hexaminecobalt(Ⅱ) ions is completed. It can be concluded that the catalytic reduction of $Co(NH_3)_6^{3+}$ is a crucial step in the simultaneous removal of SO_2 and NO with the hexaminecobalt(II) ammonia solution.

Activated carbons have been successfully put into industrial application as catalysts for their enormous surface area, porous structure, and characteristic flexibility. Bashkova et al.[11] found that activated carbon surface chemistry, particularly basic oxygen-containing groups and ash content played an important role in the catalytic oxidation of methyl mercaptan. Palm shell activated carbon (PSAC) used as a catalyst in the combined absorption of SO_2 and NO with the hexaminecobalt(II) ammonia solution has been studied[7]. Compared with PSAC, pitch-based spherical activated carbon (PBSAC) exhibits a higher mechanical strength, better adsorption performance, lower ash and symmetrical bulk density[12]. It is worthwhile investigating the performance of PBSAC as a catalyst in the process of scrubbing SO_2 and NO with the hexaminecobalt ammonia solution. A study on the catalytic characteristics of PBSAC is reported in this paper.

2 Experimental

2.1 Batch experiments

Batch experiments were performed in a stirred glass flask of 500mL to measure the $Co(NH_3)_6^{3+}$ reduction catalyzed by activated carbon. A turbine impeller of diameter 3cm was mounted on the bottom of the stirring rod. The temperature is controlled by constant temperature bath. Four hundred milliliters of 0.01mol/L $Co(NH_3)_6^{3+}$ solution was introduced into the glass flask. Appropriate dose of activated carbon was added into the $Co(NH_3)_6^{3+}$ solution as soon as the scheduled temperature attained. The pH values of the solutions were adjusted by adding appropriate amounts of H_2SO_4 or NaOH before activated carbon was introduced. The liquid samples were withdrawn every 3min to determine the change of $Co(NH_3)_6^{2+}$ concentration in the course of the experiments.

Cobalt(II) was determined spectrophotometrically at 25°C from the absorbance at 690nm of a 9mol/L solution prepared by diluting an aliquot of sample solution with concentrated HCl. This determination was made using a 10cm cell. The cobalt(II) calibration curve was obtained using standard cobalt acetate solutions ranged from 0.00 to 0.02mol/L. Least-squares fits to the data yield Eq. 8 with a correlation coefficient (r^2) 0.9999:

$$C = -0.0005 + 53.034X \quad (8)$$

where, X stands for absorbency and C for cobalt(II) concentration (mol/L).

2.2 Continuous experiments

Experiments for the simultaneous removal of SO_2 and NO were performed in a packed column with 1000mm in length and 20mm in diameter. The schematic diagram of the experimental apparatus is shown in Fig. 1. The absorber temperature was controlled using a jacket through which water from a thermostatic bath was circulated. 2% of NO in nitrogen was supplied from a cylinder, and was diluted with N_2 to the desired concentration before feeding into the absorber. SO_2 was supplied in a similar manner. Measured amount of $CoCl_2$ and ammonia solution were added into the 500mL glass circulation tank. The absorber was operated with a continuous influent gas feeding at 300mL/min from the bottom and a continuous scrubbing solution feeding, at a superficial flow rate of $5m^3/(m^2 \cdot h)$ (25mL/min) at the top. The absorbent effusing from the packed column was fed into the circulation tank. When the regeneration of $Co(NH_3)_6^{2+}$ started, the absorbent in circulation tank flew into the regeneration reactor upwardly and directly into the packed column to scrub NO and SO_2. The regeneration reactor was a fixed-bed column packed with activated carbon (20mm i.d., 800mm long). The experimental runs were carried out under atmospheric pressure. Ammonia was added in the continuous process to maintain the pH of the solution.

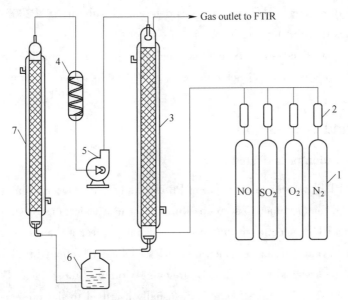

Fig. 1 Flowchart of experimental setup
1—Cylinder; 2—Massmeter; 3—Packed column; 4—Condenser;
5—Pump; 6—Circulation tanker; 7—Regeneration reactor

During the experiments, the quantitative analysis of gas compositions was achieved by an on-line Fourier transform infrared spectrometer (FTIR) (Nicolet E. S. P. 460 FT-IR) equipped with a gas cell and a quantitative package, named Quant Pad. The inlet and outlet gases were directly introduced into the gas cell of the FTIR to obtain the transient NO, N_2O, NO_2, SO_2, and H_2O concentrations in both inlet and outlet gases, as well as the transient NO conversion. This set-up was conveniently operated to monitor the absorption effect of NO and SO_2.

2.3 Characterization of carbon catalysts

The concentrations of acidic and basic groups on the carbon surface were determined by Boehm titration[13]. One gram of carbon sample was placed in 50mL of the following solutions: sodium hydroxide, sodium carbonate, sodium bicarbonate, and hydrochloric acid. The vials were sealed and shaken for 24h and then filtered. Five milliliters of the filtrate was pipetted and the excess base or acid was titrated with HCl or NaOH, respectively. The number of acidic sites was determined under the assumptions that NaOH neutralizes carboxylic, lactonic, and phenolic groups;

that Na_2CO_3 neutralizes carboxylic and lactonic groups; and that $NaHCO_3$ neutralizes only carboxylic groups. The number of basic sites was calculated from the amount of hydrochloric acid that reacted with the carbon.

pH of point of zero charge (pH_{PZC}) was determined using the mass titration method described by Noh and Schwarz[14,15]. $NaNO_3$ (0.01mol/L) solutions of pH 3, 6, and 11 were prepared using HNO_3 (0.1mol/L) and NaOH (0.1mol/L). A 100mL volume solution of different initial pH and increasing amounts of carbon sample (0.1%, 0.5%, 1.1%, 5%, 10%, 15%, and 20%) was added to 250mL Erlenmeyer flasks. The equilibrium pH was measured after 24h of shaking at (25 ± 0.1)°C.

The specific surface area (BET method) was determined with an ASAP2000 Surface Analyzer (Micromeritics Co. USA) using N_2 as the adsorbate.

Fourier transform infrared spectra were recorded with potassium bromide-pressed disks, by accumulating 32 scans at $4cm^{-1}$ resolution between 400 and $4000cm^{-1}$ using a Nicolet 5700 FTIR.

3 Results and Discussion

3.1 Characterization of activated carbon

Fig. 2 shows the FTIR transmission spectra of PBSAC and PSAC. Observation of the absorption bands are expected for the functional groups which exist in a wide range of different electronic environments. The FTIR spectra of the original samples present bands at $2918cm^{-1}$ and $2845cm^{-1}$ due to asymmetric and symmetric C—H stretching vibrations in aliphatic. The band at $1726cm^{-1}$ can be assigned to the C=O stretching vibration from lactone[16,17]. The shoulders observed at $1562cm^{-1}$ and $1455cm^{-1}$ are usually ascribed to the presence of carboxylic acid groups[18]. The band centered at $1143cm^{-1}$ is ascribed to the O—H from phenolic hydroxyl. The band at about $3500cm^{-1}$ can be assigned to the O—H stretching mode of hydroxyl functional groups. Though the bands on these two carbons are the same, the range of some bands vibration is different. For the PBSAC, there is a large increase in the intensity of the band at $1726cm^{-1}$, the C=O stretching vibration from lactone, that together with that at about $1562cm^{-1}$ and $1455cm^{-1}$, due to the C=O stretching vibration, can be assigned to carboxyl

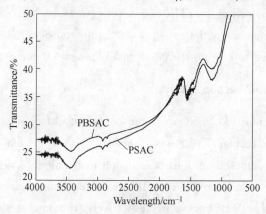

Fig. 2 FTIR spectra of PBSAC and PSAC

acid groups. There is also a predominant increase in the intensity of the band at 1143cm^{-1}, which is assigned to phenolic hydroxyl O—H stretching, while the O—H stretching mode of hydroxyl functional groups have little decrease. The results of FTIR analyses show that PBSAC possess more acid surface groups than PSAC.

Table 1 presents the surface functional groups concentration for both PBSAC and PSAC. It manifests that the amount of basic group on the PBSAC surface is nearly the same with that on the PSAC surface, but there is more acidic groups on the PBSAC surface than on the PSAC surface. The amount of acidic groups (0.2405mol/g) on PBSAC surface is approximately 1.5 times that (0.1868mol/g) on PSAC surface. The amount of carboxyl group (0.05505mol/g) on PBSAC surface is over two times that (0.02736mol/g) on PSAC surface. Much more acidic groups on the surface of PBSAC is the reason that pH_{pzc} of PBSAC is lower than that of PSAC. The data shown in Table 2 demonstrate that the specific surface area of PBSAC is larger than that of PSAC and the average pore diameter of PBSAC is bigger than that of PSAC.

Table 1 Physicochemical characteristics of carbon catalysts

Carbon type	Phenolic hydroxyl /mol·g^{-1}	Carboxyl /mol·g^{-1}	Lactonic /mol·g^{-1}	Total acid groups /mol·g^{-1}	Total basic groups /mol·g^{-1}	pH_{pzc}
PSAC	0.143	0.0271	0.0162	0.187	0.0754	9.18
PBSAC	0.162	0.0551	0.0234	0.241	0.0741	5.90

Table 2 Physical characteristics of carbon catalysts

Carbon type	BET SA/m^2·g^{-1}	Pore volume/cm^3·g^{-1}	Average pore diameter/nm
PSAC	1002	0.45	1.99
PBSAC	1232	0.67	3.50

3.2 Comparison of the catalytic ability between PBSAC and PSAC

Experiments have been performed at 60℃ with an initial pH of 5.45 to test the $Co(NH_3)_6^{3+}$ reduction catalyzed by PBSAC and PSAC, respectively. The experimental results demonstrated in Fig. 3 present that PBSAC can obtain higher $Co(NH_3)_6^{3+}$ conversion than PSAC. After a 30min operation, the $Co(NH_3)_6^{3+}$ conversion obtained under the catalysis of PBSAC is 80.0% while that under the catalysis of PSAC is 53.8%. The experimental results shows that PBSAC is a much better catalyst for $Co(NH_3)_6^{3+}$ reduction than PSAC. The catalytic performance of carbon has close relations with its chemical and physical characteristics. PSBAC holds a 20% higher specific surface area and 48.9% higher pore volume compared to PSAC. Furthermore, PBSAC has 28.9% more acidic funcitional groups on its surface than PSAC. These are the reasons that PBSAC catalyzes $Co(NH_3)_6^{3+}$ conversion much more efficiently.

3.3 Effect of stirring speed

Experiments have been performed to explore the effect of stirring speed on the reduction of $Co(NH_3)_6^{3+}$ at 60℃ with an initial pH of 5.45. The experimental results are shown in Fig. 4. It can be seen from Fig. 4 that the $Co(NH_3)_6^{3+}$ conversion may increase with the stirring speed as the stirring speed is below 300r/min. After 20min operation, the $Co(NH_3)_6^{3+}$ conversion increa-

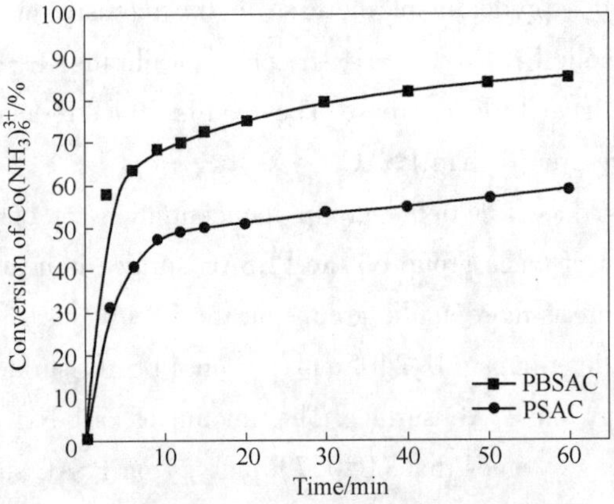

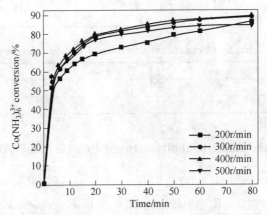

Fig. 3 Comparison of the catalytic ability between PBSAC and PSAC
(Stirring speed = 300r/min, 120mesh activated carbon = 2.5g/L, initial $Co(NH_3)_6^{3+}$ concentration = 0.01mol/L, pH = 5.45, 60℃)

ses from 69.4% to 79.7% as the stirring speed increases from 200r/min to 300r/min. The $Co(NH_3)_6^{3+}$ conversion at 300r/min is nearly equal to that at 400r/min. But the $Co(NH_3)_6^{3+}$ conversion at 500r/min is 77.6%, which is lower than that obtained at 300r/min.

Fig. 4 Effect of stirring speed on $Co(NH_3)_6^{3+}$ reduction
(0.01mol/L $Co(NH_3)_6^{3+}$, 120 mesh PBSAC = 2.5g/L, T = 60℃, pH = 5.45)

There is a viscous flow coat around the exterior of PBSAC in this liquid-solid catalytic reaction. The reaction only takes place after the reactants get through the liquid film to the surface of PBSAC. The thickness of the liquid film and the resistance of mass transfer will be reduced with the increase of stirring speed. Therefore, the reaction rate may be speeded up with the stirring speed. However, the external mass transfer limitation is eliminated as the stirring speed is above 300r/min. Therefore, the reaction rate may not increase further as the stirring speed rises. When the stirring speed is 500r/min, some PBSAC is spattered on the wall of the flask. The $Co(NH_3)_6^{3+}$ conversion rate will be reduced because the amount of PBSAC in the aqueous solution decreases.

3.4 Effect of PBSAC particle size

Experiments have been performed to explore the effect of PBSAC particle size on the reduction of $Co(NH_3)_6^{3+}$. The experimental results shown in Fig. 5 depicts that the $Co(NH_3)_6^{3+}$ conversion increases as PBSAC particle size reduces. For example, after 60min operation, the $Co(NH_3)_6^{3+}$ conversion is 71.73% catalyzed with 40-80 mesh PBSAC while the $Co(NH_3)_6^{3+}$ conversion gets to 81.67% catalyzed with 80-100 mesh PBSAC. However, the $Co(NH_3)_6^{3+}$ conversion does not increase further as the activated carbon particle size reduces to smaller than 120 mesh. The $Co(NH_3)_6^{3+}$ conversion catalyzed with 120-140 mesh activated carbon is almost equal to that catalyzed with 140-180 mesh activated carbon. It can be concluded that the internal mass transfer is negligible as the particle size is smaller than 100-120 mesh.

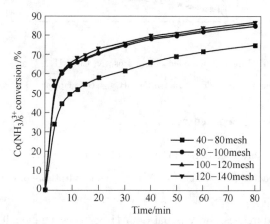

Fig. 5 Effect of activated carbon particle size on $Co(NH_3)_6^{3+}$ reduction

(0.01mol/L $Co(NH_3)_6^{3+}$, stirring speed = 300r/min, T = 60℃, pH = 5.45, PBSAC 2.5g/L)

In heterogeneous catalysis, the conversion of one reactant is very often affected by mass transfer limitations, so apparent kinetics are actually governed by external or internal mass transfer resistances. Intrinsic kinetics can only be evaluated if the above effects are minimized. According to the experimental results above, the mass transfer resistances can be eliminated under the conditions with a stirring speed greater than 300r/min and PBSAC particle size smaller than 120 mesh. The runs following are carried out under the condition without mass transfer resistances.

3.5 Effect of PBSAC dose

A set of experiments were carried out at 60℃ to investigate the effect of PBSAC dose on $Co(NH_3)_6^{3+}$ reduction. It can be seen from the experimental results shown in Fig. 6 that PBSAC greatly speeds up $Co(NH_3)_6^{3+}$ reduction. After 15min, the $Co(NH_3)_6^{3+}$ conversion reaches up to 77.5% with 5.0g/L activated carbon in the solution, while only 8.2% $Co(NH_3)_6^{3+}$ is reduced without PSAC existing in the solution. The $Co(NH_3)_6^{3+}$ conversion increases as PSAC dose increases when the PBSAC dose is below 7.5g/L. For example, after 15min operation, the $Co(NH_3)_6^{3+}$ conversion catalyzed with 1.25g/L PBSAC is 58.9% while that catalyzed with 5.0g/L is 78.8%. This can be ascribed to more carbon surface area and availability of more active sites at higher PBSAC dose. However, it can also be found from Fig. 6 that the $Co(NH_3)_6^{3+}$ conversion measured decreases as the PBSAC dose increases further above 7.5g/L. The reason is that too much dose of PSAC results in the increase of cobalt(Ⅱ) adsorption.

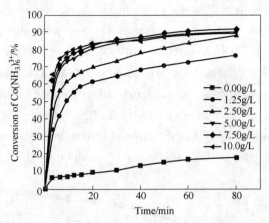

Fig. 6　Effect of activated carbon dose on $Co(NH_3)_6^{3+}$ reduction

(Stirred speed = 300r/min, 120 mesh PBSAC, initial $Co(NH_3)_6^{3+}$ concentration = 0.01mol/L, pH = 5.45, 60℃)

3.6　Effect of $Co(NH_3)_6^{3+}$ concentration

A series of experiments were conducted to test the effect of $Co(NH_3)_6^{3+}$ concentration on its reduction in the stirred reactor with initial $Co(NH_3)_6^{3+}$ concentrations of 0.005mol/L, 0.01mol/L, 0.015mol/L, and 0.02mol/L, respectively. The experimental results illustrated in Fig. 7 depict that the Co^{2+} regenerated increases with initial $Co(NH_3)_6^{3+}$ concentration. It has been found [19] that such redox reaction is first order in respect to $Co(NH_3)_6^{3+}$. $Co(NH_3)_6^{3+}$ reduction rate increases with its concentration.

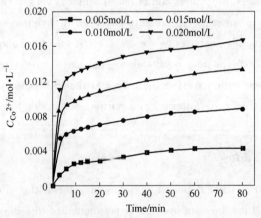

Fig. 7　Effect of initial solution concentration on $Co(NH_3)_6^{3+}$ reduction

(Stirred speed = 300r/min, 120 mesh PBSAC = 2.50g/L, pH = 5.45, T = 60℃)

3.7　Effect of temperature

Experiments were performed at pH 5.46 with temperature varied from 40 to 80℃ to investigate the effect of temperature on $Co(NH_3)_6^{3+}$ reduction. From the experimental results shown in Fig. 8, it is observed that $Co(NH_3)_6^{3+}$ reduction efficiency increases with temperature. After 20min, the $Co(NH_3)_6^{3+}$ conversion efficiency is 90% at 80℃ while that at 40℃ is 57.9%. The explanation can be given as follows: (1) high temperature is liable to make $Co(NH_3)_6^{3+}$

disintegrate easily, and conducible to produce more cobalt(Ⅲ) ions. (2) The diffusivity of solute through the external laminar layer into the micropores of PBSAC increases with temperature. (3) Dynamically, the reaction rate increases with temperature. (4) Oxygen solubility decreases with temperature, causing the oxygen produced by this reaction to stripe much more quickly from the PBSAC. All these factors are of benefit to $Co(NH_3)_6^{3+}$ reduction.

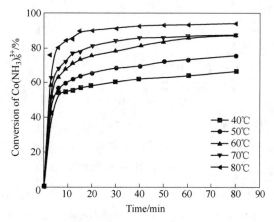

Fig. 8 Effect of temperature on $Co(NH_3)_6^{3+}$ reduction

(Stirring speed = 300r/min, 120 mesh PBSAC = 2.50g/L, initial $Co(NH_3)_6^{3+}$ concentration = 0.01mol/L, pH = 5.45)

The reaction constants can be obtained from the curves of $-\ln C_{Co(NH_3)_6^{3+}}$ versuss t plotted in terms of the experimental data shown in Fig. 8. The reaction constants for this reaction catalyzed by PBSAC are 0.0063min^{-1}, 0.0179min^{-1}, 0.0258min^{-1}, 0.0404min^{-1}, and 0.0664min^{-1} for 40℃, 50℃, 60℃, 70℃, and 80℃, respectively. An estimation of the reaction activation energy can be obtained from the Arrhenius-type plot of ln$k$ versus $1/T$. The intrinsic activation energy E for this catalytic reduction reaction by the catalysis of PBSAC has a value of E_a 51.00kJ/mol compared with E_a 56.73kJ/mol[19] by the catalysis of PSAC. It can be concluded that PBSAC greatly reduces the intrinsic activation energy for the reduction of $Co(NH_3)_6^{3+}$.

3.8 Effect of initial pH

To examine the effect of pH on the $Co(NH_3)_6^{3+}$ reduction efficiency, experiments were made at 60℃ with the pH varying from 2.0 to 8.0. The $Co(NH_3)_6^{3+}$ conversions for a reaction of 20min are shown in Fig. 9.

From Fig. 9 it is clear that PBSAC is an efficient catalyst for the reduction of $Co(NH_3)_6^{3+}$ over the pH range 2.0-6.0 and the $Co(NH_3)_6^{3+}$ conversion remains at about 80%. At lower or higher pH value the $Co(NH_3)_6^{3+}$ reduction decreases. As the pH decreasing below 2.0, there is a sharp decrease in $Co(NH_3)_6^{3+}$ reduction. For example, at pH 1.0 the $Co(NH_3)_6^{3+}$ conversion is only about 11.8% after 25min reaction as pH rises above 6.0, there is also a decrease in $Co(NH_3)_6^{3+}$ conversion with further increase in pH. For instance, after 25min run, the $Co(NH_3)_6^{3+}$ conversions are 72.7% at pH 7.0 and 60.3% at pH 8.0, respectively. The effect of pH on $Co(NH_3)_6^{3+}$ reduction can be explained as follows.

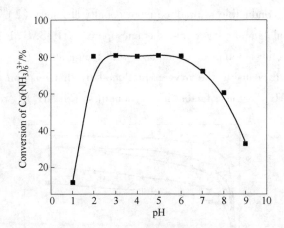

Fig. 9 Effect of initial pH on $Co(NH_3)_6^{3+}$ reduction

(Stirring speed = 300r/min, 120mesh PBSAC = 2.5g/L, initial $Co(NH_3)_6^{3+}$
concentration = 0.01mol/L, T = 60℃)

In an aqueous solution, there is equilibrium between NH_3 and NH_4^+. NH_3 become fewer with the decreasing of pH. Therefore, according to Eq. 3, disintegration of $Co(NH_3)_6^{3+}$ is enhanced as the acidity of the solution becomes stronger, which is advantageous to $Co(NH_3)_6^{3+}$ reduction. However, the OH^- concentration in the solution reduces as pH decreases, which is detrimental to the reduction of cobalt(Ⅲ) in terms of Eq. 5. Furthermore, as pH reduces, more H^+ ions will be absorbed on the surface of PBSAC and the surface of PBSAC will become more positive. It is very difficult for $Co(NH_3)_6^{3+}$ ions to be adsorbed on PBSAC surface at lower pH. Hence, the reduction of $Co(NH_3)_6^{3+}$ will be harmed while pH is lowered. Therefore, there is an optimal pH range for $Co(NH_3)_6^{3+}$ reduction catalyzed by activated carbon in aqueous solutions.

3.9 Simultaneous removal of NO and SO_2

The variation of NO removal efficiency with time coupled with the regeneration of cobalt(Ⅱ) catalyzed by PBSAC and PSAC are shown in Fig. 10. For the first 17h the operation is carried out without the regeneration of the absorbents, and then the regeneration is started and continues during the rest of the run. The experiment is performed at 50℃ with 5.0% oxygen, an initial $Co(NH_3)_6^{2+}$ concentration of 0.04mol/L, and inlet NO and SO_2 concentrations of 550 and 1500ppm, respectively. The flow rate of scrubbing solution fed into the regeneration column was 25mL/min. Before the start of regeneration, these two operations have nearly the same NO removal efficiencies. The initial NO removal efficiencies are 100%, which decline to about 53% after 17h. After activating the regeneration system, the NO removal efficiencies rise quickly. One hour after the $Co(NH_3)_6^{2+}$ regeneration catalyzed by PBSAC at the temperature of 80℃, the NO removal efficiency goes up to 92.2% and stabilizes at a level between 89.8% and 91.9% during the next 182h operation. PSAC can also enhance the removal efficiency, which raises NO removal efficiency from 53% to 83% after it catalyzes the $Co(NH_3)_6^{2+}$ regeneration for 1h at 80℃. But the NO removal efficiency decreases as the operation proceeds. After 80h operation, the NO removal efficiency reduces to 70%. Then the regeneration temperature rises to 90℃, which results in the NO removal efficiency increasing to 85%. However, the NO removal efficiency decreases to 82% 20h later. During the experiments, there is no SO_2 detected in the outlet gas streams. The SO_2 removal efficiency is 100%.

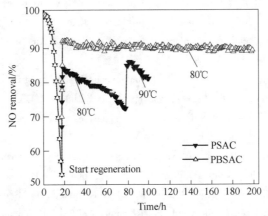

Fig. 10　NO removal efficiencies with time coupled with the regeneration catalyzed by PBSAC and PSAC
(50℃, NO = 550 × 10^{-6}, SO_2 = 1500 × 10^{-6}, O_2 = 5.2%)

The experimental results demonstrate than PBSAC can obtain a much higher NO removal efficiency than PSAC. PBSAC is a better catalyst in the simultaneous removal of SO_2 and NO with $Co(NH_3)_6^{2+}$ ammonia solution than PSAC.

4　Conclusions

The combined elimination of SO_2 and NO with the hexaminecobalt solution coupled with the regeneration of $Co(NH_3)_6^{2+}$ catalyzed by PBSAC has been studied systematically. The following specific conclusions can be drawn from the experimental results:

(1) Pitch-based spherical activated carbon is a better catalyst in the $Co(NH_3)_6^{3+}$ reduction reaction than PSAC. The $Co(NH_3)_6^{3+}$ conversion increases with the PBSAC mass when the PBSAC dose is below 7.5g/L.

(2) The $Co(NH_3)_6^{3+}$ conversion increases with its initial concentration.

(3) $Co(NH_3)_6^{3+}$ conversion increases with temperature. The intrinsic activation energy E for this catalytic reduction reaction has a value of E_a = 51.00kJ/mol.

(4) pH is a vital factor affecting this catalytic reduction reaction. The optimal pH is at the range of 2.0-6.0. The $Co(NH_3)_6^{3+}$ conversion decreases as pH is lower or higher.

(5) Pitch-based spherical activated carbon can obtain a much higher NO removal efficiency in the simultaneous absorption of NO and SO_2 coupled with the catalytic regeneration of $Co(NH_3)_6^{2+}$ than PSAC. PBSAC is a better catalyst in the simultaneous removal of SO_2 and NO with $Co(NH_3)_6^{2+}$ ammonia solution than PSAC.

Acknowledgements

The present work is supported by the Ministry of Science and Technology of China (No. 2006AA05Z307) and the state key laboratory of chemical engineering (SKL-ChE-08C05).

References

[1] C. J. Pereira, M. D. Amiridis, NO_x Control from Stationary Sources, ACS Symp. Ser. 1995, 587, 1-13.

[2] C. Mackell, B. K. Hodnett, G. Paparatto, Testing of the CuO/Al_2O_3 Calalyst-Sorbent in Extended Operation for the Simultaneous Removal of NO_x and SO_2 from Flue Gases, Ind. Eng. Chem. Res. 2000, 39, 3868.

[3] J. J. Kaczur, Oxidation Chemistry of Chloric Acid in NO_x/SO_2 and Air Toxic Metal Removal from Gas Streams, Environ. Process 1996, 15, 245.

[4] E. Sada, H. Kumazama, L. Kudo, Individual and Simultaneous Absorption of Dilute NO and SO_2 in Aqueous Slurres of $MgSO_4$ with Fe^{2+} EDTA, Ing. Eng. Chem. Proc. Dev. 1980, 19, 377.

[5] S. G. Chang, D. K. Liu, Removal of Nitrogen and Sulphur Oxides from Waste Gas Using a Phosphorus/Alkali Emulsion, Nature 1990, 343, 151.

[6] X. L. Long, W. D. Xiao, W. K. Yuan, Removal of Nitric Oxide and Sulfur Dioxide from Flue Gas Using a Hexaminecobalt(II)/Iodide Solution, Ind. Eng. Chem. Res. 2004, 43, 4048.

[7] X. L. Long, Z. L. Xin, H. X. Wang, W. D. Xiao, W. K. Yuan, Simultaneous Removal of NO and SO_2 with Hexaminecobalt(II) Solution Coupled with the Hexaminecobalt(II) Regeneration Catalyzed by Activated Carbon, Appl. Catal. B 2004, 54, 25.

[8] H. Teng, Y. T. Tu, Y. C. Lai, C. C. Lin, Reduction of NO with NH_3 over Carbon Catalysts-The Effects of Treating Carbon with H_2SO_4 and HNO_3, Carbon 2001, 39, 573.

[9] M. A. Montes-Morán, J. Angel Menéndez, E. Fuente, D. Suárez, Contribution of the Basal Planes to Carbon Basicity: An Ab Initio Study of the H_3O^+-π Interactions in Cluster Model, J Phys. Chem. B 1998, 102, 5595.

[10] L. R. Radovic, C. Moreno-Castilla, J. Rivera-Utrilla, Carbon Materials as Adsorbents in Aqueous Solutions, Chem. Phys. Carbon 2001, 27, 227.

[11] S. Bashkova, A. Bagreev, T. J. Bandosz, Catalytic Properties of Activated Carbon Surface in the Process of Adsorption/Oxidation of Methyl Mercaptan, Catal. Today 2005, 99, 323.

[12] Z. C. Liu, L. C. Ling, W. M. Qiao, L. Liu, Effect of Hydrogen on the Mesopore Development of Pitch Based Spherical Activated Carbon Containing Iron during Activation by Steam, Carbon 1999, 37, 2063.

[13] I. I. Salarne, T J. Bandosz, Surface Chemistry of Activated Carbons: Combining the Results of Temperature-prograrmned Desorption, Boehm, and Potentiometric Titrations, J. Colloid Interface Sci. 2001, 240, 252.

[14] S. S. Barton, M. J. B. Evans, E. Halliop, Acidic and Basic Site on the Surface of Porous Carbon, Carbon 1997, 35, 1361.

[15] J. S. Noh, J. A. Schwarz, Effect of HNO_3 Treatment on the Surface Acidity of Activated Carbons, Carbon 1990, 28, 675.

[16] S. Biniak, M. Pakula, G. S. Szymaanski, A. Swiatkowski, Effect of Activated Carbon Surface Oxygen- and/or Nitrogen-containing Groups on Adsorption of Copper(II) Ions from Aqueous Solution, Langmuir 1999, 15, 6117.

[17] F. Adib, A. Bagreev, T. J. Bandosz, Analysis of the Relationship between H_2S Removal Capacity and Surface Properties of Unimpregnated Activated Carbons, Environ. Sci. Technol. 2000, 34, 686.

[18] J. Zawadzki, Infrared Spectroscopy in Surface Chemistry of Carbons, in Chemistry and Physics of Carbon (Ed.: P. A. Thrower), Vol. 21, Dekker, New York 1989, pp. 147-380.

[19] H. Chen. Catalytic Reduction of Hexaminecobalt(III) by Activated Carbon, Master Dissertation, East China University of Science and Technology, Shanghai 2008.

Mechanistic Insight into Size-dependent Activity and Durability in Pt/CNT Catalyzed Hydrolytic Dehydrogenation of Ammonia Borane[*]

Abstract We report a size-dependent activity in Pt/CNT catalyzed hydrolytic dehydrogenation of ammonia borane. Kinetic study and model calculations revealed that Pt(111) facet is the dominating catalytically active surface. There is an optimized Pt particle size of ca. 1.8nm. Meanwhile, the catalyst durability was found to be highly sensitive to the Pt particle size. The smaller Pt particles appear to have lower durability, which could be related to more significant adsorption of B-containing species on Pt surfaces as well as easier changes in Pt particle size and shape. The insights reported here may pave the way for the rational design of highly active and durable Pt catalysts for hydrogen generation.

Downsizing metal particles to nanoscale in heterogeneous catalysis is often an effective method to boost the catalytic activity and thus achieve the high utilization of catalysts, which is the so-called size-dependent activity[1]. Such behavior in principle originates from the unique electronic and/or geometric properties[2], and probing the dominating properties is highly desirable for rational design of catalysts. In addition to the activity, the kinetic behavior and durability of catalysts are also important factors. However, the size-dependence of such characteristics has not been addressed very much.

Pt is one of the most investigated and industrial relevant catalysts, and its superior catalytic performance has a strong dependence on the particle size[3]. The probe reaction in the present study is hydrolytic dehydrogenation of ammonia borane:

$$NH_3BH_3 + 2H_2O \longrightarrow NH_4^+ + BO_2^- + 3H_2 \uparrow$$

It has increasingly been attracted for hydrogen production at mild conditions for fuel cell applications[4]. Previous studies revealed that Pt catalysts, especially γ-Al_2O_3 or CNT supported small Pt nanoparticles and MIL-101 confined ones, showed much higher H_2 generation rate than other catalysts[4,5]. However, in these studies on Pt catalysis, both the size effects and support effects were simultaneously involved. In other words, the intrinsic size-dependent activity is still unclear, calling for fundamental understanding under keeping supports unchanged. Additionally, effects of Pt particle size on the kinetic behavior and durability have also not been reported previously. Therefore, an attempt would be made to explore the nature of the above effects or behaviors and subsequently establish the structure-performance relationship.

In this work, we have employed the same close CNT support with the natures of highly crystalline and mesoporous[6] to load differently sized Pt nanoparticles. It makes it easy to be ac-

---

[*] Coauthors: Chen Wenyao, Ji Jian, Feng Xiang, Duan Xuezhi, Qian Gang, Li Ping, Zhou Xinggui, Chen De. Reprinted from *J. Am. Chem. Soc.* 2014, 136:16736-16739.

cessed by reactants, and it allows the imaging of almost all the particles for the accurate measurement of particle size distribution. On the resultant catalysts, intrinsic size-dependent H_2 generation activity has been investigated in hydrolytic dehydrogenation of ammonia borane. Subsequently, kinetic study and model calculations were further carried out to identify the dominating active sites as well as the contribution of the electronic and/or geometric effects to the activity. Moreover, the catalyst durability was also correlated to Pt particle size, and the possible deactivation mechanism was proposed.

Six differently sized Pt/CNT catalysts were prepared by only changing Pt loading without varying any other parameters that could affect the activity in which the procedure used has been previously reported[5c]. High angle annular dark field scanning transmission electron microscopy (HAADF-STEM) with high resolution was employed to characterize these catalysts for obtaining reliable particle size distributions[7]. Fig. 1a and Fig. S1, Supporting Information (SI), show representative images of six Pt/CNT catalysts and the corresponding particle size distributions. The mean particle sizes were calculated based on the sizes of at least 200 random particles, and the results are summarized in Table S1, SI.

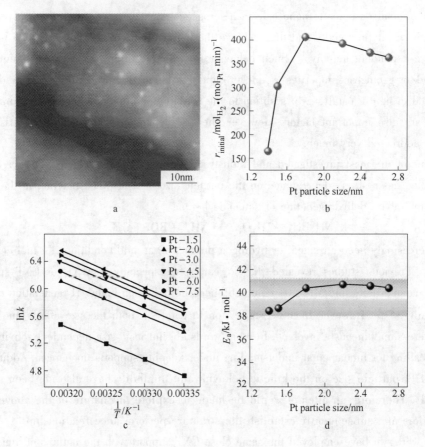

Fig. 1 Typical HAADF-STEM image of Pt/CNT with a Pt loading of 1.5wt% (i.e., Pt-1.5) (a), initial hydrogen generation rate ($r_{initial}$) as a function of Pt particle size (b), $\ln k$, derived from hydrogen generation rate versus reaction time, as a function of $1/T$ over differently sized Pt/CNT catalysts (c) and activation energy as a function of Pt particle size (d)

These catalysts were tested for hydrolytic dehydrogenation of ammonia borane in which the ratio of n_{Pt}/n_{AB} used for each reaction is maintained constant. The results of kinetic study are shown in Fig. S2, SI. It is found that the accumulated hydrogen volume almost increases linearly with the reaction time in the initial period (e. g., AB conversion being in the range of 0-50%) for all the catalysts. It suggests a zero order reaction with respect to ammonia borane. Moreover, the observed zero order reaction suggests that the reaction is not limited by external diffusion of reactants in our kinetic studies. A first reaction order is expected for an external diffusion limited reaction. The external diffusion limitation free has also been confirmed by the results in Fig. S3, SI, where the reaction rate is not influenced by the stirring speed. However, a deviation is from linear dependence of accumulated hydrogen volume with the reaction time at the late stage of the reaction or high conversions of ammonia borane. It is possibly caused by the external diffusion limitation at low concentrations. As a result, the reaction rates can be easily determined based on the slope of the hydrogen generation curves in the initial reaction period in Fig. S2, SI.

Initial rate ($r_{initial}$) of hydrogen generation over Pt/CNT catalysts was plotted as a function of the Pt particle size. As shown in Fig. 1b, the $r_{initial}$ has a strong dependence on the mean Pt particle size (d), and the optimum one appears at Pt of 1.8nm. The origin of this typical size-dependent activity would be revealed by combined kinetic study and model calculations as follows.

Considering that this reaction follows zero order kinetics over the above Pt/CNT catalysts in the initial period, the intrinsic activation energy (E_a) (SI) is easily derived by the observed overall hydrogen generation rate (i. e., the sum of the reaction rate of each type of active site) and expressed by:

$$E_a = \frac{\sum_{i=1}^{n} A_i y_i \exp[-E_i/(RT)] E_i}{\sum_{i=1}^{n} A_i y_i \exp[-E_i/(RT)]} \quad (1)$$

where, n is the number of the type of active site; A_i, y_i and E_i are the pre-exponential factor, the fraction, and the activation energy of each type of active site, respectively.

On the basis of the results of kinetic experiments at 25-40℃ and the zero reaction order characteristic in the initial period (Fig. S2, SI), reaction rate constant (k) was obtained from the slope of the fitting line. The logarithm of the as-obtained k was correlated to the reciprocal absolute temperature ($1/T$), and the results are shown in Fig. 1c. Clearly, each Pt/CNT catalyst follows a good linearity between $\ln k$ and $1/T$. The corresponding slope of the plot yields E_a (Fig. 1d). It is seen that all the E_a is in the range of 38-41kJ/mol, which are not very sensitive to Pt particle size. By combining the almost unchanged E_a for differently sized Pt/CNT catalysts with Eq. 1, it could be concluded that only one type of active site mostly contributes to the overall hydrogen generation rate. As a consequence, Eq. 1 can be simplified as:

$$E_a = \frac{A_1 y_1 \exp[-E_1/(RT)] E_1}{A_1 y_1 \exp[-E_1/(RT)]} = E_1 \quad (2)$$

It is well-known that the change in the fraction of specific active sites (e. g., corner, edge,

(111), or (100) site) with Pt particle size can be determined only when the particle shape is known. Previous studies showed that Pt nanoparticles of <5nm in size are prone to exist as truncated octahedron[8]. They mainly consist of a mixture of (100) and (111) facets in order to minimize the total interfacial free energy. Herein, we employ high-resolution transmission electron microscopy (HRTEM) to characterize the Pt particle shape and find that CNT-supported Pt nanoparticles exhibit a well-defined shape in most cases. Fig. 2a and Fig. S4, SI, show the representative shapes, which are best represented by the top slice of truncated cuboctahedron. To determine the type of the top facet of the above shape, the fast Fourier transform (FFT) of the selected area was further carried out. The top facet of (100) is observed, which is schematically shown in Fig. 2b. The numbers of corner, edge, (111), and (100) atoms of each particle were calculated by formulas in Table S2, SI[2c]. The numbers of corner, edge, (111), or (100) atoms over differently sized Pt/CNT catalyst per mole of Pt were estimated by the number of corner, edge, (111), or (100) atoms of each particle times the number of particles. The results are shown in Fig. 2c.

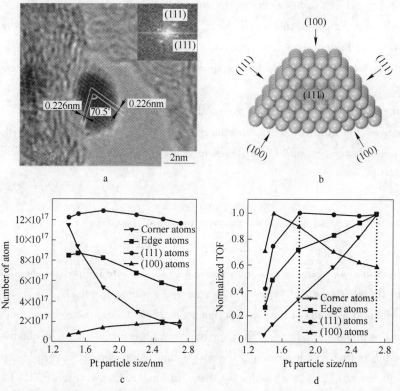

Fig. 2 Typical HRTEM image of Pt nanoparticle supported on CNT (a), schematic diagram of truncated cuboctahedron (b), plots of number of surface atoms per mole of Pt with Pt particle size of truncated cuboctahedron (c) and plots of normalized TOF with Pt particle size (d)

The activity of each specific type of active site was assumed to be uniform regardless of Pt particle size. If one type of Pt active site is responsible for the activity, the $r_{initial}$ of catalysts would increase linearly with the number of specific Pt active sites. Turnover frequency (TOF) of one type of active site, calculated from the data in Fig. 1b and Fig. 2c, should be constant. Subsequently, each TOF point was normalized to the highest TOF for each type of active site to yield normalized TOF (SI). As shown in Fig. 2d, the normalized TOF based on the number of corner or edge sites increases monotonically with increasing Pt particle size, respectively. The normalized TOF based on the number of the sites on (100) facet appears volcano-shaped as a function of Pt particle size. However, only when (111) facet atoms are considered as active sites, the normalized TOF is almost constant for Pt particles with the size of > 1.8nm. It is suggested that the (111) facets of Pt/CNT catalysts are dominating active sites for hydrolytic dehydrogenation of ammonia borane. Moreover, as shown in Fig. 2c, the number of Pt (111) atoms of differently sized Pt/CNT catalyst follows a volcano curve with Pt particle size, i.e., 1.8nm sized Pt/CNT catalyst has the maximum. This could be one reason for the volcano relationship between $r_{initial}$ and Pt particle size (Fig. 1b).

However, it can be also seen that on the smaller sized Pt/CNT catalysts (i.e., the particle size of <1.8nm) the normalized TOF increases monotonically with increasing Pt particle size (Fig. 2d). This may be dominated by electronic properties rather than geometric properties[2a]. The above results not only unravel the origin of the size-dependent activity but also reveal 1.8nm sized Pt/CNT catalyst being optimum with the highest catalytic activity and utilization of Pt. This might guide the rational design of highly active Pt catalysts.

Beside the activity and kinetic behavior measurements, the durability evaluation of catalysts is also an important issue[9]. The durability of differently sized Pt/CNT catalysts was evaluated to explore the effects of the Pt particle size on the durability. To this end, the recyclability of each catalyst was evaluated up to 5 cycles (Fig. S5, SI), where the new ammonia borane aqueous solution was added into the reactor after the completion of the last run. The deactivation function, which is defined as the $r_{initial}$ in each cycle (Fig. S5, SI) normalized by the $r_{initial}$ in the first cycle, is presented in Fig. 3a as a function of cycle number and Pt particle size. The gradual decreases in the catalytic activity are observed for all the catalysts. The results in Fig. 3a clearly indicate a size-dependent durability. In particular, the smaller the Pt particle size, the faster the deactivation. For example, the smallest sized Pt/CNT catalyst (i.e., 1.4nm) suffers from a severe deactivation, and its activity in the fifth cycle is ca. 24% of its initial activity. It is worth to mention that the AB concentration decreases with the cycle number due to the accumulation of water. However, both the molar ratio of Pt to AB and the amount of catalyst are kept constant in each cycle. Then effects of AB concentration on the hydrogen generation rate were further studied, and the typical results are shown in Fig. S6, SI. It was found that the $r_{initial}$ is almost independent of the AB concentration. The results suggest that the observed decrease in the activity with the cycle number (Fig. 3a) should be a result of catalyst deactivation.

Taking Pt/CNT catalyst with the 1.4nm of Pt particle size and the lowest durability, for example, the difference in the geometric and electronic properties between fresh catalyst and deactivated catalyst after 5 cycles was probed to gain insights into the deactivation mechanism of the catalysts. The deactivation mechanisms of pore blocking and metal leaching can be first ruled out because mesoporous close-CNT was used and no platinum species in filtrate were detected, respectively. The deactivated catalyst was characterized by HAADF-STEM. Results in Fig. 3b show that the deactivated catalyst contains some agglomerated Pt particles with irregular shape in comparison to the fresh catalyst (Fig. 1a). However, the used Pt-7.5 catalyst after 5 cycles

(Fig. S7, SI) shows similar Pt particle size distribution compared to the fresh Pt-7.5 catalyst. This may be a result of that the smaller particles have higher surface energy and thus tend to aggregate into larger particles during the reaction. Therefore, the more significant agglomeration of the smaller sized Pt particles could be one reason for the lower durability.

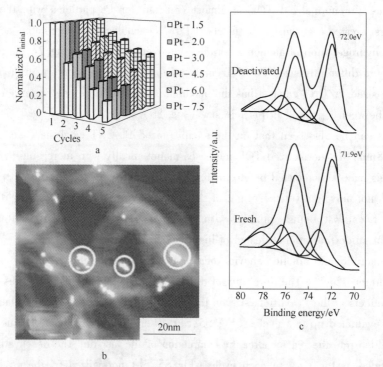

Fig. 3 Relative activities over cycles of Pt catalysts with different Pt loadings for the hydrolysis of ammonia borane (a), typical HAADF-STEM image of deactivated Pt-1.5 catalyst after 5 cycles (b) and XPS spectra of fresh Pt-1.5 catalyst and deactivated Pt-1.5 catalyst after 5 cycles (c)

Inspired by previous studies on the deactivation mechanism of $NaBH_4$ hydrolysis (i.e., strongly adsorbed B-containing species coating metal active sites)[10], we further studied whether B-containing species could strongly adsorb on the surfaces of Pt nanoparticles and thus lead to the deactivation of the catalyst. The deactivated Pt-1.5 catalyst after 5 cycles was washed using distilled water until not detecting B-containing species in the filtrate by ICP. The washed Pt-1.5 catalyst was characterized by ICP and CO chemisorption and compared to the fresh one. The results are shown in Table 1.

Table 1 Relative content of Pt species, Pt dispersion, and B content of the fresh and deactivated Pt-1.5 catalysts

Catalyst	Relative content/%			B content[①] /wt%	Pt dispersion[②] /%
	Pt^0	Pt^{2+}	Pt^{4+}		
Fresh	64	21	15	0	72
Deactivated	63	22	15	0.2	10

① Measured by ICP.
② Measured by CO chemisorption.

Inductively coupled plasma (ICP) measurements show that the washed deactivated Pt-1.5

catalyst has the 0.2wt% of the B content. CO chemisorption measurements show that the Pt dispersion of the deactivated catalyst is 10%, which is much lower than that of the fresh catalyst (72%). The estimated Pt particle size based on the measured Pt dispersion of the deactivated catalyst is ca. 11.3nm. It is much larger than that observed by HAADF-STEM (Fig. 3b). It suggests that strong adsorption of B-containing species on the Pt surfaces could be another reason for the catalyst deactivation. In addition, it is reasonable to assume that the larger fraction of Pt surfaces with low coordination number on smaller Pt particles could result in a more significant adsorption of B-containing species during the reaction. This may be the main cause for the severe deactivation of the smaller sized Pt/CNT catalyst.

Subsequently, X-ray photoelectron spectroscopy (XPS) was further employed to probe the difference in the electronic properties between fresh and deactivated Pt-1.5 catalysts. Fig. 3c shows the Pt 4f XPS spectra of fresh and deactivated catalysts, and Table 1 shows the relative contents of Pt^0, Pt^{2+}, and Pt^{4+} from the deconvolution analyses. It is observed that both catalysts have similar Pt^0, Pt^{2+}, and Pt^{4+}, contents, that is, 64%, 21%, and 15% for the fresh catalyst, while 63%, 22%, and 15% for the deactivated catalyst. Moreover, the binding energy of metallic Pt $4f_{7/2}$ for the deactivated catalyst is 72.0eV, which is slightly higher than that for the fresh catalyst (71.9eV). This could be due to the adsorption of B-containing species with higher electronegative properties creating more oxidized environment[11]. In our previous work, we have observed that the positive shift of Pt $4f_{7/2}$ binding energy led to an increase in the hydrogen generation rate[5c]. Therefore, the observed catalyst deactivation cannot be ascribed to the electronic modification of Pt/CNT catalyst.

In summary, we identify experimentally and theoretically that Pt(111) is a dominating catalytically active surface, and ca. 1.8nm sized Pt/CNT catalyst is optimum with the highest catalytic activity and utilization of Pt. Moreover, the smaller the Pt particle size, the lower the durability. The plausible mechanism of size-dependent durability has been proposed by using multiple techniques such as HAADF-STEM, XPS, ICP, and CO chemisorption. Strongly adsorbed B-containing species and the change in the Pt particle size and shape are the two main causes for the catalyst deactivation. These results demonstrate that Pt particle size plays a crucial role in the design and optimization of active and durable catalysts for hydrolytic dehydrogenation of ammonia borane.

Associated Content

Supporting information

Experimental procedures, kinetic formulas derivation, normalized TOF derivation, Table S1 and Table S2, and Fig. S1-Fig. S7. This material is available free of charge via the Internet at http://pubs.acs.org.

Authorinformation

Corresponding authors

* xzduan@ecust.edu.cn

* chen@nt.ntnu.no

Notes

The authors declare no competing financial interest.

Acknowledgements

This works was financially supported by the 111 Project of Ministry of Education of China (B08021) and the Natural Science Foundation of China (21306046 and 21276077).

References

[1] (a) Bell, A. T. Science 2003, 299, 1688. (b) Mizuno, N. ; Misono, M. Chem. Rev. 1998, 98, 199. (c) Chen, M. S. ; Cai, Y. ; Yan, Z. ; Gath, K. K. ; Axnanda, S. ; Goodman, D. W. Surf. Sci. 2007, 601, 5326. (d) Polshettiwar, V. ; Varma, R. S. Green Chem. 2010, 12, 743. (e) Li, Y. ; Somorjai, G. A. Nano Lett. 2010, 10, 2289. (f) Sanchez, A. ; Abbet, S. ; Heiz, U. ; Schneider, W. D. ; Hakkinen, H. ; Barnett, R. N. ; Landman, U. J. Phys. Chem. A 1999, 103, 9573.

[2] (a) Wilson, O. M. ; Knecht, M. R. ; Garcia-Martinez, J. C. ; Crooks, R. M. J. Am. Chem. Soc. 2006, 128, 4510. (b) Mayrhofer, K. J. J. ; Blizanac, B. B. ; Arenz, M. ; Stamenkovic, V. R. ; Ross, P. N. ; Markovic, N. M. J. Ph_ys. Chem. B 2005, 109, 14433. (c) Van Hardeveld, R. ; Hartog, F. Surf. Sci 1969, 15, 189. (d) Bond, G. C. Surf. Sci 1985, 156, 966. (e) Den Breejen, J. P. ; Radstake, P. B. ; Bezemer, G. L. ; Bitter, J. H. ; Froseth, V. ; Holmen, A. ; De Jong, K. P. J. Am. Chem. Soc. 2009, 131, 7197.

[3] (a) Van Santen, R. A. Acc. Chem. Res. 2009, 42, 57. (b) Kuhn, J. N. ; Huang, W. ; Tsung, C. K. ; Zhang, Y. ; Somorjai, G. A. J. Am. Chem. Soc. 2008, 130, 14026. (c) Tsung, C. K. ; Kuhn, J. N. ; Huang, W. Y. ; Aliaga, C. ; Hung, L. I. ; Somorjai, G. A. ; Yang, P. D. J. Am. Chem. Soc. 2009, 131, 5816. (d) Arenz, M. ; Mayrhofer, K. J. J. ; Stamenkovic, V. ; Blizanac, B. B. ; Tomoyuki, T. ; Ross, P. N. ; Markovic, N. M. J. Am. Chem. Soc. 2005, 127, 6819. (e) Allian, A. D. ; Takanabe, K. ; Fujdala, K. L. ; Hao, X. ; Truex, T. J. ; Cai, J. ; Buda, C. ; Neurock, M. ; Iglesia, E. J. Am. Chem. Soc. 2011, 133, 4498. (f) Chin, Y. H. ; Buda, C. ; Neurock, M. ; Iglesia, E. J. Am. Chem. Soc. 2011, 133, 15958. (g) Plomp, A. J. ; Vuori, H. ; Krause, A. O. ; DeJong, K. P. ; Britter, J. H. Appl. Catal. , A 2008, 351, 9.

[4] (a) Staubitz, A. ; Robertson, A. P. M. ; Manners, I. Chem. Rev. 2010, 110, 4079. (b) Jiang, H. L. ; Singh, S. K. ; Yan, J. M. ; Zhang, X. B. ; Xu, Q. ChemSusChem 2010, 3, 541. (c) Yadav, M. ; Xu, Q. Energy Environ. Sci. 2012, 5, 9698. (d) Demirci, U. B. ; Miele, P. Energy Environ. Sci. 2009, 2, 627. (e) Yan, J. M. ; Zhang, X. B. ; Akita, T. ; Haruta, M. ; Xu, Q. J. Am. Chem. Soc. 2010, 132, 5326. (f) Metin, O. ; Mazumder, V. ; Ozkar, S. ; Sun, S. J. Am. Chem. Soc. 2010, 132, 1468.

[5] (a) Chandra, M. ; Xu, Q. J. Power Sources 2006, 156, 190. (b) Chandra, M. ; Xu, Q. J. Power Sources 2007, 168, 135. (c) Chen, W. Y. ; Ji, J. ; Duan, X. Z. ; Qian, G. ; Li, P. ; Zhou, X. G. ; Chen, D. ; Yuan, W. K. Chem. Commun. 2014, 50, 2142. (d) Aijaz, A. ; Karkamkar, A. ; Choi, Y. J. ; Tsumori, N. ; Ronnebro, E. ; Autrey, T. ; Shioyama, H. ; Xu, Q. J. Am. Chem. Soc. 2012, 134, 13926. (e) Wang, X. ; Liu, D. P. ; Song, S. Y. ; Zhang, H. J. Chem. Commun. 2012, 48, 10207.

[6] (a) Zhu, J. ; Holmen, A. ; Chen, D. ChemCatChem 2013, 5, 378. (b) Serp, P. ; Corrias, M. ; Kalck, P. Appl. Catal. j A 2003, 253, 337.

[7] Zhang, B. ; Zhang, W. ; Su, D. S. ChemCatChem 2011, 3, 965.

[8] (a) Lim, B. ; Jiang, M. J. ; Camargo, P. H. C. ; Cho, E. C. ; Tao, J. ; Lu, X. M. ; Zhu, Y. M. ; Xia, Y. N. Science 2009, 324, 1302. (b) Wang, C. ; Daimon, H. ; Onodera, T. ; Koda, T. ; Sun, S. Angew. Chem. , Int. Ed. 2008, 47, 3588.

[9] (a) Bartholomew, C. H. Appl Catal A 2001, 212, 17. (b) Baylet, A. ; Royer, S. ; Marecot, P. ; Tatibouet,

J. M. ; Duprez, D. Appl. Catal. , B 2008, 77, 237. (c) Li, Y. M. ; Liu, J. H. C. ; Witham, C. A. ; Huang, W. Y. ; Marcus, M. A. ; Fakra, S. C. ; Alayoglu, P. ; Zhu, Z. W. ; Thompson, C. M. ; Arjun, A. ; Lee, K. ; Gross, E. ; Toste, F. D. ; Somorjai, G. A. J. Am. Chem. Soc. 2011, 133, 13527.

[10] Akdim, O. ; Demirci, U. B. ; Miele, P. Int. J. Hydrogen Energy 2011, 36, 13669.

[11] (a) Montilla, F. ; Morallon, E. ; De Battisti, A. ; Barison, S. ; Daolio, S. ; Vazquez, J. L. J. Phys. Chew. B 2004, 108, 15976. (b) Park, H. ; Kim, Y. K. ; Choi, W. J. Phys. Chem. C 2011, 115, 6141.

Kinetics of Gas-liquid Reaction between NO and Co(NH$_3$)$_6^{2+}$ *

Abstract Wet ammonia desulphurization process can be retrofitted for combined removal of SO$_2$ and NO from the flue gas by adding soluble cobalt (II) salts into the aqueous ammonia solutions. The Co(NH$_3$)$_6^{2+}$ formed by ammonia binding with Co^{2+} is the active constituent of scrubbing NO from the flue gas streams. A stirred vessel with a plane gas-liquid interface was used to measure the chemical absorption rates of nitric oxide into the Co(NH$_3$)$_6^{2+}$ solution under anaerobic and aerobic conditions separately. The experiments manifest that the nitric oxide absorption reaction can be regarded as instantaneous when nitric oxide concentration levels are parts per million ranges. The gas-liquid reaction becomes gas film controlling as Co(NH$_3$)$_6^{2+}$ concentration exceeds 0.02mol/L. The NO absorption rate is proportional to the nitric oxide inlet concentration. Oxygen in the gas phase is favorable to the absorption of nitric oxide. But it is of little significance to increase the oxygen concentration above 5.2%. The NO absorption rate decreases with temperature. The kinetic equation of NO absorption into the Co(NH$_3$)$_6^{2+}$ solution under aerobic condition can be written as

$$N_{NO} = \frac{K_{NO,G} k_4 K_1 (p_{NO} + \gamma C_{Co(NH_3)_6^{2+},L}) C^2_{Co(NH_3)_6^{2+},L} p_{O_2}}{k_{-3} + k_4 K_1 C^2_{Co(NH_3)_6^{2+},L} p_{O_2}}$$

Key words nitric oxide, gas-liquid reaction, kinetics, absorption

1 Introduction

The role of NO$_x$ and SO$_2$ pollutants in acid-rain formation and the destruction of lakes and forest ecosystems have been established. The removal of these contaminants to comply with environmental emission standard is of imperative necessity. The wet processes have certain economical advantages in combined NO$_x$/SO$_2$ elimination. The development of efficient processes for simultaneous SO$_2$ and NO$_x$ removal from power-plant flue gases is particularly important because fossil-fuel-fired steam boilers represent a major source of sulphur and nitrogen oxide emissions. Nitric oxide is 90%-95% of the NO$_x$ present in typical flue gas streams[1]. But existing wet flue-gas-desulphurization (FGD) scrubbers in power plants are incapable of eliminating NO from the flue-gas because of its low solubility in water. Several methods have been developed to enhance NO absorption, including the use of oxidants to oxidize NO to the more soluble NO$_2$[2-4] and the addition of various iron(II) chelates to bind and activate NO[5-7]. So far, none of these methods have been put into commercial application.

The process using an ammonia scrubber to recover sulfur dioxide from the flue gas has been developed and put into commercial application. The authors put forward a novel technique for the simultaneous elimination of NO and SO$_2$ from the flue gas by adding soluble cobalt(II)

---

* Coauthors: Long Xiangli, Xiao Wende. Reprinted from *Journal of Hazardous Materials B*, 2005, 123:210-216.

salts into the aqueous ammonia solution[8-10]. The hexamminecobalt(Ⅱ) formed by cobalt(Ⅱ) binding with ammonia can not only coordinate nitric oxide but also activate oxygen molecules in aqueous ammonia solution. The oxidant is the oxygen coexisting in the flue gas. Therefore, NO can be absorbed and oxidized simultaneously in the hexamminecobalt(Ⅱ) solution. The mechanism of NO absorption into the hexamminecobalt(Ⅱ) solution can be expressed as follows:

Nitric oxide dissolves into aqueous solution by reacting with $Co(NH_3)_6^{2+}$:

$$NO(g) \rightleftharpoons NO(aq) \quad (1)$$

$$Co(NH_3)_6^{2+}(aq) + NO(aq) \longrightarrow [Co(NH_3)_5NO]^{2+}(aq) + NH_3 \quad (2)$$

The activation of oxygen in $Co(NH_3)_6^{2+}$ solution[11,12]:

$$2Co(NH_3)_6^{2+} + O_2 \rightleftharpoons [(NH_3)_5Co\text{-}O\text{-}O\text{-}Co(NH_3)_5]^{4+} + 2NH_3 \quad (3)$$

Nitric oxide is turned into NO_2^- and NO_3^-:

$$[(NH_3)_5Co\text{-}O\text{-}O\text{-}Co(NH_3)_5]^{4+} + H_2O + 2NH_3 + [Co(NH_3)_5NO]^{2+} \longrightarrow$$
$$2Co(NH_3)_6^{3+} + 2OH^- + [Co(NH_3)_5NO_2]^{2+} \quad (4)$$

$$2[Co(NH_3)_5NO_2]^{2+} + H_2O + 4NH_3 \longrightarrow NH_4NO_2 + NH_4NO_3 + 2Co(NH_3)_6^{2+} \quad (5)$$

The net reaction can be written as follow:

$$2NO + 2O_2 + 2NH_3 + 4H^+ \longrightarrow NH_4NO_2 + NH_4NO_3 + H_2O \quad (6)$$

According to the discussion above, the oxidation and absorption of NO can be realized simultaneously in the $Co(NH_3)_6^{2+}$ solution.

The reaction between NO and $Co(NH_3)_6^{2+}$ is a heterogeneous reaction. The relation between the reactions and mass transfer is very complicated. It is imperative to investigate such relation for the scale-up of this technique and the design of industrial equipment. A study on the kinetics of NO absorption into $Co(NH_3)_6^{2+}$ solution is reported in this paper.

2 Experimental Section

The experiments were conducted in a stirred cell (as shown in Fig. 1) that has a water jacket through which water was circulated from a thermostat to maintain the desired temperature. Three turbine impellers are mounted on the same stirring rod. The upper and the bottom one provide, respectively, the mixing in the gas phase and the stirring in the liquid phase, while the middle one is used to sweep the interface between gas phase and liquid phase. The middle stirred baffle was floating on the solution. The experiments were performed at a stirred speed of 260r/min to minimize the resistance of mass transfer in the liquid. The gas-liquid interfacial areas were changed by using a stirred cell with cross-section area of 38cm^2 to substitute a stirred cell with cross-section area of 28cm^2. The absorption was carried out at atmospheric pressure. The gas phase consisted of NO in nitrogen. Two percent NO in nitrogen was supplied from a cylinder and was further diluted with N_2 to the desired concentrations before being fed to the stirred cell. The feed concentration of NO ranged from 150ppm to 900ppm. The continuous gas flow was maintained at about 0.2-0.4L/min. The solution was prepared by adding $Co(C_2H_5O_2)_2 \cdot 4H_2O$ into the aqueous ammonia solution. The chemicals of analytical reagent grade were used throughout the study.

The quantitative analysis of gas compositions was achieved by an on-line Fourier transform in-

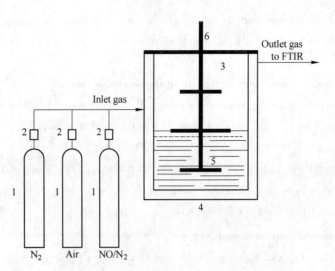

Fig. 1 Stirred cell reactor used for absorption experiments
1—Cylinder;2—Massmeter;3—Stirred cell;4—Water jacket;
5—Stirred baffle;6—Stirrer rod

frared spectrometer (Nicolet E. S. P. 460 FT-IR) equipped with a gas cell and a quantitative package, named Quant Pad. The inlet and outlet gases were directly introduced into the gas cell of the FTIR, with pipes insulated through the regulated electric coils to obtain the transient N_2O, NO, NO_2, and H_2O concentrations in both the inlet and outlet gases, as well as the transient NO conversion. This set-up is conveniently operated to monitor the nitric oxide removal efficiency.

The absorption rate can be calculated as follow[13]:

$$-N_{NO} = \frac{v_G P}{RTS}\left[\left(\frac{p_{NO}}{p_I}\right)_{in} - \left(\frac{p_{NO}}{p_I}\right)_{out}\right] \tag{7}$$

2.1 Effect of liquid volume on NO absorption rate

The experiments are performed at 50℃. The inlet NO concentration is 400ppm. The gas-liquid interfacial area is 28cm^2. Co(NH$_3$)$_6^{2+}$ concentration is 0.04mol/L. The reaction rates vary from 0.823×10^{-5} mol/(m$^2 \cdot$ s) to 0.853×10^{-5} mol/(m$^2 \cdot$ s) as liquid volumes in the stirred cell increase from 160mL to 265mL, which indicates that the nitric oxide absorption rate is almost independent of the liquid volume.

The experiments are also carried out at the same conditions as above, but with 5.2% oxygen present in the gas phase. The reaction rates vary from 1.462×10^{-5} mol/(m$^2 \cdot$ s) to 1.513×10^{-5} mol/(m$^2 \cdot$ s) as liquid volumes in the stirred cell increase from 160mL to 265mL. It can also be concluded that the nitric oxide absorption rate is independent of the liquid volume under aerobic condition.

2.2 Effect of gas-liquid interfacial area on NO absorption rate

The experiments are performed at 50℃. The inlet NO concentration is 430ppm. The liquid in the stirred cell is 265mL, 0.04mol/L Co(NH$_3$)$_6^{2+}$ solution. The reaction rates vary from

$8.530 \times 10^{-6} \text{mol}/(\text{m}^2 \cdot \text{s})$ to $11.91 \times 10^{-6} \text{mol}/(\text{m}^2 \cdot \text{s})$ as the gas-liquid interfacial areas increase from 28cm² to 38cm². A conclusion can be drawn that the absorption rate is proportional to the gas-liquid interfacial area.

The experiments were also done under the same conditions as above, but with 5.2% oxygen present in the gas phase. The reaction rates vary from $1.513 \times 10^{-4} \text{mol}/(\text{m}^2 \cdot \text{s})$ to $2.050 \times 10^{-4} \text{mol}/(\text{m}^2 \cdot \text{s})$ as the gas-liquid interfacial areas increase from 28cm² to 38cm², which manifests that the absorption rate is proportional to the gas-liquid interfacial area under aerobic condition.

According to the principle proposed by Levenspiel and Godfrey[13], the gas-liquid reaction can be regarded as an irreversible instantaneous reaction if its absorption rate is proportional to the gas-liquid interfacial area and independent of the liquid volume. A conclusion can be drawn from the experimental results that the NO absorption into $Co(NH_3)_6^{2+}$ solution can be regarded as irreversible instantaneous reaction when the NO inlet concentration is ppm ranges and the $Co(NH_3)_6^{2+}$ concentration is about 0.04mol/L. The reaction may be finished within the liquid film.

3 Theoretical

3.1 Nitric oxide reacts with $Co(NH_3)_6^{2+}$ under anaerobic condition

Under anaerobic condition, the reactions between nitric oxide and $Co(NH_3)_6^{2+}$ are reaction(1) followed with reaction (2). As discussed above, the reaction between nitric oxide and $Co(NH_3)_6^{2+}$ is instantaneous, the corresponding kinetic equation is independent of reaction rate and can be given by:

$$N_{NO} = -\frac{1}{S} \cdot \frac{dn_{NO}}{dt} = K_{NO,G}(p_{NO} + \gamma C_{Co(NH_3)_6^{2+},L}) \tag{8}$$

with

$$\gamma = \frac{1}{H_{NO}} \cdot \frac{D_{Co(NH_3)_6^{2+},L}}{D_{NO,L}} \tag{9}$$

$$K_{NO,G} = \frac{1}{\dfrac{1}{H_{NO}k_{NO,L}} + \dfrac{1}{k_{NO,G}}} \tag{10}$$

The critical concentration is expressed as follow:

$$C_{Co(NH_3)_6^{2+},C} = \frac{D_{NO,L}}{D_{Co(NH_3)_6^{2+},L}} \cdot \frac{k_{NO,G}}{k_{NO,L}} p_{NO} \tag{11}$$

when $C_{Co(NH_3)_6^{2+},L} \geq C_{Co(NH_3)_6^{2+},C}$

$$N_{NO} = k_{NO,G} p_{NO} \tag{12}$$

It can be concluded that the NO absorption rate may be increased by increasing $Co(NH_3)_6^{2+}$ concentration when $Co(NH_3)_6^{2+}$ concentration is very low. But, it is unnecessary to increase $Co(NH_3)_6^{2+}$ concentration further above its critical concentration because NO absorption has become gas film controlling. The absorption rate can not be increased further with $Co(NH_3)_6^{2+}$ concentration.

3.2 Nitric oxide reacts with $Co(NH_3)_6^{2+}$ under aerobic condition

When there is oxygen in the flue gas, reaction(2)-reaction(4) take place successively. The reaction(3) can be assumed to reach equilibrium quickly. Because NH_3 is much excessive and its concentration is nearly constant, Eq. 13 can be obtained:

$$C_D = K_1 C^2_{Co(NH_3)_6^{2+},L} p_{O_2} \tag{13}$$

where D stands for $[(NH_3)_5Co-O-O-Co(NH_3)_5]^{4+}$ and K_1 stands for the equilibrium constant of reaction (3).

According to the steady state assumption, the concentration of $[Co(NH_3)_5NO]^{2+}$ is unchangeable, Eq. 14 can be obtained from reaction(2) and reaction(3):

$$N_D = K_{NO,G}(p_{NO} + \gamma C_{Co(NH_3)_6^{2+},L}) - k_{-3}C_{[Co(NH_3)_5NO]^{2+}} - k_4 C_D C_{[Co(NH_3)_5NO]^{2+}} = 0 \tag{14}$$

Eq. 15 can be obtained from Eq. 14:

$$C_{[Co(NH_3)_6^{2+}]} = \frac{K_{NO,G}(p_{NO} + \gamma C_{Co(NH_3)_6^{2+},L})}{k_{-3} + k_4 C_D} = \frac{K_{NO,G}(p_{NO} + \gamma C_{Co(NH_3)_6^{2+},L})}{k_{-3} + k_4 K_1 C^2_{Co(NH_3)_6^{2+},L} p_{O_2}} \tag{15}$$

According to reaction (2), NO absorption rate can be expressed as Eq. 16:

$$N_{NO} = K_{NO,G}(p_{NO} + \gamma C_{Co(NH_3)_6^{2+},L}) - k_{-3} C_{[Co(NH_3)_6^{2+}]}$$
$$= \frac{K_{NO,G} k_4 K_1 (p_{NO} + \gamma C_{Co(NH_3)_6^{2+},L}) C^2_{Co(NH_3)_6^{2+},L} p_{O_2}}{k_{-3} + k_4 K_1 C^2_{Co(NH_3)_6^{2+},L} p_{O_2}} \tag{16}$$

In terms of Eq. 16, if $k_{-3} \ll k_4 K_1 C^2_{Co(NH_3)_6^{2+},L} p_{O_2}$, NO absorption rate is independent of oxygen partial pressure. If $k_{-3} \gg k_4 K_1 C^2_{Co(NH_3)_6^{2+},L} p_{O_2}$, NO absorption rate is proportional to oxygen partial pressure.

4 Results and Discussion

4.1 Effect of $Co(NH_3)_6^{2+}$ concentration on NO absorption rate

Fig. 2 shows the effect of $Co(NH_3)_6^{2+}$ concentration on NO absorption under anaerobic condition. It can be seen from Fig. 2 that NO absorption rate is proportional to $Co(NH_3)_6^{2+}$ concentration as $Co(NH_3)_6^{2+}$ concentration increases from 0.005mol/L to 0.02mol/L. For example, at 35℃ the nitric oxide absorption rate increases 6.4% (from 0.781×10^{-5} mol/($m^2 \cdot s$) to 0.831×10^{-5} mol/($m^2 \cdot s$)) as $Co(NH_3)_6^{2+}$ concentrations increase from 0.005mol/L to 0.01mol/L. But the nitric oxide absorption rates increase little as $Co(NH_3)_6^{2+}$ concentrations increase further above 0.02mol/L. For instance, at 50℃, the nitric oxide absorption rate increases only 1.1% (from 0.815×10^{-5} mol/($m^2 \cdot s$) to 0.824×10^{-5} mol/($m^2 \cdot s$)) when $Co(NH_3)_6^{2+}$ concentration increases from 0.02mol/L to 0.04mol/L. It can be concluded that the nitric oxide absorption into $Co(NH_3)_6^{2+}$ solution changes from two film controlling to gas film controlling when $Co(NH_3)_6^{2+}$ concentration exceeds 0.02mol/L. NO absorption rate is independent of $Co(NH_3)_6^{2+}$ concentration when $Co(NH_3)_6^{2+}$ concentration is over 0.02mol/L.

4.2 Effect of NO concentration on its absorption rate

Fig. 3 shows the effect of NO inlet concentration on its absorption rate into $Co(NH_3)_6^{2+}$ solu-

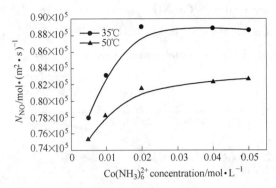

Fig. 2 Effect of Co(NH$_3$)$_6^{2+}$ concentration on reaction rate (without oxygen)
(Gas-liquid interfacial area = 28 cm^2, NO inlet concentration = 400 ppm)

tion. It can be seen from Fig. 3 that NO absorption rate is proportional to the nitric oxide inlet concentration. For instance, at 50℃, NO absorption rate into 0.05 mol/L Co(NH$_3$)$_6^{2+}$ solution increases 63.2% (from 7.784×10^{-6} mol/(m^2·s) to 12.7×10^{-6} mol/(m^2·s)) as nitric oxide inlet concentration increases from 312 ppm to 923 ppm. It can be complained by the fact that NO absorption into Co(NH$_3$)$_6^{2+}$ solution is gas film controlling when Co(NH$_3$)$_6^{2+}$ concentration is over 0.02 mol/L.

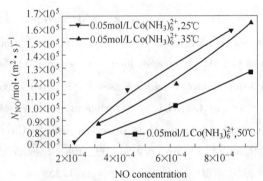

Fig. 3 Effect of NO concentration on absorption rate (without oxygen)
(Gas-liquid interfacial area = 28 cm^2)

4.3 Effect of oxygen partial pressure on NO absorption rate

Fig. 4 shows the effect of oxygen partial pressure on NO absorption rate. It can be concluded that oxygen can promote NO absorption into Co(NH$_3$)$_6^{2+}$ solution. For example, at 50℃ the NO absorption rate into 0.04 mol/L Co(NH$_3$)$_6^{2+}$ solution increases 64.6% (from 8.24×10^{-6} mol/(m^2·s) to 13.53×10^{-6} mol/(m^2·s)) as oxygen concentration increases from 0% to 5.2%. But the rate increases only 10.1% (from 13.53×10^{-6} mol/(m^2·s) to 15.29×10^{-6} mol/(m^2·s)) as oxygen concentration increases from 5.2% to 10.4%. Under aerobic conditions, the dissolved oxygen, in equilibrium with oxygen in the feed gas, activated by Co(NH$_3$)$_6^{2+}$ in aqueous ammonia solution, oxidizes nitric oxide to soluble nitric dioxide quickly. The nitric oxide oxidation rate may increase with oxygen partial pressure. Therefore, higher oxygen partial pressure in gas stream may be favorable to the formation of soluble nitrogen dioxide and enhance nitric oxide absorption into the Co(NH$_3$)$_6^{2+}$ solution. On the other hand, the formation of binuclear complex with bridging dioxygen, [(NH$_3$)$_5$Co-

O-O-Co(NH$_3$)$_5$]$^{4+}$, is connected with Co(NH$_3$)$_6^{2+}$ concentration. In other words, the activation of oxygen molecule may also be decided by Co(NH$_3$)$_6^{2+}$ concentration. Therefore, at a certain Co(NH$_3$)$_6^{2+}$ concentration, it is of little significance to increase the oxygen concentration above 5.2%.

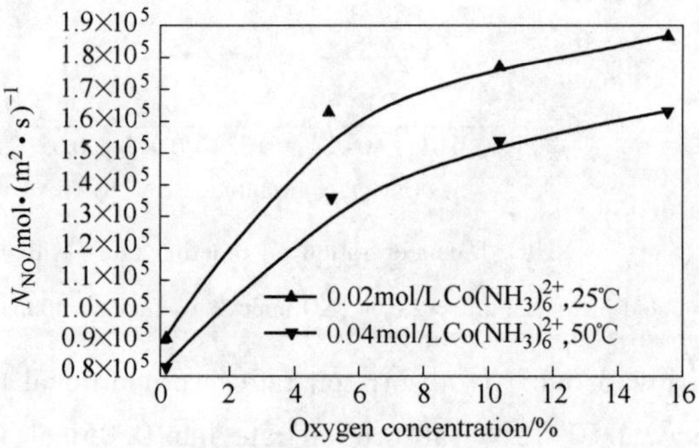

Fig. 4 Effect of oxygen concentration on absorption rate

(Gas-liquid interfacial area = 28cm^2, NO inlet concentration = 400ppm)

4.4 Effect of Co(NH$_3$)$_6^{2+}$ concentration on NO absorption rate under aerobic condition

Fig. 5 shows the effect of Co(NH$_3$)$_6^{2+}$ concentration on NO absorption rate when there is 5.2% oxygen in the flue gas. It can be seen from Fig. 5 that the NO absorption rate is proportional to Co(NH$_3$)$_6^{2+}$ concentration as Co(NH$_3$)$_6^{2+}$ concentration increases from 0.005mol/L to 0.02mol/L. For example, at 50℃ the nitric oxide absorption rate increases 4.7% (from 1.289×10^{-5} mol/(m$^2 \cdot$ s) to 1.350×10^{-5} mol/(m$^2 \cdot$ s)) as Co(NH$_3$)$_6^{2+}$ concentration increases from 0.005mol/L to 0.02mol/L. But the nitric oxide absorption rate increases little as Co(NH$_3$)$_6^{2+}$ concentrations increase further above 0.02mol/L. For instance, at 50℃, the nitric oxide absorption rate increases only 0.2% (from 1.350×10^{-5} mol/(m$^2 \cdot$ s) to 1.353×10^{-5} mol/(m$^2 \cdot$ s)) when Co(NH$_3$)$_6^{2+}$ concentration increases from 0.02mol/L to 0.04mol/L. It can be concluded that the nitric oxide absorption into the Co(NH$_3$)$_6^{2+}$ solution changes from two film controlling to gas film controlling when Co(NH$_3$)$_6^{2+}$ concentration exceeds 0.02mol/L. The NO absorption rate is independent of Co(NH$_3$)$_6^{2+}$ concentration.

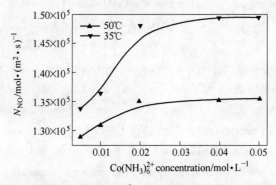

Fig. 5 Effect of Co(NH$_3$)$_6^{2+}$ concentration on reaction rate

(5.2% oxygen, gas-liquid interfacial area = 28cm^2, NO inlet concentration = 400ppm)

4.5 Effect of NO concentration on its absorption rate under aerobic condition

Fig. 6 shows the effect of nitric oxide inlet concentration on its absorption rate with 5.2% oxygen present in the gas stream. It can be concluded that, similar to the condition without oxygen, the absorption rate is also proportional to the nitric oxide inlet concentration. The reason is that NO absorption rate is gas film controlling and the reaction rate is mainly determined by NO partial pressure when $Co(NH_3)_6^{2+}$ concentration is 0.05mol/L.

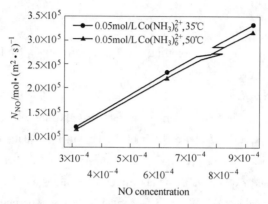

Fig. 6 Effect of NO concentration on absorption rate
(5.2% oxygen, gas-liquid interfacial area = 28cm²)

4.6 Effect of temperature on NO absorption rate

Fig. 7 shows the experimental results obtained at different temperature under anaerobic and aerobic conditions separately. It can be concluded from Fig. 7 that the nitric oxide absorption rate into $Co(NH_3)_6^{2+}$ solution decreases with temperature. The nitric oxide absorption rate decreases 10.4% (from 14.93×10^{-6} mol/(m² · s) to 13.53×10^{-6} mol/(m² · s)) as temperature increases from 35℃ to 50℃ with 5.2% oxygen present in the flue gas. The nitric oxide absorption rate decreases 23.8% (from 9.85×10^{-6} mol/(m² · s) to 7.51×10^{-6} mol/(m² · s)) as temperature increases from 21℃ to 65℃ under anaerobic condition.

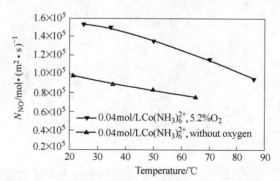

Fig. 7 Effect of temperature on reaction rate
(Gas-liquid interfacial area = 28cm²)

It is reported that at high temperature the balance of Eq. 17[14] lies to the left. High temperature is disadvantageous to the stability of $Co(NH_3)_6^{2+}$ and the coordination of nitric oxide with

$Co(NH_3)_6^{2+}$. In other words, high temperature is harmful to NO absorption into $Co(NH_3)_6^{2+}$ solution. On the other hand, the solubility of oxygen in aqueous solution decreases with temperature, which is also disadvantageous to the oxidation and absorption of NO:

$$NO + Co(II) + 5NH_3 \rightleftharpoons [Co(NH_3)_5NO]^{2+} \tag{17}$$

4.7 Kinetic equation for NO catalytic oxidation in $Co(NH_3)_6^{2+}$ solution

The kinetic Eq. 16 can be rewritten as follow:

$$\frac{1}{N_{NO}} = \frac{k_{-3}}{K_{NO,G}k_4 K_1 (p_{NO} + \gamma C_{Co(NH_3)_6^{2+},L}) C^2_{Co(NH_3)_6^{2+},L}} \cdot \frac{1}{p_{O_2}} + \frac{1}{K_{NO,G}(p_{NO} + \gamma C_{Co(NH_3)_6^{2+},L})} \tag{18}$$

A linear curve can be given by plotting $1/N_{NO}$ against $1/p_{O_2}$ (Fig. 8). The slope is $\dfrac{k_{-3}}{K_{NO,G}(p_{NO}+\gamma C_{Co(NH_3)_6^{2+},L})k_4 K_1 C^2_{Co(NH_3)_6^{2+},L}}$ and the intercept is $\dfrac{1}{K_{NO,G}(p_{NO}+\gamma C_{Co(NH_3)_6^{2+},L})}$.

In terms of the curves in Fig. 8, considering the fact that the absorption is gas film controlling when $Co(NH_3)_6^{2+}$ concentration is above 0.04mol/L, the kinetic equation of NO absorption into $Co(NH_3)_6^{2+}$ solution at 50℃ can be given as follow (correlation coefficient 0.980). The comparison between the calculated and experimental NO absorption rate values can be seen in Fig. 9. It can be seen from Fig. 9 that the calculated values are accordant with the experimental values:

$$N_{NO} = \frac{k_{NO,G} p_{NO} p_{O_2} C^2_{Co(NH_3)_6^{2+},L}}{2.664 \times 10^{-5} + p_{O_2} C^2_{Co(NH_3)_6^{2+},L}} \tag{19}$$

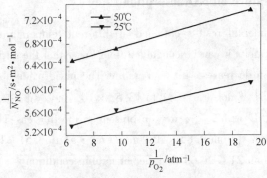

Fig. 8 Plots of $\dfrac{1}{N_{NO}}$ vs. $\dfrac{1}{p_{O_2}}$

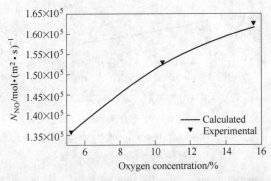

Fig. 9 Comparision between the calculated and experimental values

(Gas-liquid interfacial area = 28cm², NO inlet concentration = 400ppm, 50℃, $Co(NH_3)_6^{2+}$ = 0.04mol/L)

5 Conclusion

The experimental results in the stirred cell demonstrate that the nitric oxide absorption into $Co(NH_3)_6^{2+}$ solution can be regarded as an instantaneous reaction when nitric oxide levels are ppm ranges. The gas-liquid reaction becomes gas film controlling as $Co(NH_3)_6^{2+}$ concentration exceeds 0.02 mol/L. The rate of nitric oxide absorption into $Co(NH_3)_6^{2+}$ solution is independent of $Co(NH_3)_6^{2+}$ concentration when $Co(NH_3)_6^{2+}$ concentration is above 0.02 mol/L. The NO absorption rate is proportional to the nitric oxide inlet concentration.

Oxygen in the gas phase can promote nitric oxide absorption into $Co(NH_3)_6^{2+}$ solution. But there is little significance to increase oxygen concentration further above 5.2%. The NO absorption rate into $Co(NH_3)_6^{2+}$ solution decreases with temperature. The kinetic equation of NO absorption into the $Co(NH_3)_6^{2+}$ solution under aerobic condition can be written as:

$$N_{NO} = \frac{K_{NO,G} k_4 K_1 (p_{NO} + \gamma C_{Co(NH_3)_6^{2+},L}) \times C_{Co(NH_3)_6^{2+},L}^2 p_{O_2}}{k_{-3} + k_4 K_1 C_{Co(NH_3)_6^{2+},L}^2 p_{O_2}}$$

Nomenclature

C_{NO}	NO concentration in gas phase (mol/m^3)
$C_{Co(NH_3)_6^{2+},L}$	$Co(NH_3)_6^{2+}$ concentration in bulk liquid (mol/L)
$C_{Co(NH_3)_6^{2+},C}$	critical concentration of $Co(NH_3)_6^{2+}$ (mol/L)
$C_{[Co(NH_3)_5NO]^{2+}}$	$[Co(NH_3)_5NO]^{2+}$ concentration (mol/L)
C_{O_2}	oxygen concentration (mol%)
C_D	$[(NH_3)_5Co\text{-}O\text{-}O\text{-}Co(NH_3)_5]^{4+}$ concentration (mol/L)
D	$[(NH_3)_5Co\text{-}O\text{-}O\text{-}Co(NH_3)_5]^{4+}$
$D_{Co(NH_3)_6^{2+},L}$	diffusivity of $Co(NH_3)_6^{2+}$ in liquid (m^2/s)
$D_{NO,L}$	diffusivity of NO in liquid (m^2/s)
H_{NO}	Henry law constant of NO in water ($Pa \cdot m^3/kmol$)
$k_{NO,G}$	mass-transfer coefficient for NO in gas phase ($mol/(m^2 \cdot s \cdot Pa)$)
$K_{NO,G}$	overall mass-transfer coefficient for NO in gas phase ($mol/(m^2 \cdot s \cdot Pa)$)
k_{-3}	rate constant for reverse reaction (3) (s^{-1})
K_1	equilibrium constant for reaction (3) ($mol/(L \cdot Pa)$)
k_4	rate constant for reaction (4) ($L/(s \cdot mol)$)
$k_{NO,L}$	mass-transfer coefficient for NO in liquid phase ($mol/(m^2 \cdot s)$)
n_{NO}	NO molar number
N_D	rate of $[(NH_3)_5Co\text{-}O\text{-}O\text{-}Co(NH_3)_5]^{4+}$ reaction ($mol/(m^2 \cdot s)$)
N_{NO}	rate of NO absorption ($mol/(m^2 \cdot s)$)
P	total pressure (Pa)
p_I	partial pressure of inerts (Pa)
p_{NO}	NO partial pressure (Pa)
p_{O_2}	oxygen partial pressure (Pa)

R	gas constant (8.314 kJ/(kmol·K))
S	interfacial area (m^2)
T	absolute temperature (K)
v_G	gas volume rate (m^3/s)

Subscripts

in	inlet of the gas stream
out	outlet of the gas stream

Acknowledgements

The present work is supported by the NSFC (No. 29633030), the Ministry of Science and Technology of China (No. 2001CB711203), and the Development Project of Shanghai Priority Academic Discipline.

References

[1] C. J. Pereira, M. D. Amiridis (Eds.), NO_x Control From Stationary Sources, Washington, DC, vol. 552, 1995, p. 1.

[2] S. G. Chang, D. K. Liu, Removal of nitrogen and sulphur oxides from waste gas using a phosphorus/alkali emulsion, Nature 343 (1990) 151.

[3] Downey, G. D., Process of removing nitrogen oxides from gaseous mixtures, European Patent 0,008,488 (1980).

[4] Cooper, H. B. H., Removal and recovery of nitrogen oxides and sulfur dioxide from gaseous mixtures containing them, US Patent 4,426,364 (1984).

[5] S. G. Chang, D. Littlejohn, D. K. Liu, Use of ferrous chelates of SH-containing amino acid and peptides for the removal of NO_x and SO_2 from flue gas, Ind. Eng. Chem. Res. 27 (1988) 2156.

[6] E. K. Pham, S. G. Chang, Removal of NO from flue gases by absorption to an Iron (II) thiochelate complex and subsequent reduction to ammonia, Nature 369 (1994) 139.

[7] S. Yao, D. Littlejohn, S. G. Chang, Integrated tests for removal of nitric oxide with iron thiochelate in wet flue gas desulfurization system, Environ. Sci. Technol. 30 (1996) 3371.

[8] X. L. Long; W. D. Xiao; W. K. Yuan. A technique for the simultaneous removal of nitric oxide and sulfur dioxide from flue gases, Chinese Patent ZL 01105004.7 (2003).

[9] X. L. Long, Simultaneous removal of nitric oxide and sulfur dioxide, East China University of Science and Technology, Ph. D. Dissertation, Shanghai, the People's Republic of China, 2001.

[10] X. L. Long, W. D. Xiao, W. K. Yuan, Removal of nitric oxide and sulphur dioxide from flue gas using a hexamminecobalt (II)/iodide solution, Ind. Eng. Chem. Res. 43 (2004) 4048.

[11] M. Mori, J. A. Weil, M. Ishiguro, The formation of and interrelation between some μ-Peroxo binuclear cobalt complexes II^{1a}, J. Am. Chem. Soc. 90 (1968) 615.

[12] J. Simplicio, R. G. Wilkins, The uptake of oxygen by ammoniacal cobalt (II) solutions, J. Am. Chem. Soc. 91 (1969) 1325.

[13] O. Levenspiel, J. H. Godfrey, A gradient less contactor for experimental study of interphase mass transfer with/without reaction, Chem. Eng. Sci. 29 (1974) 1723.

[14] Gans. Peter, Reaction of nitric oxide with cobalt (II) ammine complexes and other reducing agents, J. Chem. Soc. 89 (A) (1967) 943.

DFT Studies of Dry Reforming of Methane on Ni Catalyst[*]

Abstract First-principles calculations based on density functional theory (DFT) have been used to investigate the reaction mechanism of dry methane reforming on Ni(111). The most energetically favorable adsorption configurations of the species involved in this process are identified and the transition states for all the possible elementary steps are explored by the dimer method. Then, the related thermodynamic properties at 973.15K are calculated by including the zero-point energy correction, thermal energy correction and entropic effect. It is found that CO_2 dissociates via a direct pathway to produce CO and O dominantly, and atomic O is revealed to be the primary oxidant of CH_x intermediates. Based on this information, two dominant reaction pathways are constructed as both the CH and C oxidation are found to be likely. The reaction network begins with the dissociation of CO_2 and CH_4, and then the generated CH and C are oxidized by atomic O to produce CHO and CO, followed by the CHO decomposition to finally generate CO and H_2. As for these two reaction pathways, the oxidation step is predicted to determine the overall reaction rate under the current investigated conditions, while the CH_4 dissociation is found to be the rate-limiting step at lower temperatures.

Key words density functional theory, dry reforming, nickel, carbon dioxide

1 Introduction

Dry reforming of methane has received much attention of the scientific community because it produces syngas with low H_2/CO ratio, which can be preferentially used for the synthesis of liquid hydrocarbons in the Fischer-Tropsch synthesis network and for the preparation of high-purity CO[1]. Furthermore, from an environmental perspective, a significant advantage of this process is the utilization of greenhouse gases, CH_4 and CO_2, which are leading to global warming. Finally, many potential thermodynamic heat-pipe applications such as recovery, storage and transmission of solar and other renewable energy sources have been proposed on the basis of the high enthalpy change and reversibility of dry methane reforming[2].

The cheap supported Ni catalyst is regarded as the most attractive catalyst for this process, though some noble metals such as Rh and Ru are found to be more reactive[3-7]. Nevertheless, under the same reaction conditions, the supported Ni catalyst is active for coke formation and growth of filamentous carbon as well[8-10]. Consequently, the severe catalyst deactivation becomes the main obstacle with respect to the commercialization of dry reforming of methane.

Despite numerous experimental and theoretical work devoted to investigating the reaction mechanism of dry methane reforming[6,8,11-24], some questions remain open. For example, it is well known that the role of the co-reactant CO_2 is to provide the oxidants of CH_x (x =0—3) species, i.e., OH and O. Then, which is the primary oxidant in this reaction? Bradford and Vannice[12] proposed a ki-

---

[*] Coauthors: Zhu Yian, Chen De, Zhou Xinggui. Reprinted from *Catalysis Today*, 2009, 148: 260-267.

netic model for CH_4-CO_2 reforming which correlates experimental data successfully. In their model, CO_2 participates in the reaction through the reverse water-gas shift (RWGS) reaction to produce OH, and OH group reacts with adsorbed CH_x intermediates to yield formate-type species (CH_xO). In contrast, based on the kinetic and isotopic measurements, Wei and Iglesia[6] argued that C atoms are oxidized by atomic O to form CH_xO which is taken as the most abundant intermediate on the surface.

Secondly, there is a big reaction network available for dry reforming of methane, so the confirmation of the predominant reaction pathway and rate-limiting step is challenging. Rostrup-Nielsen and Hansen[11] declared that steam reforming and dry reforming of methane on Ni catalyst proceed with the same rate, and the CH_4 dissociation and C oxidation are predicted to determine the overall reaction rate. Bradford and Vannice[12] suggested that dry reforming of methane occurs via the reversible dissociation of CH_4 and CO_2 to produce CH_xO, and the decomposition of CH_4 as well as CH_xO is the kinetically slow step. Wei and Iglesia[6] carried out some comparative experiments for the CH_4 dissociation, steam and dry reforming of methane, as well as WGS reaction. They proposed that regardless of the co-reactants, the C—H bond activation is the sole kinetically relevant step in all three reactions, and the steps involving co-reactants turns out to be quasi-equilibrated.

During the past two decades, theoretical work based on density functional theory (DFT) has also been extensively conducted to answer the aforementioned open questions[17-24]. Watwe et al.[17] performed periodic infinite plane wave slab calculations to investigate the stability and reactivity of CH_x species on Ni(111). Based on the combined DFT calculations and kinetic analyses, it was found all the CH_x species prefer threefold sites and CH as well as CO is the most abundant species on the surface. Michaelides and Hu[18-20] used DFT calculations to study the CH_3 and CH_2 adsorption on Ni(111), and declared that the C—H—Ni three-center bonding determines the chemisorption sites of CH_3 and CH_2. They also reported that the CH_3 dehydrogenation is exothermic with the reaction heat of -48 kJ/mol and the corresponding activation energy is more than 100 kJ/mol. Bengaard et al.[21] carried out first-principles calculations to suggest that both the CH_4 dissociation and nucleation of graphite are structure-sensitive, and their estimation of the barrier for the CH_4 dissociation is less than 91 kJ/mol. More recently, Blaylock et al.[22] have combined thermodynamic data with electronic activation energies to develop a microkinetic model to simulate steam reforming of methane under realistic conditions. The rate-limiting steps were found to be the CH_4 dissociative adsorption and the CH oxidation by O as well as OH. As for the CO_2 adsorption and decomposition, Wang et al.[23,24] reported that the CO_2 adsorption on Ni(111) is endothermic and the chemisorbed CO_2 molecule is negatively charged. Furthermore, the same authors suggested a mechanism for dry reforming of methane on the basis of the energy barriers obtained by DFT calculations. Their mechanism included the CO_2 decomposition to generate oxidant (atomic O), the CH_4 dissociation to surface CH, the CH oxidation by O and the CHO decomposition to produce CO, in which the CH_4 dissociation was found to be the rate-limiting step. However, in their calculations, atomic O was taken as the sole oxidant of CH and the contribution of OH was neglected.

In this contribution, the adsorption energies of the related species are firstly calculated by DFT calculations and the transition states for the involved elementary reactions are located by the dimer method[25]. Because Ni catalyst is traditionally used in dry reforming of methane at temperatures of above 973.15K[26], the corresponding Gibbs free energy barriers at this temperature are then calculated. In order to provide a comprehensive mechanistic picture, the dissociation of CO_2 and CH_4, the oxidation of CH and C, and the decomposition of CHO(H) as well as COH are investigated in detail. Consequently, the predominant reaction pathway and rate-limiting step are revealed. Finally, we conclude by discussing the implication of our results for understanding the overall mechanism of dry reforming of methane on Ni catalyst.

2 Computational Details

2.1 DFT calculations

The first-principles calculations performed in this study are based on spin-polarized DFT. The Vienna *ab initio* simulation package (VASP)[27-30] is used, in which the interactions between valence electrons and ion cores are described by pseudopotentials and the electronic wavefunctions at each k-point are expanded in terms of a discrete plane-wave basis set. Here we use Blöchl's allelectron-like projector augmented wave (PAW) method[31] as implemented by Kresse and Joubert[32], which regards the 4s 3d states as the valence configuration for Ni, 4d 5s states for Ag, 2s 2p states for C. The PAW method is necessary for accurate calculations of certain transition metals, which are sometimes poorly described by the ultrasoft pseudopotentials. Exchange and correlation of the Kohn-Sham theory are treated with the generalized gradient approximation (GGA) in the formulation of PBE functional[33]. A plane wave energy cutoff of 400eV is used in the present calculation and all geometries are optimized using a force-based conjugate gradient algorithm[34] until the forces acting on each atom are converged better than 0.1eV/nm. Brillouin zone sampling is performed using a Monkhorst-Pack grid[35] and electronic occupancies are determined according to a Methfessel-Paxton scheme[36] with an energy smearing of 0.2eV. Because there is a magnetic element (Ni) involved in the system, spin polarized effect has been considered. The calculations performed by Kresse and Hafner showed that surface magnetism is essential for the accurate quantitative description of adsorption energy[37].

The Ni(111) surface is represented as a four-layer slab with $p(3 \times 3)$ supercell and only the bottom layer of the slab is constrained to their crystal lattice positions. The neighboring slabs are separated in the direction perpendicular to the surface by a vacuum region of 1.2nm. The first Brillouin zone of the $p(3 \times 3)$ supercell is sampled with a $3 \times 3 \times 1$ k-point grid, which are evidenced to be sufficient for this cell[38,39].

The Hessian matrix for the potential energy surface is calculated using finite difference approximation, and diagonalized to find the normal modes of the investigated systems. The adsorbates and the metal atoms to which the adsorbates are attached are displaced in the direction of each Cartesian coordinate, while the other Ni atoms are kept rigid during these finite difference calculations because in the preliminary calculation, the CH_4 vibrational frequencies adopt almost the same wavenumbers with and without the relaxation of the metal atoms in the deeper layers.

The dimer method[25] is used to locate the transition state, in which the saddle point is optimized using a force-based conjugate-gradient method[34] until the maximum force in every degree of freedom is less than 0.1eV/nm. In order to obtain accurate forces, the total energy and band structure energy are converged to within 1×10^{-7} eV/atom during the electronic optimization.

2.2 Thermodynamics

2.2.1 Gas-phase species

The Gibbs free energy of a gas-phase species A(CO_2(g), CH_4(g), H_2(g), CO(g), H_2O(g), CH_3OH(g)) at temperature T and pressure P is given by:

$$G_A(T,P) = E_{\text{total},A} + E_{\text{ZPE}} + \Delta H^\circ(0 \to T) - TS(T,P)$$
$$= E_{\text{total},A} + E_{\text{ZPE}} + \Delta H^\circ(0 \to T) - TS^\circ(T) + RT\ln\frac{P}{P^\circ} \quad (1)$$

where, $E_{\text{total},A}$ is the total energy determined by DFT calculations; E_{ZPE} is the zero-point energy, which is calculated by:

$$E_{\text{ZPE}} = \sum_{i=1}^{3N-6(5)} \frac{N_A h v_i}{2} \quad (2)$$

where, N_A is Avogadro's number; h is Plank's constant; v_i is the frequency of the normal mode; N is the number of atoms involved in the system. $\Delta H^\circ(0 \to T)$ is the enthalpy change from 0K to temperature T. As for CO_2(g), H_2(g), CO(g) and H_2O(g), the enthalpy change from 0K to 298.15K can be found in Ref. [40]. The corresponding values of CH_4(g) and CH_3OH(g) are calculated by:

$$\Delta H^\circ(0 \to T) = H_{\text{trans}} + H_{\text{rot}} + (H - E_{\text{ZPE}})$$
$$= \frac{5}{2}RT + \frac{3}{2}RT + \sum_{i=1}^{3N-6} \frac{N_A h v_i e^{-hv_i/(k_B T)}}{1 - e^{-hv_i/(k_B T)}} \quad (3)$$

where, R is the universal gas constant; k_B is Boltzmann's constant. The enthalpy change of a gas-phase species A from 298.15K to temperature T and the standard entropy at temperature $T[S^\circ(T)]$ are calculated using the Shomate equation and the related parameters are taken from Ref. [40]. P° is the standard pressure which is taken to be 1 bar.

2.2.2 Weakly bound species

The Gibbs free energy for a species A adsorbed on Ni surface is calculated using a similar expression:

$$G_A(T,P) = E_{\text{total},A} + E_{\text{ZPE}} + \Delta U^\circ(0 \to T) - TS^\circ(T) \quad (4)$$

where the enthalpy change is replaced by the change of internal energy.

As for the weakly bound species, namely CH_4^* and CO_2^*, we assume to a first approximation that they behave as two-dimensional gases and maintain the full rotational and vibrational modes of the corresponding gaseous species. The standard internal energy change of adsorption is calculated by:

$$\Delta U^\circ_{\text{trans},2D}(0 \to T) - \Delta U^\circ_{\text{trans},3D}(0 \to T) = 2RT - \frac{5}{2}RT = -\frac{1}{2}RT \quad (5)$$

The standard entropy change (ΔS°) of adsorption is calculated by:

$$\Delta S^\circ = S^\circ_{\text{trans},2D} - S^\circ_{\text{trans},3D} = R\left\{\ln\left[\frac{h/(k_B T)}{(2\pi m k_B T)^{1/2}}\left(\frac{SA}{N_{\text{sat}}}\right)P^\circ\right] - \frac{1}{2}\right\} \quad (6)$$

where, SA/N_{sat} is the area occupied per adsorbed molecule at the standard state conditions. If it is assumed that the standard state is monolayer coverage, SA/N_{sat} equals the reciprocal of the surface concentration of sites.

2.2.3 Tightly bound species

As for the tightly bound species including the remaining reactants, transition states, intermediates and products, the translational and rotational modes are replaced by vibrational modes corresponding to frustrated translation and rotation on the surface. Consequently, the internal energy change and entropy of the tightly adsorbed species are given by:

$$\Delta U^\circ(0 \to T) = \sum_{i=1}^{3N} \frac{N_A h v_i e^{-hv_i/(k_B T)}}{1 - e^{-hv_i/(k_B T)}} \quad (7)$$

$$S^\circ(T) = \sum_{i=1}^{3N} \left\{ -R\ln[1 - e^{-hv_i/(k_B T)}] + \frac{N_A h v_i}{T} \cdot \frac{e^{-hv_i/(k_B T)}}{1 - e^{-hv_i/(k_B T)}} \right\} \quad (8)$$

3 Results and Discussion

Based on the previous work from other research groups[6,11,26], we propose here a more detailed mechanism for dry reforming of methane. The investigated reaction network is schematically illustrated in Fig. 1.

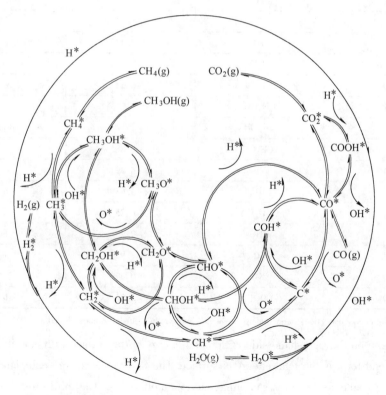

Fig. 1 Proposed mechanism for dry reforming of methane on Ni catalyst

3.1 CO_2 decomposition

As elucidated in the experimental work[6,12], two possible reaction pathways are predicted to contribute to the CO_2 decomposition. As for the first one, the adsorbed CO_2 dissociates directly to form CO and atomic O, the latter being the oxidant of CH_x. Along the second reaction pathway, CH_4 firstly dissociates to produce atomic H which subsequently activates the adsorbed CO_2 by producing the COOH intermediate, and the decomposition of COOH finally yields CO and the oxidant OH, according to:

$$CO_2^* + H^* \to COOH^* \to CO^* + OH^* \tag{9}$$

The calculated adsorption energies (ΔE_{ads}) of the species involved in dry reforming of methane are summarized in Table 1. ΔE_{ads} is calculated as:

$$\Delta E_{ads} = E_{adsorbate+surface} - E_{surface} - E_{adsorbate} \tag{10}$$

Table 1 Calculated adsorption energies (ΔE_{ads}) of the species involved in dry reforming of methane on Ni(111)

No.	Species	ΔE_{ads}[①]	Favored adsorption site
1	CH_4	-0.02	N/A
2	CH_3	-1.91	fcc
3	CH_2	-4.01	fcc
4	CH	-6.43	fcc
5	C	-6.78	hcp
6	H	-2.81	fcc
7	O	-5.67	fcc
8	OH	-3.42	fcc
9	H_2O	-0.29	N/A
10	H_2	-0.22	N/A
11	CH_3OH	-0.30	N/A
12	CH_2OH	-1.54	Bridge
13	CHOH	-3.88	fcc
14	COH	-4.39	hcp
15	CH_3O	-2.63	fcc
16	CH_2O	-0.75	fcc
17	CHO	-2.26	fcc
18	CO	-1.92	hcp
19	CO_2	-0.02	N/A
20	COOH	-2.26	fcc

① Zero-point energy correction, thermal energy correction and entropic effect are not included.

The first term on the right-hand side is the total energy of the surface with one attached molecule; the second term is the total energy of surface. The first two terms are calculated by applying the same parameters (k-point sampling, energy cutoff, etc.). The third term is the total energy of an isolated adsorbate molecule, which is estimated by putting an adsorbate molecule in a box with dimensions of 2nm × 2.05nm × 2.1nm and carrying out a spin-polarized Γ-point calculation. With this definition, a more negative value of ΔE_{ads} corresponds to stronger binding between adsorbate and surface.

The CO_2 adsorption on the Ni(111) surface has been investigated at the threefold fcc and hcp, Bridge and Atop sites. After geometry optimization, CO_2 is found to be repelled towards vacuum if it is initially placed close to Ni(111) in a parallel manner, as shown in Fig. 2a, and the adsorption energies at all binding sites take similar small values (~ -0.02eV), indicating the nature of the CO_2 physisorption on Ni(111). Moreover, in the presence of preliminarily adsorbed atomic H, CO_2 is further displaced apart from metal surface by 0.015nm. However, Wang et al.[23] performed DFT calculations to reveal that covalent bonds are formed upon the CO_2 chemisorption on Ni(111). We have reproduced the most stable adsorption configuration proposed by them, as illustrated in Fig. 2b, and found this state cannot be achieved unless the CO_2 molecule is bent artificially. Wang et al. declared that their calculated adsorption energies are positive and the CO_2 chemisorption on Ni(111) is thermodynamically unfavorable. In addition, Heiland[41] has adopted the fast ion beam and fast molecular beam techniques to investigate the CO_2 dissociation, and did not detect any chemisorbed CO_2 on Ni(111) at low temperatures. Therefore, the CO_2 physisorption on Ni(111) is verified and turns out to be the precursor state to initiate the CO_2 decomposition.

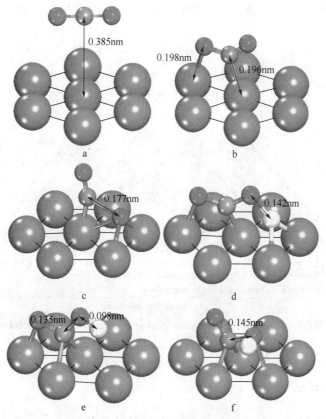

Fig. 2 Geometries of the initial states, transition states and intermediate for CO_2 decomposition on Ni(111)
a—The physisorbed CO_2 on Ni(111); b—The chemisorbed CO_2 on Ni(111);
c—The transition state for CO_2 decomposition via the direct pathway; d—The transition state for the H-inducted CO_2 decomposition;
e—The intermediate for the H-induced CO_2 decomposition; f—The transition state for COOH dissociation to generate CO and OH

By comparing the activation energy and free energy barriers at 973.15K shown in Fig. 3, it is found that CO_2 dissociates via the direct pathway dominantly owing to the relatively high reaction rate. The corresponding configuration of the saddle point is shown in Fig. 2c, in which the cleaving C—O bond is stretched from 0.118nm in gas phase to 0.177nm. To facilitate the comparison, the configurations of the two saddle points and one intermediate for the H-induced CO_2 decomposition are represented in Fig. 2d-f. In these figures, one can see that CO_2 firstly approaches to the atomic H which is preliminarily adsorbed on the surface to form the COOH species, followed by the dissociation of COOH to generate CO and OH. The adsorption energies of O and OH on Ni(111) are found to be −5.67eV and −3.42eV (see Table 1), respectively, indicating a larger site coverage of atomic O. Based on the calculated free energy barriers and adsorption energies, the primary oxidant involved in dry reforming of methane on Ni catalyst is predicted to be atomic O which is generated by the CO_2 direct decomposition. This finding agrees well with the previous kinetic and isotopic results that the CO_2 activation to produce CO and O is reversible and quasi-equilibrated under the similar reaction conditions[6]. However, we note the contribution of OH is not negligible because of the comparable barriers. Furthermore, Pan et al. [42] performed periodic DFT calculations to find on the partially hydroxylated surface of γ-Al_2O_3 which acts as the catalyst support in dry reforming of methane, the CO_2 adsorption is promoted in the vicinity of OH and simultaneously CO_2 reacts with its adjacent OH to produce a bicarbonate species.

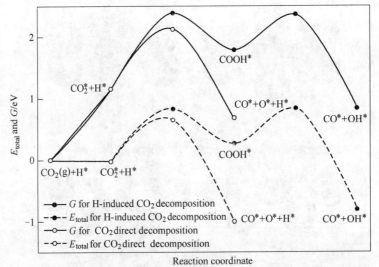

Fig. 3 Total energy and Gibbs free energy diagrams for the CO_2 direct composition and H-induced CO_2 decomposition on Ni(111)

3.2 CH_4 dissociation

Wei and Iglesia found the turnover rate of dry reforming of methane is limited solely by the C—H bond activation and unaffected by the identity as well as concentration of the coreactants[6]. Therefore, the methane consumption is expected to take place via dissociative adsorption regardless of the existence of CO_2, and the generated CH_x intermediates are subsequently oxidized by atomic O and OH to form CH_xO and CH_xOH species, respectively. The calculated free energy barriers for all the forward and reverse reactions involved in dry reforming of methane are summarized in Fig. 4, and the corresponding activation energies are listed in Table 2.

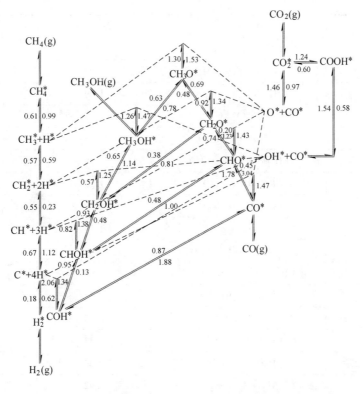

Fig. 4 Free energy barriers at 973.15K for the forward and reverse reactions involved in dry reforming of methane

Table 2 Calculated activation energies for all the forward ($E_{a,f}$) and reverse ($E_{a,r}$) elementary reactions involved in dry reforming of methane

No.	Elementary reaction	$E_{a,f}$①/eV	$E_{a,r}$②/eV
1	$CH_4(g) + {}^* \rightleftharpoons CH_4{}^*$	0	0.02
2	$CH_4{}^* \rightleftharpoons CH_3{}^* + H^*$	0.91	0.90
3	$CH_3{}^* \rightleftharpoons CH_2{}^* + H^*$	0.70	0.63
4	$CH_2{}^* \rightleftharpoons CH^* + H^*$	0.35	0.69
5	$CH^* \rightleftharpoons C^* + H^*$	1.33	0.81
6	$CO_2(g) + {}^* \rightleftharpoons CO_2{}^*$	0	0.02
7	$CO_2{}^* \rightleftharpoons CO^* + O^*$	0.67	1.65
8	$CO_2{}^* + H^* \rightleftharpoons COOH^*$	1.13	0.85
9	$COOH^* \rightleftharpoons CO^* + OH^*$	0.57	1.65
10	$CH_3{}^* + OH^* \rightleftharpoons CH_3OH^*$	2.20	1.61
11	$CH_3OH^* \rightleftharpoons CH_2OH^* + H^*$	0.88	0.69
12	$CH_2{}^* + OH^* \rightleftharpoons CH_2OH^*$	1.32	0.60
13	$CH_2OH^* \rightleftharpoons CHOH^* + H^*$	0.53	0.90
14	$CH^* + OH^* \rightleftharpoons CHOH^*$	1.48	0.80
15	$CHOH^* \rightleftharpoons COH^* + H^*$	0.15	0.86

Continued 2

No.	Elementary reaction	$E_{a,f}$[①]/eV	$E_{a,r}$[②]/eV
16	$C^* + OH^* \rightleftharpoons COH^*$	1.46	2.01
17	$COH^* \rightleftharpoons CO^* + H^*$	0.98	1.97
18	$CH_3^* + O^* \rightleftharpoons CH_3O^*$	1.59	1.31
19	$CH_3O^* \rightleftharpoons CH_2O^* + H^*$	0.93	0.64
20	$CH_2^* + O^* \rightleftharpoons CH_2O^*$	1.45	0.95
21	$CH_2O^* \rightleftharpoons CHO^* + H^*$	0.36	0.74
22	$CH^* + O^* \rightleftharpoons CHO^*$	1.53	1.08
23	$CHO^* \rightleftharpoons CO^* + H^*$	0.20	1.48
24	$C^* + O^* \rightleftharpoons CO^*$	1.59	2.94
25	$CH_3OH^* \rightleftharpoons CH_3O^* + H^*$	0.89	1.38
26	$CH_2OH^* \rightleftharpoons CH_2O^* + H^*$	0.63	1.04
27	$CHOH^* \rightleftharpoons CHO^* + H^*$	0.71	1.14
28	$O^* + H^* \rightleftharpoons OH^*$	1.35	1.16
29	$OH^* + H^* \rightleftharpoons H_2O^*$	1.33	0.92
30	$H_2O^* \rightleftharpoons H_2O(g)$	0.29	0
31	$H^* + H^* \rightleftharpoons H_2^*$	0.92	0.06
32	$H_2^* \rightleftharpoons H_2(g)$	0.22	0
33	$CO^* \rightleftharpoons CO(g)$	1.92	0
34	$CH_3OH^* \rightleftharpoons CH_3OH(g)$	0.30	0

①,② Zero-point energy correction, thermal energy correction and entropic effect are not included.

The activation energy for the first step of CH_4 dehydrogenation is calculated to be 0.91 eV and with the zero-point energy correction considered, the barrier is lowered to 0.79 eV. This is in excellent agreement with the barrier of (0.77 ± 0.1) eV obtained in the recent experiments by Egeberg et al[43]. Great care has been taken in their measurements to block defects on the surface with unreactive Au atoms in order to exclude the contribution of the step edge. The configuration of the transition state is shown in Fig. 5a, in which the cleaving C—H bond is stretched from 0.109 nm in gas phase to 0.161 nm. The detached H atom is located at a threefold site, and the remaining CH_3 fragment sits on the top of the adjacent surface Ni atom to achieve the maximum C—H—Ni three-center bonding[18,44]. After CH_4 is dissociatively adsorbed on the Ni surface, the further dehydrogenation of CH_3 and CH_2 species is quite easy and the generation of $CH_3O(H)$ as well as $CH_2O(H)$ is predicted to be unlikely because the free energy barriers for the CH_3 and CH_2 dissociation are 0.59 eV and 0.23 eV, much lower than those for the oxidation (1.47 eV and 1.53 eV for the CH_3 oxidation, 1.25 eV and 1.34 eV for the CH_2 oxidation). In contrast to the CH_4 dehydrogenation, the detached H atoms move to the top of surface Ni atoms, while the remaining CH_x fragments are adsorbed at the threefold sites with the cleaving C—H bonds stretched to 0.161 nm and 0.171 nm at the saddle points, as shown in Fig. 5b and c. Furthermore, the CH_2 dissociation is found to be exothermic with the reaction heat of −0.34eV, and therefore this process is favorable both kinetically and thermodynamically.

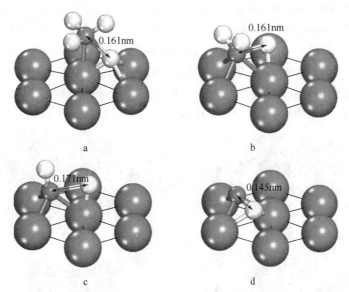

Fig. 5 Geometries of the transition states for CH$_4$ dissociation on Ni(111)

a—The transition state for CH$_4$ dehydrogenation; b—The transition state for CH$_3$ dehydrogenation;
c—The transition state for CH$_2$ dehydrogenation; d—The transition state for CH dehydrogenation

As for the CH decomposition, the free energy barrier is substantially increased to 1.12eV and remains lower than those for the CH oxidation by OH and O (1.38eV and 1.43eV). However, the preparation of both atomic C and CHO(H) is predicted to be likely because the two energy barriers are comparable and more importantly, the free energy barrier for the reverse reaction of the CH dehydrogenation (0.67eV) is much lower than that for the forward reaction. Consequently, the generated C atom recombines with atomic H to form the CH species and the CH oxidation is promoted to a certain extent. As shown in Fig. 5d, the atomic C and H are adsorbed at the threefold and Bridge sites, respectively, and the cleaving C—H bond is stretched to 0.145nm in the transition state. In fact, it is found that two saddle points are available on the potential energy surface with respect to the CH dehydrogenation step. In the other one, the atomic H moves to the top of a surface Ni atom and the C—H bond is further elongated to 0.178nm. However, this transition state is 0.09eV higher in total energy and is neglected in our subsequent calculation.

3.3 Oxidation of CH and C

As shown in section 3.2, the CH species is predicted to dissociate to generate atomic C and to react with O (or OH) simultaneously. Therefore, both CH group and atomic C are taken as the candidates to be oxidized to eventually produce CO. The transition states for the reactions of C (H) with O are shown in Fig. 6a and b, in which the atomic C and O are adsorbed at the threefold hollow and adjacent Bridge sites, respectively. The distances between the atomic C and O take almost the same values (0.184nm and 0.185nm) in the two cases, leading to the similar free energy barriers of 1.43eV and 1.47eV. The corresponding final states are shown in Fig. 7a and b, and the C—O bonds are contracted to 0.129nm and 0.120nm to form the CHO and CO species, respectively. For the most energetically favorable structure, the CHO species is adsorbed at the fcc site and the atomic C is coordinated with two surface Ni atoms besides the atomic O and H to keep the sp^3 hybridization. From the thermodynamic point of view, we find that the CH oxidation by O is endothermic with the reaction heat of 0.45eV. This disagrees with what was found by Wang et al. who suggested the reaction of CH with O is favorable both kinetically and thermodynamically[24]. The discrepancy arises from the energy reference they used, which does not refer to the sum of the total energies of separated CH and O.

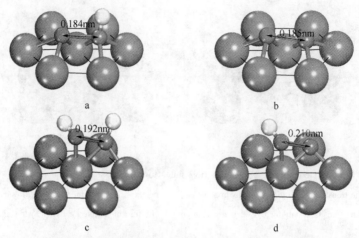

Fig. 6 Geometries of the transition states for CH and C oxidation on Ni(111)
a—The transition state for CH oxidation by O; b—The transition state for C oxidation by O;
c—The transition state for CH oxidation by OH; d—The transition state for C oxidation by O

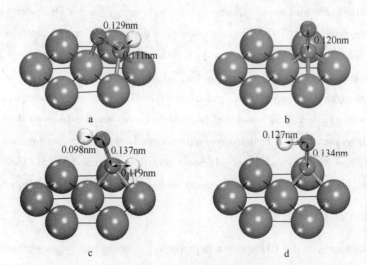

Fig. 7 Geometries of the final sates for CH and C oxidation on Ni(111)
a—CHO; b—CO; c—CHOH; d—COH

Then, the free energy barriers for the CH and C oxidation by OH have been calculated and found to be 1.38eV and 1.34eV, respectively, which are slightly lower than those for the oxidative reactions by O. However, the former pathway is not predicted to be dominant because of the low surface coverage of OH, as proposed by the previous microkinetic modeling[22]. The transition states for the CH and C oxidation by OH are shown in Fig. 6c and d, in which the OH groups move to the Atop sites adjacent to the atomic C, resulting in longer C—O bonds (0.192nm and 0.210nm in length) than those in the transition states for the oxidative reactions by O. In the final states, the CHOH and COH species are adsorbed at the fcc and hcp sites with the C—O bonds contracted to 0.137nm and 0.134nm, respectively, as shown in Fig. 7c and d. From the thermodynamic point of view, the reaction of CH with OH is endothermic, while the C oxidation by OH is exothermic. Thus, it is much more difficult for the COH species to dissociate to form OH along the reverse reaction of the C oxidation. As a result, the COH decomposition to produce CO is promoted to a certain extent.

3.4 Decomposition of CHO(H) and COH

Starting from the most energetically favorable adsorption configuration, the free energy barrier and activation energy for the CHO dissociation to generate CO are calculated to be 0.45eV and 0.20eV, respectively, the latter being consistent with the result of 0.29eV obtained by Wang et al.[24]. Furthermore, this reaction is found to be exothermic and release heat of 1.28eV. Thus, the low energy barrier and high reaction heat indicate that the CHO decomposition plays a key role in producing CO and on the other hand, this kinetically and thermodynamically favored reaction would also help to remove the thermodynamic constrain in the reaction of CH with O by continuous CHO consumption. The corresponding transition state is shown in Fig. 8a, in which the activated complex is adsorbed at the fcc site and the cleaving C—H bond is slightly stretched from 0.111nm in the CHO adsorption configuration to 0.117nm.

Similarly, the free energy barriers for the CHOH dissociation to form COH and CHO are found to be quite low (0.13eV and 0.48eV, respectively), achieving a low surface coverage of CHOH. In the transition state for the COH production, the COH species is adsorbed at the fcc site and the detached H atom is located at the adjacent Atop site with the C—H bond stretched to 0.155nm, as shown in Fig. 8b. Consequently, the angle of the C—O bond with respect to the surface normal is reduced to 12.27° as the repulsive interaction between COH and H is weakened. At the saddle point for the CHO production, the detached H atom is adsorbed at the Bridge site and the geometry of the CHO fragment is similar to that in the stable CHO adsorption configuration except that the atomic O is coordinated to the detached H atom with the O—H bond stretched from 0.098nm to 0.146nm, as shown in Fig. 8c. Furthermore, both the COH and CHO production are found to be exothermic and the reaction heats are calculated to be −0.71eV and −0.43eV, respectively. Therefore, these two species are readily to be generated for the subsequent dissociation once the CH group is oxidized by OH.

Finally, the free energy barrier for the COH decomposition to form CO is calculated to be 0.87eV. In the transition state, the C—O bond is tilted to form an angle of 29.70° with respect to the surface normal in order to coordinate H with the Ni surface, and at the same time, the distance between the atomic O and H is elongated to be 0.134nm, as shown in Fig. 8d. From the thermodynamic point of view, this reaction is energetically favorable and the reaction heat is calculated to be −0.99eV, indicating both the CH and C oxidation by OH are likely to contribute to the CO production.

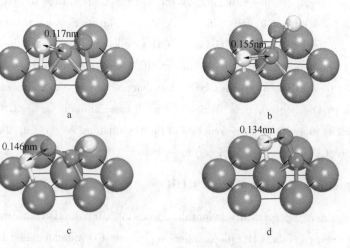

Fig. 8 Geometries of the transition states for CHO(H) and COH decomposition on Ni(111)

a—The transition state for CHO decomposition;

b—The transition state for CHOH decomposition to produce COH and H;

c—The transition state for CHOH decomposition to produce CHO and H;

d—The transition state for COH decomposition

3.5 Dominant reaction pathway and rate-limiting step

Now that the free energy barriers for the elementary steps are evaluated by the combined DFT calculations and thermodynamic analyses, the dominant reaction pathway for dry reforming of methane is predicted to involve the direct decomposition of CO_2 to form atomic O and the dissociation of methane to generate CH and C, followed by the CH and C oxidation by atomic O, and eventually the CHO decomposition to produce CO. However, the contribution of OH cannot be negligible because the free energy barrier for the H-induced CO_2 decomposition to produce OH is comparable with that for the direct CO_2 decomposition, and the CHOH and COH dissociation are found to be favorable both kinetically and thermodynamically. This prediction agrees well with the experimental data provided by Wei and Iglesia as well as their suggested model[6].

Based on the established dominant reaction pathways, we calculate the Gibbs free energies and total energies of all relevant states (including transition states) along the reaction coordinate, as shown in Fig. 9. In such a free energy diagram, the point with the highest energy defines the slowest reaction step (at standard pressures)[45]. As one can see, the oxidation step is predicted to be the rate-limiting step for both the CH and C oxidation pathways under the current investigated conditions. However, at lower temperatures, the reaction pathway including the CH oxidation is found to be dominant because of its much lower total energy barrier than that for the C oxidation pathway, and simultaneously the CH_4 dissociation turns out to determine the overall reaction rate. This shift in rate-limiting step can be used to explain a large part of the discrepancies among the various experimental studies which are summarized in Section 1.

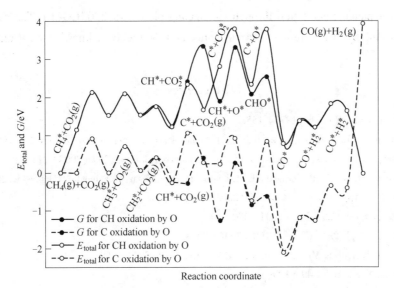

Fig. 9 Total energy and Gibbs free energy diagram at 973.15K for dominant reaction pathways

4 Conclusion

DFT calculations have been carried out to investigate the mechanism of dry reforming of methane. The most energetically favorable adsorption configurations have been identified for all the related substances listed in Table 1. Based on the interaction with the metal surface, these species are classified into three categories, namely the gas phase, weakly bound and tightly bound species. Then, the Gibbs free energies at 973.15K are calculated by including the zero-point energy correction, thermal energy correction and entropic effect. The transition states for the involved elementary reactions have been explored through the dimer method and the dominant reaction pathway is constructed on the basis of the free energy barriers.

The free energy barriers for the direct CO_2 decomposition and H-induced CO_2 decomposition are calculated to be 2.13eV and 2.40eV, respectively. Therefore, CO_2 is predicted to dissociate via a direct pathway to generate CO and O dominantly, and atomic O is taken as the primary oxidant of CH_x intermediates. However, the H-induced CO_2 decomposition contributes to dry methane reforming as well because of the comparable barriers, and the effect of the oxidant OH is not negligible. Taking atomic O as the oxidant, we construct two dominant reaction pathways as both the CH and C oxidation are found to be likely. The reaction network begins with the CO_2 decomposition to generate O and the CH_4 dehydrogenation to produce CH and C which are subsequently oxidized by atomic O to form CHO and CO, and ends with the decomposition of CHO to produce CO and H_2. As for these two dominant reaction pathways, the oxidation step is predicted to determine the overall reaction rate under the current investigated conditions, while the CH_4 dehydrogenation is found to be the rate-limiting step at lower temperatures.

Acknowledgements

This work is supported by Doctoral Fund of Ministry of Education of China (No. 200802511007), Natural Science Foundation of Shanghai (No. 08ZR1406300) and sponsored by Shanghai Educational Development Foundation through Chenguang plan (No. 2007CG41). The computational time provided by Notur project (www.notur.no) is highly acknowledged (nn2920k and nn4685k).

References

[1] D. L. Trimm, Catal. Rev. 16 (1977) 155.
[2] J. H. McCrary, G. E. McCrary, T. A. Chubb, J. J. Nemecek, D. E. Simmons, Sol. Energy 29 (1982) 141.
[3] M. Maestri, D. G. Vlachos, A. Beretta, G. Groppi, E. Tronconi, J. Catal. 259 (2008) 211.
[4] J. F. Munera, S. Irusta, L. M. Cornaglia, E. A. Lombardo, D. V. Cesar, M. Schmal, J. Catal. 245 (2007) 25.
[5] X. E. Verykios, Appl. Catal. A: Gen. 255 (2003) 101.
[6] J. Wei, E. Iglesia, J. Catal. 224 (2004) 370.
[7] J. Wei, E. Iglesia, J. Phys. Chem. B 108 (2004) 7253.
[8] U. Olsbye, T. Wurzel, L. Mleczko, Ind. Eng. Chem. Res. 36 (1997) 5180.
[9] D. L. Trimm, Catal. Today 49 (1999) 3.
[10] S. Helveg, C. Lopez-Cartes, J. Sehested, P. L. Hansen, B. S. Clausen, J. R. Rostrup-Nielsen, F. Abild-Pedersen, J. K. Norskov, Nature 427 (2004) 426.
[11] J. R. Rostrup-Nielsen, J. H. B. Hansen, J. Catal. 144 (1993) 38.
[12] M. C. J. Bradford, M. A. Vannice, Appl. Catal. A: Gen. 142 (1996) 97.
[13] Y. Nagayasu, K. Asai, A. Nakayama, S. Iwamoto, E. Yagasaki, M. Inoue, J. Jpn. Pet. Inst. 49 (2006) 186.
[14] Y. Cui, H. Zhang, H. Xu, W. Li, Appl. Catal. A: Gen. 318 (2007) 79.
[15] G. S. Gallego, C. Batiot-Dupeyrat, J. Barrault, F. Mondragon, Ind. Eng. Chem. Res. 47 (2008) 9272.
[16] J. Juan-Juan, M. C. Roman-Martinez, M. J. Illan-Gomez, Appl. Catal. A-Gen. 355 (2009) 27.
[17] R. M. Watwe, H. S. Bengaard, J. R. Rostrup-Nielsen, J. A. Dumesic, J. K. Norskov, J. Catal. 189 (2000) 16.
[18] A. Michaelides, P. Hu, Surf. Sci. 437 (1999) 362.
[19] A. Michaelides, P. Hu, J. Chem. Phys. 112 (2000) 8120.
[20] A. Michaelides, P. Hu, J. Chem. Phys. 112 (2000) 6006.
[21] H. S. Bengaard, J. K. Norskov, J. Sehested, B. S. Clausen, L. P. Nielsen, A. M. Molenbroek, J. R. Rostrup-Nielsen, J. Catal. 209 (2002) 365.
[22] D. W. Blaylock, T. Ogura, W. H. Green, G. J. O. Beran, J. Phys. Chem. C 113 (2009) 4898.
[23] S. Wang, D. Cao, Y. Li, J. Wang, H. Jiao, J. Phys. Chem. B 109 (2005) 18956.
[24] S. Wang, X. Liao, J. Hu, D. Cao, Y. Li, J. Wang, H. Jiao, Surf. Sci. 601 (2007) 1271.
[25] G. Henkelman, H. Jonsson, J. Chem. Phys. 111 (1999) 7010.
[26] X. E. Verykios, Int. J. Hydrogen Energy 28 (2003) 1045.
[27] G. Kresse, J. Hafner, Phys. Rev. B 47 (1993) 558.
[28] G. Kresse, J. Hafner, Phys. Rev. B 49 (1994) 14251.
[29] G. Kresse, J. Furthmüller, Comp. Mater. Sci. 6 (1996) 15.
[30] G. Kresse, J. Furthmüller, Phys. Rev. B 54 (1996) 11169.
[31] P. E. Blöchl, Phys. Rev. B 50 (1994) 17953.
[32] G. Kresse, D. Joubert, Phys. Rev. B 59 (1999) 1758.
[33] J. P. Perdew, K. Burke, M. Ernzerhof, Phys. Rev. Lett. 77 (1996) 3865.
[34] D. Sheppard, R. Terrell, G. Henkelman, J. Chem. Phys. 128 (2008) 10.
[35] H. J. Monkhorst, J. D. Pack, Phys. Rev. B 13 (1976) 5188.
[36] M. Methfessel, A. T. Paxton, Phys. Rev. B 40 (1989) 3616.
[37] G. Kresse, J. Hafner, Surf. Sci. 459 (2000) 287.

[38] S. Stolbov, S. Hong, A. Kara, T. S. Rahman, Phys. Rev. B 72 (2005) 155423.
[39] Y. Zhu, X. Zhou, D. Chen, W. Yuan, J. Phys. Chem. C 111 (2007) 3447.
[40] D. L. Lide, CRC Handbook of Physics and Chemistry, Academic Press, 1997.
[41] W. Heiland, Surf. Sci. 251-252 (1991) 942.
[42] Y. Pan, C. Liu, Q. Ge, Langmuir 24 (2008) 12410.
[43] R. C. Egeberg, S. Ullmann, I. Alstrup, C. B. Mullins, I. Chorkendoff, Surf. Sci. 497 (2002) 183.
[44] Y. Zhu, Y. Dai, D. Chen, W. Yuan, J. Mol. Catal. A: Chem. 264 (2007) 299.
[45] M. P. Andersson, E. Abild-Pedersen, I. N. Remediakis, T. Bligaard, G. Jones, J. Engbwk, O. Lytken, S. Horch, J. H. Nielsen, J. Sehested, J. R. Rostrup-Nielsen, J. K. Norskov, I. Chorkendorff, J. Catal. 255 (2008) 6.

First-Principles Study of C Adsorption and Diffusion on the Surfaces and in the Subsurfaces of Nonreconstructed and Reconstructed Ni(100) *

Abstract Ab initio plane wave density functional theory calculations are performed to investigate C adsorption and diffusion on the surfaces and in the subsurfaces of nonreconstructed and reconstructed Ni(100). It is observed that no reconstruction occurs on Ni(100) at a C coverage of 0.11 monolayer (ML), while the surface is driven into a well ordered Ni(100)-(2×2)$p4g$-C structure at 0.5ML. On both the nonreconstructed and reconstructed surfaces, C atoms bind most strongly to the fourfold hollow sites and the calculated geometric parameters for the C adsorption at these sites have been compared with published experimental data. Six different diffusion pathways between the fourfold hollow sites on nonreconstructed and reconstructed Ni(100) have been explored by using the nudged elastic band method. Along the most favorable pathway, a C atom hops between two fourfold hollow sites via an adjacent octahedral site in the subsurface of reconstructed Ni(100). The local lattice strain induced by the C diffusion is substantially relieved by the surface reconstruction and consequently the activation energy is decreased to 1.16eV, much lower than those for the other five diffusion pathways. Therefore, at low coverages, C atoms tend to accumulate on the surface to a higher level because of the relatively slow diffusion rate. When the critical coverage is attained, the surface reconstruction takes place and promotes the C diffusion dramatically.

1 Introduction

The interaction between C and Ni catalysts has attracted much attention during the past two decades. On the one hand, it is involved in some important chemical processes such as formation of carbon nanotubes (CNTs) and carbon nanofibers (CNFs), Fischer-Tropsch synthesis, methanation, and steam-reforming, etc.[1-4] On the other hand, C adsorption and diffusion on Ni surfaces may result in undesirable coke formation, which poisons the catalysts. During all these reactions, carbon-containing gases are mainly decomposed on the Ni(100) surface in the initial stage[5,6], and then the generated C atoms adsorb and migrate on the catalyst particles. Therefore, C adsorption and diffusion on Ni(100) are crucial in a wide range of heterogeneous Ni-catalyzed processes.

C adsorption on Ni(100) has been experimentally studied by Klink et al. through scanning tunneling microscopy (STM)[7]. The real-space information indicated that at low coverages less than 0.2 monolayer (ML), C atoms were located at the fourfold hollow sites, causing a small radial displacement ((0.015±0.015)nm) of the surrounding Ni atoms. Above a local critical coverage, the surface was driven into a (2×2)$p4g$ structure by disorder-order phase transition, which was known as "clock" reconstruction on the Ni(100) surface[8]. They suggested that the

---

* Coauthors: Zhu Yian, Zhou Xinggui, Chen De. Reprinted from *J. Phys. Chem. C*, 2007, 111: 3447-3453.

driving force for the reconstruction was the preference for C atoms to be five-coordinated. Scanned-energy mode photoelectron (PhD) study was also performed by Terborg et al. to investigate the C adsorption on Ni(100) at coverages of 0.15ML and 0.5ML[9]. It was observed that at the C coverage of 0.15ML, the adsorbed C atoms simply occupied undistorted hollow sites, while at 0.5ML, the topmost Ni atoms were displaced parallel to the surface by alternate clockwise and counterclockwise rotations around the C atoms. They concluded that the C-induced Ni—Ni repulsion was the primary driving force for the reconstruction.

Theoretical studies have also been devoted to investigating the binding between C atoms and the Ni(100) surface. Solidstate Fenske-Hall band structure calculations were used by Kirsch et al. to study the effect of C, N, and O adsorption on Ni(100) at a coverage of 0.5ML[10]. It was reported that the adsorption of C and N induced the "clock" surface reconstruction and the adsorbates sat nearly coplanar with the surface Ni atoms. In contrast, adsorbed O atoms sat slightly above the Ni(100) surface and had little effect on the overall surface structure. This was because forcing an O atom into the reconstructed surface led to the occupancy of an undesirable number of antibonding bands. Stolbov et al. conducted a first-principles study to explain the origin of the C-induced reconstruction on the Ni(100) surface[12]. At a C coverage of 0.25ML, the geometry optimization did not result in any surface reconstruction, while at 0.5ML, the system underwent a $c(2 \times 2)$-$p4g$ transformation without any energy barrier. Through the calculations of the geometric and electronic structures, they found two factors that determined the reconstruction: (1) the formation of covalent bonds between the C atoms and their nearest-neighbor (nn) Ni atoms in the second layers substantially reduced the total energy of the system; (2) the reduction of local electronic charges on the C atoms prevented the strong electron-electron repulsion. Zhang et al. performed density functional theory (DFT) calculations to study the C adsorption and diffusion on three nonreconstructed low Miller index surfaces using both cluster and slab models[11]. Their results indicated that the Ni(100) surface was the most favorable for C adsorption, while the mobility of C on Ni(100) was the lowest because of its highest energy barrier of 2.19eV. Since this barrier is also higher than the experimentally observed activation energy for the rate-limiting step of C diffusion through Ni particles (1.3-1.5eV[13,14]), the model of C diffusion on the nonreconstructed Ni(100) surface is not appropriate to simulate the real C transport scenario. However, because the C migration along Ni(100) is inevitable as mentioned above, the surface reconstruction is likely to have an effect on this process.

To our knowledge, no first-principles study has been hitherto conducted to investigate C diffusion on the surface and in the subsurface of reconstructed Ni(100). In this paper, we first calculate the adsorption energies and geometric parameters as C atoms bind to nonreconstructed and reconstructed Ni(100). The simulated results have been compared with published experimental data to evaluate the reliability of our calculation method. Then, the nudged elastic band (NEB) method is used to explore the minimum energy path (MEP) for C diffusion on the surfaces and in the subsurfaces of nonreconstructed and reconstructed Ni(100). Finally, through comparing the energy barriers for six different diffusion pathways, we conclude by discussing the implication of our results for understanding the effect of surface reconstruction on C diffusion on Ni(100).

2 Computational Details

The self-consistent total energy calculations based on DFT are performed with use of the VASP code[15-17], in which the electronic wavefunctions at each k-point in periodic systems are written as a product of a wavelike part and a cell periodic part, the latter being expanded by using a plane wave basis set. The interactions between valence electrons and ion cores are represented by Blöchl's all-electron-like projector augmented wave (PAW)[18] method, which regards the 4s 3d states as the valence configuration for Ni and 2s 2p states for C (we use the standard version of the PAW-GGA potential for Ni and the soft one for C). Exchange and correlation of the Kohn-Sham theory are treated with the generalized gradient approximation (GGA) functional of PW91[19]. A plane wave energy cutoff of 350eV is used in the present calculation and all geometries are optimized until the forces acting on each atom are converged better than 0.1eV/nm. Brillouin zone sampling is performed with use of a Monkhorst-Pack grid[20] and electronic occupancies are determined according to a Methfessel-Paxton scheme[21] with an energy smearing of 0.2eV. Because there is a magnetic element (Ni) involved in the system, spin polarized effect has been considered. The calculations performed by Kresse and Hafner showed that surface magnetism was essential for an accurate quantitative description of adsorption energy[22].

The nonreconstructed Ni surface is represented as a five-layer slab with a $p(3\times3)$ supercell, corresponding to a C coverage of 0.11ML. As for the reconstructed surface, we first use a five-layer slab with a $(2\times2)p4g$ structure to achieve a C coverage of 0.5ML, just as reported in Ref. [12]. However, it is subsequently found that this supercell is not large enough to neglect the lateral interactions of the hopping C atoms during the diffusion investigation. Therefore, in this work, we enlarge the model to a (4×4) supercell consisting of 16 Ni atoms per layer and 8 (or 7) C atoms on the surface. The bottom two layers of the slabs are constrained to their crystal lattice positions. The neighboring slabs are separated by a vacuum layer as large as 1nm to avoid periodic interactions. The first Brillouin zone is sampled with a $3\times3\times1$ and $2\times2\times1$ k-point mesh for the $p(3\times3)$ and (4×4) supercell, respectively.

As a first test of our method, the sensitivity of the C adsorption energy at the fourfold hollow sites to various computational parameters is given in Table 1. For each parameter, the adsorption energy difference is calculated with respect to the results simulated by the method described above. The most sensitive computational parameter in our calculation is k-point sampling; even so, the adsorption energy is only changed by less than 1%. From Table 1, it also can be seen that further increase of the plane wave energy cutoff and relaxation of deeper Ni layers have only a very small effect on the surface adsorption.

Table 1 Sensitivity of the C adsorption energy at the fourfold hollow site of nonreconstructed and reconstructed Ni(100) to various computational parameters

Computational parameter	Adsorption energy difference[①]/eV	
	Nonreconstructed Ni(100)	Reconstructed Ni(100)
Plane wave energy cutoff (from 350 to 400eV)	0.00	0.00
k-point sampling [from $3\times3\times1$ to $4\times4\times1$ for $p(3\times3)$ supercell, from $2\times2\times1$ to $3\times3\times1$ for (4×4) supercell]	0.06	0.04
No. of relaxed Ni layers (from three to four)	0.01	0.00

① For each computational parameter, the adsorption energy difference is calculated with respect to the results simulated by a plane wave energy cutoff of 350eV, a $3\times3\times1$ and $2\times2\times1$ Monkhorst-Pack k-point mesh for the $p(3\times3)$ and (4×4) supercell, respectively, and a five-layer slab with the bottom two Ni layers frozen.

The climbing-image NEB (CI-NEB)[23-25] method, an improved version of the NEB method, is used to locate the MEPs and transition states for C diffusion. A number of intermediate images are constructed along the reaction path between the energetically favorable reactant and product. With the two endpoints fixed, the intermediates are optimized partially by using a velocity quench algorithm[26] followed by a quasi-Newton algorithm until the maximum force in every degree of freedom is less than 0.1eV/nm. To obtain accurate forces, the total energy and band structure energy are converged to within 1×10^{-6} eV/atom during the electronic optimization.

3 Results and Discussion

3.1 C adsorption on the surface and in the subsurface of nonreconstructed Ni(100)

It was reported that the C adsorption induced no "clock" reconstruction on Ni(100) at the C coverages less than 0.3ML[7,9]. Therefore, we first investigate the interaction between C and nonreconstructed Ni(100) at the C coverage of 0.11ML. The high-symmetry sites for C binding on the surface and in the subsurface are shown in Fig. 1: a fourfold hollow site (Hollow site), a bridge site between two Ni atoms (Bridge site), an atop site above a single Ni atom (Atop site), an octahedral site in the subsurface (sub-O site), and a tetrahedral site in the subsurface (sub-T site). The adsorption energies (ΔE_{ads}) of C at these sites are calculated as:

$$\Delta E_{ads} = E(Ni_{45}C) - E(Ni_{45}) - E(C) \tag{1}$$

The first term on the right-hand side is the total energy of the slab that includes 45 Ni atoms and 1 C atom; the second term is the total energy of the slab that consists of 45 Ni atoms. The first two terms are calculated by applying the same parameters (k-point sampling, energy cutoff, etc.). The third term is the total energy of an isolated C atom in its ^3P ground state, which is estimated by putting a C atom in a cubic box with dimensions of 1nm sides and carrying out a spin-polarized Γ-point calculation. With this definition, a more negative value of ΔE_{ads} corresponds to stronger binding between C and nonreconstructed Ni(100). The adsorption energies and geometric parameters for C binding on the surface and in the subsurface of nonreconstructed Ni(100) are listed in Table 2. The fourfold hollow site is the most stable binding site for C adsorption with a binding energy of -8.26eV and Ni—C bond lengths of 0.184nm. And our prediction of the site preference agrees with published experimental and theoretical studies[11,27]. In the subsurface, C atoms prefer the sub-O site which is 1.18eV lower in adsorption energy than the hollow site. Namely the C diffusion form the surface to the subsurface is energetically unfavorable, which is opposite to C segregation into Ni(111)[28].

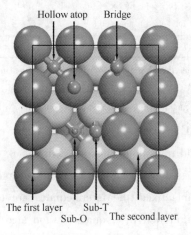

Fig. 1 Schematic representations of high-symmetry binding sites for C adsorption on the surface and in the subsurface of nonreconstructed Ni(100)

(Blue and green balls denote Ni atoms, and gray balls denote C atoms)

Table 2 Adsorption energies and geometric parameters for C binding on the surface and in the subsurface of nonreconstructed Ni(100) at the C coverage of 0.11ML

Item	Hollow	Bridge	Atop	Sub-O	Sub-T
ΔE_{ads}/eV	-8.26	-6.06	-4.54	-7.08	-6.29
d_{Ni-C}①/Å	1.84(4)	1.72(2)	1.64	1.82,1.85(4),1.90	1.74(2),1.83(2)

Note: 1Å = 0.1nm.

① d_{Ni-C} denotes the bond length(s) between the C atom and its neighboring Ni atom(s). Numbers in parentheses show the amounts of the corresponding values.

To evaluate the precision of our calculation method, we compare the calculated structural parameters to experimental observations as C atoms bind to the most favorable hollow sites. The geometric parameters are defined in Fig. 2 and the comparison is summarized in Table 3. As seen from the table, the calculated results are in qualitative agreement with the experimental data, indicating that PAW-GGA correctly predicts the adsorption structure of C on Ni(100) at low coverages. In addition, it is found that the nearest-neighbor outermost Ni atoms around the adsorbed C atom are displaced radially by 0.007nm. Simultaneously, the interlayer spacing between the nearestneighbor Ni atoms in the outermost and second layers increases by 0.01nm over the clean Ni(100) surface. As a consequence of these two kinds of displacement, the central C atom takes up a lower position to enhance the covalent bonds to the nearestneighbor outermost Ni atoms, and tends to be five-coordinated to reduce the total energy of the system.

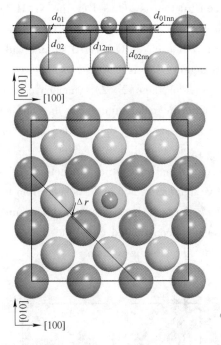

Fig. 2 Side (upper panel) and top view (lower panel) of the structural model for the C adsorption at the hollow site of nonreconstructed Ni(100)

(d denotes the interlayer spacing perpendicular to the surface. The suffixes 0,1, and 2 correspond to the adsorbate (C), outermost, and second Ni layers, respectively. Δr is the radial displacement parallel to the surface of the nearest-neighbor Ni atoms in the outermost layer)

Table 3 Comparison of the calculated geometric parameters for the C adsorption at the hollow site of nonreconstructed Ni(100) with experimental results

Item	d_{01nm}[①]/Å	d_{01}[②]/Å	d_{02nm}[③]/Å	d_{02}[④]/Å	d_{12nm}[⑤]/Å	Δz_1[⑥]/Å	Δz_2[⑦]/Å	Δr[⑧]/Å
Expt[⑨]	0.21 ±0.08	0.19 ±0.29	1.95 ±0.06	1.93 ±0.30	1.74 ±0.10	−0.02 ±0.20	−0.02 ±0.20	0.02 ±0.03
This work	0.22	0.32	2.01	2.02	1.79	0.04	−0.01	0.07

Note: 1Å = 0.1nm.

① d_{01nm} denotes the interlayer spacing perpendicular to the surface between the adsorbed C atom and its nearest-neighbor Ni atoms in the outermost layer.

② d_{01} denotes the interlayer spacing perpendicular to the surface between the adsorbed C atom and the outermost Ni layer.

③ d_{02nm} denotes the interlayer spacing perpendicular to the surface between the adsorbed C atom and its nearest-neighbor Ni atom in the second layer.

④ d_{02} denotes the interlayer spacing perpendicular to the surface between the adsorbed C atom and the second Ni layer.

⑤ d_{12nm} denotes the interlayer spacing perpendicular to the surface between the nearest-neighbor Ni atoms in the outermost and second layers.

⑥ Δz_1 denotes the relaxation of the outermost Ni layer with respect to the fully relaxed clean Ni surface.

⑦ Δz_2 denotes the relaxation of the second Ni layer with respect to the fully relaxed clean Ni surface.

⑧ Δr denotes the radial displacement parallel to the surface of the nearest-neighbor outermost Ni atoms.

⑨ Reference[9].

3.2 C adsorption on the surface and in the subsurface of reconstructed Ni(100)

Both experimental and theoretical studies have provided strong evidence that at the C coverage of 0.5ML, $p4g$-symmetry reconstruction is induced on the Ni(100) surface[7,12]. According to this information, the model of the reconstructed Ni(100) surface is established. Just as reported in Ref. [12], the simple optimization of the C-covered surface with perfect $c(2\times2)$ symmetry results in no surface reconstruction. This is because using the optimization methods such as quasi-Newton and conjugate gradient, the $c(2\times2)$ symmetry cannot be broken by the forces on Ni atoms[12]. Therefore, we start the calculation with the surface Ni atoms rotated slightly around the adsorbed C atom. After geometry optimization, the "clock" reconstruction occurs spontaneously, characterized by the lateral displacement of the outermost Ni atoms (Δxy = 0.049nm, Fig. 3). Also, the calculated geometric parameters are compared with experimental data in Table 4 and qualitative agreement is found. As seen from the table, the interlayer spacing between the adsorbed C atom and its nearest-neighbor outermost Ni atoms is slightly shortened as compared to the nonreconstructed case, and the d_{12nm} further increases to create additional room to accommodate the sinking C atom.

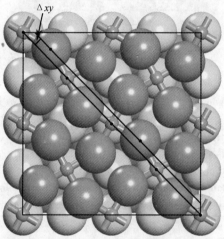

Fig. 3 Schematic representation of C-induced "clock" reconstruction on Ni(100) at 0.5ML
(Δxy is the lateral displacement of the outermost Ni atoms)

Table 4 Comparison of the calculated geometric parameters for the C adsorption at the 4-hollow site of reconstructed Ni(100) with experimental results

Item	d_{01nm}/Å	d_{02nm}/Å	d_{02}/Å	d_{12nm}/Å	Δz_2/Å	Δxy[①]/Å
Expt[②]	0.11 ± 0.04	1.94 ± 0.06	1.97 ± 0.12	1.83 ± 0.07	0.03 ± 0.13	0.41 ± 0.07
This work	0.13	1.98	2.20	1.85	−0.07	0.49

① Δxy denotes lateral displacement of the outermost Ni atoms.
② Ref. [9].

On the surface and in the subsurface of reconstructed Ni(100), seven binding sites are considered for C adsorption: a fourfold hollow site (4-hollow site), a threefold hollow site (3-hollow site), a long bridge site (LB site), a short bridge site (SB site), an atop site above a single Ni atom (Atop site), an octahedral site in the subsurface (sub-O site), and a quasitetrahedral site in the subsurface (sub-T), as shown in Fig. 4. The adsorption energies at these sites are calculated as:

$$\Delta E_{ads} = E(Ni_{80}C_8) - E(Ni_{80}C_7) - E(C) \tag{2}$$

The first term on the right-hand side is the total energy of the slab that includes 80 Ni atoms and 8 C atoms (7 C atoms adsorb at the 4-hollow sites and 1 C atom adsorbs at the investigated site); the second term is the total energy of the slab that consists of 7 C atoms that adsorb at the 4-hollow sites and 80 Ni atoms. The first two terms are calculated by applying the same parameters (k-point sampling, energy cutoff, etc.). The third term is identical with that in Eq 1. After geometry relaxation, we find that the 4-hollow, 3-hollow, and sub-O sites are stable for C adsorption, and the C atom at the LB site moves to the position between the LB and 3-hollow sites. It is also predicted that the C atoms relax to the 4-hollow, 3-hollow, and sub-O site, even if they are initially placed at the Atop, SB, and sub-T site, respectively. The calculated adsorption energies and geometric parameters are summarized in Table 5. As seen from the table, the most stable adsorption site is still the 4-hollow site, which is 0.75 eV lower in energy than the next favorable sub-O site.

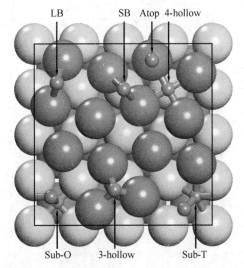

Fig. 4 Schematic representations of high-symmetry binding sites for C adsorption on the surface and in the subsurface of reconstructed Ni(100)

(For clarity, the other 7 C atoms that are coadsorbed at the 4-hollow sites are removed from the figure)

Table 5 Adsorption energies and geometric parameters for C binding on the surface and in the subsurface of reconstructed Ni(100) surface at the C coverage of 0.5ML

Item	4-hollow	3-hollow	sub-O	LB
ΔE_{ads}/eV	−7.98	−5.84	−7.23	−6.12
d_{Ni-C}/Å	1.84(4), 1.98	1.75, 1.79, 1.86	1.84(2), 1.88, 1.90, 1.91, 1.93	1.85, 1.93, 1.96

Note: 1 Å = 0.1 nm.

3.3 C diffusion on the surface and in the subsurface of nonreconstructed Ni(100)

Now that the most stable sites for C adsorption on nonreconstructed and reconstructed Ni(100) are determined, we then use the CI-NEB method to assess the diffusion rate between the fourfold hollow sites. It is assumed that the C diffusion is described by the simple version of quantum corrected harmonic transition state theory (TST)[29], which predicts the C diffusion rate constant k_{A-B} through an energy barrier E_a:

$$k_{A-B} = \frac{k_B T}{h} \exp\left(\frac{-E_a}{k_B T}\right) \qquad (3)$$

where, k_B is the Boltzmann constant; h is the Planck constant. This approach neglects contributions to the local partition functions from vibrational levels other than the ground state. These contributions can be included as yielding an effective activation energy that is slightly temperature dependent and approaches the classical result at high temperatures[30-32].

Four diffusion pathways are investigated on the surface and in the subsurface of nonreconstructed Ni(100), as shown in Fig. 5. Along PATH(Ⅰ) and PATH(Ⅱ), a C atom diffuses between the hollow sites via an adjacent Bridge and Atop site, respectively. Along PATH(Ⅲ) and PATH(Ⅳ), a C atom hops from a hollow site, passing through a sub-T and sub-O site, respectively, and finally arrives at a neighboring hollow site.

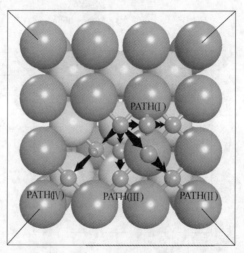

Fig. 5 Schematic representations of C diffusion on the surface and in the subsurface of nonreconstructed Ni(100) along four different pathways

The converged MEP and the configuration of the transition state for PATH(Ⅰ) are schematically represented in Fig. 6. In the transition state, the C atom is located at the Bridge site with an energy barrier of 2.20eV, consistent with the value calculated by Zhang et al. (2.19eV)[11]. In addition, it was reported by Hofmann et al. that on Ni(111), the activation energy for the C diffusion between neighboring threefold hollow sites via bridge sites was less than 0.4eV at a C coverage of 0.25ML[33]. Comparing these two diffusion barriers, we notice that the mobility of C on Ni(111) is much higher than that on Ni(100). The reason is that in the initial states, the C atoms that bind to the Ni(100) and Ni(111) surfaces are quasi-five-coordinated and three-coordinated, respectively, and it is more difficult for the C atom to leave the hollow sites on the Ni(100) surface. Fig. 7 shows the potential energy diagram for PATH(Ⅱ), together with the atomic arrangement of the corresponding transition state. The energy barrier is calculated to be 3.72eV, much higher than that for PATH(Ⅰ). This is mainly due to the instability of the transition state in which the C atom binds to the most unfavorable Atop site (Table 2).

To investigate the effect of subsurface migration on the C transport mechanism, the converged MEPs for PATH(Ⅲ) and PATH(Ⅳ) are shown in Fig. 8 and Fig. 9, respectively. In the tran-

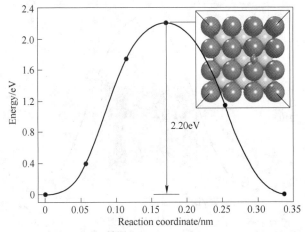

Fig. 6   MEP for the C diffusion along PATH(Ⅰ) at the C coverage of 0.11ML and the structure of the transition state

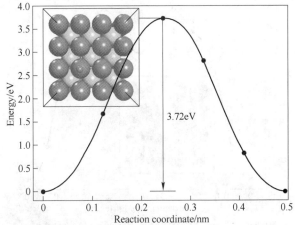

Fig. 7   MEP for the C diffusion along PATH(Ⅱ) at the C coverage of 0.11ML and the structure of the transition state

sition state of PATH(Ⅲ), the C atom is located at the sub-T site and its two adjacent surface Ni atoms are pushed apart by 0.028nm to minimize the local lattice strain. The energy barrier for the C transport along this pathway is calculated to be 1.97eV. As for PATH(Ⅳ), one can see that the MEP shows a symmetrical structure with double maxima at 1.57eV and one local minimum at 1.16eV. Therefore, two transition states and one intermediate are found during the C diffusion. At the saddle points, the C atom is five-coordinated and resides in quasi-sub-O sites, pushing the surface Ni atom (the small red ball in Fig. 9) toward vacuum by 0.054nm as compared to the fully relaxed clean surface. In the intermediate, the C atom is located at the sub-O site and the Ni atom protruding from the surface reverts approximately to its original position in the initial state. Comparing the calculated energy barriers for these four pathways, we believe that the subsurface transport mechanisms are more likely than those through surfaces. However, the activation energy for the most favorable PATH(Ⅳ) is still slightly higher than the experimentally observed results for the rate-limiting step of C transport through Ni particles (1.3-1.5eV). Therefore, C atoms tend to accumulate on the surface until a higher level to induce the "clock" reconstruction.

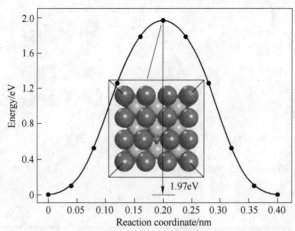

Fig. 8　MEP for the C diffusion along PATH(Ⅲ) at the C coverage of 0.11 ML and the structure of the transition state

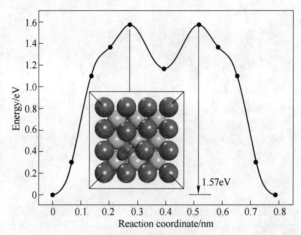

Fig. 9　MEP for the C diffusion along PATH(Ⅳ) at the C coverage of 0.11 ML and the structure of the transition state

(The small red ball denotes the Ni atom that is pushed toward vacuum)

3.4　C diffusion on the surface and in the subsurface of reconstructed Ni(100)

As calculated in Section 3.2, the surface is reconstructed to a Ni(100)-$(2 \times 2)p4g$-C structure when a critical coverage is attained. Then, we examine how the energy barrier changes as C atoms diffuse on the surface and in the subsurface of reconstructed Ni(100). In this section, the C coverage is decreased from 0.5 to 0.4375 ML to construct pathways for C transport. We find that at the coverage of 0.4375 ML, the C adsorption can also induce the "clock" reconstruction. With use of this model, the diffusion of a single C atom between two 4-hollow sites is investigated with the remaining C atoms binding to the corresponding 4-hollow sites throughout the process. Two diffusion pathways are constructed, as shown in Fig. 10. Along PATH(Ⅴ), a C atom diffuses between two 4-hollow sites via a 3-hollow site. Alternatively, along PATH(Ⅵ), a C atom hops from a 4-hollow site, passing through an adjacent sub-O site, and finally arrives at one of the nearest neighboring 4-hollow sites.

Fig. 11 and Fig. 12 show the converged MEPs and configurations of the transition states for

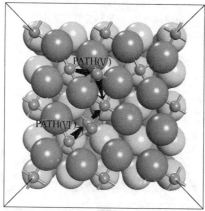

Fig. 10 Schematic representations of C diffusion on the surface and in the subsurface of reconstructed Ni(100) along two different pathways

the C diffusion along PATH(V) and PATH(VI), respectively. As for PATH(V), the calculated MEP shows a symmetrical path with two transition states and one intermediate. In the transition states, the C atom is located at the SB site with a diffusion barrier of 2.25eV. This high barrier is also due to the strong binding between the C atom and the surface in the initial state. The intermediate corresponds to the C adsorption at the 3-hollow site, which is 0.77eV lower in energy than the transition states. Among the six pathways, the most favorable one for C diffusion is PATH(VI). As seen from Fig. 12, the converged MEP is symmetrical with two maxima both at 1.16eV, much lower than those for the other five pathways. In the intermediate, the C atom binds to the sub-O site, which is 0.39eV lower in energy than the transition states. At the saddle points, the six-coordinated C atom resides at quasi-sub-O sites and the total energy of the Ni—C system is dramatically decreased by the surface reconstruction. To prove this point, we compare the structures of the transition states for PATH(VI) to those for PATH(IV) in Table 6. First, the C atom is only five-coordinated at the quasi-sub-O site of the nonreconstructed surface, while it is coordinated to 3 outermost and 3 second-layer Ni atoms at the same site after the surface reconstruction. Therefore, the transition states for PATH(VI) are stabilized by the "clock" reconstruction. Second, because of the rotation of the Ni atoms in the outermost surface, the volume of the quasi-sub-O site is substantially enlarged, e.g., the surface Ni atom denoted as a small red ball in Fig. 12 is pushed up toward vacuum by 0.058nm as compared to the fully relaxed clean surface, while the corresponding result is 0.054nm for the nonreconstructed surface. As a consequence, the local lattice strain in the transition states is dramatically relieved by the surface reconstruction.

Table 6 Comparison of the structures of the transition states for PATH(IV) and PATH(VI)

Item	d_{Ni-C}[①]/Å	Δz[②]/Å
PATH(IV)	1.78, 1.85, 1.91(2), 2.05	0.54
PATH(VI)	1.76, 1.83, 1.95(2), 2.06, 2.15	0.58

① d_{Ni-C} denotes the bond lengths between the C atom and its neighboring Ni atoms in the transition states for PATH(IV) and PATH(VI).

② Δz denotes the displaced distances of the Ni atoms which are denoted as small red balls in Fig. 9 and Fig. 11. as compared with their position on the fully relaxed clean surface.

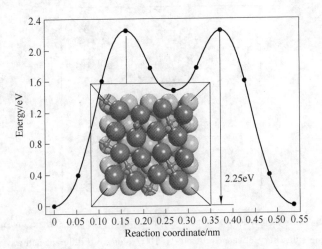

Fig. 11 MEP for the C diffusion along PATH(Ⅵ) at the C coverage of 0.4375ML and the structures of the transition states

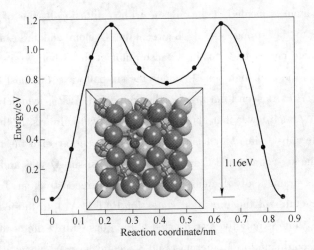

Fig. 12 MEP for the C diffusion along PATH(Ⅴ) at the C coverage of 0.4375ML and the structure of the transition state

(The small red ball denotes the Ni atom that is pushed toward vacuum)

4 Conclusions

Spin polarized DFT-GGA calculations are performed to examine the behavior of C adsorption and diffusion on the surfaces and in the subsurfaces of nonreconstructed and reconstructed Ni(100). The C adsorption is investigated at two C coverages of 0.11ML and 0.5ML. The calculated results indicate that no surface reconstruction is induced by C adsorption at 0.11ML, while the Ni atoms in the outermost layer are rotated to form a structure with $(2 \times 2)p4g$ symmetry at 0.5ML. On both the nonreconstructed and reconstructed surfaces, C atoms prefer the fourfold hollow site because higher coordination can be achieved. The calculated geometric parameters for the C adsorption at this site are in qualitative agreement with published experimental data. Then, six pathways between the fourfold hollow sites are established to investigate C diffusion on nonreconstructed and reconstructed Ni(100) with use of the CINEB method. It is found that the energy barriers for C diffusion on the surfaces are much higher than those in the subsurfaces, i. e., the subsurface migration mechanism is more likely. Among the six diffusion pathways, PATH(VI) is the most favorable because the corresponding energy barrier of 1.16eV is the lowest, and the energy barriers for the other five pathways are higher than the experimentally observed results for the ratelimiting step of C diffusion through Ni particles (1.3-1.5eV). In the transition states for PATH(VI), the six-coordinated C atom is located at the quasi-sub-O sites whose volume is enlarged by the "clock" reconstruction. Thus, the local lattice strain induced by the C diffusion is substantially relieved and the energy barrier is therefore lowered.

In summary, no reconstruction on the Ni(100) surface is induced at low C coverages. C diffusion on the surface and in the subsurface of nonreconstructed Ni(100) is very slow because of the high-energy barriers. And therefore, C atoms tend to accumulate on the surface to a higher level to induce the surface reconstruction. When the critical coverage is attained, the "clock" reconstruction takes place and substantially promotes C diffusion on Ni(100).

Acknowledgements

We thank Dr. Ping Li, Jinghong Zhou, Xiongyi Gu, Zhijun Sui, Jun Zhu, Junsheng Zheng, and Qian Zhao for valuable discussions. This work is supported by NSFC/PetroChina through a major project on multiscale methodology (No. 20490200).

References

[1] Lee, Y. H. ; Kim, S. G. ; Tománek, D. Phys. Rev. Lett. 1997, 78, 2393.
[2] De-Jong, K. P. ; Geus, J. W. Catal. Rev. Sci. Eng. 2000, 42, 481.
[3] Huber, F. ; Yu, Z. ; Logdberg, S. ; Ronning, M. ; Chen, D. ; Venvik, H. ; Holmen, A. Catal. Lett. 2006, 110, 211.
[4] Bengarrd, H. S. ; Alstrup, I. ; Chorkendorff, I. ; Ullmann, S. ; Rostrup-Nielsen, J. R. ; Nørskov, J. K. J. Catal. 1999, 187, 238.
[5] Yang, R. T. ; Chen, J. P. J. Catal. 1989, 115, 52.
[6] Zaikovskii, V. I. ; Chesnokov, V. V. ; Buyanov, R. A. Appl. Catal. 1988, 38, 41.
[7] Klink, C. ; Olesen, L. ; Besenbacher, F. ; Stensgaard, I. ; Laegsgarrd, E. ; Lang, N. D. Phys. Rev. Lett. 1993, 71, 4350.
[8] Onuferko, J. H. ; Woodruff, D. P. ; Holland, B. W. Surf. Sci. 1979, 87, 357.
[9] Terborg, R. ; Hoeft, J. T. ; Polcik, M. ; Lindsay, R. ; Schaff, O. ; Bradshaw, A. M. ; Toomes. R. L. ; Booth, N. A. ; Wooddruff, D. P. ; Rotenberg, E. ; Denlinger, J. Surf. Sci. 2000, 446, 301.
[10] Kirsch, J. E. ; Harris, S. Surf. Sci. 2003, 522, 125.
[11] Zhang, Q. ; Wells, J. C. ; Gong, X. G. ; Zhang Z. Phys. Rev. B 2004, 69, 205413.
[12] Stolbov, S. ; Hong, S. ; Kara, A; Rahman, T. S. Phys. Rev. B 2005, 72, 155423.
[13] Baker, R. T. K. ; Barber, M. A. ; Harris, P. S. ; Feates, F. S. ; Waite, R. J. J. Catal. 1972, 26, 51.
[14] Trimm, D. L. Catal. Rev. Sci. Eng. 1977, 16, 155.
[15] Kresse, G. ; Hafner, J. Phys. Rev. B 1993, 48, 13115.
[16] Kresse, G. ; Furthmüller, J. Comput. Mater. Sci. 1996, 6, 15.
[17] Kresse, G. ; Furthmüller, J. Phys. Rev. B 1996, 54, 11169.
[18] Blöchl, P. E. Phys. Rev. B 1994, 50, 17953.
[19] Perdew, J. P. ; Burke, K. ; Ernzerhof, M. Phys. Rev. Lett. 1996, 77, 3865.

[20] Monkhorst, H. J. ; Pack, J. D. Phys. Rev. B 1976, 13, 5188.
[21] Methfessel, M. ; Paxton, A. T. Phys. Rev. B 1989, 40, 3616.
[22] Kresse, G. ; Hafner, J. Surf. Sci. 2000, 459, 287.
[23] Jónsson, H. ; Mills, G. ; Jacobsen, K. W. Nudged Elastic Band Method for Finding Minimum Energy Paths of Transitions. In Classical and quantum dynamics in condensed phase simulations; Beren, B. J. , Ciccitti, G. , Coker, D. F. , Eds. ; World Scientific: Singapore, 1998; p 385.
[24] Henkelman, G. ; Uberuaga, B. P. ; Jónsson, H. J. Chem. Phys. 2000, 113, 9901.
[25] Henkelman, G. ; Jónsson. H. J. Chem. Phys. 2000, 113, 9978.
[26] Della-Valle, R. G. ; Andersen, H. C. J. Chem. Phys. 1992, 97, 2682.
[27] Isett, L. C. ; Blakely, J. M. Surf. Sci. 1976, 58, 397.
[28] Abild-Pedersen, F. ; Nørskov, J. K. ; Rostrup-Nielsen, J. R. ; Sehested, J. ; Helveg, S. Phys. Rev. B 2006, 73, 115419.
[29] Volkl, J. ; Alefeld, G. In Hydrogen in Metals; Alefeld, G. , Volkl, J. , Eds. ; Springer: Berlin, Germany, 1978; p321.
[30] Bhatia, B. ; Luo, X. ; Sholl, C. A. ; Sholl, D. S. J. Phys. : Condens. Matter 2004, 16, 8891.
[31] Kamakoti, P. ; Morreale, B. D. ; Ciocco, M. V. ; Howard, B. H. ; Killmeyer, R. P. ; Cugini, A. V. ; Sholl, D. S. Science 2005, 307, 569.
[32] Kamakoti, P. ; Sholl, D. S. Phys. Rev. B 2004, 70, 014301.
[33] Hofmann, S. ; Csányi, G. ; Ferrari, A. C. ; Payne, M. C. ; Robertson, J. Phys. Rev. Lett. 2005, 95, 036101.

Diffusion-enhanced Hierarchically Macro-mesoporous Catalyst for Selective Hydrogenation of Pyrolysis Gasoline*

Abstract A novel Pd/Al_2O_3 catalyst with the hierarchically macro-mesoporous structure was prepared and applied to the selective hydrogenation of pyrolysis gasoline. The alumina support possessed a unique structure of hierarchical mesopores and macropores. The as-prepared and calcined alumina were characterized by X-ray diffraction, N_2 adsorption-desorption and scanning electron microscopy. It showed that the hierarchically porous structure of the alumina was well preserved after calcination at 1073K, indicating high thermal stability. The 1073K calcined alumina was impregnated with palladium metal and compared with a commercial catalyst without macrochannels. Both the catalytic activity and the hydrogenation selectivity of the novel Pd/Al_2O_3 catalyst were higher than those of the commercial Pd/Al_2O_3 catalyst. In addition, apparent reaction activation energies obtained with the novel catalyst for model pyrolysis gasoline were 46%-81% higher than those with the commercial catalyst. The results adequately demonstrated the enhanced mass transfer characteristics of the novel macro-mesostructured catalyst.

Key words catalysis, diffusion, hierarchically macro-mesoporous alumina, selective hydrogenation, pyrolysis gasoline

1 Introduction

Pyrolysis gasoline(pygas), a byproduct of steam cracking of naphtha, contains a large quantity of aromatic compounds such as benzene, toluene and xylene, and a small amount of unsaturated compounds such as olefins and diolefins[1,2]. The huge amount of pygas produced annually in petrochemical plants and the high percentage of aromatics involved make pygas a potential feedstock for aromatics production. Selective hydrogenation of pygas plays a very important role in the post-treatment of pygas, which aims at converting about 90% of styrene and diolefins into ethylbenzene and monoolefins, respectively, and meanwhile, only less than 10% of monoolefins are hydrogenated into saturates[3,4].

Among various components involved in pygas, styrene and diolefins are so unstable that they are subject to polymerization through the double bond, which consequently results in gum formation[5,6]. These gums will probably cover some metal active sites of the catalyst and block some mesopores. As a result, catalyst deactivation will take place. The strategy to prevent gum formation is to reduce the residence time of diolefins inside the catalyst, in another word, to reduce the internal diffusion limitations of diolefins.

In previous work on a commercial catalyst[7] for selective hydrogenation of pygas[8], we found that the pore-size distribution of the commercial catalyst (4-20nm) was not suitable for the present reaction system from the viewpoint of long-term operation because of the heavy internal

---

* Coauthors: Zhou Zhiming, Zeng Tianying, Cheng Zhenmin. Reprinted from *AIChE Journal*, 2011, 57(8): 2198-2206.

diffusion limitations of reactants. To reduce the internal diffusion resistance, a novel catalyst support with a hierarchically macro-mesoporous structure was prepared and applied to selective hydrogenation of pygas in another previous work[9]. In that work, a preliminary test performed at a given temperature (313K) and pressure (2.0MPa) showed that the hierarchically porous catalyst exhibited higher activity and selectivity than the commercial catalyst. However, this result was not so convincing considering that only one reaction test was conducted with the novel catalyst.

In fact, hierarchically macro-mesoporous metal oxides have recently attracted considerable attention because of their potential applications[10,11]. Their unique pore structures, i.e., the monolithic macropores and the accessible mesopores, make them high-potential supports and catalysts[12]. Application examples include oxidation[13-17] and hydrogenation reactions[9,18]. Wang et al.[13,14] and Yu et al.[15] prepared hierarchically macro-mesoporous TiO_2 for photocatalytic oxidation decomposition of volatile organic compounds. Tidahy et al.[16] made use of hierarchically porous ZrO_2[19,20] as catalyst support for VOCs catalytic oxidation. Cao et al.[17] synthesized hierarchically macro-mesoporous TiO_2-supported CuO catalysts for CO oxidation. Our group prepared hierarchically porous Pd/TiO_2 and Pd/Al_2O_3 catalysts for hydrogenation of styrene[18] and selective hydrogenation of pygas[9], respectively. All these investigations reported that the uniquely bidisperse pore structure could enhance catalytic activities through improved mass transfer. However, in these studies, only few comparisons were made between the hierarchically porous catalysts and the normal catalysts, and no systematic investigation was carried out to testify the enhanced mass transfer characteristics of the novel structured catalysts.

As a part of a series of studies on selective hydrogenation of pygas, the main aim of this work is to adequately verify the diffusion-enhanced effect of the hierarchically macro-mesoporous structure of the novel Pd/Al_2O_3 catalyst. Different from our previous work[9], more detailed characterizations on the support and the catalyst are conducted in this article, and more important, the apparent reaction kinetics of selective hydrogenation of pygas over the novel Pd/Al_2O_3 catalyst is investigated, and the results are compared with those obtained with the commercial catalyst.

2 Experimental

2.1 Preparation of support and catalyst

Detailed information about preparation of the catalyst support, namely the hierarchically macro-mesoporous alumina, and the corresponding palladium-supported catalysts was presented elsewhere[9,21,22]. The hierarchically porous aluminum oxide used in this work was prepared with the aid of surfactant (cetyltrimethylammonium bromide, CTAB) under the following conditions: 2g of aluminum tri-sec-butoxide, 35mL of twice-distilled water, 15mL of absolute ethanol, and 0.4g of CTAB, pH 12, 350r/min, room temperature, and 1h of reaction. The 1073K calcined alumina with the hierarchically porous structure was sieved to a diameter range of 50-75μm and used as catalyst support. The palladium-supported catalyst with 0.3wt% (mass percent) metal loading was finally prepared by incipient-wetness impregnation of the aforementioned support with palladium chloride aqueous solution. For comparison, the commercial catalyst, supplied by the Chemical Research Institute of Lanzhou Petrochemical Corporation, was also ground and sieved to 50-75μm for the kinetic study.

2.2 Characterization

X-ray diffraction (XRD) patterns of the prepared samples were obtained on a Rigaku D/Max 2550 VB/PC diffractometer with Cu K_α radiation scanning 2θ angles ranging from 10° to 80°. Nitrogen adsorption-desorption isotherms and the corresponding pore-size distributions were acquired at 77K on a Micromeritics ASAP 2010 instrument. All the samples were degassed at 463K and 1mmHg for 6h before nitrogen adsorption measurements. The pore diameter and the pore-size distribution were determined by the BJH method. The morphology and the macroporous array of the Al_2O_3 powders were examined with a JEOL JSM 6360 LV scanning electron microscope (SEM). High-resolution transmission electron microscopy (HRTEM) investigation was performed using a JEOL JEM-2010 transmission electron microscope. The samples prepared for HRTEM investigation were first dispersed in ethanol under ultrasound and then a drop of the sample-ethanol solution was transferred onto a carbon-coated copper grid. Metal dispersions were measured by using CO pulse chemisorption on a Micromeritics AutoChem 2920 apparatus. The weighed catalysts were reduced in a mixture of 10% H_2/Ar (100mL/min) at 423K for 2h followed by a switch to helium (100mL/min) at 463K for 20min to remove adsorbed hydrogen. After the catalysts were cooled to 308K in a helium flow, carbon monoxide pulses were injected into the quartz reactor and the net volume of CO was monitored with a thermal conductivity detector. For Pd/Al_2O_3 catalysts with palladium chloride as the precursor, the mean stoichiometry of palladium metal to CO molecule can be taken as 1 according to the data summarized by Joyal and Butt[23].

2.3 Selective hydrogenation of pygas

Hydrogenation of a model pygas feed, which was composed of styrene, cyclopentadiene, 1-hexene and n-heptane (solvent), was carried out in a stirred autoclave at total pressures of 2.0-4.0MPa over a temperature range of 303-343K. During each run, the autoclave was first heated to the desired temperature under N_2 protection, and then, it was purged with preheated hydrogen for five times to exclude N_2. The autoclave was operated in a semibatch mode with hydrogen continuously entering into the autoclave to maintain the pressure. The temperature was controlled within ± 0.5K of the desired value, and the pressure within ± 0.02MPa. Liquid samples were collected for analysis at different time points.

A preliminary test on the effect of the stirring speed on the reaction rate of styrene hydrogenation showed that the stirring speed had no effect on the reaction rate when it was above 800r/min, implying the elimination of the external mass transfer limitation. In this work, all the kinetic experiments were carried out at a stirring speed of 1000r/min.

Four reaction classes were concerned with the selective hydrogenation of the model pygas, i.e.:

$$\text{(reaction 1: styrene} \rightarrow \text{ethylbenzene)} \tag{1}$$

$$\text{(reaction 2: CPD} \xrightarrow{2} \text{CPE} \xrightarrow{3} \text{CPA)} \tag{2}$$

$$\text{(reaction 3: 1-hexene} \xrightarrow{4} n\text{-hexane)} \tag{3}$$

Styrene(STY) and 1-hexene(HEX) were hydrogenated to ethylbenzene and n-hexane, respectively. Cyclopentadiene(CPD) was hydrogenated to cyclopentene(CPE), which was further hydrogenated to cyclopentane(CPA).

2.4 Analytical method

The liquid samples were analyzed with a Hewlett-Packard 6890 gas chromatograph equipped with a flame ionization detector. A HP-5 capillary column (30m × 0.32mm × 0.25μm) was used for separation. n-Octane was used as the internal standard. The oven temperature program consisted of the following segments: start at 318K (hold for 3min), ramp at 15K/min to 493K, and finally hold at 493K for 2min. The temperatures at the injector and the detector were set at 493K and 523K, respectively.

3 Results and Discussion

3.1 Physicochemical characterization

Fig. 1 shows the SEM images of the prepared aluminum oxides and the calcined alumina at different temperatures. Apparently, the macroporous channels inside the alumina are parallel to each other and perpendicular to the tangent of the outer surface. For those alumina samples calcined at a temperature range of 473-1073K (Fig. 1b-e), the macroporous structures are well

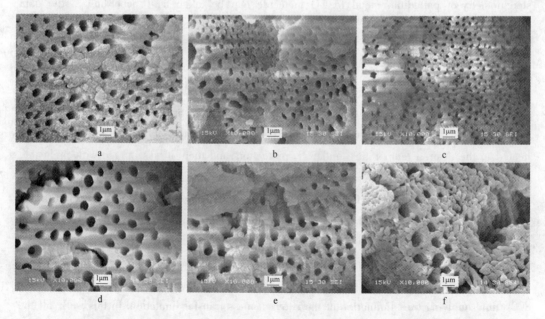

Fig. 1　SEM images of the prepared macro-mesoporous aluminum oxide(a) and the calcined alumina at different temperatures: 473K(b), 673K(c), 873K(d), 1073K(e), and 1273K(f)

preserved, and the macropore size as well as the wall thickness has less change, which indicates good thermal stability of the hierarchically porous alumina. When the calcination temperature is further up to 1273K (Fig. 1f), although wall collapse occurrs in some regions, the macroporous structure is partly preserved. The formation of the hierarchically porous structure can be explained by a spontaneous self-assembly mechanism, which is presented elsewhere[9,12,24-26].

Fig. 2 presents the XRD patterns of the prepared and the calcined alumina. The as-prepared sample (Fig. 2a) and the calcined product at 473K (Fig. 2b) exhibit diffraction peaks assigned to the boehmite phase AlOOH (JCPDS 21-1307). When the sample is calcined at 673K (Fig. 2c) and 873K (Fig. 2d), γ-Al_2O_3 phase (JCPDS 10-0425) is formed as a result of the dehydration of alumina oxyhydroxide boehmite[27]. With a further increase in the temperature to 1073K (Fig. 2e), the γ-Al_2O_3 phase transforms to δ-Al_2O_3 (JCPDS 16-0394). When the sample is calcined up to 1273K (Fig. 2f), the α-Al_2O_3 phase (JCPDS 10-0173) is observed.

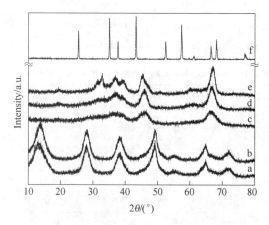

Fig. 2 X-ray diffraction patterns of the prepared macro-mesoporous aluminum oxide (a) and the calcined alumina at different temperatures: 473K(b), 673K(c), 873K(d), 1073K(e), and 1273K(f)

The N_2 adsorption-desorption isotherms of the above samples (a-f) and the pore-size distribution curves are shown in Fig. 3. The textural properties of these samples are reported in Table 1. Except for the 1273K calcined sample, all the other isotherms of the calcined samples (b-e) are of type IV with a hysteresis loop, indicating that the mesopore network is still sustained after high temperature treatment. Under calcination at 1273K, the sample shows a nonporous structure, indicating collapse of the mesopores. Sample g listed in Table 1 is a commercial Pd/δ-Al_2O_3 catalyst for selective hydrogenation of pygas, and this catalyst has no macroporous channels[7,9].

Except for the macroporous structure, the textural properties of the commercial catalyst are similar to those of sample e, i. e., the 1073K calcined Al_2O_3, and both of them exhibit the same crystalline phase, i. e., the δ-Al_2O_3 phase. In addition, as shown in Fig. 4, the pore-size distribution curves of these two catalysts almost overlap. As a result, the 1073K calcined Al_2O_3 is selected to be the support for the novel catalyst. For simplicity, the novel catalyst with the hierarchically macro-mesoporous structure and the commercial catalyst are denoted by Pd/Al_2O_3 (novel) and Pd/Al_2O_3 (com), respectively.

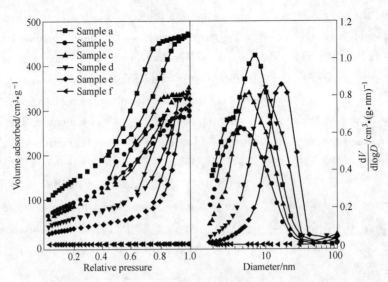

Fig. 3 N$_2$ adsorption-desorption isotherms and the pore-size distribution curves of the as-prepared and the calcined alumina samples

Table 1 Textural properties of the as-prepared and the calcined alumina samples

No.	$S_{BET}/m^2 \cdot g^{-1}$	Pore volume/$cm^3 \cdot g^{-1}$	Macropore size[①]/μm	Mesopore size[②]/nm
a	514.1	0.81	0.45	5.9
b	341.4	0.49	0.40	5.0
c	309.0	0.56	0.45	6.0
d	211.1	0.64	0.55	10.7
e	126.4	0.51	0.45	14.3
f	3.8	—	0.70	—
g[③]	98.1	0.37	—	14.9

① The average macropore diameter is obtained from analysis of the image.
② BJH pore diameter is determined from the adsorption branch.
③ The commercial Pd/δ-Al$_2$O$_3$ catalyst.

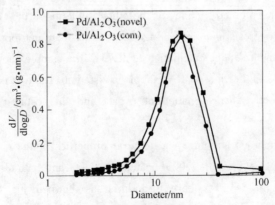

Fig. 4 Pore-size distribution curves of the novel and the commercial catalysts

Values of palladium dispersion obtained from CO chemisorption are 32.5% and 29.6% for Pd/Al$_2$O$_3$(novel) and Pd/Al$_2$O$_3$(com), respectively. Based on the dispersion values (D), the mean palladium particle size (d_P) is calculated from the following equation[28]:

$$d_P = \frac{112}{D(\%)} \quad (4)$$

The mean palladium particle sizes of Pd/Al$_2$O$_3$ (novel) and Pd/Al$_2$O$_3$ (com) catalysts are almost the same, being 3.4nm and 3.8nm, respectively.

Fig. 5 shows the HRTEM images of the novel catalyst. The metallic particles indicated by arrow are palladium metal according to the energy-dispersive X-ray analysis. The palladium particle size determined by HRTEM is about 3.6nm, which is consistent with that obtained from the CO chemisorption measurement.

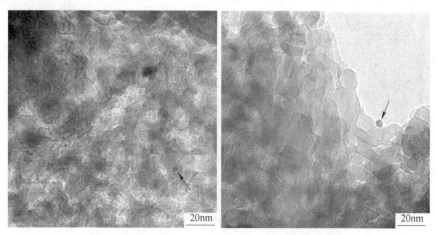

Fig. 5　HRTEM images of the Pd/Al$_2$O$_3$ (novel) catalyst

3.2　Catalytic activity

Fig. 6 presents concentration variations of various species over Pd/Al$_2$O$_3$ (novel) and Pd/Al$_2$O$_3$ (com) catalysts. It is evident that the novel catalyst exhibits much higher catalytic activity than the commercial catalyst. Taking the reaction occurring at 313K and 3.0MPa as an example (Fig. 6a), the time needed for complete conversion of cyclopentadiene and styrene over the Pd/Al$_2$O$_3$ (novel) catalyst is only 15min, but it is about 40min for the Pd/Al$_2$O$_3$ (com) catalyst.

Further observation of Fig. 6a shows that in the first 9min of reaction over the Pd/Al$_2$O$_3$ (novel) catalyst, most cyclopentadiene is conversed to cyclopentene and styrene is partly reacted, whereas 1-hexene and cyclopentene are hardly consumed. Only when cyclopentadiene and styrene are hydrogenated to a great extent do 1-hexene and cyclopentene have an obvious conversion. It implies that high selectivity of diolefin hydrogenation to monoolefin can be obtained over the novel catalyst.

To clarify which catalyst possesses the higher hydrogenation selectivity, two parameters are defined as follows:

$$S_1 = \frac{\text{Conversion of 1-hexene}}{\text{Conversion of cyclopentadiene}} \times 100\% \quad (5)$$

$$S_2 = \frac{\text{Conversion of 1-hexene}}{\text{Conversion of styrene}} \times 100\% \quad (6)$$

Obviously, the smaller S_1 and S_2, the higher the selectivity of diolefins to monoolefins. As

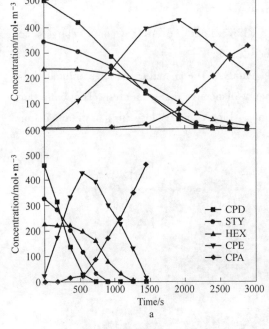
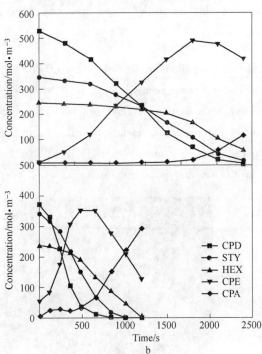

Fig. 6　Concentration-time profiles for various species over the commercial catalyst
(the above figure) and the novel catalyst (the figure below)
Reaction conditions: a—$T = 313\text{K}, P = 3.0\text{MPa}$; b—$T = 323\text{K}, P = 2.0\text{MPa}$

shown in Fig. 7, the Pd/Al$_2$O$_3$ (novel) catalyst exhibits the higher hydrogenation selectivity in terms of both cyclopentadiene and styrene. The X-axis in Fig. 7, α_{CPD} and α_{STY}, represent conversions of cyclopentadiene and styrene, respectively.

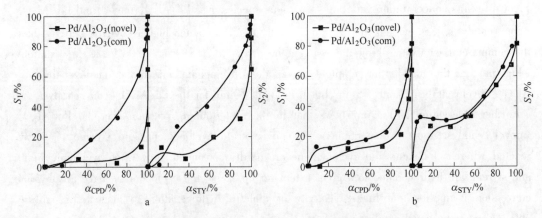

Fig. 7　Hydrogenation selectivity of different catalysts based on cyclopentadiene
(the figure on the left) and styrene (the figure on the right)
Reaction conditions: a—$T = 313\text{K}, P = 3.0\text{MPa}$; b—$T = 323\text{K}, P = 2.0\text{MPa}$

The above analyses of the activity and the selectivity of the two catalysts reveal that the novel catalyst has better catalytic performance than the commercial catalyst for the reaction system of selective hydrogenation of pygas. The main reasons lie in three aspects. First, the diffusion resistance of reactants from the outer surface to the inner surface of the catalyst can be greatly reduced in the monolithic macroporous channels[29-31]. For the catalysts with a large-pore structure, previous studies also showed that the effectiveness factors could be increased to some extent as a result of the enhancement of the effective diffusivity because of the existence of the intraparticle convection[32-35]. Second, the narrow walls separating the macropores can shorten the diffusion distance of reactants to the active sites and consequently decrease the concentration gradient of species, which in turn increases the reaction rate. Third, both macroporous channels and narrow walls can reduce the residence time of the intermediate products (monoolefins) inside the catalyst and prevent further hydrogenation of monoolefins to saturated hydrocarbons, which is favorable for the high selectivity of diolefins to monoolefins.

To further testify the advantage of the hierarchically macro-mesoporous structure of the novel catalyst in reducing the internal diffusion limitations, the apparent reaction kinetics of selective hydrogenation of pygas over the novel catalyst is investigated, and the kinetic parameters obtained with the novel catalyst and with the commercial catalyst are compared.

3.3 Reaction kinetics

The apparent kinetics of selective hydrogenation of pygas over the commercial catalyst has been systematically investigated in our previous work[3,8]. Here, we use the same kinetic model to describe the reaction system over the novel catalyst. Detailed information about the development of the kinetic model is reported elsewhere[3]. The rate expressions for reactions (Eq. 1-Eq. 3) are given by:

$$\frac{dc_{STY}}{dt} = -\frac{m}{V} \cdot \frac{k_1 b_{STY} c_{STY} \sqrt{b_H c_H}}{A_1 A_2} \tag{7}$$

$$\frac{dc_{CPD}}{dt} = -\frac{m}{V} \cdot \frac{k_2 b_{CPD} c_{CPD} \sqrt{b_H c_H}}{A_1 A_2} \tag{8}$$

$$\frac{dc_{CPE}}{dt} = \frac{m}{V} \left(\frac{k_2 b_{CPD} c_{CPD} \sqrt{b_H c_H}}{A_1 A_2} - \frac{k_3 b_{CPE} c_{CPE} \sqrt{b_H c_H}}{A_1 A_2} \right) \tag{9}$$

$$\frac{dc_{HEX}}{dt} = -\frac{m}{V} \cdot \frac{k_4 b_{HEX} c_{HEX} \sqrt{b_H c_H}}{A_1 A_2} \tag{10}$$

where, c is the molar concentration of reactant; t is the reaction time; m is the catalyst mass; V is the liquid volume; k_i is the rate constant of the ith reaction; b_j is the adsorption constant of species j. A_1 and A_2 in the denominator of Eq. 7-Eq. 10 are equal to $1 + \sqrt{b_H c_H}$ and $1 + \Sigma b_j c_j$ (j = STY, CPD, CPE, and HEX), respectively. The temperature dependences of rate constants and adsorption constants can be expressed by:

$$k_i = k_i^0 \exp[-E_i/(R_g T)], i = 1,2,3,4 \tag{11}$$

$$b_j = b_j^0 \exp[Q_j/(R_g T)], j = STY, CPD, CPE, HEX, H \tag{12}$$

where, k_i^0 is the preexponential factor of the rate constant of the ith reaction; b_j^0 is the preexponential factor of the adsorption constant of species j; E_i is the activation energy of the ith reaction; Q_j is the adsorption activation energy of species j; R_g is the universal gas constant; T is the reaction temperature.

There are 18 kinetic and adsorption parameters in the kinetic model (Eq. 7-Eq. 12), which can be estimated by fitting all the experimental data. The optimized values of these parameters are determined by using the Rosenbrock algorithm[36], which minimizes the residual sum of squares between the experimental and the calculated concentrations of all reactants. The ordinary differential equations involved in the kinetic model are integrated by the DASSL solver[37,38].

The estimated values of the 18 parameters for the novel and the commercial catalysts are summarized in Table 2. The apparent activation energies of all the four reactions (Eq. 1-Eq. 3) obtained with the novel catalyst are higher than those with the commercial catalyst, the former being 46%-81% higher than the latter. This result, on one hand indicates the heavy internal diffusion limitations of species in the commercial catalyst[39] and on the other hand, proves that the hierarchically macro-mesoporous structure does greatly reduce the influence of internal diffusion resistance.

Table 2 Estimated kinetic and adsorption parameters for Pd/Al_2O_3 (com) and Pd/Al_2O_3 (novel) catalysts

Parameter	Preexponential factor①		Apparent activation energy②	
	Pd/Al_2O_3 (com)	Pd/Al_2O_3 (novel)	Pd/Al_2O_3 (com)	Pd/Al_2O_3 (novel)
k_1	1.06×10^9	4.67×10^9	28.56	43.84
k_2	1.37×10^6	1.41×10^8	24.37	43.60
k_3	5.45×10^8	1.12×10^9	37.54	54.69
k_4	4.37×10^8	1.06×10^{10}	31.18	56.55
b_{STY}	3.67×10^{-4}	2.81×10^{-4}	12.35	11.83
b_{CPD}	2.06×10^{-3}	1.79×10^{-3}	19.32	17.61
b_{CPE}	1.53×10^{-4}	2.16×10^{-4}	10.81	20.76
b_{HEX}	2.19×10^{-4}	3.14×10^{-4}	12.59	17.67
b_H	2.39×10^{-8}	2.74×10^{-8}	14.63	16.19

① The units for k_i and b_j are $mol/(kg \cdot s)$ and m^3/mol, respectively.
② The unit for activation energy is kJ/mol.

The existence of internal mass transfer limitation in the commercial Pd/Al_2O_3 (com) catalyst can be verified by calculating from the observed reaction rate the Weisz-Prater parameter[39,40], Φ_{H_2}, as:

$$\Phi_{H_2} = (\eta\varphi^2)_{H_2} = \frac{r_H \rho_C d_P^2}{36 c_H W_S \rho_L D_{e,H}} \tag{13}$$

where, r_H is the observed reaction rate of hydrogen; ρ_C is the density of the catalyst; d_P is the diameter of the catalyst particle; W_S is the catalyst weight fraction in the liquid; ρ_L is the density of the liquid; $D_{e,H}$ is the effective diffusivity of hydrogen. Corresponding to the reactions occurring at 313K, 323K, 333K, and 343K (the pressure is 2.0MPa), the estimated Weisz-Prater parameters are 1.34, 1.45, 1.61, and 1.89, respectively. All the values are greater than 1, indicating the presence of intraparticle diffusion in the commercial catalyst.

The estimated apparent activation energy for styrene hydrogenation over the commercial catalyst in this work, 28.56kJ/mol, is very close to that reported by Nijhuis et al.[5], 27kJ/mol. Nijhuis et al.[5] used a commercial catalyst with the diameter of 40-100μm to study the reaction kinetics of styrene hydrogenation. By comparison with the values reported by Chaudhari et al.[41] (55kJ/mol) and by Jackson and Shaw[42] ((41 ± 8)kJ/mol), they found that the obtained activation energy of 27kJ/mol was greatly influenced by internal mass transfer effects. On the other hand, for the novel hierarchically structured catalyst, the apparent activation energy amounts to 43.84kJ/mol, which is in good agreement with that reported by Jackson and Shaw[42]. Therefore, the above analysis further supports that the intraparticle diffusion resistance of the novel catalyst is much smaller than that of the commercial catalyst.

Typical results of experimental and predicted concentration-time profiles for various species over the Pd/Al_2O_3(novel) catalyst are shown in Fig. 8. The agreement between the experimental and the predicted data is observed to be very well for each species, indicating the reliability and accuracy of the kinetic model to describe the reaction system of selective hydrogenation of pygas.

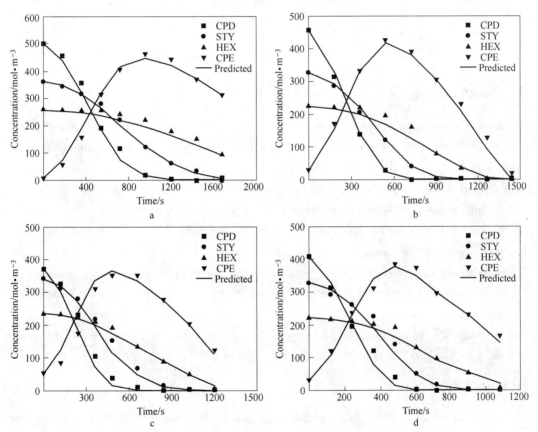

Fig. 8　Experimental and predicted concentration-time profiles over the Pd/Al_2O_3(novel) catalyst(points: experimental; lines: predicted)
Reaction conditions: a—T = 303K, P = 2.0MPa; b—T = 313K, P = 3.0MPa; c—T = 323K, P = 2.0MPa; d—T = 333K, P = 2.0MPa

After the kinetic investigation, the Pd/Al_2O_3(novel) catalyst is recovered and analyzed. As shown in Fig. 9, the pore-size distribution of the used Pd/Al_2O_3(novel) catalyst is almost the same as that of the fresh catalyst. In addition, the macrochannels are still well preserved. Neither wall collapse nor macropore blockage is observed through SEM images. Therefore, the hierarchically macro-mesoporous structure of the novel catalyst is very stable and resistant to the reaction environment of selective hydrogenation of pygas.

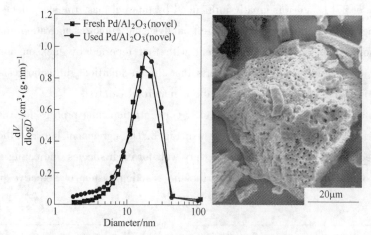

Fig. 9　Pore-size distribution curves of fresh and used Pd/Al_2O_3 (novel) catalysts as well as the SEM image of the used catalyst

4　Conclusions

Hierarchically macro-mesoporous alumina was prepared by hydrolysis and condensation of the aluminum alkoxide precursor in an aqueous solution. After calcination of the prepared aluminum oxide from 473K up to 1073K, the monolithic macropores were well preserved, indicating the high thermal stability of the alumina. A novel palladium catalyst supported on the 1073K calcined alumina was then prepared and applied to selective hydrogenation of pygas. By comparison with a commercial catalyst, the novel catalyst exhibited much higher activity and selectivity for pygas hydrogenation under various reaction conditions, which was mainly ascribed to the hierarchically porous structure of the novel catalyst. Apparent activation energies obtained with the novel catalyst for the model pygas were 46%-81% higher than those with the commercial catalyst, which further manifested the advantage of the hierarchically macro-mesoporous structure in enhancing the mass transfer of components within the catalyst. Comparison of the fresh and the used catalysts demonstrated the good stability of the novel catalyst.

Acknowledgements

The authors thank the National Natural Science Foundation of China (No. 20706018), the National High Technology Research and Development Program of China (No. 2008AA05Z405), the Program for Changjiang Scholars and Innovative Research Team in University (No. IRT0721), and the "111" Project (No. B08021) for financial support of this work.

References

[1] Zhou ZM, Cheng ZM, Yang D, Zhou X, Yuan WK. Solubility of hydrogen in pyrolysis gasoline. J Chem Eng Data. 2006;51:972-976.

[2] Mostoufi N, Sotudeh-Gharebagh R, Ahmadpour M, Eyvani, J. Simulation of an industrial pyrolysis gasoline hydrogenation unit. Chem Eng Technol. 2005;28:174-181.

[3] Zhou ZM, Cheng ZM, Cao YN, Zhang JC, Yang D, Yuan WK. Kinetics of the selective hydrogenation of pyrolysis gasoline. Chem Eng Technol. 2007;30:105-111.

[4] Yang D, Cheng ZM, Zhou ZM, Zhang JC, Yuan WK. Pyrolysis gasoline hydrogenation in the second-stage reactor: reaction kinetics and reactor simulation. Ind Eng Chem Res. 2008;47:1051-1057.

[5] Nijhuis TA, Dautzenberg FM, Moulijn JA. Modeling of monolithic and trickle-bed reactors for the hydrogenation of styrene. Chem Eng Sci. 2003;58:1113-1124.

[6] Kaminsky MP. Pyrolysis gasoline stabilization. US Patent 6,949,686,2005.

[7] Li SQ, Men XT, Liu GS, Liang SQ, Zhang XG. Selective hydrogenation catalyst for pyrolysis gasoline. US Patent 6,576,586,2003.

[8] Zhou ZM, Zeng TY, Cheng ZM, Yuan WK. Kinetics of selective hydrogenation of pyrolysis gasoline over an egg-shell catalyst. Chem Eng Sci. 2010;65:1832-1839.

[9] Zhou ZM, Zeng TY, Cheng ZM, Yuan WK. Preparation of a catalyst for selective hydrogenation of pyrolysis gasoline. Ind Eng Chem Res. 2010;49:1112-1118.

[10] Su BL, Léonard A, Yuan ZY. Highly ordered mesoporous CMI-n materials and hierarchically structured meso-macroporous compositions. C R Chimie. 2005;8:713-726.

[11] Yang XY, Li Y, Lemaire A, Yu JG, Su BL. Hierarchically structured functional materials: synthesis strategies for multimodal porous networks. Pure Appl Chem. 2009;81:2265-2307.

[12] Vantomme A, Léonard A, Yuan ZY, Su BL. Self-formation of hierarchical micro-meso-macroporous structures: generation of the new concept "hierarchical catalysis". Colloids and Surf A: Physicochem Eng Aspects. 2007;300:70-78.

[13] Wang XC, Yu JC, Ho CM, Hou YD, Fu XZ. Photocatalytic activity of a hierarchically macro/ mesoporous titania. Langmuir. 2005;21:2552-2559.

[14] Chen XF, Wang XC, Fu XZ. Hierarchical macro/mesoporous TiO_2/SiO_2 and TiO_2/ZrO_2 nanocomposites for environmental photocatalysis. Energy Environ Sci. 2009;2:872-877.

[15] Yu JG, Su YR, Cheng B. Template-free fabrication and enhanced photocatalytic activity of hierarchical macro-/mesoporous titania. Adv Funct Mater. 2007;17:1984-1990.

[16] Tidahy HL, Hosseni M, Siffert S, Cousin R, Lamonier JF, Aboukaïs A, Su BL, Giraudon JM, Leclercq G. Nanostructured macro-mesoporous zirconia impregnated by noble metal for catalytic total oxidation of toluene. Catal Today. 2008;137:335-339.

[17] Cao JL, Shao GS, Ma TY, Wang Y, Ren TZ, Wu SH, Yuan ZY. Hierarchical meso-macroporous titania-supported CuO nanocatalysts: preparation, characterization and catalytic CO oxidation. J Mater Sci. 2009;44:6717-6726.

[18] Zeng TY, Zhou ZM, Zhu J, Cheng ZM, Yuan PQ, Yuan WK. Palladium supported on hierarchically macro-mesoporous titania for styrene hydrogenation. Catal Today. 2009;147S:41-45.

[19] Yuan ZY, Vantomme A, Léonard A, Su BL. Surfactant-assisted synthesis of unprecedented hierarchical meso-macrostructured zirconia. Chem Commun. 2003;1558-1559.

[20] Blin JL, Léonard A, Yuan ZY, Gigot L, Vantomme A, Cheetham AK, Su BL. Hierarchically mesoporous/ macroporous metal oxides templated from polyethylene oxide surfactant assemblies. Angew Chem Int Ed. 2003;42:2872-2875.

[21] Deng WH, Toepke MW, Shanks BH. Surfactant-assisted synthesis of alumina with hierarchical nanopores. Adv Funct Mater. 2003;13:61-65.

[22] Zeng TY, Zhou ZM, Cheng ZM, Yuan WK. Effect of surfactant on the physical properties of hierarchically macro-mesoporous metal oxides. Chem Lett. 2010;39:680-681.

[23] Joyal CLM, Butt JB. Chemisorption and disproportionation of carbon monoxide on palladium/ silica catalysts of differing percentage metal exposed. J Chem Soc Faraday Trans 1. 1987;83:2757-2764.

[24] Collins A, Carriazo D, Davis SA, Mann S. Spontaneous template-free assembly of ordered macroporous titania. Chem Commun. 2004;568-569.

[25] Deng WH, Shanks BH. Synthesis of hierarchically structured aluminas under controlled hydrodynamic conditions. Chem Mater. 2005;17:3092-3100.

[26] Hakim SH, Shanks BH. A comparative study of macroporous metal oxides synthesized via a unified approach. Chem Mater. 2009;21:2027-2038.

[27] Ren TZ, Yuan ZY, Su BL. Microwave-assisted preparation of hierarchical mesoporous-macroporous boehmite AlOOH and γ-Al_2O_3. Langmuir. 2004;20:1531-1534.

[28] Anderson JR, Pratt KC. Introduction of Characterization and Testing of Catalysts. London: Academic Press Inc., 1985, Chapter 2.

[29] Doğu T. Diffusion and reaction in catalyst pellets with bidisperse pore size distribution. Ind Eng Chem Res. 1998;37:2158-2171.

[30] Silva VMTM, Rodrigues AE. Adsorption and diffusion in bidisperse pore structures. Ind Eng Chem Res. 1999;38:4023-4031.

[31] Delgado JA, Rodrigues AE. A Maxwell-Stefan model of bidisperse pore pressurization for Langmuir adsorption of gas mixtures. Ind Eng Chem Res. 2001;40:2289-2301.

[32] Nir A, Pismen L M. Simultaneous intraparticle forced convection, diffusion and reaction in a porous catalyst. Chem Eng Sci. 1977;32:35-41.

[33] Rodrigues AE, Ahn BJ, Zoulalian A. Intraparticle-forced convection effect in catalyst diffusivity measurements and reactor design. AIChE J. 1982;28:541-546.

[34] Leitão A, Rodrigues AE. The influence of intraparticle convection on product distribution for series-parallel catalytic reactions. Trans IChemE. 1995;73A:130-135.

[35] Leitão A, Rodrigues AE. Catalytic processes using "large-pore" materials: effects of the flow rate and operating temperature on the conversion in a plug-flow reactor for irreversible first-order reactions. Chem Eng J. 1995;60:111-116.

[36] Rosenbrock HH, Storey C. Computational Techniques for Chemical Engineers. New York: Pergamon, 1966.

[37] Petzold LR. A description of DASSL: a differential/algebraic system solver. Report SAND82-8637, Albuquerque, NM: Sandia National Laboratories, 1982.

[38] Brenan KE, Campbell SL, Petzold LR. Numerical Solution of Initial-Value Problems in Differential-Algebraic Equations. New York: North-Holland, 1989.

[39] Froment GF, Bischoff KB. Chemical Reactor Analysis and Design, 2nd ed. New York: Wiley, 1990.

[40] Fillion B, Morsi BI, Heier KR, Machado RM. Kinetics, gas liquid mass transfer, and modeling of the soybean oil hydrogenation process. Ind Eng Chem Res. 2002;41:697-709, 3052.

[41] Chaudhari RV, Jaganathan R, Kolhe DS, Emig G, Hofmann F. Kinetic modelling of a complex consecutive reaction in a slurry reactor: hydrogenation of phenyl acetylene. Chem Eng Sci. 1986;41:3073-3081.

[42] Jackson SD, Shaw LA. The liquid-phase hydrogenation of phenyl acetylene and styrene on a palladium/carbon catalyst. Appl Catal A: Gen. 1996;134:91-99.

Dryout Phenomena in a Three-phase Fixed-bed Reactor*

Abstract Understanding the mechanism of liquid-phase evaporation in a three-phase fixed-bed reactor is of practical importance, because the reaction heat is usually 7-10 times the vaporization heat of the liquid components. Evaporation, especially the liquid dryout, can largely influence the reactor performance and even safety. To predict the vanishing condition of the liquid phase, Raoult's law was applied as a preliminary approach, with the liquid vanishing temperature defined based on a liquid flow rate of zero. While providing correct trends, Raoult's law exhibits some limitation in explaining the temperature profile in the reactor. To comprehensively understand the whole process of liquid evaporation, a set of experiments on inlet temperature, catalyst activity, liquid flow rate, gas flow rate, and operation pressure were carried out. A liquid-region length-predicting equation is suggested based on these experiments and the principle of heat balance.

1 Introduction

Operation of a three-phase fixed-bed reactor is a challenging task when the liquid phase exhibits substantial volatility. In this case, evaporation of the liquid phase due to the heat of the reaction will complicate both the reaction and transport phenomena in the flow direction. Depending upon the magnitude of the reaction heat, the evaporated fraction of the liquid mixture can change from 0 to 1, and correspondingly, the reactor operation can change from a fully liquid phase to a fully gas phase, and the catalyst filling condition will evolve from a fully wetted to a partially wetted, and ultimately to a completely dry condition.

The effect of evaporation of volatile compounds on reactor performance has been considered important for a variety of systems. Singh and Carr (1983) studied the effect of solvent evaporation on oil production and total coal conversion in the solvent refined coal. Ramakrishna et al. (1985) studied the effect of vaporization of the liquid feed components on reactor performance for hydrogenation of naphthalene dissolved in a diluting solvent. Smith and Satterfield (1986) and LaVopa and Satterfield (1988) studied the effect of vapor-liquid flow ratio and vapor-liquid equilibrium on the performance of a trickle-bed reactor. More recently, a new concept on the basis of continuous phase-transitional operation has been developed for process intensification (Cheng et al., 2001a, b; 2002). The concept has been proved successful for benzene hydrogenation in the production of cyclohexane, and a pilot plant study at a capacity of 5000t/a cyclohexane is in progress.

---

* Coauthors: Cheng Zhenmin, Abdulhakeim M. Anter, Fang Xiangchen, Xiao Qiong, Suresh K. Bhatia. Reprinted from *AIChE Journal*, 2003, 49(1): 225-231.

Over the last three decades, studies on evaporation of the liquid phase in multiphase fixed-bed reactors have been confined to two basic aspects. One is the experimental work on reactor performance, as conducted by Sedriks and Kenney (1973), Hanika et al. (1975, 1976, 1977, 1986, 1999), Ruzicka and Hanika (1994), and Castellari et al. (1997). Another is the catalyst-scale study on the physiochemical aspects, as reported by Kim and Kim (1981a, b), Hu and Ho (1987), Bhatia (1988, 1989), Jaguste and Bhatia (1991), Waston and Harold (1993, 1994), as well as Hessari and Bhatia (1996). In spite of these efforts, reactor-scale description of the evaporation of the liquid is still absent, which is most likely related to the difficulty in direct observation and measurement of the liquid due to the opacity of the reactor wall and the catalyst support.

In the present work, the process of liquid evaporation has been systematically evaluated. For this purpose, Raoult's law is applied to provide a theoretical guideline, and specific experiments have been conducted to verify the validity of the theory.

2 Theoretical Development

In a previous article (Kheshgi et al., 1992), Raoult's law was introduced to account for the variations in the gas and liquid molar flow rates along the reactor, and the liquid flow rate of zero was considered identical to the disappearance of the liquid phase. However, while the validity of the approach is plausible, it was not verified experimentally. In a subsequent article by Khadilkar et al. (1999), a similar approach was followed.

In the present work, as well, we utilize Raoult's law, which has been extensively used in the description of gas-liquid equilibrium in a closed system. However, Raoult's law is extended here to a flowing system accompanied by a reaction, as described by Eq. 1:

$$A(l) + B(g) \longrightarrow C(l) \tag{1}$$

In the system chosen for the present study, A, B and C correspond to benzene, hydrogen, and cyclohexane, respectively. The balance equations are established below.

For the vapor phase:

$$y_A + y_B + y_C = 1 \tag{2}$$

For the liquid phase:

$$x_A + x_B + x_C = 1 \tag{3}$$

Eq. 1 and Eq. 2 are related through the equilibrium relations:

$$y_A = K_A x_A \quad \text{and} \quad y_C = K_C x_C \tag{4}$$

Assuming the conversion of reactant A to be α, the molar flow rates of the components are given as:

$$F_A = F_A^0 (1 - \alpha) \tag{5}$$

$$F_B = (M - \alpha) F_A^0 \tag{6}$$

$$F_C = F_C^0 + \alpha F_A^0 \tag{7}$$

where, M is the ratio of feed flow rate of hydrogen to that of benzene, that is, $F_B^0 = M F_A^0$. Eq. 5-Eq. 7 provide the total flow rate:

$$F_l + F_v = (1 + M - \alpha) F_A^0 + F_C^0 \tag{8}$$

where, F_l is the flow rate of liquid and F_v is that of the vapor. Assuming negligible solubility of hydrogen in the liquid, that is, $x_B \ll 1$, Eq. 2-Eq. 4 and the definition of y_B

$$y_B = \frac{F_B}{F_v} \tag{9}$$

provide the relation

$$F_v = \frac{(M - \alpha) F_A^0}{1 - K} \tag{10}$$

for the vapor-phase flow rate. Here the gas-liquid equilibrium constant, K, is used instead of K_A and K_C, since the physical properties of A and C are similar.

The liquid flow rate is now obtained from Eq. 8 and Eq. 10 as:

$$F_l = (1 + M - \alpha) F_A^0 + F_C^0 - \frac{(M - \alpha) F_A^0}{1 - K} \tag{11}$$

It should be noted that in benzene hydrogenation, M is always greater than 3, which is the stoichiometric ratio of hydrogen to benzene. Therefore, the parameter α can be neglected in comparison with M, since the conversion of reactant A in the liquid phase is normally less than 0.15, in view of the liquid vaporization heat being one-tenth to one-seventh of the reaction heat. Consequently, Eq. 11 can be reduced to:

$$F_l = (1 + M) F_A^0 + F_C^0 - \frac{M F_A^0}{1 - K} \tag{12}$$

Since A and C have similar physical properties, $F_A^0 + F_C^0$ can be lumped as the total liquid flow rate F_l^0, and therefore the hydrogen flow rate $M F_A^0$ can be replaced by $N F_l^0$, with N being the dimensionless flow rate of hydrogen with respect to F_l^0. Eq. 12 is therefore reduced to:

$$F_l = (1 + N) F_l^0 - \frac{N F_l^0}{1 - K} \tag{13}$$

From Eq. 13 it is seen that liquid flow rate decreases from F_l^0 to F_l in the presence of hydrogen. In addition, by increasing the system temperature, F_l can decrease to zero, since the gas-liquid equilibrium constant K increases with temperature. The vanishing condition of the liquid flow rate is therefore obtained by setting $F_l = 0$ in Eq. 13, which leads to:

$$K = \frac{1}{1 + N} \tag{14}$$

Following Raoult's law, the equilibrium constant, K, can be represented as the ratio of the saturation pressure of liquid phase at temperature, T, to the total pressure of the system, that is:

$$K = \frac{P_A^0(T)}{P_{sum}} \tag{15}$$

which has been employed by Kheshgi et al. (1992). Combining Eq. 14 and Eq. 15 provides:

$$\frac{P_A^0(T_v)}{P_{sum}} = \frac{1}{1 + N} \tag{16}$$

where, T_v is the temperature at the point at which liquid vanishes in the reactor. Eq. 16 indicates that this temperature is not only dependent on system pressure, but also on the ratio of gas-to-liquid molar flow rate, N. Assuming that the feed gas is composed of pure B, while the liquid feed is a mixture of A and B. The criterion for liquid vanishing is obtained as below:

$$\frac{F_1^0}{F_B^0} = \frac{P_A^0(T_v)}{P_{sum} - P_A^0(T_v)} \tag{17}$$

Here the saturation vapor pressure P_A^0 has been calculated with the Lee-Kesler equation as given in Reid et al (1987).

The theoretical results from Eq. 13 and Eq. 17 are plotted in Fig. 1 and Fig. 2. It is seen in Fig. 1 that the amount of liquid not vaporized is a function of the system temperature, the gas-to-liquid flow-rate ratio, as well as the system pressure. Since a zero liquid flow rate means the disappearance of the liquid, the liquid vanishing temperature, T_v can be obtained from Fig. 1 at $F_1 = 0$, and the results are given in Fig. 2. It shows that the liquid can disappear at a much lower temperature in the presence of hydrogen, which is reasonable thermodynamically.

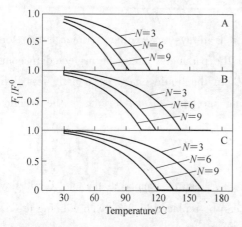

Fig. 1 Fraction of liquid not evaporated under different operating conditions
(p_{sum}: A—1.0MPa; B—2.0MPa; C—3.0MPa)

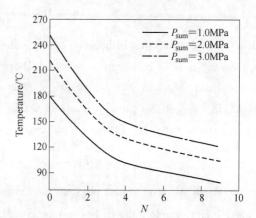

Fig. 2 Liquid vanishing temperature under different pressures and gas-to-liquid molar flow-rate ratios

3 Experimental Studies

Experiments were carried out in a three-phase fixed-bed reactor with gas and liquid flowing concurrently upward. The reactor was heated by electricity and kept adiabatic with glass-wool insulation. Temperature profiles along the reactor length were measured by means of 12 thermocouples inserted through the reactor wall. Two kinds of catalysts—Catalyst A with 0.5% Pd supported over Al_2O_3, and Catalyst B with Ni-B amorphous alloy supported over Al_2O_3—were employed. Independent experiments demonstrated that Catalyst A was less active, with an activity about one-third of Catalyst B. A detailed description for the flow diagram is available elsewhere (Cheng et al., 2001a).

4 Results and Discussion

4.1 Experimental evaluation of the liquid phase

Vaporization of the liquid phase can lead to depletion of the liquid phase and consequently to partial wetting or even dryout of the reactor. Presumably, the liquid flow rate should be zero once the vanishing temperature is attained, and therefore there will be no liquid region in the reactor once the inlet temperature is above the vanishing temperature. To identify the validity of this assumption, experiments under the inlet temperatures above T_v will provide the most rigorous demonstration.

Reactor performances under five inlet temperatures (130℃, 140℃, 150℃, 160℃, and 170℃) are shown in Fig. 3. In these experiments, the vanishing temperature, T_v is predicted to be 117℃, which is 13℃, 23℃, 33℃, 43℃, and 53℃ lower than the respective inlet temperature. It would appear that under all these conditions, the liquid can exist above the predicted vanishing temperature, T_v which could be deduced from the flat temperature profile at the entry of the reactor. The flat temperature profile in the entry region should be explained as a result of the heat balance between the evaporation of liquid and reaction heat. To clarify the boundary between the liquid- and the gas-phase regions, dashed lines are plotted by connecting the observed phase-transition points. To further verify the validity of the assumption on the corresponding relationship between the flat temperature profile and liquid region, reaction-rate dependence on benzene concentration may be explored.

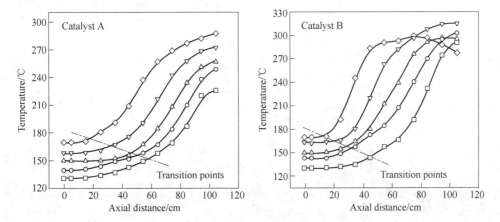

Fig. 3 Influence of inlet temperature on liquid vanishing process
(P_{sum} = 1.0MPa; C_B^0 = 16.6%; u_l = 1.46 × 10^{-3} m/s; u_g = 0.126m/s; T_v = 117℃. T_{in}, T_w: □—130℃; ○—140℃; △—150℃; ▽—160℃; ◇—170℃. Conversion: □,○,△,▽,◇—100%)

It is known that under the liquid condition the reaction rate is dominated by hydrogen pressure in view of its scant solubility in benzene, while the order with respect to the liquid reactant, benzene, will be zero. Under the gas-phase condition, the reaction-rate dependence will be different. For the present system, the reaction order for benzene will be greater than zero in the gas phase, since hydrogen is in large excess for obtaining high conversion of benzene. Therefore, if the reaction rate is found to be independent of the benzene concentration at the inlet, the reaction can be considered to be in the liquid region.

Although a direct measurement of the reaction rate is not available, temperature measurements can be used instead, since the reactor is operated almost adiabatically. In Fig. 4 it is observed that the reactor temperatures under different benzene concentrations fall into the same profile in the first 30cm, which is considered to be the liquid region. Similar temperature profiles are also observed for Catalyst B, while the only difference is that the length of the liquid region is reduced from 30cm to 10cm, which is in agreement with the difference in the activity of the two catalysts. Further downstream, the variation in temperature with inlet feed concentration indicates that the reaction rate is a function of benzene concentration. This implies that in this region the reaction occurs in the gas phase, and therefore evaporation has occurred.

The preceding experimental results indicate that the liquid phase can exist above the vanishing temperature, T_v which is in conflict with the theoretical speculation. This discrepancy can be rationalized by considering $T > T_v$ as a necessary condition, rather than a sufficient one for dryout. The existence of liquid phase over the vanishing point, T_v can be explained by considering the boiling point of the mixture of benzene and cyclohexane. At 1.0MPa, the boiling point is 180℃, which is 63℃ and 65℃ above the two T_v values shown in Fig. 4 and Fig. 5. Since the boiling point of the liquid is greater than T_v the liquid can be thermodynamically stable despite the inlet temperature being above T_v. Thus, an improved vapor-liquid equilibrium model is required for more accurate predictions of T_v in comparison to Raoult's law.

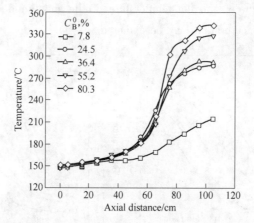

Fig. 4 Identification of liquid region from concentration experiments for Catalyst A (the less active catalyst) ($P_{sum} = 1.0$MPa; $u_l = 1.46 \times 10^{-3}$m/s; $u_g = 0.126$m/s; $T_{in} = T_w = 150℃$; $T_v = 117℃$. H_2/C_6H_6: □—33.6; ○—10.7; △—7.2; ▽—4.8; ◇—3.3. Conversion: □,○—100%; △—53.3%; ▽—44.5%; ◇—37.0%)

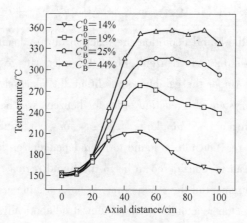

Fig. 5 Identification of liquid region from concentration experiments for Catalyst B (the more active catalyst) ($P_{sum} = 1.0$MPa; $u_l = 1.36 \times 10^{-3}$m/s; $u_g = 0.126$m/s; $T_{in} = T_w = 150℃$; $T_v = 115℃$. H_2/C_6H_6: ▽—19.9; □—11.2; ○—8.55; △—6.3. Conversion: ▽,□,○,△—100%)

4.2 Experimental dependence of liquid-region length

Verification of the existence of the liquid phase provides a qualitative description of the present subject; however, this information is insufficient for a complete understanding of the phase-transition process. For this purpose, various factors, including catalyst activity, liquid flow rate, gas flow rate, and the operation pressure, need to be studied.

The influence of catalyst activity can be observed in Fig. 4 and Fig. 5, where it is found that the liquid-region length is inversely proportional to the catalyst activity, and thereby the reaction rate. Experiments under different liquid flow rates are shown in Fig. 6. The liquid-region length is found to be approximately proportional to the liquid flow rate, which implies that the heat required for liquid vaporization under the adiabatic condition is only supplied by the reaction heat.

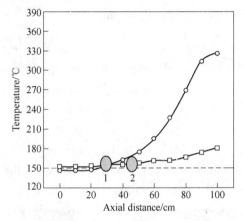

Fig. 6 Effect of liquid flow rate on phase transition for Catalyst B
($P_{sum} = 1.0$ MPa; $C_B^0 = 31.5\%$; $u_g = 0.126$ m/s; $T_{in} = T_w = 150$ ℃. u_l: ○—3.58×10^{-3} m/s; □—5.24×10^{-3} m/s. T_v: ○—142 ℃; □—152 ℃. Conversion: ○—41%; □—14%. Phase transition point: 1—30 cm; 2—45 cm)

Fig. 7 shows the experimental results for operation at different gas flow rates. The gas flow rate can greatly reduce the liquid vanishing temperature, and therefore increase the liquid evaporating rate. However, an obvious decrease in temperature profile was obtained in the liquid region

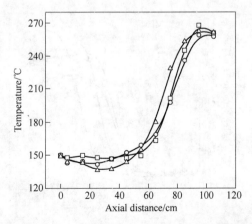

Fig. 7 Effect of gas flow rate on phase transition for Catalyst A
($C_B^0 = 21.2\%$; $P_{sum} = 1.0$ MPa; $L = 0.88 \times 10^{-3}$ m/s; $T_{in} = T_w = 150$ ℃. u_g: □—0.064 m/s; ○—0.088 m/s; △—0.126 m/s. T_v: □—123 ℃; ○—113 ℃; △—102 ℃. Conversion: □,○,△—100%)

under a gas flow rate of 0.126m/s. This can be explained by the much faster vaporization of the liquid than in the other cases, due to the much lower vanishing temperature. In spite of these details, the gas flow rate has almost no effect on the length of the liquid region.

The pressure effect on phase transition is somewhat complex, as is shown in Fig. 8. In this figure, two kinds of liquid inlet temperature, T_{in}, are employed, that are above or below T_v. When T_{in} is higher than T_v, the liquid-region lengths are the same, while when T_{in} is lower than T_v, the liquid-region lengths show some difference. This is because, in this case, the inlet liquid temperature has not reached the phase-transition temperature, so that the heat of the reaction will be used first to increase the system temperature and then for the vaporization. Therefore, enlargements of the liquid region under 2.5MPa and 3.0MPa are observed in comparison with the other low-pressure cases.

4.3 Estimating the length of the liquid region

The liquid-region length can be estimated from the heat balance between the reaction heat and the vaporization heat of the liquid fluid. Assume that the superficial liquid velocity is u_l, the reaction rate in the liquid phase is r, and the reaction heat and the vaporization heat are ΔH_r and ΔH_v, respectively. Then the liquid-region length, L, can be approximated as:

$$L = u_l \cdot \frac{l}{r} \cdot \frac{\Delta H_v}{\Delta H_r} \tag{18}$$

which assumes relative constant temperature and reaction rate in the liquid region. This is consistent with the experimental observations in Fig. 3 to Fig. 8. It should be noted that Eq. 18 applies only to the situation where liquid temperature is above the vanishing temperature. The validity of this equation in the reaction rate, r, has been verified from experiments under different catalyst activities, which shows the liquid-region length to be inversely proportional to the catalyst activity. Experiments under different liquid flow rates have verified the linear relationship between L and u_l, as can be observed from Fig. 6. It should be noted that the liquid vanishing temperature, T_v is not involved in Eq. 18, since this variable is not related to heat balance, and this consideration has been confirmed from experiments under the different gas flow rates shown in Fig. 7.

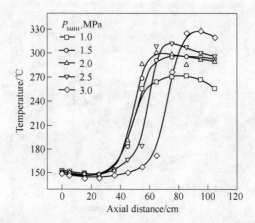

Fig. 8 Effect of pressure on phase transition for Catalyst B
($C_B^0 = 26.0\%$; $u_l = 1.16 \times 10^{-3}$m/s; $u_g = 0.126$m/s; $T_{in} = T_w = 150$℃.
T_v: □—111℃; ○—128℃; △—140℃; ▽—152℃; ◇—162℃)

5 Conclusions

The liquid vanishing process as well as the dryout phenomena in a three-phase fixed-bed reactor have been investigated. The vanishing point is defined as the point where the liquid flow rate is zero from the thermodynamic condition without considering whether or not sufficient energy is available for evaporation of the liquid phase. In the experiments, it was found that a certain entry length is needed before the liquid flow rate becomes zero, even when the inlet temperature is above the vanishing temperature, since the reaction heat in the entry region produces the required heat for liquid vanishing.

The liquid temperature can stay nearly constant in the entry section of the reactor before all the liquid has been vaporized. The length of the liquid region can be estimated from the heat balance between the reaction and vaporization, and is proportional to the liquid flow rate and the vaporization heat, but is inverse to the reaction rate and reaction heat.

Acknowledgements

The present work is supported by the Natural Science Foundation of China under Grant No. 20106005 and by SINOPEC through Contract No. 201085. Dr. Cheng is grateful to the Distinguished Visiting Scholar Program granted by the Chinese government in supporting his research in the University of Queensland, Australia.

Notation

C_B^0	inlet benzene concentration in dilution by cyclohexane, vol%
F	mass flow rate, mol/s
F_l	mass flow rate in liquid phase, mol/s
ΔH_r	reaction heat, kJ/mol
ΔH_v	vaporization heat of liquid, kJ/mol
K	equilibrium constant in Raoult's law
L	length of the liquid region, m
M, N	molar flow rate ratio of gas to the liquid
P_{sum}	total pressure of all components, MPa
$P^0(T)$	saturated vapor pressure of the liquid at temperature T, MPa
r	reaction rate, mol/(kgcat · s)
T	temperature, ℃
T_{in}	inlet temperature of the reactant mixture, ℃
T_v	vanishing temperature of the liquid, ℃
T_w	reactor wall temperature, ℃
u_g	gas-phase flow rate, m/s
u_l	liquid-phase flow rate, m/s
x	molar fraction in liquid phase
y	molar fraction in vapor phase

Greek letters

α   conversion of the reactant

Superscripts and subscripts

0   initial state
A   liquid reactant
B   gas reactant
C   liquid product
l   liquid phase
v   vapor phase

References

Bhatia, S. K., "Steady State Multiplicity and Partial Internal Wetting of Catalyst Particles," AIChE J., 34, 969 (1988).

Bhatia, S. K., "Partial Internal Wetting of Catalyst Particles with a Distribution of Pore Size," AIChE J., 35, 1337(1989).

Castellari, A. T., J. O. Cechini, L. J. Gabarain, and P. M. Haure, "Gas-Phase Reaction in a Trickle-Bed Reactor Operated at Low Liquid Flow Rates," AIChE J., 43, 1813(1997).

Cheng, Z. M., A. M. Anter, and W. K. Yuan, "Intensification of Phase Transition on Multiphase Reactions," AIChE J., 47, 1185(2001a).

Cheng, Z. M., A. M. Anter, G. M. Khalifa, J. S. Hu, Y. C. Dai, and W. K. Yuan, "An Innovative Reaction Heat Offset Operation for a Multiphase Fixed Bed Reactor Dealing with Volatile Compounds," Chem. Eng. Sci., 56, 6025(2001b).

Cheng, Z. M., and W. K. Yuan, "Influence of Hydrodynamic Parameters on Performance of a Multiphase Fixed-Bed Reactor under Phase Transition." Chem. Eng. Sci., 57, 3407(2002).

Hanika, J., K. Sporka, V. Ruzicka, and J. Krausova, "Qualitive Observations of Heat and Mass Transfer Effects on the Behaviour of a Trickle Bed Reactor," Chem. Eng. Commun., 2, 19(1975).

Hanika, J., K. Sporka, V. Ruzicka, and J. Hrstka, "Measurement of Axial Temperature Profiles in an Adiabatic Trickle Bed Reactor," Chem. Eng. J., 12, 193(1976).

Hanika, J., K. Sporka, V. Ruzicka, and R. Pistek, "Dynamic Behavior of an Adiabatic Trickle Bed Reactor," Chem. Eng. Sci., 32, 525(1977).

Hanika, J., B. N. Lukjanov, V. A. Kirillov, and V. Stanek, "Hydrogenation of 1,5-Cyclooctadiene in a Trickle Bed Reactor Accompanied by Phase Transition," Chem. Eng. Commun., 40, 183(1986).

Hanika, J., "Safe Operation and Control of Trickle-Bed Reactor," Chem. Eng. Sci., 54, 4653(1999).

Hessari, F. A., and S. K. Bhatia, "Reaction Rate Hysteresis in a Single Partially Internally Wetted Catalyst Pellet: Experiment and Modelling," Chem. Eng. Sci., 51, 1241(1996).

Hu, R., and T. C. Ho, "Steady State Multiplicity in an IncompletelyWetted Catalyst Particle," Chem. Eng. Sci., 23, 1239(1987).

Jaguste, D. N., and S. K. Bhatia, "Partial Internal Wetting of Catalyst Particles: Hysteresis Effects," AIChE J., 37, 650(1991).

Khadilkar, M. R., P. L. Mills, and M. P. Dudukovic, "Trickle-Bed Reactor Models for Systems with a Volatile Liquid Phase," Chem. Eng. Sci., 54, 2421(1999).

Kheshgi, H. S., S. C. Reyes, R. Hu, and T. C. Ho, "Phase Transition and Steady-State Multiplicity in a Trickle-

Bed Reactor," Chem. Eng. Sci. ,47,1771(1992).

Kim, D. H. , and Y. G. Kim, "An Experimental Study of Multiple Steady States in a Porous Catalyst due to Phase Transition," J. Chem. Eng. Jpn. ,14,311(1981a).

Kim, D. H. , and Y. G. Kim, "Simulation of Multiple Steady States in a Porous Catalyst due to Phase Transition," J. Chem. Eng. Jpn. ,14,318(1981b).

LaVopa, V. , and C. N. Satterfield, "Some Effects of Vapor-Liquid Equilibria on Performance of a Trickle-Bed Reactor," Chem. Eng. Sci. ,43,2175(1988).

Ramakrishna, V. N. , J. A. Guin, A. R. Tarrer, and C. W. Curtis, "Effect of Phase Behavior on Hydrotreater Performance: Simulation and Experimental Verification," Ind. Eng. Chem. Process Des. Dev,24,598(1985).

Reid, R. C. , J. M. Prausnitz, and T. K. Sherwood, The Properties of Gases and Liquids,4th ed. , McGraw-Hill, New York (1987).

Ruzicka, J. , and J. Hanika, "Partial Wetting and Forced Reaction Mixture Transition in a Model Trickle-Bed Reactor," Catal. Today,20,467(1994).

Sedriks, W. , and C. N. Kenney, "Partial Wetting in Trickle Bed Reactors—The Reduction of Crotonaldehyde over a Palladium Catalyst," Chem. Eng. Sci. ,28,559(1973).

Singh, C. P. P. , and N. L. Carr, "Process Simulation of an SRC—II Plant," Ind. Eng. Chem. Process Des. Dev, 22,104(1983).

Smith, C. M. , and C. N. Satterfield, "Some Effects of Vapor-Liquid Flow Ratio on Performance of a Trickle-Bed Reactor," Chem. Eng. Sci. ,41,839(1986).

Watson, P. C. , and M. P. Harold, "Dynamic Effects of Vaporization with Exothermic Reaction in a Porous Catalyst Pellet," AIChE J. ,39,989(1993).

Watson, P. C. , and M. P. Harold, "Rate Enhancement and Multiplicity in a Partially Wetted and Filled Pellet: Experimental Study," AIChE J. ,40,97(1994).

Influence of Hydrodynamic Parameters on Performance of a Multiphase Fixed-bed Reactor under Phase Transition*

Abstract Reactor performance of a trickle-bed reactor (TBR) concurrently under gas-liquid downward flow and that of a flooded-bed reactor (FBR) concurrently under gas-liquid upward flow was experimentally investigated under an elevated temperature and pressure (150℃ and 1.0MPa) for benzene hydrogenation. It was shown that the different hydrodynamics of TBR and FBR could result in quite different reactor behaviors, as typically observed from the temperature profiles along the reactor. The reason for this is because the present reaction is controlled by the supply of hydrogen in the liquid phase, thus external partial wetting of the catalyst pellets could increase the reaction rate. Moreover, the pronounced vapor pressure of benzene under the prescribed temperature would make the reaction remarkable over the non-wetted catalysts. Operation in the FBR is superior to the TBR, considering the operational safety. However, TBR should be considered when the catalyst partial wetting is negligible under liquid flow rates higher than 0.58cm/s as shown in this work.

Key words trickle-bed reactor, flooded-bed reactor, partial wetting, benzene hydrogenation, hydrodynamics

1 Introduction

In the authors' previous work (Cheng et al., 2001a; Cheng, Anter, & Yuan, 2001b), benzene hydrogenation into cyclohexane has been successfully performed in a three-phase fixed-bed reactor accompanied by continuous phase transition. The characteristic of that system is the great volatility of the liquid reactant benzene and the product cyclohexane in the temperature range 150-170℃. The laboratory study of this reaction was carried out in a flooded-bed reactor (FBR), with the gas and liquid phases concurrently upward flow through the catalyst bed. A pilot plant at a capacity of 5000t/a in cyclohexane is being designed in the same way. However, it is desired to know whether FBR can be further employed for large-scale industrial operations at a capacity of 100000t/a, since FBR is seldom used for commercial productions. Yet, it should be noted that the FBRs bear no difference from the TBRs (trickle-bed reactors with gas and liquid downward flow) in principle with respect to the kinetics and thermodynamics. Therefore, the influence of the different hydrodynamics on the reactor behavior will be the only reason for the reactor being operated in the upward or the downward flow fashions.

2 Theoretical Analysis

2.1 The overall considerations

The major hydrodynamic parameters affecting the performance of a three-phase fixed-bed reac-

---

* Coauthor: Cheng Zhenmin. Reprinted from Chemical Engineering Science, 2002, 57: 3407-3413.

tor could be summarized as below:

(1) The liquid-solid wetting efficiency, η_{ce}. Its influence on the overall conversion of the reactant is according to the formula introduced by Mears (1974):

$$\ln\left(\frac{1}{1-x}\right) = \frac{k\eta\eta_{ce}}{\text{LHSV}} \qquad (1)$$

where, η is the catalyst effectiveness factor in liquid phase; k is the reaction rate constant which is independent of the catalyst wetting condition.

(2) The fraction of catalyst contacted by the liquid, f. This parameter accounts for the maldistribution of the liquid, thus the averaged reaction rate of the catalyst can be expressed by the weighted contribution of the reaction rates in liquid and gas phases:

$$\bar{r} = fr_L + (1-f)r_G \qquad (2)$$

where, $\bar{r}$, r_L and r_G correspond to the above-defined reaction rates. Obviously, r_L is the reaction rate over a completely wetted catalyst, which is a function of η_{ce}.

(3) The dynamic liquid holdup, h_d. It describes the residence time 1/LHSV of the liquid reactant, therefore, the conversion of the reactant increases with h_d, as is shown in Eq. 1.

Since both η_{ce} and h_d depend on both the liquid flow rate and flow direction of the gas and liquid phases, the reactant conversion will be affected by these two factors.

2.2 Considerations for some specific cases

The above equations are generally valid for all kinds of reaction systems; however, they should be reconsidered for some special cases:

(1) When the reaction rate in the liquid phase is controlled by the gas reactant such as hydrogen. Since hydrogen is sparingly soluble in the hydrocarbon, the hydrogenation rate is usually controlled by the supply of hydrogen. Therefore the liquid reactant is in large excess, partial wetting of catalyst surface can give rise to a significant increase in the reaction rate (Funk, Harold, & Ng, 1990; Dudukovic et al., 1999).

(2) When the liquid reactant has large volatility and the reactor is operated under high temperatures. In this case, the macroscopic maldistribution of the liquid in the reactor across its diameter can lead to simultaneous occurrence of reaction on all of the completely wetted, partially wetted, and non-wetted catalyst pellets.

To illustrate the importance of liquid volatility to the gas-phase reaction, hydrogenations of α-methylstyrene (AMS) and cyclohexene has been used as an example (Watson & Harold, 1993). The normal boiling point of AMS is 165.4℃, while that for cyclohexene is 87.0℃. At 80℃, their vapor pressures are 46.0 and 743.0 Torr, respectively. Their evaporation rate and hydrogenation rate were found to be much different from the single-pellet experiment. This result is consistent with Funk et al. (1990). In their work, hydrogenation of AMS under ambient temperature with three kinds of liquid distributors was studied. It was found that uniform partial wetting of the catalyst is favorable to the gas-limited reaction, while the single-tube distributor gave the lowest conversion, since there was no adequate supply of the liquid to the catalyst. We believe this should be explained from the relatively low vapor pressure of AMS under the ambient temperature, and thus the vapor-phase reaction is negligible.

Since the present working system of benzene hydrogenation is operated at elevated temperatures, the vapor-phase reaction will be remarkable, and thus the effect of the catalyst partial wetting and liquid maldistribution will be more pronounced than in other cases.

3 Evaluation of the System Properties

3.1 The limiting reactant in the liquid- or gas-phase reactions

For a catalyzed gas-liquid reaction expressed by $A(1) + bB(g) \rightarrow C(1)$, the reaction rate is subject to both the concentration and diffusion coefficient of each reactant in the liquid phase. As a criterion, Khadilkar, Wu, Al-Dahhan, and Dudikovic (1996) proposed to use a parameter γ, which is defined as $\gamma = b(C_{Ai}D_{eA})/(C_{Be}D_{eB})$, to determine the limiting reactant of the reaction.

If $\gamma \gg 1$, the reaction will be limited by the gas reactant; if $\gamma \ll 1$, it will be limited by the liquid. In the present work, under benzene concentration of 31.5%, and the flow rates of the liquid and hydrogen at 1.32kg/h and 15nL/min, under inlet temperature of 150℃ and operation pressure of 1.0MPa, it is found that $\gamma = 38.2 \gg 1$. Therefore, in the liquid phase, the reaction rate is limited by the gas reactant hydrogen.

Under the same conditions as in the vapor phase, the value of γ is found to be 0.18. As a comparison, the vapor-phase γ value for the hydrogenation of AMS under ambient temperature is only 3.16×10^{-3}. It shows that the vapor-phase reaction on the unwetted catalysts will be very pronounced for the present work.

3.2 Determination of the hydrodynamic parameters

3.2.1 The liquid holdup

For downward flow of the gas and liquid through a bed of catalyst, four primary forms of liquid can exist external to the catalysts (Ng & Chu, 1987), and they constitute the total liquid holdup in the bed. In this work, only the dynamic liquid holdup is discussed since it determines the residence time of the liquid reactant. To estimate the dynamic liquid holdup, a correlation proposed by Ellman, Midoux, Wild, Laurent, and Charpentier (1990) is employed:

$$\beta_d = 10^\kappa \tag{3}$$

where

$$\kappa = 0.001 - \frac{R}{\xi^S} \quad \text{and} \quad \xi = \chi_L^m Re_L^m We_L^p \left(\frac{\alpha_t d_h}{1-\varepsilon}\right)^q$$

in which $R = 0.42, S = 0.48, m = 0.5, n = -0.3, p = 0$, and $q = 0.3$ for the low interaction regime.

As depicted in Table 1, Eq. 3 covers a wide range of gas and liquid properties. This correlation was recommended in the prediction of liquid saturation both in the downward and upward flow operations (Yang, Euzen, & Wild, 1992; Iliuta, Thyrion, Bolle, & Giot, 1997). However, this equation is an implicit function of the liquid flow rate, so it is difficult to establish a straightforward relationship between the conversion and liquid flow rate. Due to this reason, a different formula by Specchia and Baldi (1977) is employed, as described in Eq. 4:

$$\beta_d = 3.86 Re_L^{0.545}(Ga_L^*)^{-0.42}\left(a_s\frac{d_p}{\varepsilon}\right)^{0.65} \qquad (4)$$

Eq. 4 shows that the dependence of dynamic liquid holdup on the liquid flow rate is to the power of 0.545, which is pronounced from the normal point of view. However, it is much less according to Ellman et al. (1992) as shown in Fig. 1.

Table 1 Physical properties of the three phases from different sources

Source of data	Liquid properties			Gas properties		Catalyst properties	
	$\rho_L/kg \cdot m^{-3}$	$\mu_L/Pa \cdot s$	$\sigma_L/N \cdot m^{-1}$	$\rho_G/kg \cdot m^{-3}$	$\mu_G/Pa \cdot s$	d_p/mm	Shape
Ellman et al. (1990)	650-1146	3.1×10^{-4} - 6.63×10^{-2}	0.019-0.078	0.164-116.4	$(1.75\text{-}2.68) \times 10^{-5}$	1.16-3.06	Spherical Cylindrical
Mills and Dudukovic (1981)	650-1000	3.2×10^{-2} - 9.6×10^{-2}	—	—	—	0.54-4.1	Spherical Cylindrical
El-Hisnawi, Dudukovic, and Mills (1981)	656-1000	3.16×10^{-4} - 2.18×10^{-3}	0.0185-0.073	8.8	0.75×10^{-5}	1.3×5.6	Extrudate
This work	780	1.7×10^{-4}	0.013	56.8	1.1×10^{-5}	2-3	Spherical

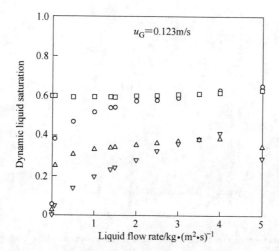

Fig. 1 Dynamic liquid holdup prediction under downward flow and upward flow fashions
(Condition: $T = 150°C$, $P = 10$bar)
Downflow: ▽—Specchia and Baldi (1977); △—Ellman et al. (1990);
Upflow: ○—Stiegel and Shah (1977); □—Yang et al. (1992)

The liquid holdup in an FBR is obviously higher than that in a TBR, as predicted from the correlations by Stiegel and Shah (1977):

$$\beta_t = 1.47 Re_L^{0.11} Re_G^{-0.14}(a_s d_p)^{-0.41} \qquad (5)$$

and Yang et al. (1992):

$$\beta_d = 1 - \frac{C_0}{\varepsilon} \cdot \frac{u_G}{u_G + u_L} \qquad (6)$$

where $C_0 = 0.16$, by assuming the benzene/cyclohexane mixture as a non-foaming system.

It should be noted that β_t in Eq. 5 is the total liquid holdup including the dynamic and static components. To get the dynamic liquid holdup, the static value of 0.12 should be subtracted.

3.2.2 The external catalyst wetting efficiency

It has been noted that the hydrodynamics in a flooded bed is similar to a trickle bed in the high interaction regimes, such as pulsing flow, etc. (Ellman, et al., 1990; Yang, et al., 1992). Following this consideration, the catalyst will be completely wetted in the upflow operation, since in the high interaction regime the liquid can fully occupy the void between adjacent particles (Cheng & Yuan, 1999). In this regard, the external wetting efficiency in an FBR can be reasonably assumed as 1.0. Nevertheless, the catalyst external surface is only partially wetted by the liquid under small liquid flow rates in a TBR. In this work, the correlation by El-Hisnawi et al. (1981) is used to predict this parameter for the benzene hydrogenation system:

$$\eta_{ce} = 1.617 Re_L^{0.146} Ga_L^{-0.071} \qquad (7)$$

In spite of its simplicity, Eq. 7 was considered to be accurate enough for its purpose (Mills & Dudukovic, 1984), which could be evidenced from the comparison shown in Fig. 2 with the equation of Mills and Dudukovic (1981):

$$\eta_{ce} = 1.0 - \exp\left[-1.35 Re_L^{0.333} Fr_L^{0.235} We_L^{-0.170} \left(\frac{a_t d_p}{\varepsilon^2} \right)^{-0.0425} \right] \qquad (8)$$

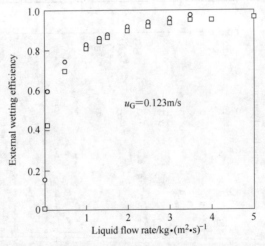

Fig. 2 External contacting efficiency of the catalyst under downward flow and upward flow fashions
(Condition: $T = 150\,°C$, $P = 10\,bar$)
Downflow: ○——El-Hisnawi et al. (1981); □——Mills and Dudukovic (1981); Upflow: $\eta_{ce} = 1.0$

4 Experimental

A fixed bed reactor of 20mm ID and 1.6m in length was used for the present work. The liquid distributor was made of a stainless steel plate with 30 uniformly distributed holes of 1mm in ID. The catalyst of Ni-B alloy over Al_2O_3 of 2-3mm in diameter was packed to a height of 1.0m between two inert sections of 30mm on both ends. The reactor was operated under a pseudo-adiabatic condition. Thirteen thermal couples, evenly distributed along the flow direction, were inserted into the reactor through the wall. The reactants, hydrogen and benzene/cyclohexane mixture, were preheated separately before flowing into the reactor. To avoid the hysteresis effect specific to the small packings, the catalyst bed was preflooded prior to each of the downflow runs.

The liquid flow rates in the present work was from 1.32 to 4.1kg/(m² · s) (0.14-0.58cm/s) and the gas flow rate was kept at 15.4L/min(0.13m/s). The hydrodynamics is predicted to fall into the trickling flow (TBR) and bubbling flow (FBR) regimes according to Charpentier and Favier (1975), Cheng and Yuan (1999), and Ramachandran and Chaudhari (1983).

5 Results and Discussion

Since the reaction in the liquid phase is limited by the mass transfer of the gas phase, the wetting degree of the catalyst will lead to different reactor performances depending on the specific hydrodynamic conditions.

5.1 Reactor performance under upflow and downflow operations

To illustrate the different behaviors of trickle- and flooded-bed reactors, the reactors were operated under different liquid flow rates. In the low liquid flow rate of 1.32kg/(m² · s)(0.14cm/s), the temperature was found to increase rapidly, as is shown in Fig. 3. In the high liquid flow rate of 4.1kg/(m² · s)(0.58cm/s), the temperature changed slowly and the reactor was in liquid phase regime in most part of the reactor, as is shown in Fig. 4.

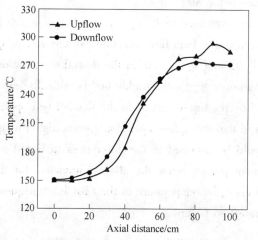

Fig. 3 Reactor performance under a low liquid flow rate
($T = 150$°C, $P = 10$bar, $L = 1.32$kg/(m² · s), $G = 15.4$nL/min, $C_B^0 = 23.8\%$)

In spite of similar trends of the two temperature profiles shown in Fig. 3, the TBR operation has a more rapid temperature increase than the FBR operation. The difference is originated from the different hydrodynamic parameters in the two flow directions. In the FBR operation, the liquid-solid contacting efficiency η_{ce} is 1.0, and the liquid saturation degree β_d is 0.53-0.60; while in the TBR operation the value of η_{ce} is estimated to be 0.83-0.85 and β_d is 0.22-0.34. Although the residence time of the reactant in the liquid region in the upflow pattern is about 2-3 times of that in the downflow, it seems its contribution to the reaction is not so large as that of the external partial wetting, and this could be explained from two points. One is the principle that has been disclosed in the literature, that is, the Biot number can be increased by

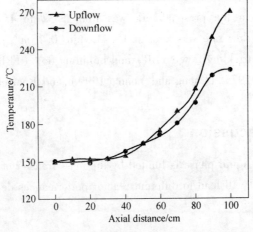

Fig. 4 Reactor performance under a high liquid flow rate
($T=150℃, P=10\text{bar}, L=4.10\text{kg}/(\text{m}^2 \cdot \text{s}), G=15.4\text{nL}/\text{min}, C_B^0=23.8\%$)

100 times due to the incomplete wetting, see Funk et al. (1990). Another is the contribution of gas-phase reaction on some unwetted catalyst, since the liquid flow rate of 0.14cm/s is not sufficient to provide uniform irrigating of the catalyst. However, this point has not been studied thoroughly in the literature since most of the research was conducted with respect to AMS and operated under temperature below 80℃.

When liquid flow rate was increased to 4.1kg/($\text{m}^2 \cdot \text{s}$), the similar hydrodynamic condition for the catalyst bed was attained in both flow directions. In the upflow operation, the value of β_d is estimated to be 0.61-0.62 and η_{ce} is 1.0; in the downflow operation, β_d is 0.38-0.41 and η_{ce} is 0.95-1.0. The variation of η_{ce} for the trickle bed is critically important, since the catalyst will be completely wetted by the liquid just as in the flooded bed operation.

From Fig. 4 it is observed that the upflow operation gives a higher temperature profile than the downflow one, and it should be ascribed to the differences in liquid residence time and mass transfer coefficients between phases. Since the above parameters for the upflow operation are larger than the downflow ones, the temperature in the FBR is consequently larger than in TBR, especially when phase transition occurs.

5.2 Reactor performance under different wetting conditions

The importance of partial wetting of the catalyst has been shown to be more significant to chemical reaction than the liquid holdup from the comparison of the TBR and FBR. In this paragraph, investigation in this direction will be continued only under downward flow condition, since the partial wetting problem will not be encountered in the upflow condition.

To generate different partial wetting degree of the catalyst, a series of liquid flow rates from low to high were employed. The lowest flow rate of 0.48kg/($\text{m}^2 \cdot \text{s}$) corresponds to a wetting degree of 0.68-0.74, and the highest one of 4.1kg/($\text{m}^2 \cdot \text{s}$) gives the wetting degree of 0.95-1.0.

The effects of partial wetting of the catalyst on performance of the reactor is exhibited from the temperature and conversion profiles shown in Fig. 5 and Fig. 6. It is found that the lower the liquid flow rate, the rapid the increase of the temperature. This result is the consequence of the long residence time and large gas-solid contacting area due to the low η_{ce} value for the low liquid flow rates. To have an understanding of the combined effect of the influence of catalyst contacting efficiency and the liquid holdup, the conversion should be plotted against the liquid flow rates.

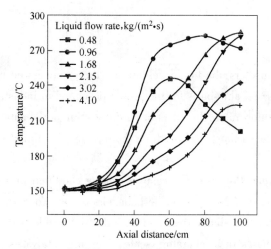

Fig. 5 Temperature profiles under different liquid flow rates for the gas-liquid downward flow

($T = 150\,\text{°C}, P = 10\,\text{bar}, G = 15.4\,\text{nL/min}, C_B^0 = 23.8\%$)

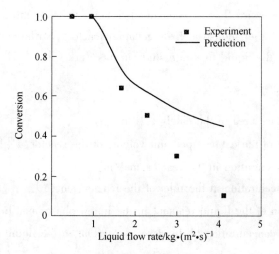

Fig. 6 Identification of the effect of external partial wetting
efficiency on the performance of a trickle-bed reactor

($T = 150\,\text{°C}, P = 10\,\text{bar}, G = 15.4\,\text{nL/min}$)

Since the reaction order to benzene in the liquid phase is zero, the conversion of benzene will be proportional to the residence time, and it can be expressed as $x \propto 1/\text{LHSV} \propto u_L^{-1}$. On the other hand, the residence time is proportional to the liquid holdup, and is therefore proportional to $u_L^{0.545}$ according to the relationship shown in Eq. 4. Accordingly, the conversion of benzene is proportional to the product of u_L^{-1} and $u_L^{0.545}$, i. e., $x \propto u_L^{-0.455}$.

Fig. 6 shows a comparison between experimental data and a predicted profile of conversion against $u_L^{-0.455}$. It is observed that the prediction gives higher conversions than the experiments, especially under high liquid flow rates. This is because the prediction is by taking liquid flow rate of 0.96kg/(m² · s) as the reference, and under which condition the reaction is not conducted completely in the liquid phase, see Fig. 5. In its application to other liquid flow rates, the same degree of catalyst wetting was assumed, and it is not true as shown in Fig. 5. Nevertheless, Fig. 6 has clearly demonstrated the impact of the vapor-phase reaction on the overall conversion of the reactant.

6 Conclusions

Two hydrodynamic parameters, the liquid holdup and the external wetting efficiency have been investigated in finding a reliable basis for the proper design of an industrial trickle-bed reactor.

(1) Under low liquid flow rates, the effect of partial wetting of the catalyst in a trickle bed can exceed the longer residence time resulting from the high level of liquid holdup in a flooded-bed reactor.

(2) Under a large liquid flow rate, the catalyst wetting condition in a trickle bed approaches that of a flooded-bed reactor. As a result, the behaviors of the two reactors are much similar.

It can be concluded that a reliable design of an industrial trickle-bed reactor should be based on upflow of the two phases in a laboratory scale, since the catalyst is completely wetted, with the liquid holdup and external mass transfer coefficients similar to the industrial trickle bed.

In considering the operation of an industrial benzene hydrogenation reactor with phase transition, the gas-liquid downward flow is recommended. Since the flow rate of liquid will be in the order of 1.0cm/s like other industrial hydrogenation reactors, partial-wetting effect will not be observed according to the liquid flow rate study in this work.

Notation

a_s specific surface area of the catalyst, 1/m

a_t surface area of the catalyst per unit volume of the reactor, $a_s(1-\varepsilon)$

C reactant concentration in the reactor, mol/m³

C_0 reactant concentration at the inlet of the reactor, mol/m³

C_{Ai} concentration of the liquid reactant in the liquid phase, mol/m³

C_{Be} equilibrium concentration of the gaseous reactant in the liquid phase, mol/m³

C_B^0 inlet concentration of benzene, wt%

d_h hydraulic diameter, m

d_p particle diameter, m

D_{eA} effective diffusivity of the liquid reactant in the catalyst, m²/s

D_{eB} effective diffusivity of the gaseous reactant in the catalyst, m²/s

f fraction of the catalyst contacted by the liquid

Fr_L Froude number for the liquid phase, $a_t u_L^2/\rho_{LG}^2$

g gravitational acceleration, m/s²

G gas flow rate, nL/min

Ga_L Galileo number, $Ga_L = d_p^3 \rho_L^2 g/\mu_L^2$

Ga_L^*	modified Galileo number, $Ga_L^* = d_p^3 \rho_L [\rho_L g + (\Delta p/\Delta z)_{LG}]/\mu_L^2$
h_d	dynamic liquid holdup
k	intrinsic reaction rate constant, 1/s
L	liquid flow rate, kg/(m² · s)
LHSV	liquid hourly space velocity, h⁻¹
P	pressure, bar
r	the overall reaction rate, mol/(m³ · s)
r_G	the gas-phase reaction rate, mol/(m³ · s)
r_L	the liquid phase reaction rate, mol/(m³ · s)
Re_L	Reynolds number of the liquid phase, $Re_L = d_p u_L \rho_L / \mu_L$
Re_G	Reynolds number of the gas phase, $Re_G = d_p u_G \rho_G / \mu_G$
T	temperature, ℃
u_G	gas superficial velocity, m/s
u_L	liquid superficial velocity, m/s
We_L	Weber number, $u_L^2 \rho_L d_p / \sigma_L$
x	conversion of the liquid reactant

Greek letters

β_d	dynamic liquid holdup with respect to the porosity of the bed
β_t	total liquid holdup with respect to the porosity of the bed
γ	criterion accounting for limitation of the reaction by gas/liquid reactant
ε	porosity of the reactor
η	catalyst effectiveness factor in liquid phase
η_{ce}	fraction of external area wetted
μ_G	gas viscosity, Pa · s
μ_L	liquid viscosity, Pa · s
ρ_G	gas density, kg/m³
ρ_L	liquid density, kg/m³
σ	surface tension of liquid, N/m
ϕ	generalized Thiele modulus
χ_L	Lockhart-Martinelli parameter

Acknowledgements

The present work was supported by the Natural Science Foundation of China under Grant No. 20106005, and co-supported by the General Petrochemical Cooperation of China (SINOPEC) through Contract No. 201085.

References

Charpentier, J. C., & Favier, M. (1975). Some liquid holdup experimental data in trickle-bed reactors for foaming and nonfoaming hydrocarbons. AIChE Journal, 21(6), 1213-1218.

Cheng, Z. M., Anter, A. M., Khalifa, G. M., Hu, J. S., Dai, Y. C., & Yuan, W. K. (2001a). An innovative reaction heat offset operation for a multiphase fixed bed reactor dealing with volatile compounds. Chemical Engineering Science,56(21-22),6025-6030.

Cheng,Z. M. ,Anter,A. M. ,& Yuan,W. K. (2001b). Intensification of phase transition on multiphase reactions. AIChE Journal,47(5),1185-1192.

Cheng,Z. M. ,& Yuan,W. K. (1999). Necessary condition for pulsing flow inception in a trickle bed. AIChE Journal,45(7),1394-1400.

Dudukovic,M. P. ,Larachi,F. ,& Mills,P. L. (1999). Multiphase reactors-revisited. Chemical Engineering Science,54,1975-1995.

El-Hisnawi,A. A. ,Dudukovic,M. P. ,& Mills,P. L. (1981). Trickle-bed reactors: Dynamic tracer tests, reaction studies,and modeling of reactor performance ACS Symposium Series,196,421-440.

Ellman,M. J. ,Midoux,N. ,Wild,G. ,Laurent,A. ,& Charpentier,J. C. (1990). A new,improved liquid holdup correlation for trickle-bed reactors. Chemical Engineering Science,45(7),1677-1684.

Funk,G. A. ,Harold,M. P. ,& Ng,K. M. (1990). A novel model for reaction in trickle beds with flow maldistribution. Industrial and Engineering Chemistry Research,29(5),738-748.

Iliuta,I. ,Thyrion,F. C. ,Bolle,L. ,& Giot,M. (1997). Comparison of hydrodynamic parameters for countercurrent and cocurrent flow through packed beds. Chemical Engineering Technology,20,171-181.

Khadilkar,M. R. ,Wu,Y. X. ,Al-Dahhan,M. H. ,& Dudukovic,M. P. (1996). Comparison of trickle-bed and upflow reactor performance at high pressures: Model predictions and experimental observations Chemical Engineering Science,51(10),2139-2148.

Mills,P. L. ,& Dudukovic,M. P. (1981). Evaluation of liquid-solid contacting in trickle-bed reactors by tracer methods. AIChE Journal,27(6),893-904.

Mills,P. L. ,& Dudukovic,M. P. (1984). A comparison of current models for isothermal trickle-bed reactorsapplication to a model reaction systems. In M. P. Dudukovic & P. L. Mills (Eds.),Chemical and catalytic reactor modeling (pp. 37-59). ACS Symposium Series,Vol. 237,Washington,DC:ACS.

Ng,K. M. ,& Chu,C. F. (1987). Trickle-bed reactors. Chemical Engineering Progress,38(11),55-63.

Ramachandran,P. A. ,& Chaudhari,R. V. (1983). Three-phase catalytic reactors. London: Gordon and Breach Science Publishers,Inc.

Specchia,V. ,& Baldi,G. (1977). Pressure drop and liquid holdup for two phase concurrent flow in packed beds. Chemical Engineering Science,32,515-523.

Stiegel,Q. J. ,& Shah,Y. T. (1977). Backmixing and liquid holdup in a gas-liquid cocurrent upflow packed column. Industrial and Engineering Chemistry Process Design Development,16(1),37-43.

Watson,P. C. ,& Harold,M. P. (1993). Dynamic effects of vaporization with exothermic reaction in a porous catalyst pellet. AIChE Journal,39(6),989-1006.

Yang,X. L. ,Euzen,J. P. ,& Wild,G. (1992). A comparison of the hydrodynamics of packed-bed reactors with cocurrent upflow and downflow of gas and liquid. Chemical Engineering Science,47(5),1323-1325.

Redistribution of Adsorbed VOCs in Activated Carbon under Electrothermal Desorption[*]

Abstract Electrothermal desorption is an electricity-promoted desorption technology developed only in the last decade. It is extremely efficient and straightforward when the adsorbent is electrically conductive, since heating can be achieved by the Joule effect. The volatile organic compound (VOC) vapors desorbing from micropores might redistribute and condense in mesopores with high concentration, which is possible since no dilution occurs. To study this problem, benzene and activated carbon were used as the working system, and a theoretical analysis was developed. In a wide temperature range up to 400℃, no VOC vapor could be condensed in mesopores with the strong micropore adsorption effect. With the weak micropore adsorption effect, however, mesopore condensation will occur, but it only takes place in mesopores smaller than 3nm in diameter, and the amount is generally negligible. To prevent any possible condensation, the desorption temperature should at least equal the liquid boiling point calculated in a 2nm capillary tube.

1 Introduction

As an established technology, adsorption has found wide applications in purification and separation of gas mixtures. For removal and recovery of the volatile organic compound (VOC) components, adsorption on activated carbon (A.C.) is most commonly used since the nonpolar substances have a relatively high affinity for the hydrophobic carbons. The adsorption capacity of VOC in A.C. is usually appreciable, for example, on the order of 0.4 to 0.5mL/g, due to the highly developed porous structure. As an example of an industrial application, a VOC recovery facility with a capacity of 125kg/h in methylene chloride, 2.7kg/h in carbon monoxide, and traces of methanol, formaldehyde, and dichloromethyl ether, has been installed in a company that produces herbicides (Thermatrix Inc., 2001). Nevertheless, the present adsorption technology is still subject to some engineering problems, for example, the long operation time and high energy required for the desorption operation, and in addition, recovery of the VOC is still a difficult task. To develop a novel compact plug-in adsorption device (CPAD), electrothermal desorption should be considered as a good candidate.

Electrothermal desorption shows some distinctive advantages over the conventional steam-heated method. In the latter process, steam is introduced into the system and then comes in contact with the adsorbent. In spite of its high heat-transfer efficiency, the negative effects are obvious (Erpelding and Bart, 1998). Since the steam is in contact with the cold adsorbent, water condensation will frequently occur on the external and internal surfaces of the adsorbent. This may seriously decrease the mass-transfer rate of VOCs from the adsorbent to the gas phase or may even destroy the structure of the adsorbent. Moreover, contamination of water by VOCs would give rise to some additional environmental pollution.

---

[*] Coauthors: Z. M. Cheng, F. D. Yu, G. Grevillot, L. Luo, D. Tondeur. Reprinted from *AIChE Journal*, 2002, 48(5): 1132-1138.

Electrothermal desorption is a promising new solution to such desorption problems when electricity is employed instead of steam (Petkovska et al., 1991; Bonnissel et al., 1998). However, one should keep in mind that VOC may be condensed in the mesopores when it is removed from the micropores. Because VOC is not diluted under electrical heating, and its concentration can be much higher than the typical relative pressure of 0.35, which has been considered as the condition of condensation in the mesopore. The purpose of this article is thus to provide an analysis of the possibility of condensation of VOC in the mesopores when it is released from the micropores during electrothermal desorption.

2 Theoretical Establishment

To establish the theoretical background for the electrothermal desorption of VOC from the adsorbent, a working system composed of benzene and activated carbon may be considered for illustration.

2.1 Phase equilibria within the micro- and mesopores

Because the VOC molecules and the micropores are similar in size, adsorption of gases within micropores usually follows a pore-filling process rather than a surface adsorption mechanism as described by the Langmuir or BET equations. Instead, the Dubinin equations, especially the Dubinin-Astakhov (DA) equation, are often employed (Dubinin, 1989; Do, 1998):

$$W = W_0 \exp\left[-\left(\frac{A}{\beta E_0}\right)^n\right] \tag{1}$$

where, W_0 is the micropore volume; A is the chemical potential defined as $R_g T \ln(p_0/p)$; E_0 is the characteristic energy and is also a measure of the interaction between the adsorbate and adsorbent; n represents the reduction in freedom of the molecular movement due to the restriction by the pore wall; and β denotes the similarity of VOC with benzene, which is measured by the ratio of their molar liquid volumes.

For the purpose of this work, the following model parameters from Do (1998) are utilized:

$$W_0 = 0.457 \text{mL/g}, E_0 = 17.61 \text{kJ/mol}, n = 1.46, \text{and } \beta = 1$$

The preceding parameter values are representative of benzene adsorption over the activated carbon, since the micropore volume of the activated carbon is normally between 0.25 and 0.5mL/g, the characteristic energy is above 15kJ/mol for the VOC, and n is from 1 to 3, depending on the relative magnitude of the adsorbate molecule diameter to the adsorbent pore size.

However, when the pore diameter is between 2 and 50nm, and especially when the reduced pressure is greater than 0.35, the DA equation is no longer valid and should be replaced by a model in describing capillary condensation. The Kelvin equation is the theory used to describe the gas-liquid equilibrium within a capillary tube, but it is questionable when the pore size is only about 2-4nm, since the pore wall potential field will influence the equilibrium. To account for the wall effect, the classic Kelvin equation should be modified (Yoshioka et al., 1997):

$$r_K = \frac{V_M \sigma \cos\theta}{R_g T \ln(p_0/p)} + t \tag{2}$$

This correction is necessary, since the diameter of the benzene molecule is estimated to be 0.595nm, which is not negligible when the pore diameter is only several nanometers.

For simplicity, the film thickness t in Eq. 2 may be taken as the molecular diameter of benzene, which implies that only one layer of molecules is adsorbed onto the pore wall prior to the capillary condensation.

The equilibrium relationships of pore filling and condensation are quite different as compared in Fig. 1 and Fig. 2. In Fig. 1, which is based on the pore-filling mechanism by Eq. 1, we see that the adsorption in the micropore is insensitive to the pressure of VOC, but is sensitive to the temperature. On the contrary, in Fig. 2a, which is obtained by Eq. 2, we see that adsorption due to condensation varies considerably with the gas-phase pressure, and the mesopore may be completely filled when the reduced pressure is over 0.95, even if the temperature is as high as 150℃. In Fig. 2b, the desorption process is not simply the reverse of the adsorption, but shows a hysteresis phenomenon. We should point out that once the VOC condensate is produced in the capillary tube, the evaporation will be very slow because the desorption will be controlled by the vapor-phase diffusion. As an illustration, the evaporation rate of benzene from a Stefan tube to a flowing air stream is 4.3mm/day (Do, 1998), which means the desorption time spent on evaporation will be on the order of 5 to 6h, if the carbon pellet diameter is about 1.0mm.

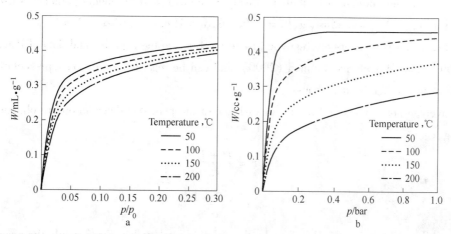

Fig. 1 Isotherm for adsorption of VOC in the micropores

a—Temperature against the reduced pressure; b—Temperature against the absolute pressure

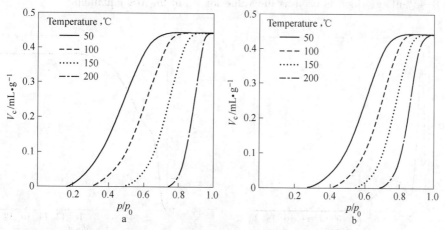

Fig. 2 Thermodynamic condition for phase transition of VOC in the mesopore

a—Evaporation; b—Condensation

Since the VOC storage in the micropore can be as high as 40% of the weight of the adsorbent, condensation will be possible in the case of electrothermal desorption. To give an accurate prediction, we need to make a theoretical analysis based on the adsorbent structure as well as the adsorption thermodynamics.

2.2 Definition of the adsorbent pore structure

To simulate the phase transition and redistribution of the VOC components within the activated carbon, physical properties of this adsorbent should be first given.

Since the macropore only acts as a transportation channel rather than the effective adsorption site, a description of its pore-size distribution is not necessary in the practical approach; therefore, a double-gamma distribution will give a satisfactory description of the joint distribution of the micro-and mesopores:

$$f(r) = V_{S,1}\frac{\alpha_1^{P_1+1}}{\Gamma(P_1+1)}r^{P_1 e - \alpha_1 r} + V_{S,2}\frac{\alpha_2^{P_2+1}}{\Gamma(P_2+1)}r^{P_2 e - \alpha_2 r} \tag{3}$$

where, $f(r)$ is the pore-size distribution density at radius r, and the model parameters used are defined as $V_{S,1} = V_{S,2} = 0.45\text{mL/g}, P_1 = 7, P_2 = 6, \alpha_1 = 15\text{nm}^{-1}$, and $\alpha_2 = 3\text{nm}^{-1}$.

The preceding parameters are regressed by fitting Eq. 3 to the experimental data of Paulsen et al. (1999) and Mazyck and Cannon (2000), and can be regarded as a good representative for a wide variety of activated carbons.

The average pore radius and the mean variance of the pore-size distribution are obtained according to:

$$\tilde{r} = \frac{P+1}{\alpha} \tag{4}$$

$$\delta = \frac{\sqrt{P+1}}{\alpha} \tag{5}$$

The overall pore-size distribution given by this relation is plotted in Fig. 3, while the accumulative pore-volume distribution is shown in Fig. 4. The total volume for pore diameters less than 2nm is 0.457mL/g, which is used as the value for W_0 in the DA equation.

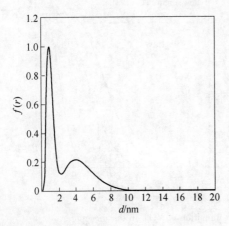
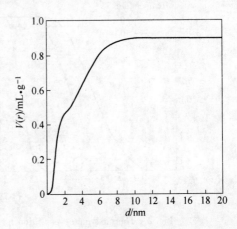

Fig. 3 Double-gamma pore-size distribution   Fig. 4 Integrated pore-volume distribution

2.3 Definition of the boundary between micro- and mesopores

To evaluate the respective volumes occupied by the micropores and the mesopores, it is important to determine the boundary between them. It should be noted that the definition given by the IUPAC is only specific to nitrogen at 77.34K, and thus cannot be applied simply to the present system.

It is known that the transition from micropore to mesopore is characterized by the occurrence of capillary condensation when the chemical potential is reduced to a relatively low value. Considering that the pore-size boundary is related to the change in the adsorption mechanism, the type of the pore can be determined from the terminal point in the adsorption-desorption hysteresis loop. Eiden and Schlünder (1990) studied this phenomenon and introduced a new parameter, A_{Gr}, to denote the chemical potential at this phase transition point. According to their study, A_{Gr} was found to depend only on the molecular size of the adsorbate, but was independent of the adsorbent property. The boundary radius from the micropore to mesopore is therefore obtained from a modified Kelvin equation:

$$r_{Gr} = \frac{2\sigma V_M}{A_{Gr}}\cos\theta + d_M \tag{6}$$

where d_M is the molecular diameter of the adsorbate.

The prediction from Eq. 6 for benzene is plotted in Fig. 5, where it is found that the r_{Gr} value decreases with an increase in temperature. This can be explained by the fact that under the same chemical potential, the VOC is easier to condense under lower temperatures, and thus the critical pore size should be correspondingly increased. It is also found that 2nm is not the exclusive upper limit for the micropore. For example, from 50℃ to 200℃, this critical diameter changes from 3.8nm to 1.9nm. This could be explained by the definition by IUPAC, for instance, the molecular diameter of nitrogen is 0.3nm, while that for benzene is 0.595nm, and correspondingly the upper limits of the micropore for these two adsorbates are 2nm and 3.8nm in diameter, respectively.

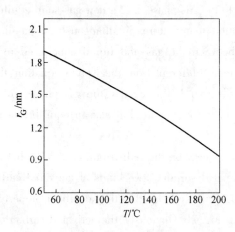

Fig. 5 Upper limit of the micropore as a function of temperature

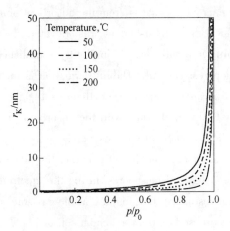

Fig. 6 Upper limit of the mesopore as a function of temperature

For the same reason, the upper limit of the mesopore domain is not 50nm either, as defined by the IUPAC. Fig. 6 shows the upper limit of mesopores in the meaning of capillary condensation denoted by the Kelvin radius, r_K, evaluated from Eq. 2 for temperatures from 50℃ to 200℃.

Since the mesopore volume is the volume contained between r_{Gr} and r_K, it can be calculated by the following equations:

$$V_{meso} = V(r_K) - V(r_{Gr}) \tag{7}$$

where

$$V(r) = V_{S,1}\frac{\Gamma(P_1+1,\alpha_1 r)}{\Gamma(P_1+1)} + V_{S,2}\frac{\Gamma(P_2+1,\alpha_2 r)}{\Gamma(P_2+1)} \tag{8}$$

and

$$\Gamma(a,b) = \int_0^b x^{a-1} e^{-x} dx \tag{9}$$

Because r_{Gr} and r_K, are not constants, the volume of VOC condensed in the mesopores not only depends on the porous structure of the adsorbent, but also on the temperature and on the specific isotherm shown in Fig. 2.

3 Results and Discussion

3.1 Redistribution of VOC among different phases

In most of the adsorption applications, the VOC concentration in the gas phase to be purified is relatively low. Under this condition, there will be no capillary condensation in the mesopore, and only the micropore is involved for adsorption, as can be seen in Fig. 2. When the adsorbent is saturated, it is believed the VOC in the gas phase is in equilibrium with the adsorbed VOC in the solid phase. To give an illustration, we assume the adsorption is performed at 20℃ and under a VOC partial pressure of 0.05bar. From Eq. 1 with the relevant parameters assumed, the adsorbed VOC amount will be 0.442mL/g, which is only a little less than the saturated value of 0.457mL/g.

As the adsorbent temperature is increased for desorption, a certain amount of VOC will leave the micropores, that is, from the adsorbent phase to the gas phase, and a new gas-solid equilibrium will be established at this elevated desorption temperature. Simultaneously, a gas-liquid equilibrium will also be implicitly established between the gas and liquid condensed in the mesopores. Since the Dubinin equation is intrinsically different from the Kelvin equation, there may be an overlap between them in the phase diagram. If this occurs, the gas phase will not only be in equilibrium with the adsorbent phase in the micropore, but also in equilibrium with the liquid phase in the mesopore.

In order to know the influence of the adsorption energy on the redistribution of VOC between micropores and mesopores during the electrothermal desorption, two kinds of activated carbons are investigated. It is found that the gas-solid equilibrium is related to the adsorbent property, as can be observed from a comparison of Fig. 7a and Fig. 7b. Obviously, the activated carbon used in Fig. 7a has a stronger adsorption effect, which is characterized by a higher characteristic energy, E_0, and a larger adsorption exponent, n.

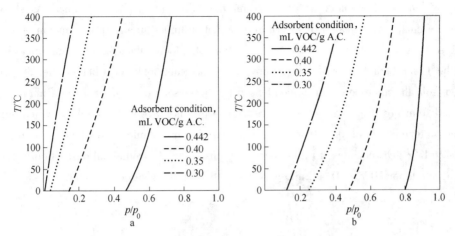

Fig. 7　Phase equilibrium between the gas and the adsorbent
Parameters in the DA equation: a—$W_0 = 0.457\text{mL/g}, E_0 = 17.61\text{kJ/mol}, n = 1.46$, b—$W_0 = 0.457\text{mL/g}, E_0 = 10.40\text{kJ/mol}, n = 1.10$

Fig. 8 shows the relationship between the thermodynamic condition and the maximum critical pore diameter available for the inception of capillary condensation. Since the upper limit for the micropore is approximately 2nm in diameter, there will be no condensation below this pore diameter, and hence the condensation is only possible in the region right to the 2nm profile shown in Fig. 8.

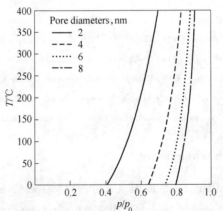

Fig. 8　Phase equilibrium between the gas and condensed liquid residing in the mesopore of different sizes

Fig. 9 shows the corresponding relationship between the temperature and relative pressure of VOC vapor under a certain amount of VOC adsorbed in the adsorbent. In Fig. 9a, the solid line indicated by the Dubinin equation means 0.442mL VOC/g A.C. can be adsorbed in the micropore if the temperature and pressure are located on the Dubinin line. However, any decrease in pressure or increase in temperature will make the VOC desorbed from the micropore space, as can be deduced from Fig. 7. The dashed line indicated by the Kelvin equation gives the temperature and pressure conditions of VOC necessary for condensation in the mesopore of 2nm in diameter. Since 2nm is the lower limit for the mesopore, any increase in temperature or decrease in pressure from the Kelvin line will make the VOC condensation impossible. Therefore, no VOC will be condensed for this kind of adsorbent. However, when a "weak" adsorbent is employed (Eiden and Schlünder, 1990), a somewhat different result is obtained. As shown in Fig. 9b, the condition indicated by the Dubinin equation under an adsorption amount of 0.40mL VOC/g A. C. is equivalent to the Kelvin equation at a pore diameter of 3nm. Since 3nm is above the lower boundary of the mesopore, it is important to know whether the VOC vapor released from the micropore can be condensed by the mesopores between 2 and 3nm. It can be calculated from Fig. 4 that the integral pore volume in this range is 0.069mL/g, which is larger than the volume of 0.042mL/g that could be released from the micropores. Therefore, there will be 0.042mL/g VOC condensed in the mesopore during the electrothermal desorption. However, it just accounts for 10% of 0.442mL/g, which is the totally adsorbed amount.

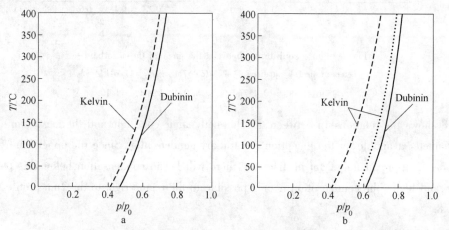

Fig. 9 Equilibrium distribution of VOC between micropore and mesopore

a—VOC in the micropores, W = 0.442cc/g (The DA equation parameters: W_0 = 0.457cc/g; E_0 = 17.61kJ/mol; n = 1.46; – – –VOC in the mesopores, the largest condensation diameter d = 2nm);

b— VOC in the micropores, W = 0.40cc/g (The DA equation parameters: W_0 = 0.457cc/g; E_0 = 10.40kJ/mol; n = 1.10; – – –VOC in the mesopores, the largest condensation diameter d = 2nm; ······VOC in the mesopores, the largest condensation diameter d = 3nm)

3.2 Minimum desorption temperature

To prevent any possible capillary condensation of the VOC species inside of the mesopore, the boiling point of liquid phase should be selected as the minimum desorption temperature.

The shape of the gas-liquid interface inside of a capillary depends on whether the outside vapor is condensed or the inside liquid is evaporated. Since the desorption process follows a similar mechanism to evaporation, the liquid boiling point should be predicted according to the Cohan equation rather than the Kelvin equation (Cheng et al., 2001):

$$\frac{p}{p_0} = \exp\left(-\frac{\sigma V_M}{R_g T} \cdot \frac{1}{r}\right) \tag{10}$$

The relationship between boiling point and the capillary size is shown in Fig. 10. As is expected, the boiling point increases rapidly when the pore diameter is decreased to 2nm. Obviously, the boiling point of 131℃ corresponding to the 2nm tube will be the minimum desorption temperature so as to avoid any possible condensation of benzene under an operation pressure of 1.0 bar.

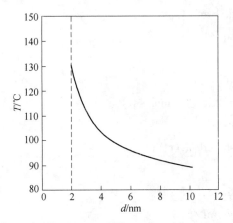

Fig. 10　Relationship between pore diameter and evaporation temperature

4　Conclusions

As a new kind of temperature swing adsorption (TSA) technology, which has been developed only in the last decade, electrothermal desorption shows distinct advantages over the traditional methods. Based on a phase-equilibrium analysis, the present work has established a method for evaluating the possible redistribution of VOC in different scales of porosities. Because of the strong interaction between the VOC vapor and the micropore as described by the Dubinin equation, the VOC vapor can hardly condense in the mesopore under the desorption temperature. It can be considered that this kind of electrically based operation can greatly intensify the VOC desorption, and the operation can also benefit from the high concentration of VOC and zero pollution to the environment.

Acknowledgements

The present work was performed in the framework of a PRA project in environmental protection (French-Chinese cooperative program). One of the authors, Zhen-Min Cheng, is deeply indebted to the French embassy in Beijing for providing the postdoctoral scholarship.

Notation

A	chemical potential, $R_g T\ln(p_0/p)$
A_{Gr}	chemical potential at the transition point from adsorption to condensation
d	pore diameter, m
d_M	diameter of a molecule, m
E_0	characteristic energy, J/mol
f	pore volume distribution density function
n	adsorption exponent in the Dubinin equation
p	gas pressure, bar
p_0	saturated gas pressure, bar

P	parameter in Eq. 4 and Eq. 5
P_1, P_2	parameters in Eq. 3
r	pore radius, m
$\tilde{r}$	average pore radius, m
R_g	universal gas constant, J/(mol·K)
r_K	the Kelvin radius, m
t	thickness of adsorbed layer, m
T	temperature, K
V_M	molar liquid volume, mL/g
V_{meso}	mesopore volume, mL/g
$V_{S,1}, V_{S,2}$	parameters in Eq. 3
W	amount adsorbed, mL/g
W_0	limiting pore volume, mL/g

Greek letters

α	parameter in Eq. 4 and Eq. 5
α_1, α_2	parameters in Eq. 3
β	similarity factor in Eq. 1
δ	distribution variance, m
Γ	gamma function
θ	angle of contact
σ	surface tension, N/m

References

Bonnissel, M., L. Luo, and D. Tondeur, "Fast Thermal Swing Adsorption Using Thermoelectric Devices and New Adsorbent," Proc. Int. Conf. on Fundamentals of Adsorption, F. Meunier, ed., Paris, p. 1065 (1998).

Cheng, Z. M., A. M. Anter, and W. K. Yuan, "Intensification of Phase Transition on Multiphase Reactions," AIChE J., 47, 1185 (2001).

Do, D. D., Adsorption Analysis: Equilibrium and Kinetics, Series on Chemical Engineering, Vol. 2, Imperial College Press, London (1998).

Dubinin, M. M., "Fundamentals of the Theory of Adsorption in Micropores of Carbon Adsorbents: Characteristics of Their Adsorption Properties and Microporous Structures," Carbon, 27, 457 (1989).

Eiden, U., and E. U. Schlünder, "Adsorption Equilibria of Pure Vapors and Their Binary Mixtures on Activated Carbon: I. Single-Component Equilibria," Chem. Eng. Process., 28, 1 (1990).

Erpelding, R., and H. J. Bart, "Modelling and Pilot-Scale Experiments of the Steam Desorption Process with Organic Solvents Loaded Activated Carbon Column," Proc. Int. Conf. on Fundamentals of Adsorption, F. Meunier, ed., Paris, p. 1029 (1998).

Mazyck, D. W., and F. S. Cannon, "Overcoming Calcium Catalysis During the Thermal Reactivation of Granular Activated Carbon Part I: Steam-Curing Plus Ramped-Temperature N_2 Treatment," Carbon, 38, 1785 (2000).

Paulsen, P. D., B. C. Moore, and F. S. Cannon, "Applicability of Adsorption Equations to Argon, Nitrogen and

Volatile Organic Compound onto Activated Carbon," Carbon,37,1843 (1999).

Petkovska, M., D. Tondeur, G. Grevillot, J. Granger, and M. Mitrovic, "Temperature-Swing Gas Separation with Electrothermal Desorption Step," Sep. Sci. Technol.,26.425 (1991).

Thermatrix Inc., http://www.thermatrix.com/(2001).

Yoshioka, T., M. Miyahara, and M. Okazaki, "Capillary Condensation Model Within Nano-Scale Pores Studied with Molecular Dynamics Simulation," J. Chem. Eng. Jpn.,30,274 (1997).

Process Flow Diagram of an Ammonia Plant as a Complex Network[*]

Abstract Complex networks have attracted increasing interests in almost all disciplines of natural and social sciences. However, few efforts have been afforded in the field of chemical engineering. An example of complex technological network, investigating the process flow of an ammonia plant (AP) is presented. The AP network is a small-world network with scale-free distribution of degrees. Adopting Newman's maximum modularity algorithm for the detection of communities in complex networks, evident modular structures are identified in the AP network, which stem from the modular sections in chemical plants. In addition, it is found that the resultant AP tree exhibits excellent allometric scaling.

Key words ammonia plant, complex network, small-world effect, scale free, modular sections

1 Introduction

Complex systems are ubiquitous in natural and social sciences. The behavior of complex system as a whole is usually richer than the sum of its parts, and it is lost if one looks at the constituents separately. Complex systems evolve in a self-adaptive manner and self-organize to form emergent behaviors due to the interactions among the constituents of a complex system at the microscopic level. The study of complexity has been witnessed in almost all disciplines of social and natural sciences (see, for instance, the special issue of *Nature* on this topic in 2001[1]). However, engineers seem a little bit indifferent as if engineering is at the edge of the science of complexity. Ottino argues that "engineering should be at the center of these developments, and contribute to the development of new theory and tools"[2], and chemical engineering is facing new opportunities[3]. The topological aspects of complex systems can be modeled by complex networks, where the constituents are viewed as vertices or nodes, and an edge is drawn between two vertices if their associated constituents interact in certain manners. In recent years, complex networks have attracted extensive interests, covering biological systems, social systems, information systems, and technological systems[4-6]. Complex networks possess many interesting properties. Most complex networks exhibit small-world traits[7], and are scale free where the distributions of degrees have powerlaw tails[8]. In addition, many real networks have modular structures or communities[9]. The fourth intriguing feature of some real networks reported recently is the self-similarity[10]. The studies of complex networks have extensively broadened and deepened our understanding of complex systems.

In the field of chemical engineering, chemical reactions and transports of mass, energy and momentum have been the traditional domains for about five decades, where the topological

---

[*] Coauthors: Jiang Zhiqiang, Zhou Weixing, Xu Bing. Reprinted from *AIChE Journal*, 2007, 53(2): 423-428.

properties are of less concerns. Amaral and Ottino have considered two examples for which the way constituents of the system are linked determines transport and the dynamics of the system, that is, food webs and cellular networks[11]. In this article, we present an example of complex technological network in traditional chemical engineering, studying the topological properties of the process flow of an ammonia plant.

The network studied here is abstracted from the process flow diagram of the Ammonia Plant of Jiujiang Chemical Fertilizer Plant (Jiangxi Province, China). The scale of the plant is 1000MT/D. In the construction of the Ammonia Plant network (AP network), towers, reactors, pumps, heat exchangers, and connection points of convergence and bifurcation of pipes are regarded as vertices. Only the equipments and pipes carrying raw materials, byproducts, and products are considered in the construction of network. The utility flows are not included in the network. The pipes connecting the vertices are treated as edges. The AP network constructed has 505 vertices and 759 edges.

2 AP Network Exhibits Small-World Effect

The average minimum path length is among the most studied quantity in complex networks[4-6]. When regarding the AP network as an undirected network, we compute the average minimum path length $\langle l \rangle = 7.76$ with a standard deviation $\sigma_1 = 2.65$. We find that the distribution of l is Gaussian. The skewness is 0.17, and the kurtosis excess is 0.01, which is close to the theoretical value 0 of a Gaussian distribution. The average minimum path length, and its fluctuation can also be estimated by a Gaussian fit to the data, which presents $\langle l \rangle = 7.85$, and $\sigma_1 = 2.74$.

In most small-world networks, the average minimum path length is somewhat larger than that for a random graph[7]. It is interesting to compare the average minimum path length of the real ammonia plant network with that of model networks. The null model is the maximally random networks with the same number of nodes and the same degree sequence as the real network. There are several methods for the generation of random graphs with prescribed degree sequences, and the chain switching method gives accurate results with acceptable computational time[12], which was used in the detection of richclub structure[13], and is the very null model in the statistical tests of network topological properties[14]. Adopting the chain switching method, we have generated 12400 random networks. The average minimum path length l_{rand} of each model network is calculated. It is found that $l_{rand} = 5.90 \pm 0.07$. What is striking is that the maximum of l_{rand} is 6.15, much smaller than $\langle l \rangle = 7.76$.

The clustering coefficient C_i of vertex i is a measure of the cluster structure indicating how much the adjacent vertices of the adjacent vertices of i are adjacent vertices of i. Mathematically, C_i is defined by:

$$C_i = \frac{E_i}{k_i(k_i - 1)/2} \quad (1)$$

where, E_i is the number of edges among the adjacent vertices of i[7]. The average clustering coefficient $C = \langle C_i \rangle$ is 0.083, which is comparable to other technological networks[5]. Using the same database of the maximally random networks, we find that $C_{rand} = 0.0075 \pm 0.0036$, and the maximum clustering coefficient of random networks is 0.025, much smaller than $C = 0.083$

for the ammonia plant network. This is the evidence supporting that the AP network is a small-world network[7].

3 AP Network is Scale-Free

The degree k of a vertex of a network is the number of edges connected to the vertex. Degree distributions of vertices are perhaps the most frequently investigated in the literature of complex networks[4-6]. The degree distributions of scale-free networks have fat tails following power laws:

$$p(k) \sim k^{-(\mu+1)} \tag{2}$$

Several mechanisms of scale-free distributions have been proposed, such as preferential attachment and its variants[4] and fitness of vertices[15,16]. In order to estimate the probability distribution of a physical variable empirically, several approaches are available. For a possible power-law distribution with fat tails, cumulative distribution or log-binning technique are usually adopted. A similar concept to the complementary distribution, called rank-ordering statistics[17], has the advantage of easy implementation, no information loss, and being less noisy.

Consider N observations of variable k sampled from a distribution whose probability density is $p(k)$. Then the complementary distribution is $P(y > k) = \int_k^\infty p(y) dy$. We sort the n observations in nonincreasing order such that $k_1 \geq k_2 \geq \cdots \geq k_i \geq \cdots \geq k_n$ where n is the rank of the observation. It follows that $NP(k \geq k_n)$ is the expected number of observations larger than or equal to k_n, that is:

$$NP(k \geq k_n) = n \tag{3}$$

If the probability density of variable k follows a power law that $p(k) \sim k^{-(\mu+1)}$, then the complementary distribution $P(k) \sim k^{-\mu}$. An intuitive relation between k_n and n follows:

$$k_n \sim n^{-1/\mu} \tag{4}$$

A rigorous expression of Eq. 4, by calculating the most probable value of k_n from the probability that the n-th value equals to k_n gives[17]:

$$k_n \sim \frac{\mu N + 1}{\mu n + 1} \tag{5}$$

When $\mu n \gg 1$ or equivalently $1 \ll n \leq N$, we retrieve Eq. 4. A plot of $\ln k_n$ as a function of $\ln n$ gives a straight line with slope $-1/\mu$ with deviations for the first a few ranks if k is distributed according to a power law of exponent μ. We note that the rank-ordering statistics is nothing but a simple generalization of Zipf's law[17-19], and has wide applications, such as in linguistics[20], the distribution of large earthquakes[21], time-occurrences of extreme floods[22], to list a few. More generally, rank-ordering statistics can be applied to probability distributions other than power laws, such as exponential or stretched exponential distributions[23], normal or log-normal distributions[17], and so on.

In Fig. 1 is shown the rank-ordering analysis of the indegree, out-degree and all-degree of the AP network in log-log plot. We see that the AP network is scale-free. Linear regression of $\ln k_n$ against $\ln n$ gives the following exponents: $1/\mu = 0.419 \pm 0.010$ for all-degree, $1/\mu = 0.407 \pm 0.009$ for in-degree, and $1/\mu = 0.4443 \pm 0.008$ for out-degree. Therefore, we have $\mu = 2.39 \pm$

0.06 for all-degree, $\mu = 2.46 \pm 0.05$ for in-degree, and $\mu = 2.31 \pm 0.04$ for out-degree.

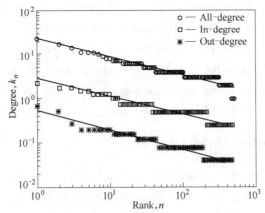

Fig. 1 Rank-ordering analysis of the in-degree, outdegree, and all-degree of the AP network.
(We have translated vertically the in-degree line by 4 and the out-degree line by 25 for better presentation.
The lines are the best fit of tail distribution to Eq. 4 (Color figure can be viewed in the online issue,
which is available at www.interscience.wiley.com))

4 Modular Structures in the AP Network

4.1 Brief review

In the recent years, much attention has been attracted to the modular clusters or community structures of real networks, such as metabolic networks[24,25], food webs[26,27], social networks[26,28-30], to list a few. There are rigorous definitions for community. A strong community is defined as a subgraph of the network requiring more connections within each community than with the rest of the network, while in a weak community the total number of connections of within-community vertices is larger than the number of connections of the vertices in the community with the rest of the network[31,32]. However, in most cases in the literature, community is only fuzzily defined in the sense that the connections within communities are denser than between communities.

Different types of algorithms have been developed for the detection of communities[9]. Sokal and Michener proposed the average-linkage method[33], which was extended to the hierarchical clustering algorithm later[34]. In 1995, Frank developed a method for direct identification of nonoverlapping subgroups[35], which was applied to detect compartments in food webs[27]. Girvan and Newman proposed a divisive algorithm that uses edge betweenness centrality to identify the boundaries of communities[26,36], which is now widely known as GN algorithm. Based on the concept of network random walking, Zhou used dissimilarity index to delimit the boundaries of communities, which was reported to outperform the algorithm based on the concept of edge betweenness centrality[37,38], An alternative divisive algorithm of Radicchi et al. is based on the edge clustering coefficient, related to the number of cycles that include a certain edge[31,32]. Another well-known algorithm is Newman's maximum modularity algorithm, which is a type of agglomerative algorithm[39,40].

Many other algorithms have been presented, for instance, the Kernighan-Lin algorim[41], the spectral method which takes into account weights and link orientations and its improvement[42,43], the resistor network approach which concerns the voltage drops[44], the information centrality algorithm that consists in finding and removing iteratively the edge with the highest information centrality[45], a fast community detection algorithm based on a q-state Potts model[46], an aggregation algorithm for finding communities of related genes[47], the maximum modularity algorithm incorporated with simulated annealing[25], the agent-based algorithm[48], the shell algorithm[49], and the algorithm based on random Ising model and maximum flow[50].

4.2 Community structure of the AP network

We apply Newman's maximum modularity algorithm[39,40] to study the community structure of the AP network. The resultant AP tree is illustrated in Fig. 2, which is not in the form of dendrogram. The shapes of the vertices represent different sections of the process flow of the AP: SGP section-oil (solid circles), rectisol section-oil (horizontal ellipses), CO-shift section (vertical ellipses), synthesis and refrigeration section (open circles), air separation section (triangles), nitrogen washing section (vertical diamonds), steam superheater unit (horizontal diamonds), ammonia storage and tank yard (rectangles), and equipments of waste treatment (squares). The maximum value of the modularity is $Q = 0.794$, which is among the largest peak modularity values reported for different networks (if not the largest), and, thus, indicates a very strong community structure in the investigated network.

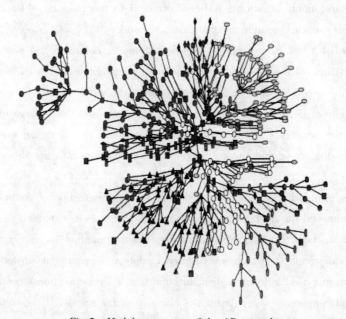

Fig. 2 Modular structure of the AP network

(The shapes of the vertices represent different sections of the process flow of the Ammonia Plant.
This figure was produced with Pajek[59] (Color figure can be viewed in the online issue,
which is available at www.interscience.wiley.com))

It has been found that random graphs and scale-free networks have modularity with analytic expressions[51], which allows us to check if the modularity observed in the AP network is mathematically significant or not. Since the modularity of a scale-free network with $S = 505$ nodes and connectivity $m = 749/505$ is:

$$Q_{SF} = \left(1 - \frac{2}{\sqrt{S}}\right)\left(a + \frac{1-a}{m}\right) = 0.6773 \quad (6)$$

which is again much smaller than $Q = 0.794$, showing that the modularity of the AP network is significant. Note that $a = 0.165$[51]. For an Erdös-Rényi random graph with $S = 505$ nodes and connection probability $p = 749/(505 \times 504/2) = 0.0059$, the maximal modularity is:

$$Q_{ER} = \left(1 - \frac{2}{\sqrt{S}}\right)\left(1 - \frac{2}{pS}\right)^{2/3} = 0.6995 \quad (7)$$

The fact that Q is greater than Q_{ER} indicates that the modular structure extracted from the AP network could not be attributed to the fluctuation of random graphs and, is, thus, still very significant.

Alternatively, we can use the same null model which employs the chain switching algorithm to generate maximally random networks with the same degree sequence of the AP network. We find that $Q_{rand} = 0.440 \pm 0.009$, and the maximum of the modularity of model networks is 0.469, which is much smaller than $Q = 0.794$. This test provides further evidence that the modular structure in the AP network is statistically significant.

The modular structures of chemical plant networks do not come out as a surprise. In a chemical plant, raw materials are fed into the process flow network and react from one section to another successively, although there are feedbacks from later sections. In general, flows are denser within a workshop section than between sections. Therefore, a section is naturally a community. In Fig. 2, most of the vertices in a given section are recognized to be members of a same community. The vertices of the storage and tank yard (rectangles) are the most dispensed in Fig. 2. This is expected since these tanks are linked from and to different sections in the process, which shows the power of Newman's maximum modularity algorithm for community detection.

4.3 Allometric scaling of the AP tree

The network shown in Fig. 2 is actually a tree. Trees exhibit intriguing intrinsic properties other than nontree networks, among which is the allometric scaling. Allometric scaling laws are ubiquitous in networking systems such as metabolism of organisms and ecosystems river networks, food webs, and so forth[52-58]. The original model of the allometric scaling on a spanning tree was developed by Banavar et al[56]. The spanning tree has one root and many branches and leaves, and can be rated as directed from root to leaves. Mathematically, each node of a tree is assigned a number 1 and two values, and are defined for each node in a recursive manner as follows:

$$A_i = \sum_j A_j + 1 \quad (8a)$$

and

$$S_i = \sum_j S_j + A_i \quad (8b)$$

where, j stands for the nodes linked from i[56]. In a food web, i is the prey and j's are its predators (thus, the nutrition flows from i to j's). The allometric scaling relation is then highlighted by the power law relation between S_i and A_i:

$$S \sim A^\eta \qquad (9)$$

For spanning trees extracted from transportation networks, the power law exponent η is a measure of transportation efficiency[56,58]. The smaller is the value of η, the more efficient is the transportation. Any spanning tree can range in principle between two extremes, that is, the chain-like trees and the star-like trees. A chain tree has one root and one leaf with no branching. Let's label leaf vertex by 1, its father by 2, and so forth. The root is labeled by n for a chain-like tree of size n. The recursive relations (Eq. 8) become $S_i = S_{i-1} + A_i$ and $A_i = A_{i-1} + 1$ with termination conditions $A_1 = S_1 = 1$. It is easy to show that $A_i = i$ and $S_i = i(i+1)/2$. Asymptotically, the exponent $\eta = 2^-$ for chain-like trees. For star-like trees of size n, there are one root and $n-1$ leaves directly connected to the root. We have $A = S = 1$ for all the leaves and $A = n$ and $S = 2n - 1$ for the root. It follows approximately that $\eta = 1^+$. Therefore, $1 < \eta < 2$ for all spanning trees.

We note that not all trees have such allometric scaling. Consider for instance the classic Cayley with n generations, where the root is the first generation. The A and S values of the vertices of the same generation are identical. If we denote A_i and S_i for the vertices of the $(n+1-i)$-th generation, the iterative equations are $A_{i+1} = 2A_i + 1$ and $S_{i+1} = 2S_i + A_{i+1}$, resulting in $A_i = 2^i - 1$, and $S_i = (i-1)2^i + 1$. This leads to $S = [\log_2(A+1) - 1]A + \log_2(A+1)$. Obviously, there is no power-law dependence between A and S.

We apply this framework on the AP tree. The calculated S is plotted in Fig. 3 as a function of A. A nice power-law relation is observed between S and A. A linear fit of $\ln S$ against $\ln A$ give $\eta = 1.21$ with regression coefficient 0.998. The trivial point ($A = 1$; $S = 1$) is excluded from the fitting[58]. This value of η is slightly larger than $\eta = 1.13 \sim 1.16$ for food webs[58], but much smaller than $\eta = 1.5$ for river networks[56]. This analysis is relevant when the flux in the pipes and reactors are considered for the investigation of the transportation efficiency, as an analogue to the river network and biological network[52,56].

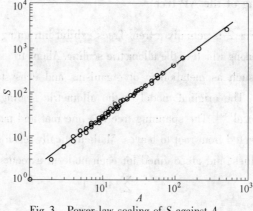

Fig. 3   Power-law scaling of S against A

(The line represents the power-law fit to the data)

5 Concluding Remarks

We have studied a complex technological network extracted from the process flow of the Ammonia Plant of Jiujiang Chemical Fertilizer Plant in Jiangxi Province of China. We have shown that the ammonia plant network is a small-world network in the sense that its minimum average path length $\langle l \rangle = 7.76$, and global clustering coefficient $C = 0.083$ are, respectively, larger than their counterparts $l_{rand} = 5.90 \pm 0.07$ and $C_{rand} = 0.0075 \pm 0.0036$ of a set of 12,400 maximally random graphs having the same degree sequences of the real AP network. We found that the shortest path lengths between two arbitrary vertices are distributed according to a Gaussian formula. The distribution of degrees follows a power law with its exponent being $\mu = 2.31 \sim 2.46$, indicating that the AP network is scale-free.

We have reviewed briefly diverse existing algorithms for the detection of community structures in complex networks, among which Newman's maximum modularity algorithm is applied to the AP network. The extracted modular structures have a very high-modularity value $Q = 0.794$ signaling the significance of the modules, which is confirmed by statistical tests. These modular structures are well explained by the workshop sections of the ammonia plant. We have constructed a spanning tree based on the community identification procedure and found that the resultant AP tree exhibits excellent allometric scaling with an exponent comparable to the universal scaling exponent of food webs.

In summary, we have studied the topological properties of the AP network from chemical engineering. More sophisticated networks can be constructed from process flows in chemical industry. There are still other open problems even in this small AP network, such as the origin of the scalefree feature, what we can learn from these topological features, robustness and sensitivity analysis on the mass flows to find out bottlenecks in the process or figure out how jamming of the nodes or cascade failure of the system can occur, to list but a few. We hope that this work will attract more affords in this direction. Further researches on complex networks containing information of transports and reactions will unveil useful properties, and benefit the field practically and theoretically.

Acknowledgements

The authors thank gratefully Hai-Feng Liu for providing the process flow diagram of Jiujiang Ammonia Plant. This work was partially supported by National Basic Research Program of China (No. 2004CB217703), the Project Sponsored by the Scientific Research Foundation for the Returned Overseas Chinese Scholars, State Education Ministry of China, and NSFC/PetroChina through a major project on multiscale methodology (No. 20490200).

References

[1] Ziemelis K. Complex systems. Nature. 2001;410;241.
[2] Ottino JM. Engineering complex systems. Nature. 2004;427;399.
[3] Ottino JM. New tools, new outlooks, new opportunities. AIChE J. 2005;51;1840-1845.
[4] Albert R, Barabási AL. Statistical mechanics of complex networks. Rev Mod Phys. 2002;74;47-97.
[5] Newman MEJ. The structure and function of complex networks. SIAM Rev. 2003;45 (2);167-256.
[6] Dorogovtsev SN, Mendes JFF. Evolution of Networks: From Biological Nets to the Internetand the WWW. Oxford: Oxford University Press;2003.
[7] Watts DJ, Strogatz SH. Collective dynamics in "small-world" networks. Nature. 1998;393;440-442.
[8] Barabási AL, Albert R. Emergence of scaling in random networks. Science. 1999;286;509-512.

[9] Newman MEJ. Detecting community structure in networks. Eur Phys J B. 2004;38:321-330.

[10] Song CM, Havlin S, Makse HA. Self-similarity of complex networks. Nature. 2005;433:392-395.

[11] Amaral LAN, Ottino JM. Complex systems and networks: Challenges and opportunities for chemical and biological engineering. Chem Eng Sci. 2004;59:1653-1666.

[12] Milo R, Kashtan N, Itzkovitz S, Newman MEJ, Alon U. Uniform generation of random graphs with arbitrary degree sequences. http://arxiv.org/abs/cond-mat/0312028.

[13] Colizza V, Flammini A, Serrano MA, Vespignani A. Detecting richclub ordering in complex networks. Nat Phys. 2006;2:110-115.

[14] Amaral LAN, Guimera R. Lies, damned lies and statistics. Nat Phys. 2006;2:75-76.

[15] Caldarelli G, Capocci A, De Los Rios P, Muňoz MA. Scale-free networks from varying vertex intrinsic fitness. Phys Rev Lett. 2002;89:258702.

[16] Servedio VDP, Caldarelli G, Butta P. Vertex intrinsic fitness: How to produce arbitrary scale-free networks. Phys Rev E. 2004;70:056126.

[17] Sornette D. Critical Phenomena in Natural Sciences-Chaos, Fractals, Self-organization and Disorder: Concepts and Tools, 2st Ed. Berlin: Springer; 2004.

[18] Zipf G. Human Behavior and the Principle of Least Effort. Massachusetts: Addison-Wesley Press; 1949.

[19] Mandelbrot BB. The Fractal Geometry of Nature. New York: W. H. Freeman; 1983.

[20] Mandelbrot BB. Structure formelle des textes et communication. Word. 1954;10:1-27.

[21] Sornette D, Knopoff L, Kagan YY, Vanneste C. Rank-ordering statistics of extreme events: Application to the distribution of large earthquakes. J Geophys Res. 1996;101:13883-13893.

[22] Mazzarella A, Rapetti F. Scale-invariance laws in the recurrence interval of extreme floods: An application to the upper Po river valley (northern Italy). J Hydrology. 2004;288:264-271.

[23] Laherrere J, Sornette D. Stretched exponential distributions in nature and economy: "Fat tails" with characteristic scales. Eur Phys J B. 1998;2:525-539.

[24] Ravasz E, Somera AL, Mongru DA, Oltvai AN, Barabási AL. Hierarchical organization of modularity in metabolic networks. Science. 2002;297:1551-1555.

[25] Guimerà R, Amaral LAN. Functional cartography of complex metabolic networks. Nature. 2005;433:895-900.

[26] Girvan M, Newman MEJ. Community structure in social and biological networks. Proc Natl Acad Sci USA. 2002;99:7821-7826.

[27] Krause AE, Frank KA, Mason DM, Ulanowicz RE, Taylor WW. Compartments revealed in food-web structure. Nature. 2003;426:282-285.

[28] Guimerà R, Danon L, Díaz-Guilera A, Giralt F, Arenas A. Self similar community structure in a network of human interactions. Phys Rev E. 2003;68:065103.

[29] Gleiser PM, Danon L. Community structure in jazz. Adv Complex Systems. 2003;6:565-573.

[30] Newman MEJ. Coauthorship networks and patterns of scientific collaboration. Proc Natl Acad Sci USA. 2004;101:5200-5205.

[31] Radicchi F, Castellano C, Cecconi F, Loreto V, Parisi D. Defining and identifying communities in networks. Proc Natl Acad Sci USA. 2004;101:2658-2663.

[32] Castellano C, Cecconi F, Loreto VD. Parisi D, Radicchi F. Self-contained algorithms to detect communities in networks. Eur Phys J B. 2004;38:311-329.

[33] Sokal RR, Michener CD. A statistical method for evaluating systematic relationships. Univ Kans Sci Bull. 1958;38:1409-1438.

[34] Eisen MB, Spellman PT, Brown PO, Botstein D. Cluster analysis and display of genomewide expression patterns. Proc Natl Acad Sci USA. 1998;85:14863-14868.

[35] Frank KA. Identifying cohesive subgroups. Soc Networks. 1995;17:27-56.

[36] Newman MEJ, Girvan M. Finding and evaluating community structure in networks. Phys Rev E. 2004;69: 026113.

[37] Zhou HJ. Network landscape from a brownian particle's perspective. Phys Rev E. 2003;67:041908.

[38] Zhou HJ. Distance, dissimilarity index, and network community structure. Phys Rev E. 2003;67:061901.

[39] Newman MEJ. Fast algorithm for detecting community structure in networks. Phys Rev E. 2004;69:066133.

[40] Clauset A, Newman MEJ, Moore C. Finding community structure in very large networks. Phys Rev E. 2004;70: 066111.

[41] Kernighan BW, Lin S. An efficient heuristic procedure for partitioning graphs. Bell System Technical J. 1970; 49:291.

[42] Capocci A, Servedio VDP, Caldarelli G, Colaiori F. Communities detection in large networks. Lect Notes Comp Sci. 2004;3243:181-187.

[43] Donetti L, Muňoz MA. Improved spectral algorithm for the detection of network communities. in:Proceedings of the 8th Granada Seminar – Computational and Statistical Physics. 2005;1-2.

[44] Wu F, Huberman BA. Finding communities in linear time:A physics approach. Eur Phys J B. 2004;38: 331-338.

[45] Fortunato S, Latora V, Marchiori M. Method to find community structures based on information centrality. Phys Rev E. 2004;70:056104.

[46] Reichardt J, Bornholdt S. Detecting fuzzy community structures in complex networks with a Potts model. Phys Rev Lett. 2004;93:218701.

[47] Wilkinson DM, Huberman BA. A method for finding communities of related genes. Proc Natl Acad Sci USA. 2004;101:5241-5248.

[48] Young M, Sager J, Csárdi G, Hága P. An agent-based algorithm for detecting community structure in networks. cond-mat/0408263.

[49] Bagrow JP, Bollt EM. A local method for detecting communities. Phys Rev E. 2005;72:046108.

[50] Son SW, Jeong H, Noh JD. Random field Ising model and community structure in complex networks. Euro Phys J B. 2006;50:431-437.

[51] Guimerà R, Sales-Pardo M, Amaral LAN. Modularity from fluctuations in random graphs and complex networks. Phys Rev E. 2004;70:025101.

[52] West GB, Brown JH, Enquist BJ. A general model for the origin of allometric scaling laws in biology. Science. 1997;276:122-126.

[53] Enquist BJ, Brown JH, West GB. Allometric scaling of plant energetics and population density. Nature. 1998; 395:163-165.

[54] West GB, Brown JH, Enquist BJ. The fourth dimension of life:Fractal geometry and allometric scaling of organisms. Science. 1999;284:1677-1679.

[55] Enquist BJ, West GB, Charnov EL, Brown JH. Allometric scaling of plant energetics and population density. Nature. 1999;401:907-911.

[56] Banavar JR, Maritan A, Rinaldo A. Size and form in efficient transportation networks. Nature. 1999;399: 130-132.

[57] Enquist BJ, Economo EP, Huxman TE, Allen AP, Ignace DD, Gillooly JF. Scaling metabolism from organisms to ecosystems. Nature. 2003;423:639-642.

[58] Garlaschelli D, Caldarelli G, Pietronero L. Universal scaling relations in food webs. Nature. 2003;423: 165-168.

[59] de Nooy W, Mrvar A, Batagelj V. Exploratory Social Network Analysis with Pajek. Cambridge:Cambridge University Press;2005.

Determination of Effectiveness Factor of a Partial Internal Wetting Catalyst from Adsorption Measurement[*]

Abstract The effect of partial internal wetting of catalyst pellets on apparent reaction kinetics at elevated temperatures and pressures is investigated experimentally and by modeling for benzene hydrogenation. A new method that combines adsorption and chemical reaction is introduced to study the kinetics influenced by capillary condensation of reagents at steady-state conditions. It is shown that the extent of liquid filling in the pellet interior has a critical effect on the global kinetics, and the current state of the catalyst depends on the history. Under certain conditions two steady states of the effectiveness factor exist. Moreover, either a decrease in temperature or increase in total pressure can increase the effectiveness factor. The model exhibits good agreement with experimental results.

Key words adsorption, capillary condensation, diffusion, partial internal wetting, effectiveness factor, modeling, benzene hydrogenation

1 Introduction

Partial internal wetting of porous catalysts often takes place in trickle bed reactors during hydrotreatment of some unsaturated hydrocarbons with substantial volatility (Kim and Kim, 1981; Bhatia, 1988; Harold, 1993; Ostrovskii et al., 1994; Watson and Harold, 1994; Kulikov et al., 2001; Fatemi et al., 2002; Koptyug et al., 2002). These reactions usually release large quantities of heat that causes the vaporization of liquid both outside and inside the catalyst. Under certain conditions, the liquid on the external surface of the catalyst may be completely evaporated, that is, the catalyst pellet is surrounded with vapor-phase reactants and products; however, the liquid still remains within some pores of the catalyst (Cheng et al., 2001a, b). Thus, the catalyst pellet is partially internally wetted.

In fact, when a porous catalyst is in contact with vapor-phase reactants containing condensable components, the larger pores of the catalyst are gas-filled, whereas the smaller pores may be filled with condensate due to capillary condensation (Kim et al., 1996; Do, 1998). Consequently, the vapor-liquid equilibrium exists in the interior pores of the catalyst pellet. When partial internal wetting of the catalyst occurs, the observed reaction rate may change significantly with the degree of partial internal wetting because the vapor-phase diffusivities are about 10^4 times larger than those of the liquid phase. Therefore, the fractional liquid filling is a key affecting the global reaction rate. Previous studies have obtained the extent of liquid filling by theoretical calculation without experimental verification (Jaguste and Bhatia, 1991; Fatemi et al., 2002). In

---

[*] Coauthors: Zhou ZhiMing, Cheng ZhenMin, Li Zhuo. Reprinted from *Chemical Engineering Science*, 2004, 59: 4305-4311.

the present work, a new method by combining adsorption and reaction is proposed to study the global kinetics. The adsorption investigation is used to establish a relationship between the fractional liquid filling and the bulk conditions while the chemical reaction is to study the effect of partial internal wetting of the catalyst on the global reaction rate. Benzene hydrogenation to cyclohexane is applied as a sample reaction on the ground that benzene and cyclohexane are volatile, and the reaction itself is exothermic. The experiments are carried out in steady-state operation and the mathematical model considering intraparticle diffusion, reaction, pore structure and vapor-liquid equilibrium within the catalyst pellet is performed.

2 Experimental

2.1 Catalyst structure

The catalyst is 0.5wt% Pd supported over γ-alumina pellet of 4mm in diameter. Pore size distribution (PSD) of the catalyst is measured by the standard nitrogen adsorption-desorption method with a Micrometrics ASAP 2010 apparatus (Fig. 1). The internal pores of the catalyst are within the mesopore range (2-50nm), according to the IUPAC classification, and the pore size follows a log-normal distribution by:

$$f(r) = \frac{V_0}{\sqrt{2\pi}r\omega}\exp\{-[\ln(r/r_a)]^2/2\omega^2\} \tag{1}$$

where, $f(r)$ is the PSD density at radius r; V_0 is the total pore volume of the catalyst. The model parameters used are regressed by fitting Eq. 1 with the experimental data, $V_0 = 0.47\text{cm}^3/\text{g}$, $\omega = 0.376$, and $r_a = 3.546\text{nm}$.

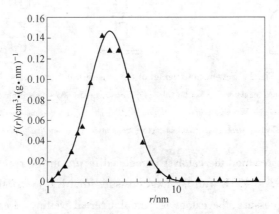

Fig. 1 Pore size distribution of catalyst pellet

2.2 Adsorption experiments

Studies on the adsorption of vapor-phase benzene and cyclohexane on porous solids under elevated temperature and pressure have not yet been reported (Gregg and Sing, 1982; Asnin et al., 2001). In this work, adsorption experiments are conducted on a fixed bed from 393 to 453K over a total pressure range of 0.3-1.0MPa. The adsorption bed is put in an oil bath where variations in temperature can be controlled within 0.2℃. The adsorbate, benzene or cyclohexane is vaporized and preheated to a desired temperature before leading to the adsorption bed. The equilibrium adsorption quantity equals the difference between the accumulative amount of inlet adsorbate and that of outlet adsorbate.

2.3 Kinetics experiments

Global reaction rate for benzene hydrogenation accompanied by capillary condensation is measured in the apparatus shown in Fig. 2. An internal recycle gradientless reactor of the Berty type is used, wherein the effect of external diffusion can be eliminated while the impeller speed exceeds 2500 r/min. The temperature difference between the inlet and outlet gas of the reactor is less than 0.5℃. The feed liquid, high purity benzene (>99.5%) is introduced by a metering pump and vaporized, mixed with hydrogen (>99.5%) and preheated to a desired temperature before entering into the reactor. The total pressure is exactly controlled by a back-pressure regulator. At the outlet, the reacted gas is cooled by a condenser and the liquid product is removed in a separator from the gas. The product is analyzed by gas chromatograph, where cyclohexane content is measured.

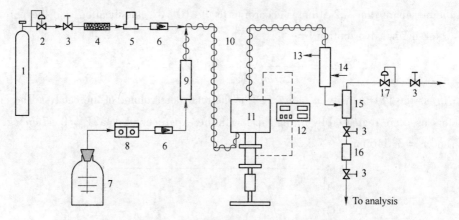

Fig. 2 Schematic diagram of the experimental apparatus

1—Hydrogen; 2—Pressure regulator; 3—Needle valve; 4—Dehydrator; 5—Mass-flow controller; 6—Check valve;
7—Benzene container; 8—Metering pump; 9—Vaporizer; 10—Heater and thermal insulator; 11—Reactor;
12—Reactor controller; 13—Cooling water; 14—Condenser; 15—Separator; 16—Sampling pipe; 17—Back-pressure regulator

Before hydrogenation reaction the catalyst is reduced *in situ* in the reactor. The reaction temperature varies from 393 to 453K with the total pressure from 1.0 to 2.0MPa. For a given reaction temperature and pressure, the extent of partial internal wetting of catalyst pellets is controlled by adjusting the mole ratio of benzene to hydrogen under certain bulk conditions.

3 Mathematical Model

3.1 Adsorption and desorption process

Gas adsorption and desorption on mesopore solids can be divided into three stages (Gregg and Sing, 1982; Do, 1998): (1) formation of a monolayer of adsorbate molecules on the pore surface at low relative pressure; (2) formation of multilayers of adsorbate on the top of the first layer with increased relative pressure; (3) capillary condensation or evaporation following stage (2). Except where specified otherwise, the interior pores of the catalyst are supposed to be cylindrical.

For a given pressure p, there exists a critical pore radius r_c. Pores with radii lower than r_c will be filled with the adsorbate, while those greater than r_c will only be covered with layers of adsorbate on their surfaces. The adsorbed amount V under a given pressure p is then:

$$V = \begin{cases} \int_{r_{min}}^{r_{max}} \dfrac{r^2 - (r-t)^2}{r^2} f(r) dr, & \text{for } r_c < r_{min} \\ \int_{r_{min}}^{r_c} f(r) dr + \int_{r_c}^{r_{max}} \dfrac{r^2 - (r-t)^2}{r^2} f(r) dr, & \text{for } r_{min} < r_c < r_{max} \end{cases} \quad (2)$$

where, t is the thickness of the adsorbed layer; r_{min} and r_{max} are the lower and upper limits of the mesopore range, i.e. $r_{min} = 1\,\text{nm}$, $r_{max} = 25\,\text{nm}$. The thickness can then be obtained from the modified Halsey equation (Androutsopoulos and Salmas, 2000):

$$t = \left(\frac{M_W}{a_m N_A \rho}\right) \left[\frac{5}{\ln\left(\frac{p_0}{p}\right)}\right]^{1/3} \left(\frac{p}{p_0}\right)^m \quad (3)$$

where, p_0 is the saturation vapor pressure of the adsorbate; m is an unknown parameter that needs to be determined by experiments. According to adsorption theories, r_c can be calculated by the following equation:

$$r_c = t + r_k \quad (4)$$

where, r_k is the Kelvin radius. It can be calculated by the classical Kelvin equation and Cohan equation, respectively:

$$r_k = \begin{cases} \dfrac{2 M_W \sigma}{R_g T \rho \ln\left(\frac{p_0}{p}\right)} & (5a) \\[2ex] \dfrac{M_W \sigma}{R_g T \rho \ln\left(\frac{p_0}{p}\right)} & (5b) \end{cases}$$

Eq. 5a is applied for desorption or evaporation process and Eq. 5b for adsorption or condensation. The above equations together with PSD of the catalyst describe adsorption of benzene and cyclohexane on the catalyst.

3.2 Reaction influenced by partial internal wetting

Benzene hydrogenation to cyclohexane on the catalyst is as follows: $C_6H_6(B) + 3H_2(H) \rightarrow C_6H_{12}(C)$. When partial internal wetting of catalyst pellets exists, interactions of reaction, diffusion and vapor-liquid equilibrium within the catalyst make the modeling complex. For simplicity, the following assumptions are made:

(1) The external diffusion is eliminated due to the internal recycle gradientless reactor used.

(2) The internal temperature gradients are negligible, which is well established for catalytic reactions, especially for steady-state conditions (Froment and Bischoff, 1990; Pan and Zhu, 1998). In addition, the low reaction rate in the range of experimental conditions is one of the reasons.

(3) Intraparticle transport is dominated by vapor-phase diffusion and liquid-phase transport is negligible, which is justifiable on the ground that the vapor-phase diffusivities are much larger

than those of the liquid phase.

Based on the adsorption studies, the fractional liquid filling of the catalyst is used to describe the vapor-liquid equilibrium, which is defined by:

$$f_w = \int_{r_{min}}^{r_c} f(r) \, dr / V_0 \tag{6}$$

For a spherical catalyst, the diffusion/reaction equation for any species i at the steady-state condition is given by:

$$\frac{1}{R^2} \cdot \frac{d}{dR}(R^2 N_i) = \alpha_i R_{Bt} \tag{7}$$

where, α_i is the stoichiometric coefficient, $\alpha_B = -1$, $\alpha_H = -3$ and $\alpha_C = 1$. The flux N_i for the catalyst with a distribution of pore size can be described by (Hessari and Bhatia, 1996):

$$N_i = \frac{1}{\tau} \int_{r_c}^{r_{max}} \left(-D_{im} c_t \frac{dy_i}{dR} \right) f(r) \, dr \tag{8}$$

where, τ is the tortuosity factor; D_{im} the mixture diffusivity. According to the dusty-gas model (Mason and Malinauskas, 1983), D_{im} is given by:

$$D_{im} = \left[\sum_{i \neq j} \frac{1}{D_{ij}} \left(y_j - \frac{N_j}{N_i} y_i \right) + \frac{1}{D_{ki}} \right]^{-1} \tag{9}$$

An implicit assumption for Eq. 9 is isobaric conditions within the catalyst and negligible surface diffusion, which proved valid in cyclohexene hydrogenation to cyclohexane (Jaguste and Bhatia, 1991), a reaction similar to benzene hydrogenation. For chemical reactions, the steady-state flux ratios are determined by the stoichiometry (Froment and Bischoff, 1990), $N_j / N_i = \alpha_j / \alpha_i$.

The overall reaction rate of a partially internally wetted catalyst can be evaluated as the weighed average of those for the wet and dry regions of the catalyst interior according to the fractional liquid filling, that is (Sedricks and Kenney, 1973; Lemcoff et al., 1988; Li et al., 1999):

$$R_{Bt} = f_w R_{Bl} + (1 - f_w) R_{Bg} \tag{10}$$

where, R_{Bl} is the liquid-phase reaction rate. Kinetics parameters for the liquid-phase reaction are estimated under conditions of complete pore filling with liquid (Zhou et al., 2004), given by:

$$R_{Bl} = 1.57 \times 10^3 e^{-\frac{43880}{R_g T}} p_H \tag{11}$$

which shows that the liquid phase reaction rate is zero-order in benzene concentration and first-order with respect to hydrogen pressure. R_{Bg} is the gas-phase reaction rate. Studies on the kinetics of gas-phase reaction are carried out in the region excluding capillary condensation (Zhou et al., 2003). The kinetics is of Langmuir-Hinshelwood type, expressed as:

$$R_{Bg} = \frac{7.17 \times 10^2 \exp\left(-\frac{28250}{R_g T}\right) p_B p_H^{0.5}}{\left[1 + 1.80 \times 10^{-4} \exp\left(\frac{41170}{R_g T}\right) p_B\right]\left[1 + 2.95 \times 10^{-2} \exp\left(-\frac{9370}{R_g T}\right) p_H^{0.5}\right]} \tag{12}$$

In the presence of local vapor-liquid equilibrium, the intrinsic rates in the vapor phase and the liquid phase should be identical (Harold and Watson, 1993). However, the two rates may be different when adsorption is not rapid (Jaguste and Bhatia, 1991; Bukhavtsova and Ostrovskii, 1998; Fatemi, et al., 2002).

The boundary conditions for the mass transfer model are given by:

$$R = 0, \frac{dy_B}{dR} = \frac{dy_H}{dR} = \frac{dy_C}{dR} = 0 \quad (13)$$

and

$$R = R_p, y_B = y_B^0, y_H = y_H^0, y_C = y_C^0 \quad (14)$$

Additionally, in the vapor phase, the following equation must be satisfied

$$\sum_i y_i = 1 \quad (15)$$

The effectiveness factor, which presents the ratio of the actual rate of reaction in the presence of diffusion resistance to that of the vapor-filled catalyst at bulk conditions, is expressed as:

$$\eta = \frac{S_p(-N_B)_{R=R_p}}{V_p(R_{Bg})_{R=R_p}} \quad (16)$$

where, S_p is external surface area of the catalyst pellet, and V_p the pellet volume. Eq. 1-Eq. 16 form the mathematical model for a diffusion-reaction process accompanied by partial internal wetting of the catalyst at steady-state conditions, whereby the mole fraction profiles y_B, y_H, y_C and the effectiveness factors at different bulk conditions can be yielded. The above equations result in a two-point boundary value problem, which is solved simultaneously by the orthogonal collocation technique together with adaptive Simpson quadrature and the Newton-Raphson method.

4 Results and Discussions

4.1 Partial internal wetting measurements

The physical properties that are used in solving the model are provided in the literature (Reid et al., 1987; Yaws, 1999). Fig. 3 shows the adsorption isotherm of benzene on the catalyst, where a well-known hysteresis loop is observed. Agreement between the calculated curves and the experimental results indicates the validity of the model for the adsorption equilibrium. Parameter m is in the range 0.10-0.15 with the temperature from 393 to 453K. It may be anticipated that with an increase in temperature, it will be more difficult for vapor-phase benzene to be con-

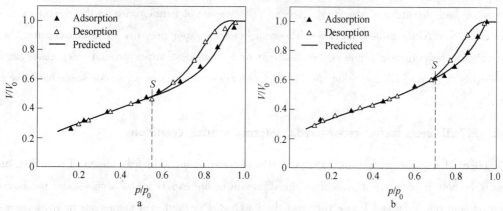

Fig. 3 The amount of benzene adsorbed on catalyst pellets versus the relative pressure
a—433K; b—453K

densed in pores of the catalyst. This is indeed observed experimentally, as seen in Fig. 3a and b. The beginning point of the hysteresis loop(shown by S) moves backwards in terms of the relative pressure with temperature variation from 433 to 453K. The relative pressures of benzene at point S are about 0.55 at 433K and 0.70 at 453K, respectively. More importantly, the size of hysteresis loop becomes narrower with increased temperature, which is justifiable because the critical pore radius above which capillary condensation can occur increases with temperature (Sonwane and Bhatia, 1999).

In addition, both experimental and calculated results show that the adsorption isotherms of benzene and cyclohexane are almost identical due to their similar properties(Fig. 4). This fact reminds us that for benzene hydrogenation, the mixture of benzene and cyclohexane within the catalyst pores can be taken as a single component.

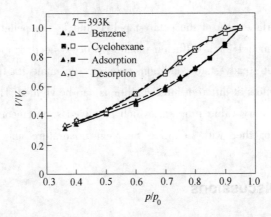

Fig. 4 Comparison of adsorption of benzene and cyclohexane on catalyst pellets

Fig. 5 gives an example of the relationship between the fractional liquid filling f_w and the mole fraction of benzene in the bulk phase at different reaction temperatures. The total pressure is 1.0MPa. This figure shows that the fractional liquid filling of the catalyst depends not only on the bulk conditions, such as mole fraction of benzene, temperature and pressure, but on the history as well. As a rule, the adsorption hysteresis results in the different fractional liquid filling at the same bulk conditions(Kim et al., 1996). The impact of temperature on the degree of liquid filling is directly caused by the variation of benzene vapor pressure with temperature, that is, with the reduction in temperature, benzene or cyclohexane vapor pressure decreases, and consequently, liquid filling within the catalyst precedes in terms of y_B at the same total pressure.

4.2 Effectiveness factor under partial internal wetting conditions

The effect of capillary condensation on the effectiveness factor is clearly observed from Fig. 6a and b. In both figures, the arrows show the direction of the experimental process, and the numbers present the fractional liquid filling of the catalyst. The predicted values are in good agreement with the experimental data. The tortuosity factor used in the model is assumed to be 4, which is reasonable for the catalyst used here(Satterfield, 1970).

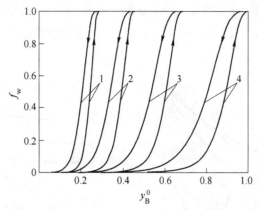

Fig. 5　Effect of mole fraction of benzene in the bulk phase on the
fractional liquid filling of catalyst pellets($p_t = 1.0$ MPa)
1—393K;2—413K;3—433K;4—453K

Increasing y_B^0 results in an increase of f_w, which leads to a decrease of the mass transfer rate into the catalyst because the liquid-phase diffusivities are much lower than those of the vapor-phase. This eventually gives rise to a decrease of the effectiveness factor directly. To be mentioned, under certain circumstances two steady states are found by way of increasing or decreasing the partial pressure of benzene in the bulk phase. The two different processes do not take place as exact reverses of each other. It means that the current state of the catalyst relies on the history, i.e., whether it is a liquid evaporation process or a vapor condensation process. As shown in Fig. 6, the increase of y_B^0 corresponds to the vapor-phase benzene condensing into liquid within pores, whereas decreasing y_B^0 leads to the condensed liquid-like benzene evaporating from some pores into vapor.

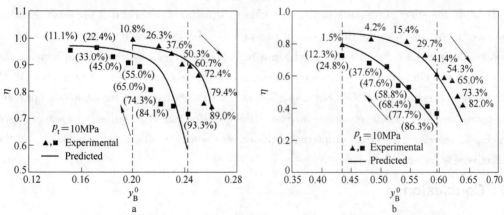

Fig. 6　Variation of effectiveness factor with mole fraction of benzene at bulk conditions
a—393K;b—433K

It is obvious that both the reaction rate and the diffusion rate increase with increase in reaction temperature. However, the effect of temperature on the former is much stronger than the latter. It is not reaction but diffusion that plays a more and more important role in determining the global reaction rate with increased temperature. Therefore, increasing temperature results in decrease of the effectiveness factor, which can be clearly seen by comparison of Fig. 6a with b. Fig. 7 presents the profiles of mole fraction of benzene, hydrogen and cyclohexane with decrease in the degree of liquid filling, corresponding to the lower branch in Fig. 6b.

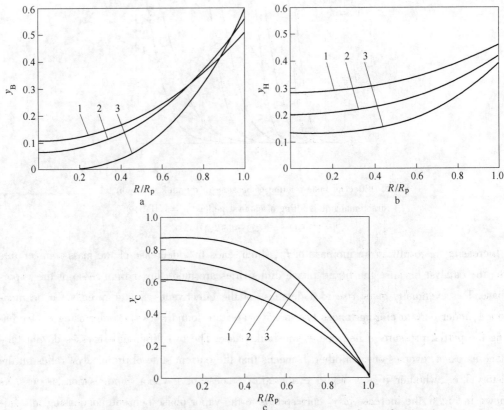

Fig. 7　Distribution of benzene, hydrogen and cyclohexane in the catalyst interior
f_w: 1—0.38; 2—0.68; 3—0.86
($T = 433$ K, $p_t = 1.0$ MPa)

To study the effect of pressure variation another experiment is performed at the same temperature as Fig. 6b, but with the different total pressure at 2.0 MPa. Fig. 8 depicts the experimental results. Comparison between Fig. 6b and Fig. 8 reveals that increasing pressure can increase the effectiveness factor. For example, when f_w is equal to 0.5, the effectiveness factor is about 0.55 for the total pressure at 1.0 MPa, but above 0.6 at 2.0 MPa. At the same conditions except reaction pressure, as the total pressure is increased, the diffusion rate is accordingly increased based on Eq. 8. Therefore, the effect of internal diffusion on the total reaction rate is reduced, and the effectiveness factor is increased.

5　Conclusions

In this work, the global kinetics of benzene hydrogenation accompanied by partial internal wetting of the mesoporous catalyst is experimentally and theoretically studied by a new method that combines adsorption and chemical reaction. An adsorption model taking account of multilayer adsorption and capillary condensation is established, by which the fractional liquid filling of the catalyst pores can be obtained. Another model for the global kinetics influenced by capillary condensation has also been developed. This model is based on intraparticle diffusion, reaction, and vapor-liquid equilibrium within the catalyst. Both models predict with a reasonable accuracy using our experimental results.

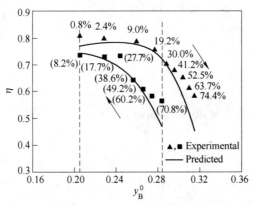

Fig. 8 Variation of effectiveness factor with mole fraction of benzene at bulk conditions
($T = 433$ K, $p_t = 2.0$ MPa)

It is shown that the adsorption hysteresis gives rise to the different fractional liquid filling of the catalyst at the same conditions, which consequently results in two steady states of the reaction rate. Reaction temperature and pressure have different effects on the effectiveness factor. Increasing temperature or decreasing pressure will reduce the effectiveness factor.

Notation

a_m molecular area of the adsorbate, $Å^2$
c_t total gas-phase concentration in the bulk phase, mol/m^3
D_{ij} binary diffusivity, m^2/s
D_{im} mixture diffusivity for component i, m^2/s
D_{ki} Knudsen diffusivity for component i, m^2/s
f_w fractional liquid filling of the catalyst
M_w molecular weight, g/mol
N_A Avogadro constant
p pressure, MPa
p_0 vapor pressure, MPa
p_t total pressure in the bulk phase, MPa
r pore radius, m
r_c critical pore radius, m
r_k Kelvin radius, m
R radial position of the catalyst, m
R_{Bg} gas-phase reaction rate for benzene, mol/(g·h)
R_{Bl} liquid-phase reaction rate for benzene, mol/(g·h)
R_{Bt} total reaction rate for benzene, mol/(g·h)
R_g universal gas constant, J/(mol·K)
R_p catalyst radius, m
t thickness of the adsorbed layer, m
T temperature, K

V_0	total pore volume of the catalyst, m^3/g
y_i	mole fraction of component i in vapor phase
y_i^0	y_i at the bulk conditions

Greek letters

α_i	stoichiometric coefficient of component i
η	effective factor
ρ	density, kg/m^3
σ	surface tension, N/m
τ	tortuosity factor

Subscripts

B	benzene
C	cyclohexane
H	hydrogen

Acknowledgements

Financial supports from the Natural Science Foundation of China (NSFC) under Grant No. 20106005 and the General Petrochemical Cooperation of China (SINOPEC) through Grant No. 201085 are gratefully acknowledged.

References

Androutsopoulos, G. P., Salmas, C. E., 2000. A new model for capillary condensation-evaporation hysteresis based on a random corrugated pore structure concept: prediction of intrinsic pore size distribution. 2. Model application. Industrial and Engineering Chemistry Research 39, 3764-3777.

Asnin, L. D., Fedorov, A. A., Chekryshkin, Yu. S., 2001. Adsorption of chlorobenzene and benzene on γ-Al_2O_3. Russian Chemical Bulletin, International Edition 50(1), 68-72.

Bhatia, S. K., 1988. Steady state multiplicity and partial internal wetting of catalyst particles. A. I. Ch. E. Journal 34(6), 969-979.

Bukhavtsova, N. M., Ostrovskii, N. M., 1998. Catalytic reaction accompanied by capillary condensation. 3. Influence on reaction kinetics and dynamics. Reaction Kinetics and Catalysis Letters 65(2), 321-329.

Cheng, Z. M., Anter, A. M., Khalifa, G. M., et al., 2001a. An innovation reaction heat offset operation for a multiphase fixed bed reactor with volatile compounds. Chemical Engineering Science 56, 6025-6030.

Cheng, Z. M., Anter, A. M., Yuan, W. K., 2001b. Intensification of phase transition on multiphase reactions. A. I. Ch. E. Journal 47(5), 1185-1192.

Do, D. D., 1998. Adsorption Analysis: Equilibrium and Kinetics. Series on Chemical Engineering, vol. 2. Imperial College Press, London.

Fatemi, S., Moosavian, M. A., Abolhamd, G., et al., 2002. Reaction rate hysteresis in the hydrotreating of thiophene in wide- and narrow-pore catalysts during temperature cycling. The Canadian Journal of Chemical Engineering 80, 231-238.

Froment, G. F., Bischoff, K. B., 1990. Chemical Reactor Analysis and Design. second ed.. Wiley, New York.

Gregg, S. J. , Sing, K. S. W. , 1982. Adsorption, Surface Area and Porosity. second ed. . Academic Press, London.

Harold, M. P. , 1993. Impact of wetting on catalyst performance in multiphase reaction systems. In: Computer Aided Design of Catalysts. Marcel Dekker, New York. pp. 391-469.

Harold, M. P. , Watson, P. C. , 1993. Bimolecular exothermic reaction with vaporization in the half-wetted slab catalyst. Chemical Engineering Science 48(5), 981-1004.

Hessari, F. A. , Bhatia, S. K. , 1996. Reaction rate hysteresis in a single partially internally wetted catalyst pellet: experimental and modelling. Chemical Engineering Science 51(8), 1241-1256.

Jaguste, D. N. , Bhatia, S. K. , 1991. Partial internal wetting of catalyst particles: hysteresis effects. A. I. Ch. E. Journal 37(5), 650-660.

Kim, D. H. , Kim, Y. G. , 1981. An experimental study of multiple steady states in a porous catalyst due to phase transition. Journal of Chemical Engineering of Japan 14(4), 311-317.

Kim, D. H. , Bhang, K. C. , Kim, J. C. , 1996. Multiple steady states induced by capillary condensation in a porous catalyst. In: Asia-Pacific Chemical Reaction Engineering Forum, Beijing, China. pp. 687-692.

Koptyug, I. V. , Kulikov, A. V. , Lysova, A. A. , et al. , 2002. NMR imaging of the distribution of the liquid phase in a catalyst pellet during α-methylstyrene evaporation accompanied by its vapor-phase hydrogenation. Journal of the American Chemical Society 124, 9684-9685.

Kulikov, A. V. , Kuzin, N. A. , Shigarov, A. B. , et al. , 2001. Experimental study of vaporization effect on steady state and dynamic behavior of catalyst pellets. Catalysis Today 66, 255-262.

Lemcoff, N. O. , Cukierman, A. L. , Martínez, O. M. , 1988. Effectiveness factor of partially wetted catalyst particles: evaluation and application to the modeling of trickle bed reactors. Catalysis Reviews Science and Engineering 30(3), 393-456.

Li, Y. X. , Cheng, Z. M. , Liu, L. H. , et al. , 1999. Catalytic oxidation of dilute SO_2 over activated carbon coupled with partial liquid phase vaporization. Chemical Engineering Science 54, 1571-1576.

Mason, E. A. , Malinauskas, A. P. , 1983. Gas Transport in Porous Media: The Dusty Gas Model. Elsevier, Amsterdam.

Ostrovskii, N. M. , Bukhavtsova, N. M. , Duplyakin, V. K. , 1994. Catalytic reactions accompanied by capillary condensation. 1. Formulation of the problems. Reaction Kinetics and Catalysis Letters 53(2), 253-259.

Pan, T. , Zhu, B. , 1998. Study on diffusion-reaction process inside a cylindrical catalyst pellet. Chemical Engineering Science 53(5), 933-946.

Reid, R. C. , Praustnitz, J. M. , Poling, B. E. , 1987. The Properties of Gases and Liquids. fourth ed. . McGraw-Hill, New York.

Satterfield, C. N. , 1970. Mass Transfer in Heterogeneous Catalysis. MIT Press, Cambridge, MA.

Sedricks, W. , Kenney, C. N. , 1973. Partial wetting in trickle-bed reactors — the reduction of crotonaldehyde over a palladium catalyst. Chemical Engineering Science 28, 559-568.

Sonwane, C. G. , Bhatia, S. K. , 1999. Analysis of criticality and isotherm reversibility in regular mesoporous materials. Langmuir 15, 5347-5354.

Watson, P. C. , Harold, M. P. , 1994. Rate enhancement and multiplicity in a partially wetted and filled pellet: experimental study. A. I. Ch. E. Journal 40(1), 97-111.

Yaws, C. L. , 1999. Chemical Properties Handbook. McGraw-Hill, New York.

Zhou, Z. M. , Li, Z. Cheng, Z. M. , et al. , 2003. Kinetics of vapor-phase benzene hydrogenation on $Pd/\gamma\text{-}Al_2O_3$. Petrochemical Technology 32(5), 392-397(in Chinese).

Zhou, Z. M. , Cheng, Z. M. , Li, Z. , et al. , 2004. Kinetics of liquid-phase benzene hydrogenation on $Pd/\gamma\text{-}Al_2O_3$. Journal of East China University of Science and Technology 30(1), 1-5(in Chinese).

Deep Removal of Sulfur and Aromatics from Diesel through Two-stage Concurrently and Countercurrently Operated Fixed-bed Reactors*

Abstract A two-stage concurrently and countercurrently operated fixed-bed reactors were investigated to remove sulphur and aromatics in producing ultra-clean diesel fuel. Under similar operating conditions, the sulfur concentration in the produced oil through countercurrent operation is much less than under concurrent flow. In a typical run with countercurrent flow at 6.0MPa and 360℃, the sulphur content could be removed from 13841μg/g to 5.7μg/g with nitrogen from 457μg/g to 1.8μg/g and the total aromatics from 41.4% to 26%. One-dimensional heterogeneous model was established to simulate the distribution of all the components involved along the reactor length in the gas and liquid phases, which revealed that the low H_2S concentration at the reactor exit region under countercurrent flow is critical to the ultra low sulphur extent in the produced oil.

Key words fixed-bed reactor, hydrodesulfurization, countercurrent flow, heterogeneous model, diesel fuel

1 Introduction

Due to the stronger competitive adsorption effect of H_2S than dibenzothiophene (DBT) compounds on the catalyst active sites, the hydrodesulfurization (HDS) reaction rate could be reduced by 70% even under low H_2S pressure of less than 0.01MPa according to a recent study of Kabe et al. (2001). It is therefore necessary to proceed the HDS reaction under H_2S concentration as low as possible, and it is needed to remove the produced H_2S from the reacting system efficiently. In this context, a countercurrent flow reactor is the ideal choice in principle, as was presented by Trambouze in 1990. Therefore, countercurrent operation is an advanced technology in HDS, and this technology has been patented by ABB Lummus Crest Inc. (Relly and Hamilton, 1993) and in commercialization for over 20 years (Reilly et al., 1973; Jackson, 2003). Moreover, it has been reported recently that the Lummus Arosat hydrogenation process could produce ultra-clean diesel with sulfur to 1μg/g and aromatics to 4 vol% (Babich and Moulijn, 2003). In spite of its technical importance of countercurrent operation in HDS, it should be noted that, detailed reaction engineering investigation has not been conducted so far, since most of the research is focused on the single-stage concurrent flow reactor (Korsten and Hoffmann, 1996; Chowdhury et al., 2002; Avraam and Vasalos, 2003; Pedernera et al., 2003). Therefore, it is the purpose of this work to extend this research in developing advanced clean fuel technology.

2 Theoretical Basis

2.1 Reaction kinetics of HDS

Petroleum feedstock includes a large number of sulfurbearing compounds, and the overall HDS

---

* Coauthors: Cheng ZhenMin, Fang Xiangchen, Zeng Ronghui, Han Baoping, Huang Lei. Reprinted from *Chemical Engineering Science*, 2004, 59: 5465-5472.

reaction is usually expressed as:

$$R - S + H_2 \longrightarrow R + H_2S \tag{1}$$

where R represents the hydrocarbon molecule which is combined with sulfur.

The HDS reaction rate equation which includes the inhibiting effect of H_2S is usually written in the Langmuir-Hinshelwood form:

$$r_{HDS} = \frac{k_{HDS} C_{H_2}^n C_S^m}{1 + K_{H_2S} C_{H_2S}} \tag{2}$$

where

$$k_{HDS} = k_{HDS}^0 \exp\left(-\frac{E_{HDS}}{RT}\right)$$

By assuming the validity of the Eq. 2, the range of reaction orders n and m are within a certain range according to the following consideration:

(1) The order on sulfur: The reaction rate of an individual sulfur compound is normally represented by a first-order kinetics, however, n^{th} order kinetics with respect to the total sulfur concentration is required to describe the global desulfurization rate (Girgis and Gates, 1991). From an analysis made by Sie (1999), the reaction order was found only varies between 1.94 and 2.18, regardless of the concentrations of sulfur in the feed.

(2) The order on H_2: The theoretical reaction order on hydrogen should be 0.5 in view of the chemical dissociation of hydrogen molecule on the catalyst surface. However, if the mass transfer rate of hydrogen is the limiting step, the order should be 1.0. Thus in the real situation, the value n should be in the range of 0.5-1.0.

In this work, two kinds of catalyst were employed in two reactors in series, and their reaction rate kinetics were measured in a Parr autoclave of 1000mL containing a catalyst holding basket in which the catalyst of commercial size could be measured. The measured reaction rates of HDS were shown as follows:

For catalyst FH-DS:

$$r_{HDS} = \frac{8.75 \times 10^{12} \exp\left(\frac{-15455}{T}\right) C_{H_2}^{0.6} C_S^2}{1 + 7 \times 10^4 C_{H_2S}} \tag{3}$$

For catalyst FH-98:

$$r_{HDS} = \frac{5.55 \times 10^{12} \exp\left(\frac{-13050}{T}\right) C_{H_2}^{0.6} C_S^2}{1 + 4 \times 10^5 C_{H_2S}} \tag{4}$$

2.2 Reaction kinetics of HDA

To simplify the modeling work, the total aromatics is conventionally classified into three groups: the mono-, the di-, and the tri-aromatics (Jaffe, 1974; Chowdhury et al., 2002), where the polyaromatics are lumped into triaromatics. The aromatics hydrogenation removal reactions (HDA) network can be expressed as follows:

$$\text{(tri-aromatic)} + H_2 \underset{K_{-tri}}{\overset{K_{tri}}{\rightleftharpoons}} \text{(di-aromatic)} \tag{5}$$

$$\text{(anthracene)} + 2H_2 \underset{K_{-di}}{\overset{K_{di}}{\rightleftharpoons}} \text{(dihydroanthracene)} \tag{6}$$

$$\text{(dihydroanthracene)} + 3H_2 \underset{K_{-mono}}{\overset{K_{mono}}{\rightleftharpoons}} \text{(tetrahydroanthracene)} \tag{7}$$

The three reactions are exothermic and reversible, and their equilibrium constants were regressed through the present work as:

$$\lg K_{mono} = 17.6 - 0.023T \tag{8}$$

$$\lg K_{di} = 11.5 - 0.017T \tag{9}$$

$$\lg K_{tri} = 5.5 - 0.007T \tag{10}$$

where, $K_{mono} = k_{mono}/k_{-mono}$, $K_{di} = k_{di}/k_{-di}$, and $K_{tri} = k_{tri}/k_{-tri}$.

The HDA reaction rates are expressed as:

$$r_{tri} = -k_{tri}C_{tri} + k_{-tri}C_{di} \tag{11}$$

$$r_{di} = k_{tri}C_{tri} - k_{-tri}C_{di} - k_{di}C_{di} + k_{-di}C_{mono} \tag{12}$$

$$r_{mono} = k_{di}C_{di} - k_{mono}C_{mono} - k_{-di}C_{mono} + k_{-mono}C_{naph} \tag{13}$$

$$r_{naph} = k_{mono}C_{mono} - k_{-mono}C_{naph} \tag{14}$$

The reaction rate constants were found as follows:

For catalyst FH-DS:

$$k_{mono} = 6.83 \times 10^4 \exp(-13471/T) \tag{15}$$

$$k_{di} = 3.52 \times 10^3 \exp(-10584/T) \tag{16}$$

$$k_{tri} = 2.82 \times 10^3 \exp(-8900/T) \tag{17}$$

For catalyst FH-98:

$$k_{mono} = 4.72 \times 10^4 \exp(-12890/T) \tag{18}$$

$$k_{di} = 3.34 \times 10^3 \exp(-9796/T) \tag{19}$$

$$k_{tri} = 9.95 \times 10^2 \exp(-8437/T) \tag{20}$$

2.3 Reactor modeling

A one-dimensional plug-flow heterogeneous model was established to simulate the concentration profiles of the reactants and products in the gas, liquid and solid phases. Under concurrent flow condition, the following equations were established:

(1) For the gas phase, the axial profile of hydrogen concentration was accounted by the mass transfer from the gas to the liquid:

$$u_G \frac{d(C_A)_G}{dz} = -(k_L a_G)_A [(C_A)_G/H_A - (C_A)_L] \tag{21}$$

A similar equation can also be written for H_2S, however, the mass transfer direction is from the liquid phase to the gas phase.

(2) For the liquid phase, the concentration profiles of hydrogen and sulfur compounds were described by Eq. 22 and Ep. 23, respectively:

$$u_L \frac{d(C_A)_L}{dz} = (k_L a_G)_A [(C_A)_G/H_A - (C_A)_L] - (k_s a_s)_A [(C_A)_L - (C_A)_S] \tag{22}$$

$$u_L \frac{d(C_B)_L}{dz} = -(k_s a_s)_B [(C_B)_L - (C_B)_S] \tag{23}$$

The axial concentration of H_2S could be described similar to Eq. 22 like H_2, however, it should be noted that H_2S is transferred from the solid to the liquid and then to gas phase. The balance of the mono-, di-, tri-aromatics and naphthene could also be described by Eq. 23, since there is no gas phase existed.

(3) For the solid phase, it is the site where the reactions take place, thus HDS was described by:

$$(k_S a_S)_B [(C_B)_L - (C_B)_S] = \rho_b r_B \qquad (24)$$

The same equation also applies to the HDA reactions.

(4) The temperature profile along the reactor was described by the heat balance:

$$\rho_L u_L c_p \frac{dT}{dz} = \rho_b (r_B \Delta H_B + r_C \Delta H_C) \qquad (25)$$

In the above equations, H_2, organic sulfur and aromatics were respectively denoted by letters A, B and C. The gas and liquid velocities u_L and u_G were the superficial values. The mass transfer coefficients $k_L a_G$ and $k_S a_S$ were defined and obtained following Smith (1981). This set of ordinary differential equations with initial boundary conditions were solved with fourth Runge-Kutta algorithm.

Similar constitute equations but with boundary conditions distributed at the two ends of the reactor could also be established for the countercurrent flow operation. The solution of theses equations was not available with conventional methods such as "shooting" method or finite differences. Instead, the "double alternate integration" method was found most reliable (Trambouze, 1990).

3 Experimental

A two-stage reactor system which is shown in Fig. 1 was installed. The two reactors were packed with different catalysts: catalyst FH-DS ($W-Mo-Ni-Co/Al_2O_3$) with good H_2S tolerance was in the first reactor to remove most of the sulfur, and catalyst FH-98 ($Ni-Mo/Al_2O_3$) was in the second one to finalize the reaction. However, when only one reactor was used, only FH-98 catalyst was used to have both high HDS and HDA activity. The two reactors were linked by a high pressure gas-liquid separator to discharge the produced H_2S. The liquid product from the first reactor flowed into the second reactor. The first reactor was always operated under concurrent flow, while the second one could be under concurrent or countercurrent flow, as shown from the flow diagram.

The reactor system used in this work was kept under isothermal condition. The reactor was 28mm in diameter inside and 500mm in length, and a thermocouple shell of 8mm was installed in the tube center. Trilobe catalyst manufactured by FRIPP was used, which has a dimension of 1.2mm in diameter and 5-8mm in length. If not specified, the catalyst loading in each reactor is 100mL, which is corresponding to a length of 31.8cm, and the total length of the two reactors was 63.6cm. Therefore, axial dispersion effect could be neglected (Tsamatsoulis and Papayannakos, 1998).

The primary crude oil properties are listed in Table 1. The mean molecular weight was estimated to be 226 from its density at 20°C as well as the middle boiling point.

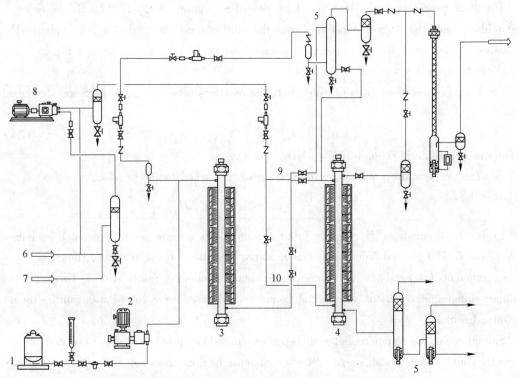

Fig. 1 Schematic flow diagram for a two-stage HDS reactor system operated in concurrent/concurrent and concurrent/countercurrent manners

1—Feedstock tank; 2—Oil pump; 3—First reactor; 4—Second reactor; 5—Gas-liquid separator;
6—Make-up hydrogen; 7—Recycling hydrogen; 8—Hydrogen recycling compressor;
9—Hydrogen inlet for concurrent operation; 10—Hydrogen inlet for countercurrent operation

Table 1 Primary properties of the raw diesel fuel

Density (20℃)/g·cm^{-3}	0.8810	Total sulfur/μg·g^{-1}	13841
Distillation/℃		Total nitrogen/μg·g^{-1}	457
IP/10%	177/241	Paraffin/%	37.8
30%/50%	271/297	Total naphthene/%	20.3
70%/90%	328/359	Total aromatics/%	41.4
95%/EP	368/375	Mono-aromatics/%	16.3
Cetane number	39.1	Di-aromatics/%	21.8
BMCI	42.7	Tri-aromatics/%	3.3

4 Results and Discussion

4.1 Influence of flow direction on reactor performance

A single-stage reactor that was operated under concurrent or countercurrent flow direction was first studied. Under the same operating condition with catalyst FH-98, i.e., at $T = 360℃$, $p = 65\,\text{bar}$, LHSV = 1.8, and H$_2$/Oil volume ratio = 500, the sulfur concentrations in the product were 341 and 245 μg/g with respect to the concurrent flow and countercurrent flow. This means countercurrent flow is superior to concurrent flow due to the simultaneous removal of H_2S from the liquid phase to the gas phase by striping of the gas phase. Nevertheless, the relatively high concentration of sulfur in the product indicates that, with only one reactor either operated concurrently or countercurrently, it is hard to reduce sulfur to the desired ultra low level of less than $30\mu g/g$.

4.2 Two-stage operation

Since a single-stage reactor can only remove sulfur to $250\text{-}350\mu g/g$, a two-stage operation was conducted. The experiments were carried out at two temperature levels, 350 and 360℃. Details of the operating condition are given in Table 2.

Table 2 Comparison of deep hydrodesulfurization under concurrent and countercurrent flow

Condition	Concurrent/concurrent			Concurrent/countercurrent		
(1) Experimental results at 350℃						
p_{H_2}/MPa	6.0	6.0	6.0	6.0	6.0	6.0
T/℃	350	350	350	350	350	350
LHSV/h^{-1}	1.0	1.5	2.0	1.0	1.5	2.0
H_2/oil volume ratio	400/400	400/400	400/400	400/300	400/300	400/300
Properties of product oil						
S content / $\mu g \cdot g^{-1}$	195	601	904	94	434	948
N content / $\mu g \cdot g^{-1}$	9.9	41.2	82.6	8	45	90
Aromatics/wt%						
Mono –	25.6	28.1	29.3	22.4	23.8	27.6
Di –	7.6	9.0	10.2	6.3	7.3	9.3
Tri –	0.7	0.7	1.0	0.5	0.5	0.7
(2) Experimental results at 360℃						
p_{H_2}/MPa	6.0	6.0	6.0	6.0	6.0	6.0
T/℃	360	360	360	360	360	360
LHSV/h^{-1}	1.0	1.5	2.0	1.0	1.5	2.0
H_2/oil volume ratio	400/400	400/400	400/400	400/300	400/300	400/300
Properties of product oil						
S content /$\mu g \cdot g^{-1}$	38.7	136	496	19	131	482
N content /$\mu g \cdot g^{-1}$	2.7	13.2	33.3	2.5	18.7	55.2
Aromatics/wt%						
Mono –	21.9	25.5	27.1	20.1	20.4	26.0
Di –	6.8	8.3	9.0	6.0	7.2	9.1
Tri –	0.6	0.7	0.8	0.5	0.6	0.7

The experimental results under above reaction conditions are also given in Table 2. It shows that countercurrent flow was superior to the concurrent one in obtaining ultra low level of sulfur even it was operated under a lower H_2/oil ratio of 300. Besides, the temperature has a significant effect on HDS since the reaction is kinetically controlled under very low sulfur concentration in view of the second-order reaction rate on sulfur. Removal of tri-, di- and mono-aromatics was better accomplished using countercurrent flow than concurrent flow, which is ascribed to the high level of hydrogen concentration at the reactor outlet. It is known HDA reactions are subjected to equilibrium limitation, however, this limitation was not obviously observed from Table 2, since higher temperature has improved removal of the aromatics. This temperature effect implies the HDA reactions are both thermodynamically and kinetically controlled.

4.3 Deep removal of sulfur and aromatics

It is known that under operating condition at 360℃ and LHSV of $1.0h^{-1}$, the sulfur from the crude diesel could be reduced to 19μg/g and the aromatics to 26.6% by the countercurrent flow. Although the product quality is fairly good, it is still to be improved to meet more stringent legislation. From the experiment under 380℃ in the first reactor and 360℃ in the second reactor with gas-liquid countercurrent flow, it was found that sulfur could be reduced to 5.6μg/g and aromatics to 26.0%. This result implies that the temperature effect on sulfur removal is significant, but only to a little degree on the aromatics, which implies that the operating temperature has reached the thermodynamic limit for HDA.

4.4 Process analysis based on numerical simulation

It shows from Fig. 2a and b that hydrogen pressure in the gas decreased gradually at the entrance of the reactor, while the hydrogen concentration in the liquid increased correspondingly, which could be explained by the gas-liquid mass transfer which is one order of magnitude less than the liquid-solid mass transfer, as was evidenced from $k_S a_S = 0.37 s^{-1}$ while $k_L a_G = 0.01 s^{-1}$.

From Fig. 2c and d, it was found over 90% of sulfur was removed in the first reactor and therefore the second reactor could be operated under low sulfur concentration. Although the H_2S concentration was rather low in the second reactor, it shows countercurrent flow operation was still superior to the concurrent one since H_2S inhibition could be reduced to the lowest level.

Removal of mono-, di-, and tri-aromatics and the production of naphthene are shown in Fig. 2e-h. In corresponding to the hydrogen profile shown in Fig. 2b, countercurrent flow has superior behavior on aromatics saturation compared with the concurrent manner.

4.5 Comparison between experiment and numerical simulation

The accuracy and reliability of numerical simulation could be evaluated directly from the comparison between experiment and calculation. As depicted in Table 3, the simulation on sulfur removal was fairly accurate, since the relative error was only 0.1%-0.15% with respect to the initial sulfur concentration. The accuracy on aromatics simulation was also acceptable except for diaromatics. The low accuracy on diaromatics is because diaromatics is the intermediate product between mono- and tri-ones, and therefore the precision of kinetic parameters should be very high.

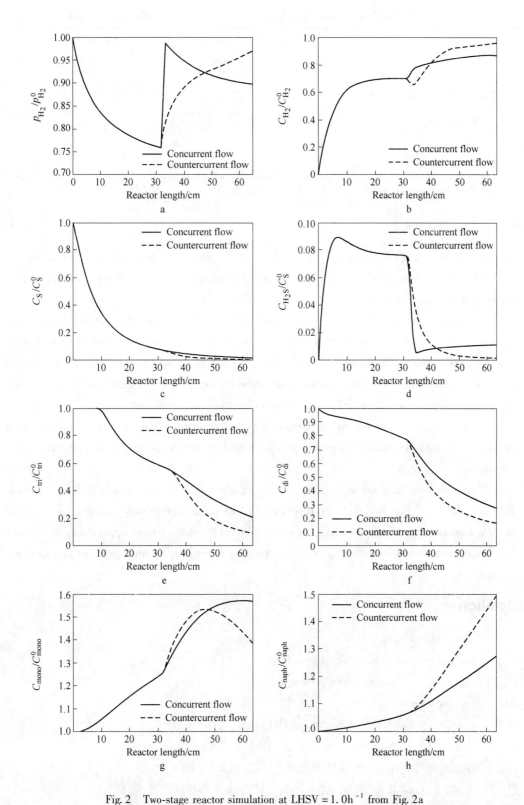

Fig. 2 Two-stage reactor simulation at LHSV = 1.0h^{-1} from Fig. 2a

a—Hydrogen pressure in the gas; b—Hydrogen concentration in the liquid; c—Sulfur concentration in the liquid;
d—H$_2$S concentration in the liquid; e—Triaromatics concentration in the liquid; f—Diaromatics concentration in the liquid;
g—Monoaromatics concentration in the liquid; h—Naphthene concentration in the liquid

Table 3 Comparison between experimental and simulation at the reactor outlet

Relative concentration	Concurrent/concurrent			Concurrent/countercurrent		
	Experiment/%	Simulation/%	Relative error	Experiment/%	Simulation/%	Relative error
S	1.41	1.51	0.1	0.68	0.50	0.18
Aromatics						
Mono-	157	157	0	137	139	2
Di-	35	28	7	29	17	12
Tri-	21	21	0	15	10	5

Note: Experiments taken at LHSV = 1.0h^{-1} according to Table 2(1).

5 Conclusions

Through the present work, a variety of variables that influence reactor performance on HDS and HDA have been investigated both experimentally and theoretically. The following considerations are suggested towards producing ultra-clean diesel fuel:

(1) A two stage-reactor system should be applied. In the first reactor, the sulfur-tolerant catalyst should be loaded since more than 90% of sulfur is removed and consequently a high H_2S concentration in the liquid. In the second reactor, the reactor should be packed with catalyst of high hydrogenation activity since the reactant concentration is very low, and on the other hand the HDS mechanism has changed from hydrogenolysis to hydrogenation.

(2) Flow direction is an important parameter to be considered. The preferred flow direction for the second-stage reactor is countercurrent flow, and by which the sulfur was reported to be reduced to 5.6μg/g.

(3) Temperature should be set at a suitable level. Since HDS is irreversible with high activation energy barrier, therefore increase of reaction temperature can greatly increase the reaction rate. On the contrary, HDA reactions are reversible and there exist an upper limit in view of thermodynamics. It was found that 380℃ was the upper temperature limit that can be practically employed.

Notation

C	concentration, mol/cm^3
E_{HDS}	activation energy for HDS, kJ/mol
H	Henry's law constant, dimensionless
ΔH	reaction heat, kJ/mol
k	forward reaction rate constant for HDA, cm^3/(kg·s)
k_-	backward reaction rate constant for HDA, cm^3/(kg·s)
k_{HDS}	reaction rate constant of HDS, (cm^3)$^{2.6}$/(mol$^{1.6}$·kg·s)
k_{HDS}^0	pre-exponential factor of HDS, (cm^3)$^{2.6}$/(mol$^{1.6}$·kg·s)
$k_L a_G$	volumetric gas-liquid mass transfer coefficient, 1/s
$k_S a_S$	volumetric liquid-solid mass transfer coefficient, 1/s
K	equilibrium constant of HDA reactions

K_{H_2S}	adsorption constant of H_2S, bar^{-1} or cm^3/mol
LHSV	liquid hourly space velocity, h^{-1}
m	reaction order in Eq. 2
n	reaction order in Eq. 2
P	pressure, MPa
R	ideal gas constant, $J/(mol \cdot K)$
T	temperature, K or ℃
u	superficial velocity, cm/s
z	axial distance of the reactor, cm

Greek letter

ρ_b	bulk density of the catalyst bed, g/cm^3

Superscripts

0	inlet condition

Subscript

A	hydrogen
B	sulfur compounds in diesel
C	aromatics
di	diaromatics
G	gas phase
H_2	hydrogen
H_2S	hydrogen sulfide
L	liquid phase
mono	monoaromatics
naph	naphthene
S	sulfur compounds in diesel
tri	triaromatics

Acknowledgements

The present work was under support by SINOPEC through Grant No. 101019 and by the Shanghai Scientific and Technical Committee through Grant No. 03JC14024.

References

Avraam, D., Vasalos, I. A., 2003. HdPro: a mathematical model of trickle-bed reactors for the catalytic hydroprocessing of oil feedstocks. Catalysis Today 79-80, 275.

Babich, I. V., Moulijn, J. A., 2003. Science and technology of novel processes for deep desulfurization of oil refinery streams: a review. Fuel 82, 607.

Chowdhury, R., Pedernera, E., Reimert, R., 2002. Trickle-bed model for desulfurization and dearomatization of diesel. A. I. Ch. E. Journal 48, 126.

Girgis, M. J. , Gates, B. C. , 1991. Reactivities, reaction networks and kinetics in high-pressure catalytic hydroprocessing. Industrial Engineering Chemical Research 30, 2021-2058.

Jackson, K. M. , 2003. Revamp technology enables deep HDS to produce ULSD. Hydrocarbon Processing 31.

Jaffe, S. B. , 1974. Kinetics of heat release in petroleum hydrogenation. Industrial Engineering Chemical Process Desta and Development 13, 34.

Kabe, T. , Aoyama, Y. , Wang, D. , Ishihara, A. , Qian, W. H. , Hosoya, M. , Zhang, Q. , 2001. Effects of H_2S on hydrodesulfurization of dibenzothiopheneand 4, 6-dimethyldibenzothiophene on aluminasupported NiMo and NiW catalysts. Applied Catalysis A General 209, 237.

Korsten, H. , Hoffmann, U. , 1996. Three-phase reactor model for hydrotreating in pilot trickle-bed reactors. A. I. Ch. E. Journal 42, 1350.

Pedernera, E. , Reimert, R. , Nguyen, N. L. , van Buren, V. , 2003. Deep desulfurization of middle distillates: process adaptation to oil fractions' compositions. Catalysis Today 79-80, 371.

Reilly, J. W. , Sze, M. C. , Saranto, U. , Schmidt, U. , Oy, N. , 1973. Aromatic reduction process is commercialized The Oil and Gas Journal 66-68.

Relly, J. W. , Hamilton, G. , 1993. Production of diesel fuel by hydrogenation of a diesel feed. USP 5, 183, 556.

Sie, S. T. , 1999. Reaction order and role of hydrogen sulfide in deep hydrodesulfurization of gas oils: consequences for industrial reactor configuration. Fuel Processing Technology 61, 149.

Smith, J. M. , 1981. External transport processes in heterogeneous reactions. Chemical Engineering Kinetics, third ed. McGraw-Hill, New York. (Chapter 10).

Trambouze, P. , 1990. Countercurrent two-phase flow fixed bed catalytic reactors. Chemical Engineering Science 45 (8), 2269-2275.

Tsamatsoulis, D. , Papayannakos, N. , 1998. Investigation of intrinsic hydrodesulphurization kinetics of a VGO in a trickle bed reactor with backmixing effects. Chemical Engineering Science 53, 3449.

A Hybrid Neural Network-first Principles Model for Fixed-bed Reactor[*]

Abstract In the present work, we combine first principles, in the form of mass and energy balance equations, with artificial neural networks (ANNs) as estimators for some of the important process parameters in modeling a wallcooled fixed-bed reactor. Experiments were carried out in a pilot wall-cooled fixed-bed reactor with benzene oxidization to maleic anhydride as a working reaction to show the performance of the proposed hybrid models. Compared with the two-dimensional pseudo-homogeneous models, the hybrid models predict equally well, but are simpler in structure and therefore easier to meet the real-time requirements of control and on-line optimization.

Key words hybrid model, fixed-bed reactor, fixed-bed reactor model, neural networks, hybrid modeling approach, benzene oxidation

1 Introduction

In chemical reaction engineering, mathematical models used for reactor design, simulation and optimization are generally mechanistic ones that are developed based on first principles. Undoubtedly, a mechanistic model is of advantage for easy analysis and reliable extrapolation. However, for fixed bed reactors, development of a precise model is usually difficult due to the existing coupled chemical/physical processes.

Determination of the heat transfer parameters is very important to guarantee the successfulness of mechanistic models because these parameters have decisive impacts on the validity and accuracy of the models. A large number of empirical correlations to estimate radial thermal conductivity, λ_{er}, and wall heat transfer coefficient. a_w, are available in the literature and textbooks. However, most of them are based on spherical or cylindrical particles, determined under "cold flow" conditions, and are valid only within certain operation ranges (Li and Finlayson, 1977; Chen, 1990). Significant errors in estimation have been reported when the operating condition changes (Lemcoff et al., 1990). Since a prior determination of transfer parameters is unreliable, experimental data are necessary to adjust the parameter values.

With these experimental data, chemical engineers can easily develop an empirical model that has a more general structure and does not require a deep understanding of the involved chemical/physical process, such as Nonlinear AutoRegressive Moving Average with exogenous inputs (NARMAX) (Thibault, 1991) and neural network(NN) model. Moreover, these models are of algebraic forms and therefore can alleviate the on-line computational load when the model are to be utilized in control and optimization of the reactor. However, being essentially black box/mod-

---

[*] Coauthors: Qi Haiyu, Zhou Xinggui, Liu Lianghong. Reprinted from *Chemical Engineering Science*, 1999, 54: 2521-2526.

els, they are of poor ability to extrapolate and difficult to interpret as well as to analyze the behaviours of the reactor. Therefore, they are not appropriate for reactor design and development.

Recently, a hybrid modeling approach, which combines a mechanistic model and an empirical one together, has been receiving increasing attention. It was used by Psichogios and Ungar (1992) in modeling a fed-batch bioreactor. Schubert et al. (1994a, b) considered a more sophisticated hybrid model, which included a mathematical model in the form of balance equations and an ANN, a fuzzy logic supervision for batch and fed-batch production of *Saccharomyces cerevisae*. Wilson and Zorzetto (1997) applied such models in on-line applications to control the liquid levels of a cascade of vessels.

In the present work, the authors combine the one-dimensional models, which are the simplest among all the mechanistic models, with neural networks in modeling a wall-cooled fixed-bed reactor, the networks being estimators for some of the important process parameters, such as the overall heat transfer coefficient.

2 The Hybrid Model for Fixed-bed Reactor

The structure of the proposed hybrid model is schematically shown in Fig. 1. It consists of a mechanistic model (denoted as MM in this figure), which can be of any kind that is developed based on first principles, and an NN. The hybrid model can be either steady state or dynamic, depending on whether the MM used is steady state or dynamic. The NN is employed to predict the parameters (kinetic parameters and/or transport parameters) in the mechanistic model. The inputs of the NN can be the operating variables (u), the axial positions (z) of the reactor, or even some of the state variables (Y) such as the central temperatures along the axis (in this case the network is recursive). Thus, the possible dependence of the parameters to the operating condition, the axial position and the state variable can be taken in to account by the NN model.

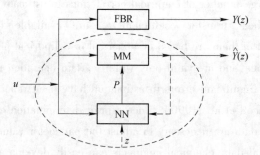

Fig. 1 Structure of the hybrid model (shown in the oval circle)

In this study, two hybrid models, one steady state and the other dynamic, have been developed for a wall-cooled fixed bed reactor in which benzene oxidation by air to maleic anhydride is carried out. The neural network is a traditional three-layered feedforward one. As will be shown below, it receives operating variables, including benzene flow rate, coolant temperature and air flow rate, as inputs and has only one output, i. e. the overall heat transfer coefficient. The mechanistic model used in the hybrid model is the simplest, i. e. the plug-flow model that ignores the radial temperature gradient. As is well known, for fixed-bed reactor with highly exothermic reactions, the one-dimensional model is no sufficient to model the reactor. However, as we will demonstrate, by taking advantage of the NN, the hybrid models, which are essentially one-dimensional ones, perform as well as the two-dimensional models that take the radial temperature gradient into consideration.

The weight factors of the neural network in the hybrid model are determined by minimizing the deviations of the model outputs from the experimental data. That is

Min
$$J = \sum_{i=1}^{N} (Y_i - \hat{Y}_i)^2$$

s. t.
$$\dot{X} = f\left(\frac{\partial X}{\partial t}, X, t, P\right)$$
$$Y = g(X)$$
$$P = h(W)$$

where, Y_i and $\hat{Y}_i$ are state variable measurements and model predictions; g denotes measurement structure; h is a symbolic representation of the relationship of the parameters in the mechanistic model to the weights in the neural network.

As can be seen, the outputs of the neural network in the hybrid model do not appear explicitly in the objective function. The observed deviations between the one-dimensional model predictions and the state variable measurements are used as error signals to the network. Since a first-principles model is included, the neural network in the hybrid model is only needed to determine some of the model parameters in the mechanistic model. Therefore, the samples used for its training can be dramatically fewer than required by conventional neural networks.

3 Experimental

To verify the feasibility and advantages of the hybrid model, experiments were carried out through a computerized pilot wall-cooled fixed-bed reactor. The reactor, shown in Fig. 2, was a stainless steel tube of 1 in diameter and 3.4m in height with the industrial V_2O_5 catalyst packed to a height of 2m and inert packings packed at both ends of the reactor. Over the catalyst, benzene was partially oxidized to maleic anhydride. This is a serial-parallel reaction in which further oxidation of benzene and maleic anhydride to CO_2 occurs simultaneously, as can be shown below:

$$C_6H_6 + 4.5O_2 \longrightarrow C_4H_2O_3 + 2CO_2 + 2H_2O$$
$$C_4H_2O_3 + 3O_2 \longrightarrow 4CO_2 + H_2O$$
$$C_6H_6 + 7.5O_2 \longrightarrow 6CO_2 + 3H_2O$$

These reactions can be assumed of first order because the oxygen is largely excessive. Ten thermal couples were installed at different axial positions (i.e., 0.0m, 0.05m, 0.1m, 0.2m, 0.3m, 0.4m, 0.5m, 0.7m, 1.2m, 2.0m from the inlet side of the packed catalyst bed) to determine the temperature profile. The outlet concentration of the reactor was determined by GC. Table 1 summarizes the operating conditions of the steady and dynamic (step-up and step-down) experiments that will be used for illustration.

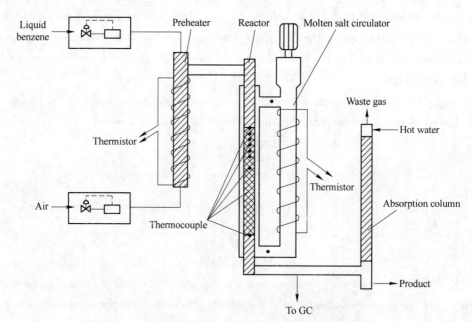

Fig. 2　Experimental setup

Table 1　Experimental conditions

No.	Steady-state experiments						Dynamic experiments					
	1	2	3	4	5	6	1	2	3	4	5	6
Coolant temperature/℃	354.3	352.1	352.5	352.1	354.1	354.1	349.4-350.1	352.3-354.8	353.6	351.0	352.9	352.1
Benzene flow rate/mL·h^{-1}	58	68	63	45	58	55	63	63	55	60-65	70-72	60-63
Air flow rate/NL·min^{-1}	27	33	33	33	30	27	33	33	30-33	33	33	33

4　Two-dimensional Models

Two two-dimensional pseudo-homogeneous models, one steady state and the other dynamic, were also developed for comparing with the proposed hybrid models. The orthogonal collocation method was used to transfer the partial differential equations into ordinary differential ones, the later being integrated by the GEAR method.

To make the comparison more convincing, the kinetic and heat transfer parameters (radial thermal conductivity λ_{er} and wall heat transfer coefficient α_w) were refined by experiments. The initial values of the kinetic and heat transfer parameters were from our previous studies (Liu, 1998). Consequently, the predictions represent the best results that the two-dimensional models can produce.

5　Results

To reduce the impacts of the possible deactivation of the catalyst, before training the hybrid model, kinetic parameters of the one-dimensional model were first adjusted to fit the steady-state experimental data. The experiments for this purpose were carried out at different wall temperatures while the inlet concentration and residence time remained unchanged. Since the overall heat transfer parameter was also uncertain, it was also estimated in the optimization. We note that kinetic parameters for the two-dimensional and the one-dimensional model are slightly different, as can be seen in Table 2. This is understandable if we take cognizance of the fact that the kinetic model is generally empirical.

Table 2 Refined parameters of the 1D and 2D models

Name	K_1/s^{-1}	$E_1/J \cdot mol^{-1}$	K_2/s^{-1}	$E_2/J \cdot mol^{-1}$	K_3/s^{-1}	$E_3/J \cdot mol^{-1}$	U /J·(m²·s·K)⁻¹	λ_{er} /J·(m·s·K)⁻¹	α_w /J·(m²·s·K)⁻¹
1D model	0.4368×10^7	0.7525×10^5	0.3689×10^8	0.1300×10^6	0.2974×10^6	0.6815×10^5	111.4		
2D model	0.4787×10^7	0.7537×10^5	0.6844×10^8	0.1026×10^6	0.1507×10^6	0.6593×10^5		6.349	120.3

Totally 14 sets of steady-state experimental data were used to verify the steady-state hybrid model. They were divided into two parts: one (eight sets) for training and the other (six sets) for verification. The Boryden-Fletcher-Goldfarb-Shanno (BFGS) method (Yuan, 1990) was used to train the network. Three hidden nodes (together with a treshold) were found necessary to produce satisfactory results.

Only part of the results is illustrated. Fig. 3 shows the steady-state hybrid model predictions compared with steady-state experiments that are used for training (Fig. 3a) and verification (Fig. 3b), respectively. Also shown in Fig. 3 are the predictions of the steady-state two-dimensional model. The predictions of the selectivity by the hybrid model are listed in Table 3 compared with the experiments. Also listed are the predictions of the two-dimensional model.

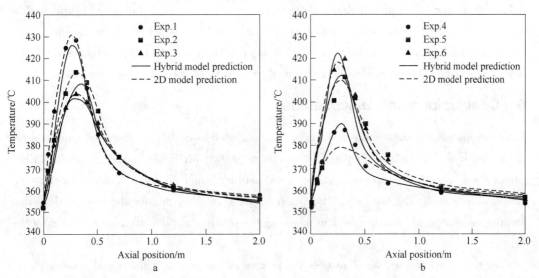

Fig. 3 Experiments compared with steady-state hybrid model and two-dimensional model predictions
a—Used for training; b—Used for verification

Table 3 Selectivity predictions of hybrid and 2D model compared with steadystate experiments

No.	1	2	3	4	5	6
Experiments	67.1	68.8	69.6	66.8	68.1	66.2
Hybrid model prediction	67.5	68.9	70.1	67.4	66.8	67.7
2D model prediction	69.2	68.9	68.6	66.9	69.3	69.5

For the dynamic hybrid model, six sets of dynamic experiments are used for training and four sets were used for verification. Three hidden nodes were also used. Fig. 4a and b illustrate the tansient responses of the hot spot temperature compared with the predictions of the dynamic hybrid model and the two-dimensional model.

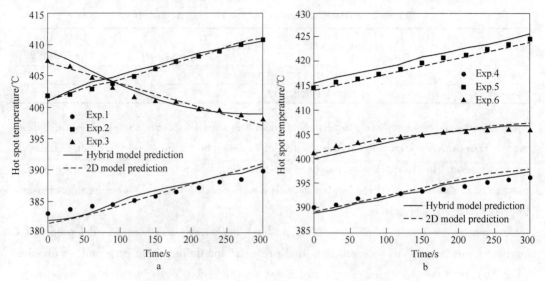

Fig. 4   Experiments compared with steady-state hybrid model and two-dimensional model predictions
a—Used for training; b—Used for verification

From these results illustrated above, we learn that the steady state and dynamic hybrid models can accurately predict the reactor behaviors, although the basic structure of the models is only one-dimensional and the reactions involved are highly exothermic. In fact, the proposed steady-state and dynamic hybrid models predict equally well with the two-dimensional models.

6 Conclusions and Discussions

A hybrid modeling approach has been presented and used to model a wall-cooled fixed-bed reactor. The hybrid model is composed of two parts: a first-principles model, which describes physical and chemical phenomena, and a neural network, which serves as a tool to estimate the process parameters, e.g. the overall heat transfer parameters in this article. Results show that the proposed approach is successful in modeling the steady state and dynamic behaviors of the reactor.

Since the first-principles model reduces the number of functions that the neural network has to choose from to approximate the process, considerably fewer data are required for training the neural network. Consequently, the hybrid model provides far better approximation accuracy, based on the same number of training samples, than the conventional neural networks. Moreover, the hybrid model is more reliable for extrapolation because it is essentially a mechanistic model.

As compared with the two-dimensional models, the hybrid models established in this article are simpler in model structure and need shorter time for numerical solutions. If the model is solved only once, the saving in computational time may be negligible. However, in control and optimization of reactors applications that require the model be solved repeatedly to compute the control moves or the optimum operating conditions, the overall computational time will be reduced significantly if the hybrid model is used. Therefore, the hybrid modeling approach proposed in this article is very promising in control and on-line optimization of the fixed-bed reactor.

Acknowledgements

This work is supported by the National Science Foundation of China (Contract No. 29676014), the China Petrochemical Corporation, and the National Science Foundation of the United States. The financial support of the State Key Laboratory of Chemical Engineering is also acknowledged.

Notation

f function relation
g function relation
h function relation
P parameters of the mechanistic model
W parameters of the neural network model
X state variable
Y observable variable

Greek letters

α_w radial thermal conductivity
λ_{er} wall heat transfer coefficient

References

Chen, Y. Z., & Wang, J. B. (1990). Estimation of heat transfer parameters in packed-bed by using orthogonal collocation. J. Chem. Ind. Engng. (Chinese), 2, 219-226.

Lemcoff, N. O., Pereira Duarte, S. I., & Martinez, O. M. (1990). Heat transfer in packed bed (Vol. 4, No. 16, pp. 1-32).

Li, C. H., & Finlayson B. A. (1977). Heat transfer in packed beds-a preevaluation. Chem. Engng Sci., 32, 1055-1066.

Liu, L. H. (1998). Control and optimization of a wall-cooled fixed-bed reactor. Ph. D. dissertation, ECUST.

Psichogios, D. C., & Ungar, L. H. (1992). A hybrid neural network-first principles approach to process modeling. A. I. Ch. E. J., 38, 1499-1511.

Schubert, J. Simutis, R., Dors, M., Havlik, I., & Lubbert, A. (1994a). Bioprocess optimization and control: application of hybrid modeling. J. Biotechnol, 35, 51-68.

Schubert, J., Simutis, R., Dors, M., Havlik, I., & Lubbert, A. (1994b). Hybrid modeling of yeast production process. Chem. Engng. Technol., 17, 10-20.

Thibault, J. (1991). Feedforward neural networks for the identification of dynamic processes, Chem. Engng Commun., 105, 109-128.

Wilson, J. A., & Zorzetto, L. F. M. (1997). A generalized approach to process state estimation using hybrid artificial neural network/mechanistic models. Comp. Chem. Engng, 21, 951-963.

Yuan, Y. (1990). Numerical methods for nonlinear programming, Shanghai, PRC: Shanghai Scientific & Technical Publishers.

Hydrodynamic Behavior of a Trickle Bed Reactor under "Forced" Pulsing Flow[*]

Abstract The hydrodynamic properties in a trickle bed reactor under forced pulsing flow are studied. It has been demonstrated by the experiments that the axial and radial liquid distributions become more uniform than those of the natural pulsing flow regime. The thickness of the liquid film over the particles is sharply thinned by the induced liquid flow. Effects of the operating conditions on liquid holdup are studied in detail. The frequency of the forced pulsing flow is found to be the most significant parameter affecting the liquid holdup.

Key words trickle bed reactor, hydrodynamic properties, forced pulsing flow, liquid holdup distribution

1 Introduction

To guarantee uniform liquid distribution in a trickle bed reactor has been a tough problem which has drawn much attention. Even in the pulsing flow regime, liquid maldistribution might occur (Beimesch & Kessler, 1971). With low liquid and gas flow rates the so-called partial irrigation occurred. Thus, ineffective catalyst wetting and sensitivity to thermal effect increase, especially for strongly exothermic reactions. The wetted portion of the catalyst surface increases with the superficial velocity of liquid till pulsing flow regime arrives (Gianetto & Specchia, 1992). However, it is generally regarded that the catalyst surface is totally wetted in the pulsing flow regime when the gas flow rate and/or the liquid flow rate are sufficiently high. But the conversion may be reduced due to the relatively high superficial velocities of the gas and liquid phases. To solve this problem the forced pulsing flow operation through a simple method is introduced. The aim of this work is to study the effect of forced gas pulsation on the hydrodynamic behaviors in a trickle bed reactor, in which a forced pulsing flow can be achieved under relatively low gas and liquid flow rates. The hydrodynamic properties under forced pulsing flow are also compared with those of the "natural" pulsing flow.

The concept of forced pulsing flow implies to make the liquid and gas phases flow through the column similar to the natural pulsing flow, i. e., liquid-rich slugs and gasrich slugs pass the reactor alternatively. Many ways can be used to operate the trickle bed reactor in a non-steady mode (Haure, Hudgins & Silveston, 1989), but a simpler one is to change the gas flow rate periodically (Jiang, Cheng & Yuan, 1997). The authors apply intermittently superimposed extra gas inputs in addition to the continuous gas stream to provide the forced gas pulsation, as shown in Fig. 1. ΔG is the forced pulse input in a time period $(0, t_0)$.

---

[*] Coauthors: Xiao Q., Cheng Z. M., Jiang Z. X., Anter A. M. Reprinted from *Chemical Engineering Science*, 2001, 56: 1189-1195.

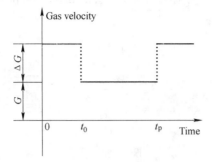

Fig. 1 Schematic diagram of the gas pulsation

In order to describe the forced pulsing flow, a parameter λ is introduced and defined as:

$$\lambda = 1 + \frac{\int_0^{t_p} \Delta G \mathrm{d}t}{G t_p} \quad (1)$$

where, t_p is the time period for a forced pulse, thus the frequency of the forced pulsing flow is $f = 1/t_p$. ΔG is a function of time and has the following form:

$$\Delta G = \begin{cases} \Delta G, & 0 \leqslant t \leqslant t_0 \\ 0, & t_0 < t \leqslant t_p \end{cases} \quad (2)$$

where, t_0 is the time period when the electromagnetic valve is open, which is controlled by a time relay. λ represents the intensity of forced input gas compared with the base gas velocity.

For a given superficial velocity of the liquid and gas, forced input frequency, forced input superficial velocity of gas and t_0, the forced pulsing flow can be distinguished uniquely. In order to describe the effect of the gas velocity on the behavior of the reactor, a new parameter G_{eff}, which is the effective superficial velocity of gas in one cycle, is defined as:

$$G_{\text{eff}} = \lambda G \quad (3)$$

where, G_{eff} is the equivalent superficial velocity of gas as the sum of the two parts of gas flow rate introduced into the reactor.

2 Experimental Setup and Schematic

Experiments are conducted in a Plexiglas reactor of an inner diameter of 0.10m, packed with 5mm glass beads at a height of 1.0m. The air-water system is applied for all the experiments. A schematic diagram of the experimental setup is shown in Fig. 2. Specially designed stainless-steel concentric-ring conductance probes, schematically shown in Fig. 3, are used to determine the liquid holdup and the axial and radial liquid distribution. Each probe includes several concentric-ring electrodes, except the one at the center. Time-dependent conductivity between each pair of electrodes can be measured online to determine the liquid holdup, and to reveal the liquid distribution as well. To investigate the properties of forced pulsing flow several probes of this kind are installed at different axial position of the column. The reactor is pre-wetted for several minutes under pulsing flow and then lowers the gas and liquid velocities to the experimental condition. The reactor is operated for a period of time to reach the steady state at the given liquid and gas flow rates, then a certain amount of gas is forced into the reactor through an electromagnetic valve. The open and shut frequency of the electromagnetic valve is adjusted through the time relays.

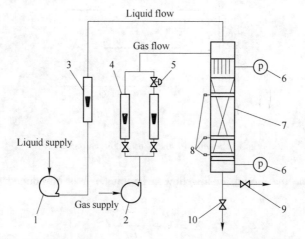

Fig. 2 Schematic diagram of the experimental setup
1—Liquid pump;2—Gas compressor;3—Liquid rotameter;4—Gas rotameter;
5—Electromagnetic valve;6—Pressure manometer;7—Trickle-bed reactor;
8—Conductance probes;9—Vent valves;10—Drainge valve

Based on electrical potential field theory, the conductance measured by the concentric-ring probe is a function of the probe's configuration, size and liquid holdup. The maximum conductance measured under a static condition for each probe, when the packed column is imbued with liquid, is used to normalize the real conductance (Tsochatzidis Karapantsios, Kostoglou & Karabelas, 1992). For the uniform liquid film, this normalized conductance presents the liquid holdup directly:

$$\beta = \frac{C}{C_{max}} \tag{4}$$

A similar formula is proposed by Tsochatzidis et al. (1992) for different ring-type electrodes, which are flush mounted onto the column wall.

3 Experiment Results

In our experiments, two probes are mounted at the lower part of the reactor with a distance of 0.12m in between. This distance is long enough not to affect each other remarkably (Tsochatzidis & Karabelas, 1995). Another is placed at the top of the packed bed. The liquid superficial velocity falls in the range from 0.354 to 0.885cm/s, and the gas superficial velocity in the range of 0.071-0.212m/s. t_0 is fixed at 1.73s in all of our experiments.

3.1 Radial liquid distribution

To measure the radial liquid distribution, pairs of electrodes as shown in Fig. 3 divide the cross-sectional area into four zones. The measured data are transformed into liquid holdup through a program based on Eq. 4 and are averaged over a period of time. Thus, the measured liquid holdup represents the area-averaged liquid holdup between the two electrodes.

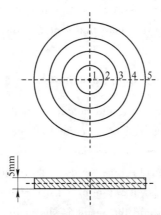

Fig. 3 Schematic of the concentric-ring conductance probe
1—node；2—2cm；3—5cm；4—8cm；5—10cm

To compare the liquid distribution under forced pulsing flow with that of the natural pulsing flow, the reactor is operated in both regimes. Experimental results in Fig. 4 show that the radial liquid distribution is more uniform under the forced pulsing flow. It should be noted that the velocities under the two flow regimes are not the same. That is because the natural pulsing flow regime never occurs below certain critical liquid and gas flow rates.

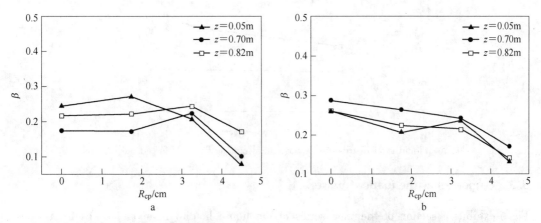

Fig. 4 Liquid radial distribution under two kinds of flow regime
a—Forced pulsing flow, $G = 0.071\text{m/s}, \Delta G = 0.212\text{m/s}, L = 0.885\text{cm/s}, \lambda = 1.74, f = 0.135\text{Hz}, 5\text{mm glass beads}$;
b—Natural pulsing flow $G = 0.212\text{m/s}, L = 1.062\text{cm/s}, 5\text{mm glass beads}$

The liquid is distributed uniformly by the forced input gas flow in radial direction except for the narrow zone adjacent to the wall of the column. This may be explained as being the result of the difference of the surface properties of the reactor wall and the packed materials. In our experiments the reactor is made of Plexiglas, which is less hydrophilic than glass. In both strongly interacting regimes, the gas and the liquid flow through the reactor competitively and alternately. The liquid holdup near the wall is very small. The voidage near the wall of the column is higher than that of the center of the reactor(Beimesch & Kessler, 1971), thus more gas preferably passes this area. Taking account of the wall of the reactor, the specific area near the wall is also higher than that of the central part of the column. While the liquid, whose viscosity is higher than that of the gas, flows through the zone near the wall, the friction acting on the liquid film is greater than that while it flows across the center of the column. That is to say, more gas goes through the area close to the wall and more liquid prefers the central part, based on the minimum energy dissipation concept.

3.2 Cross-section-averaged axial liquid distribution

To measure the cross-section-averaged liquid holdup axial distribution, the probe is composed only of the node and the outer ring, which is mounted adjacent to the column wall. It is used to measure the overall conductance of cross section. The same method as described in Eq. 4 is used to get the liquid holdup from the time series of the conductance. Liquid holdup at three different axial positions is detected. Axial liquid distribution under forced pulsing flow and natural pulsing flow are measured and the result is shown in Fig. 5. It is shown obviously that the liquid axial distribution under forced pulsing flow is more uniform than that of the natural pulsing flow. The experimental results show, while the liquid superficial velocity is greater than 0.708cm/s, that the liquid holdup decreases slightly along the flow direction. The effects of the operating condition on liquid holdup will be discussed in detail in the following section.

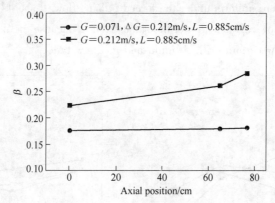

Fig. 5 Liquid axial distribution under natural pulsing flow and forced pulsing flow

3.3 Properties of the induced pulses

Fig. 6 exhibits a section of the time series of the liquid holdup in about one cycle. A short period of time ($t_0 = 1.73\text{s}$) of gas flow is superimposed onto the main gas fluid. It can be seen clearly that three sharp increases of the liquid holdup (i.e., three liquid-rich slugs passing the probe in one forced period) are triggered. The number of induced liquid pulses is strongly dependent on the base liquid holdup. It is easy to understand that the induced liquid pulses obviously increase the renewal rate of the liquid film over the particles. In view of the high velocity of the liquid pulse, the retention time of this part of liquid may be shortened. However, the difference between the average liquid holdup and the base liquid holdup is small, which is shown in Fig. 6. In this paper the base liquid holdup is defined as the averaged liquid holdup without the induced pulse. Thus, the base liquid holdup is 0.162 and the total liquid holdup is 0.175 under a condition of $G = 0.212\text{m/s}, \Delta G = 0.071\text{m/s}, L = 0.708\text{cm/s}, f = 0.135\text{Hz}$. The increment of the liquid holdup due to the induced liquid pulse is calculated as:

$$\Delta \beta_{\text{pb}} = \frac{\beta_{\text{b}} - \beta_{\text{base}}}{\beta_{\text{b}}} \times 100\% \tag{5}$$

The percentage of the liquid in the induced liquid pulse has a value of about 25% -50% of the whole liquid holdup. In Fig. 6 the liquid holdup increment is no more than 8%. Therefore, only a part of the liquid passes the reactor quickly and the average reaction time for the liquid phase will not be reduced significantly. This is different from the natural pulsing flow. Beimesch and Kessler (1971) reported that 67% of the liquid presented in the liquid-rich slug under natural pulsing flow. According to the data reported by Tsochatzidis and Karabelas (1995), this value varied from 33% to 60%.

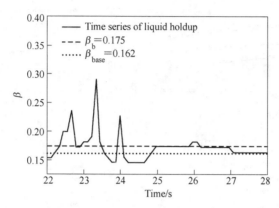

Fig. 6 Liquid holdup's time series under forced pulsing flow regime in a period
($L = 0.708 \text{cm/s}, G = 0.212 \text{m/s}, \Delta G = 0.071 \text{m/s}, F = 0.135 \text{Hz}, \lambda = 1.08$)

The ratio of β_p to β_b falls within the range of 1.1-2.0 in all our experiments. Tsochatzidis et al. (1995) reported that this ratio ranged from 1.2 to 1.5 in the natural pulsing flow regime. However, those presented by Blok, Varkevisser and Drinkenburg (1982) are 1.5-2.0. This value is very similar to our result. The length of the liquid pulse can be easily calculated from the apparent pulse velocity and the duration of the liquid pulse. In Fig. 6 the pulsing length is 0.26m. The average value of this variable is 0.30m which is similar to the data of Tsochatzidis et al. (1995). The characteristic parameters of the forced pulsing flow β_{base}, β_p and β_b depend on the liquid superficial velocity and the effective gas superficial velocity. The data are plotted in Fig. 7. All of these three parameters increase with the liquid velocity and decrease with G_{eff}.

3.4 Effects of operating conditions on the bed-averaged liquid holdup

The liquid and gas superficial velocities affect the liquid holdup in a similar way to what happens under natural pulsing flow. On increasing the liquid superficial velocity the liquid holdup increases. This can be seen clearly from Fig. 7d.

The gas flow in the system is divided into two parts, one is the base gas flow, which does not vary with time; the other is the forced input gas flow. Its properties are decided by three parameters $t_0, \Delta G$ and f. λ is an integrated parameter of the three parameters compared with the base gas velocity as expressed in Eq. 1. Then the four independent parameters $G, \Delta G, t_0$ and f determine the state of the input gas flow. Because t_0 is fixed at 1.73s in all the experiments, there are three ways to increase the effective gas superficial velocity: increase of $G, \Delta G$ and f, respectively. The experimental results are shown in Fig. 7a, b and c respectively. Increasing one of the three parameters $G, \Delta G$ and f separately always decreases the liquid holdup. Whatever the manner used, this implies that the liquid holdup decreases with increasing the effective superficial velocity of gas, which is expressed as the product of λ and G, as is shown in Eq. 3.

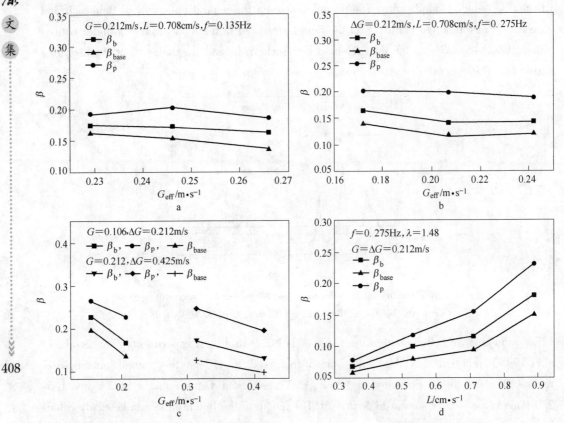

Fig. 7 Effects of the operational parameters on the characteristic properties of the induced pulse
a—Influence of ΔG on β; b—Influence of G on β; c—Influence of f on β; d—Influence of L on β

To evaluate the effects of parameters such as $G, \Delta G$ and f on liquid holdup quantitatively, the liquid holdup varying with the investigated parameter is defined as shown in the following equation:

$$R_\beta = \frac{\Delta \beta}{\Delta G_{eff}} \tag{6}$$

where, ΔG_{eff} is the increment of the total effective gas superficial velocity due to the changing of the investigated parameter. R_β is negative because the liquid holdup always decreases with gas velocity. The rate of liquid holdup is calculated and the values are 0.80 s/m, 0.371 s/m and 0.342 s/m, corresponding to the parameters $f, G, \Delta G$, respectively. Obviously, we notice that the frequency of the forced gas pulsation is the most significant parameter affecting the liquid holdup.

3.5 Velocity of induced pulses

Velocity of the induced pulses is obtained through time delay analysis. Experimental results show that the pulse velocity is actually insignificant to the liquid superficial velocity and a little significant to the gas superficial velocity. The velocity of the liquid-rich pulse is nearly 1.0m/s in our experiments. It is very similar to that of pulses under mild natural pulsing flow as reported by Tsochatzidis et al. (1995). Under natural pulsing flow Blok et al. (1982) claimed that the pulse velocity was completely determined by the gas velocity. To investigate the influence of the gas and liquid velocity on pulse velocity under forced pulsing flow, a similar way to that of natural pulsing flow is adopted. The data are plotted in Fig. 8, where the abscissa is the interstitial velocity of the gas phase calculated from the effective gas velocity. A linear correlation of the pulse velocity is proposed:

$$V_p = 0.7993 + 0.2552 u_{g_{eff}} \qquad (7)$$

The correlation coefficient of the experimental data is 0.75.

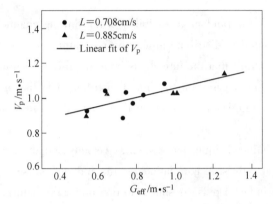

Fig. 8　Velocity of induced pulse depending on gas equivalent interstitial velocity

3.6　Comparison of the liquid holdup under forced pulsing flow and trickling flow

To investigate the effects of the forced pulsing operation mode on the liquid holdup, the liquid holdup data are compared with those of trickling operation mode. Liquid holdup under trickling flow regime at the same gas and liquid velocity is calculated using the correlation of Sato, Hirose, Takahashi and Toda (1973). The correlation proposed by Larkins, White and Jeffrey (1961) is also used but it predicts a slightly higher value of the liquid holdup than that of Sato et al. (1973), who explained that this difference was caused due to different methods used in measuring the liquid holdup. In this paper, both correlations are used to predict the liquid holdup under the same superficial velocity of liquid phase as that under forced pulsing flow, and the same gas superficial velocity as the effective gas superficial velocity under forced pulsing flow. The results are shown in Fig. 9. The abscissa shows the Lockhart-Martinelli parameter X and the ordinate expresses the liquid holdup. The forced pulsing flow operation mode can decrease the liquid holdup greatly to about half of that under trickling flow.

4　Conclusion and Discussion

Experiments show that the liquid is distributed more uniformly in the radial and axial directions by the forced gas input. Actually, in each run of forced pulsing flow experiment, at the moment of gas pulse input the system pressure shows a slight increment. The forced input gas moves down the column at a relatively high speed and pushes the liquid away in both the axial and ra-

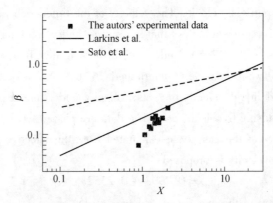

Fig. 9 Comparison of liquid holdup under two operation modes as forced pulsing flow and trickling flow

dial directions. Because of the random accessibility of the flowing channels in a packed bed, it is very difficult for the liquid to distribute uniformly in a natural way in a lower superficial velocity. Our experiments exhibit that the forced gas input improves the liquid distribution a great deal. The function of the input gases on the liquid distribution is like a special "comb". The behavior of the flowing liquid phase is combed periodically and moves down in a more regular way.

The properties of the induced liquid pulse such as velocity, length of the liquid-rich slug are similar to that of natural pulses. A correlation for the induced pulse velocity is proposed in Eq. 7. The length of the induced pulse is similar to that under the natural pulsing flow. The origin of the liquid-rich slugs under forced pulsing flow is somewhat different from that of natural pulsing flow. Under forced pulsing flow, while the reactor is operated under a steady state the liquid moves down slowly until the gas forced input. Then the liquid is pushed down and accumulated to a high liquid holdup due to the difference of velocity. At the same time, the newly formed liquid-rich slug is also accelerated to a relatively high velocity of about 1.0m/s till it moves out of the column. Before another gas pulse is inputted, the liquid holdup in the reactor is maintained at a relatively small value but increases gradually to the base liquid holdup. Under natural pulsing flow the origin of pulses is due to the waves of the liquid film which is also affected by the velocity difference between gas and liquid. Increasing the gas superficial velocity will cause the obstruction of flowing channels and liquid accumulation initially occurs at the outlet of the reactor (Sicardi, Gerhard & Hoffmann, 1979, Tsochatzidis et al., 1995; Cheng & Yuan, 1999), whereas under forced pulsing flow the liquid-rich slug appears very close to the inlet of the reactor. The frequency of the pulse under natural pulsing flow is completely governed by the real liquid velocity in excess of a critical liquid velocity at the pulsing onset (Blok et al., 1982), but the frequency of the forced pulsing flow can be adjusted in advance through the time relays. It is found that one gas pulsation input induces more than one liquid-rich slug, as shown in Fig. 6. The number of induced pulses depends on the operating conditions and especially on the base liquid holdup.

The induced liquid pulse will not reduce the liquid retention time compared with the natural pulsing flow. This is different from the natural pulsing flow, where most of the liquid quickly moves out of the reactor in the form of liquid-rich slugs due to high pulsing frequency.

The liquid holdup directly determines the thickness of the liquid film in the reactor. The liquid holdup is greatly decreased by the forced gas inputs compared with the trickling flow regime. Then the average thickness of the liquid film will also decrease. This may be the greatest advantage for this sort of operation since the thinner the liquid film over the catalyst surface, the less the resistance to mass transfer of the gas reactant. As proposed by Sie and Krishna (1998), "complete wetting" was a necessary but not sufficient condition for the ideal liquid-solid contacting, whereas "even irrigation" appeared to be a more stringent and more appropriate requirement. The periodic passing of the liquid-rich slugs will refresh the liquid film and prevent formation of hot spots. It is reasonable to consider that the liquid film over the particles is even. Changing the operation parameters can control the intensity of the forced pulsing flow. Therefore, there must be some optimal operating conditions to achieve the maximum benefit of the forced pulsing flow.

Notation

C	conductance
C_{max}	maximum conductance measured as reference value to calculate the liquid holdup using Eq. 4
L	liquid superficial velocity
G	gas superficial velocity
G_{eff}	effective gas superficial velocity
ΔG	forced input gas superficial velocity
ΔG_{eff}	increment of effective gas superficial velocity
t_0	the forced input time
t_p	time period of the forced pulsing flow
R_β	rate of liquid holdup varying with the operation parameters
V_p	induced pulse velocity
X	Lockhart-Martinelli parameter

Greek letter

β	liquid saturation
β_b	bed-averaged liquid holdup
β_p	liquid holdup of the induced pulses
β_{base}	base liquid holdup, defined as the averaged liquid holdup without the induced pulses
$\Delta\beta_{pb}$	increment of liquid saturation due to induced pulses
λ	parameter defined in Eq. 1, the intensity of the forced input gas flow compared with the base gas flow rate G

References

Beimesch, W. E., & Kessler, D. P. (1971). Liquid-gas distribution measurements in the pulsing regime of two-phase concurrent flow in packed beds. A. I. Ch. E. Journal, 17, 1160-1165.

Blok, J. R., Varkevisser, J., & Drinkenburg, A. A. H. (1982). Hydrodynamic properties of pulses in two-phase downflow operated packed columns. Chemical Engineering Journal, 25, 89-99.

Cheng, Z. M., & Yuan, W. K. (1999). Necessary condition for pulsing flow inception in a trickle bed. A. I. Ch. E. Journal, 45, 1394-1400.

Gianetto, A., & Specchia, V. (1992). Trickle-bed reactors: state of art and perspectives. Chemical Engineering Science, 47, 3197-3213.

Haure, P. M., Hudgins, R. R., & Silveston, P. L. (1989). Periodic operation of a trickle-bed reactor. A. I. Ch. E. Journal, 35, 1437-1444.

Jiang, Z. X., Cheng, Z. M., & Yuan, W. K. (1997). A kind of trickle bed reactor with gas pulsation equipment. CN 1196275A.

Larkins, R. P., White, R. R., & Jeffrey, D. W. (1961). Two-phase Concurrent Flow in Packed Beds. A. I. Ch. E. Journal, 17, 231-239.

Sato, Y., Hirose, T., Takahashi, F., & Toda, M. (1973). Pressure loss and liquid holdup in packed bed reactor with cocurrent gas-liquid down flow. Journal of Chemical Engineering of Japan, 6, 147-152.

Sicardi, S., Gerhard, H., & Hoffmann, H. (1979). Flow Regime Transition in Trickle-Bed Reactors. Chemical Engineering Journal, 18, 173-182.

Sie, S. T., & Krishna, R. (1998). Process development and scale up: III. Scale up and scale-down of trickle bed process. Review of Chemical Engineering, 14, 203-252.

Tsochatzidis, N. A., Karapantsios, T. D., Kostoglou, M. V., & Karabelas, A. J. (1992). A conductance probe for measuring liquid fraction in pipes and packed beds. International Journal of Multiphase Flow, 18, 653-667.

Tsochatzidis, N. A., & Karabelas, A. J. (1995). Properties of pulsing flow in a trickle bed. A. I. Ch. E. Journal, 41, 2371-2382.

Practical Studies of the Commercial Flow-reversed SO₂ Converter[*]

Abstract A commercial-scale flow-reversed SO_2 converter with two inter-stage heat exchangers has been studied based on the author's practices in a lead smelter. A control strategy with two control loops of cooling and cycling durations, and an special device ensuring the outlet temperature stability have been proposed and operated smoothly. Normally, the SO_2 final conversion is higher then 92%, and the maximum temperature is between 530 and 560℃ as the SO_2 composition fluctuates within 1.0% and 4.5% (vol). Problems related to the commercialization of unsteady-state SO_2 converter are also discussed.

Key words unsteady-state catalysis, fixed-bed reactor, SO_2 conversion

1 Introduction

The unsteady-state SO_2 converter with flow reversal has been commercialized to a large scale since the early 1980s by Matros and coworkers. It was in 1993, a converter of this type was established to treat the copper smelter gases in the Shenyang Smelter, Liaoling Province, China, applying the technology of Boreskov's Institute of Catalysis, with its performance basically satisfactory. On August 1, 1997, another flow-reversed converter started to be operated at the Jiyuan lead smelter, Henan Province, to treat the lead waste gases, using the UNILAB technology, designed at a gas flow rate of about 33000m³/h (STP), a superficial velocity of 0.325m/s (STP), and a SO_2 concentration of 2.93% (vol). The inner diameter of the converter is 6.5m, and the height 13.35m. According to the practical observations and experiences from the Shengyang and Jiyuan converters, however, we have found that, some practical problems are still worth studying. In this paper, some problems are discussed related to the commercialization practices of the largescale flow-reversed converter.

2 Configuration of Jiyuan Converter

The converter and the flowsheeting in Jiyuan lead smelter are programmed into a simulation package with the operating conditions and the state variables, such as temperature and pressure displaying on the computer screen on-line, as is illustrated in Fig. 1. It is a two-point heat removal unsteady-state converter with two interstage heat exchangers, each of 100m² in heat transfer surface area, hereafter referred to as the HRUS-2 converter as in Xiao et al. (1998), but is featured with five stages and three three-way valves. The converter is 6.5m in inner diameter, with an empty pillar, 2.0m in diameter, located at the bottom center, sustaining the support for each packing stage.

---

[*] Coauthors: Xiao Wende, Wang Hui. Reprinted from *Chemical Engineering Science*, 1999, 54: 4645-4652.

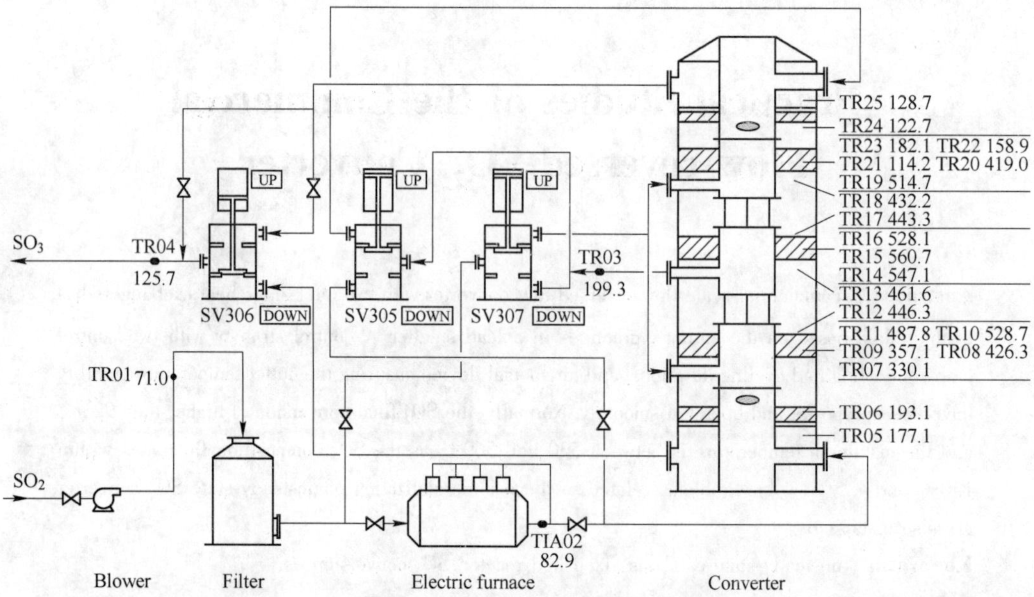

Fig. 1 Configuration and flowsheeting of the Jiyuan converter

   Among the five stages, the first and fifth ones are packed with inert ceramic pellets of 0.4m in height functioning as the output temperature buffers (OTB). The third stage is packed with a domestic catalyst, S101, 0.5m in height, sandwiched with ceramic pellets at both ends of 0.05m each in height. The second and fourth stages are packed with blended catalysts of S101 and S108 at a ratio of 1∶1, both domestically manufactured, each catalyst stage being 0.5m in height. Catalyst S101 is known more active at a temperature higher than 550℃, while S108 at lower than 500℃, their ignition temperatures being 400℃ and 360℃, respectively. This specially designed catalyst packing aims to keep a high catalyst activity in the entire reactor, based upon the primary studies shown in Xiao et al. (1998), that a temperature higher than 550℃ is generally in middle part of the reactor, and lower than 500℃ in both sides.

   Among the three three-way valves, one is for regulation of the flow rate of feed gases passing through the two inter-stage heat exchangers, denoted as SV307; the other two, denoted by SV305 and SV306, are for control of the flow reversal. The three valves are as a whole responsible for the converter operation. The valve is a piston type, its downward and upward movements driven by a computerized pneumatically controlling device. The piston positions are actually made visible on the computer screen by "UP" and "DOWN", so that whether the practical position is fit to the computer executed one can be immediately checked. All the reactant gases are introduced into the inter-stage heat exchangers as SV307 is "DOWN". The flow rate through each exchanger is made one-half of the total one by an elaborate design. In addition, when SV305 is "UP" and SV306 "DOWN", the flow is upward in the converter, and vice versa.

   Twenty one thermocouples, marked from TR05 to TR25, are mounted to define the temperature profile on-line. TR07, TR08 and TR09 are the three located at different radial positions but same axial position in the four stage to see whether the radial temperature difference exists, which undoubtedly symbols the non-uniformity of the gas flow. TR21, TR22 and TR23 are installed for the same purpose in the second stage.

Some required subsidiaries, such as the blower, the electric furnace with a power of 1200 kW for preheating the converter during start-up, and the filter for delimiting the acid fog present in the feed gases, if any, are also included in the flow sheeting.

3 SO₂ Composition in Smelter Flue Gases

The SO_2 containing gases to be treated come from a continuously sintering reactor, which is a caterpillar moving horizontally at a speed ranging from 1.0 to 1.5m/min with air blown in from the bottom. At the reactor entrance, a combustion chamber, also named as an igniter, in which heavy oil is burned, is fixed to start up the sintering reaction of the mineral lead in form of lead sulfide, i.e.:

$$PbS + 1.5O_2 = PbO + SO_2 \qquad (1)$$

The caterpillar is characterized by recycling the sintered gases and mineral and drawing the output gases from its middle part. The SO_2 content in the flue gases was first thought to be fairly stable at a design value of 2.93% (vol). However, in fact, it is quite unstable, as shown in Fig. 2, due to the unstable compositions of the feed, and to the unstable particle sizes, etc. Moreover, the process from the raw mineral pretreatment to sintering is so complicated that sudden shutdowns are very often. Therefore, there has been a common sense that, in the field of the lead smelter, the flue gases are just vented into the atmosphere without treatment. The Jiyuan converter, to the authors' knowledge, is basically the first one to recover the SO_2 from a lead smelter in China.

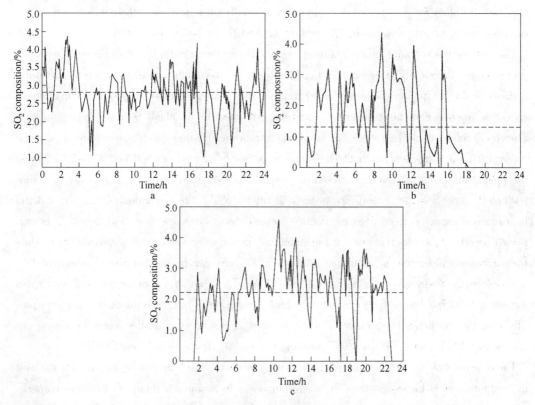

Fig. 2 SO_2 composition in the flue gases

a—Without shutdown; b—With 2.5h shutdown; c—With 7h shutdown

Solid curve—transient; Dashed—average on the domain

Fig. 2 rewrites the SO_2 composition recordings from a thermoconductive detector (TCD), that are 0.2%-0.6% lower than those by a more accurate method of iodine titration. In Fig. 2a, the average SO_2 composition is about 2.2% (vol) within 24h, with the lowest of zero from 22:30 to 1:30, meaning a shutdown period for 2h, and the highest about 4.5% (vol). Fig. 2b displays no shutdown, with the average of 2.6%, the lowest 1.0%, and the highest 4.5% (vol); Fig. 2c illustrates a shutdown period of 7h from 18:00, with the average of 2.1% from 1:00 to 18:00, the lowest zero, and the highest also 4.5%. Nevertheless, the practical SO_2 composition is generally of the order of 3% (vol) under normal operating conditions, much close to the anticipated design value.

4 Control Strategies of the Converter

As have been revealed that, the process conditions, especially the SO_2 composition in the feed gases changes with time in a broad range, and then the control conditions, such as the cycle duration of the flow reversal, and the duration of the feed gases flowing through the interstage heat exchangers should keep pace with those changes, in order to ensure an appropriate SO_2 conversion.

Two control loops are proposed and programmed into a computer package for the Jiyuan converter, with one loop for control of cooling duration introducing the cold feed gases into the exchangers using a bang-bang method, or on/off method, hereafter referred to as the cooling control, and the other for the control of cycle duration leading the gases into the catalyst bed in one direction using an adaptive method, hereafter referred to as the cycling control.

The cooling control aims at maintaining the maximum temperature of the converter within a proper scope. Practically, it is much difficult to keep the maximum temperature to a fixed value, because of the fluctuation of the SO_2 concentration, and of the long time delay of the converter response to a change of any operating condition, estimated to be about 90min (Wang, 1996). Thereby, a robust control strategy has been tried, trapping the maximum temperature in a given scope just by a bang-bang method. The scope is primarily given from 520 to 530℃ based upon the simulation results, as has been shown in Xiao et al. (1998), when the average SO_2 input composition is about 3%, the superficial velocity 0.3m/s (STP), the cycle duration 20min and the heat removal capacity about 3%. In practice, nevertheless, the scope from 530 to 560℃ is employed based on the observation that the practical final conversion hardly surpasses 90% when the maximum temperature is controlled to about 520℃, and can be improved when about 540℃.

Therefore, the three-way valve SV307 will be "DOWN", or "on", introducing the cold feed gases through the two inter-stage heat exchanges to cool the converter and to adjust the temperature profile, when the maximum temperature of the converter, also displaying on the computer screen on-line, is over 560℃, and "UP", or "off", shutting the cooling down, when below 530℃.

The objective of the cycling control is to make the maximum temperature locate near the end of the first catalyst bed when the flow is shifted, so as to produce a relatively optimal temperature profile. This concept is stemmed from our simulation studies, shown in Xiao et al. (1998), which offers an optimal cycle duration. The optimum implies, by the end of a half-cycle, the reaction wave front, denoted by the temperature profile front, is preferred to creep to the outlet boundary of the first catalyst stage. This can be accomplished by regulating the half-cycle duration. By the way, the "half-cycle duration" just means that the upward and downward durations are allowed to be unequal with each other due to the unexpected disturbance.

Two procedures have been tested for the cycling control, one being referred to as time control and the other to as temperature control. Primarily, we used the time control to set the upward and downward flow durations separately, according to the temperature changes displayed on the computer screen, and generally, the two values are equally set from 10 to 15 min. This method, though not automatic, has played an important role in the primary phase of the commercialization to find the behaviors of the newly constructed unsteady-state converter. Moreover, during the restart-up of the converter after a long period of shutdown when the converter temperature has decreased to a very low level, this method is most useful by setting different upward and downward flow durations to adjust converter quickly to establish a desirable temperature profile.

The temperature control is automatic method, and can adaptively keep pace with the changes of the operating conditions. The accurate temperature profile front is not so easy to determine in practice, so a simple method has been developed. We only use TR10 and TR20 to calculate the migration of the temperature profile fronts. When flow upward, TR10, generally behaving itself to decline, is used to track the temperature front, and when it is lower than a given value, such as 460℃ in most cases in this work, the feed direction is reversed downward. On the contrary, the downward flow will be reversed upward when TR20 shows below 460℃. Therefore, the two half-cycle durations can be automatically and adaptively regulated regardless of the frequent and extensive change of the process conditions. The resulted half-cycle duration is about 12 min when the input SO_2 concentration is of the order of 2%, and about 16 min when about 3.5%, at a flowrate of about 28000 Nm^3/h, corresponding to a superficial velocity of 0.28 Nm/s. It is in a fair agreement with the calculated results by Xiao et al. (1998).

In a word, the cooling control loop must be attempted to find a reasonable heat removal capacity for the interstage heat exchangers by regulating the cooling duration, and the cycling control loop to find a rational cycle duration. Accordingly, these loops can produce a favorite synergetic performance, as shown in Fig. 3.

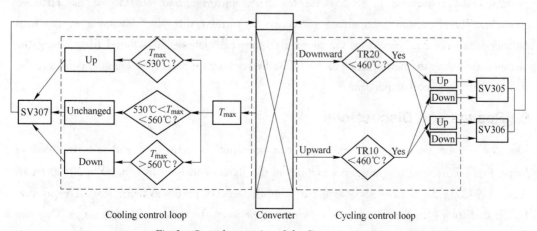

Fig. 3 Control strategies of the Jiyuan converter
Dashed frame—executed by computer

5 Buffering of Output Temperature

It is known that the flow-reversed unsteady-state fixed-bed reactor is always characterized by the periodic fluctuation of the outlet temperature (Xiao,1991;Xiao & Yuan,1996). This behavior often brings about troubles. Expansion and contraction of the outlet threeway valve, resulted from the fairly sharp heating and cooling, causes the airtightness deteriorated, and so does the final SO_2 conversion. In addition, sulfuric acid may form in the pipeline from the converter to the SO_3 absorber as the temperature is lower than the dew point of the gases (about 105℃). Finally, the acid fog is brought to the absorber and the chimney.

For the adiabatic unsteady-state SO_2 converter, hereafter referred to as AUS, the calculated results show that at $SO_2 = 3\%$, the superficial velocity = 0.3m/s, and the cycle duration = 20min, the outlet temperature stayed unchanged at the inlet one, 60℃, for the initial five minutes, but rises at a rate about 60℃/min up to 350℃, as has been calculated by Xiao (1991). For a heat removal converter, HRUS-2, with two inter-stage heat exchangers, the calculated outlet temperature is from 106 to 260℃ with a heat removal capacity of 3% and the other conditions remain same as the adiabatic converter (Xiao et al.,1998).

To resolve this problem, the authors have proposed a proprietary device, named the output temperature buffer(OTB) (Xiao & Yuan,1993). With an OTB, the output temperature fluctuates little. The calculated results are shown in Fig. 4a for the AUS converter and in Fig. 4b for the HRUS-2 converter, with $SO_2 = 3\%$ (vol), superficial velocity = 0.3Nm/s, cycle duration = 20min, and the heat removal capacity = 3%. One can notice that the OTB is very effective, which makes the output temperature much stable, almost unchanged at about 140℃ for the AUS converter, and at about 150℃ for the HRUS-2 converter. The practical results of Jiyuan converter are shown in Fig. 4c, where TR04 stands for the temperature after the OTB, and TR06 and TR08 for those before the OTB, showing a satisfactory agreement with the calculated results.

In these calculations, the inlet temperature of feed gases is assumed to be 60℃, the reaction heat of the SO_2 oxidation, $-\Delta H = 22.5$ kcal/mol and the specific heat capacity, $C_p = 7.44$ cal/mol, so the adiabatic reaction temperature rise, $\Delta T_{ad} = 90.72$℃. Moreover, the calculated average outlet conversion is 88.5% for the AUS converter, and 97.1% for the HRUS-2 converter. Therefore, the calculation confidence is verified with the difference of the heat balance less than 2℃. Based on the more elaborate calculations, it is found that, among the factors influencing the performance of the OTB, the packing heights of the ceramic pellets in the first and fifth are most important.

6 Results and Discussions

The key for the unsteady-state SO_2 converter operation is concerned with the two control loops. Fig. 4 illustrates temperature variations of the Jiyuan converter from 8:00 to 20:00 on 16 April,1997. Fig. 5a, b and c show temperature of the second, third and fourth stages, respectively. Fig. 6 displays the temperature of the chamber between the third and fourth stages. One can find that the control loops are fairly satisfactory with the maximum temperature being trapped within the scope of 530℃ and 560℃. The cycle duration was 32min.

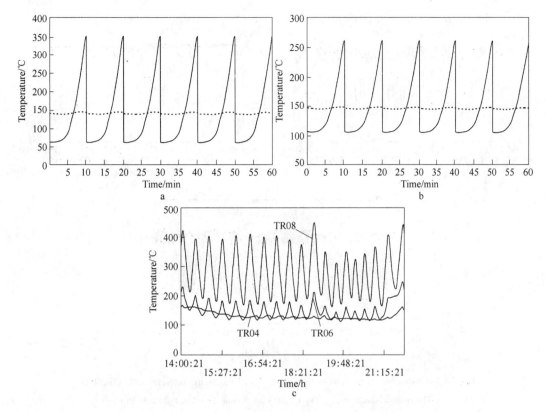

Fig. 4 The effect of OTB on output temperature
a—AUS converter: solid—without OTB, dashed—with OTB;
b—HRUS-2 converter: solid—without OTB, dashed—with OTB;
c—Jiyuan converter: TR04—after OTB, TR06—between OTBI and the first catalyst stage,
TR08—at the exit boundary of the first catalyst stage

During this period, the average input SO_2 concentration was about 2.2% (vol), based on the TCD analysis, and about 2.8% on the iodine titration, and the flow rate was about 28000Nm^3/h measured by a hole-and-plate flowmeter. The measured final conversion of the converter was from 92.2% to 93.8% by the iodine titration.

Some unexpected phenomena occurred. As can be seen that, the maximum temperatures in the three catalyst stages denoted by TR11, TR15 and TR19 are almost close, and sometimes TR15 is the highest. The temperatures in the inter-stage chambers, denoted by TR12, TR13, TR17 and TR18, are about 450℃, lower than the middle one, TR15, by about 100℃, shown more clearly in Fig. 1. According to the simulation results (Xiao et al., 1998), however, the middle stage temperature should be close to those in the chambers. We once contributed this to the maldistribution of the gas flow, but the difference between TR12 and TR13, and that between TR17 and TR18 were against this point.

The brown color of the produced sulfuric acid reminded the authors that impurities present in the feed gases, such as CO, hydrocarbons resulted from the heavy oil combustion, may affect the converter greatly. Generally, the sulfuric acid from the sulfur and pyrite roasting is clearly transparent and with no color. The brown acid, which can be decoloured by hydrogen peroxide, is

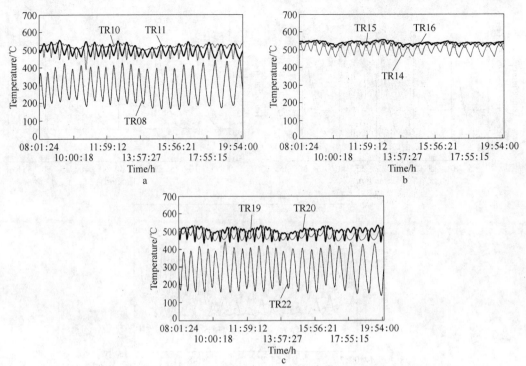

Fig. 5 Temperature variations of the catalyst bed in Jiyuan converter in 8:00-20:00
a—The first catalyst stage; b—The middle catalyst stage; c—The third catalyst stage

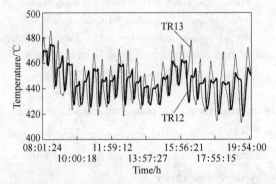

Fig. 6 Temperature variations of the chamber between the second and third stages

certainly caused by the hydrocarbons in the feed gases. As has been reported by Liu et al. (1993) that CO and hydrocarbons, particularly the latter, of only hundreds ppm in composition, may make the vanadium catalyst significantly deactivate, with the final conversion decreased by 3%-5% for the conventional converters. The lower the reaction temperature, the more considerable the deactivation, which means the deactivation mechanism may be the competitive adsorption of CO and hydrocarbons with SO_2 and O_2.

For the unsteady-state converter with flow reversal, this deactivation must be much worse, because both the front and back of the reaction zone are at a lower temperature levels. The adsorption of CO and hydrocarbons is understandably much considerable, and so is the catalyst deactivation.

Simulation was done to conclude that the reaction rate could be reduced to 40% of the basic rate, calculated by the correlation of Ge et al. (1984) for the S101 catalyst. The middle stage temperature, represented by the middle point temperature, increases from 450 to 540℃, and the average outlet conversion decreases from 97.1% to 93.5%, when $SO_2 = 3\%$, $u = 0.3 Nm/s$, cycle duration = 20min, and the heat removal capacity equals to 3%. This indeed provides a satisfactory evidence to the practical observations.

7 Problems to be Studied

7.1 Flow distribution in the converter

There has been well established that the superficial velocity has great effect on the maximum temperature and the final SO_2 conversion, (Matros, 1989; Xiao, 1991), so the well distributed flow in the flow direction is necessary. As the velocity increases from 0.1 to 0.4m/s with the input SO_2 concentration fixed to 3% (vol), for example, the maximum temperature rises from 564 to 643℃, and the final conversion decreases from 96.8% to 85.5% in an AUS converter. However, the situation is better for the HRUS-2 converter as has been studied by Xiao et al. (1998).

Another reason for a uniform flow distribution is that the catalyst bed may subside nearby the reactor wall due to the upward and downward blowing. Higher voidages bring short-cuts of the gases, and make the conversion deteriorate considerably. We have regrettably noticed this fact from the Jiyuan converter, though it may not happen to the conventional steady-state converter in which the gas flow direction maintains unchanged and is generally downward.

For compensation, an elaborately designed distributor has been developed and applied to revamping of the Jiyuan converter.

7.2 The three-way valve

The three-way valves, 1.2m in diameter, are the critical devices for the flow-reversed SO_2 converters. In the Jiyuan converter, two out of the three valves are used for cycling control and the rest for cooling control. One for cycling for the inlet gases with SO_2, and the other for the outlet gases with SO_3. To our experiences, the SO_3 valve requires much care.

In practice, the three-way valve is of the piston type with its stem moving up and down. When moving upwards, the part which contacts with SO_3 is exposed to the moisture surroundings, and thereby concentrated sulfuric acid forms. On condition of concentrated sulfuric acid and relatively high temperature up to about 180℃, few kinds of steel can endure the corrosion except for the Hastelloy alloys. In the Jiyuan converter, the 317 steel stem was first used, corrosion occurred obviously after half a year operation. Therefore, an improvement on the valves was proposed to isolate the stem for the moisture surroundings.

7.3 Special catalysts to the unsteady-state SO_2 conversion

There has been no catalysts specially for the flowreversed SO_2 converters, to the authors' knowledge. The strength of the catalyst seems to be a major concern, but in the Jiyuan converter, little pulverization was found after half a year operation, although the catalysts had to undergo the periodical gas blowing downward and upward. Sulfuric acid condensed in the converter, especially in the second and fourth catalyst stages, being responsible for the catalyst pulverization, has not been found in the Jiyuan converter.

The particle size of the catalyst has been reported affecting the performance of the unsteady-state SO_2 converters by Matros (1989). The maximum temperature decreases, and then the conversion increases with increase in the catalyst diameter. Therefore, coarser catalyst pellets should be employed for the relatively concentrated SO_2 gases, and but for the lean gases, based on the author's calculations, small-sized catalysts are preferred.

8 Conclusions

This contribution presents the observations and some problems to be studied in a commercialized HRUS-2 converter. This converter has been commercialized based upon the authors' pilot-scale (0.8m in diameter) experiments from 1990 to 1993 (Xiao et al., 1995; Xiao & Yuan, 1996) and on the computer simulation studies. A 60-fold scale-up was finally realized.

The results are basically satisfactory with the SO_2 conversion more than 92% and the recovered sulfuric acid (93%, wt) about 80 tons/day, provided the upstream sintering supplies the desired gases. The HRUS-2 converter, employing the domestic S101 and S108 catalysts, has been flexible to fluctuation of the SO_2 composition. It can be operated as an AUS converter as the SO_2 composition is low. The restart-up of this HRUS-2 converter is convenient, too, provided the temperature level of the converter is above the ignition temperature of the catalysts. The highest score of shutdown period is up to 36h for the Jiyuan converter before restart-up.

The proposed two control loops are effective to the converter operation, with the maximum temperature trapped in a given range and the cycle duration of flow direction resulting in a preferred temperature profile. The control strategies have been assembled into a package, which is very convenient for the operators and supervisors. The OTB is effective, too.

The Jiyuan converter has been being operated for half and a year with about 30000 tones of sulfuric acid of 93% (wt) produced. As for the economics of the installation, the cost of sulphuric acid is about 280 RMB (33.5 $) per ton, less than those from the sulfur or pyrite roasting plants, where it is over 400 RMB (48.3 $) per ton. Among the cost, the electric power consumption and the investment depreciation are responsible for 85%, as the feedstock SO_2 is free of charge.

Acknowledgements

The present work is supported by the National Science Foundation of China, the State Key Laboratory of Chemical Engineering of China, and by the State Education Commission of China.

References

Ge, H. X., Han, Z. H., & Xie, K. C. (1984). Mechanism and kinetics of SO_2 oxidation on "K-V" and "K-Na-V" Catalysts. Journal of Chemistry and Industry Engineering (China), 37, 244-256.

Liu, S. W., Qi, Y., Zhao, S. Q., & Ding, R. B. (1993). Technologies for sulfuric acid production. Dongnan University Press, (in Chinese).

Matros, Yu. Sh. (1989). Catalytic processes under unsteady-state conditions. Amsterdam: Elsevier.

Wang, H. (1996). Study on the stability and control of the flow-reversed SO_2 converter. M. S. thesis, East China University of Science and Technology.

Xiao, W. D. (1991). Study on the behaviours and control of unsteadystate fixed-bed reactor with flow reversal. Ph. D. dissertation. East China University of Science and Technology.

Xiao, W. D., Wang, H., & Yuan, W. K. (1998). Unsteady-state SO_2 converter with inter-stage heat removals. Chemical Engineering Science, submitted for publication.

Xiao, W. D., & Yuan, W. K. (1996). Modelling and experimental studies on unsteady-state SO_2 converters. The Canadian Journal of Chemistry Engineering, 74, 772.

Xiao, W. D., Yuan, W. K., Ma, J., & Zheng, Q. (1995). Pilot-scale research on unsteady-state SO_2 conversion. Sulfuric Acid Industry No. 1, 3-13; No. 3, 3-18 (in Chinese).

Xiao, W. D., & Yuan, W. K. (1993) Integrated unsteady-state SO_2 converters with inter-stage cooling, Chinese Patent. CNP: 91107321. 3.

Nanomaterials Synthesized by Gas Combustion Flames: Morphology and Structure*

Abstract The flame technology has been employed broadly for large-scale manufacture of carbon blacks, fumed silica, pigmentary titania, and also ceramic commodities such as SiO_2, TiO_2, and Al_2O_3. A deeper understanding of the process also made it possible for production of novel nanomaterials with high functionality-various novel nanomaterials such as nanorods, nanowires, nanotubes, nanocoils, and nanocomposites with core/shell, hollow and ball-in-shell structures, have been synthesized recently via gas combustion technology, while the mechanisms of the material formation were investigated based on the nucleation-growth and chemical engineering principles. Studies of the fluid flow and mass mixing, supported by principles of chemical reaction engineering, could provide knowledge for better understanding of the process, and thus make rational manipulation of the products possible.

Key words flame, nanomaterials, synthesis, chemical engineering, structure

1 Introduction

The gas combustion flame technology refers to the formation of nanomaterials from gases in flames. This technology has been initially developed for preparation of carbon black since prehistoric times as depicted in Chinese ink artwork. In the late 19th century, fine carbon black used as fillers in rubbers was commercially produced through pyrolyzing the undesirable by-product natural gas from oil fields (Stark & Pratsinis, 2002). Industrial laboratories led the research in this field in the mid-20th century, motivated from the industrial importance of fumed silica which was first made in the 1940s and was used mainly as fillers for silicon rubbers. In the late 1970s, the flame technology contributed decisively to the manufacture of ultrapure silica fiber that became the basic material for lightguides in telecommunications. Nowadays, the flame technology is employed routinely in large scale manufacturing of carbon blacks and ceramic commodities such as fumed silica and pigmentary titania and, to a lesser extent, for specialty chemicals such as zinc oxide and alumina powders. The flame technology appears to possess many advantages over the conventional wet processes because the product can be collected easily by filters, so the postheating treatment is no longer required, especially due to its continuous nature and the ability for scale up. Therefore, this technology has been regarded as a well-established method for production of nanomaterials with high purity and controllable structures.

Ulrich (1984) pioneered the investigation of flame synthesis of ceramic powders by making SiO_2 powders through $SiCl_4$ oxidation in premixed flames. Since the early 1990s, the pace of research has been further intensified with a renewed interest in flame technology for manufactur-

---

* Coauthors: Li Chunzhong, Hu Yanjie. Reprinted from *Particuology*, 2010, 8:556-562.

ing advanced materials mainly for nanosized particles (Pratsinis, 1998). In 1996, fullerenic nanostructures were synthesized in flames (Das Chowdhury, Howard, & Vandersande, 1996), and single-wall carbon nanotubes were prepared in the $C_2H_2/O_2/Ar$ flames (Richter et al., 1996). More recently, an array of nanoscaled, sophisticated products such as catalysts, sensors, dental and bone replacement composites, phosphors, fuel cell and battery materials, and even nutritional supplements were made with flames (Strobel & Pratsinis, 2007). In addition, many novel metal oxide nanomaterials with potential application in electronics, biotechnology and catalysis are also synthesized by gas flame combustion (Li, 2010). However, it is still difficult to produce multicomponent materials with controlled morphology and structures, because of the nonuniformity of temperature and gas composition in the flame.

Flame synthesis includes combustion of volatile precursors in a hydrocarbon, hydrogen or halide flame (Strobel & Pratsinis, 2007). Precursor is injected into the burner as the gas phase, or in the form of droplets or even solid particles, that reacts to form intermediate and product molecules and clusters. Products form via nucleation, surface reaction and/or coagulation and subsequent coalescence into larger aggregates. As the aerosol stream leaves the high temperature zone, particle growth continues mostly by coagulation to the complete particles. The flame synthesis for the nanomaterials involves processes such as rapid high-temperature vapor reaction, nucleation, particle growth, and agglomeration. Meanwhile these processes correlate and interact, leading to complexity of the nanomaterial formation. Therefore, preparation of nanomaterials and corresponding growth mechanism in this complex flame environment have become the focus of the study in the past years. For such a complex system, it would be prudent to study the system only to set up the ties between the so-called dominating-scale structure, and certain holistic performance of the system that interests people, and then to manipulate the found dominating-scale structure to achieve our target (Yuan, 2007). For synthesis of nanomaterials by the flame technology, the key target is to control the micro/meso-structures of the products, because the application of nanomaterials is determined mainly by their chemical composition, together with their micro/meso-structures. Aided by the chemical engineering principles and developed methods, many novel nanomaterials have been synthesized by our laboratory, using the flame technology. Equipments for SiO_2 and TiO_2 nanoparticles have been scaled up successfully.

2 Nanoparticles

The gas combustion flame synthesis is most suitable for synthesizing nanoparticles of high-purity, high specific surface area and controlled particle size distribution. Many kinds of nanoparticles have been synthesized commercially for applications in electronics, biotechnology and catalysis, etc. (Pratsinis, 1998). Recently, nanosized Al_2O_3 and SnO_2 were prepared by oxidation of doped metalorganic precursors like $Al(CH_3)_3$ or $Sn(CH_3)_3$ in low pressure premixed $H_2/O_2/Ar$ flames (Lindackers, Janzen, Rellinghaus, Wassermann, & Roth, 1998). Amorphous spherical Al_2O_3 particles were found in a size range from 4.7 to 8.44nm, while SnO_2 particles ranged between 2.7 and 8.3nm. A theoretical model was developed including the full H_2/O_2 reaction kinetics as well as the transport properties of a burner stabilized flame. With an external electric field, needle and plate electrodes could be used to synthesize fumed silica with controlled particle size distribution from HMDS in a coflow double diffusion flame at atmospheric pressure (Kammler & Pratsinis,2000). The average primary particle diameter could be reduced when the applied electric field strength between the two needle electrodes was increased from 0 to 1.5 kV/cm. Based on the design of a multiport diffusion type burner composed of five concentric tubes, Jan and Kim synthesized TiO_2 nanoparticles ranging from 10 to 30nm by oxidation of titanium tetrachloride ($TiCl_4$) in a diffusion flame reactor (Jang & Kim, 2001), and the mass fraction of the synthesized anatase reached 40% to 80% in their experiments.

Among various photocatalysts, TiO_2 has outstanding photocatalytic properties and broad applications in many fields due to its proper bandgap energy, large surface area, high stability, low cost, and non-toxicity. However, when taking TiO_2 as a photocatalyst, the low photon quantum efficiency and the utilization of UV light as the excitation source would limit its application. By using a multijet flame to control the distribution of temperature and residence time, and also the mixing of reactants, a modified diffusion flame method was established to fabricate TiO_2 nanoparticles by introducing diluting gas in the central tube (Zhao, Li, Liu, & Gu, 2007; Zhao, Li, Liu, Gu, et al, 2007; Zhao, Li, Liu, Gu, & Du, 2008). This method partially overcomes the disadvantages of the traditional methods, which would result in the formation of TiO_2 with larger particle sizes and inhomogeneous distribution. The as-prepared TiO_2 nanoparticles show much better photocatalytic activity (three times) compared to the P-25 TiO_2 when degrading rhodamine B under visible light irradiation (Fig. 1), that can be attributed to well-dispersion characteristics of the TiO_2 nanoparticles prepared by modified diffusion flame process (Zhao, Li, Liu, & Gu, 2007). For the Zn^{2+} doped TiO_2 nanoparticles prepared by the modified diffusion flame method, most part of dopant Zn^{2+} ions supposedly locate on the surface of TiO_2 nanoparticles to form small ZnO nuclei randomly dispersed on the anatase TiO_2 surface. The doped TiO_2 shows better photocatalytic properties than the pure TiO_2, because of the discrepancy of the energy band position of TiO_2 and ZnO (Zhao, Li, & Gu, 2008). The properly Fe^{3+} doped TiO_2 nanoparticles exhibit higher photocatalytic activity under UV excitation than undoped TiO_2, since Fe^{3+} helps to separate photogenerated electrons and holes by trapping them temporarily and shallowly (Zhao, Li, & Liu, 2007). However, high-level Fe^{3+} doping would lead to charge pair recombination and thus decreasing the photocatalytic activities of TiO_2 nanoparticles. Tian et al. prepared V-doped TiO_2 nanoparticles by a simple one-step flame spray pyrolysis (FSP) technique(Tian, Li, Gu, Jiang, Hu, et al., 2009). Benefiting from the short residence time and high quenching rate during the flame spray process, V^{4+} ions are successfully incorporated into the TiO_2 crystal lattice. It reveals that V-doping favors the primary particle size growth as well as the increase of rutile content in the products, and thus V-doping enhances the photocatalytic activity under both UV and visible light irradiation.

Based on self-assembly of TiO_2 nanoparticles synthesized by H_2/O_2 combustion flames, self-cleaning films were prepared from multilayer deposition of poly(sodium 4-styrene sulfonate) on the treated TiO_2 nanoparticles and SiO_2 nanoparticles with electrostatic interaction by adsorbing positively charged poly (diallyldimethylammonium chloride) via layer-bylayer assembly processes. The films of TiO_2/SiO_2 assembled for 10 cycles provide effective photocatalytic

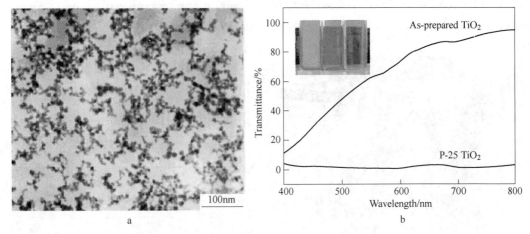

Fig. 1 The morphology (a) and the transmittance of as-prepared TiO_2 and TiO_2 P-25 (b)
(The insets show the images of RhB/TiO_2 P-25 system (left), RhB/as-prepared
TiO_2 system (middle), and RhB solution (right))

properties with a maximum transmittance of 99.3%, and can shorten the water droplet spreading time down to 0.29s. The multilayer films assembled for ten cycles are four times more active than films assembled with five cycles, indicating that flame-synthesized TiO_2 with good crystallinity can be used to fabricate high transparent self-cleaning films under proper assembly conditions (Wang, Hu, Zhang, & Li, 2010).

3 Nanorods and Nanowires

Flame-made metal oxides are always of spherical particles and chainlike agglomerates instead of the nanostructures with different shapes, particularly one-dimensional semiconductor nanomaterials with unique electronic and optical properties for nanodevice applications. ZnO nanorods were prepared via a flame spray pyrolysis by introducing indium and tin dopants which selectively affected a specific ZnO crystal plane (Height, Mädler, & Pratsinis, 2006). SnO_2 nanorods were prepared via the flame at atmospheric pressure using a multi-element diffusion flame burner with a gas-phase precursor for SnO_2 and solid-phase precursor for metal additives (Bakrania, Perez, & Wooldridge, 2007). Cobalt nanowires were prepared with magnetic fields as an effective structure directing agents from a metal ferromagnetic nanoparticle gas stream (Athanassiou, Grossmann, Grass, & Stark, 2007). This template free, continuous and rapid production method afforded nanowires with an aspect ratio of over 1000 (length/diameter) at a production rate of over 30g/h. Therefore, it is still a challenge, via the flame technology, to realize morphology-controlled synthesis of one-dimensional nanostructures for further promoting their applications. Recently a continuous and scalable iron-assisted flame approach was developed (Liu, Gu, Hu, & Li, 2010), and well-crystalline SnO_2 nanorods were primarily synthesized with a production rate up to 50g/h in the laboratory-scale (Fig. 2). The as-prepared SnO_2 nanorods with uniform length up to 200nm and diameter around 20nm are of single crystal rutile structures, growing along the [001] direction. The morphology and structure can be easily controlled by introducing Fe dopant and adjusting the flame residence time. Meanwhile, the photoluminescence (PL) spectrum of SnO_2 nanorods exhibits a broad, stronger orange-emission peak around 620nm, suggesting their potential applications in optoelectronics.

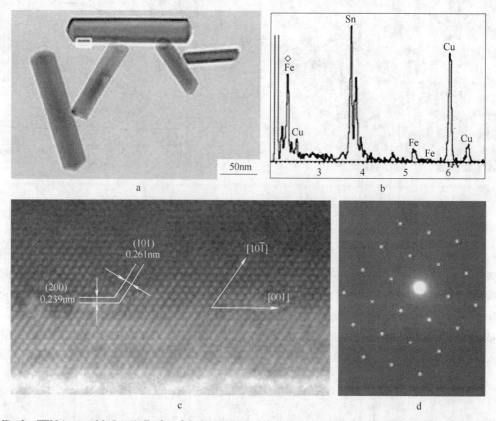

Fig. 2 TEM image of 2.5 at% Fe-doped SnO_2 nanorods (a), and their EDS analysis (b), HRTEM image (c) and corresponding SAED pattern (d) taken from the white box in Fig. 2a showing the preferred [001] orientation

A large-scale composite method, premixed atmospheric flat flame deposition, combining advantages of both flame synthesis and thermal evaporation, has been successfully developed to construct SnO_2 nanowires with novel arrow-like tips by controlling the reaction conditions. SnO_2 nanowires with special tips are different from the previously mentioned nanowires. They are structurally uniform single crystals, growing along the [001] direction, lengthing up to 4μm (Fig. 3). The arrow-like nanowire of this kind exhibits a much stronger emission peak at 620nm, allowing for potential applications in optoelectronics. The synthesis of complex, arrow-like nanowires will provide building blocks for the future architecture of functional nanodevices.

4 Nanotubes and Nanocoils

Since the discovery of carbon nanotubes, their unique structure and physical properties have attracted much attention. Flame synthesis is considered to be a very energy efficient process: a portion of the fuel serves as the heating source to leave the rest as the reactant. Therefore, flame synthesis was tentatively used for synthesizing carbon nanotubes. However, due to the complex chemical and temperature environment, flame synthesis of carbon nanotubes has been regarded as unsuccessful. Yuan et al. (Yuan, Saito, Hu, & Chen, 2001; Yuan, Saito, Pan, Williams, & Gordon, 2001) prepared multi-walled carbon nanotubes (MWCNTs) by immersing metallic sub-

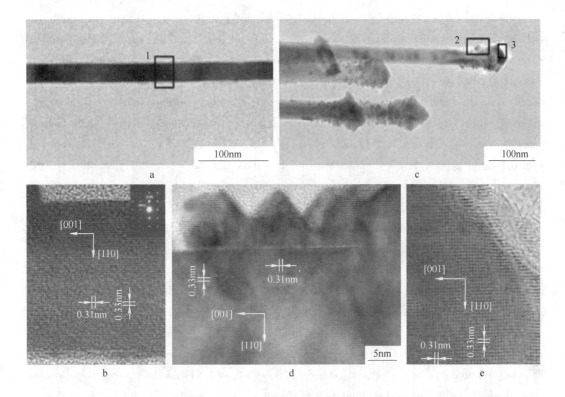

Fig. 3 Low magnification TEM image and its HRTEM image and the corresponding SAED pattern
a—TEM image of the middle part of a randomly selected nanowire; b—HRTEM image taken from Box1 in Fig. 3a, showing the preferential growth direction is [001]; c—TEM image of nanowires tips; d,e—HRTEM images taken, respectively, from Boxes 2 and 3 in Fig. 3c, displaying the structure characteristic of the wire tips

strates in a coflow diffusion flame composed of methane and ethylene. Single-walled carbon nanotubes (SWNTs) were prepared in a hydrocarbon (acetylene or ethylene)/air diffusion flame (Vander Wal, Ticich, & Curtis, 2000). Merchán-Merchán, Saveliev, and Nguyen (2009) prepared carbon and metal-oxide nanostructures on molybdenum probes inserted in a counter-flow oxy-fuel flame. Flame position and probe diameter were varied to achieve a controlled growth of carbon and metal-oxide nanostructures in the fuel and oxygen-rich flame zones. Molybdenum probes of 1mm diameter were introduced in the flame at various heights, starting from the upper hydrocarbon-rich zone on the fuel side of the flame to the oxygen-rich zone on the oxidizer side. High density layers of carbon nanocoils (CNCs) and filamentous structures containing ribbon shape and straight nanofibers were formed in the upper hydrocarbon-rich flame zone. MoO_2 microchannel structures were formed on the oxidizer side in the vicinity of the flame front. The micro-channels appeared as rectangular and square-framed shapes; they were completely hollow, closed, and semi-open with a small circular cavity at their tips. However, these approaches are very complicated, and the gaseous fuels cannot be safely and carefully controlled. Therefore, it still remains a challenge to develop simpler and safer approaches to synthesize CNTs.

For preparing carbon nanotubes, a methane diffusion flame was established in a tube-like burner with a fuel tube in the centre (Zhou, Li, Gu, & Du, 2008). With a high yield of carbon nanotubes, less carbon impurities were formed. Yields and purities of carbon nanotubes could be enhanced obviously in a suitable flame environment and over a proper catalyst. Multi-walled carbon nanotubes grew directly on the stainless steel mesh in a controllable methane diffusion flame. On a HCl pre-etched mesh, high density carbon nanotubes were synthesized with uniform outer diameter of about 60nm and a large inner diameter of about 50nm (Fig. 4). Carbon nanotubes were also formed in the methane diffusion flame by using of flower-shaped NiO architectures as catalysts (Zhou, Gu, & Li, 2009; Zhou, Li, Gu, & Wang, 2008). The NiO 3D architecture exhibited good catalytic characteristics for carbon nanotube formation; nanotubes grew along the surface of nanosheets, resulting in patterning growth. Most CNTs are estimated to be of 15nm outer diameter and 7nm inner diameter. The graphite crystal structure has a characteristic peak at $1580 cm^{-1}$, while multicrystal or noncrystal carbon materials have a peak at $1345 cm^{-1}$. The low value of I_{1345}/I_{1580} indicates a good graphitization degree of multiwall carbon nanotubes. Carbon nanotubes (CNTs) with ultrafine inner diameter have also been synthesized successfully through a simple ethanol flame method (Wang, Li, Gu, & Zhou, 2008; Wang, Li, Zhou, & Gu, 2007). The inner diameter as well as the crystallinity of the CNTs can be altered greatly by controlling the experimental conditions. The whole process has been divided into three steps: (1) pyrolysis of ethanol and formation of catalyst particles; (2) surface adsorption of pyrolyzed products on the catalyst particles; (3) growth of CNTs on the catalyst particle surface (including diffusion and precipitation of carbon through catalyst particles and formation of CNTs at the other side of the particles). During the formation of MWCNTs, it should be noted that the morphology, especially the inner diameter of the tubes, is strongly dependent on the burning rate.

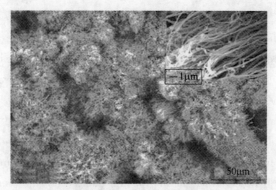

Fig. 4 Typical SEM image of the carbon nanotubes on HCl-etched mesh
(Inset shows a high-magnification SEM image of the carbon nanotubes)

Carbon nanocoils (CNCs) were first synthesized by using tin oxide nanoparticles as catalyst formed *in situ* from stannic chloride precursor in an ethanol flame (Wang, Li, Gu, & Zhang, 2009). The obtained CNCs of a mesoporous character have tight coil pitches, and the average fiber and coil diameters are approximately 50-80nm and 80-100nm, respectively (Fig. 5). The anisotropic deposition rates of carbon among tin dioxide crystal planes give rise to the driving force for the coiling of carbon fibers. The crystal plane (101) of SnO_2, which is the most favorable face for carbon precipitation, is situated on the outer side of the CNCs, while the crystal plane (110) with the lowest carbon extrusion speed is situated on the inner side. The CNCs present excellent specific capacitance of ca. 40F/g while used as polarized electrodes, considerably higher than that of the micro-coiled carbon fibers or carbon nanofibers, and possibly a candidate for super capacitors.

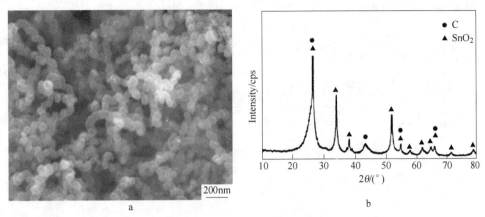

Fig. 5 SEM image (a) and XRD pattern (b) of the CNCs synthesized by using ethanol flame over SnO_2

5 Novel Nanostructures

In the recent years, flame synthesis has been chosen to prepare many novel nanostructures, such as nanoflakes, nanoneedles, core/shell, hollow and ball-in-shell structures. Single and bicrystal α-Fe_2O_3 nanoflakes and CuO nanoneedles were grown in the postflame region by a solid diffusion mechanism. The α-Fe_2O_3 nanoflakes reached lengths exceeding 20μm after only 20min of growth (Rao & Zheng, 2009). This rapid growth rate is attributed to a large initial heating rate of the metal substrate in the flame and to the presence of water vapor and carbon dioxide in the gas phase that together generate thin and porous oxide layers which greatly enhance the diffusion of the deficient metal to the nanostructure growth site and enable growth at higher temperatures than previously demonstrated. Athanassiou, Mensing, and Stark (2007) prepared carbon-coated copper nanoparticles with different carbon layers; the core/shell geometry of these carbon/metal composites afforded two distinctly different electrical behaviors depending on the carbon layer properties. Graphene layers with a predominant sp^2 character showed an ill-defined bandgap structure as evidenced by UV-vis diffuse reflectance spectroscopy.

Based on controlling the mixing state of reactants of $TiCl_4$ and $SiCl_4$, TiO_2/SiO_2 nanocomposite particles were prepared in a premixed flame in form of dispersing structure of TiO_2 nanoparticles depositing in amorphous SiO_2 matrix and core/shell structure of TiO_2 nanoparticles encapsulated by amorphous SiO_2 (Hu, Li, Cong, Jiang, & Zhao, 2006; Hu, Li, Cong, Jiang, & Zhao, 2007; Hu, Li, Gu, & Zhao, 2007). In the dispersing structure, the size of the crystalline TiO_2 particles was about 1-2nm, and would be changed along with the Ti/Si ratio in the composites. Formation of such a dispersing structure is attributed to the heterogeneous nucleation of SiO_2 on the TiO_2 surface and the short residence time in the high temperature region (Fig. 6) (Hu et al., 2006). Thickness of the SiO_2 layer in the core/shell structure was about 3-5nm and can be tuned by changing the processing parameters. The formation of core/shell TiO_2/SiO_2 nanostructures is explained by heterogeneous nucleation and growth of SiO_2 on the surface of TiO$_2$ particles (Fig. 7) (Hu, Li, Gu, & Zhao, 2007). With addition of SiO$_2$ into the matrix, the transformation from anatase to rutile has been hindered due to the surface modification by SiO$_2$. The luminescence has been enhanced remarkably by adding SiO$_2$, mostly due to surface modification of SiO$_2$ around TiO$_2$ nanoparticles.

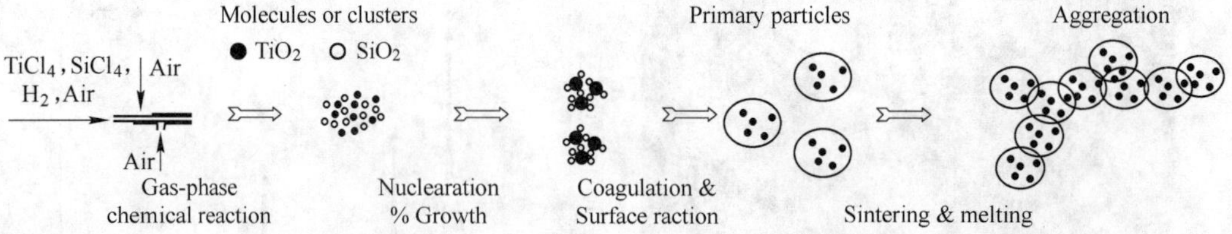

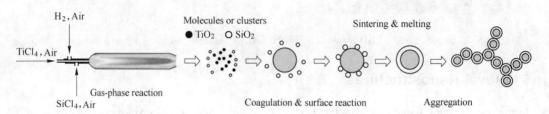

Fig. 6 Illustration for the formation mechanism of the dispersing TiO$_2$/SiO$_2$ nanostructures

Fig. 7 Illustration for the formation mechanism of the core/shell TiO$_2$/SiO$_2$ nanostructures

When the Joule-Thomson throttle cooling and high speed jet entrainment phenomena in the multi-jet reactor were taken advantage of, a novel γ-Al$_2$O$_3$ hollow nanospheres as well as the Al$_2$O$_3$/TiO$_2$ and Al$_2$O$_3$/SiO$_2$ hollow nanocomposites were successfully prepared (Hu, Li, Gu, & Ma, 2007). The particle size of these hollow nanostructures was in a range of 100-200nm and a shell thickness of 10-30nm. The shell was constructed by small particles of 5-10nm. It is desirable to state that, even for the very small spheres existing in the samples, hollow interior still exists with the shell thickness of 5nm, which is rare when prepared with conventional methods. The formation mechanism of such hollow nanostructures conforms to the One-Droplet-to-One-Particle theory (ODOP), where hydrolysis and nucleation occur on the surface of droplet when the surface concentration of droplet is greater than critical degree of super-saturation and its inner concentration is less than the equilibrium concentration. Introducing TiO$_2$ can improve Al$_2$O$_3$ crystallization to form Al$_2$TiO$_5$, whose better crystallinity can reduce formation of defects such as oxygen vacancies, making the luminescent intensity weaker.

Ball-in-shell structured TiO$_2$ nanospheres with good crystalline nature and thermal stability were formed by feeding a mixture of titanium tetrachloride and alcohol vapor to a facile diffusion flame (Liu, Hu, Gu, & Li, 2009). The resultant ball-in-shell spheres were composed of nanocrystallites, with their shell thickness and void space width of 30-50nm and 10-30nm, respectively (Fig. 8). The formation mechanism of ball-in-shell spheres depended on the relative rate between chemical reaction and diffusion. This structure results in the increased light absorbency of the spheres, and thus offers increased potential for designing and preparing novel materials with enhanced photocatalytic activities.

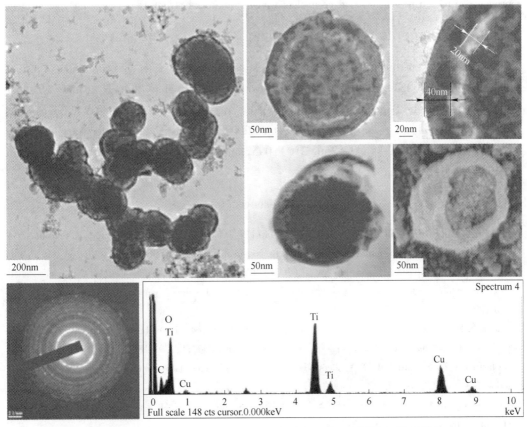

Fig. 8 TEM and SEM images, SAED pattern, and EDS analysis of ball-in-shell TiO_2 spheres

6 Summary

The flame technology has been employed widely for large-scale manufacture of powdery products, such as fumed silica, pigmentary titania, and also ceramic commodities lick SiO_2, TiO_2, and Al_2O_3. An in-depth understanding of the process makes it possible to play its role in synthesis of simple oxides nanoparticles as well as more complex, functional novel nanomaterials. Recently, many kinds of novel nanomaterials such as nanorods, nanowires, nanotubes, nanocoils, and nanocomposites with core/shell structures, hollow structure and ball-in-shell structures, were successfully synthesized by gas combustion flames, and the formation mechanism were investigated based on the nucleation-growth and chemical engineering principles.

Controlling gas concentration, as well as temperature distribution, is very important for synthesis of nanomaterials because their characteristics are often determined by temperature and concentration in the micro-scale regions. Temperature distribution, and concentration distribution, together with the residence time distribution in the flames often offer us alternatives to manipulate complex and functional nanomaterials. Micromixing of the reactants is a key to control flame temperature, reaction and nucleation-growth rates, and subsequently the morphology, composition and structure of the products. Fluid flow and mixing studies based on principles of chemical reaction engineering provide the tools for deeper understanding and better controlling of product characteristics. Also helpful are the chemical engineering principles and methods that give us more ideas in the preparation and control of the nanomaterial structure.

Acknowledgements

This paper is written under the support of the National Natural Science Foundation of China (20925621, 20906027, 20706015), the Program of Shanghai Subject Chief Scientist (08XD1401500), the Shanghai Shuguang Scholars Tracking Program (08GG09), the Special Projects for Key Laboratories in Shanghai (09DZ2202000), the Special Projects for Nanotechnology of Shanghai (0852nm02000, 0952nm02100, 0952nm02100), the Shanghai Pujiang Program (09PJ1403200).

References

Athanassiou, E. K., Grossmann, P., Grass, R. N., & Stark, W. J. (2007). Template free, large scale synthesis of cobalt nanowires using magnetic fields for alignment. Nanotechnology, 18, 165606.

Athanassiou, E. K., Mensing, C., & Stark, W. J. (2007). Insulator coated metal nanoparticles with a core/shell geometry exhibit a temperature sensitivity similar to advanced spinels. Sensors and Actuators A, 138, 120-129.

Bakrania, S. D., Perez, C., & Wooldridge, M. S. (2007). Methane-assisted combustion synthesis of nanocomposite tin dioxide materials. Proceedings of the Combustion Institute, 31, 1797-1804.

Das Chowdhury, K., Howard, J. B., & Vandersande, J. B. (1996). Fullerenic nanostructures in flames. Journal of Materials Research, 11, 341-347.

Height, M. J., Mädler, L., & Pratsinis, S. E. (2006). Nanorods of ZnO made by flame spray pyrolysis. Chemistry of Materials, 18, 572-578.

Hu, Y., Li, C., Cong, D., Jiang, H., & Zhao, Y. (2006). Morphology and structure of TiO_2/SiO_2 nanocomposites prepared by muti-jet flame reactor. Chinese Journal of Inorganic Chemistry, 22(12), 2253-2257 (in Chinese).

Hu, Y., Li, C., Cong, D., Jiang, H., & Zhao, Y. (2007). Mechanism analysis and preparation of core-shell TiO_2/SiO_2 nanoparticles by H_2/Air flame combustions. Journal of Inorganic Materials, 22(2), 205-208 (in Chinese).

Hu, Y., Li, C., Gu, F., & Ma, J. (2007). Preparation and formation mechanism of alumina hollow nanostructures via high speed jet flame combustion. Industrial & Engineering Chemistry Research, 46(24), 8004-8008.

Hu, Y., Li, C., Gu, F., & Zhao, Y. (2007). Facile flame synthesis and photoluminescent properties of core-shell TiO_2-SiO_2 nanoparticles. Journal of Alloys and Compounds, 432(1-2), L5-L9.

Jang, H. J., & Kim, S. K. (2001). Controlled synthesis of titanium dioxide nanoparticles in a modified diffusion flame reactor. Materials Research Bulletin, 36, 627-637.

Kammler, H. K., & Pratsinis, S. E. (2000). Electrically-assisted flame aerosol synthesis of fumed silica at high production rates. Chemical Engineering and Processing, 39, 219-227.

Li, C. (2010). Structure controlling and process scale-up in the fabrication of nanomaterials. Frontiers of Chemical Engineering in China, 4(1), 18-25.

Lindackers, D., Janzen, C., Rellinghaus, B., Wassermann, E. F., & Roth, P. (1998). Synthesis of Al_2O_3 and SnO_2 particles by oxidation of metalorganic precursors in premixed H_2/O_2/Ar low pressure flames. Nanostructured Materials, 10(8), 1247-1270.

Liu, J., Gu, F., Hu, Y., & Li, C. (2010). Flame synthesis of tin oxide nanorods: A continuous and scalable approach. Journal of Physics Chemistry C, 114, 5867-5870.

Liu, J., Hu, Y., Gu, F., & Li, C. (2009). Flame synthesis of ball-in-shell-structured TiO_2 nanospheres. Industrial & Engineering Chemistry Research, 48, 735-739.

Merchán-Merchán, W., Saveliev, A. V., & Nguyen, V. (2009). Opposed flow oxyflame synthesis of carbon and oxide nanostructures on molybdenum probes. Proceedings of the Combustion Institute, 32, 1879-1886.

Pratsinis, S. E. (1998). Flame aerosol synthesis of ceramic powders. Progress in Energy and Combustion Science, 24(3), 197-219.

Rao, P. M., & Zheng, X. L. (2009). Rapid catalyst-free flame synthesis of dense, aligned α-Fe_2O_3 nanoflake and CuO nanoneedle arrays. Nano Letter, 9(8), 3001-3006.

Richter, H., Hemadi, K., Caudano, R., Fonseca, A., Migeon, H. N., Nagy, B. J., et al. (1996). Formation of nanotubes in low pressure hydrocarbon flames. Carbon, 34, 427-429.

Stark, W. J., & Pratsinis, S. E. (2002). Aerosol flame reactors for manufacture of nanoparticles. Powder Technology, 126, 103-108.

Strobel, R., & Pratsinis, S. E. (2007). Flame aerosol synthesis of smart nanostructured materials. Journal of Materials Chemistry, 17, 4743-4756.

Tian, B., Li, C., Gu, F., Jiang, H., Hu, Y., & Zhang, J. (2009). Flame sprayed V-doped TiO_2 nanoparticles with enhanced photocatalytic activity under visible light irradiation. Chemical Engineering Journal, 151(1-3), 220-227.

Ulrich, G. D. (1984). Flame synthesis of fine particles. Chemical & Engineering News, 62(32), 22-29.

Vander Wal, R. L., Ticich, T. M., & Curtis, V. E. (2000). Diffusion flame synthesis of single-walled carbon nanotubes. Chemical Physics Letters, 323(3-4), 217-223.

Wang, H., Hu, Y., Zhang, Z., & Li, C. (2010). Self-cleaning films with high transparency based on TiO_2 nanoparticles synthesized via flame combustion. Industrial & Engineering Chemistry Research, 49(8), 3654-3662.

Wang, L., Li, C., Gu, F., & Zhang, C. (2009). Facile flame synthesis and electrochemical properties of carbon nanocoils. Journal of Alloys and Compounds, 473(1-2), 351-355.

Wang, L., Li, C., Gu, F., & Zhou, Q. (2008). Morphology and structure of carbon nanocoils synthesized via the flame combustion of ethanol. Journal of Inorganic Materials, 23(6), 1179-1183 (in Chinese).

Wang, L., Li, C., Zhou, Q., & Gu, F. (2007). Controllable synthesis of carbon nanotubes with ultrafine inner diameters in ethanol flame. Physica B, 398(1), 18-22.

Yuan, L., Saito, K., Hu, W., & Chen, Z. (2001). Ethylene flame synthesis of well-aligned multi-walled carbon nanotubes. Chemical Physics Letters, 346(1-2), 23-28.

Yuan, L., Saito, K., Pan, C., Williams, F. A., & Gordon, A. S. (2001). Nanotubes from methane flames. Chemical Physics Letters, 340(3-4), 237-241.

Yuan, W. K. (2007). Targeting the dominating-scale structure of a multiscale complex system: A methodological problem. Chemical Engineering Science, 62, 3335-3345.

Zhao, Y., Li, C., & Gu, F. (2008). Zn-doped TiO_2 nanoparticles with high photocatalysis activity synthesised by hydrogen-oxygen diffusion flames. Applied Catalysis B: Environmental, 79(3), 208-215.

Zhao, Y., Li, C., & Liu, X. (2007). Photosensitized degradation activity of dye by Fe-doped TiO_2 nanocrystals synthesize via gas combustion flames. Journal of Inorganic Materials, 22(6), 1070-1074 (in Chinese).

Zhao, Y., Li, C., Liu, X., & Gu, F. (2007). Highly enhanced degradation of dye with well-dispersed TiO_2 nanoparticles under visible irradiation. Journal of Alloys and Compounds, 440(1-2), 281-286.

Zhao, Y., Li, C., Liu, X., Gu, F., & Du, H. (2008). Surface characteristics and microstructure of dispersed TiO_2 nanoparticles prepared by diffusion flame combustion. Materials Chemistry and Physics, 107(2-3), 344-349.

Zhao, Y., Li, C., Liu, X., Gu, F., Jiang, H., Shao, W., et al. (2007). Synthesis and optical properties of TiO_2 nanoparticles. Materials Letters, 61(1), 79-83.

Zhou,Q. ,Gu,F. ,& Li,C. (2009). Self-organized NiO architectures:Synthesis and catalytic properties for growth of carbon nanotubes. Journal of Alloys and Compounds,474(1-2),358-363.

Zhou,Q. ,Li,C. ,Gu,F. ,& Du,H. (2008). Flame synthesis of carbon nanotubes with high density on stainless steel mesh. Journal of Alloys and Compounds,463(1-2),317-322.

Zhou,Q. ,Li,C. ,Gu,F. ,& Wang,L. (2008). Effects of Ni coated cordierite catalyst on flame synthesis of carbon nanotubes. Journal of Inorganic Materials,23(4),805-810 (in Chinese).

Co-pyrolysis of Residual Oil and Polyethylene in Sub- and Supercritical Water*

Abstract At the temperature of 693K and water densities from 0.10 to 0.30g/cm^3, co-pyrolysis of residual oil and polyethylene in sub- and supercritical water (sub-CW and SCW) was experimentally investigated. With the increase in water density, the phase structure of the co-pyrolysis system may evolve from a liquid/liquid/solid three-phase structure to a liquid/solid two-phase one. By co-pyrolysis of polyethylene with residual oil, H-rich paraffins, the main pyrolysis product of polyethylene, are released continuously into the reaction system. At higher water densities with the favorable liquid/solid two-phase structure, the contact of aromatic radicals from the pyrolysis of residual oil and paraffins from that of polyethylene is promoted, ensuring the coupling between pyrolysis networks of residual oil and polyethylene. Consequently, dealkylation of aromatic radicals may follow the desired mechanism by which the production of coke-inducing components is effectively depressed. A significantly reduced coke yield is observed in the co-pyrolysis of residual oil and polyethylene in sub-CW and SCW.

Key words Co-pyrolysis, supercritical water, polyethylene, residual oil

1 Introduction

Nowadays, nearly 70% of the reserved fossil resources in the world are heavy oil. With the development of enhanced oil recovery methods such as steam flooding and carbon dioxide injection, the refineries are facing serious challenge to process unconventional extra heavy oil. Since 1990s, upgrading of heavy oil using sub-CW or SCW as the reaction media has been attracting increasing attention in literature[1-7].

Upgrading of heavy oil in SCW is simple in operation, having no specific requirements for the quality of raw materials. Due to the acid/base nature of SCW in the vicinity of the critical point, heteroatoms and heavy metals contained in heavy oil can be partly removed by hydrolysis. Besides, SCW might be a H-donor for the upgrading through the ion-mechanism based addition of SCW to olefins[8,9]. Despite of these advantages, upgrading of heavy oil in SCW however has never been attempted in industrial scale applications. The essential problem lies in the fact that the limited improvement to upgrading performance can hardly compensate for the rapidly rising operation cost.

It is gradually realized that the introduction of external H sources should be vital to the feasibility of upgrading of heavy oil in SCW. Cheng et al. performed catalytic hydrocracking of

---

* Coauthors: Bai Fan, Zhu Chunchun, Liu Yin, Yuan Peiqing, Cheng Zhenmin. Reprinted from *Fuel Processing Technology*, 2013, 160: 267-274.

Gudao residue using SCW-syngas as an alternative H source[10]. With the aid of catalyst phosphomolybdic acid, the yield of middle distillates was elevated together with depressed asphaltene formation. The *in-situ* produced hydrogen through water-gas shift reaction (WGSR) was found to be more active than molecular hydrogen. Sato et al. investigated the upgrading of asphalt with and without the partial oxidation in SCW, trying to obtain the H source through the partial oxidation of hydrocarbons and WGSR in series[11]. However, in the absence of hydrogenation or WGSR related catalysts no significant improvement to the product distribution and degradation rate of asphalt was observed. No matter the molecular or *in-situ* produced hydrogen was used, the involvement of catalysts was confirmed to be indispensable. In the presence of various explosive gases and catalysts, the originally simple process becomes extremely complicated, which makes it infeasible for the practical implement.

Reactions of heavy oil and polyolefins in sub-CW and SCW both are composed mainly of pyrolysis which is based on free radical mechanism[12-16]. By the consideration that the H source might be re-distributed through the coupling between the pyrolysis networks of polyolefins and heavy oil, the authors thus proposed the co-pyrolysis of heavy oil and polyolefins in the presence of sub-CW and SCW. Hopefully, the H source is to be transferred from the pyrolysis network of polyolefins to that of heavy oil so as to improve the upgrading product distribution of heavy oil and to depress the formation of coke.

Hereby, the co-pyrolysis of residual oil and low density polyethylene (LDPE) in sub-CW and SCW was experimentally investigated. For comparison, the pyrolysis of residual oil alone under the same hydrothermal environments was also examined. By detailed characterization on the co-pyrolysis and pyrolysis products, the possible mechanism of the coupling between pyrolysis networks was proposed. Besides, the essential roles of sub-CW and SCW in co-pyrolysis were discussed simultaneously. In what follows, "co-pyrolysis" means in particular the thermal cracking of residual oil accompanied with LDPE and "pyrolysis" is the thermal cracking alone of residual oil or LDPE.

2 Experimental

2.1 Apparatus and reaction run

LDPE used in this work has the density of 0.92g/cm^3 and a melting point between 378 and 388K. The weight mean molecular weight of LDPE is 1.25×10^5. The detailed properties of the raw residual oil are listed in Table 1. According to thermogravimetry analysis, the raw residual oil and LDPE both start thermal cracking at temperatures near 643K.

Table 1 Properties of raw residual oil

$\rho/g \cdot cm^{-3}$		H/C ratio		SARA fractions/wt%			
				Saturates	Aromatics	Resins	Asphaltenes
0.98		1.1		37.1	18.2	20.9	23.8
H distribution/%				C distribution/%			H_m
H_{Ar}	H_α	H_β	H_γ	C_{Ar}	C_N	C_P	
6.5	10.3	64.5	18.7	15.0	37.4	47.6	0.48

All experiments were applied in a Parr 4598-HPHT autoclave made of SS316L steel with the capacity of 0.1L. The autoclave is equipped with flat paddles whose stirring rate was kept at 1000 rpm during reaction. The temperature and pressure were measured with a K-type thermocouple and a pressure gauge. Normally, the fluctuation of the reaction temperature could be controlled within ±3K.

In a typical run for co-pyrolysis, 0.5g of LDPE and 10g of water were loaded into the autoclave first. After purging with high purity N_2, the reactor was sealed and pre-heated to 633K. Then, 20g of residual oil and a certain amount of water were charged into the reactor with two metering pumps in 5min, by which the water density in the reactor was adjusted between 0.10 and 0.30g/cm³. Subsequently, at a slope of 15K/min the reactor was heated to 693K. After 20 to 60min's reaction, the reactor was subjected to forced air cooling to rapidly terminate the reaction. In general, the process for pyrolysis was similar to that of co-pyrolysis except that no LDPE was loaded into the autoclave.

2.2 Product separation and analytical procedures

After the reaction, the autoclave was washed thoroughly with tetrahydrofuran to collect co-pyrolysis or pyrolysis products. The obtained product was separated according to a procedure as shown in Fig. 1. Solid coke was first separated by vacuum filtration. Then, tetrahydrofuran and water were removed from the product by rotary evaporation. The following separation of the obtained liquid oil product was based on the Industrial Standard of Chinese Petrochemical NB/SH/T 0509—2010. Basically, n-heptane insoluble precipitates were defined as asphaltenes. The remaining maltenes were further divided into saturates, aromatics, and resin by washing in an activated γ-Al_2O_3 chromatographic column with different solvents.

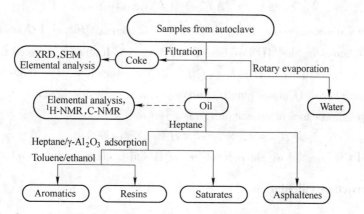

Fig. 1 Separation and analytical procedures for raw residual oil and upgrading products

The yield of liquid oil fractions (Y_i) or coke (Y_c) was evaluated by:

$$Y_i = [m_i/(m_{residual} + m_{LDPE})] \times 100\% \quad (1)$$

$$Y_c = [m_{coke}/(m_{residual} + m_{LDPE})] \times 100\% \quad (2)$$

where, m_i denotes the weight of any liquid oil fraction; m_{coke} is that of coke; $m_{residual}$ and m_{LDPE} are the weight of the loaded raw residual oil and LDPE.

The mass balance (M-B), hydrogen balance (H-B), and carbon balance (C-B) in each run were evaluated by:

$$M\text{-}B = \frac{\sum m_i + m_{\text{coke}}}{m_{\text{residual}} + m_{\text{LDPE}}} \times 100\% \tag{3}$$

$$H\text{-}B = \frac{\sum m_i \cdot \dfrac{H/C_{\text{oil}}}{H/C_{\text{oil}} + 12} + m_{\text{coke}} \cdot \dfrac{H/C_{\text{coke}}}{H/C_{\text{coke}} + 12}}{m_{\text{residual}} \cdot \dfrac{H/C_{\text{residual}}}{H/C_{\text{residual}} + 12} + m_{\text{LDPE}} \cdot \dfrac{1}{7}} \times 100\% \tag{4}$$

$$C\text{-}B = \frac{\sum m_i \dfrac{12}{H/C_{\text{oil}} + 12} + m_{\text{coke}} \dfrac{12}{H/C_{\text{coke}} + 12}}{m_{\text{residual}} \dfrac{12}{H/C_{\text{residual}} + 12} + m_{\text{LDPE}} \cdot \dfrac{6}{7}} \times 100\% \tag{5}$$

where, H/C_{oil} and H/C_{coke} are the H/C ratio of the liquid oil product and that of coke; H/C_{residual} is the H/C ratio of the raw residual oil. It was assumed that the H/C ratio of LDPE is 2.

The elemental analysis of oil and coke samples was applied on a vario EL Ⅲ element analyzer. The XRD pattern of coke was recorded on a Bruker D8 advance X-ray diffractometer using Ni-filtered Cu K_α radiation, the morphology of coke being observed on a NanoScope Ⅲa Multi-Mode AFM microscope with the accelerating voltage of 15kV. The ^{1}H-NMR and C-NMR spectra of oil samples were analyzed on a Bruker AVANCE 500MHz NMR Spectrometer. CDCl$_3$ and TMS were used as the solvent and the internal standard for calibrating chemical shift, respectively. Protons in oil were classified as aromatic H (H_{Ar}), α-alkyl H (H_α), γ-alkyl H (H_γ), and other alkyl H (H_β). Carbon atoms were classified as aromatic C (C_{Ar}), alicyclic C (C_{N}), and paraffinic C (C_{P})[17-19].

The distributions of H and C of different definition in the oil samples were evaluated by:

$$\% H_i = (A_i H / \sum A_i H) \times 100\% \tag{6}$$

$$\% C_i = (A_i C / \sum A_i C) \times 100\% \tag{7}$$

where, subscript i denotes anyone of the defined H or C atoms; $A_i H$ and $A_i C$ are peak areas of various H and C atoms on the ^{1}H-NMR and C-NMR spectra. It should be noted that $\% H_{\text{Ar}}$ and $\% C_{\text{Ar}}$ are also the aromaticity of H atoms ($f_a H$) and the aromaticity of C atoms ($f_a C$).

The average number of H atoms bonding directly with an aromatic C atom (H_m), an indication of the polycyclic degree of oil samples, was given by:

$$H_m = 12 \times (f_a H \times H\% / f_a C \times C\%) \tag{8}$$

where, $H\%$ and $C\%$ are the weight percentages of H and C elements in the oil sample.

2.3 Phase structure calculation

The phase structure of residual oil in the presence of sub-CW and SCW was calculated using Aspen Engineering Suite 2006. The calculation applied a flash operation in which a rigorous VLL phase equilibrium of residual oil and water under a hydrothermal condition was performed. The Soave-Redlich-Kwong (SRK) cubic equation of state was implemented, and this method used Kabadi-Danner mixing rule to deal with the water-hydrocarbon system. Based on the built in property method, the raw residual oil was divided into a series of pseudo-components whose characteristics were represented by the assay data of true boiling points listed in Table 2 and API gravity.

Table 2 True boiling points of raw residual oil

Volume/%	0	5	10	30	50	70	90	95	100
Temperature/K	708	737	755	793	823	858	904	922	939

3 Results and Discussion

3.1 Pyrolysis of residual oil in sub-CW and SCW

Prior to discussing the co-pyrolysis of residual oil and LDPE, it is necessary to survey their individual reaction behavior in sub-CW and SCW. So far, a comprehensive knowledge about the pyrolysis of LDPE in SCW has already been established in literature. Before complete decomposition, LDPE suspended in SCW releases PE macromolecules into the water phase continuously. The products of the homogeneous pyrolysis of free PE macromolecules in SCW consist primarily of C_7 to C_{24} alkanes and α-alkenes, suggesting that the reaction follows FSS mechanism derived from the pyrolysis of hexadecane[20-22]. Refer to Eq. 9 to Eq. 11. Usually, a PE macromolecule R first produces two primary carbon radicals $R_1 \cdot$ and $R_2 \cdot$ by C-C cleavage. Then, radical $R_1 \cdot$ is saturated through H-abstraction from another PE macromolecules R', leaving a secondary carbon radical $R' \cdot$ at random position along the main chain. Further β-scission of $R' \cdot$ produces an α-olefin $O=$ and a primary carbon radical $R_3 \cdot$:

$$R \longrightarrow R_1 \cdot + R_2 \cdot \tag{9}$$

$$R_1 \cdot + R' \longrightarrow R_1 H + R' \cdot \tag{10}$$

$$R' \longrightarrow O= + R_3 \cdot \tag{11}$$

As for the pyrolysis of residual oil in sub-CW and SCW, no consistent literature results can be found till now. By this consideration, pyrolysis of residual oil alone was carried out under a typical hydrothermal condition, 693K and a water density of 0.20g/cm³. The product distribution at varied reaction time is illustrated in Fig. 2. For convenience, the corresponding data of mass balance, hydrogen balance, carbon balance, and H/C ratios of oil products and coke are also labeled on the figure. Besides, the related NMR results of the liquid products are listed in Table 3 (detailed NMR spectra can be seen in the supporting materials).

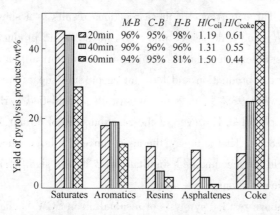

Fig. 2 Yield of pyrolysis products of residual oil vs. reaction time;
693K and water density of 0.20g/cm³

Table 3 NMR results of liquid pyrolysis products at varied reaction time (693K and ρ_{water} of 0.20g/cm^3)

Reaction time/min	H distribution/%				C distribution/%			H_m
	H_{Ar}	H_α	H_β	H_γ	C_{Ar}	C_N	C_P	
20	8.1	12.7	60.3	18.9	14.2	38.0	47.8	0.68
40	11.9	14.3	54.4	19.4	23.0	37.0	40.0	0.68
60	12.2	15.5	50.8	21.5	28.4	32.6	39.0	0.64

At the early reaction stage, the pyrolysis of residual oil is composed of the decomposition of asphaltenes and resins to saturates and aromatics. After a 20min's run, the conversions of asphaltenes and resins have reached 54% and 43%. At the same time, the fractions of saturates and aromatics in the liquid products increase from the initial values of 37.1wt% and 18.2wt% to 52.3wt% and 20.9wt%, respectively. The increased proportions of C_N and C_P, and an increase in H_m from 0.48 to 0.68 all suggest that the quality of the liquid pyrolysis products is significantly improved. It is noteworthy that a coke yield of 10wt%, resulted from the direct condensation of asphaltenes, can already be obtained. Essentially the pyrolysis of hydrocarbons is a process of decarbonization, so the appearance of coke is accompanied by the increase in the H/C ratio of the liquid products. Asphaltenes are macromolecules with polycondensed aromatic benzene units. From the point view of reaction barrier, decomposition of asphaltenes occurs preferentially at the aliphatic substituents. The formed aromatic radicals may further decompose to saturates and aromatics through dealkylation; accordingly, an increased proportion of H_α and a decreased one of H_β in the liquid products can be observed.

As the reaction goes on, the yields of asphaltenes and resins decrease continuously. Although the fractions of saturates and aromatics in the liquid products still keep a rising tendency, their yields actually decrease in varying degree. What is more important, the yield of coke increases drastically since then. At the reaction time of 60min a high H/C ratio of the liquid product up to 1.50 can be obtained, but it is at the cost of the undesired coke yield of 48wt%. The continuously decreased proportions of C_P and C_N as well as the increased one of C_{Ar} indicate that condensation should be dominant at the middle and later pyrolysis stages. Generally, condensation may include reactions as follows: deep thermal cracking of saturates to short olefins and diolefins; dehydrogenation of cycloparaffins to aromatics; and condensation of aromatics to resins and asphaltenes. The occurrence of these reactions results in a simultaneously increased proportions of H_γ and H_{Ar} and a decreased one of H_β in the liquid products. Besides, the rising proportion of H_α suggests the accumulation of aromatics with short aliphatic substituents.

Due to the neglected gas production and loss during product separation, the mass balance of pyrolysis may vary between 93% and 96%. Meanwhile, a satisfied carbon balance between 95% and 96% can be obtained. However, at the reaction time of 60min, the hydrogen balance decreases drastically to 81%. That is to say, the appearance of a large amount of coke at the latter pyrolysis stage should be accompanied simultaneously by the significant production of H_2 as well as hydrogen rich lower alkanes and alkenes.

On the basis of the above results, pyrolysis of residual oil in SCW is characterized by dealkylation of aromatic radicals at the early reaction stage and by condensation at the middle and later reaction stages. Detailed analysis shows that dealkylation of aromatic radicals can follow several mechanisms as shown in Fig. 3, which in turn affects the coking behavior and product distribution at the subsequent reaction stages.

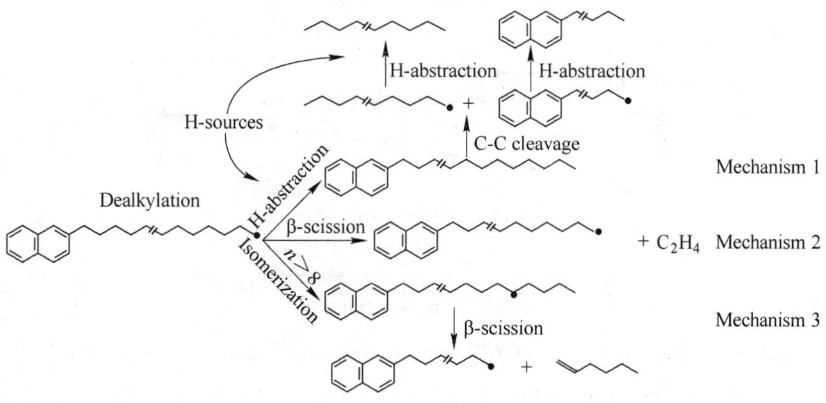

Fig. 3 Possible dealkylation mechanisms of aromatic radicals involved in early pyrolysis stage

On condition that the aromatic radical is surrounded by H-sources, the radical can be readily saturated through mechanism 1 during which H-abstraction occurs. On the contrary, dealkylation of aromatic radicals has to follow mechanism 2 or 3 in terms of the length of the aliphatic substituents. Theoretical calculations suggested that the ultimate dealkylation products through mechanism 2, ethylene and methylated aromatics, are active coke-inducing components[23-30]. The reaction barriers of the involved C-C cleavage, H-abstraction, isomerization, and β-scission are about 340kJ/mol, 40kJ/mol, 40kJ/mol, and 120kJ/mol. Detailed dealkylation pathway of aromatic radicals depends not only on the intrinsic kinetics of the elementary reactions, but also closely on the mass transfer environment between aromatic radicals and H-sources. In practice, it can be alternative among three mechanisms.

3.2 Phase structure of co-pyrolysis system

To understand the phase behavior of the co-pyrolysis system, the phase structure of sub-CW or SCW/residual oil was investigated. The calculated phase diagrams at the mass ratios of water/oil of 1∶4 and 1∶2 are illustrated in Fig. 4 and Fig. 5, together with the applicable pyrolysis temperature range labeled.

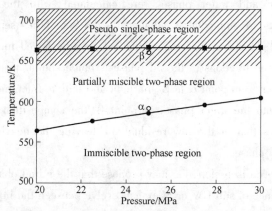

Fig. 4 Phase structure of sub-CW or SCW/residual oil at the mass ratio of 1∶4; shadow region means the applicable pyrolysis temperature range

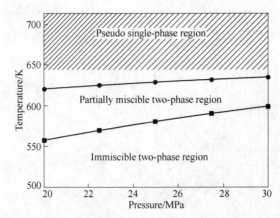

Fig. 5 Phase structure of sub-CW or SCW/residual oil at the mass ratio of 1:2;
shadow region means the applicable pyrolysis temperature range

The phase diagrams of sub-CW or SCW/residual oil are divided by two boundaries into three regions, that is, immiscible two-phase region, partially miscible two-phase region, and pseudo single-phase region. Usually, pyrolysis of hydrocarbons occurs spontaneously at temperatures higher than 643 K. At the mass ratio of 1:4, the applicable pyrolysis temperature range not only covers the pseudo single-phase region, but also extends into the partially miscible two-phase region. At that time, a slight change in reaction temperature near the upper boundary may result in different phase structures. With the increase in the mass ratio of water to oil, the upper and lower boundaries on the phase diagram both shift downwards, by which the area of the partially miscible region is significantly reduced. Consequently, the overlap between the applicable pyrolysis temperature range and the partially miscible two-phase region disappears. The pyrolysis of residual oil now can only be applied in the pseudo single-phase region. It should be noted that the upgrading of residual oil also results in the shift of both two boundaries towards the low temperature region. That is to say, at the middle and later reaction stages the phase structure of sub-CW or SCW/residual oil will inevitably migrate into the single-phase region regardless of the initial situation.

At the phase points α and β indicated in Fig. 4, the TBP curves of oil fractions in the partially miscible oil phase and water phase were calculated, with the results illustrated in Fig. 6. These two points are immediately adjacent to the boundaries separating the phase diagram. At point α, the lightest fractions in residual oil are transferred into the water phase, and the composition of the oil fractions in the oil phase is much similar to that of the raw residual oil. During the transition from point α to β, the light and middle fractions of residual oil are extracted continuously into the water phase. At point β, the composition of oil fractions in the water phase approaches that of the raw residual oil, leaving the heaviest oil fractions highly concentrated in the oil phase.

According to the above discussion, one may propose that the initial phase structure of residual oil/LDPE in the presence of sub-CW or SCW can evolve between the liquid/liquid/solid three-phase structure and liquid/solid two-phase one as shown in Fig. 7. The increase in temperature, elevation of water density, and upgrading of residual oil all can promote phase transition. In the

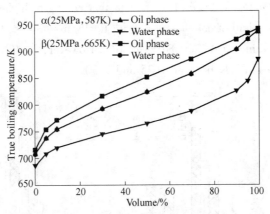

Fig. 6　TBP curves of oil fractions in the partially miscible water phase and oil phase at points α and β indicated in Fig. 4

former phase structure, the partially miscible oil phase and water phase are in dynamic equilibrium, and LDPE particles are mainly suspended in the oil phase with concentrated heavier fractions. In the latter phase structure, the non-polar sub-CW or SCW dissolved with saturates, aromatics, and resins consists of the continuous medium of the liquid phase in which the polar asphaltene macromolecules are highly dispersed. Meanwhile, LDPE particles are also suspended in the liquid phase.

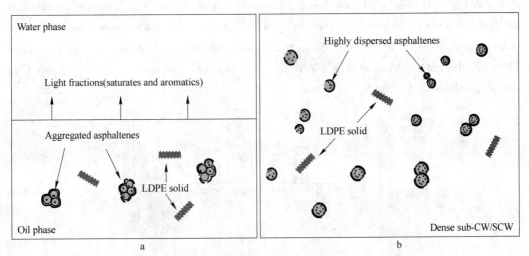

Fig. 7　Initial phase structure of residual oil/LDPE/sub-CW or SCW
a—Liquid/liquid/solid three-phase structure; b—Liquid/solid two-phase structure

3.3　Co-pyrolysis of residual oil and LDPE in sub-CW and SCW

On the basis of the available information, co-pyrolysis of residual oil and LDEP with the mass ratio of 40:1 was surveyed at 693K and water densities ranging from 0.10 to 0.30g/cm³, with the results at the reaction time of 1h illustrated in Fig. 8.

With the elevation of water density from 0.10 to 0.30g/cm³, the yield of saturates increases from 21.0wt% to 32.9wt% and that of aromatics increases slightly from 18.4wt% to 19.4wt%. Meanwhile, the yield of asphaltenes presents a decreasing tendency but that of resins increases from 10.8wt% to 13.9wt%. It is evident that coking at higher water densities is effectively depressed. The yield of coke decreases drastically from 36.0wt% to 25.8wt%, whereas the hydrogen balance increases from 84% to 96%. However, the H/C ratio of the liquid co-pyrolysis products varies merely between 1.25 and 1.30.

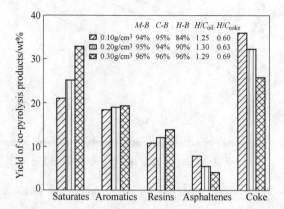

Fig. 8   Yields of co-pyrolysis products at increased water density
(reaction time: 1h, 693K)

By applying the pyrolysis of residual oil alone under the same hydrothermal conditions, one may find some differences in the product distribution between co-pyrolysis and pyrolysis as shown in Fig. 9. On the whole, the yields of aromatics, resins, and asphaltenes in co-pyrolysis are higher than those in pyrolysis. Only the yield of saturates shows an opposite tendency. It is noteworthy that coking in pyrolysis is drastically depressed after introducing a small amount of LDPE. At the water density of 0.30g/cm³, the yield of coke in co-pyrolysis is much lower than that in pyrolysis by 16.5wt%.

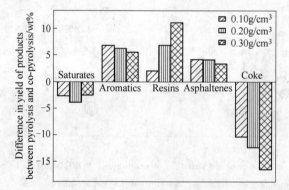

Fig. 9   Difference in yield of products between co-pyrolysis and pyrolysis at increased water density
(reaction time: 1h, 693K)

With the elevation of water density, the distribution of H and C in the liquid products of both co-pyrolysis and pyrolysis shows some similarities. Refer to Table 4. Firstly, the proportions of C_P and H_β increase, which is coincident with the simultaneously increased fraction of saturates in the liquid products. Secondary, the proportion of C_{Ar} decreases, and it agrees with the depressed condensation tendency at higher water densities. The monotonic decrease in the proportion of H_α suggests the reduction in the aromatics with short aliphatic substituents. Apart from the above similarities, there appear some differences in the H and C distributions between co-pyrolysis and pyrolysis. The proportion of C_{Ar} in the liquid products of co-pyrolysis is higher than that of pyrolysis. However, the polycyclic degree of the co-pyrolysis products is lower than that of the pyrolysis products, which is evidenced by the higher values of H_m of the co-pyrolysis products.

Table 4 NMR results of liquid products of pyrolysis and co-pyrolysis (693K, reaction time: 1 h)

	ρ_{water}/g·cm^{-3}	H distribution/%				C distribution/%			H_m
		H_{Ar}	H_α	H_β	H_γ	C_{Ar}	C_N	C_P	
Pyrolysis	0.10	11.5	18.0	50.8	19.7	30.2	30.9	38.9	0.56
	0.20	12.2	15.5	50.8	21.5	28.4	32.6	39.0	0.64
	0.30	11.0	14.2	53.1	21.7	25.3	31.6	43.1	0.66
Co-pyrolysis	0.10	14.5	17.5	49.9	18.1	31.3	30.9	37.8	0.58
	0.20	15.5	16.2	47.5	20.8	30.0	33.8	36.2	0.68
	0.30	15.2	15.7	50.2	18.9	27.8	32.0	40.2	0.71

3.4 Coupling between pyrolysis networks of residual oil and LDPE

After introducing a small amount of LDPE, coupling between the pyrolysis networks of residual oil and LDPE is experimentally confirmed, together with the significant change in the product distribution. The degree of coupling is found to be sensitive to the applied density of sub-CW or SCW. It can be reasonably proposed that the decomposition of asphaltenes and resins should be independent of the release of PE macromolecules from the surface of LDPE solid. Furthermore, coupling between the pyrolysis networks only occurs in the continuous medium whose properties can be adjusted by the applied reaction temperature and water density. Accordingly, one may suggest the possible coupling mechanism as shown in Fig. 10.

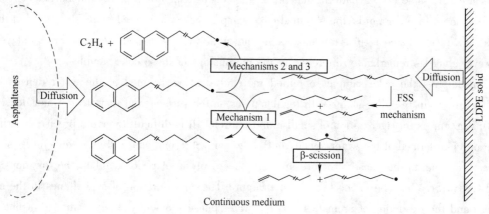

Fig. 10 Proposed mechanism of coupling between pyrolysis networks of residual oil and LDPE

At low water densities with the liquid/liquid/solid three-phase structure, the continuous medium of the oil phase is composed mainly of concentrated heavier fractions of residual oil. Although LDPE is also suspended in the oil phase, the contact between the homogeneous pyrolysis product of PE and the aromatic radicals released from asphaltenes is retarded by the high viscosity of the oil phase. Without appropriate H-sources, dealkylation of localized aromatic radicals may follow mechanisms 2 and 3 shown in Fig. 3 with priority, resulting in the production of aromatics with short aliphatic substituents, ethylene, and α-olefins falling within the definition of saturates. With the increase in water density, the phase structure of co-pyrolysis transits to the liquid/solid two-phase structure in which the continuous medium consists of dense sub-CW or SCW dissolved with saturates, aromatics, and resins. Upon being released from asphaltenes into the continuous medium, aromatic radicals can contact promptly with the homogeneous pyrolysis products of PE and be readily saturated further. The production of coke-inducing components thus is depressed, together with the preferential production of aromatics with longer aliphatic substituents. Nevertheless, the production of saturates through dealkylation mechanism 3 is also restrained.

Along with the development of the reaction, the phase structure of the co-pyrolysis system will inevitably evolve to the liquid/solid two-phase structure. However, the difference in the initial phase structure has already established distinct concentration profiles of the coke-inducing components in the reaction system, which may be responsible largely for the different condensation and coking behavior at the middle and later reaction stages. Since the release of PE macromolecules into the co-pyrolysis system is an independent process, the presence of LDPE forms a sustaining H-source for the pyrolysis of residual oil.

The mechanism suggested above cannot account for the fact that the yields of asphaltenes and resins in co-pyrolysis are higher than those in pyrolysis. In order to get a comprehensive understanding of co-pyrolysis, coke samples formed in co-pyrolysis and pyrolysis were further characterized with XRD and SEM.

3.5 Coking behavior in pyrolysis and co-pyrolysis

At 693K and water densities from 0.1 to 0.3g/cm^3, coke formed in both pyrolysis and co-pyrolysis all shares the same crystal pattern, that is, only 2-dimensional (01) plane can be detected on XRD spectra. In other words, the obtained coke samples have an ordered plane structure but the arrangement of planes is just random. As is shown in Fig. 11, the subsequent SEM analysis however suggests remarkable difference in the morphology of these coke samples.

Consider first the coke samples obtained in pyrolysis. Coke formed at the water density of 0.10g/cm^3 is in the plate-shape with regular pores on the surface, and that at the water density of 0.30g/cm^3 is the aggregation of small round particles. It is difficult to give a precise description on coke formed at the water density of 0.20g/cm^3. Still, one can find tiny pores on its surface. According to the phase structure analysis, the pyrolysis of residual oil in the low density sub-CW or SCW can be in the two-phase structure. The concentration of asphaltenes in the oil phase and the extraction of aromatics into the water phase promote the aggregation of asphaltenes. Since coke precursors are first resulted from the direct condensation of asphaltenes, the aggregation of asphaltenes should favor the interaction of coke precursors to form coke plane even at the early pyrolysis stage. The subsequent transverse of the produced gas, such as methane and hydrogen, leaves pores on the surface of the coke plane. As for the pyrolysis of residual oil in the high density sub-CW or SCW with the single-phase structure, asphaltenes are highly dispersed in the continuous phase. The formation of coke precursors and the growth of coke occur independently in each asphaltene molecule; consequently, small coke particles are obtained finally.

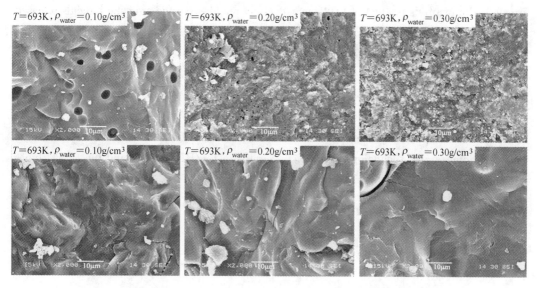

Fig. 11　SEM images of coke formed in pyrolysis (up) and co-pyrolysis (down) at increased water density from 0.1 to 0.3 g/cm^3
(reaction time: 1h, 693K)

Regardless of the variation of water density, coke samples obtained in co-pyrolysis are all in the plate-shape. In terms of the coking behavior in pyrolysis, the dispersion state of asphaltenes under different phase structures plays a vital role on the morphology of coke. The same morphology of coke indicates that the dispersion state of asphaltenes in co-pyrolysis should be irrelevant to the applied phase structure. At higher water densities with the liquid/solid two-phase structure, asphaltenes, which are supposed to be highly dispersed, are virtually in the aggregated form. Aggregation of asphaltenes in co-pyrolysis here is proposed to be attributed to the high flexibility of the free PE macromolecules. No matter in the liquid/liquid/solid three-phase structure or in the liquid/solid two-phase one, PE macromolecules released into the continuous medium may curl rapidly and entrain adjacent asphaltene molecules, resulting in the aggregation of asphaltenes especially at higher water densities. Apparently, the aggregation is unfavorable for the decomposition of asphaltenes, and those incompletely decomposed asphaltenes may fall within the definition of resins. Accordingly, the yields of asphaltenes and resins in co-pyrolysis are always higher than those in pyrolysis. Meanwhile, coke samples with the same plane morphology can be observed at varied water density.

4　Conclusions

In sub-CW or SCW, the pyrolysis of residual oil is characterized by the dealkylation of aromatic radicals at the early reaction stage and by the condensation at the middle and later reaction stages, while pyrolysis of LDPE mainly follows FSS mechanism in which H-rich paraffins are released. In the co-pyrolysis of residual oil and LDPE, the presence of LDPE establishes a potential H-source for the pyrolysis network of residual oil. Further introduction of sub-CW or SCW medium into the co-pyrolysis system guarantees the coupling between the pyrolysis networks of residual oil and LDPE. At higher water density with the liquid/solid two-phase structure, the pyrolysis product of LDPE may contract readily with the aromatic radicals released from the decomposition of asphaltenes. Consequently, the dealkylation of aromatic radicals can follow the favorable mechanism by which the production of coke-inducing components is effectively depressed.

Acknowledgements

This work was supported by the Fundamental Research Funds for the Central Universities, the Shanghai Science and Technology Committee (Grant No. 09ZR1407500) and the CNPC Innovation Foundation (2011D-5006-0406).

Appendix A. Supplementary Data

Supplementary data to this article can be found online at http://dx.doi.org/10.1016/j.fuproc.2012.07.031.

References

[1] I. V. Kozhevnikov, A. L. Nuzhdin, O. N. Martyanov. Transformation of petroleum asphaltenes in supercritical water, Journal of Supercritical Fluids 55 (2010) 217-222.

[2] M. Morimoto, S. Sato, T. Takanohashi. Conditions of supercritical water for good miscibility with heavy oils, Journal of the Japan Petroleum Institute 53 (2010) 61-62.

[3] S. Kokubo, K. Nishida, A. Hayashi, H. Takahashi, O. Yokota, S. I. Inage. Effective demetalization and suppression of coke formation using supercritical water technology for heavy oil upgrading, Journal of the Japan Petroleum Institute 51 (2008) 309-314.

[4] Wahyudiono, M. Sasaki, M. Goto. Kinetic study for liquefaction of tar in sub- and supercritical water, Polymer Degradation and Stability 93 (2008) 1194-1204.

[5] M. Meng, H. Q. Hu, Q. M. Zhang, M. Ding. Extraction of Tumuji oil sand with sub- and supercritical water, Energy & Fuels 20 (2006) 1157-1160.

[6] H. Luik, L. Luik. Extraction of fossil fuels with sub- and supercritical water, Energy Sources 23 (2001) 449-459.

[7] H. Q. Hu, J. Zhang, S. C. Guo, G. H. Chen. Extraction of Huadian oil shale with water in sub- and supercritical states, Fuel 78 (1999) 645-651.

[8] T. Moriya, H. Enomoto. Role of water in conversion of polyethylene to oils through supercritical water cracking, Kagaku Kogaku Ronbunshu 25 (1999) 940-946. (in Japanese)

[9] T. Moriya, H. Enomoto. Conversion of polyethylene to oil using supercritical water and donation of hydrogen in supercritical water, Kobunshi Ronbunshu 58 (2001) 661-673. (in Japanese)

[10] J. N. Cheng, Y. H. Liu, Y. H. Lou, G. H. Que. Hydrocracking of Gudao residual Oil with dispersed catalysts using supercritical water-syngas as a hydrogen source, Petroleum Science and Technology 23 (2005) 1453-1462.

[11] T. Sato, T. Adschiri, K. Arai, G. L. Rempel, F. T. T. Ng. Upgrading of asphalt with and without partial oxidation in supercritical water, Fuel 82 (2003) 1231-1239.

[12] X. L. Su, Y. L. Zhao, R. Zhang, J. C. Bi. Investigation on degradation of polyethylene to oils in supercritical

water, Fuel Processing Technology 85 (2004) 1249-1258.

[13] Z. Fang, Jr. R. L. Smith, H. Inomata, K. Arai. Phase behavior and reaction of polyethylene in supercritical water at pressures up to 2.6 GPa and temperatures up to 670℃, Journal of Supercritical Fluids 16 (2000) 207-216.

[14] M. Watanabe, M. Mochiduki, S. Sawamoto, T. Adschiri, K. Arai. Partial oxidation of n-hexadecane and polyethylene in supercritical water, Journal of Supercritical Fluids 20 (2001) 257-266.

[15] T. Moriya, H. Enomoto. Characteristics of polyethylene cracking in supercritical water compared to thermal cracking, Polymer Degradation and Stability 65 (1999) 373-386.

[16] H. F. Zhang, X. L. Su, D. K. Sun, R. Zhang, J. C. Bi. Investigation on degradation of polyethylene to oil in a continuous supercritical water reactor, Journal of Fuel Chemistry and Technology 35 (2007) 487-491. (in Chinese)

[17] S. W. Lee, B. Glavincevski. NMR method for determination of aromatics in middle distillate oils, Fuel Processing Technology 60 (1999) 60.

[18] Y. Yang, B. Liu, H. T. Xi, X. Q. Sun, T. Zhang. Study on relationship between the concentration of hydrocarbon groups in heavy oils and their structural parameter from ^{1}H NMR spectra, Fuel 82 (2003) 721-727.

[19] N. Tsuzuki, N. Takeda, M. Suzuki, K. Yokoi. The kinetic modeling of oil cracking by hydrothermal pyrolysis experiments, International Journal of Coal Geology 39 (1999) 227-250.

[20] D. Depeyre, C. Flicoteaux. Modeling of thermal steam cracking of n-hexadecane, Industrial and Engineering Chemistry Research 30 (1991) 1116-1130.

[21] T. J. Ford. Liquid-phase thermal decomposition of hexadecane: reaction mechanisms, Industrial and Engineering Chemistry Fundamentals 25 (1986) 240-243.

[22] F. Khorasheh, M. R. Gray. High-pressure thermal cracking of n-hexadecane, Industrial and Engineering Chemistry Research 32 (1993) 1853-1863.

[23] K. Hemelsoet, V. V. Speybroeck, M. Waroquier. A DFT-based investigation of hydrogen abstraction reactions from methylated polycyclic aromatic hydrocarbons, ChemPhysChem 9 (2008) 2349-2358.

[24] V. V. Speybroeck, D. V. Neck, M. Waroquier, S. Wauters, M. Saeys, G. B. Marin. Ab initio study on elementary radical reactions in coke formation, International Journal of Quantum Chemistry 91 (2003) 384-388.

[25] V. V. Speybroeck, K. Hemelsoet, M. Waroquier, G. B. Marine. Reactivity and aromaticity of polyaromatics in radical cyclization reactions, International Journal of Quantum Chemistry 96 (2004) 568-576.

[26] Y. T. Xiao, J. M. Longo, G. B. Hieshima, R. J. Hill. Understanding the kinetics and mechanisms of hydrocarbon thermal cracking: an ab initio approach, Industrial and Engineering Chemistry Research 36 (1997) 4033-4040.

[27] V. V. Speybroeck, D. V. Neck, M. Waroquier. Ab initio study of radical addition reactions: addition of a primary ethylbenzene radical to ethene (I), Journal of Physical Chemistry A 104 (2000) 10939-10950.

[28] M. K. Sabbe, A. G. Vandeputte, M. F. Reyniers, V. V. Speybroeck, M. Waroquier, G. B. Marin. Ab initio thermochemistry and kinetics for carbon-centered radical addition and β-scission reactions, Journal of Physical Chemistry A 111 (2007) 8416-8428.

[29] A. Comandini, K. Brezinsky. Theoretical study of the formation of naphthalene from the radical/π-bond addition between single-ring aromatic hydrocarbons, Journal of Physical Chemistry A 115 (2011) 5547-5559.

[30] J. P. Leininger, C. Minot, F. Lorant, F. Behar. Density functional theory investigation of competitive free-radical processes during the thermal cracking of methylated polyaromatics: estimation of kinetic parameters, Journal of Physical Chemistry A 111 (2007) 3082-3090.

Experimental Measurements and Modeling of Solubility and Diffusivity of CO_2 in Polypropylene/Micro- and Nanocalcium Carbonate Composites[*]

Abstract The effects of the filler size and concentration on the solubility and diffusivity of CO_2 in polypropylene (PP)/calcium carbonate ($CaCO_3$) composites were investigated in this work. The apparent solubility of CO_2 in PP and its composites containing 5wt% and 10wt% micro- or nano-$CaCO_3$ was measured by using a magnetic suspension balance (MSB) at temperatures of 200 and 220℃ and CO_2 pressures up to 22MPa. Meanwhile, the swelling volume of the PP composites/CO_2 solutions was experimentally measured at the same conditions by using a high-pressure view cell with direct visual observation. It was then used to correct the gas buoyancy acting on the PP composites in the MSB measurement so that the real solubility of CO_2 in the PP composites was determined. Meanwhile, the diffusion coefficient of CO_2 in the PP composites was estimated from the sorption lines at gas pressures ranged form 5 to 10MPa. It was found that the experimental solubility and diffusivity of CO_2 decrease with increasing micro-$CaCO_3$ concentration whereas they increase with increasing nano-$CaCO_3$ loading in the PP composites. Two new models based on free volume theory considering the effects of micro-$CaCO_3$ on the free volume and the diffusion path and the lubricant effect of nano-$CaCO_3$ were proposed and used to well correlate the experimental diffusion coefficient of CO_2 in PP/micro-$CaCO_3$ and PP/nano-$CaCO_3$ composites, respectively.

1 Introduction

The concept of the microcellular foaming process was proposed in the 1980s by Suh[1] at MIT to produce novel high-performance polymeric foams. It was aimed at reducing the material consumption while not or slightly deteriorating the mechanical properties. In the past 30 years, intensive studies on polymeric foaming especially using the environmental benign blowing agent CO_2 have been conducted to realize commercial application through industrial processes such as the microcellular injection molding[2,3], foaming extrusion[4-6], and so forth. However, maintaining the mechanical properties of foams still remains a major challenge. Recently, the polymer composite foams have received tremendous interest from both theoretical and industrial areas as they are advantageous over neat polymer foams in terms of enhanced mechanical properties, better thermal stability, and improved barrier and flame retardancy[7-9] due to the fact that the fillers can not only improve the mechanical properties of polymer itself but also serve as the nucleation agent to increase the bubble size and reduce the bubble density which in turn enhances the mechanical strength of the foams. For example, PP/$CaCO_3$ composites are widely used due to the excellent property of PP and low price of $CaCO_3$[10]. Incorporation of

---

[*] Coauthors: Chen Jie, Liu Tao, Zhao Ling. Reprinted from *Ind. Eng. Chem. Res.*, 2013, 52: 5100-5110.

CaCO$_3$ in PP not only would increase the tensile strength of composites but also acts as an excellent β crystal nucleation agent to enhance impact strength[11], which also makes PP/CaCO$_3$ composites a promising material to overcome the mechanical weakness of their foams for industrial application. Many studies have been done on polymer composites foams[12-17]. Chen et al.[14] investigated the effect of the filler size on the morphology of HDPE/CaCO$_3$ composites foamed with high pressure CO$_2$. The results showed that the smaller bubble size and higher bubble density could be obtained in PE/nano-CaCO$_3$ composites foams. Javni et al.[15] and Saint-Michel et al.[16] also found the similar results in polyurethane composites with micro- and nanosilica fillers. All of these works attribute the smaller bubble size only to the better nucleation effect of nanofiller compared to that of microfiller.

However, as the gas solubility and diffusivity play a crucial role in controlling the bubble nucleation and growth of the foaming process, the investigation of the effect of micro- and nanosize fillers on gas solubility and diffusivity in polymer composites is also very important. The accurate measurement of these data can not only provide economic guidance into design and manufacturing of foaming parts but also provide insights into to the mechanism of the thermodynamic behavior for the theoretical study of polymer composites/high pressure gas solution. When gas dissolves into the polymer matrix, its volume would expand. As a result, the swelling ratio indicates the special PVT behavior of polymer/gas solution. Gas solubility, namely, the maximum amount of gas that can be dissolved into the polymer under certain temperature and gas pressure, would affect the nucleation rate during foaming and largely determine the final bubble density. The diffusivity of gas controls the bubble growth rate and affects the final bubble size. Much work has been done on the swelling of the polymer/CO$_2$ solution[18-21] and the solubility and diffusivity[22-26] of CO$_2$ in neat polymers. However, the knowledge of those in polymer composites is very limited. It is necessary to investigate the thermodynamic property of sorption, diffusion, and the relating sweling of polymer composites/CO$_2$ system.

The free volume plays a significant role on the thermodynamic properties (such as gas solubility and related swelling) of the polymer/gas system and gas diffusion process. It is considered as the unoccupied volume inside polymers that consist of lots of single and interconnected nanospaces between polymeric chains[27,28]. Generally, the free volume is consisted of the statistic and dynamic free volumes. The statistic free volume plays a major role on the gas sorption, and the dynamic one mainly contributes to the diffusion[29]. For the gas diffusion process, the free volume based models are widely used to describe the coefficient of gas in molten polymers. It was first proposed by Cohen and Turnbull et al[30]. who expressed the diffusion coefficient as an exponential function of inversed free volume fraction. Thereafter, Fujida[31] modified the model and used a comprehensive expression of free volume fraction including the temperature and gas concentration dependence. Maeda and Paul et al.[32] gave a very simple form of free volume as a function of inversed free volume. Combining Fujida's and Maeda's models, Areerat et al.[33] proposed a new model that use the simple form of free volume expression but maintain the temperature dependence. It is considered simple but effective to correlate the diffusion coefficients of a variety of gases in polymers[34].

It is generally known that the size, shape, and concentration of the fillers as well as the interface bonding of filler and polymer can affect the properties of polymer composites and gas solubility and diffusivity as well. The effect of the surface bonding had been investigated in our previous work by comparing the gas sorption and diffusion in PP/micro-$CaCO_3$ composites without and with interface modification[35]. This work aims to study the two other effects, i. e., the filler size and concentration, on the gas solubility and diffusivity in polymer composites with interface modification. The swelling, solubility, and diffusion data in PP and its composites containing 5% and 10% micro- and nano-$CaCO_3$ were determined precisely under high CO_2 gas pressure. The swelling ratio of PP/$CaCO_3$ composites was determined in a high pressure view cell at 200 and 220 ℃ and CO_2 pressures up to 22MPa. The solubility was then precisely estimated form the apparent solubility that was measured by using magnetic suspension balance (MSB) with the experimental swelling data correction at the same experimental temperature and gas pressure. Meanwhile, the diffusion coefficient was estimated from the sorption lines in a gas pressure range of 5-10MPa. The free volume models based on the modification of Areerat's free volume model were proposed and employed to correlate the diffusion coefficient of CO_2 for both PP/micro- and nano-$CaCO_3$ composites.

2 Experimental Section

2.1 Materials

Isotactic polypropylene (Y1600) with a weight-average molar mass (M_m) of 197000g/mol and an average melt flow index (MFI) of 16.0g/10min was purchased from Shanghai Petrochemical Co., China. The melting temperature and the crystallinity of iPP are 167.8℃ and 47.8%, respectively, measured by differential scanning calorimetry (DSC) in atmospheric nitrogen. micro-$CaCO_3$ with the density of 2.73g/cm^3 and the particle size of 30-50μm was provided by Shanghai LingFeng Chemical Co., China. Nano-$CaCO_3$ with the density of 2.68g/cm^3 and the particle size of 100-150nm was obtained form Shanghai Juqian Chemical Co., China. The compatibilizer of the composites used in this work, polypropylene graft maleic anhydride (PP-g-MA), was prepared in our laboratory using scCO_2-assisted solid-state grafting technique[36], with the maleic anhydride content of 0.6wt%. CO_2 (purity:99.99% w/w) was supplied by shanghai Chenggong Gases Co., China. All the materials were used as received.

2.2 Preparation of PP composites with micro- and nano-$CaCO_3$

Both PP/$CaCO_3$ composites with micro- and nano-$CaCO_3$ fillers were melt-compounded in Haake Minilab (Thermo Electron, Germany). The compatibilizer was added into the composites to enhance the interface interaction between the polymer matrix and inorganic fillers. The detail formulation of the composites is given in Table 1. Before compounding, PP, PP-g-MA pellets, and the $CaCO_3$ powder were dried under vacuum at 80℃ for 12h to remove moisture, respectively. After that, all the components were thoroughly dry-mixed and then fed simultaneously into Haake Minilab. Melt mixing was carried out at a screw rotating speed of 50r/min for 10min under 0.6MPa nitrogen atmosphere. The mixing temperature was set at 190℃. The rod-like samples collected at the die exit were pelletized and then hot pressed into cylinders with 10mm in height and 3mm in diameter for swelling and sorption measurements. PP with 10wt% PP-g-MA blend was also prepared at the same condition for comparison. The morphology of PP microcomposites and nanocomposites was characterized by using scanning electron microscopy (SEM) (JSM-6360LV, JEOL Ltd. Tokyo, Japan) and transmission electron microscopy (TEM) (JEOL-2010, Japan). The dispersion of the element, calcium (Ca), was also analyzed by using energy dispersive spectrometer (EDS) (EDAX TEAM Apollo, USA) to reflect the dispersion of the filler, $CaCO_3$, in the polymer matrix. As shown in Fig. 1 and Fig. 2, micro- or nano-$CaCO_3$ fillers were dispersed uniformly in the polymer composites.

Table 1 Composition of various PP/$CaCO_3$ composites based on PP, PP-g-MA, and $CaCO_3$

Sample	PP blend	PP/wt%	PP-g-MA/wt%	$CaCO_3$/wt%
0	PP/PP-g-MA	90	10	
1	PP/PP-g-MA/micro-$CaCO_3$	85.5	9.5	5
2	PP/PP-g-MA/micro-$CaCO_3$	81	9	10
3	PP/PP-g-MA/nano-$CaCO_3$	85.5	9.5	5
4	PP/PP-g-MA/nano-$CaCO_3$	81	9	10

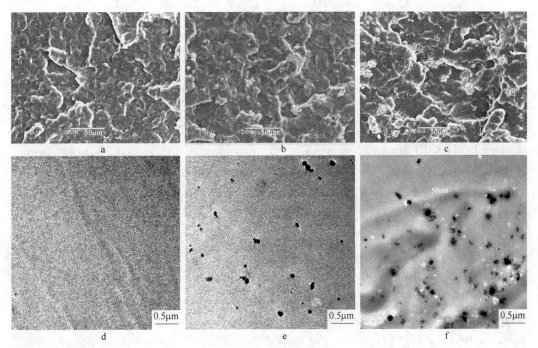

Fig. 1 SEM and TEM pictures of PP/nano-$CaCO_3$ composites

a, d—Sample 0; b—Sample 1; c—Sample 2; e—Sample 3; f—Sample 4

2.3 Swelling measurement

The swelling measurements on both PP/micro- and nano-$CaCO_3$ composites were conducted at temperatures of 200 and 220℃ and CO_2 pressure up to 22MPa, using a self-designed high-temperature and-pressure view cell. The details of the view cell system together with the swelling measurement procedure had been described in our previous work[37]. The swelling ratio of the

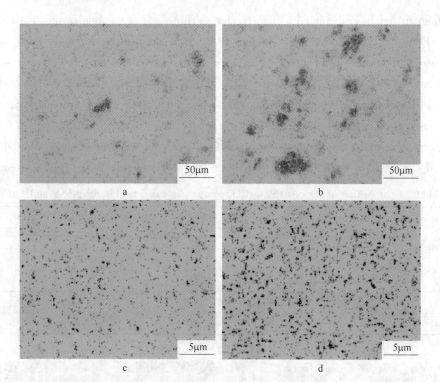

Fig. 2 EDS mapping graphs of the element of calcium (Ca) in PP/micro-CaCO$_3$ composites of sample 1(a) and sample 2(b), and PP/nano-CaCO$_3$ composites of sample 3(c) and sample 4(d)

(The green areas in (a) and (b) indicated Ca in PP/micro-CaCO$_3$ composites. The red areas in (c) and (d) indicated Ca in PP/nano-CaCO$_3$ composites)

samples can be calculated by the volume change in the glass cylinder container which has been captured by the CCD camera before and after CO_2 dissolution. For PP composites, as $CaCO_3$ filler is the solid particle and its volume cannot change at the experimental conditions, the volume expansion is resulted only from the swelling of the polymer matrix. The volume swelling ratio is defined as:

$$S_{swell}(P,T) = \frac{V_{PP}(T,P,w_{CO_2})}{V_0(T,P)} = \frac{s \times h - m_{comp} \times w_f/\rho_f}{m_{comp} \times (1-w_f)/\rho_0(T,P)} \quad (1)$$

where, $V_0(T,P)$ and $V_{PP}(T,P,w_{CO_2})$ are the volumes of the polymer matrix of the samples before and after swelling. For PP/PP-g-MA, $V(T,P,w_{CO_2})$ could be directly calculated from the height of sample/CO_2 solution, h, and the inner sectional area of the glass container, s, in the image captured by the CCD camera. For the PP composites, the volume of the polymer matrix $V(T,P,w_{CO_2})$ could be obtained from the total volume in the images subtracted by the volume of the fillers that was estimated from the mass of the PP/$CaCO_3$ composites, m_{comp}, the weight percentage, w_f, and the fillers density, ρ_f, $V_0(T,P)$ was calculated from the sample mass and the density of polymer matrix $\rho_0(T,P)$, which was obtained from the Tait equation. As the concentration of MA in PP/PP-g-MA was small, the density of PP/PP-g-MA was considered to be that of neat PP and thus was estimated from the parameters for neat PP[21]:

$$\rho_0(P,T) = 1/[7.46 \times 10^6/(6.45 \times 10^9 + P) + 1.06 \times 10^2 \times T/(9.86 \times 10^7 + P)] \quad (2)$$

where, the units of the density, the temperature, and the pressure are m^3/kg, ℃, and Pa, respectively.

2.4 CO_2 solubility and diffusivity measurements

The apparent CO_2 uptake in the samples was measured by using a magnetic suspension balance (MSB, Rubotherm Prazisionsmesstechnik GmbH, Germany) at temperatures of 200 and 220℃ and pressure up to 22MPa. The real solubility of CO_2 with respect to the mass of the polymer matrix, $S_{CO_2}(P,T)$, can be calculated from the apparent solubility, W_{app}, with the swelling correction, as given by Eq. 3:

$$S_{CO_2}(P,T,w_{CO_2}) = \frac{W_{app} + \rho_{CO_2}(P,T) \times [V_{comp}(P,T,w_f) + V_B] + \rho_{CO_2}(P,T) \times \Delta V_{swell}(P,T)}{m_{PP}}$$

$$= \frac{W_t(P,T) - W_0(0,T) + \rho_{CO_2}(P,T) \times [V_{comp}(P,T) + V_B] + \rho_{CO_2}(P,T) \times [V_{comp}(P,T) - m_{comp} \times w_f/\rho_f \times S_{swell}(P,T)]}{m_{comp} \times (1 - w_f)}$$

(3)

where, $W_0(0,T)$ and $W_t(P,T)$ are the readouts of the micro-balance before and after gas sorption; $\rho_{CO_2}(P,T)$, $V_{comp}(P,T)$, and V_B are the density of carbon dioxide, the volume of the polymer composites, and the volume of the sample basket and relative attachments, respectively. They can be determined by MSB in a separate blank test. The last term in Eq. 3 represents the swelling correction and can be evaluated from the experimental swelling ratio, $S_{swell}(P,T)$, weight percentage, and fillers density.

The diffusion coefficient of CO_2 in $PP/CaCO_3$ composites melts was simultaneously determined from the sorption line by recording the weight change of sample versus sorption time t. Fick's second law of one-dimensional diffusion was employed to describe the diffusion behavior with the assumption that the thickness of the sample remains constant during the diffusion. To minimize the effect of polymer swelling on the thickness of samples, the diffusion experiments were conducted at CO_2 pressures below 10MPa with a stepwise increase of CO_2 pressure at 1MPa/step. The one dimensional diffusion equation in Fick's second law could be derived as[38]:

$$\frac{M_t}{M_{eq}} = 1 - \frac{8}{\pi^2} \sum_{n=0}^{\infty} \frac{1}{(2n+1)^2} \exp\left[\frac{-(2n+1)^2 \pi^2 Dt}{4L^2}\right] \quad (4)$$

where, M_t and M_{eq} are the amount of CO_2 dissolved into the samples at time t and sorption equilibrium under a certain stepwise pressure change, respectively. The sample thickness L was calculated from the volume of the sample with swelling correction divided by the area of the basket bottom.

3 Results and Discussion

3.1 Swelling ratio of PP/micro- and nano-$CaCO_3$ composites under CO_2

The swelling ratio of the PP composites/CO_2 solution was measured by using a high-pressure view cell with direct visual observation. The apparatus and technique had been checked in our previous work[37]. To eliminate the interface gap due to the uncompleted wetting and weak interface bonding between the fillers and polymer matrix, the interface compatibilizer (PP-g-MA) was incorporated in both PP composites. Fig. 3 illustrates the dependence of the experimental swelling ratio of both PP/micro- and nano-CaCO$_3$ composites under CO$_2$ on the CO$_2$ pressure at temperatures of 200 and 220℃, respectively, as well as PP blended with 10% PP-g-MA for comparison. Each experimental point was repeated three times, and their average was taken as the final value. The deviation of the swelling ratio was usually no more than ±5%. As shown in Fig. 3, the swelling ratio of both PP composites/CO$_2$ solutions increased with CO$_2$ pressure and decreased with increasing temperature as that of other polymers and composites did due to the dissolution of CO$_2$[18,37]. It also shows that the swelling ratio of PP/micro-CaCO$_3$ composites decreased with increasing micro-CaCO$_3$ concentration, while that of PP/nano-CaCO$_3$ composites increased with increasing nano-CaCO$_3$ concentration. It indicates that the fillers with different sizes place a different effluence on the properties of polymer composites, which result in the different dependences of swelling ratio on the filler concentration.

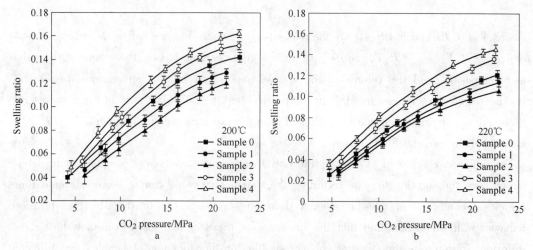

Fig. 3　Dependence of the swelling ratio of PP and PP composites on CO$_2$ pressure at temperatures of 200℃ (a) and 220℃ (b)

(The lines are 2nd-order polynomial fitting of swelling degree data)

Rheological measurements were conducted on both PP composites at the low shear rates ranging from 0 to 100s^{-1} and 190℃ by AR2000 (TA Instruments, New Castle, DE). The profiles of the viscosity versus the shear rate are illustrated in Fig. 4. It is found that the viscosity curves also show different dependences on the filler concentration. For PP/micro-CaCO$_3$ composites, the viscosity increased with increasing the micro-CaCO$_3$ concentration, whereas the viscosity decreased with increasing the nano-CaCO$_3$ loading for PP/nano-CaCO$_3$ composites. The viscosity of PP/micro-CaCO$_3$ composites was found increased in most polymer composites[37-41]. However, for polymer nanocomposites, the viscosity was found either increased or decreased with the incorporation of nanofillers, depending on the shape, concentration, and surface treatment of the filler. Xie et al.[42] reported that the viscosity decreased in PVC/CaCO$_3$ nanocomposites with the addition of nano-CaCO$_3$ filler. Mackay et al.[43] also found the similar viscosity reduction in polystyrene solution filler with highly cross-linked spherical polystyrene nanoparticles. Therefore, the viscosity of nanocomposites would be reduced with the addition of a spherically shape filler like nano-CaCO$_3$ at low concentration. As the free volume is closely related to the viscosity and plays a dominant role on the gas transport properties in polymer, it is believed that the presence of the large size micro-$CaCO_3$ filler would block the movement of polymer chains and reduce the free volume of the polymer matrix, resulting in a depression of swelling ratio in PP/micro-$CaCO_3$ composites and an increase in viscosity. However, as the nano-$CaCO_3$ spherical particle is small, it would otherwise increase the free volume and thus reduce the viscosity of the PP composites. The increase of the free volume by the addition of other nanofiller was also found in the study of the permeability of the polymer nanocomposite membrane, in which the free volume was experimentally determined by the position annihilation lifetime spectroscopy (PALS)[44-46]. In this work, the reduction of the free volume for PP/micro-$CaCO_3$ composites and the increase on that of PP/nano-$CaCO_3$ composites were also quantitatively verified by the experimental free volume determined from the swelling and sorption data and will be discussed below.

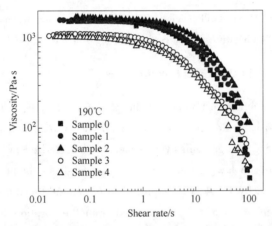

Fig. 4 Viscosity of PP and PP/$CaCO_3$ composites versus shear rate

3.2 Solubility of CO_2 in PP/$CaCO_3$ composites with the experimental swelling correction

Fig. 5 illustrates the experimental solubility of CO_2 in PP/micro-$CaCO_3$ and PP/nano-$CaCO_3$ composites melts, respectively, which are determined from the apparent sorption data by MSB combined with the experimental swelling correction. Similar to the experimental solubility in neat PP melt determined by the same approach in our previous work[37], the CO_2 sorption data in both PP composites increased almost linearly with respect to an increase of gas pressure, indicating that after eliminating the uncompleted wetting, the CO_2 gas dissolves only in the composites matrix, resulting in a sorption behavior that also follows Henry's law.

Fig. 5 also shows that the CO_2 solubility in PP/micro-$CaCO_3$ composites at the same temperature and CO_2 pressure remains almost unchanged with increasing micro-$CaCO_3$ loading compared to that in neat PP. However, that in PP/nano-$CaCO_3$ composites increased with increasing nano-$CaCO_3$ concentration. As the static free volume is unoccupied volume inside polymers, it relates directly to the quantity of gas that can dissolve in polymer and its composites. The swelling and rheological behaviors suggested that the free volume decreased with the addition of micro-$CaCO_3$ while increased with the addition of nano-$CaCO_3$. Therefore, the different dependence of the solubility on the filler concentration could be attributed to the different free volume change induced by the fillers with different sizes.

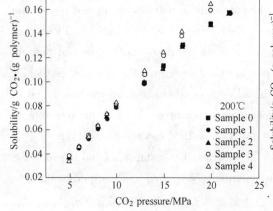
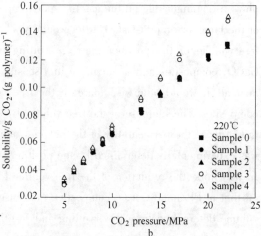

Fig. 5 Solubility of CO_2 in PP, PP/micro-$CaCO_3$, and PP/nano-$CaCO_3$ composites at 200 ℃ (a) and 220 ℃ (b) with swelling correction using experimental swelling ratio

3.3 Diffusivity of CO_2 in PP/$CaCO_3$ composites

The gas diffusivity is an important property of the gas/polymer system that dictates the gas diffusion rate and helps to estimate the time needed for the system to reach equilibrium in gas assisted polymer processing. The gas diffusivity is usually characterized by the diffusion coefficient derived from Fick's diffusion law. In this work, Fick's second law for unidirection gas diffusion into a sheet, as expressed by Eq. 4, was employed to determine the diffusion coefficient. Fig. 6 shows the diffusion coefficients of CO_2 determined from the sorption lines in the PP/micro-$CaCO_3$ and PP/nano-$CaCO_3$ composites, respectively. The diffusion coefficient of CO_2 in neat PP obtained in this work was of the same order of magnitude as that in other works and was found in good agreement with the data reported by Areerat et al.[33] Similar to the neat PP, the diffusion coefficients of CO_2 in both PP composites increased with increasing temperature and fluctuated with increasing gas pressure. The diffusion coefficient of CO_2 in PP/micro-$CaCO_3$ composites decreased with increasing micro-$CaCO_3$ loading, whereas that in PP/nano-$CaCO_3$ composites increased with increasing nano-$CaCO_3$ concentration. It should be attributed to the different free volume change induced by the fillers with different sizes.

3.4 Diffusion modeling using free volume theory

3.4.1 Modeling of CO_2 diffusion coefficient in neat PP

For polymer/gas solution, the experimental diffusion coefficients, namely, the mutual diffusion coefficients, relate to the self-diffusion coefficient of gas D_1 as:

$$D_{\text{mutual}} = \frac{x_2 D_1}{RT}\left(\frac{\partial \mu_1^P}{\partial \ln x_1}\right)_{T,P} \tag{5}$$

where, subscripts of 1 and 2 stand for CO_2 and the polymer, respectively; x is the mole fraction of the component; μ_1^P is the chemical potential of CO_2 in the polymer.

The free volume theory based models are well established statistical thermodynamic ones that have been widely used to predict the self-diffusion coefficient of gases in polymers[47,48]. There

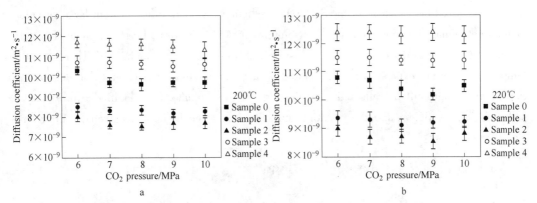

Fig. 6 Experimental diffusion coefficient of CO_2 in PP, PP/micro-$CaCO_3$, and PP/nano-$CaCO_3$ composites at 200℃ (a) and 220℃ (b)

are several versions of the free volume models. Although these models vary in reflecting temperature and concentration dependence of the penetrant, they are all related with the free volume of polymer/penetrant solutions by an exponential function of the inversed free volume or free volume fraction. Among these free volume models, Areerat's model is a simple but effective model that combines the term of temperature contribution in Fujita's model[31] and the term of free volume in Maeda and Paul's model[32]. Areerat et al. applied this model to correlate the diffusion coefficients of CO_2 in a variety of molten polymers and found it could well predict the gas diffusion behavior of polymers in the molten state. In the model, the self-diffusion coefficient, D_{free}, is expressed as:

$$D_{\text{self}} = RTA\exp\left(\frac{-B}{V_{\text{free}}}\right) = RTA\exp\left[\frac{-B}{\hat{V}_{\text{mix}}(P,T,w_{CO_2}) - \hat{V}^0_{\text{mix}}}\right] \quad (6)$$

where, D_{self} is an exponential function of the inversed free volume, V_{free}. It is proportional to the temperature. A and B are the parameters for correlation and are assumed to be dependent on the gas types but invariant to polymer types. The free volume is defined as the difference of specific volume of polymer/CO_2 mixture, $\hat{V}_{\text{mix}}(P,T,w_{CO_2})$, and the occupied specific volume of the mixture, $\hat{V}^0_{\text{mix}}$. In the model, $\hat{V}_{\text{mix}}(P,T,w_{CO_2})$ is evaluated from the S-L EOS[49,50]. $\hat{V}^0_{\text{mix}}$ is estimated from the occupied specific volumes of CO_2, $\hat{V}^0_{CO_2}$, and that of polymers at absolute zero, $\hat{V}^0_{\text{poly}}$, by using the weight average mixing law as follows:

$$\hat{V}^0_{\text{mix}} = w_{CO_2} \hat{V}^0_{CO_2} + (1 - w_{CO_2}) \hat{V}^0_{\text{poly}} \quad (7)$$

Using the proposed model, they correlated the diffusion data in different polymers that have a large difference of free volume at a certain temperature. Although the obtained parameters in their study are universal and can be used to estimate the diffusion coefficient of CO_2 in various polymer types, the accuracy of prediction will be reduced for each certain polymer.

As Areerat's model can well describe gas diffusion behavior in molten polymer, a new model was proposed based on it to correlate the diffusion coefficient of CO_2 in PP/PP-g-MA at different temperatures. It briefly follows assumptions and formulation of Areerat's model but with two modifications. First, as both the gas solubility and volume of gas/polymer solution were determined experimentally in this work, the experimental specific volume of polymer/CO_2

mixture was used here instead of the S-L EOS predictions. Second, different from large variation of the free volume of various polymer/gas solutions in Areerat's work, the free volume of certain polymers (PP/PP-g-MA) under the measurement were basically unchanged. Therefore, the temperature was not expressed as a separated term in the modified model, and the temperature dependence could still be reflected from the term of free volume. The modified expressions of self-diffusion coefficient and the resulting mutual diffusion coefficient are given by Eq. 8 and Eq. 9, respectively:

$$D_{self} = A\exp\left(\frac{-B}{V_{free}}\right) = A\exp\left[\frac{-B}{\hat{V}_{mix}(P,T,w_{CO_2}) - \hat{V}^0_{mix}}\right] \quad (8)$$

$$D_{mutual} = \frac{x_2 A}{RT}\exp\left[\frac{-B}{\hat{V}_{mix}(P,T,w_{CO_2}) - \hat{V}^0_{mix}}\right]\left(\frac{\partial \mu_1^P}{\partial \ln x_1}\right)_{T,P} \quad (9)$$

The parameters in the model are also assumed to be soely associated with the gas type and independent of the polymers type, which makes it applicable to predict diffusion coefficients of CO_2 in PP composites in the following part. Using the modified model, the new parameters A and B were correlated from experimental CO_2 diffusion data of PP/PP-g-MA at temperature from 160 to 240℃ and CO_2 pressure ranging from 5 to 10MPa. The values of $\hat{V}_{mix}(P,T,w_{CO_2})$ were calculated from the swelling and solubility data determined experimentally in this work by Eq. 10:

$$\hat{V}_{mix}(P,T,w_{CO_2}) = \frac{1 + S_{swell}(P,T)}{\rho_0(T,P)(1 + S_{CO_2}(P,T,w_{CO_2}))} \quad (10)$$

The occupied specific volume of the polymer/CO_2 mixture, $\hat{V}^0_{mix}$, was estimated by the same weight average mixing law with Eq. 6 as used in Areerat et al.'s work, where $\hat{V}^0_{CO_2}$ and $\hat{V}^0_{poly}$ are the occupied specific volume of CO_2 and that of polymer at absolute zero. The values of $\hat{V}^0_{CO_2}$ and $\hat{V}^0_{poly}$ are fixed to be 0.589cm³/g[51] and 39.975cm³/mol[32], respectively. The free volume, V_{free}, was obtained from the difference of $\hat{V}_{mix}(P,T,w_{CO_2})$ and $\hat{V}^0_{mix}$ and summarized in Table 2.

Table 2 Calculated free volume for PP/CO_2 solution

CO_2 pressure/bar	Free volume/cm³·g⁻¹				
	160℃	180℃	200℃	220℃	240℃
60	0.4126	0.4235	0.4318	0.4340	0.4463
70	0.4132	0.4242	0.4325	0.4347	0.4460
80	0.4125	0.4227	0.4324	0.4347	0.4473
90	0.4107	0.4223	0.4317	0.4359	0.4480
100	0.4070	0.4195	0.4296	0.4351	0.4459

Note: 1bar = 10⁵Pa.

The value of $(\partial \mu_1^P/(\partial \ln x_1))_{T,P}$ at a given temperature and CO_2 pressure for the PP/PP-g-MA/CO_2 mixture was estimated from the S-L EOS with a mixing rule:

$$\tilde{P} = -\tilde{\rho}^2 - \tilde{T}\left[\ln(1-\tilde{\rho}) + \left(1 - \frac{1}{r}\right)\tilde{\rho}\right] \quad (11)$$

$$\tilde{P} = \frac{P}{P^*}, \tilde{\rho} = \frac{\rho}{\rho^*}, \tilde{T} = \frac{T}{T^*}, r = \frac{M_w P^*}{RT^* \rho^*}$$

where, M_W is the molecular weight of polymers; P^*, ρ^*, and T^* are the characteristic parameters of the S-L EOS; r is the number of sites occupied by a polymer chain. The mixing rules of the S-L EOS could be found in Sato et al. 's work[52].

The chemical potential of CO_2 in the polymer, μ_1^P, is derived from S-L EOS[53] as Eq. 12:

$$\frac{\mu_1^P}{RT} = \ln\phi_1 + \left(1 - \frac{r_1^0 T_1^* P_2^*}{r_2^0 T_2^* P_1^*}\right)\phi_2 + \frac{r_1^0 \rho T_1^* \phi_2^2 [P_1^* + P_2^* - 2(1-k_{12})(P_1^* \times P_2^*)^{0.5}]}{P_1^* \rho^* T} +$$

$$r_1^0 \left[-\frac{\rho T_1^*}{\rho^* T} + \frac{P\rho^* T_1^*}{P_1^* \rho T} + \left(\frac{\rho^*}{\rho} - 1\right)\ln\left(1 - \frac{\rho}{\rho^*}\right) + \frac{1}{r_1^0}\ln\left(\frac{\rho}{\rho^*}\right) \right] = \frac{\mu_1^G}{RT}$$

$$\frac{\mu_1^G}{RT} = r_1^0 \left[-\frac{\rho_1 T_1^*}{\rho_1^* T} + \frac{P\rho_1^* T_1^*}{P_1^* \rho_1 T} + \left(\frac{\rho_1^*}{\rho_1} - 1\right)\ln\left(1 - \frac{\rho_1}{\rho_1^*}\right) + \frac{1}{r_1^0}\ln\left(\frac{\rho_1}{\rho_1^*}\right) \right] \quad (12)$$

Then, $[\partial \mu_1^P / (\partial \ln x_1)]_{T,P}$, can be calculated from Eq. 13 as[33]:

$$\left(\frac{\partial \mu_1^P}{\partial \ln x_1}\right)_{T,P} = \left(\frac{\partial \mu_1^P}{\partial \ln \phi_1}\right)_{T,P}\left(\frac{\partial \phi_1}{\partial \ln x_1}\right)_{T,P} = \left[1 + \phi_1\left(\frac{M_{w_2}\rho_1^* - M_{w_1}\rho_2^*}{M_{w_1}\rho_2^*}\right)\right]\left(\frac{\partial \mu_1^P}{\partial \ln \phi_1}\right)_{T,P} \quad (13)$$

where, M_{w_1} and M_{w_2} are the molecular weights of CO_2 and polymer, respectively. The values of $(\partial \mu_1^P / (\partial \ln \phi_1))_{T,P}$ are calculated numerically by plotting μ_1^P against $\ln\phi_1$ at temperature T and gas pressure of P. Then, the term $(\partial \mu_1^P / (\partial \ln x_1))_{T,P}$ was calculated by Eq. 13.

Once $(\partial \mu_1^P / (\partial \ln x_1))_{T,P}$, $\hat{V}_{mix}(P, T, w_{CO_2})$, and $\hat{V}_{mix}(P, T, w_{CO_2})$ were known, the self-diffusion coefficient D_{self} can be estimated from Eq. 8 at a given temperature and gas pressure. By taking the logarithm of D_{self} in Eq. 8, the parameters of A and B could be estimated from the slope and intercept values. The fitted A and B are $4.10 \times 10^{-8} m^2/s$ and $0.56 cm^3/g$, respectively. Using the obtained A and B, the predicted mutual diffusion coefficients of PP/PP-g-MA at various temperatures were plotted in Fig. 7. The predicted values showed good agreement with the experimental ones.

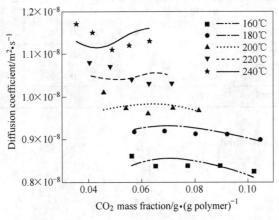

Fig. 7 Fitting results of free volume model for mutual diffusion coefficient of CO_2 in PP/PP-g-MA melts

3.4.2 Modeling of the diffusion coefficient of CO_2 in PP/micro-$CaCO_3$ based on free volume theory

As shown in Fig. 6, for PP/micro-$CaCO_3$ composites, the diffusion coefficient decreased with increasing micro-$CaCO_3$ concentration. The negative effect of microsize $CaCO_3$ fillers on the gas diffusion can be mainly attributed to two reasons. First, the increased viscosity of PP/micro-

CaCO₃ composites, as show in Fig. 4, indicated that the polymer chains movement would be blocked by the microsize fillers, which would reduce the free volume of the polymer matrix. As shown in Table 3, the free volume of CO_2/PP/micro-$CaCO_3$ is also calculated from experimental swelling and solubility data in this work by using Eq. 10, which quantitatively proved the free volume was reduced by the addition of micro-$CaCO_3$. Meanwhile, the micro-$CaCO_3$ particles act as the impermeable barriers by increasing the length of the diffusion path. In this work, a new model considering both of these two facts was proposed to describe the CO_2 diffusion behavior in PP/micro-$CaCO_3$ composites. The free volume model used for PP/PP-g-MA above was employed to describe the diffusion behavior in the composites matrix, while a diffusion tortuosity factor τ was used to account for block effect of micro-$CaCO_3$ in the diffusion path. The model is expressed in Eq. 14 and Eq. 15, respectively:

$$D = \frac{D_0}{\tau} \qquad (14)$$

$$D_0 = \frac{Ax_2}{RT}\exp\left(\frac{-B}{\hat{V}_{mix} - \hat{V}^0_{mix}}\right)\left(\frac{\partial \mu_1}{\partial \ln x_1}\right)_{T,P} \qquad (15)$$

where, D_0 is the CO_2 diffusion coefficient in the matrix of PP/micro-$CaCO_3$ composites. The free volume in the matrix was estimated from the solubility and swelling data of PP/micro-$CaCO_3$/CO_2 solution so that the effect of the filler on the free volume of matrix was included. The term $[\partial \mu_1/(\partial \ln x_1)]_{T,P}$ was estimated from the S-L EOS using the apparent solubility of CO_2 in the composites.

Table 3 Calculated free volume for PP/micro-$CaCO_3$/CO_2 solution

CO_2 pressure/bar	Free volume/cm³·g⁻¹					
	Sample 0		Sample 1		Sample 2	
	200℃	220℃	200℃	220℃	200℃	220℃
60	0.4318	0.4340	0.4145	0.4284	0.4029	0.4249
70	0.4325	0.4347	0.4155	0.4290	0.4023	0.4253
80	0.4324	0.4347	0.4148	0.4285	0.4033	0.4245
90	0.4317	0.4359	0.4143	0.4296	0.4028	0.4254
100	0.4296	0.4351	0.4121	0.4288	0.4008	0.4246

Note: 1 bar = 10⁵ Pa.

The tortuosity factor τ is defined as:

$$L' = L + \left(\frac{\pi}{2} - 1\right)L\phi^{1/3} \qquad (16)$$

$$\tau = 1 + \left(\frac{\pi}{2} - 1\right)\phi^{1/3} \qquad (17)$$

So,

$$\tau = \frac{L'}{L} \qquad (18)$$

In this work, the tortuosity factor τ is theoretically derived by assuming that identical spherical micro-$CaCO_3$ uniformly dispersed in the matrix. As shown in Fig. 8, the blue line is the normal gas diffusion path and the red one is the real path in PP/micro-$CaCO_3$ composites. The pass length will increase by $\pi\tau - 2r$ each time when the gas molecule encounters the filler. L' and L are the diffusion length with and without the impermeable barriers, respectively. If N is the average number of the fillers that the gas will meet during the diffusion, the prolonged path length L can be expressed by:

$$L' = L + (\pi r - 2r)N \tag{19}$$

Since N can be estimated from the filler volume ϕ:

$$N = \frac{L\phi^{1/3}}{2r} \tag{20}$$

τ becomes:

$$\tau = 1 + \left(\frac{\pi}{2} - 1\right)\phi^{1/3} \tag{21}$$

Combining Eq. 14 and Eq. 21, the CO_2 diffusion coefficient in PP/micro-$CaCO_3$ composites can be written as:

$$D = \frac{D_0}{1 + \left(\frac{\pi}{2} - 1\right)\phi^{1/3}} = \frac{Ax_2}{RT}\exp\left(\frac{-B}{\hat{V}_{mix} - \hat{V}^0_{mix}}\right)\left(\frac{\partial \mu_1}{\partial \ln x_1}\right)_{T,P} \frac{1}{1 + \left(\frac{\pi}{2} - 1\right)\phi^{1/3}} \tag{22}$$

The fitting results are illustrated in Fig. 9, indicating that the proposed model could correlate the mutual diffusion coefficients of CO_2 in PP/micro-$CaCO_3$ composites within an average relative deviation of 10%.

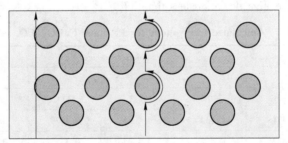

Fig. 8 Diffusion pass of gas in PP/micro-$CaCO_3$ composites

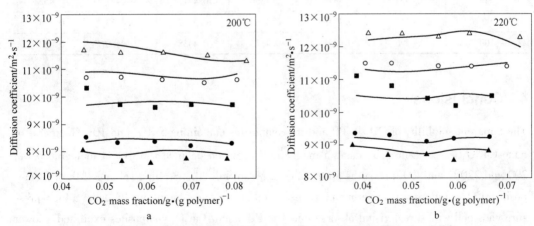

Fig. 9 Fitting results of mutual diffusion coefficient of CO_2 in PP/micro-$CaCO_3$ and PP/nano-$CaCO_3$ composites melts

3.4.3 Modeling of CO_2 diffusion coefficient in PP/nano-$CaCO_3$ composites

The diffusion coefficients of CO_2 in PP/nano-$CaCO_3$ are also illustrated in Fig. 6. Different from those in PP/micro-CaCO$_3$ composites, the diffusion coefficients of CO$_2$ in PP/nano-CaCO$_3$ increased with increasing the filler concentration. The free volume of CO$_2$/PP/nano-CaCO$_3$ solution was also calculated by Eq. 10 and collected in Table 4. It showed that the free volume increases with increasing the nano-CaCO$_3$ concentration, as suggested in the viscosity test of PP/nano-CaCO$_3$ composites in Fig. 4. However, the diffusion coefficients of CO$_2$ predicted from Eq. 9 with the calculated free volume are underestimated as compared to the experimental ones, suggesting that the nanofillers have other effects on the gas diffusion process. Xie et al. [42] and Mackay et al. [43] found that the lubricant effect (or "ball bearing" effect) of nano-CaCO$_3$ filler in polymers would ease the movement of polymer chains. In that case, the dynamic free volume will also increase with the addition of nano-CaCO$_3$. As a result, besides the increase of the static free volume, the increase of dynamic free volume would also contribute to the increase of the diffusion coefficient. In this work, a parameter β is included to account for the lubricant effect. The final free volume diffusion coefficient model for PP/nano-CaCO$_3$ is then expressed as:

$$D = D_0\beta = \frac{Ax_2\beta}{RT}\exp\left(\frac{-B}{\hat{V}_{mix} - \hat{V}^0_{mix}}\right)\left(\frac{\partial \mu_1}{\partial \ln x_1}\right)_{T,P} \quad (23)$$

The resulting β is 1.09 and 1.15 for 5% and 10% PP/nano-CaCO$_3$ composites, respectively. The fitting results by the proposed model, as illustrated in Fig. 9, show a reasonable correlation with an average relative deviation less than 3%.

Table 4 Calculated free volume for PP/nano-CaCO$_3$/CO$_2$ solution

CO$_2$ pressure/bar	Free volume/cm^3 · g^{-1}					
	Sample 0		Sample 3		Sample 4	
	200℃	220℃	200℃	220℃	200℃	220℃
60	0.4318	0.4340	0.4351	0.4379	0.4453	0.4479
70	0.4325	0.4347	0.4341	0.4385	0.4472	0.4508
80	0.4324	0.4347	0.4366	0.4416	0.4471	0.4523
90	0.4317	0.4359	0.4363	0.4432	0.4463	0.4547
100	0.4296	0.4351	0.4373	0.4438	0.4460	0.4522

Note: 1 bar = 10^5 Pa.

4 Conclusions

The apparent solubility of CO$_2$ in PP and its composites containing 5wt% and 10wt% micro- or nano-CaCO$_3$ was measured by using a magnetic suspension balance (MSB) at temperatures of 200 and 220℃ and CO$_2$ pressures up to 22MPa. Meanwhile, the swelling volume of the PP composites/CO$_2$ solutions was experimentally measured at the same conditions by using a high-pressure view cell with direct visual observation. The PP/micro-CaCO$_3$ composites exhibited a lower swelling ratio than that of PP/nano-CaCO$_3$. Moreover, the swelling ratio of the PP/micro-CaCO$_3$ composites decreased with increasing the micro-CaCO$_3$ concentration whereas that of the PP/micro-CaCO$_3$ composites increased with increasing nano-CaCO$_3$ loading. The rheological tests on the both PP composites indicated it can be attributed to the different effects of CaCO$_3$ with different filler size on the static free volume. The addition of micro-$CaCO_3$ resulted in a decrease in the static free volume while the nano-$CaCO_3$ induced an increase in the static free volume. The experimental swelling ratio was then used to correct the gas buoyancy acting on the PP composites in the MSB measurement so that the real solubility of CO_2 in the PP composites was determined. The experimental results showed that the solubility of CO_2 in PP/nano-$CaCO_3$ composites was higher than that in PP/micro-$CaCO_3$. Moreover, the solubility of CO_2 decreased with increasing micro-$CaCO_3$ concentration whereas increased with increasing nano-$CaCO_3$ loading in the PP composites. The different dependence of the solubility in PP/micro-$CaCO_3$ and PP/nano-$CaCO_3$ composites on the filler concentration could also be attributed to the different static free volume change induced by the fillers with different sizes.

The diffusion coefficient of CO_2 in the PP composites was also estimated from the sorption lines at gas pressures ranged from 5 to 10MPa. It decreased with increasing micro-$CaCO_3$ loading in the PP/micro-$CaCO_3$ composites while increased with increasing the nano-$CaCO_3$ concentration in the PP/nano-$CaCO_3$ composites. In the PP/micro-$CaCO_3$ composites, besides the reduced free volume by the micro-$CaCO_3$ filler, it also acts as an impermeable barrier, forcing a longer diffusion pass, whereas in PP/nano-$CaCO_3$ composites the nanoparticle induced an increase in both static and dynamic free volume. These results indicated that the smaller bubble size and higher bubble density obtained in polymer nanocomposite foams might also be attributed to the higher solubility and diffusivity of CO_2 in polymer nanocomposites than those in polymer microcomposites besides the better nucleation effect of nanofiller than that of microfiller. The free volume model incorporated with a diffusion tortuosity factor τ accounting for the block effect of micro-$CaCO_3$ in the diffusion path was proposed and employed to correlate the experimental diffusion coefficient of CO_2 in PP/micro-$CaCO_3$ composites within an average relative deviation of 10%. For the PP/nano-$CaCO_3$ composites, the free volume model incorporated with a parameter β accounting for the lubricant effect of nano-$CaCO_3$ was proposed and used to correlate the experimental diffusion coefficient of CO_2 in the nanocomposites within an average relative deviation of 3%.

Associated Content

Supporting information

Tables showing the solubility of CO_2 and CO_2 diffusion coefficients. This material is available free of charge via the Internet at http://pubs.acs.org.

Author Information

Corresponding author

* Tel.: +86-21-64253175. Fax: +86-21-64253528. E-mail: zhaoling@ecust.edu.cn.

Notes

The authors declare no competing financial interest.

Acknowledgements

The authors are grateful to the National Natural Science Foundation of China (Grant No. 20976045), National Programs for High Technology Research and Development of China (863 Project, 2012AA040211), the joint research project for Yangtze River Delta (12195810900), Program for New Century Excellent Talents in University (NCET-09-0348), Shanghai Shuguang Project (08SG28), Fundamental Research Funds for the Central Universities, and the 111 Project (B08021).

References

[1] Suh, N. P. Innovation in Polymer Processing; Hanser/Gardner Publications: New York, 1996.

[2] Pierick, D.; Jacobsen, K. Injection Molding Innovation: The Microcellular Foam Process. Plast. Eng. 2001, 57, 46.

[3] Xu, J.; Kishbaugh, L. Simple Modeling of the Mechanical Properties with Part Weight Reduction for Microcellular Foam Plastic. J. Cell. Plast. 2003, 39, 29.

[4] Baldwin, D. F.; Park, C. B.; Suh, N. P. An Extrusion System for the Processing of Microcellular Polymer Sheets: Shaping and Cell Growth Control. Polym. Eng. Sci. 1996, 36, 1425.

[5] Gale, M. Carbon Dioxide Extrusion Foaming of Engineering Thermoplastics; Technomic: Lancaster, 2000.

[6] Leaversuch, R. D. Carbon Dioxide Foaming is Explored for Thin Extrusions. Mod. Plast. 2000, 77, 27.

[7] Kiliaris, P.; Papaspyrides, C. D. Polymer/Layered Silicate (Clay) Nanocomposites: An overview of Flame Retardancy. Prog. Polym. Sci. 2010, 35, 902.

[8] Spitalsky, Z.; Tasis, D.; Papagelis, K.; Galiotis, C. Carbon Nanotube-Polymer Composites: Chemistry, Processing, Mechanical and Electrical Properties. Prog. Polym. Sci. 2010, 35, 357.

[9] Yuan, M. J.; Turng, L. S. Microstructure and Mechanical Properties of Microcellular Injection Molded Polyamide-6 Nanocomposites. Polymer 2005, 46, 7273-7292.

[10] Feng, J.; Chen, M.; Huang, Z. Assessment of efficacy of trivalent lanthanum complex as surface modifier of calcium carbonate. J. Appl. Polym. Sci. 2001, 82, 1339-1345.

[11] Chan, C. M.; Wu, J.; Li, J. X.; Cheung, Y. K. Polypropylene/Calcium Carbonate Nanocomposites. Polymer 2002, 43, 2981-2992.

[12] Abbasi, M.; Khorasani, S. N.; Bagheri, R.; Esfahani, J. M. Microcellular Foaming of Low-Density Polyethylene using Nano-$CaCO_3$ as a Nucleating Agent. Polym. Compos. 2011, 32, 1718.

[13] Gilbert-Tremblay, H.; Mighri, F.; Rodrigue, D. Morphology Development of Polypropylene Cellular Films for Piezoelectric Applications. J. Cell. Plast. 2012, 48, 341-354.

[14] Chen, L.; Blizard, K.; Straff, R.; Wang, X. Effect of Filler Size on Cell Nucleation During Foaming Process. J. Cell. Plast. 2002, 38, 139-148.

[15] Javni, I.; Zhang, W.; Karajkov, V.; Petrovic, Z. S.; Divjakovic, V. Effect of Nano- and Micro-Silica Fillers on Polyurethane Foam Properties. J. Cell. Plast. 2002, 38, 229-239.

[16] Saint-Michel, F.; Chazeau, L.; Cavaille, J. Y. Mechanical Properties of High Density Polyurethane Foams: II Effect of the Filler Size. Compos. Sci. Technol. 2006, 66, 2709.

[17] Yoon, J. D.; Kim, J. H.; Cha, S. W. The Effect of Control Factors and the Effect of $CaCO_3$ on the Microcellular Foam Morphology. Polym-Plast. Technol. Eng. 2005, 44, 805.

[18] Royer, J. R.; DeSimone, J. M.; Khan, S. A. Carbon Dioxide-Induced Swelling of Poly (dimethylsiloxane). Macromolecules 1999, 32, 8965-8973.

[19] Liu, D. H.; Li, H. B.; Noon, M. S.; Tomasko, D. L. CO_2-Induced PMMA Swelling and Multiple Thermody-

namic Property Analysis using Sanchez-Lacombe EOS. Macromolecules 2005,38,4416.

[20] Pantoula,M.;von Schnitzler,J.;Eggers,R.;Panayiotou,C. Sorption and Swelling in Glassy Polymer/Carbon Dioxide Systems-Part II -Swelling. J. Supercrit. Fluids 2007,39,426.

[21] Li,Y. G.;Park,C. B.;Li,H. B.;Wang,J. Measurement of the PVT Property of PP/CO_2 Solution. Fluid Phase Equilib. 2008,270,15-22.

[22] Sato,Y.;Fujiwara,K.;Takikawa,T.;Sumarno;Takishima,S.;Masuoka,H. Solubilities and Diffusion Coefficients of Carbon Dioxide and Nitrogen in Polypropylene, High-Density Polyethylene, and Polystyrene under High Pressures and Temperatures. Fluid Phase Equilib. 1999,162,261.

[23] Sato,Y.;Takikawa,T.;Sorakubo,A.;Takishima,S.;Masuoka,H.;Imaizumi,M. Solubility and Diffusion Coefficient of Carbon Dioxide in Biodegradable Polymers. Ind. Eng. Chem. Res. 2000,39,4813.

[24] Li,G.;Li,H.;Turng,L. S.;Gong,S.;Zhang,C. Measurement of Gas Solubility and Diffusivity in Polylactide. Fluid Phase Equilib. 2006,246,158.

[25] Carbone, M. G. P.;Di Maio, E.;Iannace, S.;Mensitieri, G. Simultaneous Experimental Evaluation of Solubility, Diffusivity, Interfacial Tension and Specific Volume of Polymer/Gas Solutions. Polym. Test. 2011,30,303.

[26] Handa,Y. P.;Zhang,Z. Y. Sorption, Diffusion, and Dilation in Linear and Branched Polycarbonate-CO_2 Systems in Relation to Solid State Processing of Microcellular Foams. Cell. Polym. 2002,21,221.

[27] Park, H. B.;Jung, C. H.;Lee, Y. M.;Hill, A. J.;Pas, S. J.;Mudie, S. T.;Van Wagner, E.;Freeman, B. D.;Cookson,D. J. Polymers with Cavities Tuned for Fast Selective Transport of Small Molecules and Ions. Science 2007,318,254.

[28] Yave,W.;Car,A.;Peinemann,K. V.;Shaikh,M. Q.;Ratzke,K.;Faupel,F. Gas Permeability and Free Volume in Poly (amide-b-ethylene oxide)/Polyethylene Glycol Blend Membranes. J. Membr. Sci. 2009, 339,177.

[29] Forsyth,M.;Meakin,P.;MacFarlane,D. R.;Hill,A. J. Free Volume and Conductivity of Plasticized Polyether-Urethane Solid Polymer Electrolytes. J. Phys.:Condens. Matter 1995,7,7601.

[30] Cohen,M. H.;Turnbull,D. Molecular Transport in Liquids and Glasses. J. Chem. Phys. 1959,31,1164.

[31] Fujita,H. Diffusion in Polymer-Diluent Systems. Adv. Polym. Sci. 1961,3,1.

[32] Maeda,Y.;Paul,D. R. Effect of Anti-plasticization on Gas Sorption and Transport. III. Free Volume Interpretation. J. Polym. Sci. ,Part B:Polym. Phys. 1987,25,1005.

[33] Areerat,S.;Funami,E.;Hayata,Y.;Nakagawa,D.;Ohshima,M. Measurement and Prediction of Diffusion Coefficients of Supercritical CO_2 in Molten Polymers. Polym. Eng. Sci. 2004,44,1915.

[34] Chen, K. H. J.;Rizvi, S. S. H. Measurement and Prediction of Solubilities and Diffusion Coefficients of Carbon Dioxide in Starch-Water Mixtures at Elevated Pressures. J. Polym. Sci. , Part B:Polym. Phys. 2006, 44, 607.

[35] Chen,J.;Liu,T.;Zhao,L.;Yuan,W. K. Solubility and Diffusivity of Carbon Dioxide in Polypropylene/Micro- Calcium Carbonate Composites. Proceedings of the 70th Annual Technical Conference of the Society of Plastics Engineers;Orlando,FL,2012.

[36] Tong,G.;Liu,T.;Zhao,L.;Yuan,W. K. Supercritical Carbon Dioxide-Assisted Preparation of Polypropylene Grafed Acrylic Acid with High Grafted Content and Small Gel Percent. J. Supercrit. Fluid. 2009, 48,261.

[37] Chen,J.;Liu,T.;Zhao,L.;Yuan,W. K. Determination of CO_2 Solubility in Isotactic Polypropylene Melts with Different Polydispersities using Magnetic Suspension Balance Combined with Swelling Correction. Thermochim. Acta 2012,530,79.

[38] Crank. J. The Mathematics of Diffusion,2nd ed. ;Clarendon Press:Oxford,1976.

[39] Poslinski, A. J.;Ryan, M. E.;Gupta, R. K.;Seshadri, S. G.;Frechette, F. J. Rheological Behavior of

Filled Polymeric Systems I. Yield Stress and Shear-Thinning Effects. J. Rheol. 1988, 32, 703.

[40] Maiti, S. N.; Hassan, M. R. Melt Rheological Properties of Polypropylene-Wood Flour Composites. J. Appl. Polym. Sci. 1989, 37, 2019.

[41] Da Silva, A. L. N.; Rocha, M. C. G.; Moraes, M. A. R.; Valente, C. A. R.; Coutinho, F. M. B. Mechanical and Rheological Properties of Composites Based on Polyolefin and Mineral Additives. Polym. Test. 2002, 21, 57.

[42] Xie, X. L.; Liu, Q. X.; Li, R. K. Y.; Zhou, X. P.; Zhang, Q. X.; Yu, Z. Z.; Mai, Y. W. Rheological and Mechanical Properties of Pvc/$CaCO_3$ Nanocomposites Prepared by *in Situ* Polymerization. Polymer 2004, 45, 6665.

[43] Mackay, M. E.; Dao, T. T.; Tuteja, A.; Ho, D. L.; Van Horn, B.; Kim, H. C.; Hawker, C. J. Nanoscale Effects Leading to Non-Einstein-Like Decrease in Viscosity. Nat. Mater. 2003, 2, 762.

[44] Song, Q. L.; Nataraj, S. K.; Roussenova, M. V.; Tan, J. C.; Hughes, D. J.; Li, W.; Bourgoin, P.; Alam, M. A.; Cheetham, A. K.; Al-Muhtaseb, S. A.; Sivaniah, E. Zeolitic Imidazolate Framework (Zif-8) Based Polymer Nanocomposite Membranes for Gas Separation. Energy Environ. Sci. 2012, 5, 8359.

[45] Lue, S. J.; Lee, D. T.; Chen, J. Y.; Chiu, C. H.; Hu, C. C.; Jean, Y. C.; Lai, J. Y. Diffusivity Enhancement of Water Vapor In Poly (Vinyl Alcohol)-Fumed Silica Nano-Composite Membranes: Correlation with Polymer Crystallinity and Free-Volume Properties. J. Membr. Sci. 2008, 325, 831.

[46] Winberg, P.; Eldrup, M.; Pedersen, N. J.; van Es, M. A.; Maurer, F. H. J. Free Volume Sizes in Intercalated Polyamide 6/Clay Nanocomposites. Polymer 2005, 46, 8239.

[47] Ramesh, N.; Davis, P. K.; Zielinski, J. M.; Danner, R. P.; Duda, J. L. Application of Free-Volume Theory to Self Diffusion of Solvents in Polymers Below the Glass Transition Temperature: A Review. J. Polym. Sci. Part B: Polym. Phys. 2011, 49, 1629.

[48] Zielinski, J. M.; Duda, J. L. Predicting Polymer/Solvent Diffusion Coefficients using Free-Volume Theory. AIChE J. 1992, 38, 405.

[49] Sanchez, I. C.; Lacombe, R. H. An elementary molecular theory of classical fluids. Pure fluids. J. Phys. Chem. 1976, 80, 2352-2362.

[50] Sanchez, I. C.; Lacombe, R. H. Statistical thermodynamics of polymer solutions. Macromolecules 1978, 11, 1145-1156.

[51] Bondi, A. Van Der Waals Volumes and Radius. J. Phys. Chem. 1964, 68, 441.

[52] Sato, Y.; Takikawa, T.; Yamane, M.; Takishima, S.; Masuoka, H. Solubility of Carbon Dioxide In PPo And PPo/Ps Blends. Fluid Phase Equilib. 2002, 194-197, 847-858.

[53] Li, G.; Li, H.; Wang, J.; Park, C. B. Investigating the solubility of CO_2 In Polypropylene Using Various EOS Models. Cell. Polym. 2006, 25, 237.

Effect of Supercritical Carbon Dioxide-assisted Nano-scale Dispersion of Nucleating Agents on the Crystallization Behavior and Properties of Polypropylene[*]

Abstract This work aims at using supercritical carbon dioxide ($scCO_2$) to disperse a nucleating agent, sodium 2,2-methylene-bis (4,6-di-*tert*-butylphenyl) phosphate, denoted as NA40, in an isotactic polypropylene (PP) on the nanometer scale and at studying the nucleating efficiency of the nano-dispersed NA40 by differential scanning calorimeter (DSC) and polarized optical microscope (POM). The Avrami equation and a model combining Avrami equation and Ozawa equation were used to describe the isothermal and non-isothermal crystallization kinetics of the nucleated PP, respectively. The results showed a consistent trend: both the isothermal crystallization rate and the non-isothermal crystallization temperature of the PP in which NA40 was dispersed under $scCO_2$ were higher than those of the virgin PP and the PP in which the same amount of NA40 was incorporated by a classical extruder compounding process. Moreover, the size of the spherulites and the haze of the former PP were smaller than those of the virgin PP and the latter. The trend was reversed in terms of flexural and tensile strengths.

Key words supercritical carbon dioxide, polypropylene, nucleating agent, nano-dispersion, crystallization kinetics

1 Introduction

Polypropylene (PP) is one of the most versatile commodity polymers. Nucleating agents are often added to it to enhance mechanical and optical properties[1-4].

PP is a semi-crystalline polymer, namely, its chains in the amorphous phase are in a random coil conformation, while those in the crystalline phase are in a helical conformation[5]. It starts crystallizing at nucleation sites[5]. Nucleating agents increase the number of nucleation sites, resulting in an increase in the overall crystallization rate and a decrease in the spherulite size[6]. Both the number and size of nucleation sites may affect the crystallization process and therefore the product performance. Molecular interactions between the polymer and the surface of a nucleating agent must match so as to increase the rate of crystallization or temperature at which the maximum rate occurs upon cooling from the molten state[7-9]. Smaller size and more uniform distribution of nucleating particles reduce the interfacial free energy barrier for spontaneous nucleation and growth of a birefringent dendritic phase in molten polymers[10].

It is common practice to use a screw extruder to incorporate a nucleating agent in polymers. A much finer scale of dispersion of the nucleating agent may be possible under a supercritical fluid[11,12]. Carbon dioxide is the most frequently used one in polymer processing[13-15] for its

---

[*] Coauthors: Li Bin, Hu Guohua, Cao Guiping, Liu Tao, Zhao Ling. Reprinted from *The Journal of Supercritical Fluids*, 2008, 44:446-456.

unique properties: nonflammable, nontoxic, relatively inexpensive and relatively easy to reach supercritical conditions.

A previous work showed the feasibility of penetrating sodium benzoate (NaBz) as a nucleating agent to a PP under supercritical carbon dioxide ($scCO_2$)[12]. This work aims at comparing: (1) the isothermal and non-isothermal crystallization between the PP in which NA40 was dispersed under $scCO_2$ and the virgin PP as well as the PP in which the same amount of sodium 2,2-methylene-bis(4,6-di-*tert*-butylphenyl) phosphate (NA40) was incorporated by classical extruder compounding; (2) their mechanical and optical properties; (3) the nucleation efficiency between NA40 and NaBz.

2 Experimental

2.1 Materials

Commercial isotactic polypropylene pellets with an average size of 3-4mm were kindly provided by Shanghai Petrochemical Company, China, with a mass-average molar mass (M_w) of 188700g/mol and a polydispersity index of 5.1. The nucleating agent, sodium 2,2-methylene-bis(4,6-di-*tert*-butylphenyl) phosphate (NA40) was synthesized in the laboratory. NA40 was in the form of powder with an average diameter of 20μm and had a melting point of 410℃. Fig. 1 shows the chemical formula of NA40. Carbon dioxide (purity: 99.9%) was obtained from Air Product Co. and used as received. Acetone (purity: >99.8%) was purchased from Shanghai Reagent Company.

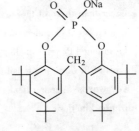

Fig. 1 Chemical formula of NA40

2.2 Solubility of NA40 in $scCO_2$

The solubility of NA40 in $scCO_2$ was measured with a dynamic flow technique. Fig. 2 is a schematic of the apparatus used. It consisted of a CO_2 cylinder, a compressor, a pre-mixer, a heat exchanger, an equilibrium vessel, a solute collector, a wet gas meter, a co-solvent reservoir, and tubing, valves and fittings of various types.

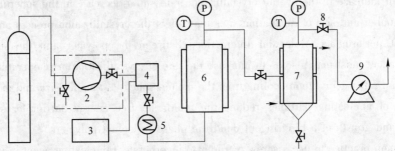

Fig. 2 Schematic of the apparatus used for measuring solubilities of the NA40 in $scCO_2$
1—CO_2 cylinder; 2—Diaphragm type compressor; 3—Co-solvent reservoir; 4—Pre-mixer; 5—Heat exchanger;
6—Equilibrium vessel; 7—Solute collector; 8—Back-pressure-valve; 9—Wet gas meter

The CO_2 was pumped and mixed with the co-solvent (acetone) via a preheater to allow it to reach the desired supercritical conditions and to ensure complete mixing of the CO_2 and acetone. The equilibrium vessel was loaded with alternate layers of NA40 powder (~5g) and glass beads of 1.0-1.5mm in diameter. The saturated CO_2 + NA40 solution was depressurized downstream to the ambient atmosphere through a back-pressure-valve. NA40 was then precipitated and collected in methanol. The gas was released through a wet gas meter that measured its volume. The collected NA40 was dissolved in methanol and its concentration was measured by a UV spectrophotometer. The wave number used was 260nm. Fig. 3 shows a calibration curve of NA40 concentration in methanol. The solubility of NA40 in CO_2 was calculated based on the measured solute mass and gas volume.

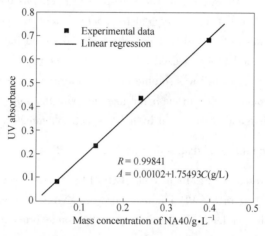

Fig. 3 UV calibration curve of the NA40 concentration in methanol

2.3 Incorporation of NA40 in PP

Three types of PP samples were studied in this work and are designated as PP0, PP1 and PP2, respectively. Table 1 shows their preparation conditions and the NA40 contents in those samples. The first one (PP0) was the virgin PP that was extruded and pelletized using a conventional twin-screw extruder without $scCO_2$. The second one (PP1) was the virgin PP in which 0.36wt% of NA40 was incorporated. That amount of NA40 was incorporated in PP by first blending PP and NA40 in a high-speed mixer without $scCO_2$. The resulting mixture was then extruded and pelletized following the same procedure as PP0. The third one (PP2) contained the same amount of NA40 as PP1 except that NA40 was first incorporated in the virgin PP in a high-pressure cell charged with $scCO_2$ and the resulting product was then extruded and pelletized following the same procedure as PP0 or PP1.

Table 1 PP samples studied in this work

Sample	Preparation technique	Mass percentage of NA40 in PP/%
PP0	Virgin PP but extruded and pelletized	0
PP1	PP0 blended with NA40 in a high-speed mixer, then extruded by a twin-screw extruder (SJSH-30) through a standard die and pelletized	0.36
PP2	PP0 penetrated with NA40 under $scCO_2$ (60℃, 10MPa, 4h) in a 1l autoclave, then extruded by a twin-screw extruder (SJSH-30) through a standard die and pelletized	0.36

The scCO$_2$-assisted incorporation of NA40 in the virgin PP for PP2 was carried out in an apparatus described elsewhere[11,12] except that the volume of the high-pressure cell was larger, 2l. About 100g of the PP pellets were placed in the high-pressure cell together with 1wt% of NA40. When a desired temperature (60℃, an optimal one for the incorporation of NA40[11,12]) was reached, the CO$_2$ was charged to the high-pressure cell till a desired pressure was attained. Acetone (0.05wt% with respect to the PP) was then charged to the system with a high-pressure advection plump (LB-10, Beijing Satellite Instrument Co., China) to improve the mass uptake of NA40 in the PP[11,12]. The incorporation process lasted 4h, the time necessary for the mass uptake to reach equilibrium. The high-pressure cell was then cooled down, opened up and the PP pellets taken out for subsequent treatment and/or analyses.

Acetone was used to wash off NA40 that was physically absorbed but not impregnated on the PP pellet surfaces while methanol was used to extract the penetrated nucleating agent from inside the PP pellets[16]. Since NA40 was soluble in methanol, its mass uptake in the PP could be calculated through its concentration in methanol together with the solution volume. The concentration of NA40 in methanol was measured by a UV spectrophotometer.

2.4 Assessment of the state of dispersion of NA40 in PP

Scanning electrical microscopy (SEM) of type JEOL/EO JSM-6360 was used to visualize the state of dispersion of NA40 in the PP. For that purposes, NA40 was incorporated in PP films (1mm × 3mm × 0.3mm) under scCO$_2$ according to the same procedure as for PP2. The PP films were then fractured in liquid nitrogen. Fractured surfaces were analyzed under SEM.

2.5 Crystallization kinetics of the PP by differential scanning calorimetry (DSC)

Isothermal and non-isothermal crystallization kinetics of PP0, PP1 and PP2 were carded out using a Perkin-Elmer Diamond DSC (Perkin-Elmer Company, USA). The instrument was calibrated with In and Zn before the measurements. All the measurements were conducted under high-purity (>99.99%) nitrogen atmosphere. The mass of samples was about 5mg.

The isothermal crystallization kinetics was performed as follows. A sample sealed in an aluminum pan with a diameter of 8mm was heated up from room temperature to 210℃ and held at that temperature for 5min to erase any thermal and/or mechanical history. It was then cooled down to a prescribed crystallization temperature at a cooling rate of 200℃/min. The heat flux as a function of time was recorded. As for the non-isothermal crystallization kinetics, a sample was also heated up to 210℃ and held at that temperature for 5min to eliminate any thermal and/or mechanical history. It was then cooled down to room temperature at different cooling rates. The later were 5℃/min, 10℃/min 15℃/min, 20℃/min and 30℃/min, respectively. The corresponding heat flux was recorded as a function of time.

2.6 Spherulitic morphologies by polarized optical microscopy (POM)

The spherulitic morphologies of PP0-PP2 were observed with an optical polarized microscope (BX51, Olympus, Japan) equipped with a hot stage (THMS 600, Linkam, Great Britain). Prior to the observation, a PP sample was sandwiched between two microscope cover glasses. The whole system was heated up to 210℃ and held at that temperature for 5min to erase any thermal and/or mechanical history. It was then cooled down to 140℃ with a cooling rate of 100℃/min.

2.7 Characterization of mechanical and optical properties

An injection-molding machine (CJ-80E) was used to mold PP0, PP1 and PP2 pellets into test specimens. The mold temperature, injection time, injection pressure, melt temperature, holding time, holding pressure and cooling time were 50℃, 8s, 6.5MPa, 230℃, 15s, 5.0MPa and 20s, respectively. The flexural modulus and tensile strength of PP0, PP1 and PP2 were measured according to ASTM D-790 and D-638 using a universal testing machine (DXLL-20000). Their haze was measured using a hazemeter according to ASTM D-1003 standard. Haze is the percentage of transmitted light that, in passing through a specimen, deviates from the incident beam by the forward scattering. Only is the light flux deviating more than 2.5° on the average considered to be haze. Fig. 4 is a schematic of the hazemeter (WGT-S, Jinyi Instruments, Shanghai, China).

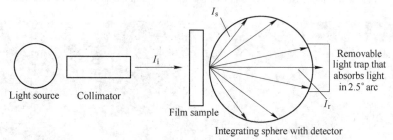

Fig. 4 Schematic diagram of hazemeter. I_i, I_r and I_s are the intensities of the incident light beam, regular transmitted light and scattered transmitted light, respectively

The haze value of a film sample is calculated by:

$$\% \text{Haze} = \frac{I_s > 2.5°}{I_s + I_r} \times 100\% \quad (1)$$

where, I_r and I_s are the intensities of the regular transmitted light and scattered transmitted light, respectively.

3 Results and Discussion

3.1 Partition of NA40 between scCO$_2$ and PP

Fig. 5 depicts the basic principle of the scCO$_2$-assisted incorporation of NA40 in PP in the high-pressure cell. NA40 is first dissolved in scCO$_2$ and then penetrates PP under scCO$_2$. After a certain period of time, NA40 is partitioned between a liquid phase (NA40 + scCO$_2$) and a solid phase (PP + NA40 + scCO$_2$). Upon release of scCO$_2$, it is expected that NA40 that has penetrated PP will remain and crystallize as very fine particles.

Fig. 6 shows the SEM micrographs of NA40 alone (Fig. 6a) and its state of dispersion in PP2 after the scCO$_2$-assisted incorporation step of NA40 but before the extrusion and pelletizing step (Fig. 6b). No SEM micrograph on PP1 is shown here because unlike NA40 particles in PP0 or PP2, those in PP1 already acted as nucleating sites and were incorporated in PP spherulites. The small white dots in the micrographs correspond to NA40. As expected, they were uniformly and finely dispersed in PP2 with a diameter on the order of 100nm.

The partition coefficient of NA40 between PP0 and scCO$_2$, Y, was defined as follows:

$$Y = \frac{F_{NA40}^{PP}}{F_{NA40}^{CO_2}} \quad (2)$$

where, F_{NA40}^{PP} and $F_{NA40}^{CO_2}$ are the mass fractions of NA40 in PP0 and in $scCO_2$, respectively.

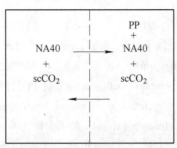

Fig. 5 Schematic diagram of the basic principle of $scCO_2$-assisted
incorporation of NA40 in PP in the high-pressure cell
(NA40 is first dissolved in $scCO_2$ and then penetrates PP with the help of $scCO_2$ at 60℃)

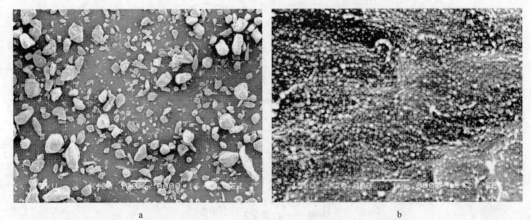

Fig. 6 SEM micrographs of the NA40 alone (a, scale bar = 100μm) and its
state of dispersion in the PP2 (b, scale bar = 1μm)
(The penetration process was carried out under the following conditions: temperature = 60℃, CO_2
pressure = 8MPa and time = 4h. The amount of acetone added was 0.05wt% of the PP)

Fig. 7 shows, respectively, the solubility of NA40 in $scCO_2$, its mass uptake in PP0 and its partition coefficient between PP0 and $scCO_2$ at 60℃ as a function of the $scCO_2$ pressure in the presence of acetone as a co-solvent. The solubility of NA40 in the $scCO_2$ at 60℃ increased almost linearly with increasing $scCO_2$ pressure. The mass uptake of NA40 in PP0 at 60℃ first increased with increasing $scCO_2$ pressure, reached a maximum when the $scCO_2$ pressure was about 10MPa. It then decreased with further increasing $scCO_2$ pressure. As for the partition coefficient of NA40 between PP0 and $scCO_2$, it decreased with increasing $scCO_2$ pressure, implying that the solubility of NA40 in $scCO_2$ increased more than in PP as the $scCO_2$ pressure increased. The existence of a maximum in the mass uptake of NA40 in PP0 was related to the fact that as the $scCO_2$ pressure increased, the solubility of NA40 in the CO_2 increased and whereas the partition coefficient of NA40 between PP0 and $scCO_2$ decreased.

3.2 Isothermal crystallization kinetics

3.2.1 Analysis based on Avrami equation

Fig. 8 compares the isothermal crystallization rate among PP0, PP1 and PP2. As expected, for all the three samples, an increase in the crystallization temperature led to a later start of crystallization and a decrease in the crystallization rate.

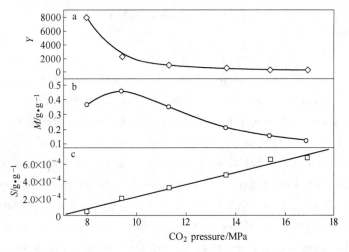

Fig. 7 Solubility of the NA40 in the scCO$_2$, its mass uptake in PP0 and its partition coefficient between PP0 and scCO$_2$ at 60℃ as a function of the scCO$_2$ pressure in the presence of 0.05wt% of acetone with respect to PP0

a—Y, the partition coefficient of NA40 between PP and scCO$_2$; b—M, the mass uptake of NA40 in PP, g/g; c—S, the solubility of NA40 in scCO$_2$, g/g

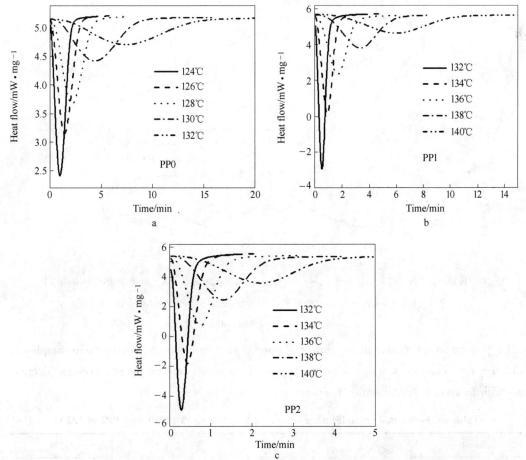

Fig. 8 Heat flow as a function of time during the isothermal crystallization at different crystallization temperatures

a—PP0 (virgin PP); b—PP1 (PP blend with NA40); c—PP2 (PP penetrated with NA40 by scCO$_2$)

The Avrami equation[17,18] was used to quantitatively describe the effect of the nucleating agent on the isothermal crystallization kinetics as follows:

$$X(t) = 1 - \exp(-Kt^n) \tag{3}$$

or

$$\ln\{-\ln[1 - X(t)]\} = n\ln t + \ln K \tag{4}$$

where, $X(t), t, K, n$ are the relative degree of crystallinity, crystallization time, overall kinetic constant and Avrami exponent, respectively. The latter is related to the nucleation mechanism and growth dimension. Based on the original assumptions of the theory, the value of n should be an integer, ranging from 1 to 4, corresponding to various growth geometries from rod-like to sphere-like.

The isothermal crystallization half time, $t_{1/2}$, characterizes the time at which 50% of the total crystallization process has completed. A smaller value of $t_{1/2}$ corresponds to a more rapid crystallization process. Table 2 gives the values of n and K determined from the initial linear sections in Fig. 9. It also shows the values of the time at which the crystallization rate reached a maximum, t_{max}. The latter was obtained from the heat flow curves in Fig. 8. The value of $t_{1/2}$ follows the order: PP0≫PP1 > PP2, while that of the crystallization rate constant K follows the opposite order: PP0≪PP1 < PP2. This indicates that the crystallization rate of the PP is obviously increased because of the addition of NA40, especially when it is uniformly dispersed in the PP matrix through the scCO$_2$ penetration process.

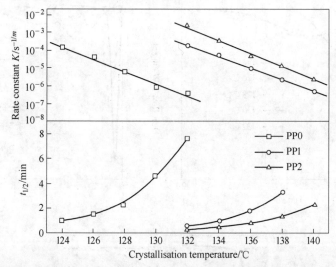

Fig. 9 Isothermal crystallization half time $t_{1/2}$, and crystallization rate constant K of PP0 (virgin PP), PP1 (PP blend with NA40) and PP2 (PP penetrated with NA40 by scCO$_2$) as a function of the crystallization temperature

Fig. 9 compares the value of $t_{1/2}$ and crystallization rate constant, K, for the three samples at different crystallization temperatures. The $t_{1/2}$ increases and crystallization rate constant decreases with increasing crystallization temperature.

Table 2 Isothermal crystallization kinetic parameters of PP0, PP1 and PP2 at 132°C

Sample	$t_{1/2}$/s	n	$K/\text{s}^{-1/n}$	t_{max}/s
PP0	452.6	2.82	7.25×10^{-7}	481.9
PP1	32.5	2.45	1.78×10^{-4}	57.4
PP2	17.8	2.20	2.47×10^{-3}	37.9

3.2.2 Surface free energy (σ_e)

According to the Lauritzen-Hoffman secondary nucleation theory[19], the crystal growth rate G of a crystalline aggregate (e.g. spherulite) in each regime depends on the degree of undercooling ΔT, and is defined by the following equation:

$$G = G_0 \exp\left[-\frac{U^*}{R(T_c - T_\infty)}\right] \exp\left(-\frac{K_g}{T_c \Delta T f}\right) \quad (5)$$

$$K_g = \frac{nb_0 \sigma \sigma_e T_m^0}{\Delta h_f^\circ k} \quad (6)$$

where, G_0 is the pre-exponential factor; U^* the transport activation energy and is usually equal to 6300J/mol; T_∞ a hypothetical temperature below which all viscous flow ceases and is usually equal to $T_g - 30$; T_c the crystallization temperature; ΔT the degree of supercooling and is equal to $T_m^0 - T_c$, where T_m^0 is the equilibrium melting temperature; f a correction term to account for the variation in the bulk enthalpy of fusion per unit volume with temperature and is equal to $2T_c/(T_m^0 + T_c)$; K_g the nucleation parameter and can be defined by Eq. 6; $n = 4$ for a nucleated crystallization of PP corresponding to regime III; k the Boltzmann constant; Δh_f° the bulk melting enthalpy per unit volume for a fully crystalline polymer ($\Delta h_f^\circ = H_m^\circ \rho_c$), for iPP $\Delta h_f^\circ = 1.34 \times 10^8 J/m^3$; b_0 the thickness of a monomolecular layer in the crystal, for iPP $b_0 = 6.56 \times 10^{-10}$m; σ_e the folding surface free energy; and σ is the lateral surface free energy, and it can be evaluated through $\sigma = ab_0 \Delta h_f^\circ$, $a(\approx 1)$ is an empirical constant. With the above assumptions, K_g can be expressed as:

$$K_g = \frac{nT_m^0}{k} \sigma_e b_0^2 \quad (7)$$

Generally, the crystal growth rate can be calculated through $G = 1/t_{1/2}$[19], and then Eq. 5 can be expressed as follows:

$$\ln(t_{1/2}^{-1}) + \frac{U^*}{R(T_c - T_\infty)} = \ln G_0 - \frac{K_g}{T_c \Delta T f} \quad (8)$$

Fig. 10 plots $\ln(t_{1/2}^{-1}) + U^*/R(T_c - T_\infty)$ versus $1/(T_c \Delta T f)$ and the values of K_g for PP0-PP2 are obtained. The K_g values for PP0-PP2 are $7.86 \times 10^5 K^2$, $5.68 \times 10^5 K^2$ and $4.85 \times 10^5 K^2$, respectively. The chain folding free energy for PP0-PP2 can therefore be calculated with K_g values based on Eq. 6, which is $0.131 J/m^2$, $0.094 J/m^2$ and $0.081 J/m^2$, respectively. The lower the folding free energy is, the less the energy consumed for the polymer macromolecule to fold on the nuclei surface to form spherulite. The smallest value of σ_e for PP2 shows that a uniformly and nano-scale dispersed nucleating agent reduces the folding free energy and accelerate the crystallization process.

3.3 Non-isothermal crystallization kinetics

3.3.1 Non-isothermal crystallization behaviors of PP

Fig. 11 shows the DSC curves of PP cooled from the melt at various cooling rates. In all cases, the exothermic trace becomes wider and the crystallization temperature shifts to a lower temperature when the cooling rate is increased. Fig. 12 summarizes the crystallization temperatures of PP0, PP1 and PP2 at various cooling rates. At a given cooling rate, the crystallization temperature follows the order: PP0 < PP1 < PP2, indicating that PP2 has higher crystallization rate compared with PP0 and PP1. This result is similar to that of the isothermal crystallization.

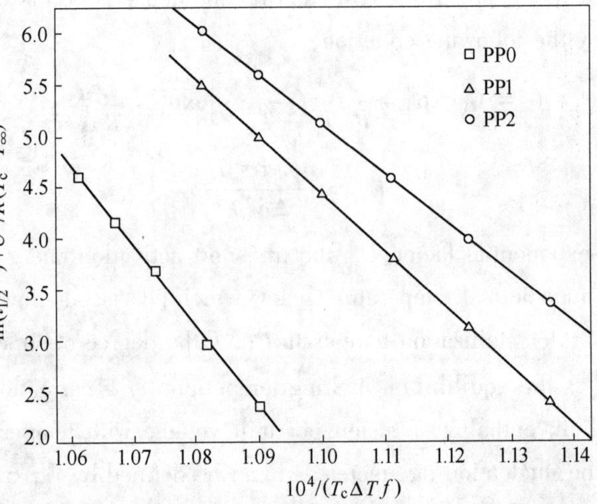

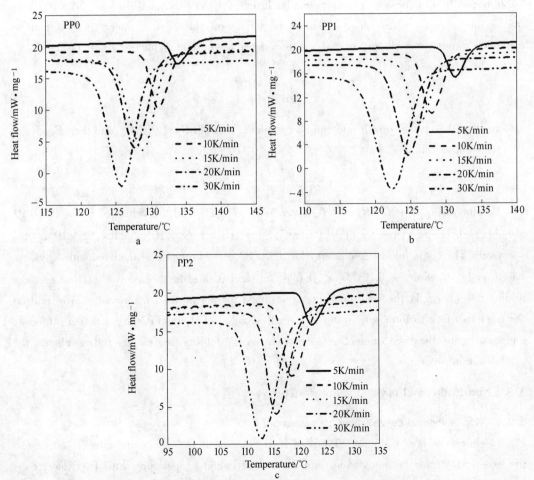

Fig. 10 Analysis of the isothermal growth rate of PP0, PP1 and PP2 as a function of the crystallization temperature based on the Lauritzen-Hoffman secondary nucleation theory

Fig. 11 Effect of the cooling rate on the crystallization of PP0 (virgin PP) (a), PP1 (PP blend with NA40) (b) and PP2 (PP penetrated with NA40 by $scCO_2$) (c)

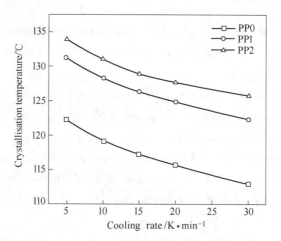

Fig. 12 Effect of the cooling rate on the crystallization temperature corresponding to the maximum crystallization rate of PP0 (virgin PP), PP1 (PP blend with NA40) and PP2 (PP penetrated with NA40 by scCO$_2$)

3.3.2 A model combining the Avrami and Ozawa models

As the non-isothermal crystallization is a rate-dependent process, assuming that the polymer melt be cooled at a constanl rate and the mathematical derivation of Evans[20] be valid, and accounting for the effect of cooling rate on crystallization, Ozawa[21] modified the Avrami equation by replacing t in Eq. 1 with T/R_c, as follows:

$$1 - X(t) = \exp\left[-\frac{k(T)}{R_c^m}\right] \quad (9)$$

where, $k(T)$ is cooling function; m is Ozawa exponent and depends on the crystal growth and nucleation mechanism.

While the Ozawa equation describes the non-isothermal crystallization behavior of polymers to a certain extent, it is proven to fail to adequately describe the non-isothermal crystallization kinetics for many polymers, such as PE[22], PEEK[23] and PEEKK[24], in which a large portion of crystallization is attributed to the secondary process. In addition, as the crystallinity is related to the cooling rate R_c and crystallization time t (or temperature T), a relationship between R_c and t could be built up at a given crystallinity. This work adopts a convenient kinetic model by Liu and Mo[24,25] to deal with non-isothermal data by combining the Avrami Eq. 3 with the Ozawa Eq. 9, as follows:

$$n\ln t + \ln K = \ln k(t) - m\ln R_c \quad (10)$$

$$\ln R_c = \frac{1}{m}\ln\frac{k(t)}{K} - \frac{n}{m\ln t} \quad (11)$$

Given

$$F(T) = \left[\frac{k(t)}{K}\right]^{1/m} \quad (12)$$

and

$$a = \frac{n}{m} \quad (13)$$

The final form of Eq. 11 becomes:

$$\ln R_c = \ln F(T) - a\ln t \quad (14)$$

The smaller the value of $F(T)$, the higher the crystallization rate becomes. At a given degree of crystallization, on plotting $\ln R_c$ versus $\ln t$, the values of $F(T)$ and a can be obtained from the intercepts and slopes of these lines.

Generally, the value of $F(T)$ increased with an increase in the relative crystallinity for PP0, PP1 and PP2, respectively, meaning that to obtain a higher degree of crystallinity, at an unit crystallization time, a higher cooling rate is required. Table 3 summarizes the value of $F(T)$ and a with different degrees of crystallization $X(t)$. Fig. 13 shows the plot of $\ln R_c$ versus $\ln t$ for $X(t) = 50\%$. At the same $X(t)$ value, the value of $F(T)$ follows the order: PP0 > PP1 > PP2, which indicates again that the crystallization rate of PP2 is higher than that of PP0 and PP1.

Table 3　Non-isothermal crystallization kinetics for PP0, PP1 and PP2 based on a model combining the Avrami and Ozawa models (Eq. 14)

$X(t)/\%$	$F(T)$			a		
	PP0	PP1	PP2	PP0	PP1	PP2
30	101.9	70.8	68.5	1.058	0.941	0.942
40	102.7	70.9	68.6	1.059	0.937	0.941
50	103.6	70.9	68.8	1.060	0.938	0.939
60	104.5	71.0	68.9	1.061	0.936	0.938
70	105.3	71.1	69.1	1.062	0.934	0.936

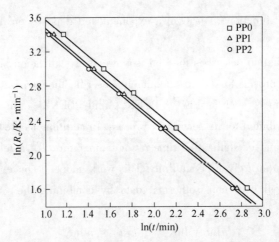

Fig. 13　Plot of $\ln R_c$ vs. $\ln t$ for PP0 (virgin PP), PP1 (PP blend with NA40) and PP2 (PP penetrated with NA40 by $scCO_2$) for $X(t) = 50\%$

3.3.3　Nucleation efficiency

The nucleation efficiency is based on the calorimetric efficiency scale proposed by Lotz and co-workers[26] and is calibrated according to their procedure based on the "ideally" nucleated polymer attainable via self-nucleation. Two limits need to be defined: the lower limit refers to the virgin (non-nucleated) PP and the upper limit to an optimally self-nucleated PP. The nucleation efficiency is a percentage of the range defined by the two limits and is expressed as:

$$N_E = \frac{T_{c,\text{lower}} - T_c}{T_{c,\text{upper}} - T_c} \times 100 \qquad (15)$$

where, $T_{c,\text{lower}}$ and $T_{c,\text{upper}}$ are the crystallization temperatures of the virgin and best self-nucleated PP, respectively. T_c is the crystallization temperature obtained in the presence of a nucleating agent (NA40). N_E is thus equal to 0 for a non-nucleating action and 100 for the maximum efficiency.

Jain et al.[27] determined the values of $T_{c,\text{lower}}$ and $T_{c,\text{upper}}$ of an iPP by DSC. They were 110.6℃ and 138.6℃, respectively. Fig. 14 shows that the nucleation efficiency for crystallization of PP0, PP1 and PP2 decreased with increasing cooling rate. At a given cooling rate, the nucleating efficiency followed the order: PP0≪PP1 < PP2, which indicates that the nano-scale dispersed nucleating agent provided the PP with more nucleating sites and promoted the overall nucleating efficiency.

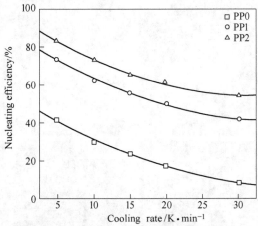

Fig. 14　Nucleating efficiency of PP0 (virgin PP), PP1 (PP blend with NA40) and PP2 (PP penetrated with NA40 by scCO$_2$) as a function of the cooling rate

3.4　Spherulite morphologies of PP under POM

Fig. 15 is the spherulitic morphologies of PP0, PP1 and PP2 isothermally crystallized at 140℃. For all the PP samples, the size of the spherulite increases as the crystallization proceeds. It can be seen that the size of spherulite follows the order: PP0≫PP1 > PP2. These results are in agreement with the state of NA40 dispersion in the PP described above (see Fig. 6). Spherulites in PP2 grow much faster in size than PP0 and PP1.

3.5　Mechanical and optical properties

The presence of a nucleating agent enhances the crystallization rate and promotes the formation of intercrystalline links[6]. The latter are bridges between and within spherulites generated by a PP chain that has one segment crystallized in a spherulite and another segment crystallized in another spherulite or another part of the same spherulite. Thus, a PP chain can "link" two spherulites together. A nucleating agent, through its ability to create intercrystalline links and smaller spherulites, improves the mechanical properties of the PP Fig. 16 compares mechanical properties of PP0, PP1 and PP2 at room temperature. PP2 is the best in terms of the flexural and tensile strengths.

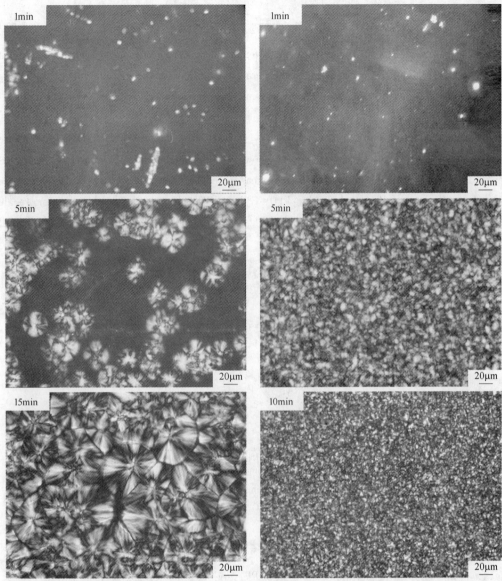

Fig. 15 Evolution of the spherulites of PP0 (virgin PP), left pictures, PP1 (PP blend with NA40) and PP2 (PP penetrated with NA40 by scCO$_2$) right pictures, during the isothermal crystallization at 140 ℃

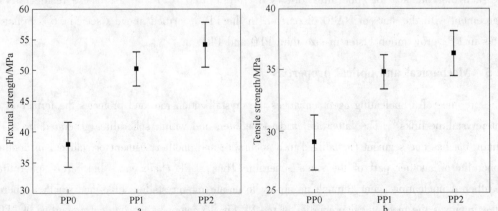

Fig. 16 Comparison of flexural and tensile strength among PP0 (virgin PP), PP1 (PP blend with NA40) and PP2 (PP penetrated with NA40 by scCO$_2$)

Fig. 17 compares the haze values of PP0, PP1 and PP2. They follow the order: PP0≫PP1 > PP2, indicating that PP2 was more transparent than PP1 and the latter was much more transparent than PP0. This can be attributed to the spherulites that induced light scattering. More light is scattered when the spherulites are large.

From the above results, the superiority of PP2 over PP1 in terms of mechanical and optical properties was not as big as expected from the difference in the state of dispersion of NA40 between PP1 and PP2 shown in Fig. 6. The similarity in properties between PP1 and PP2 might be explained by the difference of the concentration of NA40 in the PP, because the concentration of NA40 in PP2 might not be estimated with good accuracy. Nevertheless, aggregation of NA40 particles during the extrusion in the twin-screw extruder was mostly likely responsible for. This raises an issue of how to stablize finely dispersed NA40 particles in PP during the extrusion step.

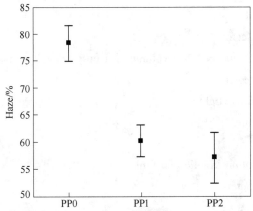

Fig. 17 Haze value of PP0 (virgin PP), PP1 (PP blend with NA40) and PP2 (PP penetrated with NA40 by $scCO_2$)

4 Conclusions

This work aims at: (1) dispersing a nucleating agent, sodium 2,2-methylene-bis(4,6-di-*tert*-butylphenyl) phosphate denoted as NA40, in an isotactic polypropylene (PP) on the nanometer scale under supercritical carbon dioxide ($scCO_2$); (2) studying the nucleating efficiency of the nano-dispersed NA40 by differential scanning calorimeter (DSC) and polarized optical microscope (POM); (3) mechanical properties of such nucleated PP.

Under $scCO_2$ and in the presence of acetone as a co-solvent, NA40 was uniformly dispersed in the PP with a diameter of the order of 100nm. The isothermal and non-isothermal crystallization kinetics of the PP in which NA40 was dispersed on the nanometer scale were much more rapid than those of the virgin PP and its blend with same amount of NA40 but dispersed by a conventional extruder compounding process. The PP with NA40 dispersed on the nanometer scale also showed superior flexural and tensile strengths as well as lower haze value indicating higher transparency.

Acknowledgements

The authors gracefully thank Prof. Xin Zhong (East China University of Science and Technology) for providing NA40 sample. This work was supported by a NSFC/PetroChina joint project for multiscale methodology and a Nano-technology fund of Shanghai Science and Technology Commission (0552nm 039). They also thank the reviewers for their pertinent questions and constructive comments.

Nomenclature

a	empirical constant
b_0	thickness of a monomolecular layer in the crystal (m)
f	correction term
F_{NA40}^{PP}	mass fraction of NA40 in PP0 (g/g)
$F_{NA40}^{CO_2}$	mass fraction of NA40 in scCO$_2$ (g/g)
$F(T)$	crystallization variable (see Eq. 12)
G	crystal growth rate (s^{-1})
G_0	pre-exponential factor
Δh_f°	bulk melting enthalpy per unit volume of a fully crystalline polymer (kJ/mol)
I_r	regular transmitted light
I_s	scattered transmitted light
k	Boltzmann constant
$k(T)$	cooling function
K	overall crystallization kinetic constant
K_g	nucleation parameter
m	Ozawa exponent
n	Avrami exponent
N_E	nucleating efficiency (%)
R_c	cooling rate (°C/min)
t	crystallization time (s)
$t_{1/2}$	half crystallization time (s)
T	temperature
T_c	crystallization temperature
$T_{c,lower}$	crystallization temperatures of the virgin PP
$T_{c,upper}$	crystallization temperatures of best self-nucleated PP
T_∞	hypothetical temperature below which all viscous flow ceases
T_m^0	equilibrium melting temperature
ΔT	degree of supercooling
U^*	transport activation energy (J/mol)
$X(t)$	relative degree of crystallinity (%)
Y	partition coefficient

Greek letters

σ	lateral surface free energy (J/m^2)
σ_e	folding surface free energy (J/m^2)

References

[1] T. Xu, H. Lei, C. S. Xie, The effect of nucleating agent on the crystalline morphology of polypropylene, Mater. Design 24 (2003) 227-230.

[2] Y. Feng, X. Jin, J. N. Hay, Effect of nucleating agent addition on crystallization of isotactic polypropylene, J. Appl. Polym. Sci. 69 (1998) 2089-2095.

[3] R. Phillips, J. E. Manson, Prediction and analysis of nonisothermal crystallization of polymers, J. Polym. Sci. Part B: Polym. Phys. 35 (1997) 875-878.

[4] C. Y. Kim, Y. C. Kim, S. C. Kim, Temperature dependence of the nucleation effect of sorbitol derivatives on polypropylene crystallization, Polym. Eng. Sci. 33 (1993) 1445-1451.

[5] B. Wunderlich, Macromolecular Physics, Academic Press, New York, 1976.

[6] H. D. Keith, F. J. Padden, R. G. Vadimsky, Intercrystalline links in polyethylene crystallized from the melt, J. Polym. Sci. Part A2-4: Polym. Phys. 4 (1966) 267-281.

[7] J. C. Wittmann, B. Lotz, Epitaxial crystallization of polymers on organic and polymeric substrates, Prog. Polym. Sci. 15 (1990) 909-948.

[8] B. J. Chisholm, P. M. Fong, J. G. Zimmer, R. Hendrix, Properties of glassfilled thermoplastic polyesters, J. Appl. Polym. Sci. 74 (1999) 889-899.

[9] J. P. Mercier, Nucleation in polymer crystallization: a physical or a chemical mechanism, Polym. Eng. Sci. 30 (1990) 270-278.

[10] J. G. Tang, Y. Wang, H. Y. Liu, L. A. Belfiore, Effects of organic nucleating agents and zinc oxide nanoparticles on isotactic polypropylene crystallization, Polymer 45 (2004) 2081-2091.

[11] B. Li, G. P Cao, T. Liu, L. Zhao, W. K. Yuan, G. H. Hu, Preliminary study on the characteristics of isotactic polypropylene with nucleating agent swollen by supercritical carbon dioxide, Chin. J. Chem. Eng. 13 (2005) 673-677.

[12] B. Li, G. H. Hu, G. P Cao, T. Liu, L. Zhao, W. K. Yuan, Supercritical carbon dioxide-assisted dispersion of sodium benzoate in polypropylene and crystallization behavior of the resulting polypropylene, J. Appl. Polym, Sci. 102 (2006) 3212-3220.

[13] D. L. Tomasko, H. B. Li, D. H. Liu, X. M. Han, M. J. Wingert, L. J. Lee, K. W. Koelling, A review of CO_2 applications in the processing of polymers, Ind. Eng. Chem. Res. 42 (2003) 6431-6456.

[14] T. Liu, G. H. Hu, G. S. Tong, L. Zhao, G. P. Cao, W. K. Yuan, Supercritical carbon dioxide assisted solid-state grafting of maleic anhydride onto polypropylene, Ind. Eng. Chem. Res. 44 (2005) 4292-4299.

[15] Z. M. Xu, X. L. Jiang, T. Liu, G. H. Hu, L. Zhao, Z. N. Zhu, W. K. Yuan, Foaming of polypropylene with supercritical carbon dioxide, J. Supercrit. Fluids 41 (2007) 299-310.

[16] Y. Q. Wang, Polymer impregnation and surface modification using supercritical carbon dioxide, Doctor Dissertation, The Ohio State University, 2001.

[17] M. J. Avrami, Kinetics of phase change Ⅰ, J. Chem. Phys. 7 (1939) 1103-1112.

[18] M. J. Avrami, Kinetics of phase change Ⅱ, J. Chem. Phys. 8 (1940) 212-214.

[19] J. I. Lauritzen, J. D. Hoffman, Extension of theory of growth of chainfolded polymer crystals to larger undercoolings, J. Appl. Phys. 44 (1973) 4340-4352.

[20] U. R. Evans, The laws of expanding circles and spheres in relation to the lateral growth of surface films and the grain size of metals, Trans. Faraday Soc. 41 (1945) 365-372.

[21] T. Ozawa, Kinetics of non-isothermal crystallization, Polymer 12 (1971) 150-158.

[22] M. Eder, A. Wlochowicz, Kinetics of non-isothermal crystallization of polyethylene and polypropylene, Polymer 24 (1983) 1593-1595.

[23] P. Cebe, S. D. Hong, Crystallization behavior of poly(ether-ether-ketone), Polymer 27 (1986) 1183-1192.

[24] T. X. Liu, Z. S. Mo, S. E. Wang, H. F. Zhang, Nonisothermal melt and cold crystallization kinetics of poly (aryl ether ether ketone), Polym. Eng. Sci. 37 (1997) 568-573.

[25] T. X. Liu, Z. S. Mo, H. F. Zhang, Nonisothermal crystallization behavior of a novel poly (aryl ether ketone): PEDEKmK, J. Appl. Polym. Sci. 67 (1998) 815-821.

[26] B. Fillon, A. Thierry, B. Lotz, J. C. Wittman, Efficiency scale for polymer nucleating agents, J, Therm. Anal. 42 (1994) 721-731.

[27] S. Jain, H. Goossens, M. van, P. Duin, Lemstra, Effect of *in situ* prepared silica nano-particles on non-isothermal crystallization of polypropylene, Polymer 46 (2005) 8805-8818.

CO_2-induced Phase Transition of Isotactic Poly-1-butene with Form III upon Heating[*]

Abstract This work is aimed at studying the effect of CO_2 on the phase transition of isotactic poly-1-butene (iPB-1) with form III upon heating. The melting behaviors of form III under atmospheric N_2 and compressed CO_2 at different heating rates ranging from 1 to 20℃/min were investigated using high-pressure differential scanning calorimetry (DSC). The results showed that the plasticization effect of CO_2 promoted melting of form III and inhibited the phase transition of form III to II as a whole. By analyzing the melting parameters obtained from the DSC measurements, we deduced that the phase transition of form III to II might comprise another transition process besides the melt-recrystallization mechanism. In-situ wide-angle X-ray diffraction (WAXD) measurement on form III under atmospheric N_2 at a heating rate of 0.25℃/min verified that the phase transition of form III to II passed through the solid-solid phase transition before melt-recrystallization. In-situ high-pressure Fourier transform infrared (FTIR) was then used to detect the phase transition of form III under atmospheric N_2 and compressed CO_2 at the heating rate of 1℃/min. It was also shown that the phase transition of form III to II passed through the solid-solid phase transition and melt-recrystallization under atmospheric N_2, 1 and 2MPa CO_2. However, form II formed completely through the melt-recrystallization under 3MPa CO_2, and could not generate with further increasing CO_2 pressure to 4MPa. Moreover, more form I′ generated during heating through the solid-solid phase transition with increasing CO_2 pressure. Besides carbon tetrachloride solution prepared form III, the other two solutions, i. e., dilute toluene and o-xylene, cast form III also exhibited the similar generation processes of form II upon heating under atmospheric N_2 and compressed CO_2 as measured by *in-situ* high-pressure FTIR.

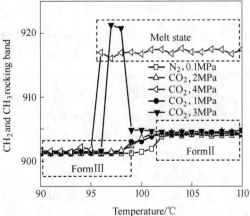

1 Introduction

Isotactic poly-1-butene (iPB-1), first synthesized by Natta in the 1950s[1], is a polymorphous semicrystal polyolefin with outstanding properties, such as high creep resistance, low stiffness,

---

[*] Coauthors: Li Lei, Liu Tao, Zhao Ling. Reprinted from *Macromolecules*, 2011, 44: 4836-4844.

good temperature and chemical resistances[2-5]. iPB-1 may exit in four different crystal structures, designated as forms Ⅰ, Ⅱ, Ⅲ and Ⅰ'[6,7]. Forms Ⅰ and Ⅰ' have the same 3/1 helix conformation with trigonal and untwined hexagonal crystal structure respectively[8,9]. Form Ⅱ has the tetragonal unit cell packed by 11/3 helix conformation, and form Ⅲ with 4/1 helix chain conformation has the orthorhombic unit cell[10-13]. The routes for preparing the iPB-1 in various crystal forms have been investigated in detail[14,15]. Crystallized from melt at atmospheric pressure, metastable form Ⅱ can be obtained. However, form Ⅱ is unstable under atmospheric condition, and will transform into form Ⅰ, a stable crystal form[7,16-22]. Forms Ⅲ and Ⅰ' are usually formed by crystallization from certain dilute solutions[23,24]. Form Ⅰ can also be obtained through crystallization from melt under high hydrostatic pressure[25-27]. In recent studies, forms Ⅰ and Ⅰ' are directly crystallized from the iPB-1 melt containing different concentration of stereodefects (rr triads defects)[28,29].

Compressed or supercritical carbon dioxide (CO_2) is well established for using as a promising alternative to organic and other toxic or harmful solvents in polymer processing, such as polymer modification, microcellular foaming, polymer blending, particle production and polymerization[30-33]. Dissolving CO_2 into the polymers increases the free volume of polymers and leads to an acceleration of polymer chains relaxation[34,35]. The plasticization effect depresses the glass transition temperature and melting temperature and also leads to lowering of the energy barriers making the phase transition possible at a much reduced temperarure[36-39]. The phase transitions of poly(L-lactide) (PLLA) and syndiotactic polystyrene (sPS) under CO_2 had shown significant difference from those at atmospheric conditions[40,41]. The phase transition of iPB-1 form Ⅱ to Ⅰ also exhibited difference from that without CO_2[42]. Whereas there are numerous investigations on CO_2-induced polymer phase transition at a constant temperature[41,43-48], CO_2-induced polymer phase transition upon heating has been seldom studied. The phase transition of iPB-1 with form Ⅲ upon heating under atmospheric condition had been wildly studied, and complicated phase transformations were involved[49,50]. It was found that form Ⅱ generated during heating and form Ⅰ' could also be obtained through a solid-solid transition from form Ⅲ[51]. The polymorphous transformations of form Ⅲ arising during heating will provide us a good example to investigate the effect of CO_2 on the complicated phase transition upon heating.

In this work, the melting behaviors of iPB-1 with form Ⅲ under atmospheric N_2 and compressed CO_2 at different heating rates were investigated using high-pressure differential scanning calorimetry (DSC). It was found that the phase transition of form Ⅲ to Ⅱ might comprise another transition process besides the melt-recrystallization mechanism. In-situ wide-angle X-ray diffraction (WAXD) measurement on form Ⅲ under atmospheric N_2 at a heating rate of 0.25℃/min verified the hypothesis. In-situ high-pressure Fourier transform infrared spectroscopy (FTIR) was then used to detect the influence of compressed CO_2 on the phase transition of form Ⅲ and the effect of form Ⅲ preparations on the form Ⅱ generation process upon heating at a heating rate of 1℃/min.

2 Experimental Section

2.1 Materials and sample preparations

iPB-1 pellets (PB 0110M) were kindly provided by Basell Polyolefins. Before used, they were purified by Soxhlet extraction in acetone for at least 24h and then dried in a vacuum oven at 40℃ for 2 days. Then, they were dissolved in a 3wt% solution of carbon tetrachloride at the solvent boiling temperature of 78℃ for 2h. The iPB-1 film with form Ⅲ was obtained by evaporating the solvent completely under atmospheric conditions. The film thickness measured by micrometer caliper was (32 ± 2) μm. CO_2 (purity: 99.9% w/w) was purchased from Air Products Co., Shanghai, China.

2.2 Wide-angle X-ray diffraction

WAXD of the type Rigaku D/max 2550 VB/PC X-ray diffractometer (Cu K_α Ni-filtered radiation) was used to study the modification of the prepared iPB-1 films. The scan rate was 1° (θ)/min, and the diffraction angular range was between 3° and 50° 2θ. In-situ WAXD measurement on the phase transition of form Ⅲ upon heating was also performed in Rigaku D/max 2550 VB/PC X-ray diffractometer with a Paar Physica TCU 750 temperature control unit under atmospheric N_2. The operation conditions were 20kV, 200mA, 0.02° 2θ step^{-1} from 5° to 30°, and scanning speed 8°(θ)/min. The heating rate was controlled at 0.25℃/min.

2.3 Differential scanning calorimetry

DSC (NETZSCH DSC 204 HP, Germany) was used to characterize the melting process of iPB-1 with form Ⅲ under atmospheric N_2 and compressed CO_2[52]. The calorimeter was calibrated by carrying out the measurement of the melting points and the heat of fusion of In, Bi, Sn, Pb, and Zn at atmospheric N_2 and high CO_2 pressures, respectively. Under compressed CO_2 conditions, iPB-1 films were held at 30℃ for 2h to ensure CO_2 completely diffusing into the films before heating. For each DSC measurement, about (5 ± 1) mg of the iPB-1 was heated from 30℃ to 150℃ at a constant heating rate. We employed several heating rates ranging from 1 to 20℃/min.

2.4 Fourier transform infrared spectroscopy

The melting process of form Ⅲ was also investigated using *in-situ* FTIR of type Bruker Equinox-55 equipped with a Harrick high-pressure demountable liquid cell, the details of which had been described elsewhere[53]. Before being heated from 30 to 150℃ at a heating rate of 1℃/min, the films were held at 30℃ for 2h. FTIR spectra were recorded at a resolution of 4.0cm^{-1} and a rate of 1 spectrum per 32s. The IR intensities refer to the peak height. The scanned wavenumber was in the range of 4000-400cm^{-1}.

3 Results and Discussion

3.1 Melting behavior of iPB-1 with form Ⅲ under atmospheric N_2 and Compressed CO_2

The solvent and casting conditions had significant influences on the final crystal form of the obtained iPB-1 specimens[54,55]. As shown in Fig. 1, the WAXD profile of the prepared iPB-1 film was the typical crystal structure of form Ⅲ. It shows three strong (101), (111), and (201) reflections at 2θ = 12.2°, 17.1°, and 18.6° and two weak (200) and (120) reflections at 2θ = 14.2° and 21.3°[56,57]. Further DSC and FTIR measurements will also confirm that the iPB-1 film was in form Ⅲ.

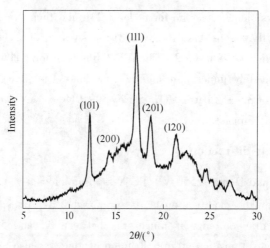

Fig. 1　WAXD profile of the prepared iPB-1 film

The melting behaviors of iPB-1 with form Ⅲ under atmospheric N_2 at heating rates ranging from 1 to 20℃/min, characterized by DSC, are shown in Fig. 2. Two endothermic peaks were observed at all those heating rates. The first melting peak at the lower temperature was attributed to the melting of form Ⅲ and the other at the higher temperature to the melting of the generated form Ⅱ. Moreover, a sharp exothermic peak corresponding to recrystallization of form Ⅱ was detected between the two endothermic melting peaks. With increasing heating rate, the form Ⅱ recrystallization peak moves to the higher temperatures. The results were consisted with other previous work[58,59]. Lee et al. reported that the exothermic peak disappeared when the heating rate was above 20℃/min and the endothermic peak of form Ⅱ existed only as a slight shoulder at heating scan of 100℃/min[59]. It is well-known that the DSC curves changes its shape with changing the heating rate if the phase transition is governed by the recrystallization mechanism, since the recrystallization process requires suitable time for crystallizing into another crystalline form[60,61].

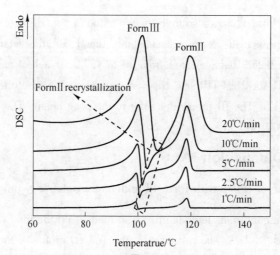

Fig. 2　DSC traces of the iPB-1 with form Ⅲ under atmospheric N_2 at various heating rates

The melting behaviors of iPB-1 with form Ⅲ at different CO_2 pressures ranging from 1 to 4MPa and heating rates ranging from 1 to 20℃/min, characterized by DSC, are shown in Fig. 3. The fluctuation in the DSC curves at 2MPa, 3MPa and 4MPa was caused by the compressed CO_2. With increasing CO_2 pressure, the exothermic peak and endothermic peak of form Ⅱ tended to disappear at the high heating rates. At the CO_2 pressure of 2MPa, the exothermic peak disappeared and the endothermic peak of form Ⅱ was hardly detected at the heating rate of 20℃/min. With increasing CO_2 pressure to 3MPa, no melting peak of form Ⅱ was detected at the heating rates of 20℃/min, 10℃/min and 5℃/min. However, at 2MPa and 3MPa, with decreasing heating rate, the exothermic peak of form Ⅱ recrystallization and endothermic peak of form Ⅱ melting were observed at heating rates lower than 10℃/min and 2.5℃/min, respectively. It indicated more time was needed for form Ⅱ recrystallization upon heating at high CO_2 pressures. When melting at 4MPa, no recrystallization and melting peaks of form Ⅱ were detected even at the heating rate of 1℃/min.

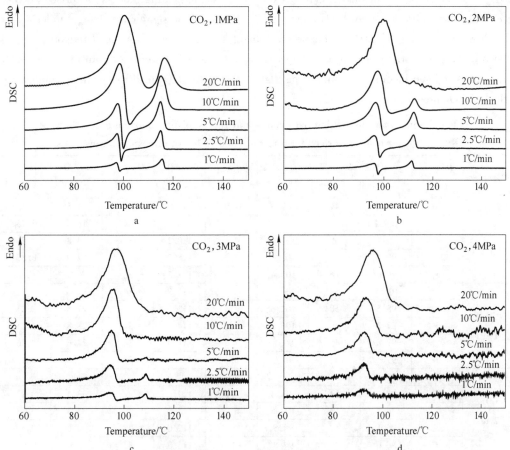

Fig. 3 High-pressure DSC diagrams of the iPB-1 with form Ⅲ scanned under 1MPa(a), 2MPa(b), 3MPa(c) and 4MPa(d) CO_2 at various heating rates

For each condition, DSC measurements were conducted at least three times, and the average melting parameters and corresponded maximum deviations were obtained. Table 1 collects the melting temperatures ($T_{mⅢ}$ and $T_{mⅡ}$) of forms Ⅲ and Ⅱ, recrystallization temperature ($T_{rⅡ}$) of form Ⅱ, fusion enthalpies ($\Delta H_{fⅢ}$ and $\Delta H_{fⅡ}$) of forms Ⅲ and Ⅱ, and recrystallization enthalpies ($\Delta H_{rⅡ}$) and half-recrystallization time of form Ⅱ ($t_{1/2}$) of form Ⅱ corresponding to the peaks and areas under the DSC curves in Fig. 2 and Fig. 3. The definition of $\Delta H_{fⅢ}$, $\Delta H_{rⅡ}$ and $\Delta H_{fⅡ}$ is exhibited in Fig. S1 of the Supporting Information. $T_{mⅢ}$ increased with increasing heating rate under atmospheric N_2 and compressed CO_2, which was ascribed to increase of the superheat of form Ⅲ. Moreover, as shown in Fig. S2 of the Supporting Information, $T_{mⅢ}$ decreased linearly with increasing CO_2 pressure at a certain heating rate with a similar value of $dT_{mⅢ}/dP = -1.8℃/MPa$ for various heating rates. Linear decrease of $T_{mⅢ}$ as a function of CO_2 pressure was due to the plasticization effect of CO_2 on the iPB-1 and could be explained on the basis of the Flory-Huggins theory[62,63]. Another interesting observation in Table 1 was that $\Delta H_{fⅢ}$ decreased with decreasing heating rate under atmospheric N_2, 1MPa, 2MPa and 3MPa CO_2, which indicated less form Ⅲ melted and more form Ⅲ might transform into the intermediate form Ⅱ without passing through melt-recrystallization with increasing relaxation time of polymer chain in form Ⅲ. It was also evidenced by the gradual decrease in the melt-recrystallization rate of form Ⅱ with decreasing the heating rate as demonstrated by the change of $t_{1/2}$. $\Delta H_{fⅢ}$ should not be the full fusion enthalpy of form Ⅲ if the heating process comprised another phase transition of form Ⅲ to Ⅱ. However, $\Delta H_{fⅢ}$ did not show significant change under 4MPa CO_2 because the form Ⅲ melted completely without arranging into form Ⅱ. Therefore, the full fusion enthalpy of form Ⅲ could be obtained. Furthermore, $\Delta H_{fⅢ}$ increased with increasing CO_2 pressure at a given heating rate, indicating that dissolution of CO_2 in iPB-1 or the plasticization effect of CO_2 promoted melting of form Ⅲ.

Table 1 Nonisothermal melting parameters and corresponding maximum deviations of iPB-1 with form Ⅲ at various heating rates and different atmosphere conditions

Atmosphere	R/℃·min^{-1}	$T_{mⅢ} \pm 0.2$ /℃	$\Delta H_{fⅢ} \pm 2.0$ /J·g^{-1}	$T_{rⅡ} \pm 0.2$ /℃	$\Delta H_{rⅡ} \pm 0.5$ /J·g^{-1}	$T_{mⅡ} \pm 0.2$ /℃	$\Delta H_{fⅡ} \pm 1.0$ /J·g^{-1}	$\Delta H_{rⅡ}/\Delta H_{fⅡ}$ /%	$t_{1/2}$ /min
N_2, 0.1MPa	20	101.5	41	107.3	4.9	119.4	28	17.5	0.13
	10	100.1	39	103.5	4.8	118.3	29	16.6	0.15
	5	99.7	38	101.6	4.6	118.2	31	14.8	0.21
	2.5	99.2	37	100.6	4.2	118.2	35	12.0	0.37
	1	99.0	36	100.4	2.7	118.3	36	7.5	1.06
CO_2, 1MPa	20	100.4	52			116.2	14		
	10	98.5	50	102.6	6.4	115.1	23	27.8	0.21
	5	97.8	49	100.1	9.2	114.9	32	28.8	0.25
	2.5	97.5	46	99.1	7.7	114.9	34	22.6	0.36
	1	97.1	44	98.6	5.1	115.5	35	14.6	0.99
CO_2, 2MPa	20	99.8	56						
	10	97.6	54	103.3	2.0	112.4	6	33.3	0.23
	5	96.7	53	100.5	6.7	112.2	19	35.3	0.38
	2.5	96.1	52	98.5	11.2	112.0	28	40.0	0.50
	1	96.0	48	97.8	14.6	111.3	32	45.6	0.98
CO_2, 3MPa	20	97.0	62						
	10	95.8	61						
	5	94.8	60						
	2.5	94.4	59	97.9	3.8	108.5	7	54.3	0.67
	1	94.2	57	97.0	9.1	108.5	15	60.7	1.75
CO_2, 4MPa	20	95.9	60						
	10	92.8	61						
	5	92.5	61						
	2.5	92.3	60						
	1	91.8	61						

$\Delta H_{fⅡ}$ increased with decreasing the heating rate at a certain atmosphere condition, implying that more form Ⅱ had generated with increasing relaxation time of polymer chain. Meanwhile, $\Delta H_{fⅡ}$ decreased with increasing CO_2 pressure at a given heating rate, implying that less form Ⅱ had generated due to the increased plasticization effect of CO_2. $\Delta H_{rⅡ}/\Delta H_{fⅡ}$ characterized qualitatively the ratio of the content of form Ⅱ recrystallized from melt to the total content of form Ⅱ. Under atmospheric N_2 and 1MPa CO_2, $\Delta H_{rⅡ}$ was much lower than $\Delta H_{fⅡ}$, which might verify the existence of the other phase transition process of form Ⅲ to Ⅱ before the melt-recrystallization during heating. Meanwhile, $\Delta H_{rⅡ}/\Delta H_{fⅡ}$ decreased with decreasing the heating rate, indicating less form Ⅱ formed through the melt-recrystallization with increasing relaxation time of polymer chain. As CO_2 pressure increased to 2MPa, however, both $\Delta H_{rⅡ}$ and $\Delta H_{rⅡ}/\Delta H_{fⅡ}$ increased with decreasing the heating rate. Especially at the low heating rates, i.e., 2.5℃/min and 1℃/min, $\Delta H_{rⅡ}$ was 11.2J/g and 14.6J/g, respectively, much higher than that at 1MPa CO_2 pressure. It indicated that the plasticization effect of 2MPa CO_2 significantly promoted the melt-recrystallization of form Ⅱ and more form Ⅱ generated through the melt-recrystallization with decreasing the heating rate. At CO_2 pressure of 3MPa, $\Delta H_{fⅡ}$ decreased to a very low level, e.g., 15J/g even at the heating rate of 1℃, and $\Delta H_{rⅡ}/\Delta H_{fⅡ}$ increased to a high level, e.g., 60.7%. It confirmed that the plasticization effect of 3MPa CO_2 significantly inhibited the other generation process of form Ⅱ before melt-crystallization, and most of the form Ⅱ generated through the melt-recrystallization. Meanwhile, there was a transition that $\Delta H_{rⅡ}$ decreased in comparison with that at the CO_2 pressure of 2MPa at the given heating rates, implying that the plasticization effect at the CO_2 pressure of 3MPa preferred to promote melting instead of recrystallization. At the CO_2 pressure of 4MPa, the plasticization effect of CO_2 was so strong that the iPB-1 with form Ⅲ melted completely and could not recrystallize into form Ⅱ even at the heating rate of 1℃/min.

3.2 In-situ WAXD measurement on the phase transition of form Ⅲ during heating under atmospheric N_2

In order to verify the existence of another generation process of form Ⅱ before melt-recrystallization, in-situ WAXD was applied to study the phase transition of form Ⅲ upon the heating under atmospheric N_2 at a relative low heating rate of 0.25℃/min for more form Ⅱ might formed through the phase transition before melt-recrystallization as stated above. Fig. 4 illustrated the variation of WAXD profiles of form Ⅲ heated under atmospheric N_2 at a heating rate of 0.25℃/min as a function of temperature. The DSC curve of form Ⅲ scanned at the same heating rate under atmospheric N_2 was also exhibited at upper right corner of Fig. 4. The crystal structure changes of form Ⅲ during heating can be well identified by WAXD[56]. The WAXD pattern of form Ⅰ′ exhibits characteristic intensity peak at $2\theta = 9.9°$ corresponding to the diffraction of crystal reflection from planes (110). Form Ⅱ presents the (200) and (301) reflections at $2\theta = 11.9°$ and 18.4°. It was evident that the form Ⅲ underwent a comprehensive crystal phase transition during heating. The intensities of (101) and (111) reflections at $2\theta = 12.2°$ and 17.1° of form Ⅲ gradually decreased with increasing temperature. Form Ⅰ′ generated at 80℃, as demonstrated by the presence of the (110) reflection of form Ⅰ′ at 9.9°, and the intensity of (110) reflection increased with increasing the temperature until to 98℃, indicating the solid-solid phase transition of form Ⅲ to Ⅰ′. At 96℃, a weak (200) reflection of form Ⅱ at $2\theta = 11.9°$ was detected before form Ⅲ completely melted and its intensity increased with increasing temperature, which verified the existence of another generation process of form Ⅱ before from Ⅲ completely melt. With the temperature further increasing, form Ⅲ completely melted at 101℃ as indicated by the vanish of the (111) reflection of form Ⅲ at 17.1°, which agreed well with the DSC measurement.

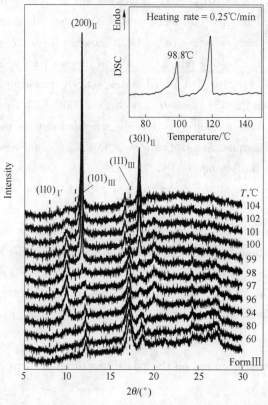

Fig. 4 WAXD profiles of form Ⅲ meting process under atmospheric
N_2 as a function of temperature at a heating rate of 0.25℃/min
(At the upper right corner is the DSC curve of form Ⅲ scanned under atmospheric N_2 at the same heating rate)

To have a clear insight into the phase transition of form Ⅲ to Ⅱ, the intensities of the (111) reflection of form Ⅲ, (200) reflection of form Ⅱ, and (110) reflection of form Ⅰ′ as a function of temperature are normalized in Fig. 5. It was apparent that the intensity of the (200) reflection of form Ⅱ gradually increased from 96 to 99℃ and then increased abruptly at temperature above 99℃. As shown in the DSC curve of form Ⅲ in Fig. 4, iPB-1 located in the endothermic peak in the temperature region of 96-99℃ and shifted to the exothermic peak at temperature above 100℃. It demonstrated that form Ⅱ generated through the melt-recrystallization since iPB-1 entered an isotropic liquid phase or mesophase at temperature above 100℃. In the temperature region of 96-100℃ before form Ⅲ completely melted, form Ⅱ should formed through a solid-solid transition without entering the isotropic liquid phase or mesophase[64]. Especially at 96 and 97℃, the majority of form Ⅲ was in the solid state.

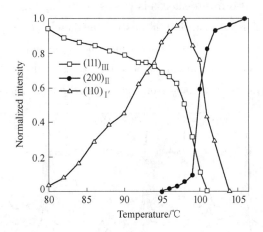

Fig. 5　Normalized intensity of the (111) reflection of form Ⅲ, (200) reflection of form Ⅱ, and (110) reflection of form Ⅰ′ as a function of temperature

3.3　In-situ FTIR measurements on phase transition of form Ⅲ upon heating under atmospheric N_2 and compressed CO_2

To further confirm the changes in the form Ⅱ generation process, *in-situ* high-pressure FTIR was also employed to characterize the melting behavior of iPB-1 with form Ⅲ under atmospheric N_2 and compressed CO_2. Because of the rapid phase transition at a fast heating rate and the relative long time during the FTIR scan period, the form Ⅲ melting behavior was only studied at the heating rate of 1℃/min. Fig. 6 illustrates the FTIR spectra of iPB-1 film with form Ⅲ at different temperatures. Previous work had shown there were distinct differences of form Ⅲ, Ⅱ and Ⅰ′ among the infrared spectrum range in 800-950 cm^{-1}. The band at 901 cm^{-1} is known to be the characteristic band of form Ⅲ, while the bands at 904 cm^{-1} and 924 cm^{-1} are corresponding to the characteristic of form Ⅱ and Ⅰ (or Ⅰ′), respectively. They all correspond to the CH_2 and CH_3 rocking vibrations[59,65,66].

As shown in Fig. 6a, form Ⅲ underwent a comprehensive crystal structure transition during heating under atmospheric N_2. The intensity of 901 cm^{-1} band decreased slightly with increasing temperature from 80℃ to 100℃ due to the melting of form Ⅲ and solid-solid transition to form Ⅰ′ as demonstrated by the appearance of a weak shoulder at 924 cm^{-1} band. Meanwhile, the intensity of 924 cm^{-1} band increased slightly with increasing the temperature, indicating the increase of form Ⅰ′ content during heating. When the temperature reached 102℃, the band at 924 cm^{-1} disappeared and the band at 901 cm^{-1} moved to 904 cm^{-1}, indicating disappearance of the forms Ⅲ and Ⅰ′ and generation of the form Ⅱ. At this temperature, the IR spectrum gave no evidence for the appearance of the completely amorphous phase. And the band at 904 cm^{-1} kept absorbance intensity until 114℃ and vanished at 120℃, at which a completely amorphous IR spectrum was obtained. From the DSC curve of form Ⅲ melted under atmospheric N_2 at the heating rate of 1℃/min, as shown in Fig. 2, iPB-1 also melted completely at 120℃. Jang et al.[50] had studied the form Ⅲ melting process by using *in-situ* synchrotron small- and wide-angle X-ray scattering and found no completely amorphous pattern was recorded before form Ⅱ formed. Miyoshi et al.[49] also investigated this process by using high-resolution solid-state ^{13}C NMR spectroscopy, and claimed that form Ⅱ phase immediately grown after melting of form Ⅲ and the content of the amorphous phase continuously increased only from 47.3% at 100℃ to 59.3% at 107℃. The *in-situ* FTIR measurements also revealed that the generation of form Ⅱ under atmospheric N_2 comprised the solid-solid phase transition and melt-recrystallization.

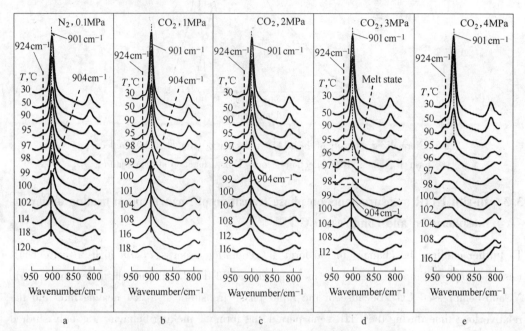

Fig. 6 IR spectra of from Ⅲ melting process under atmospheric N_2 (a), 1MPa CO_2 (b), 2MPa CO_2 (c), 3MPa CO_2 (d), and 4MPa CO_2 (e)

The IR spectra of form Ⅲ melting processes under 1MPa and 2MPa CO_2, as shown in Fig. 6b, c, exhibited similar phase transition behaviors in comparison with that under atmospheric N_2. The 901cm^{-1} gradually reduced with increasing temperature. The disappearance temperatures of forms Ⅲ and Ⅰ′ decreased with increasing CO_2 pressure. During the generation of the form Ⅱ, no completely amorphous IR spectra were detected. Amorphous IR spectra were obtained at 118℃ under 1MPa CO_2 and at 116℃ under 2MPa CO_2, which agreed well with the DSC results in Fig. 3a, b. Meanwhile, CO_2 also promoted the generation of the form Ⅱ at a lower temperature. In addition, the intensity of the 924cm^{-1} band in IR spectra at 2MPa was stronger than that at 1MPa, which indicated that more form Ⅰ′ generated at higher CO_2 pressure.

It should be noted that generation process of form Ⅱ at 3MPa exhibited very different from that under atmospheric N_2, 1MPa and 2MPa CO_2. As shown in Fig. 6d, when the temperature increased from 75 to 96℃, the intensity of the 901cm^{-1} band decreased slightly and that of 924cm^{-1} band increased. However, a completely amorphous IR spectrum was obtained at 97℃ before the IR spectra of form Ⅱ appeared. The band of 904cm^{-1} began to emerge at 98℃. The intensity of 904cm^{-1} band increased from 100 to 104℃, which confirmed that form Ⅱ generated totally through the melt-recrystallization from form Ⅲ. Moreover, the form Ⅱ completely melted at 112℃, which also corresponded to the completely melting of form Ⅱ in the DSC result. Fig. 6e illustrates that iPB-1 completely melted without crystallizing into form Ⅱ when heated up under 4MPa CO_2.

Fig. 7 shows the spectra shifts of the 901cm^{-1} band as a function of temperature under atmospheric N_2 and compressed CO_2. It demonstrated clearly that the coexistence of the solid-solid transition and melt-recrystallization of form III to II under atmospheric N_2, 1MPa and 2MPa CO_2. When the pressure reached 3MPa, the form II generated completely through the melt-recrystallization from the melt or mesophase. The application of high pressure CO_2 had changed the generation process of form II. Upon further increase of the CO_2 pressure to 4MPa, as shown in Fig. 6e and Fig. 7, the forms III and I′ melted directly without recrystallizing into the form II.

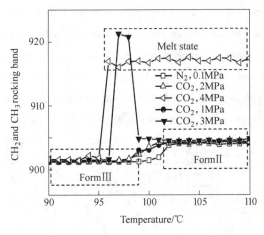

Fig. 7 Change of the 901cm^{-1} band as a function of temperature at different atmosphere conditions

The changes in the form II generation process can be explained by the plasticization effect of CO_2 on the polymer chain's motion in the amorphous phase. Handa et al.[43] had investigated the effect of compressed CO_2 on the phase transitions and polymorphism in syndiotactic polystyrene (sPS), and found sPS underwent planar mesophase to β, α to β, and γ to β transitions at different conditions of temperature and CO_2 pressure. They claimed that the extent of plasticization played an important role in determining the resulting crystal modifications. The extent of plasticization had a significant influence on the polymer chain motion and flexibility in the amorphous region[30,34,67]. The increase of polymer's chain motion influenced the polymer crystallization and phase transition behaviors. De Rosa et al. had obtained forms I and I directly from the melt iPB-1 prepared with metallocene catalysts. They ascribed this effect to the presence of rr defects in the iPB-1, which increased the flexibility of the chains and the crystallization rate of form I or I′[28,29]. The studies on the phase transition of form II to I in iPB-1 also revealed that normal stress and molecular mobility within noncrystalline regions play a crucial role in the phase transformation[42,68-70]. In the present work, the extent of CO_2 plasticization effect enhanced with increasing CO_2 pressure. The increased polymer chain motion and flexibility substantially changed the generation process of form II. This effect was also presented by the intermediate melt state at 3MPa before recrystallization occurred. The increase in the motion and flexibility of polymer chains could make more form III melt before recrystallization. Moreover, further increased

CO_2 plasticization effect, i. e., at the CO_2 pressure of 4MPa, made the polymer chains motion and flexibility even more intense and polymer chains unable to reorganize into form II.

3.4 CO_2-induced phase transition of form III to I′ upon heating

Another interesting observation from Fig. 8 was that the intensity of 924cm^{-1} band also changed with CO_2 pressure. The transformed fraction of form I′ was derived from the ratio between the absorbance band area at 924cm^{-1} (A_{924}) and the sum of that at 901cm^{-1} and 924cm^{-1} ($A_{901} + A_{924}$). The superposition of two bands was analyzed by a PEAK-FIT V4.12 program that was usually applied to deconvolute complex IR spectra. The fraction of the form I′, with respect to total fraction of crystalline phase, at various pressures as a function of temperature is shown Fig. 8. It is apparent that the fraction of form I′ increased with temperature at a given pressure. The form I′ content also increased with CO_2 pressure at the same temperature. It was ascribed to the plasticization effect of CO_2, which allowed the form III to I′ transition occur at a lower temperature by increasing the polymer chain motion. That was to say more form I′ generated at a higher CO_2 pressure through elongation of the transition time.

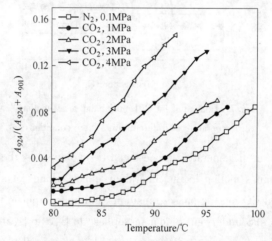

Fig. 8 Dependence of form I′ relatively content on the temperature under atmospheric N_2 and compressed CO_2

3.5 Influence of form III preparations on the phase transition of form III upon heating

The DSC thermograms of dilute toluene and o-xylene solutions cast form III at a heating rate of 10℃/min are shown in Fig. S3 of the Supporting Information. It was apparent the solvents influenced the thermal properties of prepared form III. The form III prepared by toluene and o-xylene solutions has smaller ΔH_{rII} and ΔH_{fII} than that prepared by carbon tetrachloride solution.

Fig. 9 illustrates the FTIR spectra of toluene solution prepared form III during heating at a heating rate of 1℃/min under atmospheric N_2 and 2-4MPa CO_2. As shown in Fig. 9, no completely amorphous iPB-1 spectrum was detected before the 904cm^{-1} band existed during heating, indicating the recrystallization of form II under these two conditions were not pass through the completely melt or mesophase. However, when heated under 2.7MPa CO_2, a completely amorphous IR spectrum of iPB-1 was observed at 100℃, as exhibited in Fig. 9c. Then, the intensity of the 904cm^{-1} band increased with increasing the temperature until to 108℃, which revealed the recrystallization of form II from the melt or mesophase. Completely amorphous IR spectra of iPB-1 were also detected from 99 to 101℃ during heating under 3MPa CO_2, as shown in Fig. 9d. After that, the 904cm^{-1} band was also observed with increasing temperature. Compared with the melting process of form III under 2.7MPa CO_2, the recrystallization of form II under 3MPa underwent at a higher temperature and the relative intensity of form II absorbance band decreased indicating the recrystallization of form II was gradually inhibited. Heated under 4MPa CO_2, form III melt directly without recrystallizing into form II, as shown in Fig. 9e.

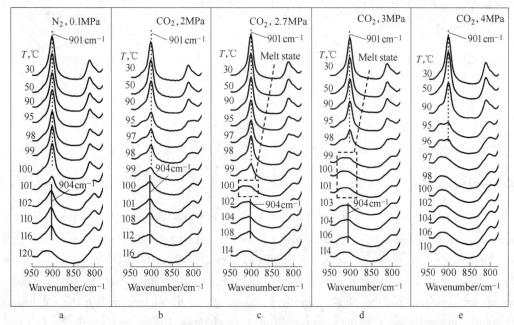

Fig. 9 FTIR spectra of toluene solution prepared form III melting process under atmospheric N_2(a), 2MPa CO_2(b), 2.7MPa CO_2(c), 3MPa CO_2(d), and 4MPa CO_2(e)

Shown in Fig. 10 are the FTIR spectra of o-xylene solution prepared form III heated under atmospheric N_2 and 1-3MPa CO_2. For melting of the form III under atmospheric N_2 and 1-1.5MPa CO_2, as shown in Fig. 10a-c, no completely amorphous IR spectra of iPB-1 were detected before 904cm^{-1} band appeared, which revealed the form II was not generated totally through melt-recrystallization. Completely amorphous IR spectra of iPB-1 were detected at 99℃ and 100℃ before the generation of form II during heating under 2MPa CO_2, as shown in Fig. 10d, which verified that form II was recrystallized from the melt or mesophase. As shown in Fig. 10e, form III directly melted under 3MPa CO_2.

Those results above showed that the dilute toluene and o-xylene solutions prepared form III exhibited the similar generation processes of form II upon heating with the form III prepared by carbon tetrachloride solution under atmospheric N_2 and compressed CO_2. Moreover, high-pressure CO_2 also changed the generation process of form II into to a total melt-recrystallization.

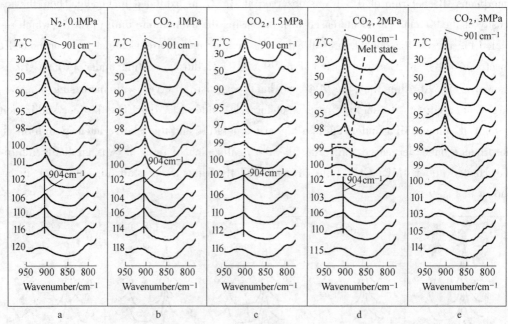

Fig. 10 FTIR spectra of o-xylene solution prepared form Ⅲ melting process under atmospheric N_2(a), 1MPa CO_2(b), 1.5MPa CO_2(c), 2MPa CO_2(d), and 3MPa CO_2(e)

4　Conclusion

The melting behaviors of form Ⅲ under atmospheric N_2 and compressed CO_2 at different heating rates ranging from 1 to 20℃/min were investigated using high-pressure DSC. It was shown that the plasticization effect of CO_2 promoted melting of form Ⅲ and inhibited the phase transition of form Ⅲ to Ⅱ as a whole. By analyzing the melting parameters obtained from the DSC measurements, we deduced that the phase transition of form Ⅲ to Ⅱ might comprise the solid-solid transition process besides the melt-recrystallization mechanism. Under atmospheric N_2, less and less form Ⅱ generated through the melt-recrystallization with decreasing the heating rate. The plasticization effect of 2MPa and 3MPa CO_2 significantly promoted the melt-recrystallization of form Ⅱ and more form Ⅱ generated through the melt-recrystallization with decreasing the heating rate. Under 4MPa CO_2, the plasticization effect of CO_2 was so strong that the form Ⅲ was completely molten and could not recrystallize into form Ⅱ, which indicated that the full fusion enthalpy of form Ⅲ could be measured. In-situ WAXD measurement on form Ⅲ under atmospheric N_2 at a heating rate of 0.25℃ verified that the phase transition of form Ⅲ to Ⅱ passed through the solid-solid phase transition before melt-recrystallization.

In-situ high-pressure FTIR was then applied to detect the phase transition of form Ⅲ under atmospheric N_2 and compressed CO_2 at the heating rate of 1℃/min. It was also shown that the phase transition of form Ⅲ to Ⅱ passed through the solid-solid phase transition and melt-recrystallization under atmospheric N_2, 1MPa and 2MPa CO_2. However, form Ⅱ generated completely through the melt-recrystallization from the melt or mesophase under 3MPa CO_2, and could not generate with further increasing CO_2 pressure to 4MPa. Moreover, more form Ⅰ′ generated during heating through the solid-solid phase transition with increasing CO_2 pressure. Besides carbon tetrachloride solution prepared form Ⅲ, the other two solutions, i. e., dilute toluene and o-xylene, cast form Ⅲ also exhibited the similar generation processes of form Ⅱ upon heating under atmospheric N_2 and compressed CO_2 as measured by *in-situ* high-pressure FTIR. CO_2 changed the phase transition of dilute toluene prepared form Ⅲ to Ⅱ to the total melt-recrystallization at 2.7MPa, while that of dilute o-xylene form Ⅲ to Ⅱ at 2MPa.

5 Associated Content

Supporting Information: Text giving experimental details on the preparation of form Ⅲ from dilute toluene and o-xylene solutions and the definition of $\Delta H_{fⅢ}$, $\Delta H_{rⅡ}$ and $\Delta H_{fⅡ}$, including figures showing definition of $\Delta H_{fⅢ}$, $\Delta H_{rⅡ}$ and $\Delta H_{fⅡ}$, plots of $T_{mⅢ}$ vs CO_2 pressure, and DSC curves of the different solutions prepared form Ⅲ. This material is available free of charge via the Internet at http://pubs.acs.org.

Author Information

Corresponding Author

* Tel +86-21-64253175; Fax +86-21-64253528; e-mail liutao@ecust.edu.cn (T. L.), zhaoling@ecust.edu.cn(L. Z.).

Acknowledgements

The authors are grateful to the National Natural Science Foundation of China (Grants 20976045 and 20976046), Shanghai Shuguang Project (08SG28), Program for New Century Excellent Talents in University (NCET-09-0348), Program for Changjiang Scholars and Innovative Research Team in University (IRT0721) and the 111 Project (B08021).

References

[1] Natta, G. Makromol. Chem. 1960, 35, 94-131.
[2] Tosaka, M.; Kamijo, T.; Tsuji, M.; Kohjiya, S.; Ogawa, T.; Isoda, S.; Kobayashi, T. Macromolecules 2000, 33, 9666-9672.
[3] Azzurri, F.; Flores, A.; Alfonso, G. C.; Calleja, F. J. B. Macromolecules 2002, 35, 9069-9073.
[4] Men, Y.; Rieger, J.; Homeyer, J. Macromolecules 2004, 37, 9481-9488.
[5] Di Lorenzo, M. L.; Righetti, M. C. Polymer 2008, 49, 1323-1331.
[6] Natta, G.; Corradini, P.; Bassi, I. Nuovo Cimento (Suppl.) 1960, 15, 52-67.
[7] Danusso, F.; Gianotti, G. Makromol. Chem. 1963, 61, 139-156.
[8] Turner-Jones, A. J. Polym. Sci., Part B: Polym. Lett. 1965, 3, 591-600.
[9] Turner-Jones, A. Polymer 1966, 7, 23-59.
[10] Turner-Jones, A. J. Polym. Sci., Part B: Polym. Lett. 1963, 1, 455-456.
[11] Cojazzi, G.; Malta, V.; Celotti, G.; Zannetti, R. Makromol. Chem. 1976, 177, 915-926.
[12] Petraccone, V.; Pirozzi, B.; Frasci, A.; Corradini, P. Eur. Polym. J. 1976, 12, 323-327.
[13] Dorset, D. L.; McCourt, M. P.; Kopp, S.; Wittmann, J. C.; Lotz, B. Acta Crystallogr., Sect B 1994, 50, 201-208.
[14] Holland, V. F.; Miller, R. L. J. Appl. Phys. 1964, 35, 3241-3248.
[15] Luciani, L.; Seppala, J.; Lofgren, B. Prog. Polym. Sci. 1988, 13, 37-62.

[16] Boor, J. ; Mitchell, J. C. J. Polym. Sci. , Part A: Gen. Pap. 1963, 1, 59-84.

[17] Danusso, F. ; Gianotti, G. Makromol. Chem. 1965, 88, 149-158.

[18] Powers, J. ; Hoffman, J. D. ; Weeks, J. J. ; Quinn, F. A. , Jr. J. Res. Natl. Bur. Stand. 1965, 69A, 335-345.

[19] Schaffhauser, R. J. J. Polym. Sci. , Part B: Polym. Lett. 1967, 5, 839-841.

[20] Foglia, A. J. J. Appl. Polym. Sci. Appl. Polym. Symp. 1969, 11, 1-18.

[21] Gohil, R. M. ; Miles, M. J. ; Petermann, J. J. Macromol. Sci. , Part B: Phys. 1982, 21, 189-201.

[22] Fujiwara, Y. Polym. Bull. 1985, 13, 253-258.

[23] Miller, R. L. ; Holland, V. F. J. Polym. Sci. , Part B: Polym. Lett. 1964, 2, 519-521.

[24] Mathieu, C. ; Stocker, W. ; Thierry, A. ; Wittmann, J. C. ; Lotz, B. Polymer 2001, 42, 7033-7047.

[25] Armeniades, C. D. ; Baer, E. J. Macromol. Sci. , Part B: Phys. 1967, 1, 309-334.

[26] Nakafuku, C. ; Miyaki, T. Polymer 1983, 24, 141-148.

[27] Kalay, G. ; Kalay, C. R. J. Appl. Polym. Sci. 2003, 88, 814-824.

[28] De Rosa, C. ; Auriemma, F. ; Resconi, L. Angew. Chem. Int. Ed. 2009, 48, 9871-9874.

[29] De Rosa, C. ; Auriemma, F. ; Ruiz de Ballesteros, O. ; Esposito, F. ; Laguzza, D. ; Di Girolamo, R. ; Resconi, L. Macromolecules 2009, 42, 8286-8297.

[30] Kazarian, S. G. ; Vincent, M. F. ; Bright, F. V. ; Liotta, C. L. ; Eckert, C. A. J. Am. Chem. Soc. 1996, 118, 1729-1736.

[31] Nalawade, S. P. ; Picchioni, F. ; Janssen, L. P. B. M. Prog. Polym. Sci. 2006, 31, 19-43.

[32] Hua, C. ; Chen, Z. ; Xu, Q. ; He, L. J. Polym. Sci. , Part B: Polym. Phys. 2009, 47, 784-792.

[33] Kiran, E. J. Supercrit. Fluids 2009, 47, 466-483.

[34] Pasquali, I. ; Comi, L. ; Pucciarelli, F. ; Bettini, R. Int. J. Pharm. 2008, 356, 76-81.

[35] Kikic, I. J. Supercrit. Fluids 2009, 47, 458-465.

[36] Chow, T. S. Macromolecules 1980, 13, 362-364.

[37] Condo, P. D. ; Johnston, K. P. Macromolecules 1992, 25, 6730-6732.

[38] Asai, S. ; Shimada, Y. ; Tominaga, Y. ; Sumita, M. Macromolecules 2005, 38, 6544-6550.

[39] Shieh, Y. T. ; Hsiao, T. T. J. Supercrit. Fluids 2009, 48, 64-71.

[40] Ma, W. ; Yu, J. ; He, J. Polymer 2005, 46, 11104-11111.

[41] Marubayashi, H. ; Akaishi, S. ; Akasaka, S. ; Asai, S. ; Sumita, M. Macromolecules 2008, 41, 9192-9203.

[42] Li, L. ; Liu, T. ; Zhao, L. ; Yuan, W. K. Macromolecules 2009, 42, 2286-2290.

[43] Handa, Y. P. ; Zhang, Z. ; Wong, B. Macromolecules 1997, 30, 8499-8504.

[44] Zhang, Z. ; Handa, Y. P. Macromolecules 1997, 30, 8505-8507.

[45] Teramoto, G. ; Oda, T. ; Saito, H. ; Sano, H. ; Fujita, Y. J. Polym. Sci. , Part B: Polym. Phys. 2004, 42, 2738-2746.

[46] Shieh, Y. T. ; Hsiao, T. T. ; Chang, S. K. Polymer 2006, 47, 5929-5937.

[47] Li, L. ; Liu, T. ; Zhao, L. Macromol. Symp. 2010, 296, 517-525.

[48] Li, L. ; Liu, T. ; Zhao, L. ; Yuan, W. K. Asic-Pac. J. Chem. Eng. 2009, 4, 800-806.

[49] Miyoshi, T. ; Hayashi, S. ; Imashiro, F. ; Kaito, A. Macromolecules 2002, 35, 2624-2632.

[50] Jiang, Z. ; Sun, Y. ; Tang, Y. ; Lai, Y. ; Funari, S. r. S. ; Gehrke, R. ; Men, Y. J. Phys. Chem. B 2010, 114, 6001-6005.

[51] Jiang, T. ; Liu, M. ; Fu, P. ; Wang, Y. ; Fang, Y. ; Zhao, Q. Polym. Eng. Sci. 2009, 49, 1366-1374.

[52] Liu, T. ; Hu, G. H. ; Tong, G. S. ; Zhao, L. ; Cao, G. P. ; Yuan, W. K. Ind. Eng. Chem. Res. 2005, 44, 4292-4299.

[53] Li, B. ; Li, L. ; Zhao, L. ; Yuan, W. K. Eur. Polym. J. 2008, 44, 2619-2624.

[54] Kaszonyiova, M. ; Rybnikar, F. ; Geil, P. H. J. Macromol. Sci. , Part B: Phys. 2004, B43, 1095-1114.

[55] Kaszonyiova, M. ; Rybnikar, K. ; Geil, P. H. J. Macromol. Sci. , Part B: Phys. 2005, B44, 377-396.

[56] Nakamura, K. ; Aoike, T. ; Usaka, K. ; Kanamoto, T. Macromolecules 1999, 32, 4975-4982.
[57] Rusa, C. C. ; Wei, M. ; Bullions, T. A. ; Rusa, M. ; Gomez, M. A. ; Porbeni, F. E. ; Wang, X. ; Shin, I. D. ; Balik, C. M. ; White, J. L. ; Tonelli, A. E. Cryst. Growth Des. 2004, 4, 1431-1441.
[58] Geacintov, C. ; Miles, R. B. ; Schuubmans, H. J. L. J. Polym. Sci. , Part C: Polym. Symp. 1966, 14, 283-290.
[59] Lee, K. H. ; Snively, C. M. ; Givens, S. ; Chase, D. B. ; Rabolt, J. F. Macromolecules 2007, 40, 2590-2595.
[60] Geacintov, C. ; Schotl, R. S. ; Miles, R. B. J. Polym. Sci. , Part C: Polym. Symp. 1964, 6, 197-207.
[61] Kawai, T. ; Rahman, N. ; Matsuba, G. ; Nishida, K. ; Kanaya, T. ; Nakano, M. ; Okamoto, H. ; Kawada, J. ; Usuki, A. ; Honma, N. ; Nakajima, K. ; Matsuda, M. Macromolecules 2007, 40, 9463-9469.
[62] Kishimoto, Y. ; Ishii, R. Polymer 2000, 41, 3483-3485.
[63] Li, B. ; Zhu, X. ; Hu, G. H. ; Liu, T. ; Cao, G. ; Zhao, L. ; Yuan, W. K. Polym. Eng. Sci. 2008, 48, 1608-1614.
[64] Cheng, S. Z. D. Phase transitions in polymers: The role of metastable states; Elsevier Science: Amsterdam, 2008, p 24.
[65] Luongo, J. P. ; Salovey, R. J. Polym. Sci. , Part A2: Polym. Phys. 1966, 4, 997-1008.
[66] Goldbach, G. ; Peitscher, G. J. Polym. Sci. , Part B: Polym. Lett. 1968, 6, 783-788.
[67] Kazarian, S. G. ; Brantley, N. H. ; Eckert, C. A. Vib. Spectrosc. 1999, 19, 277-283.
[68] Goldbach, G. Angew. Makromol. Chem. 1973, 29, 213-227.
[69] Goldbach, G. Angew. Makromol. Chem. 1974, 39, 175-188.
[70] Weynant, E. ; Haudin, J. M. ; G'Sell, C. J. Mater. Sci. 1982, 17, 1017-1035.

CO_2-induced Polymorphous Phase Transition of Isotactic Poly-1-butene with Form III upon Annealing[*]

Abstract In-situ high-pressure FTIR was used to investigate the polymorphous phase transition of isotactic poly-1-butene (iPB-1) with form III upon annealing at temperatures ranging from 75 to 100℃ and CO_2 pressures ranging from 2 to 12MPa. It was shown that the phase transition of form III changed from form III to II not through form III to I′ with increasing temperature and application of CO_2 increased the content of generated form I′. Wide-angle X-ray diffraction (WAXD) measurement on the annealed iPB-1 with form III verified the phase transition of form III. The crystalline morphology of the annealed iPB-1 films was investigated using polarized optical microscopy (POM). The results implied that the phase transition of form III to I might process via a solid-solid transition, which did not affect the orientation of the lamellar stacks. The orientation of form II lamellar stacks depended strongly on the formation process. To obtain strong orientation, the formation process displayed the following order: melt crystallization at ambient condition > melt recrystallization under CO_2 > phase transition upon annealing at ambient condition. Avrami equation could be well established to describe the phase transition of form III to I′ through a solid-solid phase transition.

Key words CO_2, phase transition, iPB-1

1 Introduction

Isotactic poly-1-butene (iPB-1), as a polymorphous semicrystal polyolefin, has high creep resistance, low stiffness, good temperature and chemical resistances, so that it has been found many applications in pressure tanks, piping and tubing, molding, films and sheets, and composites and blends[1-3]. It can develop to four crystal modifications with different helix conformations and crystal unit-cell dimensions, designated as form I, II, III and I′, depending on the preparation conditions[4,5]. Forms I and I′ have the same 3/1 helix conformation with twined hexagonal and untwined hexagonal crystal structure respectively[6,7]. Form II has the tetragonal crystal lattice packed by 11/3 helix conformation, and form III with 4/1 helix chain conformation has the orthorhombic unit cell[8]. Crystallized from the melt at ambient condition, metastable form II can be obtained and transforms into form I via a solid-solid phase transition[9,10]. Form III and I′ are usually formed through crystallization from certain dilute solution[11,12]. Form I′ with low content can also be obtained through crystallization under high hydrostatic pressure or through solid-solid phase transition from form III above certain temperature during heating at ambient condition[13,14], while much more form I′ can be generated during heating form III under high pressure CO_2[15]. In the recent studies, forms I and I′ are directly

---

[*] Coauthors: Li Lei, Liu Tao, Zhao Ling. Reprinted from *Polymer*, 2011, 52: 3488-3495.

crystallized from the iPB-1 melt containing different concentration of stereodefects (rr triads defects)[16,17].

Compressed or supercritical CO_2 has attracted much attention because of its unique properties such as nonflammable, nontoxic and relatively inexpensive[18,19]. It can be used as processing solvents or plasticizers in polymer applications including polymer modification, formation of polymer composites, polymer blending, microcellular foaming, particle production and polymerization[20]. Dissolution of CO_2 in polymer can increase the inter-chain distance as well as the chains-segmental mobility, and result in the decrease in the glass transition temperature (T_g), melting temperature (T_m) and crystallization temperature (T_c)[21-23]. The plasticization effect of CO_2 also leads to lowering of the energy barriers, thereby making the crystal phase transition possible at much lower temperatures[24]. Meanwhile, the thermodynamic activity of CO_2 varies continuously with simply changing its pressure and temperature[25]. Thus, the phase transition in the polymorphous polymers under CO_2 might be manipulated through changing the pressure and temperature of CO_2. Handa et al. found that the syndiotactic polystyrene (sPS) underwent planar mesophase to β, α to β, and γ to β transitions at different CO_2 pressures and temperatures[26]. Asia et al. also reported the crystal modification of poly(L-lactide) (PLLA) changed continuously from the disorder α (α″) to α forms not through the α′ one with increasing temperature under CO_2[27]. The CO_2-induced phase transition of iPB-1 from form II to I also showed strong dependence on the temperature and CO_2 pressure[28].

The crystal phase transition had been extensively investigated in many research areas[29-34]. The polymorphic phase transition of iPB-1 with form III upon heating and isothermal annealing at atmospheric pressure had also been widely studied by various methods[35-37]. In our previous work, we discussed briefly the phase transition of form III after CO_2 treatment and compared with that after annealing under ambient N_2[38,39]. The results showed that the temperature and CO_2 pressure had significant influence on final crystal structure of the annealed sample. However, a detailed investigation on the effect of CO_2 on the phase transition process of form III has not been conducted yet. This work reports the detailed characterization of the phase transition under compressed CO_2 by using in-situ high-pressure Fourier transform infrared (FT-IR). The crystal structure was also investigated by using wide-angle X-ray diffraction (WAXD). Meanwhile, the crystal morphology was studied by polarized optical microscopy (POM). It was shown that the phase transition, and crystal form and morphology of the annealed iPB-1 with form III depended significantly on the temperature and CO_2 pressure of the annealing condition.

2 Experimental Section

2.1 Materials and sample preparation

The iPB-1 pellets (PB 0110M) were kindly provided by Basell Polyolefins. Before used, they were purified by Soxhlet extraction in acetone for at least 24h and dried in a vacuum oven at 40℃ for 2 days. Then, they were dissolved in 3wt% solution of carbon tetrachloride at the solvent boiling temperature of 78℃ for 2h. The iPB-1 film with form III was obtained by evaporating the solvent completely at ambient condition. The film thickness measured by micrometer caliper was (32 ± 2) μm. CO_2 (purity: 99.9% w/w) was purchased from Air Products Co., Shanghai, China.

2.2 Characterization of polymer

2.2.1 Wide-angle X-ray diffraction (WAXD)

WAXD of the type Rigaku D/max 2550 VB/PC X-Ray Diffractometer (Cu Kα Ni-filtered radiation) was used to study the crystal modification of the iPB-1 film. The scan rate was $10°$ (θ)/min and the diffraction angular range was between $3°$ and $50°$ 2θ.

2.2.2 Differential scanning calorimeter (DSC)

DSC (NETZSCH DSC 204 HP, Germany) was used to characterize the melting behaviors of iPB-1 with form Ⅲ under atmospheric N_2 and compressed CO_2. The calorimeter was calibrated by carrying out the measurement of the melting points and the heat of fusion of In, Bi, Sn, Pb, and Zn at ambient and high CO_2 pressures, respectively[40]. Under compressed CO_2, iPB-1 film was held at 30℃ for 2h to ensure CO_2 completely diffuse into the film before heating. For each DSC measurement, about (5 ± 1) mg of iPB-1 film with form Ⅲ was heated from 30 to 170℃ at a rate of 10℃/min.

2.2.3 In-situ high-pressure Fourier transform infrared (FTIR)

The isothermal phase transition of form Ⅲ upon annealing under compressed CO_2 was investigated by using *in-situ* high-pressure FTIR of type Bruker Equinox-55 equipped with a Harrick high-pressure demountable liquid cell, the details of which had been described elsewhere[41]. IPB-1 film in the IR cell was heated to the desired temperature at the heating rate of 20℃/min and kept at it for 2min. Thereafter, CO_2 at the desired temperature was injected into the IR cell to the desired pressure ranging from 2 to 12MPa. Annealing of form Ⅲ under atmospheric N_2 was also conducted for comparison. FTIR spectra were recorded at a resolution of $4 cm^{-1}$ and a rate of 1 spectrum per 32s. The IR intensities referred to the peak height. The scanned wave number was in the range of $4000-400 cm^{-1}$.

2.2.4 Polarized optical microscopy (POM)

After FTIR measurement, the annealed iPB-1 film was depressurized and cooled to ambient condition at a natural cooling rate. The crystal modification was detected by WAXD. Meanwhile, the crystal morphology of the annealed film was investigated by Olympus BX 51 POM attached with a DP70 digital camera.

3 Results and Discussions

3.1 Plasticization effect of CO_2 on form Ⅲ

The solvent and casting conditions had significant influences on the final crystal structure of the obtained iPB-1 film[42]. Before further experiments, the obtained iPB-1 film was characterized by WAXD and DSC measurements, respectively, as shown in Fig. S1A and B in Supplementary Content. The WAXD pattern exhibits a sequence of reflections of $12.2°, 14.2°, 17.1°, 18.6°$ and $21.3°$, corresponding to the (101), (200), (111), (201) and (120) lattice planes of form Ⅲ crystal[14,43]. The DSC thermogram also presents the typical melting curve of form Ⅲ.

The endothermal peaks at 99℃ and 118℃ were related to the melting of form Ⅲ and recrystallized form Ⅱ during the heating, and the exothermal peak at 103℃ to the recrystallization of form Ⅱ[36]. From the WAXD and DSC measurements, it was confirmed that the iPB-1 film was in form Ⅲ.

The melting behaviors of iPB-1 with form Ⅲ at various CO_2 pressures, characterized by DSC, are shown in Fig. 1. The fluctuation in the DSC curves at CO_2 pressure above 2MPa was caused by the high pressure CO_2. At low CO_2 pressure, i.e., 0.1MPa, 0.5MPa or 2MPa, the DSC curve shows two melting peaks of form Ⅲ and the recrystallized form Ⅱ, respectively. However, the melting peak of form Ⅱ could not be detected at above 2MPa, indicating recrystallization of form Ⅱ should be inhibited by the high pressure CO_2 during heating. Due to the plasticization effect of CO_2 on the polymer, the melting peaks of form Ⅲ shifted to lower temperatures with increasing CO_2 pressure. The melting temperature of form Ⅲ ($T_{mⅢ}$) is plotted against the CO_2 pressure in Fig. S2 in Supplementary Content. It was apparent that $T_{mⅢ}$ decreased linearly with increasing CO_2 pressure with a slope of -1.6℃/MPa, which could be explained by Flory-Huggins theory[44].

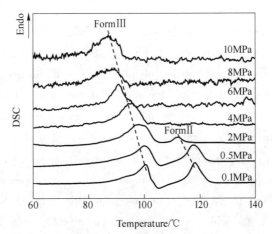

Fig. 1 High-pressure DSC diagrams of iPB-1 with form Ⅲ at various CO_2 pressures during the melting process

3.2 *In-situ* FTIR measurements on phase transition of form Ⅲ upon annealing under atmospheric N_2 and compressed CO_2

3.2.1 Phase transition of form Ⅲ to Ⅱ

The crystal modification of forms Ⅲ, Ⅱ and Ⅰ′ can be well distinguished by the infrared spectrum, for there are distinct differences among the IR spectrum in the range of 800-950cm^{-1}. The band at 901cm^{-1} is known to be the characteristic band of solvent cast form Ⅲ, while the bands at 904cm^{-1} and 924cm^{-1} are corresponding to the characteristic of forms Ⅱ and Ⅰ′, respectively[36,45,46]. Annealing of iPB-1 with form Ⅲ was conducted at temperatures ranging from 75 to 100℃ and CO_2 pressures ranging from 2 to 12MPa. Fig. 2 shows the spectra of iPB-1 films with form Ⅲ immersed in atmospheric N_2 or compressed CO_2 at different temperatures for different times. When the film was immersed in 6MPa CO_2 at 90℃, see Fig. 2a, form Ⅲ melted immediately. With elongation of the annealing time, the form Ⅱ characteristics IR band at 904cm^{-1} and the form Ⅰ′ at 924cm^{-1} appeared and increased intensity. It was claimed that forms Ⅱ and Ⅰ' generated by crystallization from a transition state (mesophase). Dependence of the normalized intensity of the two bands on the annealing time is shown in Fig. 3. The band at 904cm^{-1} appeared earlier and increased with the annealing time much more abruptly than the band at 924cm^{-1} did, indicating that form Ⅱ generated more easily than form Ⅰ' did. At ambient condition, the growth rate of form Ⅰ crystals (form Ⅰ' can be regarded as an imperfect form Ⅰ with many defects) was about 1/100 that of form Ⅱ crystals[17,47], so that no form Ⅰ' can generate when cooled from the melt. Form Ⅰ' can generate from the melt when annealing at 90℃ and 6MPa, which should be ascribed to the plasticization effect of CO_2 on the mobility of iPB-1 polymer chains. De Rosa et al. had reported that increase in the polymer chains motion by introducing *rr* stereodefects in iPB-1 increased the crystallization rate of form Ⅰ' and made the melted iPB-1 directly crystallize into form Ⅰ' at high stereodefect content[16,17]. Dissolution of CO_2 in polymer could also increase its free volume, and its chains motion as well[22], which should subsequently increase the crystallization rate of form Ⅰ'.

When immersed in 4MPa CO_2 at 95℃, see Fig. 2b, form Ⅲ also melted immediately. The intensity of the band at 904cm^{-1} increased with the annealing time, and the band at 924cm^{-1} was not detected during the annealing process. When annealed under atmospheric N_2 at 100℃, see Fig. 2c, the form Ⅲ characteristic band at 901cm^{-1} gradually shifted to form Ⅱ characteristic band at 904cm^{-1} with increasing the annealing time (see Fig. S3 in Supplementary Content). It took about 500min for the phase transition of form Ⅲ into Ⅱ. Moreover, the intensity of the characteristic band decreased, indicating that the iPB-1 film was partially melted. Completely amorphous phase IR spectra of iPB-1 were not detected during the annealing process. Similar results were also reported during the melting process of form Ⅲ under atmospheric N_2. Miyoshi et al. had studied the form Ⅲ melting process by high-resolution solid-state ^{13}C nuclear magnetic resonance (NMR) spectroscopy and found no evidence for appearance of the amorphous phase after melting of form Ⅲ[35]. Men et al. investigated the form Ⅲ heating process through *in-situ* synchrotron small- and wide-angle X-ray scattering (SAXS and WAXS), and they also found no completely amorphous pattern during the formation of form Ⅱ[37]. When immersed in 2 and 4MPa CO_2 at 100℃, see Fig. 2d-e, form Ⅲ transited into Ⅱ through melt recrystallization, as evidenced by the rapid increase of the intensity of the band at 904cm^{-1} from the melt state. The crystallization of form Ⅱ completed in a few minutes.

3.2.2 Phase transition of form Ⅲ to Ⅰ'

Fig. 4 shows the spectra of iPB-1 with form Ⅲ annealed at CO_2 pressures ranging from 2 to 12MPa and temperatures ranging from 75 to 95℃ for different times. At these annealing conditions, the form Ⅲ characteristic band at 901cm^{-1} gradually shifted to form Ⅰ' characteristic band at 924cm^{-1} with increasing the annealing time. The absorbance ratio of the bands at 901cm^{-1} and 924cm^{-1} depended significantly on the annealing time, temperature and CO_2 pressure. The content of form Ⅰ' in the annealed iPB-1 changed with the annealing conditions. However, completely amorphous phase IR spectra of iPB-1 were not detected during the annealing process.

3.2.3 Effect of CO_2 on the phase transition of form Ⅲ upon annealing

As shown in Fig. 2 and Fig. 4, the IR spectra tend to be unchanged after annealing for 560min,

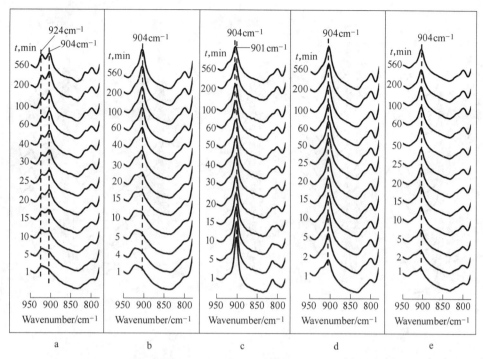

Fig. 2 FTIR spectrum of form Ⅲ annealed at 90℃ and 6MPa(a), 95℃ and 4MPa(b), 100℃ and ambient N_2(c), 100℃ and 2MPa(d), and 100℃ and 4MPa(e) for different times

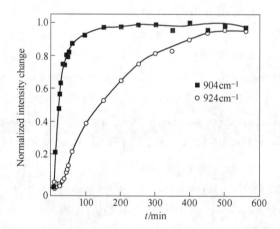

Fig. 3 Normallized bands intensity at 924cm^{-1} and 904cm^{-1} as a function of annealing time

indicating completion of the phase transition. Fig. 5 illustrates the IR spectra of form Ⅲ after annealed at temperatures ranging from 75 to 100℃ under atmospheric N_2 and compressed CO_2 for 560min. For form Ⅲ annealed at 75℃ under atmospheric N_2 and 2MPa CO_2, see Fig. 5a, the band at 901cm^{-1} showed no change after annealing. Whereas, it gradually weakened with further increasing CO_2 pressure, and the band at 924cm^{-1} appeared and strengthened. It indicated that the phase transition could only occur at above 2MPa and the content of form Ⅲ transformed into form Ⅰ′ increased with increasing CO_2 pressure at 75℃. For form Ⅲ annealed at 80℃, see Fig. 5b, the band at 924cm^{-1} appeared at 2MPa and strengthened with increasing CO_2 pressure. Meanwhile, at a given CO_2 pressure, the intensity of the IR band at 924cm^{-1} after

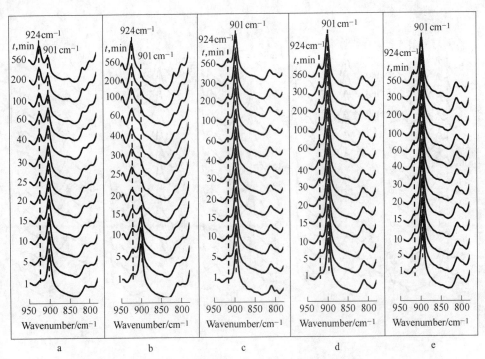

Fig. 4 FTIR spectrum of form Ⅲ annealed at 75℃ and 12MPa(a), 80℃ and 12MPa(b), 85℃ and 6MPa(c), 90℃ and 4MPa(d), and 95℃ and 2MPa(e) for different times

annealing at 80℃ was stronger than that at 75℃, which meant that more form Ⅰ′ generated at the higher temperature. When the temperature reached 85℃, form Ⅲ could transform into form Ⅰ′ under atmospheric N_2 as evidenced by the presence of the band at 924cm^{-1} as shown in Fig. 5c. The intensity of the band at 924cm^{-1} also increased with increasing CO_2 pressure until to 6MPa. However, completely melted IR spectra of iPB-1 were obtained at 8MPa and 12MPa. Annealed at 90℃, form Ⅲ transformed into form Ⅰ′ under atmospheric N_2, 2MPa and 4MPa CO_2, as presented in Fig. 5d that the intensity of the band at 901cm^{-1} decreased and that of the band at 924cm^{-1} increased. However, the band at 901cm^{-1} disappeared in the spectra of the film after annealing at 6MPa, and the bands at 904cm^{-1} and 924cm^{-1} appeared, indicating that form Ⅲ transformed completely into form Ⅰ′ and Ⅱ. With further increasing CO_2 pressure to 8 and 12MPa, the completely melted iPB-1 spectra were also detected. For form Ⅲ annealed at 95℃, see Fig. 5e, the IR spectra showed the band at 924cm^{-1} under atmospheric N_2 and 2MPa CO_2, whereas showed only the band at 904cm^{-1} at 4MPa, indicating only form Ⅱ was formed. With increasing CO_2 pressure to 6MPa, 8MPa and 12MPa, the completely melted iPB-1 IR spectra were detected. The iPB-1 films with form Ⅲ annealed at 100℃, see Fig. 5f, exhibited only one IR intense absorbance band at 904cm^{-1} under ambient N_2, 2 and 4MPa CO_2, implying that only form Ⅱ generated. Annealed under CO_2 pressures above 6MPa, the films were completely melted, as evidenced by the amorphous iPB-1 IR spectra.

The crystal structures of form Ⅲ annealed at different conditions are summarized in Fig. 6. The relative contents of form Ⅰ′ after annealing are also listed. Due to the fluctuation of WAXD patterns and the possible phase transition of form Ⅱ to Ⅰ before WAXD measurement,

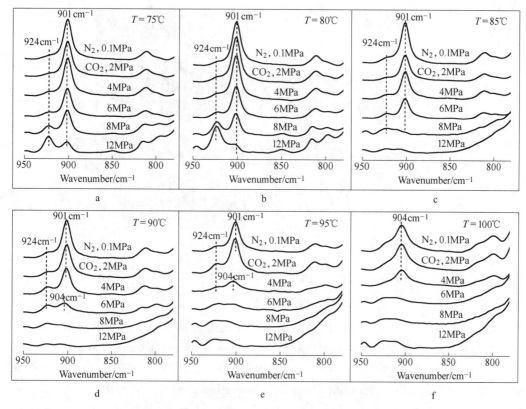

Fig. 5 IR spectrum of form Ⅲ annealed at ambient N_2 and various CO_2 pressure at 75℃ (a), 80℃ (b), 85℃ (c), 90℃ (d), 95℃ (e) and 100℃ (f) for 560min

the content of form I′ in the iPB-1 films was calculated from the IR data. The relative percentage of form I′ in the total crystal was derived from the ratio between the absorbance band area at $924cm^{-1}$ (A_{924}) and the sum of that at $901cm^{-1}$ (or $904cm^{-1}$) and $924cm^{-1}$ (A_{901} (or A_{904}) + A_{924}). The superposition of the IR bands was analyzed by a PEAK-FIT V4.12 program that is usually applied to deconvolute complex IR spectra. It was clarified that the phase transition of form Ⅲ depended strongly on the annealing temperature and CO_2 pressure. At 75-85℃, high pressure CO_2-induced and promoted the phase transition of form Ⅲ to I′, and the relative content of form I′ increased with CO_2 pressure at a given temperature. At 90-95℃ and in the range of low CO_2 pressure, CO_2 also induced the phase transition of form Ⅲ into form I′, whereas form Ⅲ turned to transit into form Ⅱ with further increasing CO_2 pressure. Especially at 90℃ and 6MPa, form Ⅲ completely transformed into form Ⅱ and I′. At 100℃, form Ⅲ transformed into form Ⅱ at low CO_2 pressures, and melted at high CO_2 pressures.

3.3 WAXD measurement on the crystal modification of form Ⅲ after annealing

To further identify the crystal modification in the iPB-1 after annealing under atmospheric N_2 and compressed CO_2, WAXD measurement was employed and the results were shown in Fig. 7. As shown in Fig. 7a, b, the WAXD profiles exhibited no changes for the samples annealed at 75℃ and 80℃ under atmospheric N_2 and at 75℃ under 2MPa CO_2, indicating that no form transition occurred. The WAXD pattern of form I′ might exhibit three major peaks, at

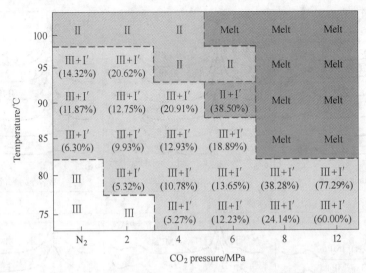

Fig. 6 Phase diagram of crystalline structure of the form III annealed at ambient N_2 and various CO_2 pressure from 75 to 100 ℃ for 560min

$2\theta = 9.9°, 17.3°$ and $20.1°$, corresponding to the diffraction of crystal reflections from planes (110), (300) and (220), respectively. In Fig. 7a-c, the diffraction peaks of form I′ appeared and increased with CO_2 pressure from 4 to 12MPa at 75℃, 2-12MPa at 80℃ and 2-6MPa at 85℃. The WAXD profile of the iPB-1 film after annealing at 85℃ and 8MPa exhibited three major peaks, at 11.9°, 16.9°, and 18.4°, corresponding to the diffraction of form II crystal reflections from planes (200), (220), and (301) respectively[3]. This was attributed to the crystallization of form II from the iPB-1 melt as evidenced by the *in-situ* FTIR measurements. The form I′ diffraction peaks also appeared at 90℃ and 2MPa and 4MPa, and 95℃ and 2MPa, as shown in Fig. 7d-e. At 90℃ and 6MPa, the forms II and I′ diffraction peaks both existed, implying the coexistence of forms II and I′ in the film. For the film treated at 90℃ and 8MPa, a WAXD pattern of complete form II was obtained. It should be noted that the weak diffraction peak of form I characteristic diffraction peak at $2\theta = 9.9°$ in the WAXD profile was ascribed to the transition of form II into I after the treatment before WAXD measurement, for there was no 924cm^{-1} band was observed at the condition from the *in situ* FTIR measurement. For the films annealed at 100℃, the WAXD curves also showed the form II characteristic diffraction peaks in Fig. 7f. It was apparent that the WAXD measurements consisted well with the *in-situ* FTIR results.

3.4 Crystalline morphology of the iPB-1 films with form III after annealing

Fig. 8 shows the polarized optical micrographs of the iPB-1 films with form III after annealing under atmospheric N_2 and compressed CO_2. In the upper right corner, the crystalline morphology of form III before annealing exhibits no Maltese cross pattern, which means the orientation of the solution cast form III lamellar stacks was randomness distributed in the crystal[48]. For the films annealed under CO_2, where only did crystal phase transition of form III to I′ occur, the crystalline morphology of the films showed no significant change compared with that of original form III. It implied that the phase transition of form III to I′ might process via a solid-solid

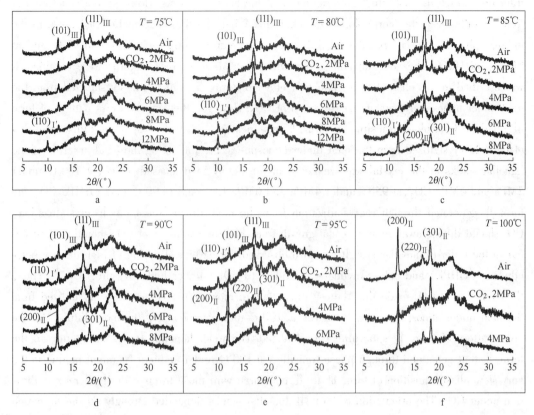

Fig. 7 WAXD profiles of form Ⅲ annealed at ambient N_2 and various CO_2 pressure at 75℃(a), 80℃(b), 85℃(c), 90℃(d), 95℃(e) and 100℃(f) for 560min

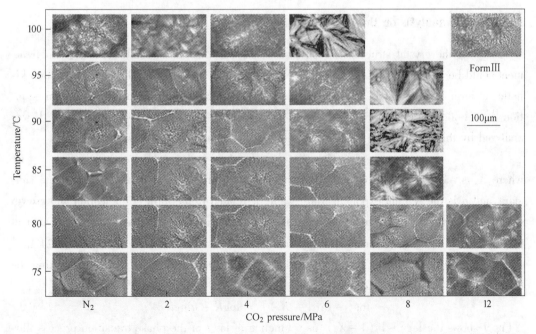

Fig. 8 Polarized optical micrographs of form Ⅲ annealed at various temperatures under ambient N_2 and compressed CO_2

transition, which did not affect the orientation of the lamellar stacks. However, the generation of form Ⅱ influenced the morphologies of spherulite. When the films had melted, as shown in the micrographs of form Ⅲ annealed at 8MPa and temperatures ranging from 85 to 95℃ and at 6MPa and 100℃, the compact spherulites with clear Maltese cross patterns were obtained. It was ascribed to the clear double reflection effect of form Ⅱ, which was crystallized from the melt under atmospheric air after depression of CO_2. However, the Maltese cross pattern of form Ⅱ obtained from phase transition of form Ⅲ by annealing under CO_2 was more diffuse than that of form Ⅱ crystallized from the melt under atmospheric air, indicating the local ordering of the lamellar stacks in the crystal was weaker than the latter. As discussed above, at 100℃ under 2MPa and 4MPa CO_2, at 95℃ under 4MPa and 6MPa CO_2, and at 90℃ under 6MPa CO_2, the form Ⅱ generated through recrystallization from the melt. During the form Ⅱ crystallization, CO_2 should diffuse away from a crystal growth front, and the exclusion of CO_2 should at least occur in the order of lamellar size. It was also observed during crystallization of high-density polyethylene (HDPE) and poly(vinylidene fluoride) (PVDF) under high pressure CO_2[49]. Thus, the exclusion of CO_2 made the Maltese cross pattern more diffuse than those crystallized from the melt under atmospheric air. Moreover, at 100℃ under ambient N_2, the most diffuse Maltese cross pattern of form Ⅱ spherulite was detected. The weak lamellar stacks orientation in the form Ⅱ that generated through phase transition at 100℃ under ambient N_2 might be due to the very slow phase transition of form Ⅲ to Ⅱ compared with those through the melt recrystallization under CO_2. The orientation of form Ⅱ lamellar stacks depended strongly on the formation process. To obtain the strong orientation, the formation process displayed the following order: melt crystallization at ambient condition > melt recrystallization under CO_2 > phase transition upon annealing at ambient condition.

3.5 Kinetic analysis on the phase transition of form Ⅲ to Ⅰ′

By assuming the crystal structure transition kinetics was nucleation limited, the Avrami treatment could be used to describe the overall phase transition kinetics[50]. The phase transition kinetics of form Ⅱ to Ⅰ under ambient condition had been well analyzed by using Avrami equation[51,52]. Following this approach, the solid-solid phase transition kinetics of form Ⅲ to Ⅰ′ was analyzed by this method. The Avrami equation can be expressed as follows:

$$1 - X_t = \exp(-Kt^n) \tag{1}$$

where, X_t is relative content of form Ⅰ′ in the sample at annealing time of t, K is the rate constant, and n is the Avrami exponent whose value is related to the form of nucleation and growth of the new phase. X_t is derived from the ratio between the form Ⅰ′ absorbance band area at 924cm^{-1} (A_{924}) and the sum of form Ⅲ and Ⅰ′ absorbance band area at 901 and 924cm^{-1} ($A_{901} + A_{924}$) at aging time t. To facilitate the evaluation of the Avrami parameters, taking the logarithm of this equation twice gives:

$$\log[-\ln(1 - X_t)] = \log K + n\log t \tag{2}$$

Fig. 9 shows the $\log[-\ln(1 - X_t)]$ as a function of $\log t$ of the phase transition process illustrated in Fig. 4a-e. It was apparent that the plot of $\log[-\ln(1 - X_t)]$ against $\log t$ varied linearly in a relatively large range of phase transition, though a slight deviation from the straight line derived in the middle of the curves was also observed for the transition processes at 75℃ and 12MPa, and at 80℃ and 12MPa. Geil et al. had also observed the two transition processes during the phase transition from form Ⅱ to Ⅰ, and they believed this effect was similar to that of secondary crystallization during initial crystallization[52]. Hong and Spruiell found the two stages phase transition of iPB-1 form Ⅱ to Ⅰ, and they ascribed the change in the Avrami plot to a change in the phase transition mechanism[51]. It should be noted that the second phase transition process was only observed at the late stage of the phase transition. This was the reason why only one straight line was observed for the films annealed at 85℃ and 6MPa, 90℃ and 4MPa, and 95℃ and 2MPa. The crystal phase transition at these conditions was so slow that the phase transition was in the first phase transition process during annealing, thus only one straight line was detected. From these results, it was concluded that the Avrami equation could be well established to describe the phase transition of form Ⅲ to Ⅰ′ through a solid-solid phase transition.

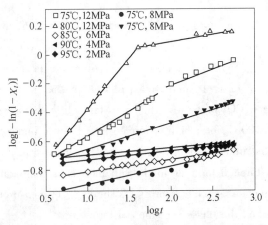

Fig. 9 Avrami plots for form Ⅲ transition process at various CO_2 pressures and temperatures

The Avrami exponents and the rate constants could be obtained from the slope and the intercept of the straight lines of $\log[-\ln(1-X_t)]$ vs $\log t$ plots. Due to the slow crystal phase transition rate of form Ⅲ into Ⅰ′ at certain CO_2 pressure and temperature, it was not appropriate to obtain the reasonable Avrami exponents when the transition fraction of form Ⅰ′ was too low (relative percentage of form Ⅰ′ was lower than 10%). The Avrami exponents and rate constants of the first phase transition process are listed in Table 1. The Avrami exponent increased with CO_2 pressure at 75℃ and 80℃ and showed no significant change at 85℃, 90℃ and 95℃. The changes in the Avrami exponent were related to a possible change in the phase transition process. Similar results were also found for isothermal crystallized from polymer melts under high pressure CO_2[26,53]. Meanwhile, it should also be noted that the Avrami exponent of first stage of form Ⅱ to Ⅰ transition was nearly unity[52]. However, the phase transition of form Ⅲ to Ⅰ′ under CO_2 was only a fraction of unity. Then, the physical meaning of this exponent was lost. The low exponents might be due to a non-negligible volume fraction of pre-formed unclei or the changed isothermal phase transition rate with time for reasons such as the existence of a large interfacial region[50].

Table 1 Values of Avrami exponent, n and the rate constant, K, at various temperature and CO_2 pressures

Conditions	Avrami analysis	
	n	K
75℃, 12MPa	0.36	0.11
75℃, 8MPa	0.18	0.076
80℃, 12MPa	0.67	0.093
80℃, 8MPa	0.21	0.13
85℃, 6MPa	0.072	0.13
90℃, 4MPa	0.046	0.18
95℃, 2MPa	0.059	0.16

4 Conclusions

In-situ high-pressure FTIR was used to investigate the polymorphous phase transition of iPB-1 with form Ⅲ upon annealing at temperatures ranging from 75 to 100℃ and CO_2 pressures ranging from 2 to 12MPa. At 75℃, 80℃ and 85℃, CO_2-induced the phase transition of form Ⅲ into Ⅰ′ and the content of form Ⅰ′ increased with CO_2 pressure. At 90℃ and 95℃, form Ⅲ transformed into Ⅰ′ at low CO_2 pressures, while into form Ⅱ with increasing CO_2 pressure before form Ⅲ melted. At 100℃, form Ⅲ transformed into Ⅱ even under atmospheric N_2 and low pressure CO_2, and melted under high pressure CO_2. The phase transition of form Ⅲ changed from form Ⅲ to Ⅱ not through form Ⅲ to Ⅰ′ with increasing temperature, and application of CO_2 increased the content of generated form Ⅰ′. The WAXD measurements on the annealed iPB-1 with form Ⅲ verified the phase transition of form Ⅲ.

It should be noted that forms Ⅱ and Ⅰ′ can be obtained by crystallization through a transition mesophase at 90℃ under 6MPa CO_2 and at 95℃ under 4MPa CO_2. The transition mesophase in iPB-1 may be in an analogy similar to the mesphase, proposed by Welsh et al.[54,55] and Abou-Kandil et al.[56-58], during annealing or drawing polyethylene terephthalate (PET), polyethylene naphthoate (PEN) and random copolymer of PET and PEN. The transition mesophase in iPB-1 with form Ⅲ during annealing under CO_2 appears to act as a precursor to crystallization of forms Ⅱ and Ⅰ′.

The crystalline morphology of the annealed iPB-1 films was investigated using the POM. The results showed that the phase transition of form Ⅲ into Ⅰ′ had no significant influence on the crystalline morphology, which implied the phase transition might process via the solid-solid transition. The orientation of form Ⅱ lamellar stacks was strongly depended on the formation process. More diffuse Maltese cross patterns were observed for the form Ⅱ generated under high pressure CO_2 during the annealing compared with those of melt crystallized form Ⅱ under ambient air, which was ascribed to exclusion effect of CO_2 on the crystallization process of the iPB-1. To obtain the strong orientation, the formation process displayed the following order: melt crystallization at ambient condition > melt recrystallization under CO_2 > phase transition upon annealing at ambient condition. Avrami equation could be well established to describe the phase transition of form Ⅲ to Ⅰ′ through a solid-solid phase transition.

Acknowledgements

The authors are grateful to the National Natural Science Foundation of China (Grant No. 20976045 and 20976046). Shanghai Shuguang Project (08SG28), Program for New Century Excellent Talents in University (NCET-09-0348), Program for Changjiang Scholars and Innovative Research Team in University (IRT0721) and the 111 Project (B08021).

Appendix. Supplementary Data

The supplementary data associated with this article can be found in the online version at doi: 10.1016/j. polymer. 2011. 05. 042.

References

[1] Fu Q, Heck B, Strobl G, Thomann Y. Macromolecules 2001;34:2502-2511.
[2] Jiang S, Duan Y, Li L, Yan D, Chen E, Yan S. Polymer 2004;45:6365-6374.
[3] Causin V, Marega C, Marigo A, Ferrara G, Idiyatullina G, Fantinel F. Polymer 2006;47:4773-4780.
[4] Danusso F, Gianotti G. Makromol Chem 1963;61:139-156.
[5] Luciani L, Seppala J, Lofgren B. Prog Polym Sci 1988;13:37-62.
[6] Natta G, Corradini P, Bassi I. Nuovo Cimento 1960;15(Suppl.):52-67.
[7] Maring D, Spiess MWHW, Meurer B, Weill G. J Polym Sci B Polym Phys 2000;38:2611-2624.
[8] Samon JM, Schultz JM, Hsiao BS, Wu J, Khot S. J Polym Sci B Polym Phys 2000;38:1872-1882.
[9] Fujiwara Y. Polym Bull 1985;13:253-258.
[10] Di Lorenzo ML, Righetti MC. Polymer 2008;49:1323-1331.
[11] Dorset DL, McCourt MP, Kopp S, Wittmann JC, Lotz B. Acta Crystallogr B 1994;50:201-208.
[12] Mathieu C, Stocker W, Thierry A, Wittmann JC, Lotz B. Polymer 2001;42:7033-7047.
[13] Nakafuku C, Miyaki T. Polymer 1983;24:141-148.
[14] Nakamura K, Aoike T, Usaka K, Kanamoto T. Macromolecules 1999;32:4975-4982.
[15] Li L, Liu T, Zhao L, Yuan W-k. Macromolecules 2011, doi:10.1021/ma200988y.
[16] De Rosa C, Auriemma F, Resconi L. Angew Chem Int Edit 2009;48:9871-9874.
[17] De Rosa C, Auriemma F, Ruiz de Ballesteros O, Esposito F, Laguzza D, Di Girolamo R, et al. Macromolecules 2009;42:8286-8297.
[18] Kikic I. J Supercrit Fluids 2009;47:458-465.
[19] Kiran E. J Supercrit Fluids 2009;47:466-483.
[20] Nalawade SP, Picchioni F, Janssen LPBM. Prog Polym Sci 2006;31:19-43.
[21] Condo PD, Johnston KP. Macromolecules 1992;25:6730-6732.
[22] Kazarian SG, Vincent MF, Bright FV, Liotta CL, Eckert CA. J Am Chem Soc 1996;118:1729-1736.
[23] Takada M, Hasegawa S, Ohshima M. Polym Eng Sci 2004;44:186-196.
[24] Shieh YT, Hsiao TT, Chang SK. Polymer 2006;47:5929-5937.
[25] Zhang Z, Handa YP. Macromolecules 1997;30:8505-8507.
[26] Handa YP, Zhang Z, Wong B. Macromolecules 1997;30:8499-8504.
[27] Marubayashi H, Akaishi S, Akasaka S, Asai S, Sumita M. Macromolecules 2008;41:9192-9203.
[28] Li L, Liu T, Zhao L, Yuan W-k. Macromolecules 2009;42:2286-2290.
[29] Tanaka R, Tashiro K, Kobayashi M. Polymer 1999;40:3855-3865.
[30] Egorov EA, Zhizhenkov VV, Gorshkova IA, Savitsky AV. Polymer 1999;40:3891-3894.

[31] Rosas G, Perez R. Mater Lett 1998;36:229-234.

[32] Aoki K, Suzuki T, Minoda H, Tanishiro Y, Yagi K. Surf Sci 1998;408:101-111.

[33] Ceelen WCAN, Denier van der Gon AW, Reijme MA, Brongersma HH, Spolveri I, Atrei A, et al. Surf Sci 1998;406:264-278.

[34] Inkson BJ, Clemens H, Marien J. Scripta Mater 1998;38:1377-1382.

[35] Miyoshi T, Hayashi S, Imashiro F, Kaito A. Macromolecules 2002;35:2624-2632.

[36] Lee KH, Snively CM, Givens S, Chase DB, Rabolt JF. Macromolecules 2007;40:2590-2595.

[37] Jiang Z, Sun Y, Tang Y, Lai Y, SrS Funari, Gehrke R, et al. J Phys Chem B 2010;114:6001-6005.

[38] Li L, Liu T, Zhao L, Yuan W-k. Asia Pac J Chem Eng; 2009:800-806.

[39] Li L, Liu T, Zhao L. Macromol Symp 2010;296:517-525.

[40] Liu T, Hu G-H, Tong G-s, Zhao L, Cao G-p, Yuan W-k. Ind Eng Chem Res 2005;44:4292-4299.

[41] Li B, Li L, Zhao L, Yuan W-k. Eur Polym J 2008;44:2619-2624.

[42] Kaszonyiova M, Rybnikar K, Geil PH. J Macromol Sci B Phys 2005;B44:377-396.

[43] Shieh YT, Lee MS, Chen SA. Polymer 2001;42:4439-4448.

[44] Kishimoto Y, Ishii R. Polymer 2000;41:3483-3485.

[45] Luongo JP, Salovey R. J Polym Sci A-Polym Phys 1966;4:997-1008.

[46] Goldbach G, Peitscher G. J Polym Sci B Polym Lett 1968;6:783-788.

[47] Yamashita M, Hoshino A, Kato M. J Polym Sci B Polym Phys 2007;45:684-697.

[48] Wunderlich B. Thermal analysis of polymeric materials. Berlin:Springer;2005.

[49] Koga Y, Saito H. Polymer 2006;47:7564-7571.

[50] Cheng SZD. Phase transitions in polymers:the role of metastable states. Amsterdam:Elsevier Science;2008.

[51] Hong K-B, Spruiell JE. J Appl Polym Sci 1985;30:3163-3188.

[52] Chau KW, Yang YC, Geil PH. J Mater Sci 1986;21:3002-3014.

[53] Varma-Nair M, Handa PY, Mehta AK, Agarwal P. Therm Acta 2003;396:57-65.

[54] Welsh GE, Blundell DJ, Windle AH. Macromolecules 1998;31:7562-7565.

[55] Welsh GE, Blundell DJ, Windle AH. J Mater Sci 2000;35:5225-5240.

[56] Abou-Kandil AI, Goldbeck-Wood G, Windle AH. Macromolecules 2007;40:6448-6453.

[57] Abou-Kandil AI, Windle AH. Polymer 2007;48:4824-4836.

[58] Abou-Kandil AI, Windle AH. Polymer 2007;48:5069-5079.

Supercritical Carbon Dioxide-assisted Dispersion of Sodium Benzoate in Polypropylene and Crystallization Behavior of the Resulting Polypropylene*

Abstract This work aimed at studying the efficiency of the use of supercritical carbon dioxide (scCO$_2$) as a swelling agent to disperse sodium benzoate (NaBz), a nucleating agent, in polypropylene (PP), on the one hand; and the crystallization behavior of the resulting PP, on the other hand. Under scCO$_2$, the NaBz was uniformly dispersed in the PP at a nanometer scale. The use of ethanol or acetone as a cosolvent further increased its state of dispersion and its mass uptake in the PP. Isothermal and nonisothermal crystallization kinetics indicated that the PP with the NaBz being dispersed in it under scCO$_2$ had a much higher crystallization rate than that of the pure PP or the PP with the NaBz being dispersed in it by a conventional melt compounding process. The size of the PP crystallites was also much smaller when the NaBz was dispersed at a nanometer scale.

Key words supercritical carbon dioxide, polypropylene, sodium benzoate, dispersion

1 Introduction

As a general-purpose polymer, polypropylene (PP) has various excellent properties such as mechanical rigidity, thermal and chemical resistance, ease of processing, and recycling. Many properties of PP depend very much on its crystallinity and crystalline morphology. Nowadays processing techniques require shorter and shorter cycle times. The addition of nucleating agents to PP accelerates the crystallization process and therefore shortens the cycle time[1-4].

Sodium benzoate (NaBz) has frequently been used as a flavoring or antimicrobial agent[5]. It may also be used as a nucleating agent for PP[6-9]. This is because its melting point is high (439℃), is insoluble in the PP melt at typical processing temperatures, and crystallizes much earlier than PP[10]. A nucleating agent like NaBz is usually added to PP by extrusion compounding. The driving force for dispersion is a mechanical force the extruder imparts to the NaBz particles. Under such conditions, the state of dispersion of the NaBz is not optimal.

Supercritical fluids (SCFs) have been gaining popularity in polymer processing[11-17]. Carbon dioxide has by far been the most frequently used SCF candidate owing to its unique properties: nonflammable, nontoxic, relatively inexpensive, relatively easy to reach supercritical conditions, etc. The aim of this work is two-folds. First, we study the extent to which the NaBz can be dissolved in supercritical carbon dioxide (scCO$_2$) and then in PP under scCO$_2$. Since the NaBz is a salt bearing a polar group, we test the efficiency of using a cosolvent at enhancing its solubility and therefore its mass uptake in the apolar PP. Second, we compare the isothermal and nonisother-

---

* Coauthors: Li Bin, Hu Guohua, Gao Guiping, Liu Tao, Zhao Ling. Reprinted from Journal of Applied Polymer Science, 2006, 102: 3212-3220.

mal crystallization kinetics of PP in which the NaBz is dispersed by different processes: conventional melt compounding, scCO$_2$-assisted dispersion, and scCO$_2$/cosolvent-assisted dispersion.

2 Experimental

2.1 Penetration of NaBz to PP

A commercial isotactic polypropylene (PP0) was used in this study. It was provided by Shanghai Petrochemical Company, China, in a powdery form under the tradename Y1600. About 15g of the PP0 powder was first molten in an oven at 200℃ and then pressed to a film of 0.3mm thickness using a press under 10MPa. The film was then cut into pieces of 1mm × 3mm × 0.3mm. The latter were refluxed in acetone for 24h to remove impurities and then dried at 100℃ for 2h. The NaBz was purchased from Shanghai Chemical Reagent Company, China. Table 1 gathers some of the characteristics of the PP0 and NaBz.

Table 1 Characteristics of the PP and NaBz used in this work

Material	Property	Value
PP0	Melt flow index(230℃)/g·(10min)$^{-1}$	16
	Polydispersity index	5.1
	M_w/g·mol^{-1}	188700
	T_g/℃	−5
NaBz	Density/g·cm^{-3}	1.46
	Formula mass/g·mol^{-1}	144.1
	Average particle size/μm	200
	Melting point/℃	439

Fig. 1 shows the experimental apparatus used for the penetration of the NaBz to the PP0 under scCO$_2$. It was mainly composed of a gas cylinder, a highpressure reactor (Parr Instrument, Co.), a gas booster (Haskel International), a digital pressure gauge, an electrical heating bath, and valves and fittings of different kinds. All the metallic parts in contact with chemicals were made of stainless steel. The apparatus was tested up to 40MPa. It had a volume capacity of 500mL.

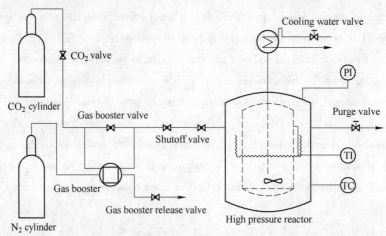

Fig. 1 Schematic of the apparatus used for the scCO$_2$-assisted penetration of the NaBz in the PP0

PP films were placed in the high-pressure cell together with 1wt% of the NaBz. The system was then purged with CO_2. When it reached a desired temperature of 60℃ (an optimal one for the penetration of nucleating agent according to a previous work[17]), CO_2 was charged till a desired pressure was attained. The penetration process lasted 4h. Thereafter the high-pressure cell was rapidly depressurized. It was then cooled down, opened up, and the PP films were taken out. A gravimetric technique[18] was used to measure the mass uptake of the NaBz in the PP films. A high-precision electrical balance (BP211D, Sartorius, Germany) was used for that purpose. It had an accuracy of within ±0.01mg. A scanning electron microscope (SEM) of type JEOL/EO JSM-6360 was used to visualize the state of dispersion of the NaBz in the PP.

Ethanol and acetone were used as cosolvents for the penetration of the NaBz in the PP0. They were purchased from Shanghai Reagent Company and their purities were above 99.8%. When prescribed pressure and temperature were reached, 20mL of ethanol or acetone (0.05wt% with respect to the PP0) was added to the system using a high-pressure advection pump (LB-10, Beijing Satellite Instrument Co., China).

2.2 Morphology of the PP crystalline phases

The morphology of the PP crystalline phases was studied using an optical polarized microscope (BX51, Olympus, Japan) equipped with a hot stage (THMS 600, Linkam, Great Britain). A PP sample was sandwiched between two microscope cover glasses, heated to 200℃, kept at that temperature for 5min, and then cooled down to 140℃ with a cooling rate of 100℃/min for isothermal crystallization and crystalline morphology observations.

2.3 Crystallization analysis

A differential scanning calorimeter of type PerkinElmer Pyris Diamond DSC was used to study the crystallization behavior of PP samples. Five types of samples (PP0-PP4) were studied, as shown in Table 2. All of them were extruded and pelletized with a twinscrew extruder (SJSH-30) for DSC analysis.

Table 2 PP samples used in this work

Sample	Preparation technique	Mass percentage of NaBz in PP
PP0	Virgin	0
PP1	NaBz was dispersed in the PP0 by a conventional compounding technique	0.12
PP2	NaBz was dispersed in the PP0 by a conventional compounding technique	0.85
PP3	NaBz was incorporated in the PP0 under $scCO_2$	0.12
PP4	NaBz was incorporated in the PP0 under $scCO_2$ and in the presence of ethanol as a co-solvent (0.05wt% with respect to the PP0)	0.85
PP5	NaBz was incorporated in the PP0 under $scCO_2$ and in the presence of acetone as a co-solvent (0.05wt% with respect to the PP0)	1.65

In a typical isothermal crystallization experiment, about 5mg of a PP sample in the form of pellets was heated up to 200℃ and kept at that temperature for 10min to remove any thermal history. It was then cooled down to a prescribed crystallization temperature. For the nonisothermal crystallization, the crystallization thermograms were obtained by cooling the sample at different cooling rates ranging from 1 to 20℃/min. For the isothermal crystallization, the crystallization temperature ranged from 110 to 135℃ and the cooling rate was 100℃/min.

A nonisothermal crystallization thermogram showed the variation of the enthalpy of crystallization as a function of temperature under a given cooling rate. On the other hand, an isothermal crystallization thermogram showed the variation of the enthalpy of crystallization as a function of time at a given crystallization temperature.

The crystallinity of the PP samples was calculated by the following equation[3,19]:

$$X_c = \frac{\Delta H(t)}{\Delta H_f} \qquad (1)$$

where, $\Delta H(t)$ is the enthalpy of crystallization at time t, ΔH_f is the enthalpy of fusion of a 100% crystalline PP(209J/g). The relative crystallinity was calculated by[20-22]:

$$V_c(t) = \frac{X(t)}{X_{c \to \infty}} = \int_0^t (dH/dt) dt / \int_0^\infty (dH/dt) dt \qquad (2)$$

where $X_{c \to \infty}$ is the crystallinity at equilibrium and dH/dt is the enthalpy variation rate.

3 Results and Discussion

3.1 Penetration and mass uptake of the NaBz in the PP0

Penetration of the NaBz into the PP0 matrix proceeded under $scCO_2$ with and without the presence of a cosolvent. The SEM images in Fig. 2 show the morphology of the NaBz alone and its state of dispersion in PP3, PP4, and PP5. It is seen that the size of the NaBz particles was greatly reduced when they penetrated into PP matrix under $scCO_2$, especially when ethanol was added. Moreover, the addition of the cosolvent seemed to lead to more homogeneous dispersion of the NaBz particles in the PP.

Fig. 3 compares the mass uptake of the NaBz in the PP0 obtained under three different penetration conditions. The mass uptake was always very low (of the order of 0.1wt% only) when the NaBz penetrated the PP0 under $scCO_2$ without a cosolvent, irrespective of the $scCO_2$ pressure (4-18MPa). When a cosolvent (acetone or ethanol) was used, the mass uptake became much higher, reaching a maximum of 1.65wt%. This significant increase in the mass uptake could be explained by an enhanced solubility of the NaBz in the $scCO_2$ when a cosolvent was added. Acetone was clearly much more efficient in increasing the mass uptake of the NaBz than ethanol. Moreover, the mass uptake depended very much on the $scCO_2$ pressure too. Apparently there was an optimum for the $scCO_2$ pressure corresponding to a maximum of the mass uptake of the NaBz in the PP0. It was somewhere between 7 and 9MPa for both cosolvents. That optimum could be related to the partitioning of the NaBz between the $scCO_2$ and PP0 matrix. The solubility of the NaBz in the $scCO_2$ increased rapidly with increasing $scCO_2$ pressure. When the latter was very high, it might be that the NaBz was partitioned more in the $scCO_2$ than in the PP0 matrix. Moreover, during the rapid CO_2 depressurization process when the initial CO_2 was high, some of the NaBz might leach out from the PP0 matrix further contributing to a decrease in the mass uptake of the NaBz.

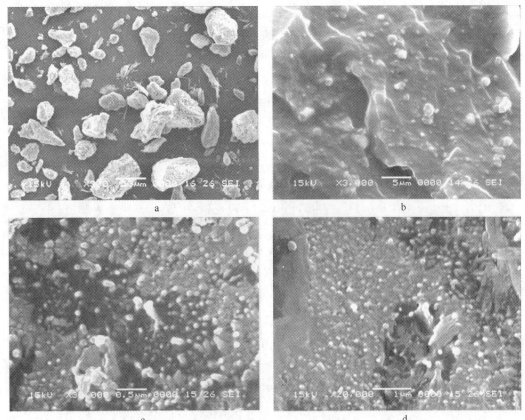

Fig. 2 SEM images showing the morphology of the NaBz particles alone and the state of dispersion of the latter in PP3, PP4, and PP5

(The penetration process was carried out under the following conditions: temperature = 60℃, CO_2 pressure = 8.3MPa, and time = 4h. When a cosolvent was used, its amount was 0.05wt% of the PP)

a—NaBz alone; b—PP3; c—PP4; d—PP5

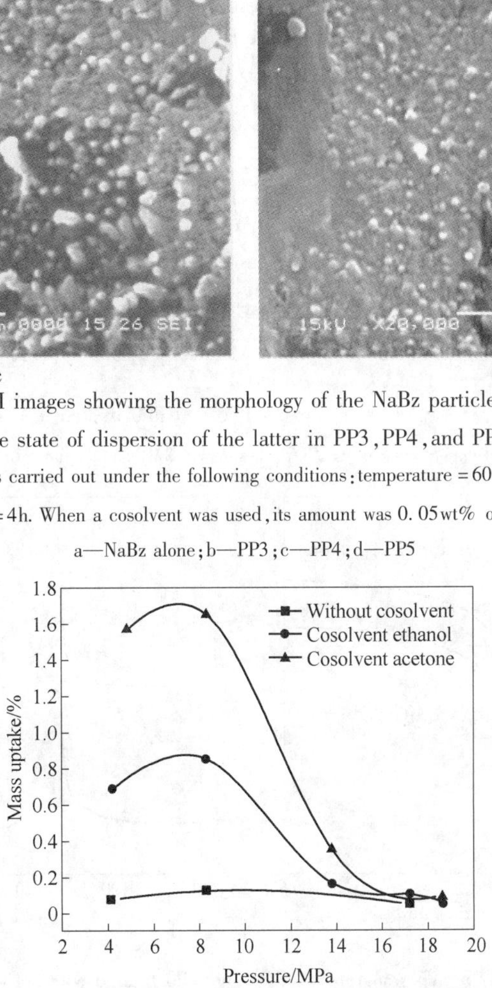

Fig. 3 Effect of the scCO$_2$ pressure on the mass uptake of the NaBz in the PP0

(The penetration process was carried out under the following conditions: temperature = 60℃, CO_2 pressure = 8.3MPa, and time = 4h)

3.2 Morphology of the PP crystalline phases

Before we will discuss on the effect of the NaBz on the crystallization behavior and crystalline structures of the PP0, it is important to check whether or not scCO$_2$ had an effect on them. Fig. 4 compares the X-ray diffractograms between the PP0 and the PP0 treated with scCO$_2$ without NaBz and under the following conditions: temperature = 60℃, CO$_2$ pressure = 8.3 MPa, and time = 4h. The two diffractograms were virtually the same, indicating that the crystalline structures of the PP0 did not change after the treatment with scCO$_2$ under the above conditions. This is further confirmed by their isothermal and nonisothermal DSC traces in Fig. 5. Again, both the PP0 and the PP0 treated with scCO$_2$ under the above conditions had the same isothermal and nonisothermal crystallization curves, and thus the same crystallization behavior. In short, under the above conditions, scCO$_2$ alone did not impart any noticeable effect on the crystalline structures of the PP0.

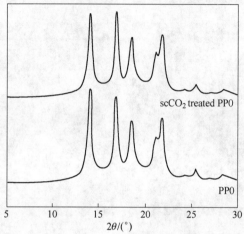

Fig. 4 X-ray diffractograms of PP0 and PP0 treated with scCO$_2$

(Temperature = 60℃, CO$_2$ pressure = 8.3 MPa, and time = 4h)

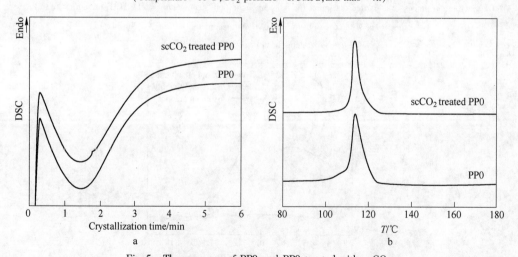

Fig. 5 Thermograms of PP0 and PP0 treated with scCO$_2$

(Temperature = 60℃, CO$_2$ pressure = 8.3 MPa, and time = 4h)

a—Isothermal (crystallization temperature, 120℃); b—Nonisothermal crystallization (cooling rate, 10℃/min)

Fig. 6 compares, in a qualitative manner, the crystalline morphologies of PP0, PP1, PP2, PP3, and PP4. As expected, the size of the PP crystallites followed the order: PP0 > PP1 > PP2 > PP3 > PP4. Those results were in agreement with both the state of dispersion and the amount of the NaBz in the PP0 described above (see Table 2 and Fig. 2). It is interesting to recall that the

PP1 and PP3 contained both the same amount of the NaBz, i. e. ,0.12wt%. The only difference was that in the former case the NaBz was dispersed in the PP0 by a conventional melt compounding process whereas in the latter case, it was done under $scCO_2$. Thus the fact that the size of crystallites of the PP3 was smaller than that of the PP1 by more than an order of magnitude was a strong piece of evidence supporting the advantage of using $scCO_2$ for finely dispersing the NaBz in the PP0. This is further confirmed by comparing PP2 and PP4. They also contained the same but larger amount of NaBz, i. e. ,0.85wt%.

Fig. 6 Optical microscopic photographs of the PP samples crystallized at 140℃ for an hour after having been kept at 200℃ for 10min and then cooled down to 140℃ with a cooling rate of 100℃/min
a—PP0; b—PP1; c—PP2; d—PP3; e—PP4

3.3 Isothermal crystallization kinetics

Fig. 7 and Fig. 8 show the variation of the heat flow and that of the relative crystallinity as a function of the crystallization time, respectively. The results show that the presence of the NaBz greatly accelerated the crystallization process of the PP0, confirming that the NaBz did act as a nucleating agent for the PP0. The crystallization rate followed the order: PP0 < PP1 < PP2 < PP3 < PP4, as expected.

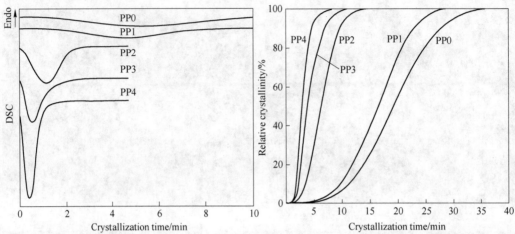

Fig. 7　Heat flow versus time for the isothermal crystallization at 130℃

Fig. 8　Relative crystallinity versus time during the isothermal crystallization at 130℃

The isothermal crystallization half-time, $t_{1/2}$, characterizes the time at which 50% of the total crystallization process has completed. A smaller value of $t_{1/2}$ corresponds to a more rapid crystallization process. Fig. 9 compares the values of $t_{1/2}$ for the four PP samples. Again as expected, the value of $t_{1/2}$ followed the order: PP0 > PP1 > PP2 > PP3 > PP4.

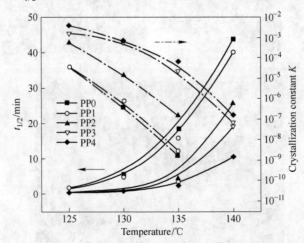

Fig. 9　Isothermal crystallization half-time and crystallization constant of the PP samples as a function of the crystallization temperature

The kinetics of the isothermal crystallization is often described by the modified Avrami equation[3-5,10,23-28]:

$$1 - X_t = \exp(-kt^n) \tag{3}$$

where, X_t is the relative volume-fraction crystallinity at the crystallization time t; k is a crystallization kinetic constant; n is the Avrami exponent whose value depend on the nucleation mechanism and the form of the growing crystals.

The above equation can also be written in the following double logarithm form:

$$\ln[-\ln(1-X_t)] = \ln k + n\ln t \qquad (4)$$

A plot of $\ln[-\ln(1-X_t)]$ versus $\ln t$ yields both the value of the Avrami exponent n (slope of the straight line) and that of the crystallization kinetic constant k (intersection with the coordinate). In principle, the Avrami parameter n is 3 for a spherical three-dimensional crystallization process with a heterogeneous nucleation mechanism. Its exact value depends on combined nucleation modes, secondary crystallization, intermediate dimensionality of crystal growth, etc. Table 3 gathers the values of the crystallization kinetic parameters (n, k, and $t_{1/2}$) for PP0, PP1, PP2, PP3, and PP4 in Table 2. The values of n followed the order: PP0 ≈ PP1 ≈ PP2 > PP3 ≈ PP4. The fact that the values of n of the PP3 and PP4 were smaller than those of PP0, PP1, and PP2 could be explained as follows. When the NaBz was dispersed at a nanometer scale, the growth of spherulites could be more constrained because of a much larger number of nucleating sites and consequently a much smaller spacing between the growing spherulites. Thus the values of n were smaller. Fig. 9 shows that at a given crystallization temperature the value of the crystallization rate constant k followed the logic order: PP0 < PP1 < PP2 < PP3 < PP4.

Table 3 Isothermal crystallization kinetic parameters of PP0, PP1, PP2, PP3, and PP4

Sample	Temperature/℃	n	K	$t_{1/2}$/min
PP0	120	2.5	1.2×10^{-4}	0.55
	125	2.2	3.7×10^{-5}	1.67
	130	2.5	4.1×10^{-7}	5.34
	135	2.9	1.6×10^{-9}	18.38
PP1	120	2.5	2.4×10^{-4}	0.42
	125	2.3	3.8×10^{-5}	1.34
	130	2.5	7.6×10^{-7}	4.57
	135	2.8	2.7×10^{-9}	15.73
PP2	120	2.4	1.8×10^{-3}	0.224
	125	2.2	5.8×10^{-4}	0.451
	130	2.5	1.6×10^{-5}	1.20
	135	2.7	1.6×10^{-7}	4.19
PP3	125	2.3	1.6×10^{-3}	0.26
	130	1.8	7.6×10^{-4}	0.66
	135	2.0	2.2×10^{-5}	3.29
	140	2.4	6.6×10^{-8}	15.92
PP4	125	2.0	4.1×10^{-3}	0.24
	130	2.1	7.1×10^{-4}	0.45
	135	1.9	6.6×10^{-5}	2.24
	140	2.5	1.5×10^{-7}	10.42

3.4 Nonisothermal crystallization kinetics

The results of the nonisothermal crystallization kinetics of PP0, PP1, PP2, PP3, and PP4 were in line with those of their isothermal crystallization kinetics. For example, Fig. 10 and Fig. 11 show the variation of the heat flow and relative crystallinity as a function of temperature at a cooling rate of 5℃/min, respectively. As expected, the temperature at which crystallization started to proceed was the highest for the PP4 and the lowest for the PP0. The crystallization temperature, T_c, defined as a temperature at which the crystallization rate was the maximum, followed the order: PP0 < PP1 < PP2 < PP3 < PP4.

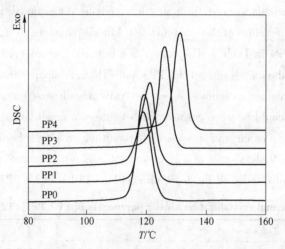

Fig. 10 Heat flow versus temperature during the nonisothermal crystallization of the PP samples at a cooling rate of 5℃/min

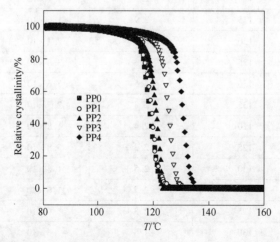

Fig. 11 Relative crystallinity of the PP samples at a cooling rate of 5℃/min

Fig. 12 shows the effect of the cooling rate on the recrystallization temperature and crystallinity, respectively, for PP0, PP1, PP2, PP3, and PP4. Over the entire cooling rate range, both the recrystallization temperature and crystallinity followed the order: PP0 < PP1 < PP2 < PP3 < PP4. Moreover, they both decreased with increasing cooling rate. This is because a higher cooling rate corresponded to a shorter time available for polymer chains to crystallize.

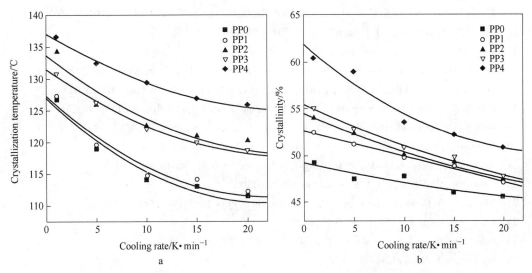

Fig. 12 Nonisothermal crystallization behavior of the PP samples as a function of the cooling rate
a—Crystallization temperature; b—Crystallinity

4 Conclusions

This paper has reported on the feasibility and potential advantages of using $scCO_2$ as a swelling agent for assisting in the penetration and dispersion of NaBz, a nucleating agent, in PP. Under $scCO_2$, the NaBz successfully penetrated and was then finely dispersed in the PP, especially when a cosolvent like acetone or ethanol was used. In the latter case, the dispersion of the NaBz in the PP reached a nanometer scale, which was not possible otherwise by a conventional melt compounding process. Moreover, the mass uptake of the NaBz in the PP was much higher. Isothermal and nonisothermal crystallization kinetics showed that the NaBz acted as a very good nucleating agent. Its nucleating performance was much better when it was dispersed in the PP at a nanometer scale. The crystallization rate and crystallization temperature were much higher and the size of crystallites much smaller.

References

[1] Xu, T.; Lei, H.; Xie, C. S. Mater Des 2003, 24, 227.

[2] Feng, Y.; Jin, X.; Hay, J. N. J Appl Polym Sci 1998, 69, 2089.

[3] Phillips, R.; Manson, J. E. J Polym Sci Part B: Polym Phys 1997, 35, 875.

[4] Kim, C. Y.; Kim, Y. C.; Kim, S. C. Polym Eng Sci 1993, 33, 1445.

[5] Oestergaard, E. Acta Odontol Scand 1994, 52, 335.

[6] Jang, G. S.; Cho, W. J.; Ha, C. S. J Polym Sci Part B: Polym Phys 2001, 39, 1001.

[7] Zhu, P. W.; Edward, G. Macromolecules 2004, 37, 2658.

[8] Zhang, J. L. J Appl Polym Sci 2004, 93, 590.

[9] Nagarajan, K.; Levon, K.; Myerson, A. S. J Therm Anal Calorim 2000, 59, 497.

[10] Wunderlich, B. Macromolecular Physics, Vol. 2; Academic Press: New York, 1976.

[11] Tomasko, D. L.; Li, H. B.; Liu, D. H.; Han, X. M.; Wingert, M. J.; Lee, L. J.; Koelling, K. W. Ind Eng Chem Res 2003, 42, 6431.

[12] Alsoy, S.; Duda, J. L. Chem Eng Tech 1999, 22, 971.

[13] Woods, H. M. ; Silva, M. M. ; Nouvel, C. ; Shakesheff, K. M. ; Howdle, S. M. J Mater Chem 2004, 14, 1663.

[14] Garcia, L. M. ; Lesser, A. J. J Appl Polym Sci 2004, 93, 1501.

[15] Beckman, E. J. Ind Eng Chem Res 2003, 42, 1598.

[16] Liu, T. ; Hu, G. H. ; Tong, G. Sh. ; Zhao, L. ; Cao, G. P. ; Yuan, W. K. Ind Eng Chem Res 2005, 44, 4292.

[17] Li, B. ; Cao, G. P. ; Liu, T. ; Zhao, L. ; Yuan, W. K. ; Hu, G. H. Chin J Chem Eng 2005, 13, 673.

[18] Berens, A. R. ; Huvard, G. S. ; Korsmeyer, R. W. ; Kunig, F. W. J Appl Polym Sci 1992, 46, 231.

[19] Dorazio, L. ; Mancarella, C. ; Martuscelli, E. ; Sticotti, G. J Mater Sci 1991, 26, 4033.

[20] Dogopolsky, I. ; Silberman, A. ; Kenig, S. Polym Adv Technol 1995, 6, 653.

[21] Avalos, F. ; Lopez-Manchado, M. A. ; Arroyo, M. Polymer 1996, 37, 5681.

[22] Yokoyama, Y. ; Ricco, T. J Appl Polym Sci 1997, 66, 1007.

[23] Miteva, T. ; Minkova, L. ; Magagnini, P. Macromol Chem Phys 1998, 199, 1519.

[24] Tjong, S. C. ; Xu, S. A. Polym Int 1997, 44, 95.

[25] Iroh, J. O. ; Berry, J. P. Polymer 1993, 34, 4747.

[26] Binsbergen, F. L. ; de Lange, B. G. M. Polymer 1968, 9, 23.

[27] Progelhof, R. C. ; Throne, J. L. Polymer Engineering Principles; Academic Press: New York, 1993; pp 95-150.

[28] Seo, Y. S. ; Kim, J. H. ; Kim, K. U. Polymer 2000, 41, 2639.

Controlling Crystal Phase Transition from Form II to I in Isotactic Poly-1-butene Using CO_2 *

Abstract Controlling crystal phase transition from modification II to I in isotactic Poly-1-butene (iPB-1) using CO_2 is presented in this article. The intrinsic kinetics of CO_2-induced phase transition from modification II to I in iPB-1 at 40℃ and different CO_2 pressures were detected using *in situ* high-pressure Fourier transform infrared spectroscopy (FTIR) and correlated by Avrami equation. Sorption of CO_2 in iPB-1 matrix was measured at 40℃ and different CO_2 pressures using both FTIR and magnetic suspension balance (MSB) and the diffusivity was determined by Fick's second law. An algorithm combining the CO_2 diffusion and induced phase transition was subsequently proposed to calculate the CO_2 concentration as well as the phase transition degree in the iPB-1 matrix with different thickness at different saturation time. The calculated phase transition degree in the iPB-1 agreed well with the FTIR results. In addition, the yield stresses of iPB-1 specimens annealed in the air and 6MPa CO_2 at 40℃ with different durations were also experimentally investigated. The algorithm was applied to predict the phase transition degree of the iPB-1 specimens. The appropriate CO_2 treatment time for getting high yield stress was in consistent with that predicted by the algorithm.

Key words iPB-1, crystal phase transition, mathematical modeling

1 Introduction

Since isotactic poly-1-butene (iPB-1) was synthesized by Natta in 1954[1], it has been extensively investigated and applied due to its polymorphism and many outstanding properties[2]. iPB-1 may exist in four different crystal structures, designated as modifications I, II, III and I′[3-5]. Crystallized from melt under atmospheric pressure, the kinetically favored tetragonal crystalline modification II can be obtained[3,6]. However, modification II is unstable under ambient condition, and will transform into the trigonal (twinned hexagonal) crystalline modification I, a stable crystal form[3,7,8]. The transition of modification II into I substantially enhances the mechanical, thermal and physical properties so that iPB-1 with modification I is most widely used. However, completion of the transition under ambient condition requires several days or weeks[7]. De Rosa et al.[9,10] made a breakthrough: modifications I and I′ crystallized from the melt of the metallocene catalysts prepared iPB-1 containing different concentration of stereodefects (*rr* triads defects). Finding solutions to directly obtain form I from the melt still stands in the way of the commercial development of the Ziegler-Natta catalysts prepared iPB-1. The phase transition of modification II into I has been widely researched by density measurements[11,12], differential scanning calorimetry (DSC)[13], infrared spectroscopy (IR)[6,7,14-16], transmission electron microscopy (TEM)[17], X-ray diffractometry (XRD)[16,18], electron diffraction[19], small-angle X-ray scattering (SAXS)[20], microindentation hardness[21] and atomic

---

* Coauthors: Xu Yang, Liu Tao, Li Lei, Li Dachao, Zhao Ling. Reprinted from *Polymer*, 2012, 53: 6102-6111.

force microscopy (AFM)[22]. The rate of crystal transformation was affected by various chemical, physical and mechanical factors: (1) temperature[11]; (2) mechanical deformation, hydrostatic pressure, orientation and local thermal stresses[12,22,23]; (3) X-ray radiation and ultrasonic treatment[7]; (4) molecular weight and tacticity[15]; (5) additives, blends and nucleating agents[12]; (6) copolymerization[24,25]. It was found a number of methods such as application of uniaxial, shear orientation and hydrostatic pressure and addition of additives or other polymers[26] could accelerate the phase transition.

Supercritical carbon dioxide (scCO$_2$) has been increasingly considered and used as a promising alternative to organic and other toxic or harmful solvents for applications in polymer processing, such as grafting, foaming, and impregnation of additives[27,28]. On the one hand, the sorption of CO_2 in polymers can plasticize the materials and thus decrease the glass transition temperature and melting temperature[29,30]. On the other hand, the crystallization[31-35], crystal form transformation[35-38] or conformational transition[39] may be induced by CO_2 treatment. In both cases, the structure and morphology of polymers have been changed significantly. Li et al.[40] found the deformation of iPB-1 matrix, during the CO_2-assisted foaming process, makes the iPB-1 melt crystallize into modification I rather than modification I′, which crystallizes after annealing under high-pressure CO_2 without foaming. Through high-pressure DSC, wide-angle X-ray diffraction (WAXD) and *in situ* high-pressure Fourier transform infrared spectroscopy (FTIR) measurements, they also found that modification II melt-crystallized under 0.5-8MPa CO_2 and modification I′ melt-crystallized directly at CO_2 pressures higher than 10MPa[35]. Application of pressurized CO_2 was capable of promoting the phase transition of modification II to I in iPB-1, which was much more effective in comparison with hydrostatic pressure[37]. However, the rates of the CO_2-induced phase transition have not been quantitatively given. In fact, the CO_2-induced phase transition from modification II to I was affected by CO_2 diffusion so that the phase transition rate should depend on the combined effects of the intrinsic kinetics of the CO_2-induced phase transition and CO_2 diffusion[41].

In this work, the intrinsic kinetics of CO_2-induced phase transition from modification II to I in iPB-1 were detected at 40°C and different CO_2 pressures using *in situ* high-pressure FTIR. Both FTIR and magnetic suspension balance (MSB) were used to determine the solubility and diffusivity of CO_2 in iPB-1 matrix. An algorithm combing the CO_2 diffusion and induced phase transition was subsequently proposed to predict the phase transition degree in the iPB-1 matrix with different thickness at different saturation time. In addition, the yield stresses of iPB-1 specimens annealed in the air and pressurized CO_2 with different durations were also investigated. The appropriate CO_2 treatment time for getting high yield stress was well predicted by the algorithm.

2 Experimental

2.1 Materials and sample preparations

iPB-1 pellets (PB 0110 M) were kindly provided by Basell Polyolefins. The weight-average molecular weight (M_w) is 4.39×10^5 g/mol, and the melt index is 0.4g/10min (190℃/2.16kg). Before used, they were dried in a vacuum oven at 40℃ for 2 days. CO_2 (purity: 99.9% w/w) was purchased from Air Products Co., Shanghai, China.

The iPB-1 films for FTIR characterization were prepared from pellets using a hot press at 160℃ and 10MPa between two terylene slides for 10min, rapidly quenched by plunging into liquid nitrogen or ice water and removed from the slide using a spatula in the liquid nitrogen or ice water to minimize the deformation. The films with thickness ranging from 20 to 30μm or 0.2-1.2mm were prepared. The exact thickness was measured by micrometer caliper. As shown in Fig. 1, the melting behaviors of the prepared iPB-1 films were immediately characterized by DSC (NETZSCH DSC 204 HP, Germany). Table 1 collects the melting temperatures and fusion enthalpies of the iPB-1 films. The indistinct shoulder on the high-temperature side of melting peak in each DSC curve and the crystallinities of modifications Ⅱ and Ⅰ shown in Table 1 indicated that tiny modification Ⅰ existed in the films. The modification Ⅰ formed supposedly in two ways: Direct crystallization from the melt at the existence of nucleating agent[42]; Solid-solid transformation from modification Ⅱ accelerated by contraction induced local strains during fast cooling process[17] and some additives[13], such as heat stabilizer and nucleating agent.

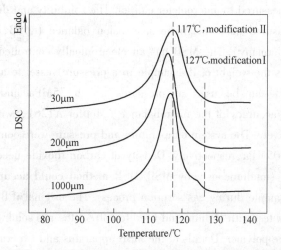

Fig. 1 DSC curves of the prepared iPB-1 films with different thicknesses

Table 1 DSC results for iPB-1 films as soon as prepared with different thicknesses

Thickness/μm	$T_{mⅡ}$/℃	$\Delta H_Ⅱ$/J·g^{-1}	$\Delta H_Ⅰ$/J·g^{-1}	$\alpha_Ⅱ$/%	$\alpha_Ⅰ$/%	α_{DSC}/%	$X_Ⅰ$/%
30	117.3	34.26	0.85	55.3	0.6	55.9	1.1
200	115.8	33.91	1.07	54.7	0.8	55.5	1.4
1000	116.6	33.89	2.20	54.6	1.6	56.2	2.8

$T_{mⅡ}$ stands for the melting temperatures of modification Ⅱ, $\Delta H_Ⅱ$ and $\Delta H_Ⅰ$ for the melting enthalpies of modifications Ⅱ and Ⅰ, $\alpha_Ⅱ$ and $\alpha_Ⅰ$ for the crystallinities of modifications Ⅱ and Ⅰ, α_{DSC} for the total crystallinity obtained by the DSC measurement and $X_Ⅰ$ for the mass ratio of modification Ⅰ to the total crystals. The heat fusion values of the two ideal infinite crystals are $\Delta H°_{mⅡ} = 62$J/g and $\Delta H°_{mⅠ} = 141$J/g for the crystal modifications Ⅱ and Ⅰ, respectively[21].

2.2 *In situ* high-pressure FTIR

In situ FTIR was widely used to the diffusion of solvent in polymers[43,44] and the polymer crystals[35,39,45]. In this work, CO_2 solubility and diffusivity in iPB-1 and the relative contents of modifications II and I were measured by using *in situ* high-pressure FTIR of type Bruker Equinox-55 equipped with a Harrick high-pressure demountable cell. Its detail was described elsewhere[37]. Before the FTIR measurement, the temperature of the high-pressure cell and the buffer vessel containing high-pressure CO_2 was controlled at 40℃. The iPB-1 film as soon as prepared with thickness of 20-30μm or 0.2-1.0mm was transferred in the cell. FTIR spectra collection was started up. Then, CO_2 in the buffer vessel filled into the cell through a valve and the desired pressure reached in less than 2s. FTIR spectra were recorded at a resolution of $4.0 cm^{-1}$ and a rate of one spectrum per 32s. The IR intensities refer to the peak height. The scanned wavenumber was in the range of $4000-400 cm^{-1}$.

2.3 Magnetic suspension balance

iPB-1 sheets with about 1.0mm thickness was cut into a rectangle of 23.5mm × 12.5mm. The exact thickness was measured by micrometer caliper. The solubility and diffusivity of CO_2 in the iPB-1 films were measured using magnetic suspension balance (MSB, Rubotherm Prazisions messtechnik GmbH, Germany). The MSB has an electronically controlled magnetic suspension coupling that transmits the weight of the sample in a pressure vessel to a microbalance outside of the cell. The MSB can be used at pressures up to 35MPa and temperatures up to 523K. Resolution and accuracy of the microbalance (Mettler AT261, Switzerland) are 0.01mg and 0.002%, respectively. The system temperature and pressure were controlled at the accuracy of ±0.2℃ and ±0.05MPa, respectively. Density of carbon dioxide needed for buoyancy correction was measured simultaneously by MSB. MSB method could accurately detect the mass variation of polymer sample during gas sorption process. The original iPB-1 sheet's volume was determined by a blank test with Helium and used to correct the gas solubility by considering gas buoyancy acting on the polymer. Details of the MSB apparatus and experimental procedure used in this work have been described in previous publications[46,47].

2.4 Differential scanning calorimeter (DSC)

DSC (NETZSCH DSC 204 HP, Germany) was used to confirm the modification of the iPB-1 crystal and to measure the crystallinity of the iPB-1 sample before and after the phase transition. For each DSC measurement, about 10mg of the iPB-1 was heated from 30 to 170℃ at a heating rate of 10℃/min.

2.5 CO_2 treatment and tensile testing

Treatments of the iPB-1 as soon as prepared with thickness of 1.2mm using 6MPa CO_2 were performed in a high-pressure vessel placed in a homemade water bath with a temperature controller and its accuracy was ±0.5℃. The temperature of the water bath was controlled at 40℃. After a preset period of time, the CO_2 in the vessel was slowly released. The iPB-1 sheets were immediately die cut to a tensile specimen, which conformed in dimensions to ASTM D638 Type V. The iPB-1 specimens with form Ⅱ were also annealed at ambient condition for comparison. Instron 3327 Series tensile testing machine with a video extensometer was used for the tensile test at the crosshead speed of 10mm/min and temperature of 23℃.

3 Results and Discussion

3.1 Intrinsic kinetics of CO_2-induced phase transition from modification Ⅱ to Ⅰ in iPB-1

The infrared spectra and characteristic absorption bands of iPB-1 polymorphs were reported by Luongo and Salovey[6,7] early in 1965. The band at 925cm^{-1} (CH_2 and CH_3 rocking) is known to be characteristic of form Ⅰ while 905cm^{-1} (CH_2 and CH_3 rocking) is characteristic of modification Ⅱ[15,37]. Moreover, the bands at 905cm^{-1} and 925cm^{-1} do not appear in the spectrum of pure CO_2. Therefore, the infrared spectra can be used to characterize the crystal phase transition of iPB-1 when the iPB-1 with form Ⅱ is immersed into CO_2 in the high-pressure IR cell.

Fig. 2 shows the spectra of iPB-1 films with modification Ⅱ and thickness of 20-30 μm immersed in CO_2 at pressures ranging from 1 to 6MPa and 40℃ for different CO_2 treatment times. Compared with the phase transition time, the CO_2 diffusion time (or saturation time) in such thin film could be negligible. Therefore, it was assumed that the film was saturated as soon as the high-pressure CO_2 was applied, and the obtained kinetics of CO_2-induced phase transition from modification Ⅱ to Ⅰ in iPB-1 should be intrinsic. As the CO_2 treatment time was prolonged, the absorbance of the 925cm^{-1} band grew, whereas the absorbance of the 905cm^{-1} band decreased, indicating the phase transition from modification Ⅱ to Ⅰ in iPB-1. With increasing the CO_2 pressure, the phase transition was obviously accelerated. At CO_2 pressure of 1MPa, the peak at 905cm^{-1} still existed obviously even after CO_2 treatment for 350min. Whereas at CO_2 pressure of 6MPa, the peak at 905cm^{-1} disappeared only after CO_2 treatment for 50min. The higher CO_2 pressure, the faster the band at 905cm^{-1} disappeared.

A weak and broad band at 905cm^{-1} in the melt iPB-1 spectrum was also observed and considered the characteristic of amorphous region as reported by Luongo and Salovey[7]. The absorbance of the amorphous regions in the iPB-1 film was subtracted first from the spectrum before the Gaussian function and the least squares method were used to separate the overlapped peaks at 905cm^{-1} and 925cm^{-1}. The peak areas, A_{905} and A_{925}, were then obtained. The absorptivity ratio of the 925cm^{-1} and 905cm^{-1} bands, α, was calculated through the linear regression analysis[48]. Fig. 3 shows the regression results with average $R^2 = 0.99$ at CO_2 pressures ranging from 0.01 to 6.01MPa. α varies between 1.20 and 1.35, which was supposedly attributed to the slight variation of crystallinity in the iPB-1. The relative content of modification Ⅰ, at each CO_2 pressure, was derived from Eq. 1 using the corresponding absorptivity ratio, α. Because the total crystallinity of modifications Ⅰ and Ⅱ remained constant during the transformation, which had been confirmed by both Li[37] and Azzurri[21] through DSC and WAXD study, it was reasonable to use the relative content, X_t, to stand for the absolute quantity of modification Ⅰ in the iPB-1 film at CO_2 treatment time of t. In addition, the relative content wasn't influenced by the film

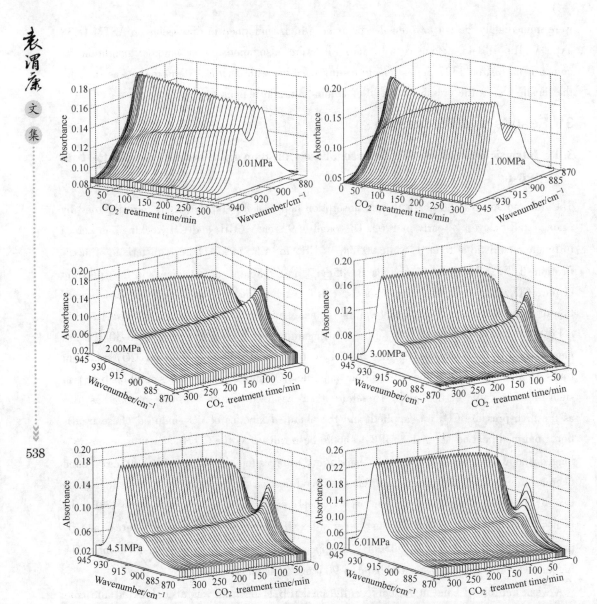

Fig. 2 IR absorption peaks at 905cm^{-1} and 925cm^{-1} of iPB-1 at 40℃ and different CO_2 pressures for various times

thickness so that the internal standard peak[7,14,15], indicating the variation of film thickness, was not necessary to be considered in the experiment:

$$X_t = \frac{A_{925}}{A_{925} + \alpha A_{905}} \quad (1)$$

Fig. 4 shows the relative content of modification I, X_t, as a function of CO_2 treatment time at various CO_2 pressures. With increasing CO_2 pressure, X_t increases at the initial stage and then tends to level off. The relative phase transition degree at 316min and half transition time at each CO_2 pressure were listed in Table 2. The results indicated that the phase transition from modification II to I in the iPB-1 was significantly accelerated by the dissolved CO_2 in the polymer.

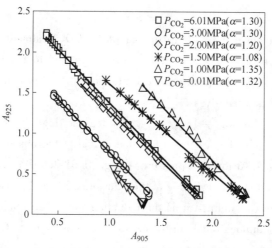

Fig. 3 Plots of the integral absorbance of 925cm^{-1} band vs that of 905cm^{-1} band for iPB-1 samples with different transformation degrees

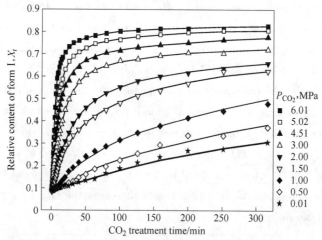

Fig. 4 The relative content of form I, X_t, in the iPB-1 films as a function of CO_2 treatment time at 40℃ and different CO_2 pressures

Table 2 The relative transition degree at 316min and half transition time at each CO_2 pressure

CO_2 pressure/MPa	$X_t=316$	$t_{1/2}$/min
6.01	0.826	5.9
5.02	0.806	10.0
4.51	0.774	15.0
3.00	0.723	25.3
2.00	0.659	63.5
1.50	0.624	113.0
1.00	0.499	316.5
0.50	0.384	631.0①
0.00	0.308	1047.0①

① Extrapolated from the Avrami plot.

The Avrami equation, deduced from analyzing the growth of random distributed crystal nuclei[49,50], was employed here for a deep analysis of the transformation process. It had been well used to analyze the phase transition kinetics of form II to I under ambient condition[12,18]. It can be expressed as follows:

$$1 - X_t = \exp(-kt^n) \tag{2}$$

where X_t is relative content of form I in the sample at CO_2 treatment time of t, k is the rate constant, and n is the avrami exponent whose value is related to the form of nucleation and growth of the new phase[51]. To facilitate the evaluation of the Avrami parameters, taking the logarithm of the equation twice gives:

$$\log[-\ln(1 - X_t)] = \log k + n\log t \tag{3}$$

Fig. 5 shows the $\log[-\ln(1 - X_t)]$ as a function of $\log t$ of the phase transition process illustrated in Fig. 2 and Fig. 4. At low CO_2 pressures (e.g., 0.01-2.0MPa), a distinct period with a small slope was observed at the beginning of the transformation and substantially shortened with increasing CO_2 pressure. We defined it as the induction stage. The whole transformation should consist of three stages although the third one was not observed at CO_2 pressures of 0.01-1.0MPa due to the shortage of the transformation time. Chau et al.[18] studied the phase transformation of modification II to I in iPB-1 at atmospheric condition using XRD. They observed the posterior two transformation stages: (1) the instantaneous nucleation of form I, during which growth of the stable phase is the rate-determining step; (2) the exhaustion of active nuclei. Jiang's research[22] further pointed out that local thermal stresses played a very important role in generating the nuclei of the iPB-1 form I crystals. At high CO_2 pressures (e.g., 3-6MPa), the first stage of the transformation disappeared and the whole transformation consisted of only two stages. Since irregular stacking of lamellae, local variations of lamellar thicknesses, and bending of lamellae were common features in semicrystalline polymers, it was speculated that the lamellar distortion and elongation, both of which were proved to be accelerators of the form transformation[17,20,52,53], were enhanced by the CO_2-induced swelling of the heterogeneously distributed amorphous regions. As the CO_2 pressure increases, the larger the polymer swells and the faster the crystal transforms.

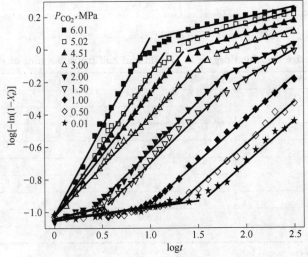

Fig. 5 Avrami plots for the phase transformation from modification II to I in iPB-1 films at 40℃ and different CO_2 pressures

From the apparent Avrami analysis, three stages, namely induction, nucleation controlled and growth controlled periods, were depicted during the phase transformation from form II to I in iPB-1. As shown in Table 3, the obtained Avrami parameters n and k were strongly affected by the CO_2 pressure, i.e., the local CO_2 concentration in the iPB-1, and transition time. The corresponding R^2 was quite close to 1, indicating that the intrinsic kinetics of the CO_2-induced transformation in iPB-1 could be described by Avrami equation.

Table 3 Avrami parameters of each phase transition stage at different CO_2 pressures

P_{CO_2}/MPa		6.01	5.02	4.51	3.00	2.00	1.50	1.00	0.50	0.01
Stage 1	n	—	—	—	—	0.151	0.148	0.074	0.090	0.079
	k	—	—	—	—	0.091	0.090	0.090	0.088	0.089
	R^2	—	—	—	—	0.892	0.937	0.938	0.925	0.944
Stage 2	n	1.093	0.872	0.745	0.611	0.599	0.576	0.526	0.517	0.519
	k	0.095	0.093	0.095	0.094	0.060	0.053	0.033	0.024	0.018
	R^2	0.997	0.998	0.998	0.999	0.998	0.998	0.999	0.999	0.998
Stage 3	n	0.132	0.158	0.173	0.172	0.282	0.351	—	—	—
	k	0.855	0.689	0.558	0.486	0.217	0.132	—	—	—
	R^2	0.945	0.959	0.977	0.981	0.989	0.995	—	—	—

Note: R^2 is the square of the Pearson product moment correlation coefficient.

3.2 Solubility and diffusivity of CO_2 in the iPB-1

The FTIR and MSB methods were both adopted to investigate the solubility and diffusivity of CO_2 in the iPB-1 sheet at 40℃ and CO_2 pressures ranging from 1 to 6MPa for mutual corroboration. Fig. 6 shows that CO_2 has three absorbance bands in the near-IR at 4837cm^{-1}, 4966cm^{-1}, and 5088cm^{-1} which don't overlap with the iPB-1 spectrum. Similar to Brantley's work[44], the absorbance band area at 4966cm^{-1} was utilized to determine the concentration of CO_2 in the iPB-1 sheet through the Beer-Lambert law:

$$A = \varepsilon c l \tag{4}$$

where, A was the absorbance area; ε was the molar absorptivity; c was the CO_2 concentration and l was the path length. The integrated molar absorptivity can be considered independent of the CO_2 density in the range of temperatures and pressures studied[54].

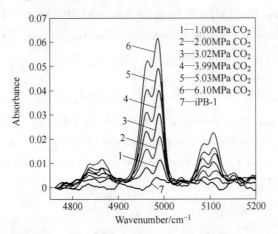

Fig.6 Absorbances of pure CO_2 at 40℃ and pressures ranging from 1.00 to 6.10MPa, as well as that of pure iPB-1, in the near-IR region

The molar absorptivity was calibrated by relating the absorbance band area at 4966cm^{-1} of pure CO_2 through a path length of 0.995mm at 40℃ and pressures ranging from 1.00 to 6.10MPa to the CO_2 concentration. The latter, C_{CO_2}, was calculated by the BWR equation of state[55]. The plot of A_{4966} versus C_{CO_2} is shown in Fig. 7 with a slope of 0.882 ($R^2 = 0.972$). The molar absorptivity of the absorbance band was then obtained to be $\varepsilon = 8.864$L/(mol·cm) lying between the values measured by Brantley[44] and Buback[56], 6.4L/(mol·cm) and 10L/(mol·cm), respectively.

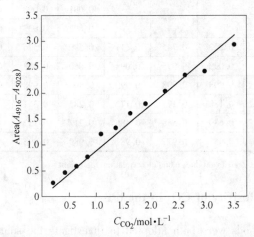

Fig. 7 The IR absorbances of pure CO_2 with different densities at 40℃
(The solid line across the origin is the linear fitting of the experimental points)

On the one hand, for the semicrystalline polymers, it is generally accepted that the sorption of gas occurs mainly in the amorphous regions[57] and causes a slight swelling of solid samples[58]. The CO_2-induced phase transition from form II to I takes place in the crystal regions and results in a shrinkage of ca. 2% in one dimension[59] and a volume contraction of 5.7%[11] on the other hand. Accordingly, it is acceptable to ignore the thickness change. Moreover, the solid-solid phase transformation from form II to I doesn't alter the crystal morphology[18,22] and induces only a small change in the fraction of amorphous material[11].

The absorbance of dissolved CO_2 in polymer at different CO_2 treatment time, $A_{t,4966}$, was calculated by subtracting the absorbance of CO_2 in space. The solubility of CO_2 in the amorphous regions of the iPB-1 sheet was then obtained by:

$$S_{CO_2,\text{FTIR}} = \frac{\overline{A}_{4966}}{\varepsilon \times h \times (1 - \alpha_{\text{DSC}})} \quad (5)$$

where, $\overline{A}_{4966}$ was the $A_{t,4966}$ at equilibrium; h was the thickness of the iPB-1 sheet. The solubility of CO_2 in the iPB-1 as a function of CO_2 pressure was obtained and shown in Fig. 8.

It is also reasonable to presume that the diffusion was independent on the phase transition and could be expressed by Fick's second law:

$$\frac{\partial C}{\partial t} = D \frac{\partial^2 C}{\partial x^2} \quad (6)$$

Because the other 2-D of the iPB-1 sheet were more than 8 times larger than the thickness, CO_2 entered effectively through the plane faces and a negligible amount through the edges. If the polymer

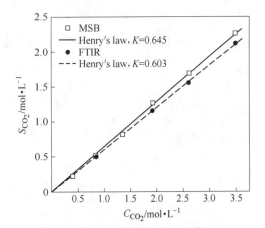

Fig. 8 Sorption isotherms of CO_2 in iPB-1 measured by MSB (□) and FTIR (●) at 40℃ and CO_2 pressures ranging from 1 to 6MPa

(The solid and dashed lines were linear fittings of the MSB and FTIR results via Eq. 10, respectively)

region $-l < x < l$ was initially at a uniform concentration C_0, and the surfaces $x = \pm l$ are kept at a constant concentration C_1, the solution[60] of one-dimensional diffusion equation becomes:

$$\frac{C - C_0}{C_1 - C_0} = 1 - \frac{4}{\pi}\sum_{n=0}^{\infty}\frac{(-1)^n}{2n+1}\exp\left[\frac{-D(2n+1)^2\pi^2 t}{4l^2}\right]\cos\frac{(2n+1)\pi x}{2l} \qquad (7)$$

If M_t denotes the total amount of diffusing gas which has entered the sheet at time t, and M_∞ the corresponding quantity after infinite time, then:

$$\frac{M_t}{M_\infty} = 1 - \sum_{n=0}^{\infty}\frac{8}{(2n+1)^2\pi^2}\exp\left[\frac{-D(2n+1)^2\pi^2 t}{4l^2}\right] \qquad (8)$$

where, M_t and M_∞ are the mass of dissolved gas in the polymer at $t = t$ and $t = \infty$, respectively. The relative mass of absorbed CO_2, M_t/M_∞ in Eq. 8, could be replaced by the relative absorbance band area of dissolved CO_2, $A_{t,4966}/\bar{A}_{4966}$. As shown in Fig. 9, the relative amount of dissolved CO_2 in the iPB-1 obtained by FTIR was then adopted to correlate the diffusion coefficient of CO_2 in the iPB-1 with Eq. 8. The diffusivities as a function of CO_2 pressure were compared with those measured by MSB in Fig. 10.

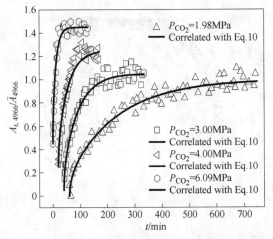

Fig. 9 The relative CO_2 uptake profiles measured by FTIR spectra (points) and the regression results (lines) via Eq. 8 at 40℃ and various CO_2 pressures

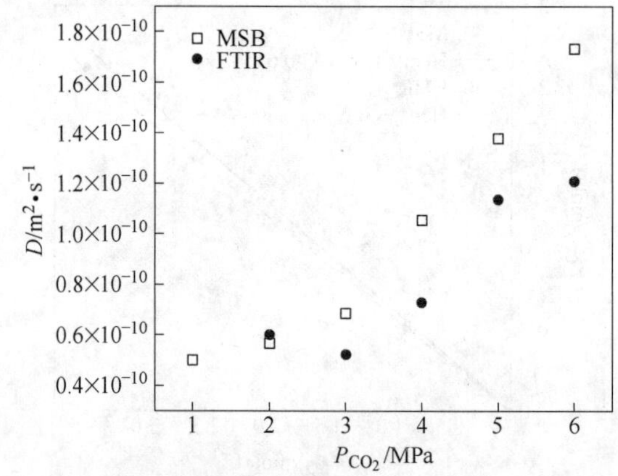

Fig. 10 Diffusivities of CO_2 in iPB-1 sheets at 40℃ and CO_2 pressures ranging from 1 to 6MPa measured by MSB (□) and FTIR (●)

The MSB method was also adopted to investigate the solubility and diffusivity of CO_2 in the iPB-1 sheet at the same condition for comparison. In the measurement, the CO_2 pressure was increased from 1 to 6MPa step by step. In each step, the saturation time was set to 180min that is long enough for the system to reach the sorption equilibrium. The solubility of CO_2 in the amorphous region of iPB-1 was calculated from the equation as follows:

$$S_{CO_2,MSB} = \frac{m_{CO_2}^{\infty}}{m_{PB} \times (1 - \alpha_{DSC}) \times \rho_a} \quad (9)$$

where, $m_{CO_2}^{\infty}$ was the saturated CO_2 mass in polymer; m_{PB} was the real mass of the iPB-1 sample obtained in the reference gas measurement; α_{DSC} was the crystallinity of the iPB-1 sample obtained through DSC; ρ_a was the density of the amorphous region of iPB-1. $\rho_a = 0.868g/cm^3$ was adopted[11]. The solubility of CO_2 in the iPB-1 as a function of CO_2 pressure was obtained and shown in Fig. 8.

As shown in Fig. 11, the normalized mass uptake profiles of each section obtained by MSB was adopted to correlate the diffusion coefficient of CO_2 in the iPB-1 with Eq. 8. The diffusivities as a function of CO_2 pressure were compared with those measured by FTIR in Fig. 10.

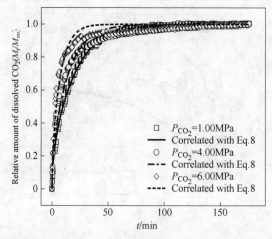

Fig. 11 Sorption profiles for CO_2 in iPB-1 plates as soon as prepared measured by MSB at 40℃ and different CO_2 pressures

The solubilities and diffusivities of CO_2 in iPB-1 at 40℃ and various pressures, both measured by the MSB and FTIR, were collected and compared in Fig. 8 and Fig. 10, respectively. The solubilities of CO_2, S_{CO_2}, were correlated with the concentration of CO_2 in the fluid phase, C_{CO_2}, by Henry's law:

$$S_{CO_2} = K \times C_{CO_2} \qquad (10)$$

Fig. 8 shows the correlation results with Henry coefficients 0.645 ($R^2 = 0.999$) and 0.603 ($R^2 = 0.998$) for MSB and FTIR, respectively. As can be seen from these two figures, the solubility and diffusivity measured by MSB are systematically bigger than those by FTIR, which may result from the step-by-step procedure in MSB measurement. In other words, the already dilated and relaxed polymer matrix in the preceding step was more favorable for CO_2 diffusion.

3.3 Effect of CO_2 diffusion on phase transition from form II to I in thick iPB-1 sheets

The intrinsic kinetics had shown that dissolved CO_2 in the iPB-1 could significantly accelerate the phase transition from modification II to I. However, dissolution of CO_2 into thick iPB-1 products, such as hot water pipes and moldings[2], depended on the solubility and diffusivity of CO_2 in the polymer. Combining the intrinsic kinetics of the phase transition and CO_2 diffusion, an algorithm designed for a plane sheet was proposed and applied to predict the phase transition from form II to I in the iPB-1 sheet with different thickness at different CO_2 pressures. The phase transition of the iPB-1 sheets as soon as prepared with modification II and the thickness of 0.2mm and 0.4mm under CO_2 was experimentally investigated using *in situ* FTIR for verification.

The schematic of the algorithm was given in Fig. 12a. At the beginning, the solubility and diffusivity of CO_2 were obtained through Henry's law and interpolation, respectively. The iPB-1 sheet was then equally divided into $2n + 1$ layers parallel with the surface. The thickness of each layer was so small that the CO_2 concentration distribution was uniform in each individual layer. A few assumptions were further made: (1) The diffusion of CO_2 in the iPB-1 sheet could be expressed by Fick's second law; (2) The CO_2 concentration distribution was uniform in each individual thin layer; (3) The CO_2 concentration and the rate of CO_2-induced crystallization in an individual thin layer were constant in a short time; (4) Crystallinity of iPB-1 sheet had no effect on the diffusion coefficient. In such case, the concentration of CO_2 at each layer could be determined from Eq. 7, where C_1 was the corresponding solubility. The intrinsic kinetic curve of each layer was assumed being only dependent on the local CO_2 concentration. Therefore, the Avrami parameters at any CO_2 concentration could be interpolated from the experiment data in Table 3. After that, the transformation degree $X_{t,y}$ at the next step, was predicted following the procedure shown in Fig. 12b. Averaging calculated $X_{t,y}$ throughout the sheet, the relative transformation degree, X_t, was obtained. It could also be experimentally monitored by the transmission FTIR spectrum.

On the basis of the algorithm, the modeled evolution of the phase transformation degree in thick iPB-1 sheets at different CO_2 pressures were treated with the Avrami method and shown in Fig. 13. The pressure of CO_2 was the predominant factor determining the ultimate content of form I, while the thickness affected mainly the first two stages of the phase transition. The ex-

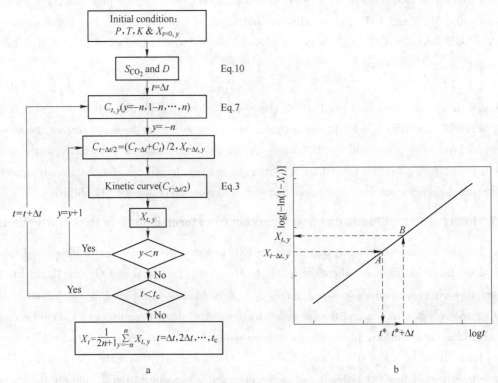

Fig. 12 Calculation flow chart of the algorithm (a) and details of the X_t prediction at each time step (b)

perimental results (dots) were also compared with the predictions (lines) at the CO_2 pressure of 4 MPa and 6 MPa. The experimental values are evenly distributed around the model curve, except at the third stage, where all of the data lies below the model line. The algorithm combining the intrinsic kinetics of the phase transition and CO_2 diffusion could well predict the phase transition degree in iPB-1 with different thickness.

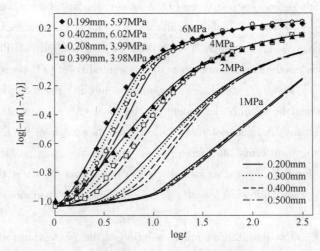

Fig. 13 Comparison between simulated Avrami curves and experimentally measured results (points) for the phase transition from modification Ⅱ to Ⅰ in iPB-1 sheets with different thicknesses at 40 ℃ and different CO_2 pressures

3.4 Tensile property of the iPB-1 with different phase transition degree

The yield stresses and the melting behaviors of iPB-1 specimens with thickness of 1.2mm annealed in the air and 6MPa CO_2 with different durations were investigated. Fig. 14a shows the nominal stress-strain curves of the iPB-1 specimens with different phase transition degree, obtained directly from the load-elongation charts. None of the samples exhibits yield drop, indicating a fully homogeneous deformation (no necking process). During the tensile process, the iPB-1 specimen underwent a viscoelastic deformation, a gradual yielding and a progressive strain hardening up to rupture. Revealed by Fig. 14a, the Young's moduli of the specimens annealed in the air seemed not increase obviously, but the yield stresses were substantially improved due to the phase transformation, which was also observed by T. Asada[14]. To determine the crystal forms in the specimen, DSC analysis was simultaneously performed. Fig. 14b and c show the DSC curves of the iPB-1 specimens annealed in the air and pressurized CO_2, respectively. As the annealing time was prolonged, the melting peaks in Fig. 14b and c moved gradually to the higher temperature side, indicating the modification II - I transition during the annealing processes. However, the peaks width in Fig. 14b remained almost the same while those in Fig. 14c got wider. It implied that the crystal was broken by the dissolved CO_2 and the distribution of lamellar thicknesses was broadened. This speculation, in some degree, explained the decrease of yield stresses of CO_2 annealed iPB-1 samples and was consistent with the molecular mechanism previously proposed that the lamellar distortion and elongation, both of which were proved to be accelerators of the form transformation[17,20,52,53], were enhanced by the CO_2-induced swelling of the heterogeneously distributed amorphous regions.

The transformation curve of the iPB-1 sheet treated with 6MPa CO_2 at 40℃ was also calculated from the algorithm and displayed in Fig. 15. It was known that the increase of modification I actually became very slow in the third stage. Therefore, the suitable treatment time should lie between the cross point of stages two and three and the bifurcate point in stage three, designated as A and B in Fig. 15, respectively. Tensile test of the specimens, annealed in CO_2 at 40℃ and 6MPa for different times, were performed and the results were also displayed in Fig. 14a. The same trend was observed here as revealed by the air-annealed samples that the more modification I, the larger the yield stress. However, the yield stress didn't increase any more when the treatment time was longer than 1h. Namely, 1h was the proper treatment time lying between 0.66h (point A) and 1.26h (point B).

Comparing yield stresses of specimens annealed in the air and pressurized CO_2, it was observed that the yield stresses of CO_2 treated samples with higher content of modification I were even lower than those of air-annealed samples with lower content of modification I. It was proposed by T. Asada[14] and E. Weynant[61,62] that the orientation of modification II and the phase transformation both took place in the viscoelastic range, and that the reversible extension of molecules in the amorphous phase and particularly of tie molecules played a pivotal role to the elastic strain. Therefore, the relaxation effect of CO_2 on the amorphous region of iPB-1 might be the reason for the decrease of the yield stress with respect to the samples annealed in the air.

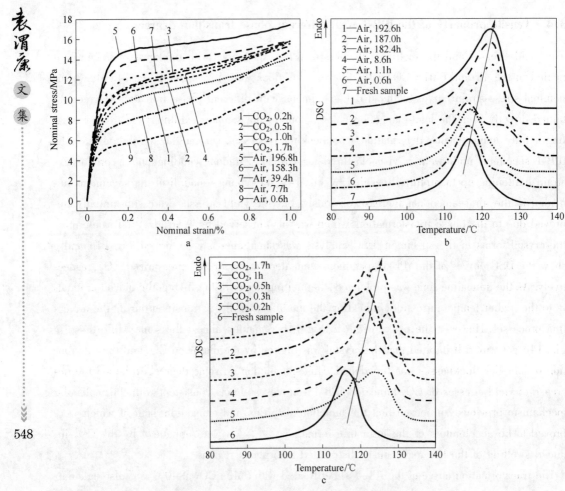

Fig. 14 Nominal stress-strain curves of iPB-1 specimens annealed in the air (black lines) and 6MPa CO_2 (colored lines) at 40℃ for different time (a), DSC curves of iPB-1 specimens annealed in the air for different time (b), and DSC results of iPB-1 specimens annealed under 6MPa CO_2 at 40℃ for different time (c)

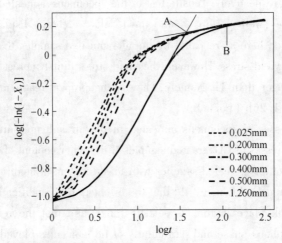

Fig. 15 Simulated Avrami curves of phase transformation from form Ⅱ to Ⅰ in iPB-1 sheets with different thickness treated with CO_2 at 40℃ and 6MPa

(The appropriate treating time for the 1.26mm thick iPB-1 sheet was speculated to be between points A and B)

4 Conclusions

In this work, the intrinsic kinetics of CO_2-induced phase transition from modification II to I in iPB-1 at 40℃ and different CO_2 pressures were detected using *in situ* high-pressure FTIR and correlated by Avrami equation. Three stages, namely induction, nucleation controlled and growth controlled periods, were discovered during the CO_2-induced phase transformation. Application of CO_2 substantially shortened the induction stage and the transition rate increased with increasing CO_2 pressure. When the CO_2 pressure was higher than 3MPa, the induction stage disappeared. The molecular mechanism of this acceleration effect was supposed to be the lamellar distortion and elongation resulted from the CO_2-induced swelling of the heterogeneously distributed amorphous regions.

Sorption of CO_2 in iPB-1 matrix was measured at 40℃ and different CO_2 pressures using both FTIR and MSB and the diffusivity determined by Fick's second law. The solubility and diffusivity measured by MSB were systematically bigger than those by FTIR, which should result from the different procedure in the two methods. The solubility could be well correlated by Henry's law. The diffusion coefficient had an order of magnitude of 10^{-11}-10^{-10} m^2/s in the iPB-1 at 40℃.

An algorithm combining the CO_2 diffusion and induced phase transition was subsequently proposed to calculate the CO_2 concentration as well as the phase transition degree in the iPB-1 matrix with different thickness at different saturation time. The calculated phase transition degree in the iPB-1 agreed well with the FTIR results. In addition, the yield stresses of iPB-1 specimens with thickness of 1.2mm annealed in the air and 6MPa CO_2 at 40℃ with different durations were also investigated. The yield stress of the iPB-1 specimens annealed in CO_2 increased with the annealing time and reached the maximum at about 60min. However, the iPB-1 specimen annealed in CO_2 had a lower yield stress than that annealed in the air for more than 39h, which was supposedly attributed to the plasticization effect of CO_2 on the amorphous regions of iPB-1. The algorithm was applied to predict the phase transition degree of the iPB-1 specimens. The appropriate CO_2 treatment time for getting high yield stress was in consistent with that predicted by the algorithm.

Acknowledgements

The authors are grateful to the National Natural Science Foundation of China (Grant No. 20976046), National High-tech R&D Program of China (863 Program) (Grant No. 2012AA040211), Innovation Program of Shanghai Municipal Education Commission (11CXY23), Fundamental Research Funds for the Central Universities, State Key Laboratory for Modification of Chemical Fibers and Polymer Materials (Dong Hua University) and Program for New Century Excellent Talents in University (NCET-09-0348).

References

[1] Natta C. Journal of Polymer Science 1960;48:219-239.
[2] Luciani L, Seppälä J, Löfgren B. Progress in Polymer Science 1988;13:37-62.

[3] Natta G, Corradini P, Bassi I. Il Nuovo Cimento (1955-1965) 1960;15:52-67.

[4] Danusso F, Gianotti G. Die Makromolekulare Chemie 1963;61:139-156.

[5] Belfiore LA, Schilling FC, Tonelli AE, Lovinger AJ, Bovey FA. Macromolecules 1984;17:2561-2565.

[6] Luongo JP, Salovey R. Journal of Polymer Science Part B: Polymer Letters 1965;3:513-515.

[7] Luongo JP, Salovey R. Journal of Polymer Science Part A-2: Polymer Physics 1966;4:997-1008.

[8] Boor Jr J, Youngman E. Journal of Polymer Science Part B: Polymer Letters 1964;2:903-907.

[9] De Rosa C, Auriemma F, Ruiz de Ballesteros O, Esposito F, Laguzza D, Di Girolamo R, et al. Macromolecules 2009;42:8286-8297.

[10] De Rosa C, Auriemma F, Resconi L. Angewandte Chemie 2009;121:10055-10058.

[11] Boor J, Mitchell JC. Journal of Polymer Science Part A: General Papers 1963;1:59-84.

[12] Hong K-B, Spruiell JE. Journal of Applied Polymer Science 1985;30:3163-3188.

[13] Clampitt BH, Hughes RH. Journal of Polymer Science Part C: Polymer Symposia 1964;6:43-51.

[14] Asada T, Sasada J, Onogi S. Polymer Journal 1972;3:350-356.

[15] He A, Xu C, Shao H, Yao W, Huang B. Polymer Degradation and Stability 2010;95:1443-1448.

[16] Lee KH, Snively CM, Givens S, Chase DB, Rabolt JF. Macromolecules 2007;40:2590-2595.

[17] Gohil RM, Miles MJ, Petermann J. Journal of Macromolecular Science, Part B 1982;21:189-201.

[18] Chau K, Yang Y, Geil P. Journal of Materials Science 1986;21:3002-3014.

[19] Kopp S, Wittmann J, Lotz B. Journal of Materials Science 1994;29:6159-6166.

[20] Marigo A, Marega C, Cecchin G, Collina G, Ferrara G. European Polymer Journal 2000;36:131-136.

[21] Azzurri F, Flores A, Alfonso GC, Baltá Calleja FJ. Macromolecules 2002;35:9069-9073.

[22] Jiang S, Duan Y, Li L, Yan D, Chen E, Yan S. Polymer 2004;45:6365-6374.

[23] Fujiwara Y. Polymer Bulletin 1985;13:253-258.

[24] Jones AT. Polymer 1966;7:23-59.

[25] Jones AT. Journal of Polymer Science Part C: Polymer Symposia 1967;16:393-404.

[26] Shieh YT, Lee MS, Chen SA. Polymer 2001;42:4439-4448.

[27] Cooper AI. Journal of Materials Chemistry 2000;10:207-234.

[28] Li B, Hu GH, Cao GP, Liu T, Zhao L, Yuan WK. The Journal of Supercritical Fluids 2008;44:446-456.

[29] Chiou JS, Barlow JW, Paul DR. Journal of Applied Polymer Science 1985;30:2633-2642.

[30] Zhang Z, Handa YP. Macromolecules 1997;30:8505-8507.

[31] Bao JB, Liu T, Zhao L, Hu GH. Industrial and Engineering Chemistry Research 2011;50:9632-9641.

[32] Chiou JS, Barlow JW, Paul DR. Journal of Applied Polymer Science 1985;30:3911-3924.

[33] Takada M, Tanigaki M, Ohshima M. Polymer Engineering and Science 2001;41:1938-1946.

[34] Li D, Liu T, Zhao L, Yuan W. AIChE Journal 2011;58:2512-2523.

[35] Li L, Liu T, Zhao L. Polymer 2011;52:5659-5668.

[36] Handa YP, Zhang Z, Wong B. Macromolecules 1997;30:8499-8504.

[37] Li L, Liu T, Zhao L, Yuan Wk. Macromolecules 2009;42:2286-2290.

[38] Li L, Liu T, Zhao L, Yuan Wk. Macromolecules 2011;44:4836-4844.

[39] Li B, Li L, Zhao L, Yuan W. European Polymer Journal 2008;44:2619-2624.

[40] Li L, Liu T, Zhao L. Macromolecular Rapid Communications 2011;32:1834-1838.

[41] Kalospiros NS, Astarita G, Paulaitis ME. Chemical Engineering Science 1993;48:23-40.

[42] Burns JR, Turnbull D. Journal of Polymer Science Part A-2: Polymer Physics 1968;6:775-782.

[43] Davis EM, Benetatos NM, Regnault WF, Winey KI, Elabd YA. Polymer 2011;52:5378-5386.

[44] Brantley NH, Kazarian SG, Eckert CA. Journal of Applied Polymer Science 2000;77:764-775.

[45] Li L, Liu T, Zhao L, Yuan W. Polymer 2011;52:3488-3495.

[46] Sato Y, Takikawa T, Takishima S, Masuoka H. The Journal of Supercritical Fluids 2001;19:187-198.

[47] Lei Z, Ohyabu H, Sato Y, Inomata H, Smith RL. The Journal of Supercritical Fluids 2007;40:452-461.
[48] Qian R, Shen D, Sun F, Wu L. Macromolecular Chemistry and Physics 1996;197:1485-1493.
[49] Avrami M. Journal of Chemical Physics 1939;7:1103-1112.
[50] Avrami M. Journal of Chemical Physics 1940;8:212-224.
[51] Wunderlich B. Macromolecular physics, Crystal nucleation, growth, annealing, Vol. 2. New York: Academic, 1976.
[52] Shi J, Wu P, Li L, Liu T, Zhao L. Polymer 2009;50:5598-5604.
[53] Maruyama M, Sakamoto Y, Nozaki K, Yamamoto T, Kajioka H, Toda A, et al. Polymer 2010;51:5532-5538.
[54] Duarte ARC, Anderson LE, Duarte CMM, Kazarian SG. The Journal of Supercritical Fluids 2005;36:160-165.
[55] Li D, Liu T, Zhao L, Yuan W. Industrial and Engineering Chemistry Research 2009;48:7117-7124.
[56] Buback M, Schweer J, Tups H. Zeitschrift Naturforschung Teil A 1986;41:505.
[57] Tomasko DL, Li H, Liu D, Han X, Wingert MJ, Lee LJ, et al. Industrial and Engineering Chemistry Research 2003;42:6431-6456.
[58] Wissinger RG, Paulaitis ME. Journal of Polymer Science Part B: Polymer Physics 1987;25:2497-2510.
[59] Guo M, Bowman J. Journal of Applied Polymer Science 1983;28:2341-2362.
[60] Crank J. The mathematics of diffusion, 2nd ed. Oxford: Clarendon; 1975.
[61] Weynant E, Haudin JM, G'Sell C. Journal of Materials Science 1980;15:2677-2692.
[62] Weynant E, Haudin JM, G'Sell C. Journal of Materials Science 1982;17:1017-1035.

Effects of Crystal Structure on the Foaming of Isotactic Polypropylene Using Supercritical Carbon Dioxide as a Foaming Agent[*]

Abstract This paper aims to study, for the first time, the effect of crystal structure on the cell formation in an isotactic polypropylene (iPP) during a solid-state foaming process using supercritical carbon dioxide ($scCO_2$) as a foaming agent. Results show that the spherulite structure exerted a significant impact on the cell morphology of foamed iPP. Very interestingly under a relatively low pressure, microcells could appear at the centers of spherulites of iPP where the melting started proceeding first. They also appeared in the amorphous domains located in between spherulites and the interlamellar regions of spherulites of iPP. The larger the size of an amorphous area, the lower the CO_2 saturation pressure needed to induce cell formation. When microcells were generated in the interlamellar regions, tie fibrils bridging lamellae could be stretched. γ-Crystals were formed at very high CO_2 saturation pressure.

Key words supercritical carbon dioxide, foaming, polypropylene, crystal structure, cell structure, semi-crystalline polymer

1 Introduction

Semicrystalline polymers possess wide applications due to desirable mechanical properties they offer, such as high stiffness and high strength. Microcellular semicrystalline polymers have attracted increasing attention in recent years since they can reduce the cost and the density of products while retaining desired mechanical properties. However, the microcellular foaming behavior of semicrystalline polymers can be different from that of amorphous polymers in that crystal structures may have a large impact on the foaming behavior and foam structure.

The effect of crystal structure on the foaming behavior of semicrystalline polymers may depend on the type of the foaming process. Foaming conditions may not all be the same for batch foaming, injection molding foaming and extrusion foaming processes. In a solid-state batch foaming process, due to the structurally heterogeneous nature of semicrystalline polymers and the concomitantly non-uniform dispersion of the foaming agent in the polymer matrix, it is conceivable that the crystal structure affects both the cell nucleation and growth. Doroudiani et al.[1] observed uniform cells with fine cell size in low-crystallinity polymers while non-uniform structures were developed in highcrystallinity polymers. They attributed the above phenomena to differences in gas solubility and stiffness of the material in different regions of polymers. Baldwin et al.[2,3] observed a higher cell density in semicrystalline polymers than in amorphous polymers and believed that interfaces between the crystalline and amorphous regions could be

---

[*] Coauthors: Jiang Xiulei, Liu Tao, Xu Zhimei, Zhao Ling, Hu Guohua. Reprinted from *The Journal of Supercritical Fluids*, 2009, 48: 167-175.

the preferential cell nucleation sites during the microcellular foaming process. Itoh and Kabumoto[4] indicated that when the foaming temperature of polyphenylene sulfide was above its crystallization temperature, high cell nucleation would be achieved in the surroundings of crystallized areas because the gas in the vicinity of crystallized areas was pushed out into the surroundings as crystallization proceeded. When studying the microcellular foaming behavior of polyamide-6 (PA6)/clay nanocomposites in an injection molding process, Yuan et al. [5,6] found that the cell wall structure and smoothness were determined by the size of the crystal structure. Meanwhile, nanoclays in the microcellular injection molding process promoted the γ-form and suppressed the α-form crystal structure of the PA6. To obtain polypropylene foams with a large volume expansion ratio in a continuous extrusion foaming process, a strategy that Naguib et al. [7] proposed was to optimize processing conditions in the die to avoid too-rapid crystallization.

This work concentrated primarily on the effect of crystal structure on the foaming of isotactic polypropylene (iPP) in a solid-state foaming process using supercritical carbon dioxide (sc-CO_2) as a foaming agent. The so-called "solid-state" means that the foaming temperature is chosen to be below the melting temperature of the iPP. Since the iPP is not completely molten during the foaming process, its crystal structure would possibly exert an impact on the cell nucleation, cell growth and cell morphology. The microstructure of iPP is complex since it may crystallize in several distinct forms (α-monoclinic, β-pseudohexagonal and γ-orthorhombic), depending on crystallization conditions and additives. The mostcommon crystal obtained is the α-monoclinic, with β-pseudohexagonal form and γ-orthorhombic formed under special conditions[8]. The α-form of iPP tends to build spherulites upon crystallization in the melt. Spherulites themselves are semicrystal structures that consist of single-crystal lamellae radiating from a common central nucleus and interlamellar amorphous matter[9]. The spherical symmetry of α-spherulite of iPP is developed with a so-called central multi-directional growth mechanism, instead of the sheaf-like unidirectional growth mechanism[10]. As a unique feature of iPP, in addition to dominant radially growing lamellae, spherulites of iPP include a second set of tangential lamellae which are oriented in a direction perpendicular to the former[11]. Accordingly, it is necessary to study the effects of such local structures of spherulites of iPP on its foaming behavior in a solid-state foaming process. On the other hand, the aforementioned features of iPP might, in return, be revealed by the cell morphology in different parts of spherulites.

It is generally accepted that crystallites are impenetrable for most non-reactive molecules including CO_2. Small molecules can diffuse into the amorphous regions but not the crystalline regions[12]. With respect to spherulites, small molecules can only enter the amorphous interlayers between lamellae but not the radiating stacks of crystal lamellae. However, those considerations are largely based on experimental sorption and diffusion data with the crystallinity of semicrystalline polymers, in which crystalline regions are often represented by the dispersed impermeable spheres in a permeable matrix[13-18]. Since cell nucleation and growth only occur in the amorphous regions where the CO_2 are dissolved, the state of dispersion of $scCO_2$ in iPP might be revealed by the foam structure: existence of cells in the amorphous regions and the absence of cells in crystalline regions.

In summary, to the best of the authors' knowledge, this paper is the first report on the effects of the crystal structure of a semicrystalline polymer on its foaming behavior and foam structure and provides better insight into the structure of spherulites of the iPP and the state of dispersion of scCO$_2$ in the iPP.

2 Experimental

2.1 Materials

An iPP with a melt flow index of 16.0g/10min was purchased from Shanghai Petrochemical Co., China. The mass-average molar mass of the iPP was 188000g/mol. The crystallinity and melting temperature of the iPP was 47% and 169℃, respectively. The iPP was used as received. The CO$_2$(purity 99%) was purchased from Airproduct Co., Shanghai, China.

2.2 Foaming process

A depressurization batch foaming process was used to foam the iPP (Fig. 1). The central piece of the apparatus was a high-pressure vessel with an internal volume of 80mL. A pressure transducer of type P31 from Beijing Endress & Hauser Ripenss Instrumentation Co., Ltd., was used to measure the pressure with an accuracy of ±0.01MPa and a valve of type Swagelok SS-1RS8MM to release the CO$_2$ gas. A computer installed with a PCI bus data acquisition system was connected to the above pressure transducer to record the pressure decay during the depressurization process. The highpressure vessel was loaded with about 10PP pellets with an average diameter of 3-4mm. It was sealed and then immersed in a silicone oil bath with a laboratory-made temperature controller with an accuracy of ±0.2℃. After the high-pressure vessel was purged with low-pressure CO$_2$, a given amount of CO$_2$ was pumped into the vessel. The CO$_2$ loading was achieved by a DZB-1A syringe pump of Beijing Satellite Instrument Co., China, with an accuracy of 0.01cm^3. Then the samples were saturated with CO$_2$ under given pressure and temperature for 40min during which the dissolution of CO$_2$ in the PP reached equilibrium[19]. Thereafter, the valve was rapidly opened to release the CO$_2$ in the high-pressure vessel in order to induce cell nucleation. Then the high-pressure vessel was opened up quickly to take out the foamed samples for subsequent analyses. The maximum depressurization rate was controlled at about 90MPa/s. Fig. 2 shows the pressure change during a typical depressurization process. The cell morphologies of the foamed iPP were characterized by a JSM-6360LV scanning electron microscopy (SEM).

3 DSC Measurement

It is well-known that a small molecule like scCO$_2$ induces the melt temperature depression of semi-crystalline polymers. A highpressure differential scanning calorimeter of type Netzsch 204 HP, Germany, was used to determine the scCO$_2$ induced melting temperature depression of the iPP used in this work. Prior to the DSC measurement, the iPP pellets were saturated with scCO$_2$ under 10MPa and at 155℃ for an hour using the same high-pressure vessel in which the foaming experiments were conducted. Unlike the foaming process, after the saturation process the high-pressure vessel was taken out from the oil bath and cooled in an ice-water bath. Then the CO$_2$ in the

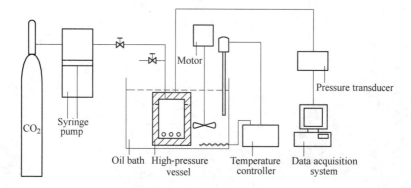

Fig. 1　Schematic of the foaming apparatus

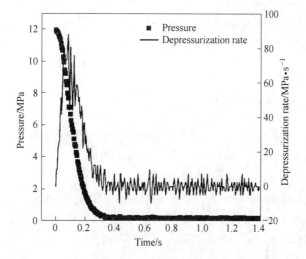

Fig. 2　Typical changes in pressure and depressurization rate
with time during a depressurization process

vessel was released very slowly to avoid foaming. The DSC measurement was conducted under a prescribed CO_2 pressure. The iPP was heated up from 20 to 200℃ at a rate of 10℃/min.

The melting of the iPP did not take place at a single temperature. Rather it started at a temperature denoted as $T_{m,s}$ and ended at a higher temperature denoted as $T_{m,e}$. The temperature corresponding to the maximum of the DSC curve is often considered as the melting temperature and is denoted as T_m. Fig. 3 shows the starting melting temperature ($T_{m,s}$), melting temperature (T_m) and ending melting temperature ($T_{m,e}$) of the iPP as a function of the applied CO_2 pressure. The foaming conditions were also shown (see open circles). Under those conditions, the PP either barely started melting or was partly molten.

The thermophysical properties of foamed iPP were measured with differential scanning calorimetry (DSC) of type NETZSCH 204 HP, Germany, under an ambient nitrogen atmosphere. 5-10mg of a sample was used for each DSC measurement. It was heated from 20 to 200℃ with a heating rate of 10℃/min.

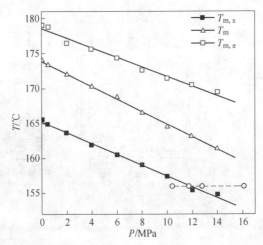

Fig. 3 Evolution of the starting melting temperature ($T_{m,s}$), melting temperature (T_m) and ending melting temperature ($T_{m,e}$) of the iPP as a function of CO_2 pressure

($T_{m,s}$(℃) = 165.3 − 0.78P(MPa); T_m = 173.9 − 0.90P; $T_{m,e}$ = 178.0 − 0.68P.

The open circles correspond to the foaming conditions in this work)

4 Results and Discussion

A previous work showed that at a given foaming temperature, there was a lower limit for the sc-CO_2 saturation pressure below which the iPP could not foam because it was too stiff[19]. In this work, the foaming temperature was fixed at 156℃, unless stated otherwise. Fig. 4 shows the cell morphology of a foamed iPP obtained at 156℃ and 10.4MPa (according to Fig. 3, under those conditions the iPP barely started melting). Cells started to form under the above foaming conditions. From Fig. 4a, there were only sporadic microcells generated. A higher magnification allowed observing a number of submicro-sized cells around a microcell (Fig. 4b and c). The texture around the microcell, characterized by the radially distributed submicro-sized cells, revealed the architecture of the corresponding spherulite. It also infers that the microcells were formed from the centers of spherulites of iPP.

The fact that microcells started to form from the centers of spherulites is related to the structural difference between the central area and the dominant radial region of a spherulite of iPP. Okada et al.[20] indicated that the isotropic embryo, developed at the early stage of crystallization in iPP, had highly disordered crystalline arrangement due to the fast crystallization at the early stage of crystallization in PP. They also noticed that the isotropic precursor was highly disordered even when it had grown up to 0.5μm in radius. Weng et al.[21,22] pointed out that the crosshatched centers of the spherulites of iPP might have become totally molten at a temperature near the melting temperature while the dominant radial lamellae remained unaffected. In view of those considerations and the experimental observations of this work, it is believed that under the chosen foaming conditions (10.4MPa and 156℃) the centers of spherulites of iPP likely begun to melt while the dominant radial lamellae remained almost unaffected. Once the centers of spherulites became molten, cell nucleation and growth would proceed there.

On the other hand, the radially distributed submicro-sized cells around the microcells indicate

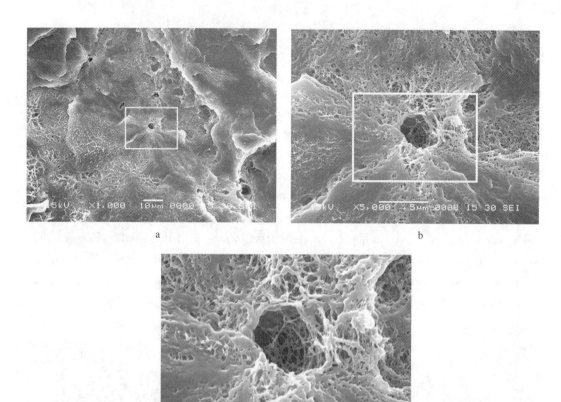

Fig. 4　Cell morphology of a foamed iPP prepared under the foaming conditions of 156℃ and 10.4MPa

a—Low magnification；b—A magnification of the white rectangle in Fig. 4a；c—A magnification of the white rectangle in Fig. 4b

that under the aforementioned foaming conditions, the cell nucleation could also occur in the interlamellar amorphous regions in which CO_2 could be dissolved. Additionally, some thin lamellae would also have molten under the foaming conditions. Consequently the interlamellar regions could be deformable to a certain extent when the centers of spherulites were molten. However, the cells formed in the interlamellar regions were unable to grow large, because of the constraint of neighboring unmolten lamellae and the high viscosity of the interlamellar materials. This is shown schematically in Fig. 5.

Fig. 5　Schematic of cell formation in the interlamellar amorphous regions under a relatively low CO_2 saturation pressure

Fig. 6 shows the geometry of the spherulites on different levels of the structural hierarchy of the foamed iPP obtained at 156℃ with a higher $scCO_2$ saturation pressure (11.7MPa). Under those conditions, the melting of the iPP was still very limited. Like the case with 156℃ and 10.4MPa, there were also microcells in the centers of spherulites (Fig. 6a and b). Nevertheless, they were larger in size than those in Fig. 4. Moreover, instead of submicro-sized cells, open cells were found in the interlamellar regions (Fig. 6c and d). At a given foaming temperature, an increase in the scCO$_2$ saturation pressure brought about an increase in the amount of CO$_2$ dissolved in the iPP and a depression in the melting temperature. Consequently, the centers of spherulites had further molten, leading to larger expansion of the cells. The interlamellar amorphous materials including the thin crystal lamellae might also have become molten because of an increased scCO$_2$ saturation pressure. In general, besides the tie fibrils consisting of tie molecules passing through several lamellae and amorphous interlayers, the interlamellar amorphous materials are mainly composed of chains of low molar mass or stereo-irregular chains that do not exhibit the required symmetry for crystallization and consequently are rejected out of lamellae during crystallization[23]. Accordingly, the elongational viscosity of the interlamellar materials would be very low when they are molten. Thus cell rupture would occur during cell growth and the cells generated in the interlamellar regions would become open because the interlamellar materials were in the molten state.

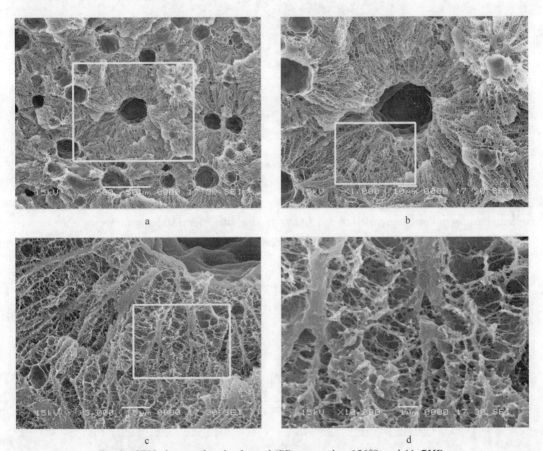

Fig. 6 SEM photographs of a foamed iPP prepared at 156℃ and 11.7MPa
a—Low magnification; b—Magnification of the white rectangle in Fig. 6a;
c—Magnification of the white rectangle in Fig. 6b; d—Magnification of the white rectangle in Fig. 6c

From Fig. 6c and d, the separated dendritic piled-lamellae can be clearly seen after foaming. The cell expansion in the molten interlamellar regions led to the split of adjoining lamellae from each other. However, similar to the deformation caused by external drawing[24-29], piled-lamellae, consisting of tens of lamellae (estimated according to the thickness of the radial piled-lamellae block), behaved as a single block with no separation of single lamella during the stretching of lamellae, as illustrated in Fig. 7. This result implies that cell nucleation only occurred in the interlamellar regions between the piled-lamellae units instead of each individual lamella. Because of the split of the piled-lamellae, tie fibrils bridging lamellae in the interlamellar regions were stretched, as revealed by the tangential fibrils between piled-lamellae (Fig. 6d).

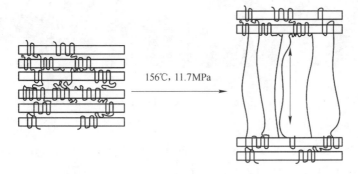

Fig. 7 Schematic of the split of piled-lamellae blocks due to cell expansion in the molten interlamellar regions under a moderate CO_2 saturation pressure

The fact that cell expansion could only occur in the interlamellar amorphous regions between the piled-lamellae blocks while the dominant radial lamellae remained unaffected suggests that CO_2 gas could only diffuse into the amorphous regions of the iPP and could not enter the crystalline lamellae. Unlike the literature on the gas sorption and diffusion, this work provides, for the first time, straightforward evidence that gas can only diffuse into the amorphous regions of a semicrystalline polymer.

In addition to microcells formed in the centers of spherulites, closed microcells also begun to form at the boundaries of spherulites of the iPP under those foaming conditions. The white rectangle in Fig. 8 shows closed microcells formed at the conjunct boundaries of three neighboring spherulites. Since the lamellae of neighboring spherulites can penetrate into or impinge each other during the later stage of crystallization, "island-like" areas containing amorphous materials were formed at the boundaries of spherulites and their sizes depended upon the contact of the "edgeon" lamellae of the neighboring spherulites[30]. The impingement of the "edge-on" lamellae created more amorphous areas at the conjunct boundaries. Therefore, the amorphous areas at the conjunct boundaries of three spherulites had relatively large sizes. More CO_2 would be dissolved in those areas than other areas of the boundaries of spherulites, increasing the probability of cell formation in those areas. However, similar to the dominant radial regions, no closed microcells were formed in the boundaries because the "edge-on" lamellae of neighboring spherulites deeply penetrated into each other. For example, the left side of the white rectangle in Fig. 8a shows the deep penetration of the "edge-on" lamellae of neighboring spherulites and the absence of microcells over there. Those results suggest that at a given foaming temperature, microcells can be formed at the boundaries of spherulites only when the saturation pressure is high enough.

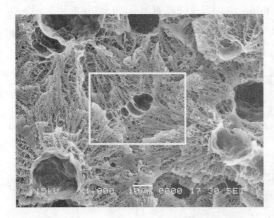
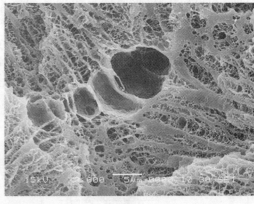

a　　　　　　　　　　　　　　　b

Fig. 8　Cell morphology in the boundaries of spherulites obtained at 156℃ and 11.7MPa (a)
and magnification of the white rectangle in Fig. 8a (b)

Fig. 9 shows the cell morphology of the foamed iPP obtained at an even higher saturation pressure, namely, 12.8MPa. The foaming temperature remained the same (156℃). More closed cells were formed along the boundaries of spherulites under such foaming conditions. Since the saturation pressure was increased, amorphous areas with relatively small sizes at the boundaries could also be foamed. However, it should be noted that there were still no closed cells formed at the interlamellar amorphous regions in the dominant radial areas under such foaming conditions.

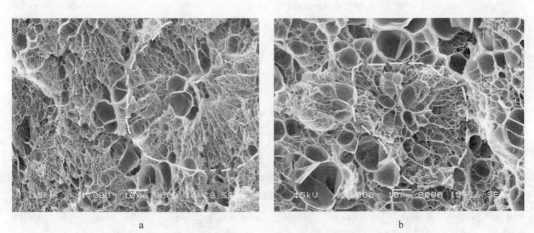

a　　　　　　　　　　　　　　　b

Fig. 9　Cell morphology at the boundaries of spherulites
(Foaming conditions: 156℃ and 12.8MPa, the dashed lines mark the boundaries of a spherulite)

Fig. 10 shows the cell morphology of foamed iPP obtained at 156℃ and 16.1MPa. Even more closed microcells were generated in the iPP. Unlike the foamed iPP obtained under low saturation pressure, the spherulite structures were not discernable any more. With regardto the microcells, the walls of some of them were empty and those of others were filled with stretched fibrils. The latter were shown in a clearer manner in Fig. 10c and d. Based on the above discussion about the cell morphology obtained under low pressures, we speculate that the empty cells with smooth walls were formed either in the centers of spherulites or the boundaries of spherulites. As the tie fibrils consisting of bundles of tie molecules only existed in the interlamellar amorphous regions of the dominant radial areas, it is safe to conclude that the cells filled with stretched tie fibrils were formed between adjacent piled-lamellae blocks. It is seen that the cells generated at the centers and boundaries of spherulites had large cell sizes owing to the large growing space, whereas the cells generated at the interlamellar regions were relatively small because of the constraint of neighboring lamellae. This result indicates that the pressure of 16.1 MPa was high enough to induce microcell formation in the dominant radial regions when the foaming temperature was 156℃. Due to the growth of microcells in the interlamellar amorphous regions, the neighboring piled-lamellae were split from each other. The walls with stretched fibrils were the bended piled-lamellae, as shown in Fig. 11.

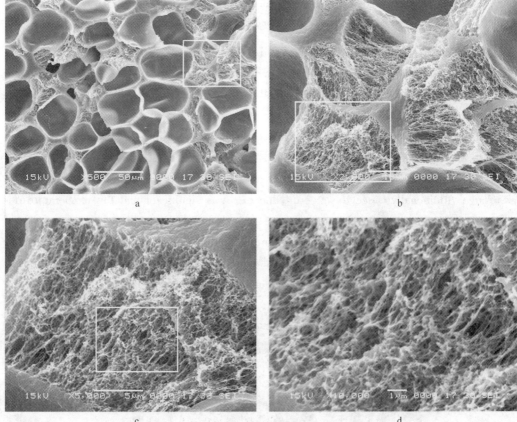

Fig. 10 Cell morphology of a foamed iPP prepared at 156℃ and 16.1 MPa

a—Low magnification; b—Magnification of the white rectangle in Fig. 10a;

c—Magnification of the white rectangle in Fig. 10b; d—Magnification of the white rectangle in Fig. 10c

According to the foregoing discussion, the effect of spherulite structure of iPP on the microcell formation depended on the local properties of spheruites of iPP. When the foaming temperature was close to the melting temperature of the iPP, the centers of spherulites melted first and consequently microcells started forming from there. In addition to the centers of spherulites, microcells could also be formed in the amorphous areas of the spherulites, including those in the boundaries and the interlamellar regions of spherulites of the iPP. Also, the larger the amorphous area, the larger the amount of CO_2 dissolved in that area. Therefore, the CO_2 saturation pressure needed for the cell generation in the amorphous areas of the boundaries of spherulites was lower than that in the interlamellar amorphous regions.

Fig. 11　Schematic of the split and bending of piled-lamellae blocks due to strong gas expansion during the cell growth under a very high CO_2 saturation pressure

Fig. 12 shows the cell morphology of a foamed iPP obtained at 25.0MPa and 152℃. Note that the CO_2 saturation pressure was much higher and the foaming temperature was much lower than the above cases. Under those conditions, the cell density was much higher and the cell uniformity was much improved. Moreover, it seemed the cells grew in a similar manner in the centers of spherulites, the interlamellar regions and the boundaries of spherulites. This may be because a decrease in the foaming temperature and an increase in the saturation pressure led to an increase in the amount of CO_2 dissolved in the amorphous regions, allowing cells to be able to grow large. Additionally, since the pressure increase was significant and the temperature decrease was marginal, the foaming temperature was closer to the melting temperature of the iPP. Thus more thin crystals would have molten, facilitating the cell growth in the interlamellar regions. Consequently, the cells generated in the interlamellar regions could grow as large as those generated in the centers and boundaries of spherulites of the iPP.

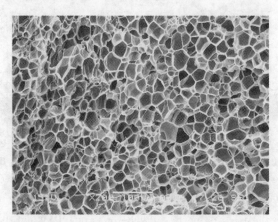

Fig. 12　Cell morphology of a foamed iPP obtained at 25MPa and 152℃
(White bar = 100μm)

5　Thermophysical Properties of Foamed iPP

Fig. 13 compares the DSC thermograms among the pure iPP, the iPP annealed at 156℃ and ambient pressure for 40min (the same time as that for the $scCO_2$ saturation of the iPP before foaming) and foamed iPP in Fig. 4, Fig. 6 and Fig. 9. The annealing effect at 156℃ and ambient pressure brought about a small increase in the melting temperature of the iPP. The melting temperature of the pure iPP was 169℃ and that of the annealed one 172℃. On the other hand, the iPP that had been subjected to the scCO$_2$ saturation at the same temperature and for the same period of time showed two melting peaks. WAXD showed that no new crystals were formed. Previous studies identified the origin of the two peaks of iPP isothermally crystallized at high temperature[11,31,32]. The lower-temperature endotherm corresponds to the melting of thin tangential lamellae while the higher temperature endotherm the melting of thick radial lamellae. The pure iPP and the iPP annealed under ambient pressure had each one endotherm, suggesting that the radial lamellae and the tangential lamellae had almost the same thickness. After the annealing with scCO$_2$, the radial lamellae were significantly thickened, as revealed by the large increase in the melting temperature. Meanwhile, the lower temperature endotherm did not shift much, implying that the thickness of tangential lamellae did not change much during the scCO$_2$ saturation process. Moreover, the intensity of the lower-temperature endotherm increased with increasing CO$_2$ saturation pressure. This infers that the content of tangential lamellae increased with increasing CO$_2$ saturation pressure.

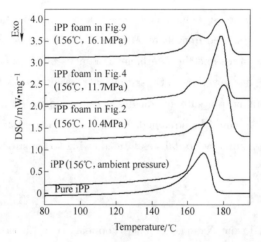

Fig. 13　DSC endotherms of the pure iPP
(iPP annealed at 156℃ without scCO$_2$ for 40min, the foamed iPP obtained at 156℃ and different saturation pressures)

Fig. 14 shows the WAXD and DSC of the foamed iPP obtained at 152℃ and 25MPa. As shown in Fig. 14a, the diffraction peaks of the foamed iPP are sharper than those of the pure iPP because of the scCO$_2$-induced crystallization. The new peak at 20.1 is characteristic of γ-crystal, indicating that γ-crystal was formed when iPP was saturated at 152℃ and 25MPa. The DSC also detected the existence of γ-crystal. Fig. 14b shows three melting peaks. The lowest one corresponds to the melting of γ-crystal. The other two correspond to the melting of tangential lamellae and radial lamellae, respectively.

6　Conclusion

This work studied, for the first time, the effect of crystal structure on the cell formation in an iPP during a solid-state foaming process using scCO$_2$ as a foaming agent. The results showed that the spherulite structure exerted a significant impact on the cell morphology of the resulting foamed

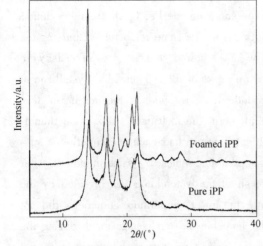
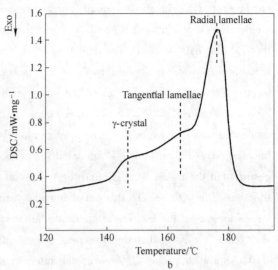

Fig. 14 WAXD of the pure iPP and a foamed iPP obtained at 152℃ and 25MPa (a) and DSC of the foamed iPP obtained at 152℃ and 25MPa (b)

iPP. Under a relatively low pressure, microcells could be formed from the centers of spherulites of iPP because the melting started from there. Microcells could also be formed in the amorphous areas located in the boundaries and the interlamellar regions of the spherulites of iPP. The larger the amorphous area, the lower the CO_2 saturation pressure needed to induce cell formation. At higher CO_2 saturation pressure, microcells were formed in the interlamellar regions and tie fibrils bridging lamellae could be stretched. When the iPP were foamed at a very high pressure and a very low temperature, γ-crystal was formed owing to the strong $scCO_2$-induced plasticization effect.

Acknowledgements

The authors are grateful to the National Science Foundation of China and PetroChina for the support of a joint project on multiscale methodologies (20490204), the National Science Foundation of China (50703011), Program for Changjiang Scholars and Innovative Research Team in University (IRT0721) and the 111 Project (B08021).

References

[1] S. Doroudiani, C. B. Park, M. T. Kortschot, Effect of the crystallinity and morphology on the microcellular foam structure of semicrystalline polymers, Polym. Eng. Sci. 36 (1996) 2645-2662.

[2] D. F. Baldwin, C. B. Park, N. P. Suh, A microcellular processing study of poly(ethylene terephthalate) in the amorphous and semicrystalline states. Part I: microcellular nucleation, Polym. Eng. Sci. 36 (1996) 1437-1445.

[3] D. F. Baldwin, C. B. Park, N. P. Suh, A microcellular processing study of poly(ethylene terephthalate) in the amorphous and semicrystalline states. Part II: cell growth and process design, Polym. Eng. Sci. 36 (1996) 1446-1453.

[4] M. Itoh, A. Kabumoto, Effects of crystallization on cell morphology in microcellular polyphenylene sulfide, Furukawa Rev. 28 (2005) 32-38.

[5] M. Yuan, L. -S. Turng, Microstructure and mechanical properties of microcellular injection molded polyamide-6 nanocomposites, Polymer 46 (2005) 7273-7292.

[6] M. Yuan, L. -S. Turng, S. Gong, A. Winardi, Crystallization behavior of polyamide-6 microcellular nanocomposites, J. Cell. Plast. 40 (2004) 397-409.

[7] H. E. Naguib, C. B. Park, U. Panzer, N. Reichelt, Strategies for achieving ultra lowdensity polypropylene foams, Polym. Eng. Sci. 42 (2002) 1481-1492.

[8] E. B. Bond, J. E. Spruiell, J. S. Lin, AWAXD/SAXS/DSC study on the melting behavior of Ziegler-Natta and Metallocene catalyzed isotactic polypropylene, J. Polym. Sci. Part B: Polym. Phys. 37 (1999) 3050-3064.

[9] K. G. Gatos, C. Minogianni, C. Galiotis, Quantifying crystalline fraction within polymer spherulites, Macromolecules 40 (2007) 786-789.

[10] D. R. Norton, A. Keller, The spherulitic and lamellar morphology of meltcrystallized isotactic polypropylene, Polymer 26 (1985) 704-716.

[11] R. G. Alamo, G. M. Brown, L. Mandelkern, A. Lehtinen, R. Paukkeri, A morphological study of a highly structurally regular isotactic poly(propylene) fraction, Polymer 40 (1999) 3933-3944.

[12] M. Hedenqvist, U. W. Gedde, Diffusion of small-molecule penetrants in semicrystalline polymers, Prog. Polym. Sci. 21 (1996) 299-333.

[13] J. S. Chiou, J. W. Barlow, D. R. Paul, Polymer crystallization induced by sorption of CO_2 gas, J. Appl. Polym. Sci. 30 (1985) 3911-3924.

[14] E. Beckman, R. S. Porter, Crystallization of bisphenol A polycarbonate induced by supercritical carbon dioxide, J. Polym. Sci. Part B: Polym. Phys. 25 (1987) 1511-1517.

[15] N. H. Brantley, S. G. Kazarian, C. A. Eckert, *In situ* FTIR measurement of carbon dioxide sorption into poly(ethylene terephthalate) at elevated pressures, J. Appl. Polym. Sci. 77 (2000) 764-775.

[16] M. Aguilar-Vega, D. R. Paul, Gas transport properties of poly (2, 2, 4, 4-tetramethyl cyclobutane carbonate), J. Polym. Sci. Part B: Polym. Phys. 31 (1993) 991-1004.

[17] P. Fossati, A. Sanguineti, M. G. De Angelis, M. G. Baschetti, F. Doghieri, G. C. Sarti, Gas solubility and permeability in MFA, J. Polym. Sci. Part B: Polym. Phys. 45 (2007) 1637-1652.

[18] L. Hardy, E. Espuche, G. Seytre, I. Stevenson, Gas transport properties of poly(ethylene-2,6-naphtalene dicarboxylate) films: influence of crystallinity and orientation, J. Appl. Polym. Sci. 89 (2003) 1849-1857.

[19] Z. -M. Xu, X. -L. Jiang, T. Liu, G. -H. Hu, L. Zhao, Z. -N. Zhu, W. -K. Yuan, Foaming of polypropylene with supercritical carbon dioxide, J. Supercrit. Fluids 41 (2007) 299-310.

[20] T. Okada, H. Saito, T. Inoue, Time-resolved light scattering studies on the early stage of crystallization in isotactic polypropylene, Macromolecules 25 (1992) 1908-1911.

[21] J. Weng, R. H. Olley, D. C. Bassett, P. Jääskeläinen, On morphology and multiple melting in polypropylene, J. Macromol. Sci. Part B: Phys. 4-6 (2002) 891-908.

[22] J. Weng, R. H. Olley, D. C. Bassett, P. Jääskeläinen, Changes in the melting behavior with the radial distance in isotactic polypropylene spherulites, J. Polym. Sci. Part B: Polym. Phys. 41 (2003) 2342-2354.

[23] S. Swaminarayan, C. Charbon, A multiscale model for polymer crystallization. I : growth of individual spherulites, Polym. Eng. Sci. 38 (1998) 634-643.

[24] A. Peterlin, Molecular model of drawing polyethylene and polypropylene, J. Mater. Sci. 6 (1971) 490-508.

[25] K. -H. Nitta, M. Takayanagi, Role of tie molecules in the yielding deformation of isotactic polypropylene, J. Polym. Sci. Part B: Polym. Phys. 37 (1999) 357-368.

[26] Y. Nozue, Y. Shinohara, Y. Ogawa, T. Sakurai, H. Hori, T. Kasahara, N. Yamaguchi, N. Yagi, Y. Amemiya, Deformation behavior of isotactic polypropylene spheruite during hot drawing investigated by simultaneous

microbeam SAXS-WAXS and POM measurement, Macromolecules 40 (2007) 2036-2045.

[27] T. Sakurai, Y. Nozue, T. Kasahara, K. Mizunuma, N. Yamaguchi, K. Tashiro, Y. Amemiya, Structural deformation behavior of isotactic polypropylene with different molecular characteristic during hot drawing process, Polymer 46 (2005) 8846-8858.

[28] K.-H. Nitta, Structural factors controlling tensile yield deformation of semicrystalline polymers, Macromol. Symp. 170 (2001) 311-319.

[29] Y. Koike, M. Cakmak, Atomic force microscopy observation on the structure development during the uniaxial stretching of PP from partially molten state: effect of isotacticity, Macromolecules 37 (2004) 2171-2181.

[30] Y.-H. Luo, Y. Jiang, X.-G. Jin, L. Li, C.-M. Chan, Real-time AFM study of lamellar growth of semicrystalline polymers, Macromol. Symp. 192 (2003) 271-279.

[31] P. S. Dai, P. Cebe, M. Capel, R. G. Alamo, L. Mandelkern, *In situ* wide- and small-angle scattering study of melting kinetics of isotactic poly(propylene), Macromolecules 36 (2003) 4042-4050.

[32] T. W. Huang, R. G. Alamo, Fusion of isotactic poly(propylene), Macromolecules 32 (1999) 6374-6376.

Foaming of Linear Isotactic Polypropylene Based on its Non-isothermal Crystallization Behaviors under Compressed CO_2 *

Abstract The non-isothermal crystallization behaviors of isotactic polypropylene (iPP) under ambient N_2 and compressed CO_2 (5-50 bar) at cooling rates of 0.2-5.0℃/min were carefully studied using high-pressure differential scanning calorimeter. The presence of compressed CO_2 had strong plasticization effect on the iPP matrix and retarded the formation of critical size nuclei, which effectively postponed the crystallization peak to lower temperature region. On the basis of these findings, a new foaming strategy was utilized to fabricate iPP foams using the ordinary unmodified linear iPP with supercritical CO_2 as the foaming agent. The foaming temperature range of this strategy was determined to be as wide as 40℃ and the upper and lower temperature limits were 155℃ and 105℃, which were determined by the melt strength and crystallization temperature of the iPP specimen under supercritical CO_2, respectively. Due to the acute depression of CO_2 solubility in the iPP matrix during the foaming process, the iPP foams with the bi-modal cell structure were fabricated.

Key words linear isotactic polypropylene, compressed CO_2, bi-modal, foam, foaming structure

1 Introduction

Isotactic polypropylene (iPP), as one of the most widely used commercial polymers, has many desirable and beneficial properties, such as high melting point, high tensile modulus, low density, excellent chemical resistance and easy recycling[1]. These outstanding properties as well as a low material cost have made iPP more competitive in producing polymer foams than other polyolefins or thermoplastics such as polyethylene (PE) and polystyrene (PS) in various industrial applications[2,3]. iPP foams, like PE foams, can offer good thermal stability and high chemical resistance, furthermore, they can be used at higher temperatures than PE and PS foams, which have made iPP foams a potential substitute for these polymer foams.

Foaming of iPP was generally carried out by dissolving a blowing agent into the polymer matrix. Thereafter, the solubility of the blowing agent was reduced rapidly by producing thermodynamic instability in the structure (e.g. by increasing temperature or decreasing pressure), to induce nucleation and growth of bubbles. On the basis of the state of polymer matrix during foaming, strategies for fabricating iPP foams could be classified as the solid-state[4,5] foaming process, melt foaming process and polymer solution extrusion[6,7]. The so-called "solid-state" meant that the foaming temperature was chosen to be lower than the melting temperature of iPP. Since CO_2 has many unique properties such as nonflammable, nontoxic, relatively inexpensive, and relatively large solubility in polymers[8,9], it has been utilized as a popular

---

* Coauthors: Li Dachao, Liu Tao, Zhao Ling. Reprinted from *The Journal of Supercritical Fluids*, 2011, 60: 89-97.

physical foaming agent in many foaming applications[4,5,10-15]. The existed crystalline regions in iPP matrix not only depress the solubility of CO_2 in the polymer matrix, but also affect dramatically cell nucleation and bubble growth[5,10,11,16,17]. In the bubble nucleation step, the interface between lamellar and amorphous domains is a high energy region resulting from the surface effects. In these regions, the Gibbs free energy necessary for nucleating a stable cell is less than that for homogeneous nucleation, resulting in the preferential nucleation of cells at the interface[16,17]. While in the bubble growth step, because of less mobility of molecule chains in crystalline regions, the formed cells is constrained by the neighboring lamellar[5,10,11]. Xu et al.[4] studied the foaming behavior of linear iPP during a solid-state foaming process with CO_2. It was found the maximum foaming temperature range of linear iPP in a solid state foaming process was very narrow, only about 4℃. Doroudiani et al.[18] observed uniform cells with fine cell size were developed in low-crystallinity polymers while non-uniform structures in high-crystallinity polymers. The high crystallinity of iPP is adverse to fabricate fine iPP foams with uniform bubble size and high expansion ratio.

The melt process for fabricating polymer foams, such as the foam extrusion[3,14,19] and injection molding[20-23], can certainly avoid the drawbacks raised by the crystal regions. Owing to the linear structure of iPP molecular chain, the melt strength and melt elasticity is too weak to prevent the formed bubble from coalesce and collapse during the foam processing. Consequently, the formed iPP foams usually have the open-cell morphology and non-uniform cell size distribution. Various methods have been adopted to improve iPP's melt strength and reduce iPP foams density, such as long-chain branching[24-27], chemical or radiation cross-linking[28,29], and polymer compounding. For example, Zhai et al.[27] synthesized PS and poly(methyl methacrylate) (PMMA) grafted iPP copolymers by atom transfer radical polymerization using a branched iPP as polymerization precursor. The foaming behaviors of linear as well as the branched and grafted iPP were studied by using CO_2 as the blowing agent in a batch foaming method. Blending with PE or PS[30-32] has also been used to improve the cell morphology of PP foam. Introduction of nano-particles could also increase the polymer's melt strength and induce strain hardening in the melt. Zheng et al.[19] presented the foaming behaviors of linear iPP and iPP/clay nanocomposites blown with supercritical CO_2 and indicated that the nano-particles have a positive impact on improving the cell morphology, the cell density and the expansion ratio of the linear PP foams. Although it is acceptable to fabricate iPP foams with a modified material which has stronger melt strength, it may be more valuable to develop a new method, for our current purpose, which could prepare iPP foams using ordinary unmodified linear materials with low melt strength.

Dissolution of CO_2 in polymer will affect the polymer properties at both melt and solid states. It depresses the glass-transition temperature[19,33-35], alters the crystallization temperature[36] and changes the crystallization kinetics of several semi-crystalline polymers[36-40]. Takada et al.[40] investigated the effect of CO_2 on the isothermal crystallization kinetics of iPP using high-pressure differential scanning calorimeter (DSC). It was found that the crystallization rate was accelerated by CO_2 at the temperature in the crystal-growth rate controlled region, and depressed in the nucleation controlled region. However, the non-isothermal crystallization kinetics of iPP in contact with compressed CO_2 was still unreported. One important feature of iPP was the onset temperature of crystallization peak could be depressed significantly by the compressed CO_2, especially at high pressures. Thus, presence of compressed CO_2 may provide us an opportunity to conduct the melt foaming process at much lower temperatures while keeping iPP from crystallizing. Since the operating temperature decreases, the iPP melt is of higher melt strength to keep the formed bubbles from rupturing during foaming to produce iPP foams with high expansion ratio.

In the past decade, efforts have been made to investigate the influence of processing conditions and the nature of polymers on the final cell morphologies of polymer foams. Many studies have focused on preparing high-performance polymer foams by increasing the number of cells and decreasing cell size[8,41,42]. However, little research has been reported on fabricating polymer foams with a bi-modal distribution of cell sizes[43,44]. It is reported that the bi-modal cell structure has outstanding heat insulation property[44]. In an effort to produce the bi-modal foam structure, a combination of two different blowing agents is often used in order to induce two different nucleating mechanisms rather than a single blowing agent. Otherwise, a two-step depressurization process is required to produce bubble nucleation twice in order to generate bi-modal cell structure[45]. A method utilizing just one blowing agent through a one-step depressurization foaming process is still unreported.

In this work, it is shown for the first time that iPP foams with either uniform or bi-modal distribution of cell size are controllably produced with compressed CO_2 using the ordinary unmodified linear iPP. The non-isothermal crystallization behaviors of the iPP melts under compressed CO_2 were first studied in detail using high-pressure DSC, based on which a new foaming strategy combining foaming and crystallization of iPP melts with compressed CO_2 was utilized. In the foaming process, supercritical CO_2 did not only serve as a physical foaming agent but also a powerful tool to control the crystallization behavior of iPP melts, i. e. , decrease the crystallization temperature and prevent the iPP melts from crystallization prior to foaming.

2 Experimental

2.1 Materials

Linear iPP with melt flow index of 16. 0g/10min was purchased from Shanghai Petrochemical Co. , China. The mass-average and number-average molar masses of the PP were 197000 and 42000g/mol, respectively. Its crystallinity and melting temperature were 47% and 169℃ in nitrogen at ambient pressure. The barlike iPP specimen with a diameter of 2-2. 5mm were extruded using Haake Minilab system (Thermo Electron Co.) under 0. 6MPa N_2 at 190℃ and a screw speed of 50 rotations/min. They were cut into circular sheets with a mass of approximately 15mg, and then used for DSC experiments. CO_2 (purity: 99. 9% w/w) was purchased from Air Products Co. , Shanghai, China and used as received.

2.2 Foaming apparatus

The batch foaming process was performed in a high-pressure vessel[4,5,32]. The internal volume of the vessel was 40cm³, calibrated with distilled water by a syringe pump. The vessel was placed in a homemade electronically controlled temperature oil bath. The temperature of the latter was measured with a calibrated mercury thermometer at an accuracy of ±0.02℃ and was controlled at an accuracy of ±0.2℃. The vessel pressure was measured at an accuracy of ±0.01MPa by a pressure transducer of type P31 from Beijing Endress & Hauser Ripenss Instrumentation Co., Ltd. A valve of type Swagelok SS-1RS8MM was used to release CO_2 in the vessel. A computer installed with a PCI bus data acquisition system was connected to the pressure transducer to record the pressure decay during depressurization. The CO_2 loading was achieved by a DZB-1A syringe pump of Beijing Satellite Instrument Co., China, with an accuracy of 0.01cm³.

2.3 Characterization

The high-pressure DSC (NETZSCH 204 HP, Germany) was used to characterize the melting and crystallization behaviors of the iPP in contact with atmospheric N_2 (1 bar) and compressed CO_2. The maximum operating pressure of the DSC equipment was 14MPa. The calorimeter was calibrated by carrying out the measurement of the heat of fusion of In, Bi, Sn, Pb, and Zn under ambient and high CO_2 pressures, respectively[46].

To measure the CO_2-induced depressions in T_m, an extruded iPP specimen with a mass of 15mg was loaded in the DSC high-pressure chamber, and the chamber was pressurized by CO_2 up to a desired level. Subsequently, it was kept at 35℃ for 3 hours to dissolve CO_2. Then, it was heated to 200℃ at the rate of 10℃/min. T_m was determined from the measured DSC profile. The effects of CO_2 on T_m were investigated by varying the CO_2 pressure from 1 to 50 bar. For a reference, T_m was also measured in contact with atmospheric nitrogen.

To investigate the effect of CO_2 on the non-isothermal crystallization behavior, after the iPP sample was loaded in the chamber of the high-pressure DSC, CO_2 with desired pressure was applied to the sample, then the sample was heated to 200℃ at a rate of 10℃/min, and kept at this temperature for 10min to melt all the crystals and dissolve CO_2 into the sample. After that, the system was cooled to 60℃ and the exothermic heat flow was measured at different cooling rates (i.e., 0.2℃/min, 0.5℃/min, 1.0℃/min, 2.0℃/min, and 5.0℃/min) under 1 bar N_2 and compressed CO_2 at pressures ranging from 5 to 50 bar. To minimize the effect of the noise induced by the compressed CO_2 environment, at least three DSC measurements were performed for each cooling rate and gas environment.

The crystallinity, X_c, of the obtained iPP foams was estimated from DSC results under atmospheric N_2 at a heating rate of 10℃/min by using the relation $X_c = \Delta H_m / \Delta H_m^o$, where ΔH_m was the enthalpy of crystallization per gram of the sample and ΔH_m^o the enthalpy of crystallization per gram of 100% crystalline iPP. The latter was 209J/g[47].

The cell morphologies of the iPP foams were characterized by a JSM-6360LV (JEOL Ltd. Tokyo, Japan) scanning electron microscopy (SEM). The samples were immersed in liquid nitrogen for 10min and then fractured. The SEM scanned fractured surfaces with Pd (palladium) coating. The average cell size was obtained through the analysis of the SEM photographs by the software of Image-Pro Plus (Media Cybernetics, Silver Spring, Maryland, USA). The number average diameter of all the cells in the micrograph, D, was calculated using the following equation:

$$D = \frac{\sum d_i n_i}{\sum n_i} \quad (1)$$

where, n_i was the number of cells with a perimeter-equivalent diameter of d_i.

The volume expansion ratio of the iPP foams, R_v, defined as the ratio of the bulk density of the virgin iPP (ρ_0) to that of the foamed one (ρ_f), could be calculated as follows:

$$R_v = \frac{\rho_0}{\rho_f} \quad (2)$$

The mass densities of foamed iPP samples ρ_f were measured according to ASTM D792-00 involving weighing polymer foam in water using a sinker. ρ_f was calculated as follows:

$$\rho_f = \frac{a}{a + w - b} \rho_{water} \quad (3)$$

where, a was the apparent mass of specimen in air without sinker; b the apparent mass of specimen and sinker completely immersed in water; w was the apparent mass of the totally immersed sinker.

The cell density (N_0), the number of cells per cubic centimeter of unfoamed iPP was determined from the equation:

$$N_0 = \left(\frac{nM^2}{A}\right)^{3/2} R_v \quad (4)$$

where, n was the number of cells in the SEM micrograph; M was the magnification factor; A was the area of the micrograph (cm^2); R_v was the volume expansion ratio.

3 Results and Discussion

3.1 Plasticization effect of CO_2 on the iPP

It was known that dissolution of CO_2 in polymer would affect polymer properties at both melt and solid states. The dissolved CO_2 depressed the glass transition temperature[33-35], altered the crystallization temperature[36] and changed the crystallization kinetics of several semi-crystalline polymers[36-40]. The depression in T_g and T_m of iPP was plotted against CO_2 pressure in Fig. 1. ΔT_g stood for the magnitude of depression and was given by $\Delta T_g = T_{g,CO_2} - T_{g,nitrogen}$[36], where $T_{g,nitrogen}$ was T_g measured under atmospheric N_2. Similarly, ΔT_m denoted $\Delta T_m = T_{m,CO_2} - T_{m,nitrogen}$, where $T_{m,nitrogen}$ was T_m measured under atmospheric N_2. With the same definition of ΔT_g and ΔT_m, the data of Takada et al.[36,40] were also shown in Fig. 1. ΔT_m decreased almost linearly with increasing CO_2 pressure, while the depression magnitude of T_g was almost the same as that of T_m. The linear relationship between ΔT_m and CO_2 pressure was given: $\Delta T_m = -1.30 \times P_{CO_2}$.

3.2 Non-isothermal crystallization behavior of the iPP under compressed CO_2

The non-isothermal crystallization behaviors of iPP were studied using high-pressure DSC at different cooling rates (i.e. 0.2℃/min, 0.5℃/min, 1.0℃/min, 2.0℃/min, and 5.0℃/min)

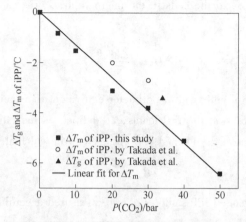

Fig. 1 Depression of T_g and T_m by compressed CO_2

and CO_2 pressures ranging from 5 to 50 bar. Fig. 2 shows the typical high-pressure DSC curves of the iPP at cooling rate of 0.5 ℃/min and different CO_2 pressures. As expected, the DSC curve shifted to the low-temperature side with increasing CO_2 pressure, which indicated that the dissolution of CO_2 in iPP matrix effectively postponed the crystallization of iPP melts. In addition, the crystallization enthalpy, i. e., the integral area of the crystallization peak, also decreased with increasing CO_2 pressure.

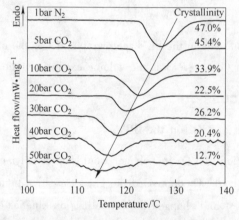

Fig. 2 Representative DSC cooling traces of the iPP melts at a cooling rate of 0.5 ℃/min under different gas environments

The effects of CO_2 on the peak crystallization temperature of the iPP at different cooling rates are shown in Fig. 3. It was observed that CO_2 effectively postponed the crystallization peak of the iPP to lower temperature region. In fact, there would be two aspects that affecting the crystallization behavior of polymer melt when high pressure gas was employed, i. e., external pressure and dissolved gas[48]. The crystallization kinetics would be affected by the hydraulic pressure applied externally by the gas on the polymer melt. Especially when the solubility of the gas in the polymer is very low or negligible, the kinetics of crystallization will be governed by the hydraulic pressure and the crystallization rate will be enhanced. In the case of iPP/CO_2 system, it was believed that the large magnitude of change in the crystallization peak of iPP was dominated by the relatively high solubility of CO_2 in iPP matrix. The dissolved CO_2 would induce iPP swelling and increase the free volume so as to enhance the mobility of molecular chain. Thus, the crystallization temperature of the iPP was depressed. This mechanism is the same as that of the so-called "plasticization effect", which could explain the depression of the glass transition temperature and melting temperature of iPP by the dissolved CO_2. However, as shown in Fig. 4, the magnitude of change in the crystallization temperature was larger than that of T_m. ΔT_c had the similar definition with ΔT_m and ΔT_g, i. e., the magnitude of depression, and was given by $\Delta T_c = T_{c,CO_2} - T_{c,nitrogen}$ [38], where T_{c,CO_2} and $T_{c,nitrogen}$ were T_c measured under high-pressure CO_2 and atmospheric N_2, respectively. It indicated that beside plasticization effect, the dissolved CO_2 had other effect on the crystallization behavior of iPP. Takada et al. [36,40] investigated the effect of dissolved CO_2 on the isothermal crystallization behaviors of PP and concluded that the dissolved CO_2 would prevent the formation of critical size nuclei, i. e., decrease the crystallization rate within the nucleation dominated temperature region. It was believed that the dissolved CO_2 would also prevent the formation of critical size nuclei of iPP in a non-isothermal crystallization process so as to depress the crystallization temperature. In summary, the postpone or depression effect of compressed CO_2 on the crystallization peak generated in a non-isothermal crystallization process was dominated by the plasticization effect and retardation of the formation of critical size nuclei, while the effect of external pressure was negligible.

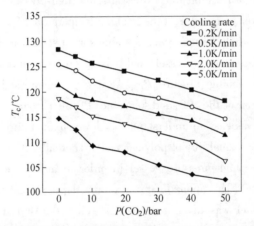

Fig. 3 Effects of cooling rate and CO_2 pressure on the crystallization temperature of the iPP melts

3.3 Foaming strategy and determination of foaming conditions

On the basis of the postpone or depression effects of compressed CO_2 on the crystallization peak generated in both non-isothermal and isothermal crystallization processes of iPP, the foaming strategy proposed by Li et al. [49] was utilized to fabricate iPP foams using the unmodified linear iPP with low melt strength. In this strategy, CO_2 did not only serve as a physical foaming agent but also a powerful tool to control the crystallization behaviors of iPP melts.

As shown in Fig. 5, the process was started by placing the iPP specimen in the high-pressure vessel (state 1 in Fig. 5). After a given amount of CO_2 was charged, the vessel was put into an oil-bath at a temperature of 200 ℃ (the saturation temperature, T_{sat}). It usually took 15-20 min for the temperature in the vessel to reach 200 ℃ and the CO_2 pressure to reach a desired value

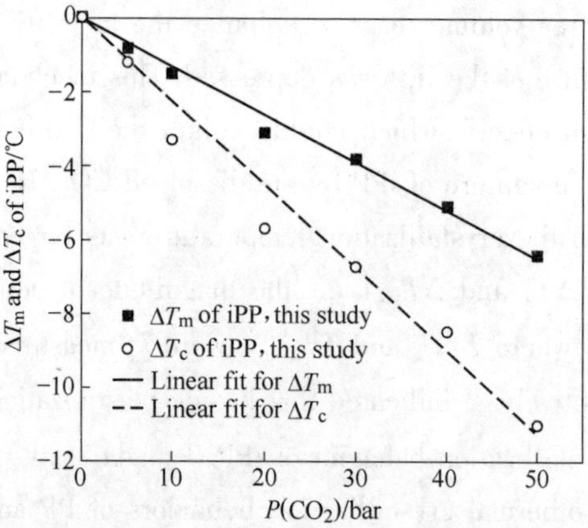

Fig. 4 Effects of CO_2 pressure on the melting temperature and crystallization temperature of the iPP

(The heating rate and cooling rate are 10 ℃/min and 0.5 ℃/min, respectively)

(the saturation pressure, P_{sat}). The vessel was retained there for another 20 min to completely melt iPP crystals and dissolve CO_2 into the sample (state 2 in Fig. 5). Selection of 200 ℃ as the saturation temperature was to guarantee the completed melting of iPP matrix, while the saturation pressure of 30 MPa was pretty close to the maximum operation pressure of our high-pressure vessel. We utilized such a high saturation pressure because we want to achieve high CO_2 solubility and also high depressurization rate. Afterward, the high-pressure vessel was quenched in another oil-bath at a lower temperature (the foaming temperature, T_f), and retained for another 25 min to make sure the whole system reach a stable state (state 3 in Fig. 5). During the cooling process, the CO_2 pressure in the vessel would also decrease with decreasing temperature and reach a stable one (the foaming pressure, P_f). At state 3, compressed CO_2 served as a tool to postpone the crystallization of the iPP, and keep the iPP melt from crystallizing at the lowered foaming temperature. It was easy to understand that lower temperature would increase the melt strength and enhance the foamability of polymer melts. Thereafter, the valve was rapidly opened to release the CO_2 in the high-pressure vessel in order to induce cell nucleation and bubble growth (state 4 in Fig. 5). The approximate maximum depressurization rate was between 300 and 400 MPa/s. The vessel was taken out from the oil-bath. After a natural cooling process (state 5 in Fig. 5), the vessel was opened and the foamed samples were taken out for subsequent analyses.

Although the foaming strategy was designed based on the non-isothermal behaviors of the iPP, the information provided by the non-isothermal behaviors under compressed CO_2 was not enough to explain the foaming process. That was because the whole process was a combination of both an isothermal (state 3 to state 4) and non-isothermal crystallization (state 4 to state 5) processes. To make sure that barely any crystals had formed at state 3, the high-pressure DSC was utilized to simulate the foaming process from state 1 to state 3 as following. Under 10 MPa CO_2, the iPP sample was heated to and kept at 200 ℃ for 20 min to melt crystals and dissolve CO_2, which simulated state 2 of the process. Afterward, the temperature was decreased quickly to a lower temperature (i.e., T_f) and retained for another 25 min to simulate state 3. Thereafter, a heating rate of 10 ℃/min was utilized and Fig. 6 shows the DSC thermograms of the iPP annealed under different T_fs.

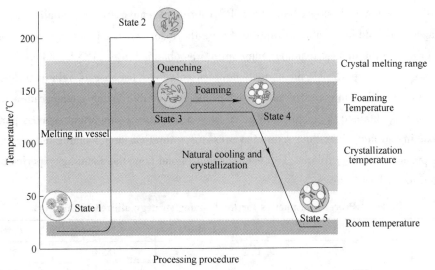

Fig. 5 Schematic of the foaming strategy utilized in this work[49]

As shown in Fig. 6, there was barely any melt peak detected for the iPP melt annealed at 120℃, indicating no crystal regions had formed during the annealing process. The melt peak did not obviously show up until the annealing temperature was decreased to 115℃, at which an unconspicuous peak was observed, indicating some incomplete and thin lamellar had formed during the 25min annealing period. For comparison, the DSC heating race of the original iPP measured under 10MPa CO_2 at a heating rate of 10℃/min is also given in Fig. 6. The crystallinities for the three specimens, i. e., the original iPP pellet, iPP melt annealed at 115℃ and 120℃, were 38.3%, 15.9% and 5.8%, respectively. As mentioned above, the purpose of keeping temperature at 200℃ was to eliminate the iPP crystal regions that would reduce the bubble size and foreclose formation of uniform cell structures. Thus, the foaming temperature must be higher than 115℃ which could prevent the iPP matrix from crystallizing at experimental conditions.

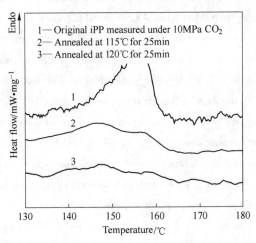

Fig. 6 High-pressure DSC heating traces of the iPP melts after annealed at different temperatures and under 10MPa CO_2 for 25min

(For comparison, the DSC heating races of the original iPP measured under 10MPa CO_2 with a heating rate of 10℃/min was also given. The crystallinities of the three specimens (i. e., the original iPP pellet, iPP melt annealed at 115℃ and 120℃) were 38.3%, 15.9%, and 5.8%, respectively)

Because of the low melt strength of linear iPP, exorbitant temperature essentially leads to bubble collapse and bubble-membrane rupture during the foam expansion. After a series of exploring experiments, the highest foaming temperature was obtained, i. e. ,155℃. When the foaming temperature was above 155℃, the formed bubbles collapsed because of relatively lower melt strength and no iPP foam could be fabricated. Thus, the suitable foaming temperature range for the process was obtained, i. e. ,115-155℃. As shown in Table 1, important process parameters of the foaming strategy, including saturation pressure, saturation time, foaming temperatures and foaming pressures, were explored. For each foaming condition, the foaming experiments were conducted for at least 3 times.

Table 1 Process parameters for the foaming strategy utilized in this work

T_{sat}/℃	t_{sat}/min	P_{sat}/MPa	T_f/℃	P_f/MPa
200.0	20-25	30.0	155.0	23.9
			150.0	23.3
			145.0	22.7
			140.0	22.0
			135.0	21.4
			130.0	20.7
			125.0	20.0
			120.0	19.3
			115.0	18.6

3.4 Characterization of the iPP foams

As shown in Fig. 7, iPP foams could be fabricated using the unmodified linear iPP in a wide temperature range between 115 and 155℃. It indicated that the foaming temperature range of the iPP was tremendously extended from 4℃ [4] to 40℃ using this foaming strategy. The upper limit of the foaming temperature was determined by the melt strength of the linear iPP. If the foaming temperature was higher than 155℃, the formed bubbles would collapse because of the lower melt strength and no iPP foam could be fabricated. The lower limit of the foaming temperature was determined by the crystallization temperature of the iPP melt under compressed CO_2, as shown in Fig. 6. Uniform cell size distribution could be obtained at the foaming temperature of 155℃. While in the foaming temperature range of 150-130℃, the bi-modal cell morphology was observed. When the foaming temperature was lower than 130℃, the obtained cell morphology trended to be anomalous and the formed bubbles seemed to be stretched, which might due to the effect of the formed crystals on the bubble growth.

The average bubble size and cell density of the bi-modal iPP foams are illustrated in Fig. 8a-d, respectively. Fig. 8 also shows the error bars to represent the overall distribution of the experimental data. The standard error was utilized to obtain the error bars:

$$\text{Standard error} = \frac{\text{standard deviation}}{\sqrt{N}} \qquad (5)$$

where, N was the number of replicates of certain data.

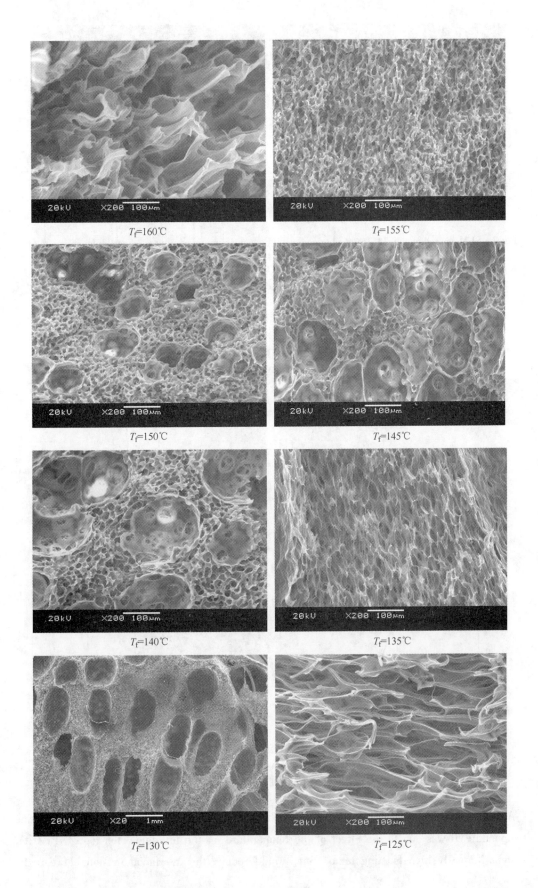

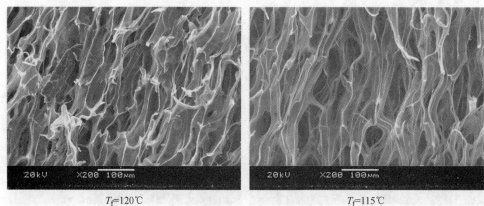

T_f=120℃　　　　　　　　　　　T_f=115℃

Fig. 7　SEM images of the iPP foams obtained at different foaming temperatures using the foaming strategy

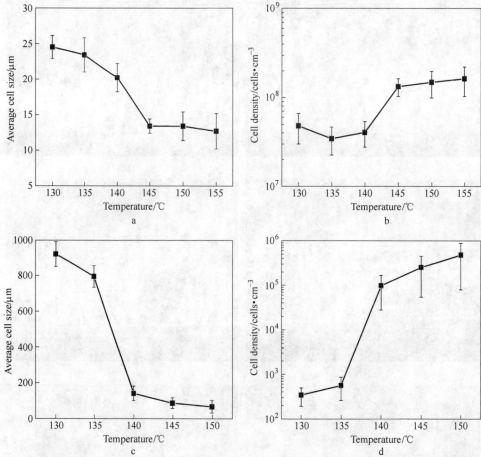

Fig. 8　Average cell size and cell density of the bi-modal iPP foams
a—Average cell size for small cells; b—Cell density for small cells;
c—Average cell size for large cells; d—Cell density for large cells

As shown in Fig. 8a and b, for the small cells, the cell density increased with the foaming temperature. It was due to the effect of the temperature on the interfacial tension of iPP/CO_2 system. Normally, high foaming temperature would reduce the interfacial tension between the gas phase and polymer melts, which was favorable for the nucleation of gas bubbles. Because of the increase of the bubble nucleation, more gas was utilized to induce formation of the bubble nucleus rather than growth of the formed bubbles. Thus, the obtained average cell diameter decreased with increasing temperature. Although variation of the average cell size and cell density of the large cells with the foaming temperature was the same as those of the small cells, we thought they had different formation mechanism. The large cells were generated during a relatively long time (20-25min) cooling process, which could be considered as a depressurization process at a very slow depressurization rate. During this depressurization process, a lower foaming temperature would induce greater super-saturation degree and driven more CO_2 out from the polymer matrix. Because of the very slow depressurization rate, the excluded CO_2 would prefer blowing the already existed bubble even bigger and making the adjacent bubble combine with each other but not forming new bubble nucleus. Thus, a tremendous decrease of cell density and increase of average cell size were observed with decreasing the foaming temperature, as shown in Fig. 8c and d.

3.5 Discussion on the iPP foams

Normally, in an effort to produce the bi-modal foam structure, a combination of two different blowing agents was often used in order to induce two different nucleating mechanisms rather than a single blowing agent. Otherwise, a two-step depressurization process was required to produce bubble nucleation twice so as to generate bi-modal cell structure. Thus, it could be concluded that, in order to produce the bi-modal foam structure from a single phase polymer matrix, the bubble nucleation should be induced twice. The first bubble nucleation would generate the former of the relatively large bubbles, and the second nucleation would induce the already existed bubble growing bigger and generating new and small bubbles.

A schematic diagram showing the pressure and temperature evolution versus time in the foaming strategy is exhibited in Fig. 9. During the cooling step of state 2 to states 3 and 4, the vessel was sealed and no more CO_2 was added into the system so that the CO_2 pressure in the vessel would decrease with decreasing temperature. The final CO_2 pressure (i.e., the pressure in the vessel before foaming took place) of each experiment was recorded and shown in Table 1. Although the temperature and pressure in the vessel decreased simultaneously, we supposed the solubility of CO_2 in the iPP matrix would evidently decrease and the discharged CO_2 lead to the first nucleation.

To demonstrate our assumption, we adopt the solubility data of Li et al.[50], who investigated the solubilities of CO_2 and N_2 in the linear iPP melt at temperatures from 180 to 220℃ and pressures up to 27.6MPa using a magnetic suspension balance (MSB). Then, we extrapolated their data utilizing the correlated interaction parameter k_{12} of Sanchez-Lacombe equation of state and obtained the solubility at the saturation pressure and temperature of 30MPa and 200℃, i.e., 370.5g CO_2/kg iPP. We also adopted the solubility data from our previous work[9], in which the solubility and diffusivity of CO_2 in the amorphous phase of solid-state iPP at temperatures of 100℃, 125℃ and 150℃ were measured. The experiment data were also extrapolated to the foaming pressure as shown in Fig. 10. During the cooling process of state 2 to states 3 and 4 as shown in Fig. 5 or Fig. 9, the CO_2 pressure in the vessel would decrease from 30MPa to 23.9-18.6MPa as the temperature decreased from 200℃ to 155-115℃, and the corresponding solubility of CO_2 in the iPP melts would decreased in the shadow area in Fig. 10, which was depressed at least 100g CO_2/kg polymer comparing to that at state 2. Driven by the thermodynamic instability, phase separation occurred between the polymer and gas phase, which would induce the first nucleation to form cell nuclei during the cooling process. Thus, prior to the opening of the valve, there would be a certain number of existed bubbles in the iPP/CO_2 matrix. When the CO_2 pressure in the vessel was released at state 4 in Fig. 5, the over saturated CO_2 would induce the already existed bubble growing bigger and generating new and small bubbles and the bi-modal iPP foams were obtained.

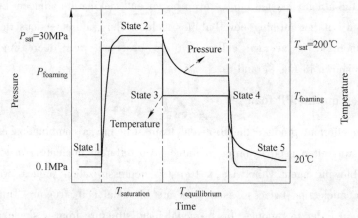

Fig. 9 Schematic diagram of the pressure and temperature evolution versus time in the foaming strategy

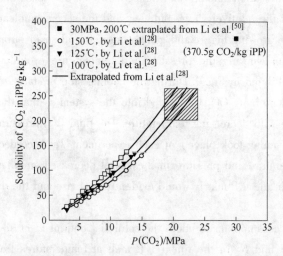

Fig. 10 Solubility of CO_2 in iPP at different conditions

In our previous work[49] concerning about batch foaming of PLA using the same strategy, PLA foams with inter-connected structures, porosity of 67.9%-91.4% and expansion ratio of 15-30 times are controllably produced. However, no bi-modal PLA foam were obtained, which was because the solubility variation of CO_2 in PLA matrix from state 2 to state 4 was not as obvious as that of CO_2 in iPP. Besides, the foaming temperature range of PLA was not as wide as that of iPP, which could be attributing to the relatively low molecular weight and corresponding low melt strength of the PLA.

4 Conclusions

This work aimed at fabricating porous iPP materials using a linear unmodified raw material, which had a melt flow index of 16.00g/10min and low melt strength. The non-isothermal crystallization behaviors of the iPP under ambient N_2 and compressed CO_2 (5-50 bar) at cooling rates of 0.2-5.0℃/min were carefully studied using high-pressure DSC. The results suggested that the presence of compressed CO_2 had strong plasticization effect on the iPP matrix and retarded the formation of critical size nuclei, which effectively postponed the crystallization peak to lower temperature region. On the basis of these findings, a new foaming strategy combining a batch foaming process and crystallization of iPP melts with supercritical CO_2 was utilized to fabricate iPP foams using the unmodified linear iPP. The foaming temperature range of the strategy was determined to be as wide as 40℃, and the upper and lower temperature limits were 155℃ and 115℃, which were determined by the melt strength and crystallization temperature of the iPP specimen under compressed CO_2, respectively. In this foaming strategy, during the temperature of the high-pressure vessel was decreased from saturation temperature to foaming temperature, the acute depression of CO_2 solubility in iPP matrix induced the first nucleation. When the CO_2 pressure in the vessel was released rapidly by opening the valve, the over saturated CO_2 would induce the already existed bubble growing bigger and generating new and small bubbles. Consequently, the iPP foams with the bi-modal cell structure were obtained.

Acknowledgements

The authors are grateful to the National Natural Science Foundation of China (Grant Nos. 50703011, 20976045, 20976046), Shanghai Shuguang Project (08SG28), Program for New Century Excellent Talents in University (NCET-09-0348), Program for Changjiang Scholars, Innovative Research Team in University, the 111 Project (B08021) and the support of "the Fundamental Research Funds for the Central Universities".

References

[1] C. Vasile, Handbook of Polyolefins, 2nd ed., Marcel Dekker, New York, 2000.

[2] J. S. Colton, The nucleation of microcellular foams in semicrystalline thermoplastics, Materials & Manufacturing Processes 4 (1989) 253-262.

[3] W. T. Zhai, T. Kuboki, L. L. Wang, C. B. Park, E. K. Lee, H. E. Naguib, Cell structure evolution and the crystallization behavior of polypropylene/clay nanocomposites foams blown in continuous extrusion, Industrial & Engineering Chemistry Research 49 (2010) 9834-9845.

[4] Z. M. Xu, X. L. Jiang, T. Liu, G. H. Hu, L. Zhao, Z. N. Zhu, W. K. Yuan, Foaming of polypropylene with supercritical carbon dioxide, J Supercritical Fluids 41 (2007) 299-310.

[5] X. L. Jiang, T. Liu, Z. M. Xu, L. Zhao, G. H. Hu, W. K. Yuan, Effects of crystal structure on the foaming of isotactic polypropylene using supercritical carbon dioxide as a foaming agent, J Supercritical Fluids 48 (2009) 167-175.

[6] E. Kiran, Foaming strategies for bioabsorbable polymers in supercritical fluid mixtures. Part I. Miscibility and foaming of poly(L-lactic acid) in carbon dioxide plus acetone binary fluid mixtures, J Supercritical

Fluids 54 (2010) 296-307.

[7] E. Kiran, Foaming strategies for bioabsorbable polymers in supercritical fluid mixtures. Part Ⅱ. Foaming of poly(epsilon-caprolactone-co-lactide) in carbon dioxide and carbon dioxide plus acetone fluid mixtures and formation of tubular foams via solution extrusion, J Supercritical Fluids 54 (2010) 308-319.

[8] D. L. Tomasko, H. B. Li, D. H. Liu, X. M. Han, M. J. Wingert, L. J. Lee, K. W. Koelling, A review of CO_2 applications in the processing of polymers, Industrial & Engineering Chemistry Research 42 (2003) 6431-6456.

[9] D. C. Li, T. Liu, L. Zhao, W. K. Yuan, Solubility and diffusivity of carbon dioxide in solid-state isotactic polypropylene by the pressure-decay method, Industrial & Engineering Chemistry Research 48 (2009) 7117-7124.

[10] D. F. Baldwin, C. B. Park, N. P. Suh, A microcellular processing study of poly(ethylene terephthalate) in the amorphous and semicrystalline states. 1. Microcell nucleation, Polymer Engineering & Science 36 (1996) 1437-1445.

[11] D. F. Baldwin, C. B. Park, N. P. Suh, A microcellular processing study of poly(ethylene terephthalate) in the amorphous and semicrystalline states. 2. Cell growth and process design, Polymer Engineering & Science 36 (1996) 1446-1453.

[12] Y. W. Di, S. Iannace, E. Di Maio, L. Nicolais, Reactively modified poly(lactic acid): properties and foam processing, Macromolecular Materials and Engineering 290 (2005) 1083-1090.

[13] J. Reignier, R. Gendron, M. F. Champagne, Extrusion foaming of poly(lactic acid) blown with CO_2: toward 100% green material, Cellular Polymers 26 (2007) 83-115.

[14] S. T. Lee, L. Kareko, J. Jun, Study of thermoplastic PLA foam extrusion, J. Cellular Plastics 44 (2008) 293-305.

[15] S. Pilla, S. G. Kim, G. K. Auer, S. Q. Gong, C. B. Park, Microcellular extrusion-foaming of polylactide with chain-extender, Polymer Engineering & Science 49 (2009) 1653-1660.

[16] J. S. Colton, N. P. Suh, Nucleation of microcellular foam-theory and practice, Polymer Engineering & Science 27 (1987) 500-503.

[17] W. T. Zhai, J. Yu, L. C. Wu, W. M. Ma, J. S. He, Heterogeneous nucleation uniformizing cell size distribution in microcellular nanocomposites foams, Polymer 47 (2006) 7580-7589.

[18] S. Doroudiani, C. B. Park, M. T. Kortschot, Effect of the crystallinity and morphology on the microcellular foam structure of semicrystalline polymers, Polymer Engineering & Science 36 (1996) 2645-2662.

[19] W. G. Zheng, Y. H. Lee, C. B. Park, Use of nanoparticles for improving the foaming behaviors of linear PP, J. Applied Polymer Science 117 (2010) 2972-2979.

[20] K. Shelesh-Nezhad, A. Taghizadeh, Shrinkage behavior and mechanical performances of injection molded polypropylene/talc composites, Polymer Engineering & Science 47 (2007) 2124-2128.

[21] G. J. Zhong, Z. M. Li, L. B. Li, E. Mendes, Crystalline morphology of isotactic polypropylene (iPP) in injection molded poly(ethylene terephthalate) (PET)/iPP microfibrillar blends, Polymer 48 (2007) 1729-1740.

[22] G. W. Beckermann, K. L. Pickering, Engineering and evaluation of hemp fibre reinforced polypropylene composites: fibre treatment and matrix modification, Composites Part A: Applied Science and Manufacturing 39 (2008) 979-988.

[23] G. Q. Zheng, W. Yang, M. B. Yang, J. B. Chen, Q. Li, C. Y. Shen, Gas-assisted injection molded polypropylene: the skin-core structure, Polymer Engineering & Science 48 (2008) 976-986.

[24] H. E. Naguib, C. B. Park, Strategies for achieving ultra low-density polypropylene foams, Polymer Engineering & Science 42 (2002) 1481-1492.

[25] P. Spitael, C. W. Macosko, Strain hardening in polypropylenes and its role in extrusion foaming, Polymer

Engineering & Science 44 (2004) 2090-2100.

[26] M. A. Rodriguez-Perez, Crosslinked polyolefin foams: Production, structure, properties, and applications, Advances in Polymer Science 184 (2005) 97-126.

[27] W. T. Zhai, H. Wang, J. Yu, J. Y. Dong, J. S. He, Foaming behavior of isotactic polypropylene in supercritical CO_2 influenced by phase morphology via chain grafting, Polymer 49 (2008) 3146-3156.

[28] D. H. Han, J. H. Jang, H. Y. Kim, B. N. Kim, B. Y. Shin, Manufacturing and foaming of high melt viscosity of polypropylene by using electron beam radiation technology, Polymer Engineering & Science 46 (2006) 431-437.

[29] C. S. Liu, D. F. Wei, A. Zheng, Y. Li, H. N. Xiao, Improving foamability of polypropylene by grafting modification, J. Applied Polymer Science 101 (2006) 4114-4123.

[30] M. H. Lee, C. Tzoganakis, C. B. Park, Extrusion of PE/PS blends with supercritical carbon dioxide, Polymer Engineering & Science 38 (1998) 1112-1120.

[31] P. C. Lee, J. Wang, C. B. Park, Extruded open-cell foams using two semicrystalline polymers with different crystallization temperatures, Industrial & Engineering Chemistry Research 45 (2006) 175-181.

[32] X. L. Jiang, T. Liu, L. Zhao, Z. M. Xu, W. K. Yuan, Effects of blend morphology on the foaming of polypropylene/low-density polyethylene blends during a batch foaming process, J. Cellular Plastics 45 (2009) 225-241.

[33] T. S. Chow, Molecular interpretation of the glass-transition temperature of polymer-diluent systems, Macromolecules 13 (1980) 362-364.

[34] Z. Y. Zhang, Y. P. Handa, An *in situ* study of plasticization of polymers by high-pressure gases, J. Polymer Science B: Polymer Physics 36 (1998) 977-982.

[35] Y. P. Handa, Z. Zhang, B. Wong, Solubility, diffusivity, and retrograde vitrification in PMMA-CO_2, and development of sub-micron cellular structures, Cellular Polymers 20 (2001) 1-16.

[36] M. Takada, S. Hasegawa, M. Ohshima, Crystallization kinetics of poly(L-lactide) in contact with pressurized CO_2, Polymer Engineering & Science 44 (2004) 186-196.

[37] K. Mizoguchi, T. Hirose, Y. Naito, Y. Kamiya, CO_2-induced crystallization of poly(ethylene-terephthalate), polymer 28 (1987) 1298-1302.

[38] S. M. Lambert, M. E. Paulaitis, Crystallization of poly(ethylene terephthalate) induced by carbon dioxide sorption at elevated pressures, J. Supercritical Fluids 4 (1991) 15-23.

[39] Y. P. Handa, Z. Y. Zhang, B. Wong, Effect of compressed CO_2 on phase transitions and polymorphism in syndiotactic polystyrene, Macromolecules 30 (1997) 8499-8504.

[40] M. Takada, M. Tanigaki, M. Ohshima, Effects of CO_2 on crystallization kinetics of polypropylene, Polymer Engineering & Science 41 (2001) 1938-1946.

[41] C. B. Park, A. H. Behravesh, R. D. Venter, A strategy for the suppression of cell coalescence in the extrusion of microcellular high-impact polystyrene foams, ACS symposium Series 669 (1997) 115-129.

[42] X. Xu, C. B. Park, D. L. Xu, R. Pop-Iliev, Effects of die geometry on cell nucleation of PS foams blown with CO_2, Polymer Engineering & Science 43 (2003) 1378-1390.

[43] K. A. Arora, A. J. Lesser, T. J. McCarthy, Preparation and characterization of microcellular polystyrene foams processed in supercritical carbon dioxide, Macromolecules 31 (1998) 4614-4620.

[44] K. M. Lee, E. K. Lee, S. G. Kim, C. B. Park, H. E. Naguib, Bi-cellular foam structure of polystyrene from extrusion foaming process, J. Cellular Plastics 45 (2009) 539-553.

[45] J. B. Bao, T. Liu, L. Zhao, G. H. Hu, A two-step depressurization batch process for the formation of bi-modal cell structure polystyrene foams using scCO_2, J. Supercritical Fluids 55 (2011) 1104-1114.

[46] T. Liu, G. H. Hu, G. S. Tong, L. Zhao, G. P. Cao, W. K. Yuan, Supercritical carbon dioxide assisted solid-state grafting process of maleic anhydride onto polypropylene, Industrial & Engineering Chemistry Re-

search 44 (2005) 4292-4299.

[47] S. I. Brandrup, E. M. , Polymer Hand Book, Interscience, New York, 1975.

[48] H. E. Naguib, C. B. Park, S. W. Song, Effect of supercritical gas on crystallization of linear and branched polypropylene resins with foaming additives, Industrial & Engineering Chemistry Research 44 (2005) 6685-6691.

[49] D. C. Li, T. Liu, L. Zhao, X. S. Lian, W. K. Yuan, Foaming of poly(lactic acid) based on its nonisothermal crystallization behavior under compressed carbon dioxide, Industrial & Engineering Chemistry Research 50 (2011) 1997-2007.

[50] G. Li, J. Wang, C. B. Park, R. Simha, Measurement of gas solubility in linear/branched PP melts, J. Polymer Science B: Polymer Physics 45 (2007) 2497-2508.